Lecture Notes in Computer Science   12114

More information about this series at http://www.springer.com/series/7409

Yunmook Nah · Bin Cui ·
Sang-Won Lee · Jeffrey Xu Yu ·
Yang-Sae Moon · Steven Euijong Whang (Eds.)

Database Systems
for Advanced Applications

25th International Conference, DASFAA 2020
Jeju, South Korea, September 24–27, 2020
Proceedings, Part III

 Springer

Editors
Yunmook Nah
Dankook University
Yongin, Korea (Republic of)

Sang-Won Lee
Sungkyunkwan University
Suwon, Korea (Republic of)

Yang-Sae Moon 🄳
Kangwon National University
Chunchon, Korea (Republic of)

Bin Cui
Peking University
Haidian, China

Jeffrey Xu Yu
Department of System Engineering
and Engineering Management
The Chinese University of Hong Kong
Hong Kong, Hong Kong

Steven Euijong Whang 🄳
Korea Advanced Institute of Science
and Technology
Daejeon, Korea (Republic of)

ISSN 0302-9743   ISSN 1611-3349   (electronic)
Lecture Notes in Computer Science
ISBN 978-3-030-59418-3   ISBN 978-3-030-59419-0   (eBook)
https://doi.org/10.1007/978-3-030-59419-0

LNCS Sublibrary: SL3 – Information Systems and Applications, incl. Internet/Web, and HCI

This Springer imprint is published by the registered company Springer Nature Switzerland AG
The registered company address is: Gewerbestrasse 11, 6330 Cham, Switzerland

Preface

It is our great pleasure to introduce the proceedings of the 25th International Conference on Database Systems for Advanced Applications (DASFAA 2020), held during September 24–27, 2020, in Jeju, Korea. The conference was originally scheduled for May 21–24, 2020, but inevitably postponed due to the outbreak of COVID-19 and its continual spreading all over the world. DASFAA provides a leading international forum for discussing the latest research on database systems and advanced applications. The conference's long history has established the event as the premier research conference in the database area.

To rigorously review the 487 research paper submissions, we conducted a double-blind review following the tradition of DASFAA and constructed the large committee consisting of 16 Senior Program Committee (SPC) members and 212 Program Committee (PC) members. Each valid submission was reviewed by three PC members and meta-reviewed by one SPC member who also led the discussion with the PC members. We, the PC co-chairs, considered the recommendations from the SPC members and looked into each submission as well as its reviews to make the final decisions. As a result, 119 full papers (acceptance ratio of 24.4%) and 23 short papers were accepted. The review process was supported by the EasyChair system. During the three main conference days, these 142 papers were presented in 27 research sessions. The dominant keywords for the accepted papers included neural network, knowledge graph, time series, social networks, and attention mechanism. In addition, we included 4 industrial papers, 15 demo papers, and 3 tutorials in the program. Last but not least, to shed the light on the direction where the database field is headed to, the conference program included four invited keynote presentations by Amr El Abbadi (University of California, Santa Barbara, USA), Kian-Lee Tan (National University of Singapore, Singapore), Wolfgang Lehner (TU Dresden, Germany), and Sang Kyun Cha (Seoul National University, South Korea).

Five workshops were selected by the workshop co-chairs to be held in conjunction with DASFAA 2020: the 7th Big Data Management and Service (BDMS 2020); the 6th International Symposium on Semantic Computing and Personalization (SeCoP 2020); the 5th Big Data Quality Management (BDQM 2020); the 4th International Workshop on Graph Data Management and Analysis (GDMA 2020); and the First International Workshop on Artificial Intelligence for Data Engineering (AIDE 2020). The workshop papers are included in a separate volume of the proceedings also published by Springer in its *Lecture Notes in Computer Science* series.

We would like to thank all SPC members, PC members, and external reviewers for their hard work to provide us with thoughtful and comprehensive reviews and recommendations. Many thanks to the authors who submitted their papers to the conference. In addition, we are grateful to all the members of the Organizing Committee, and many volunteers, for their great support in the conference organization. Also, we would like to express our sincere thanks to Yang-Sae Moon for compiling all accepted papers and for working with the Springer team to produce the proceedings. Lastly, we acknowledge the generous financial support from IITP[1], Dankook University SW Centric University Project Office, DKU RICT, OKESTRO, SUNJESOFT, KISTI, LG CNS, INZENT, Begas, SK Broadband, MTDATA, WAVUS, SELIMTSG, and Springer.

We hope that the readers of the proceedings find the content interesting, rewarding, and beneficial to their research.

September 2020   Bin Cui
                                                                          Sang-Won Lee
                                                                          Jeffrey Xu Yu

[1] Institute of Information & communications Technology Planning & Evaluation (IITP) grant funded by the Korea government (MSIT) (No. 2020-0-01356, 25th International Conference on Database Systems for Advanced Applications (DASFAA)).

istration Chair

-Soo Kim   DGIST, South Korea

lication Co-chairs

g-Sae Moon   Kangwon National University, South Korea
en Euijong Whang   KAIST, South Korea

icity Co-chairs

xia Shao   Beijing University of Posts and Telecommunications,
                              China
yung Wang   California State University Northridge, USA
oon Chun   Myongji University, South Korea

Chair

o Song   Pukyong National University, South Korea

ce Chair

eop Kwon   Myongji University, South Korea

r Chair

Choi   Sunjesoft Inc., South Korea

A Steering Committee Liaison

k Shim   Seoul National University, South Korea

am Committee

Program Committee Members

k Candan   Arizona State University, USA
                            The Hong Kong University of Science
                              and Technology, Hong Kong
in Han   POSTECH, South Korea
S. Jensen   Aalborg University, Denmark
                            University of Utah, USA
Liu   Swinburne University of Technology, Australia
utt   Free University of Bozen-Bolzano, Italy
nizuka   Osaka University, Japan
Shim   Seoul National University, South Korea
Tong   Beihang University, China
iao   National University of Singapore, Singapore
                            East China Normal University, China
in   The University of Queensland, Australia
                            ETH Zurich, Switzerland

Organization

Organizing Committee

General Chair

Yunmook Nah — Dankook University, South Korea

Program Co-chairs

Bin Cui — Peking University, China
Sang-Won Lee — Sungkyunkwan University, South
Jeffrey Xu Yu — The Chinese University of Hong I

Industry Program Co-chairs

Jinyang Gao — Alibaba Group, China
Sangjun Lee — Soongsil University, South Kore
Eenjun Hwang — Korea University, South Korea

Demo Co-chairs

Makoto P. Kato — Kyoto University, Japan
Hwanjo Yu — POSTECH, South Korea

Tutorial Chair

U. Kang — Seoul National University, So

Workshop Co-chairs

Chulyun Kim — Sookmyung Women's Unive
Seon Ho Kim — USC, USA

Panel Chair

Wook-Shin Han — POSTECH, South Korea

Organizing Committee Chair

Jinseok Chae — Incheon National Universi

Local Arrangement Co-chairs

Jun-Ki Min — Koreatec, South Korea
Haejin Chung — Dankook University, Sou

Re
Mir

Pul
Yar
Stev

Pub
Ying

Taeh
Jong

Web
Ha-Jo

Finan
Dongs

Spons
Junho

DASF
Kyuseo

Progr

Senior
K. Selcu
Lei Che

Wook-Si
Christian
Feifei Li
Chengfei
Werner N
Makoto (
Kyuseok
Yongxin
Xiaokui X
Junjie Ya
Hongzhi
Ce Zhang

Organization

Organizing Committee

General Chair

Yunmook Nah — Dankook University, South Korea

Program Co-chairs

Bin Cui — Peking University, China
Sang-Won Lee — Sungkyunkwan University, South Korea
Jeffrey Xu Yu — The Chinese University of Hong Kong, Hong Kong

Industry Program Co-chairs

Jinyang Gao — Alibaba Group, China
Sangjun Lee — Soongsil University, South Korea
Eenjun Hwang — Korea University, South Korea

Demo Co-chairs

Makoto P. Kato — Kyoto University, Japan
Hwanjo Yu — POSTECH, South Korea

Tutorial Chair

U. Kang — Seoul National University, South Korea

Workshop Co-chairs

Chulyun Kim — Sookmyung Women's University, South Korea
Seon Ho Kim — USC, USA

Panel Chair

Wook-Shin Han — POSTECH, South Korea

Organizing Committee Chair

Jinseok Chae — Incheon National University, South Korea

Local Arrangement Co-chairs

Jun-Ki Min — Koreatec, South Korea
Haejin Chung — Dankook University, South Korea

Registration Chair

Min-Soo Kim   DGIST, South Korea

Publication Co-chairs

Yang-Sae Moon   Kangwon National University, South Korea
Steven Euijong Whang   KAIST, South Korea

Publicity Co-chairs

Yingxia Shao   Beijing University of Posts and Telecommunications,
                               China
Taehyung Wang   California State University Northridge, USA
Jonghoon Chun   Myongji University, South Korea

Web Chair

Ha-Joo Song   Pukyong National University, South Korea

Finance Chair

Dongseop Kwon   Myongji University, South Korea

Sponsor Chair

Junho Choi   Sunjesoft Inc., South Korea

DASFAA Steering Committee Liaison

Kyuseok Shim   Seoul National University, South Korea

Program Committee

Senior Program Committee Members

K. Selcuk Candan   Arizona State University, USA
Lei Chen   The Hong Kong University of Science
                               and Technology, Hong Kong
Wook-Shin Han   POSTECH, South Korea
Christian S. Jensen   Aalborg University, Denmark
Feifei Li   University of Utah, USA
Chengfei Liu   Swinburne University of Technology, Australia
Werner Nutt   Free University of Bozen-Bolzano, Italy
Makoto Onizuka   Osaka University, Japan
Kyuseok Shim   Seoul National University, South Korea
Yongxin Tong   Beihang University, China
Xiaokui Xiao   National University of Singapore, Singapore
Junjie Yao   East China Normal University, China
Hongzhi Yin   The University of Queensland, Australia
Ce Zhang   ETH Zurich, Switzerland

Qiang Zhu	University of Michigan, USA
Eenjun Hwang	Korea University, South Korea

Program Committee Members

Alberto Abello	Universitat Politècnica de Catalunya, Spain
Marco Aldinucci	University of Turin, Italy
Akhil Arora	Ecole Polytechnique Fédérale de Lausanne, Switzerland
Jie Bao	JD Finance, China
Zhifeng Bao	RMIT University, Australia
Ladjel Bellatreche	LIAS, ENSMA, France
Andrea Calì	University of London, Birkbeck College, UK
Xin Cao	The University of New South Wales, Australia
Yang Cao	Kyoto University, Japan
Yang Cao	The University of Edinburgh, UK
Barbara Catania	DIBRIS, University of Genoa, Italy
Chengliang Chai	Tsinghua University, China
Lijun Chang	The University of Sydney, Australia
Chen Chen	Arizona State University, USA
Cindy Chen	University of Massachusetts Lowell, USA
Huiyuan Chen	Case Western Reserve University, USA
Shimin Chen	ICT CAS, China
Wei Chen	Soochow University, China
Yang Chen	Fudan University, China
Peng Cheng	East China Normal University, China
Reynold Cheng	The University of Hong Kong, Hong Kong
Theodoros Chondrogiannis	University of Konstanz, Germany
Jaegul Choo	Korea University, South Korea
Lingyang Chu	Simon Fraser University, Canada
Gao Cong	Nanyang Technological University, Singapore
Antonio Corral	University of Almeria, Spain
Lizhen Cui	Shandong University, China
Lars Dannecker	SAP SE, Germany
Ernesto Damiani	University of Milan, Italy
Sabrina De Capitani	University of Milan, Italy
Dong Den	Rutgers University, USA
Anton Dignös	Free University of Bozen-Bolzano, Italy
Lei Duan	Sichuan University, China
Amr Ebaid	Google, USA
Ju Fan Renmin	University of China, China
Yanjie Fu	University of Central Florida, USA
Hong Gao	Harbin Institute of Technology, China
Xiaofeng Gao	Shanghai Jiao Tong University, China
Yunjun Gao	Zhejiang University, China
Tingjian Ge	University of Massachusetts Lowell, USA

Zheng Liu	Nanjing University of Posts and Telecommunications, China
Chunbin Lin	Amazon Web Services, USA
Guanfeng Liu	Macquarie University, Australia
Hailong Liu	Northwestern Polytechnical University, China
Qing Liu	CSIRO, Australia
Qingyun Liu	Facebook, USA
Eric Lo	The Chinese University of Hong Kong, Hong Kong
Cheng Long	Nanyang Technological University, Singapore
Guodong Long	University of Technology Sydney, Australia
Hua Lu	Aalborg University, Denmark
Wei Lu	Renmin University of China, China
Shuai Ma	Beihang University, China
Yannis Manolopoulos	Open University of Cyprus, Cyprus
Jun-Ki Min	Korea University of Technology and Education, South Korea
Yang-Sae Moon	Kangwon National University, South Korea
Mikolaj Morzy	Poznan University of Technology, Poland
Parth Nagarkar	New Mexico State University, USA
Liqiang Nie	Shandong University, China
Baoning Niu	Taiyuan University of Technology, China
Kjetil Nørvåg	Norwegian University of Science and Technology, Norway
Vincent Oria	New Jersey Institute of Technology, USA
Noseong Park	George Mason University, USA
Dhaval Patel	IBM, USA
Wen-Chih Peng	National Chiao Tung University, Taiwan
Ruggero G. Pensa	University of Turin, Italy
Dieter Pfoser	George Mason University, USA
Silvestro R. Poccia	Polytechnic of Turin, Italy
Shaojie Qiao	Chengdu University of Information Technology, China
Lu Qin	University of Technology Sydney, Australia
Weixiong Rao	Tongji University, China
Oscar Romero	Universitat Politènica de Catalunya, Spain
Olivier Ruas	Peking University, China
Babak Salimi	University of Washington, USA
Maria Luisa Sapino	University of Turin, Italy
Claudio Schifanella	University of Turin, Italy
Shuo Shang	Inception Institute of Artificial Intelligence, UAE
Xuequn Shang	Northwestern Polytechnical University, China
Zechao Shang	The University of Chicago, USA
Jie Shao	University of Electronic Science and Technology of China, China
Yingxia Shao	Beijing University of Posts and Telecommunications, China
Wei Shen	Nankai University, China

Jianqiu Xu	Nanjing University of Aeronautics and Astronautics, China
Quanqing Xu	A*STAR, Singapore
Tong Yang	Peking University, China
Yu Yang	City University of Hong Kong, Hong Kong
Zhi Yang	Peking University, China
Bin Yao	Shanghai Jiao Tong University, China
Lina Yao	The University of New South Wales, Australia
Man Lung Yiu	The Hong Kong Polytechnic University, Hong Kong
Ge Yu	Northeastern University, China
Lele Yu	Tencent, China
Minghe Yu	Northeastern University, China
Ye Yuan	Northeastern University, China
Dongxiang Zhang	Zhejiang University, China
Jilian Zhang	Jinan University, China
Rui Zhang	The University of Melbourne, Australia
Tieying Zhang	Alibaba Group, USA
Wei Zhang	East China Normal University, China
Xiaofei Zhang	The University of Memphis, USA
Xiaowang Zhang	Tianjin University, China
Ying Zhang	University of Technology Sydney, Australia
Yong Zhang	Tsinghua University, China
Zhenjie Zhang	Yitu Technology, Singapore
Zhipeng Zhang	Peking University, China
Jun Zhao	Nanyang Technological University, Singapore
Kangfei Zhao	The Chinese University of Hong Kong, Hong Kong
Pengpeng Zhao	Soochow University, China
Xiang Zhao	National University of Defense Technology, China
Bolong Zheng	Huazhong University of Science and Technology, China
Kai Zheng	University of Electronic Science and Technology of China, China
Weiguo Zheng	Fudan University, China
Yudian Zheng	Twitter, USA
Chang Zhou	Alibaba Group, China
Rui Zhou	Swinburne University of Technology, Australia
Xiangmin Zhou	RMIT University, Australia
Xuan Zhou	East China Normal University, China
Yongluan Zhou	University of Copenhagen, Denmark
Zimu Zhou	Singapore Management University, Singapore
Yuanyuan Zhu	Wuhan University, China
Lei Zou	Peking University, China
Zhaonian Zou	Harbin Institute of Technology, China
Andreas Züfle	George Mason University, USA

External Reviewers

Ahmed Al-Baghdadi
Alberto R. Martinelli
Anastasios Gounaris
Antonio Corral
Antonio Jesus
Baozhu Liu
Barbara Cantalupo
Bayu Distiawan
Besim Bilalli
Bing Tian
Caihua Shan
Chen Li
Chengkun He
Chenhao Ma
Chris Liu
Chuanwen Feng
Conghui Tan
Davide Colla
Deyu Kong
Dimitrios Rafailidis
Dingyuan Shi
Dominique Laurent
Dong Wen
Eleftherios Tiakas
Elena Battaglia
Feng Yuan
Francisco Garcia-Garcia
Fuxiang Zhang
Gang Qian
Gianluca Mittone
Hans Behrens
Hanyuan Zhang
Huajun He
Huan Li
Huaqiang Xu
Huasha Zhao
Iacopo Colonnelli
Jiaojiao Jiang
Jiejie Zhao
Jiliang Tang
Jing Nathan Yan
Jinglin Peng
Jithin Vachery

Joon-Seok Kim
Junhua Zhang
Kostas Tsichlas
Liang Li
Lin Sun
Livio Bioglio
Lu Liu
Luigi Di Caro
Mahmoud Mohammadi
Massimo Torquati
Mengmeng Yang
Michael Vassilakopoulos
Moditha Hewasinghage
Mushfiq Islam
Nhi N.Y. Vo
Niccolo Meneghetti
Niranjan Rai
Panayiotis Bozanis
Peilun Yang
Pengfei Li
Petar Jovanovic
Pietro Galliani
Qian Li
Qian Tao
Qiang Fu
Qianhao Cong
Qianren Mao
Qinyong Wang
Qize Jiang
Ran Gao
Rongzhong Lian
Rosni Lumbantoruan
Ruixuan Liu
Ruiyuan Li
Saket Gurukar
San Kim
Seokki Lee
Sergi Nadal
Shaowu Liu
Shiquan Yang
Shuyuan Li
Sicong Dong
Sicong Liu

Sijie Ruan
Sizhuo Li
Tao Shen
Teng Wang
Tianfu He
Tiantian Liu
Tianyu Zhao
Tong Chen
Waqar Ali
Weilong Ren
Weiwei Zhao
Weixue Chen
Wentao Li
Wenya Sun
Xia Hu
Xiang Li
Xiang Yu
Xiang Zhang
Xiangguo Sun
Xianzhe Wu
Xiao He
Xiaocong Chen
Xiaocui Li
Xiaodong Li
Xiaojie Wang
Xiaolin Han
Xiaoqi Li
Xiaoshuang Chen
Xing Niu
Xinting Huang
Xinyi Zhang
Xinyu Zhang
Yang He
Yang Zhao
Yao Wan
Yaohua Tang
Yash Garg
Yasir Arfat
Yijian Liu
Yilun Huang
Yingjun Wu
Yixin Su
Yu Yang

Yuan Liang
Yuanfeng Song
Yuanhang Yu
Yukun Cao
Yuming Huang
Yuwei Wang

Yuxing Han
Yuxuan Qiu
Yuyu Luo
Zelei Cheng
Zhangqing Shan
Zhuo Ma

Zicun Cong
Zili Zhou
Zisheng Yu
Zizhe Wang
Zonghan Wu

Financial Sponsors

Academic Sponsors

Contents – Part III

Query Processing

Social Network

Sequential Multi-fusion Network for Multi-channel Video CTR Prediction

Wen Wang[1], Wei Zhang[1,2](✉), Wei Feng[3], and Hongyuan Zha[4]

[1] School of Computer Science and Technology,
East China Normal University, Shanghai, China
51164500120@stu.ecnu.edu.cn, zhangwei.thu2011@gmail.com
[2] Key Laboratory of Artificial Intelligence, Ministry of Education, Shanghai, China
[3] Facebook, Menlo Park, USA
whitepapers824@gmail.com
[4] Georgia Institute of Technology, Atlanta, USA
zha@cc.gatech.edu

Abstract. In this work, we study video click-through rate (CTR) prediction, crucial for the refinement of video recommendation and the revenue of video advertising. Existing studies have verified the importance of modeling users' clicked items as their latent preference for general click-through rate prediction. However, all of the clicked ones are equally treated in the input stage, which is not the case in online video platforms. This is because each video is attributed to one of the multiple channels (e.g., TV and MOVIES), thus having different impacts on the prediction of candidate videos from a certain channel. To this end, we propose a novel Sequential Multi-Fusion Network (SMFN) by classifying all the channels into two categories: (1) target channel which current candidate videos belong to, and (2) context channel which includes all the left channels. For each category, SMFN leverages a recurrent neural network to model the corresponding clicked video sequence. The hidden interactions between the two categories are characterized by correlating each video of a sequence with the overall representation of another sequence through a simple but effective fusion unit. The experimental results on the real datasets collected from a commercial online video platform demonstrate the proposed model outperforms some strong alternative methods.

Keywords: Click-through rate prediction · Sequential recommendation · Recurrent neural networks

1 Introduction

Click-Through Rate (CTR) prediction is a fundamental task in computational advertising and recommender system. On the one hand, the predicted CTRs,

This work was partially conducted while Wen Wang and Wei Feng were with Hulu. It was supported in part by the National Key Research and Development Prograssm (2019YFB2102600), NSFC (61702190), Shanghai Sailing Program (17YF1404500), and the foundation of Key Laboratory of Artificial Intelligence, Ministry of Education, P.R. China.

with bid price, jointly determine the ranking of candidate advertisements in the common cost-per-click scenario. They have a direct impact on the total revenue of ad publishers [23]. On the other hand, CTR could be leveraged for the refinement of candidate recommendations [8]. As such, the recommendations are more promising to receive users' positive feedback (e.g., click). Since the candidate advertisements or recommendations are indispensable for performing CTR prediction, two-stage frameworks are commonly adopted by current large business platforms, wherein the candidate generation corresponds to the first stage and the CTR prediction constitutes the second stage [8,30]. In this work, we concentrate on video CTR prediction, which has the aforementioned benefits to the more and more popular online video platforms, such as Hulu[1], Netflix[2], iQiyi[3], to name a few.

To achieve accurate CTR prediction for different types of items in a general manner, some existing methods pursue to leverage users' clicked items to characterize their latent preference [4,12,30]. Recent advances in recurrent neural networks (RNNs) [15,20] have promoted the development of the sequential user preference modeling [14], although only a few studies have been found in the literature of CTR prediction [28,29]. A common procedure shared in these methods is to model the past clicked items as a single sequence, mapping each item into a low dimensional embedding and feeding these embeddings to an RNN for obtaining a fixed-length hidden representation. As such, each item is equally treated in the input stage and temporal dependency between different clicked items could be partially captured.

However, in many scenarios of online video services, all the videos are divided into different channels. To consider the multi-channel video CTR prediction, an initial and natural attempt is to merge the clicked videos from a certain user into a single sequence, no matter which channels they belong to. Then previous non-sequential models like [30] or RNN based models [29] might be utilized for prediction. However, these approaches regard all the clicked videos equally in the input stage, which is sub-optimal because each channel has its own characteristics, making clicked videos from different channels have different impacts on the prediction of candidate videos from a certain channel. Hence, users' clicked videos on different channels might reflect diverse user preferences. As a result, simply treating videos from different channels equally is not reasonable, possibly limiting the predictive performance.

A simple illustration of the multi-channel video CTR prediction situation is shown in Fig. 1 where the TV and MOVIES channels are chosen for display. If we only look at the user's past clicked videos from the TV channel, we may feel that "Friends" is the one the user most wanted, according to "2 Broke Girls" and "The Big Bang Theory". If the user's behaviors from both channels are considered and videos from different channels are treated equally, it might be that "Agents of S.H.I.E.L.D." or "Inhumans" is what the user most wanted, according to

[1] https://www.hulu.com/.
[2] https://www.netflix.com/.
[3] https://www.iqiyi.com/.

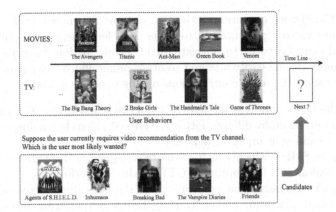

Fig. 1. Illustration of video CTR prediction using the TV and MOVIES channels. The aim is to predict which video a target user most likely wants to watch in next, given the historical behaviors from both channels.

"The Avengers" and "Ant-Man". However, both of the above perspectives have some limitations: (i) only considering user behavior from the current channel will lose some potentially useful information from other channels for CTR prediction; (ii) treating videos from different channels equally will ignore the different characteristics of channels. Another consideration is to treat the clicked videos from each channel separately, building a channel-dependent video sequence for different modeling parts. However, there might exist hidden and complex relations between the different video sequences, which might not be easy to be acquired by the above methods. As a result, to improve multi-channel video CTR prediction, an essential challenge raises: how to effectively fuse video sequences from different channels in a principled manner?

In this paper, we take Hulu as the testbed to study the multi-channel video CTR prediction problem. All the channels are first divided into two categories: target channel and context channel. The target channel denotes the video channel which a user is staying on. And recommendations should be generated based on video candidates belonging to the channel. The context channel includes all the left channels, excepting for the target channel. Based on this, we propose a novel Sequential Multi-Fusion Network (short for SMFN, see Fig. 2) to differentiate the roles of the clicked videos from different channels for better multi-channel video CTR prediction. Concretely, we regard the currently deployed recommender system in the store-shelf scenario of Hulu as the candidate generator in the first stage [25]. SMFN involves four main modules. At first, two GRU networks [7] are adopted to model the clicked videos in each channel category, resulting in an overall representation for each behavior sequence. Afterwards, to characterize the hidden interactions between the two clicked video sequences, a simple but effective fusion unit is proposed to correlate each video of a sequence with the overall representation of another sequence. Then sum pooling operation is conducted on each updated sequence, to get the final representations of the two sequences. Finally, we concatenate the two channel-dependent representations and candidate video representation together, and feed them into a multi-layer feed-forward neural network to generate final CTR prediction. Worth noting our approach could also be easily adapted to other online video platforms.

We summarize the main contributions of this paper as follows:

- We are the first to address the multi-channel issue existing in the video CTR prediction problem, by differentiating the clicked videos belonging to different channels as different input sequences.
- We propose SMFN which can fuse the two sequences through a simple but effective fusion unit for improving CTR prediction performance.
- We conduct comprehensive experiments on datasets from a real-world commercial video platform, demonstrating SMFN outperforms several strong competitors, validating the benefit of the fusion unit.

2   Related Work

2.1   CTR Prediction

CTR prediction is developed in the scenario of sponsored search [23]. The traditional pipeline is to first construct tailored handcrafted features and then build statistical models for prediction. Despite the early widely adopted logistic regression model [19] in industrial applications, some other models are also considered, including the dynamic Gamma-Poisson model [1], Bayesian linear regression model [11], and tree-based method [13]. CTR prediction could be used to determine which advertisements are displayed in sponsored search, and benefit the practical usage of recommender systems. For example, Agarwal et al. [1] refined the recommendation results based on the predicted CTR. Many online platforms deploy recommender systems to generate candidate recommendations for users, making the reranking of recommendations based on CTR to be feasible.

The flourish of deep learning methods in the last decade has witnessed the growth trend of their application in the CTR prediction scenarios. Chen et al. [3] studied image CTR prediction by leveraging convolutional neural networks [16] to learn image representation, which is fed to several fully connected layers. DeepFM [12] integrates factorization machines [22], originally proposed for recommendation, with deep learning, hoping to inherit both of their advantages in a unified model. Deep interest network (DIN) [30] overcomes the limitation of using only a fixed-length vector to represent users' dynamic preferences. It employs the attention mechanism [2] to adaptively select the past interacted items of users to construct their representations. Due to the sequential nature of user behaviors, sequential modeling approaches have been largely investigated in recommendation [14]. However, only a few studies have been seen in the literature of CTR prediction. Zhang et al. [29] first adopted RNNs to characterize the temporal dependency between users' click items. DeepIntent further incorporates the attention mechanism into the RNN modeling process [28]. However, none of the above studies investigate the issue of taking multiple item sequences at one time for CTR prediction, just as the problem studied in this paper. In the experiments, we also consider two simple extensions of deep learning based CTR models for fusing multi-channel data. Nevertheless, their performance is not very satisfactory, showing the necessity of pursuing a more powerful multi-channel model.

2.2 Video CTR Prediction and Recommendation

The methods for video CTR prediction and recommendation share some common aspects and could be used interchangeably with tailored modifications, such as revising loss functions. Therefore, we provide a brief discussion of their relevant studies together. Davidson et al. [10] considered several categories of video features, capturing video quality, user specificity, and diversification. Those features are integrated by an empirical linear combination and CTR is adopted as the primary metric to evaluate the recommendation quality. Cui et al. [9] learned a common space for users and videos. However, it relied on the social attributes of users and video content features, which might not be easy to be obtained. Li et al. [17] investigated the role of user groups in TV recommendation and utilized the traditional matrix factorization method. Yang et al. [27] focused on users' multi-site behaviors and preference to specific sites was additionally captured. Recently, deep learning methods have also been applied in this scenario. Youtube recommender system [8] has benefited from deep learning approaches by fusing multi-types of feature embeddings. A simple feed-forward neural network is adopted, without characterizing the temporal dependency between different clicked videos of a user. Moreover, a deep hierarchical attention network is designed to perform CTR for cold-start micro-videos by considering video content features [4]. Chen et al. [5] expanded DIN with multi-view learning to incorporate visual features of videos. Nevertheless, the temporal relation between videos is only handled by splitting a whole user behavior sequence into several subsequences. In summary, all the above studies have treated all the videos equally in the input stage, ignoring the fact that videos from different channels would have different impacts on the CTR prediction of candidate videos from a certain channel. This issue motivates our investigation of multi-channel CTR prediction in this paper.

3   Problem Definition

As aforementioned, we take the currently deployed recommender system in Hulu as the candidate generator, and our CTR prediction approach is performed on these candidate recommendations. Suppose the whole video set is V and the current candidate recommendation is $v_c \in V$. For a given target user, its clicked video sequence is split into two sequences, according to the introduction of the target channel and the context channel. To be specific, we denote $P = [p_1, p_2, \ldots, p_{N_p}]$ for target channel based sequence and $Q = [q_1, q_2, \ldots, q_{N_q}]$ as the context channel based sequence, where N_p and N_q correspond to the numbers of clicked videos belonging to the two channels. All the videos in P and Q

are ordered by their clicked time, respectively. With the above notations, we can formally define the studied problem as below:

Problem 1 (Multi-Channel Click-Through Rate Prediction). Given the candidate video v_c and the target user's clicked video sequences P and Q, the problem aims to predict the probability that the user will click the video.

It is worth noting that our model does not use any user profile features. On one hand, our focus is on effectively fusing multiple user click behavior sequences, and video representation itself could somewhat represent the users' latent interest. On the other hand, our CTR prediction method can be regarded as a wrapper for the existing video recommendation system, which has fully considered to model user profile features for candidate video generation.

4   The SMFN Method

Model Overview. Our model SMFN is an end-to-end learning framework that takes a candidate video and a user's historical behaviors in two categories of channels as input, and output the CTR score prediction. A high-level graphical illustration of SMFN is shown in Fig. 2. Essentially, it is composed of four core modules: a) *Sequential Representation Learning*, which learns an overall representation of each sequence. b) *Sequential Elementwise Fusion*, which fuses each video representation in a sequence with the sequence-level representation of another video sequence via a fusion unit. c) *Sequential Representation Reconstruction*, which reconstructs the representation of the new sequence obtained in the previous step. And d) *Learning for CTR Prediction*, which outputs the CTR score. In what follows, we will go deeper into the details of each module with necessary mathematical formulations.

4.1   Input Representation

As aforementioned, we represent the two video sequences as $P = [p_1, p_2, \ldots, p_{N_p}]$ and $Q = [q_1, q_2, \ldots, q_{N_q}]$, respectively, and the candidate video as v_c. We assume that the original representation of each video in $\mathcal{V}$ is based on one-hot encoding, which is a binary vector where only one position takes the value one and all the left take 0. Concretely, we use $\bar{p}_i \in \mathbb{R}^{|\mathcal{V}|}$ to denote the one-hot encoding of p_i and it is similar for $\bar{q}_j$ and $\bar{v}_c$. Because the one-hot representation is a high-dimensional sparse vector, we adopt the commonly used lookup table operation in representation learning to map each one-hot representation to its low-dimensional dense vector by a shared video embedding matrix E:

$$\hat{p}_i = E^T \bar{p}_i, \hat{q}_j = E^T \bar{q}_j, \hat{v}_c = E^T \bar{v}_c, \tag{1}$$

where $E \in \mathbb{R}^{|\mathcal{V}| \times K}$ is the embedding matrix, and K is the size of embedding. We omit the lookup procedure of the model in Fig. 2. Thus the inputs in the figure are two embedding sequences ($\hat{P} = [\hat{p}_1, \hat{p}_2, \ldots, \hat{p}_{N_p}]$ and $\hat{Q} = [\hat{q}_1, \hat{q}_2, \ldots, \hat{q}_{N_q}]$), and an embedding vector $\hat{v}_c$.

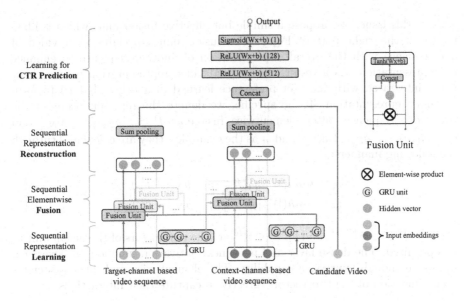

Fig. 2. The architecture of the proposed SMFN.

4.2 Sequential Representation Learning

In order to learn the sequential user behavior to characterize the temporal relation, we adopt two GRUs to model the target channel and context channel, respectively. Take the clicked video sequence $\hat{P}$ as an example. At each time step t, a GRU unit has an update gate z_t and a reset gate r_t. The corresponding hidden state h_t is calculated through the following equations:

$$z_t = \sigma(W_z \hat{p}_t + U_z h_{t-1} + b_z),\tag{2}$$

$$r_t = \sigma(W_r \hat{p}_t + U_r h_{t-1} + b_r),\tag{3}$$

$$\bar{h}_t = tanh(W_h \hat{p}_t + U_h(r_t \circ h_{t-1}) + b_h),\tag{4}$$

$$h_t = (1 - z_t) \circ h_{t-1} + z_t \circ \bar{h}_t,\tag{5}$$

where $\circ$ is the element-wise multiplier. $W_{\{z,r,h\}}$, $U_{\{z,r,h\}}$ and $b_{\{z,r,h\}}$ are GRU's parameters to be learned. σ is the sigmoid activation function. From the above four formulas, we can see that h_t is calculated by h_{t-1} and $\hat{p}_t$. h_t could be regarded as the overall representation of first t elements of the sequence. After recurrent computation for each time step, we get the last hidden state h_{N_p}, which is the overall representation of the whole sequence. Similarly, we could obtain h_{N_q} to represent the whole video sequence of the context channel. These two representations are exploited for later correlation computation.

4.3 Sequential Elementwise Fusion

There might exist hidden and complex relations between the two video sequences, which are not easy to be acquired by the above computational formulas. To address this issue, we propose a simple but effective fusion unit, which is illustrated in the right part of Fig. 2. The fusion unit correlates each video of one sequence with the overall representation of another sequence to enhance its representation. As a result, the original video representation in one channel is integrated with the user preference learned in another channel to form the new representation. To be specific, we denote the representations of the two video sequences after elementwise fusion as $\tilde{P} = [\tilde{p}_1, \tilde{p}_2, \ldots, \tilde{p}_{N_p}]$ and $\tilde{Q} = [\tilde{q}_1, \tilde{q}_2, \ldots, \tilde{q}_{N_q}]$. Take $\tilde{p}_t$ and $\tilde{q}_t$ as the examples, they are calculated through the following manners:

$$\tilde{p}_t = tanh(W_F^p[\hat{p}_t; h_{N_q}; \hat{p}_t \circ h_{N_q}] + b_F^p), \tag{6}$$

$$\tilde{q}_t = tanh(W_F^q[\hat{q}_t; h_{N_p}; \hat{q}_t \circ h_{N_p}] + b_F^q), \tag{7}$$

where $W_F^{\{p,q\}} \in \mathbb{R}^{K \times 3K}$ and $b_F^{\{p,q\}} \in \mathbb{R}^K$ are the parameters of the fusion unit to be optimized. The dense layer and non-linear activation function are leveraged to ensure enough expressive capability. The element-wise multiplier associates each dimension of the two vectors directly to capture their interactions.

4.4 Sequential Representation Reconstruction

After obtaining the two new sequences by the previous module, a fixed-length sequence-level representation for each new sequence is acquired via this module through the pooling operation. In order to preserve the overall situation of each element in the sequence, the sum pooling operation is adopted for each new sequence, which is defined as follows:

$$\dot{p} = \sum_{t=1}^{N_p} \tilde{p}_t, \dot{q} = \sum_{t=1}^{N_q} \tilde{q}_t, \tag{8}$$

where $\dot{p}$ and $\dot{q}$ are the final representations to denote each channel-based sequence. Through the above manner, they contain not only the original representation of each video in its own channel, but also the hidden relations between different channels.

4.5 Learning for CTR Prediction

Finally, we concatenate these two final channel-based representations and candidate video embedding together to form a unified representation. This representation is then fed into two fully-connected dense layers, each of which is followed by the rectified linear unit (ReLU) activation function. The above layers provide more nonlinear representation ability. To be specific, the overall representation r is defined as follows:

$$r = ReLU(W_2 ReLU(W_1[\dot{p}; \dot{q}; \hat{v}] + b_1) + b_2), \tag{9}$$

where ReLU represents the rectified linear unit, with the form, $\text{ReLU}(x) = max(0, x)$. $W_1 \in \mathbb{R}^{512 \times 3K}$, $b_1 \in \mathbb{R}^{512}$, $W_2 \in \mathbb{R}^{128 \times 512}$, $b_2 \in \mathbb{R}^{128}$ are parameters of these layers to be learned.

Afterwards, since the CTR score ranges from 0 and 1, we adopt a sigmoid layer to get it according to the overall representation r. More specifically, we define the computational formula as follows:

$$o = \sigma(W_3 r + b_3), \tag{10}$$

where $W_3 \in \mathbb{R}^{128}$ and $b_3 \in \mathbb{R}$ are the parameters of the layers to be learned. And o indicates the predicted CTR we strive to generate.

4.6 Loss Function

Our model is trained in an end-to-end fashion by minimizing the loss $\mathcal{L}$. Following [30], we optimize our model parameters by minimizing the cross-entropy loss, which is defined as:

$$\mathcal{L} = -\frac{1}{|\mathcal{S}|} \sum_{(x,y) \in \mathcal{S}} (y \log f(x) + (1 - y) \log(1 - f(x))), \tag{11}$$

where $\mathcal{S}$ is the training data set. The input x includes users' past clicked videos and candidate videos. $y \in \{0, 1\}$ is the ground-truth label of x. Since the anonymous users' log could reflect whether they click the recommended videos or not, we determine the labels to be 0 or 1 easily, unlike the implicit feedback scenario where only the label 1 could be known. $f(x)$ is the output of our SMFN model, representing the predicted CTR score o.

5 Experimental Setup

This section clarifies some necessary experimental settings for testing the performance of various methods on CTR prediction. First, we introduce the datasets adopted in the experiments, including their basic statistics. Then we specify the evaluation metrics for comparing different models. Moreover, the baselines used in the experiments are explained. Ablation study is further adopted to validate the rationality of the proposed model. Finally, we describe the training details, including the hyper-parameter settings.

5.1 Datasets

As aforementioned, we conduct experiments on real datasets collected in Hulu. All the used users are anonymized for privacy protection. The deployed recommender system in Hulu has generated a large number of user logs to record the interactions between users and recommended videos. If a video is clicked by a user and at the same time, watched by the user for more than two minutes, we regard this as a valid positive click.

Table 1. Basic statistics of the datasets.

Data	One Month	Three Months
#User	47,053	21,132
#Video	15,428	14,940
#Train	985,225	1,392,576
#Validation	123,152	174,072
#Test	123,152	174,072

In Hulu, most of videos come from two channels, TV and MOVIES, thus we collect user logs in these two channels for testing. Two datasets with different time spans are built. The first dataset is based on the user logs from December 1st to December 30th in 2018 (One Month) while the second dataset is from October 1st to December 30th in 2018 (Three Months). In order to make the volumes of the two datasets roughly the same, we sample about 1/3 users from the whole user set to build the Three Months dataset. Since the volume of the videos not clicked by users is exceedingly huge, we randomly sample 10% of them. The ratios of positive and negative instances in our two datasets are about 1:10. Finally, we remove users with less than 100 records in each dataset and consider all the videos that have appeared. We sort the datasets in a chronological order and divide them into training, validation and test sets with a ratio of 8:1:1. The summarization of the experimental datasets is shown in Table 1.

5.2 Competitors

To validate the advantages of our proposed SMFN model, we compare it with several alternative baselines, some of which have strong performance.

- **MF.** Matrix factorization (MF) [24] is a model based collaborative filtering algorithm widely used in recommendation systems. It learns latent factors for each user and item identifier. The inner product of a pair of user and item factors is regarded as the corresponding rating score. In our implementation, we use the average embedding of user clicked videos as the user latent factor. This could be regarded as a weak baseline for comparison.
- **DeepFM.** DeepFM [12], a factorization-machine [22] based neural network for CTR prediction, combines the power of factorization machines and deep learning in a unified architecture. It is widely used in recommendation system and CTR prediction task.
- **Wide&Deep.** Wide&Deep [6] consists of two parts: cross-product feature transformations (wide) and deep neural networks (deep). It jointly trains wide linear models and deep neural networks to combine the benefits of memorization and generalization for recommendation systems. We follow the practice in [30] to take cross-product of user behaviors and candidates as the input of the "wide" module.

- **DIN.** Deep Interest Network (DIN) [30] designs a local activation unit to adaptively learn the representation of user interest from historical behaviors with respect to a certain advertisement. Because our model uses the history of multiple channels as two sequences, for a fair comparison, we regard both of the two sequences as the user historical behaviors.
- **PaGRU.** Since [14], GRU is widely used in recommendation system. In order to compare with them, we present PaGRU, which is a very intuitive method that uses two GRU networks to model the two video sequences separately, without a deep fusion as our model.
- **DAN.** Dual Attention Networks (DAN) [21] are proposed to jointly leverage visual and textual attention mechanisms to capture fine-grained interplay between vision and language. Considering that it also models both sequences simultaneously, we implement it to compare with our model. The model DAN shows two architectures in its paper, reasoning-DAN (r-DAN) and matching-DAN (m-DAN). We use the former one for comparison, which is more suitable for our problem setting.
- **CoAtt.** Co-Attention (CoAtt) [18] is similar to DAN, except the difference that the interactive attention calculation of two sequences performs in parallel or alternatively. We implement the alternating form for comparison.

In the experiments, we test and analyze the effect of different processing ways of users' historical behavior sequences in two channels. In addition, we will also discuss some variants of our proposed model, including ablation study and different sequence lengths.

5.3 Metrics

The major goal of the studied problem is to predict the CTR score. Since the ratio of positive and negative instances in many CTR prediction tasks, including ours, is very unbalanced, some of the common metrics for binary classification problems such as accuracy, recall, and F1-score cannot reveal the performance well. AUC (Area Under the Receiver Operating Characteristics Curve) is another common metric for binary classification problems, which is insensitive to ratio of positive and negative instances, and widely used in the CTR prediction field [10, 30]. Hence we adopt AUC to evaluate our model.

Besides, due to that random guessing without any prior information gets 0.5 in AUC, we introduce the RelaImpr metric to measure relative improvements over models, following [26,30]. Specifically, RelaImpr is defined as below:

$$RelaImpr = \left(\frac{\text{AUC}_{\text{measured model}} - 0.5}{\text{AUC}_{\text{base model}} - 0.5} - 1 \right) \times 100\%, \tag{12}$$

5.4 Implementation Details

The main hyper-parameters of all the adopted methods are tuned on the validation sets and then performance evaluations are conducted on the test sets. To

Table 2. Performance of SMFN and other adopted competitors by AUC and RelaImpr. And RelaImpr is calculated by comparing with DeepFM.

Methods	One Month		Three Months	
	AUC	RelaImpr	AUC	RelaImpr
MF	0.6553	−27.70%	0.6580	−25.65%
cre DeepFM	0.7148	0.00%	0.7125	0.00%
Wide&Deep	0.7149	0.05%	0.7137	0.56%
DIN	0.7169	0.98%	0.7161	1.69%
PaGRU	0.7174	1.21%	0.7185	2.82%
DAN	0.7112	−1.68%	0.7110	−0.71%
CoAtt	0.7152	0.19%	0.7135	0.47%
SMFN	**0.7268**	**5.59%**	**0.7313**	**8.85%**

reduce the variance of the experimental results, all the experiments are repeated 10 times with different random seeds and the averaged results are reported. We implement our proposed SMFN based on TensorFlow, and initialize video embedding through a uniform distribution $U(-0.01, 0.01)$ consistently for all relevant methods to ensure fair comparison. Without loss of generality, we set the sequence length to 10 (the effect of length is analyzed in Fig. 3). And the size of video embedding and hidden state in GRU are both set to 64. We train all the deep learning approaches using Adam with a learning rate of 0.0001, mini-batch size of 32, and exponential decay rates of 0.9 and 0.999. The model learning process is terminated with an early stopping strategy, by referring to the model performance on the validation dataset every 3000 batches. When the best performance keeps unchanged for more than 10 iterations, the learning process will be stopped.

6   Experimental Results and Discussion

In this section, we present the detailed experimental results and some intuitive analysis to first answer the following core research questions:

Q1 How does SMFN compare with the adopted competitors?
Q2 Is there any performance gain by considering the characteristics of a multi-channel video sequence? And what is the benefit of the fusion unit?

Besides, we further provide experimental discussions about the influence of different sequence lengths and different methods of sequential modeling.

6.1   Model Comparison

Table 2 shows the performance comparison between SMFN and all the competitors. First of all, we can see that all the deep network methods with users' historical behavior sequences beat MF significantly, which demonstrates the power

Table 3. Performance comparison of several representative methods w.r.t. different sequence processing cases. All the other lines calculate RelaImpr by comparing with the first line.

Sequence Processing Settings	Methods	One Month		Three Months	
		AUC	RelaImpr	AUC	RelaImpr
Case 1: One sequence (without other channels)	DeepFM	0.7090	0.00%	0.7035	0.00%
	DIN	0.7102	0.57%	0.7036	0.05%
	GRU	0.7135	2.15%	0.7140	5.16%
Case 2: One merged sequence (fusing multiple channels)	DeepFM	0.7148	2.78%	0.7125	4.42%
	DIN	0.7169	3.78%	0.7161	6.19%
	GRU	0.7151	2.92%	0.7122	4.28%
Case 3: Two sequences (one for other channels)	DeepFM	0.7153	3.01%	0.7137	5.01%
	PaDIN	0.7157	3.21%	0.7141	5.21%
	PaGRU	0.7174	4.02%	0.7185	7.37%
	SMFN	**0.7268**	**8.52%**	**0.7313**	**13.66%**

of deep networks and the importance of user's historical behavior sequences in this question. DIN with specially designed local activation unit performs better than DeepFM and Wide&Deep. But it performs worse than PaGRU, which illustrates that the architecture of GRU is more suitable than the architecture of the adopted attention mechanism for our problem. A noteworthy phenomenon is that PaGRU performs better than DeepFM, Wide&Deep and DIN, which do not consider the sequence order. It indeed reflects the importance of considering the order of clicked video sequence for CTR prediction, and also verifies that GRU is appropriate for sequence-based recommendations. By comparing the two methods of considering the association between the sequences, DAN and CoAtt, with other deep network methods, we observe no improvements, indicating that methods perform well in other fields might not necessarily solve our problem well. SMFN stands out significantly among all the competitors, which answers the question Q1. This proves that SMFN is not only better than other models in the same field, but also better than similar methods in other fields. We attribute the success to the specially designed fusion unit, which integrates the representation of one sequence with the overall representation of another sequence learned by GRU. This is a property that no other methods have.

6.2 Combinatorial Analysis of Two-Channels

In order to test the effect of different sequence processing methods, we consider the following three cases:

- Case 1: Only using the user clicked videos belonging to the target channel as one sequence.
- Case 2: Merging user clicked videos in the target channel and other channels as a whole sequence.
- Case 3: Treating user clicked videos in the target channel and other channels as two independent sequences.

We choose three representative methods, i.e., DeepFM, DIN and GRU, to test their performances in these 3 cases. In case 1 and case 2, GRU means modeling a

Table 4. Ablation study of SMFN. The metric AUC is used for evaluation.

Methods	One Month	Three Months
SMFN w/o fusion unit	0.7191	0.7188
SMFN	**0.7268**	**0.7313**

single sequence using a GRU network. And in case 3, since the input to DIN is a single sequence, we develop PaDIN (Paired DIN), a modification of DIN, so as to accept two sequences as input. The results are shown in Table 3. First, we focus on case 1 and case 2. Obviously, both of DeepFM and DIN in case 2 are much better than those in case 1. It shows that users' behaviors in other channels are indeed beneficial for CTR prediction in the target channel. But GRU in case 2 is not much better than that in case 1 for the One Month dataset, and even slightly worse than that in case 1 with Three Months dataset. It might be attributed to the fact that GRU is not very suitable for modeling sequences containing multiple categories.

We further compare their performance in case 2 and case 3. In both cases, the results of DeepFM show no obvious difference. It might be explained by the fact that the special architecture of DeepFM makes it less sensitive to the input data form. As the best method in case 2, DIN performs even slightly better than PaDIN, showing that it is hard for DIN to differentiate the role of different sequences for prediction. Moreover, GRU is better than DIN in case 1, but it worse than DIN in case 2, and PaGRU is better than PaDIN again in case 3. The reason might be that GRU is better for processing sequences that only contain the same category. In contrast, DIN is good at extracting useful information from sequences containing multiple categories. Finally, SMFN is significantly better than all of these methods in all of the three cases. It demonstrates that using a well-designed method can be better in case 3 than in case 2. Based on the above illustration, we can answer the first part of question Q2 that differentiating users' past behaviors based on target channel and context channel is indeed meaningful. Besides, compared with the alternative multi-channel methods, SMFN does not incur much additional computational cost since the fusion unit is simple.

6.3   Model Analysis

Ablation Study. To validate the effectiveness of the fusion part in SMFN, we conduct ablation experiments and show the results in Table 4. "SMFN w/o Fusion Unit" denotes the method of removing the fusion unit from SMFN, leaving GRU networks and sum pooling operation to model the two sequences. By comparing with this method, it is very obvious that the fusion unit plays a crucial role in SMFN. Based on the results, we can see the contribution of the fusion unit and can answer the second part of question Q2.

Influence of Sequence Length. We visualize the performance variations with the increase of the modeled sequence length in Fig. 3. Two representative methods,

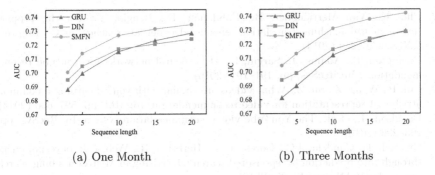

(a) One Month   (b) Three Months

Fig. 3. Results for different sequence lengths. Noting that both channels have the same sequence length.

GRU and DIN, are also tested for comparison. As expected, the performance becomes better with larger sequence length, and the variation trends turn to be stable. Regardless of the sequence length, SMFN outperforms others consistently.

7   Conclusion

In this paper, we have studied the problem of multi-channel video CTR prediction. By innovating the manner to differentiate the role of video sequences from different channels, we propose a novel Sequential Multi-Fusion Network (SMFN). The major novelty lies in the deep fusion of two sequences through a simple but effective mechanism. We conduct comprehensive experiments on several datasets collected from a real-world commercial video platform, demonstrating SMFN outperforms several strong competitors, and validating the fusion unit is beneficial for improving the CTR prediction performance.

References

1. Agarwal, D., Chen, B., Elango, P.: Spatio-temporal models for estimating click-through rate. In: WWW, pp. 21–30 (2009)
2. Bahdanau, D., Cho, K., Bengio, Y.: Neural machine translation by jointly learning to align and translate. In: ICLR (2015)
3. Chen, J., Sun, B., Li, H., Lu, H., Hua, X.: Deep CTR prediction in display advertising. In: MM, pp. 811–820 (2016)
4. Chen, X., Liu, D., Zha, Z., Zhou, W., Xiong, Z., Li, Y.: Temporal hierarchical attention at category- and item-level for micro-video click-through prediction. In: MM, pp. 1146–1153 (2018)
5. Chen, Z., Xu, K., Zhang, W.: Content-based video relevance prediction with multi-view multi-level deep interest network. In: MM, pp. 2607–2611 (2019)
6. Cheng, H., et al.: Wide & deep learning for recommender systems. In: DLRS, pp. 7–10 (2016)

7. Cho, K., Van Merriënboer, B., Bahdanau, D., Bengio, Y.: On the properties of neural machine translation: encoder-decoder approaches. arXiv preprint arXiv:1409.1259 (2014)
8. Covington, P., Adams, J., Sargin, E.: Deep neural networks for Youtube recommendations. In: RecSys, pp. 191–198 (2016)
9. Cui, P., Wang, Z., Su, Z.: What videos are similar with you?: Learning a common attributed representation for video recommendation. In: MM, pp. 597–606 (2014)
10. Davidson, J., et al.: The YouTube video recommendation system. In: RecSys, pp. 293–296 (2010)
11. Graepel, T., Candela, J.Q., Borchert, T., Herbrich, R.: Web-scale bayesian click-through rate prediction for sponsored search advertising in Microsoft's bing search engine. In: ICML, pp. 13–20 (2010)
12. Guo, H., Tang, R., Ye, Y., Li, Z., He, X.: DeepFM: a factorization-machine based neural network for CTR prediction. In: IJCAI, pp. 1725–1731 (2017)
13. He, X., et al.: Practical lessons from predicting clicks on ads at Facebook. In: DMOA, pp. 1–9 (2014)
14. Hidasi, B., Karatzoglou, A., Baltrunas, L., Tikk, D.: Session-based recommendations with recurrent neural networks. In: ICLR (2016)
15. Hochreiter, S., Schmidhuber, J.: Long short-term memory. Neural Comput. 9(8), 1735–1780 (1997)
16. Krizhevsky, A., Sutskever, I., Hinton, G.E.: ImageNet classification with deep convolutional neural networks. In: NIPS, pp. 1097–1105 (2012)
17. Li, H., Zhu, H., Ge, Y., Fu, Y., Ge, Y.: Personalized TV recommendation with mixture probabilistic matrix factorization. In: SDM, pp. 352–360 (2015)
18. Lu, J., Yang, J., Batra, D., Parikh, D.: Hierarchical question-image co-attention for visual question answering. In: NIPS, pp. 289–297 (2016)
19. McMahan, H.B., et al.: Ad click prediction: a view from the trenches. In: SIGKDD, pp. 1222–1230 (2013)
20. Mikolov, T., Karafiát, M., Burget, L., Černocký, J., Khudanpur, S.: Recurrent neural network based language model. In: ISCA (2010)
21. Nam, H., Ha, J.W., Kim, J.: Dual attention networks for multimodal reasoning and matching. In: CVPR, pp. 299–307 (2017)
22. Rendle, S.: Factorization machines. In: ICDM, pp. 995–1000 (2010)
23. Richardson, M., Dominowska, E., Ragno, R.: Predicting clicks: estimating the click-through rate for new ads. In: WWW, pp. 521–530 (2007)
24. Salakhutdinov, R., Mnih, A.: Probabilistic matrix factorization. In: NIP, pp. 1257–1264 (2007)
25. Xu, X., Chen, L., Zu, S., Zhou, H.: Hulu video recommendation: from relevance to reasoning. In: RecSys, p. 482 (2018)
26. Yan, L., Li, W.J., Xue, G.R., Han, D.: Coupled group lasso for web-scale CTR prediction in display advertising. In: ICML, pp. 802–810 (2014)
27. Yang, C., Yan, H., Yu, D., Li, Y., Chiu, D.M.: Multi-site user behavior modeling and its application in video recommendation. In: SIGIR, pp. 175–184 (2017)
28. Zhai, S., Chang, K., Zhang, R., Zhang, Z.M.: Deepintent: learning attentions for online advertising with recurrent neural networks. In: SIGKDD, pp. 1295–1304 (2016)
29. Zhang, Y., et al.: Sequential click prediction for sponsored search with recurrent neural networks. In: AAAI (2014)
30. Zhou, G., et al.: Deep interest network for click-through rate prediction. In: SIGKDD, pp. 1059–1068 (2018)

Finding Attribute Diversified Communities in Complex Networks

Afzal Azeem Chowdhary$^{(\boxtimes)}$ (iD), Chengfei Liu$^{(\boxtimes)}$, Lu Chen$^{(\boxtimes)}$, Rui Zhou$^{(\boxtimes)}$, and Yun Yang

Department of Computer Science and Software Engineering,
Faculty of Science Engineering and Technology,
Swinburne University of Technology, Melbourne, Australia
{achowdhary,cliu,luchen,rzhou,yyang}@swin.edu.au

Abstract. Recently, finding communities by considering both structure cohesiveness and attribute cohesiveness has begun to generate considerable interest. However, existing works only consider attribute cohesiveness from the perspective of attribute similarity. No work has considered finding communities with attribute diversity, which has good use in many applications. In this paper, we study the problem of searching *attribute diversified communities* in complex networks. We propose a model for attribute diversified communities and investigate the problem of *attribute diversified community search* based on k-core. We first prove the NP-hardness of the problem, and then propose efficient branch and bound algorithms with novel effective bounds. The experiments performed on various complex network datasets demonstrate the efficiency and effectiveness of our algorithms for finding attribute diversified communities and entitle the significance of our study.

1 Introduction

In the current environment, complex networks such as *Facebook, Twitter, IMDb, DBLP, etc..* have garnered widespread attention by acting as a bridge between physical entities and virtual web space. These new interactions lead to a vast amount of big and rich data. Making sense of such big and rich data finds many applications, including science, engineering, environment, public health, business, etc. One of the most interesting and challenging problems in complex network data analysis is to find different kinds of communities of users with close structural connections with each other for different purposes.

A network can be modelled as a graph that consists of a set of nodes representing the individuals/entities, and a set of the edges corresponding to the interactions among them. Then a community can be defined as an induced subgraph of the graph, where the set of nodes inside the subgraph are densely interconnected. While community search has attracted widespread studies, the majority of research in this area has focused on pure structure-based community search without considering attributes, e.g., k-core [2,20], k-truss [9], k-edge-connected graph [5,6,27] and k-vertex-connected graph [22].

© Springer Nature Switzerland AG 2020
Y. Nah et al. (Eds.): DASFAA 2020, LNCS 12114, pp. 19–35, 2020.
https://doi.org/10.1007/978-3-030-59419-0_2

Recently, there has been a significant interest in finding communities which are both structurally cohesive and attribute cohesive. It is motivated by the need to make the resultant communities more meaningful, i.e., the communities shall be not only structurally dense but also attribute cohesive with regard to user-provided queries. In [26], Zhang et al. find (k,r)-core community such that socially the vertices in (k,r)-core is a k-core and from similarity perspective pairwise vertex similarity is more than a threshold r. Chen et al. [8] finds co-located communities by taking distance similarity into account based on k-truss as the cohesiveness measure. Li et al. [16] propose a skyline community model for searching communities in attributed graphs.

However, all these works only consider attribute cohesiveness from the perspective of attribute similarity. To the best of our knowledge, no work has considered finding communities with diversified attributes, i.e., diversity with regard to attributes. We call this kind of communities *attribute diversified communities*.

By an attribute diversified community, we mean that the connected nodes inside the community exhibit difference in terms of interested attributes. Actually, searching attribute diversified communities has many applications. For example, gathering socially connected experts of different marketing fields to brainstorm a marketing session of different new products, selecting a panel of concerted engineers with different technological expertise for reviewing and testing different products to show the collective information pool of the panel, etc. Finding these attribute diversified communities allows individuals inside a community to express different opinions in order to achieve an innovative outcome.

To model such an attribute diversified community, we need to explore the attributes of users to derive the associated diversity values. These values will further express the attribute diversity of the community. Simultaneously, we also need to maintain the structure cohesiveness of the community, e.g., each user is connected to at least k other users, which implicitly incorporate a large number of diversified relationships. By doing so, we can end up with a community that is both structurally cohesive and attribute diversified.

We look at the following motivating example to see how the attributes of connected users will play an important role in determining the diversity of the community.

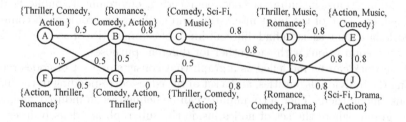

Fig. 1. Motivating example

Figure 1 shows an IMDb graph where each node represents a person who belongs to Hollywood and each relationship represents the work the pair of persons have collaborated on. Each person has a set of keywords to describe different movie genres he/she has worked in. Now, let's consider a scenario where we want to organize a film festival which needs a panel of acquainted jurors to judge the wide variety of movies to be premiered. The acquaintance of jurors is to ensure the decision making process smooth because conflicts often exist implicitly in unestablished relations. Consequently, we need to find a community whose diversity is maximized along with it being structurally cohesive. The attribute diversified community model can be used to find a set of connected jurors who have diverse attributes to express the vast expanse of the industry. From Fig. 1, we see that nodes such as {A,F,G,H} are not quite diverse enough w.r.t. each other in terms of attributes. Here, each person is connected to at least 2 other persons which makes the graph structurally cohesive.

In this scenario, we are keen to find a community where each person inside this community would contribute significantly to both attribute diversity and structural cohesiveness of the community. We find that the community with maximum diversity is the subgraph induced by {B,C,D,E,I,J}, and is of both high attribute diversity and structural cohesiveness. Even though the attribute diversity between {B,I} is moderate, they still contribute to the community to increase its overall attribute diversity. While nodes {A,F,G} are excluded as they contribute less to attribute diversity. Node {H} gets deleted by not satisfying the degree constraint.

This paper is devoted to the study on how to model and search attribute diversified communities. In particular, we investigate the problem with a purpose of diversification as illustrated above. To do this, we first define a diversity measure to capture difference between users of a community through their relationship. For this purpose of diversification, we define avgDiv diversity measure, where avgDiv is designed for finding a community by maximizing the overall average attribute diversity of the community. To guarantee the structure cohesiveness of a community, we use k-core to ensure that each user has at least k neighbours in the community. Based on the diversity measure and the k-core structure cohesiveness constraint, we model an attribute diversified community based on avgDiv diversity measure problem. In this paper, we develop efficient algorithms to solve the avgDiv diversity measure problem and conduct extensive experiments to evaluate the performance of the proposed algorithms.

The main contributions of this work are summarised as follows:

- To the best of our knowledge, this is the first work to study the problem of finding attribute diversified communities in a complex network, which is significant and complementary research to this field. We propose a model for attribute diversified communities and investigate the *attribute diversified community search* problem for finding attribute diversified communities.
- For the attribute diversified community search problem, we first prove the NP-hardness of the problem, and then propose efficient branch and bound algorithms with novel effective bounds.

– We conduct extensive experiments on real datasets, and the results demonstrate the efficiency and effectiveness of the proposed algorithms for the problem.

The remainder of this paper is organised as follows. We first review the related work in Sect. 2 and then formally define the attribute diversified community model and present the attribute diversified community search problem in Sect. 3. Section 4 first proves the hardness of the problem, and then presents a baseline approach and some optimisation strategies to the problem. We evaluate the performance of our proposed methods in Sect. 5 and finally conclude the paper in Sect. 6.

2 Related Work

Community Search in Attributed Graph. In [16], Li et al. propose a skyline community model for searching communities in attributed graph. Zhang et al. propose (k, r)-core community model that considers k-core and pairwise vertex similarity [26]. Fang et al., propose a community model that is sensitive to query attributes. In [14], an attributed community model is proposed by using k-truss for capturing social cohesiveness and the resultant community shall contain attributes similar with query attributes. In [11,21] community models considering spatial closeness are studied. The works above would find communities with users having similar attributes while our work find communities with users having diversified attributes. In [7], a parameter-free contextual community model is studied. Community models considering influence are studied in [3,15,17]. In [3,17], they use max-min objective function. They aim to find influential communities where scores are defined on vertices. In our work, we study an attribute diversified community problem and scores are defined on relationships. Additionally, we explore diversifying community users by using avgDiv objective function.

Community Detection in Attributed Graph. Works including [18] consider graph structure with LDA model to detect attributed communities. Unified distance [28] is also considered for detecting attributed communities. In [28], attributed communities are detected by using proposed structural/attribute clustering methods, in which structural distance is unified by attribute weighted edges. Xu et al. [24] propose a Bayesian based model. Ruan et al. [19] propose an attributed community detection method that links edges and content, filters some edges that are loosely connected from content perspective, and partitions the remaining graphs into attributed communities. In [13], Huang et al. propose a community model considering attributes based on an entropy-based model. Recently, Wu et al. propose an attributed community model [23] based on an attributed refined fitness model. Yang et al. [25] propose a model using probabilistic generative model. Compared to our work, these works do not have explicit search specification and they focus on finding communities with users having similar attributes whereas we explore how to find communities with diversified users.

Table 1. Notations

Notation	Definition
G	An attributed graph
H, H'	A subgraph of G
k	The coreness of a subgraph
n	The total number of vertices in G
m	The total number of edges in G
$E(H)$	The edges in H
$V(H)$	The vertices in H
u, v	The vertices in the attributed graph
W_u	The attributes of u in G
(u, v)	An edge in the attributed graph
$div((u, v))$	The diversity of edge (u, v)
$deg(u, H)$	The degree of u in H
$N(u, H)$	The neighbour of u in H
$avgDiv(H)$	The average edge diversity of H
H^*	The optimum result for an algorithm
h	A connected k-core

3   Problem Formulation

In this section, we discuss some preliminaries and then formulate the problem. The notations frequently used in this paper are summarised in Table 1.

3.1   Preliminaries

Attributed Graph. An attributed graph $G = (V, E, W)$ contains a set of vertices $V(G)$, edges $E(G)$ and attributes W. Each vertex $v \in V(G)$ is attached with a set of attributes $W_v \subseteq W$. Given $v \in V(G)$, $deg(v, G)$ denotes the degree of v in G and $N(v, G)$ denotes the neighbours of v in G.

3.2   Problem Formulation

We model a complex network as an attributed graph $G = (V, E, W)$.

We consider attribute diversification as well as structure cohesiveness over users simultaneously to model a community. Before proposing the community formally, we discuss and define structure cohesiveness and attribute diversity as follows.

Structure Cohesiveness. We use k minimum acquaintance to measure structure cohesiveness. Given a subgraph H, H is considered as structure cohesive if it is a k-core subgraph defined as follows.

Definition 1 k-core subgraph. *Given a subgraph $H \subseteq G$, an integer k, H is called k-core subgraph if for every $v \in V(H)$, $deg(v, H) \geq k$ and such maximum k is called the coreness of H.*

Intuitively, a k-core is a subgraph in which vertex has at least k neighbours. A k-core with a large value k indicates strong internal connections over users. A k-core is maximal if it cannot be extended.

Subgraph Diversity. We explore the diversity of a subgraph by considering the attribute diversity of an edge first.

Definition 2 Edge diversity. *Given an edge, $(u, v) \in E(G)$ with attributes W_u and W_v, we propose an edge diversity function $div(e)$ based on Jaccard similarity as follows,*

$$div((u, v)) = 1 - \frac{|W_u \cap W_v|}{|W_u \cup W_v|}.$$

In **Definition** 2, $\frac{|W_u \cap W_v|}{|W_u \cup W_v|}$ is Jaccard similarity. The idea of transformation from similarity measure to diversity measure using $1 -$ similarity has been widely accepted such as [1]. Please note that we can use different distance metrics such as *cosine distance, edit distance, Pearson coefficient, etc.* to define edge diversity. The edge diversity values can be pre-computed.

Next we propose our diversity function to measure the overall attribute diversity for a subgraph.

Average Attribute Diversity. We measure the overall diversity of a subgraph by using the average edge diversity. It is denoted as $avgDiv$ and is defined formally as follows.

Definition 3 avgDiv. *Given a subgraph $H \subseteq G$, we measure its diversity as follows,*

$$avgDiv(H) = \frac{\sum_{(u,v) \in H} div((u, v))}{|V(H)|}.$$

Note that, although we only define diversities on directly connected vertices. Actually, our solutions also work if diversity is generalized on any pair of vertices.

Attribute Diversified Community. Now we are ready to propose the attribute diversified community model.

Definition 4 Attribute diversified community. *Given a subgraph $H \subseteq G$, an integer k, H is defined as an attribute diversified community if H satisfies the following constraints simultaneously:*

- *Connectivity: H is connected;*
- *Structure Cohesiveness: H is a k-core subgraph;*
- *Maximizing Average Edge Diversity: For $avgDiv(H)$, H is $argmax_{H'}$ $\{avgDiv(H')|H' \subseteq G\}$;*

Accordingly, given G and an integer k, the research problem we focus on in this paper is as follows.

Attribute Diversified Community Search Problem. Find the subgraph $H \subseteq G$ that maximises $avgDiv(H)$.

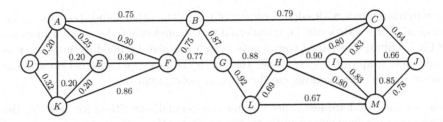

Fig. 2. Graph with edge diversity

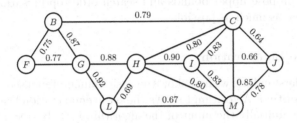

Fig. 3. Result for $k = 2$

To briefly show the results of the above problem we will discuss the example shown in Fig. 2. For the attribute diversified community search problem with $k = 2$, the result is the subgraph {B, C, F, G, H, I, J, L, M} with diversity of 1.44 as shown in Fig. 3.

4   Attribute Diversified Community Search Problem

Before showing efficient algorithms for solving the problem of finding attribute diversified community maximising average edge diversity, i.e., the attribute diversified community search problem, we first study the hardness of the problem.

Lemma 1. *The attribute diversified community search problem is NP-hard.*

Proof. We reduce the well known NP-hard problem, densest at least size ρ problem, denoted as *dalsρ*, to an instance of the attribute diversified community search problem. Given a graph G, a size constraint ρ, *dalsρ* problem finds the subgraph H with the highest density, defined as $\frac{|E(H)|}{|V(H)|}$, among all subgraphs that have sizes no less than ρ. Given any instance of *dalsρ* problem with G and ρ, we can construct an instance of attribute diversified community search problem with k and G' as follows. k is set to be $\rho - 1$. G' is created based on G together with its attributes using the similar approach in [10]. This makes G' become a complete graph with $V(G') = V(G)$. Besides, the edge diversity of $e \in E(G')$ is 1 if $e \in E(G)$, otherwise 0. With such a construction, every subgraph in G' with size greater than k is a k-core. The answer for the attribute diversified community search problem in G' if it exists can be used to derive the answer for *dalsρ* by removing edges with edge diversity of 0. On the other hand, the answer for $dals\rho$ in G if it exists can be transferred to the answer for the created instance of the attribute diversified community search problem by making the answer as a complete graph. It is easy to see that both the instance construction and the answer equivalence checking can be done in polynomial time.

Due to the NP-hardness, there is no polynomial algorithm for solving the attribute diversified community search problem. We will propose branch and bound algorithm for solving attribute diversified community search problem, in which we will propose upper bounds and search order optimisations to reduce the search space as much as possible.

4.1 Baseline Enumeration

For ease of understanding, we first show the basic enumeration used in the branch and bound algorithm. Algorithm 1 shows the basic enumeration that derives the optimum result. Initially the input of the algorithm is G. By recursively calling itself, Algorithm 1 tries all possible subgraphs of G if the subgraphs may contain the optimum result (lines 12 to 16) and checks if there is a feasible solution in the current recursion (lines 8 to 10). If there is a feasible solution h in the recursion and the feasible solution is greater than the current optimum one H^*, H^* is updated to h (line 9).

Algorithm 1: basicADC(G)

1 $H^* \leftarrow \phi$;
2 basicEnum (G);
3 return H^*;
4 **Procedure** basicEnum (H)
5 $\quad$ $H' \leftarrow k\text{-core(H)}$;
6 $\quad$ let $\mathcal{H}'$ be the set of connected component in H';
7 $\quad$ **foreach** $h \in \mathcal{H}'$ **do**
8 $\quad\quad$ **if** $avgDiv(h) > avgDiv(H^*)$ **then**
9 $\quad\quad\quad$ $H^* \leftarrow h$;
10 $\quad\quad$ **end**
11 $\quad$ **end**
12 $\quad$ **foreach** $h \in \mathcal{H}'$ **do**
13 $\quad\quad$ **foreach** $v \in V(h)$ **do**
14 $\quad\quad\quad$ basicEnum $(h \setminus \{v\})$;
15 $\quad\quad$ **end**
16 $\quad$ **end**

Search Space Reduction. Algorithm 1 also applies space reduction optimisations based on the observations as follows.

Observation 1. *The optimum result can only be contained in a connected k-core of G if it exists when the enumeration starts.*

Observation 2. *During the recursion with an input H, the optimum result can only be contained in a connected k-core of H.*

With the observations, when a recursion starts, Algorithm 1 first reduces the input to the maximal k-core, which would make the input become a set of maximal connected k-cores. Algorithm 1 only tries combinations in each connected k-core. As such, the search space can be reduced significantly.

Running Example. Consider the graph in Fig. 4(a). To find the result for $k = 2$, Algorithm 1 enumerates all the possible outcomes in the search space to get the subgraph with the maximum avgDiv score. In this case, the subgraph induced by $\{A, B, C, D, E\}$ with avgDiv score of 1.304 is the maximum attribute diversified community as shown in Fig. 4(b).

In Algorithm 1, the time complexity for each recursion is $O(m)$, i.e, finding connected components takes $O(m)$ time, while the overall time complexity of Algorithm 1 is exponential. Thus, it is infeasible to use this approach to find attribute diversified communities for large scale complex networks.

The correctness of Algorithm 1 is clear. This is because all combinations of vertices that potentially lead to the optimum result are explored by the algorithm. However, it cannot scale to even medium-size datasets. In the following sub-section, we will propose upper bounds and study upper bound based prunings to improve the search performance. To make the prunings more effective, we also propose heuristic rules to find subgraphs with large diversities as early as possible. Those optimisations improve the performance substantially.

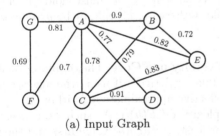   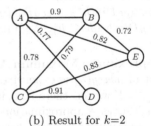

      (a) Input Graph   (b) Result for $k=2$

Fig. 4. Running example with result

4.2 Optimisations

Upper Bound Based Pruning. The idea is that we estimate the upper bound of the average edge diversity of the current search branch. If the upper bound is smaller than the diversity of the optimum result found so far, we terminate the search branch.

Next, we will propose three upper bounds.

Upper Bound Based on Core Property. We firstly show an upper bound for a connected k-core based on core property. The upper bound for h is defined as follows.

$$ubcore(h) = \frac{\sum_{(u,v)\in E(h)} div((u,v))}{k+1} \tag{1}$$

The correctness of the upper bound is clear. Given a connected k-core h, there will be no subgraph of h with aggregated edge diversity greater than the numerator of $ubcore(h)$. At the same time, there are no feasible answers contained in h with size less than $k+1$ since a feasible answer must be a k-core graph and by definition a k-core shall contain as least $k+1$ number of vertices. As such $ubcore(h)$ is clearly an upper bound, i.e., the best result contained in h has average edge density no greater than $ubcore(h)$.

The upper bound based on core property would only be tight when h contains an optimum result with size close to $k+1$. However, it has limited pruning effectiveness when h contains large-size results. Next we study tight bounds for arbitrary h.

Maximum Average Diversity in a Core. Given a connected k-core h, this bound is defined as follows.

$$ubavg(h) = max\{avgDiv(h')|h' \subseteq h\} \tag{2}$$

Lemma 2. *ubavg(h) is an upper bound for h.*

Proof. Let h^* be the attribute diversified community contained in h and h' be $argmax_{h'}\{avgDiv(h')|h' \subseteq h\}$. By definition, since h^* must be a k-core while h' relaxes the k-core constraint. As such $h^* \subseteq h'$ must hold. Therefore, $avgDiv(h^*) \leq avgDiv(h')$ must hold. Otherwise, h' will not be $argmax_{h'}$ $\{avgDiv(h')|h' \subseteq h\}$. We finish the proof.

Approximate Maximum Average Diversity in a Core. The computational cost of $ubavg(h)$ is expensive. It would take $O(|V(h)|^3)$ if using the algorithm in [12]. However, there is a simple but effective approximate algorithm [4] that can achieve $\frac{1}{2}$-approximation with complexity $O(|E(h)|)$. As such we can use the approximation algorithm to get an at least $\frac{1}{2}$ $ubavg(h)$ value first and then multiple it by 2 to derive a slightly loose bound, denoted as $apxubavg(h)$.

In implementation, $ubcore(h)$ and $apxubavg(h)$ are prioritised as they are cheap.

Search Order. For each connected k-cores that cannot be pruned, we sort them in non-increasing order according to their upper bounds. By doing this, we can heuristically find communities with large average diversity as early as possible. This would make the upper bound based pruning more effective.

The Advanced Algorithm. Algorithm 2 shows the algorithm with the discussed optimisations. Different from Algorithm 1, it has an extra loop (lines 8

Algorithm 2: advADC(G)

---

1 $H^* \leftarrow \phi$;
2 advEnum (G);
3 **return** H^*;
4 **Procedure** advEnum (H)
5      $H' \leftarrow k\text{-core}(H)$;
6   let $\mathcal{H}'$ be the set of connected component in H';
7   **foreach** $h \in \mathcal{H}'$ **do**
8   **if** H^* *not empty* **then**
9 | calculate necessary upper bounds of h and prune h based on H^* ;
10   **end**
11   **if** h *is pruned* **then**
12 | *continue*;
13   **end**
14   **if** $avgDiv(h) > avgDiv(H^*)$ **then**
15 | $H^* \leftarrow h$;
16   **end**
17   **end**
18   sort h in the remaining $\mathcal{H}'$;
19   **foreach** $h \in \mathcal{H}'$ **do**
20   **foreach** $v \in V(h)$ **do**
21 | advEnum $(h \setminus \{v\})$;
22   **end**
23   **end**

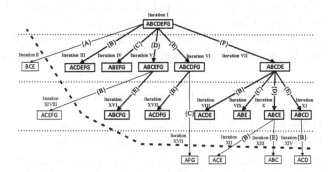

Fig. 5. Search space explored

to 13) to efficiently prune connected k-cores with least cost; after that, search order optimisations is applied (line 18).

The correctness of Algorithm 2 is immediate given the correctness of Algorithm 1 and the correctness of upper bounds that we have discussed.

Running Example. Consider the previous example in Fig. 4(a). The whole search space is shown in Fig. 5 for $k = 2$. The *normal arrows* with node id's represent the order of nodes deleted during the enumeration whereas iteration id's are shown to represent the order of search space enumeration. To see the efficiency of Algorithm 2, the search space with *bold arrows* are only enumerated to find the optimum result. The search space that is below the bold dotted line is pruned out. These *normal arrow* results were previously enumerated by Algorithm 1 to find the optimum result. This clearly shows the pruning power of Algorithm 2.

Compared to Algorithm 1, the total number of recursions in Algorithm 2 is substantially reduced with the help of the optimizations that include heuristics and pruning strategies presented above, even though the time complexity for each recursion remains $O(m)$. As shown in the experiments later, Algorithm 2 is able to find results much faster than Algorithm 1 on real datasets.

5   Experiments

We evaluated the performance of the proposed algorithms on four real datasets. All the experiments were conducted on a PC with 3.9 GHz Intel Core i7 (6-core) processor, 16 GB RAM, running Windows 10.

Implemented Algorithms. We implemented all algorithms proposed in this paper. Algorithms 1 and 2 are denoted as Alg-1 and Alg-2 correspondingly in this section. All the algorithms have found the most attribute diversified community and have correspondingly reported the performance. All the algorithms were implemented in *C++* and were compiled with *MinGW Compiler 64-bit ver 6.3*.

Datasets. The four real datasets include Facebook, Brightkite, DBLP and ACM that are obtained from *snap.standford.edu* and *konect.uni-koblenz.de*. The data were cleaned by removing all self-loops. The edge diversity values are pre-computed using Jaccard similarity. Table 2 presents the statistics for all datasets.

Table 2. Real world datasets

Datasets	Vertices	Edges	d_{max}	d_{avg}	Range of k
Facebook	3k	17k	62	4.43	52, 53, 54, 55, 56
Brightkite	58k	214k	1,134	7.30	33, 36, 39, 42, 45
DBLP	317k	1M	343	6.62	25, 30, 35, 40, 45
ACM	359k	1.06M	5720	2.95	10, 12, 14, 16, 18

Parameter Settings. The experiments are evaluated using different settings of query parameters: k (the minimum core number) to detect results, i.e most attribute diversified community. The range of the parameter k is shown in the last column Table 2. All the experiments were conducted at least 5 times and their averages were taken. The experiments that ran for more than an hour are shown with INF label.

5.1   Efficiency Evaluation

Varying k. We vary the value of k for detecting results. The range of k for every dataset is chosen to select a diversified community with high cohesiveness. We report the results in Figs. 6. For all the dataset, we see that Alg-2 outperforms Alg-1 for different values of k. This is because of the pruning effectiveness over the enumeration search space is much higher for smaller values of k compared with large values of k. Because, the size of maximal k-core is greater for small values of k which makes bound checking in Alg-2 more effective than Alg-1. But for larger values of k, the bound in Alg-2 becomes tighter by following Definition 3. For DBLP and ACM dataset, Alg-2 slightly betters Alg-1 by a magnitude margin of around 20%. Although, Alg-2 performs better, the graph itself is very dense, which makes the bound checking moderately effective for those values of k.

Scalability. To see the scalability of the proposed Alg-1 and Alg-2 on varying the size of graph. We show the results for attribute diversified communities for different datasets with default k set at 55, 39, 40, 14 in order presented in Table 2. Figs. 7 shows that, Alg-2 performs better than Alg-1 in all the scenarios when varying graph sizes, with the main reason being the effective upper bound check and search order optimization for Alg-2. Also, since the density of the graph varies with different ratios, the pruning effectiveness of Alg-2 clearly outperforms Alg-1.

Pruning Effectiveness Evaluation. We also evaluate the pruning effectiveness when varying k. This is evaluated by two measurements: the number of recursions used by Alg-2 over the number of recursions used by Alg-1, denoted by Alg-2/Alg-1, and the percentage of the total search space explored by Alg-2 for finding the attribute diversified community results, denoted by Alg-2/theo. The results are shown in Figs. 8 respectively. The Alg-2/Alg-1 ratio for Facebook

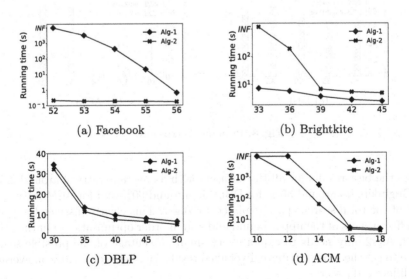

(a) Facebook

(b) Brightkite

(c) DBLP

(d) ACM

Fig. 6. Varying k

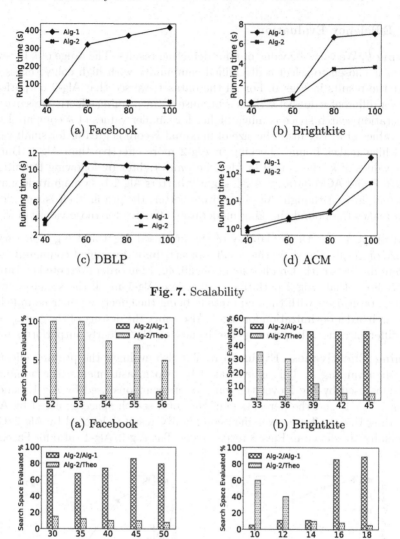

Fig. 7. Scalability

Fig. 8. Pruning effectiveness

shows the best results for all the datasets with an average ratio of around 2.5%, for Brightkite its around 35%, for DBLP its around 60% and for ACM its around 15% of the total search space explored by Alg-1. This clearly justifies the pruning effectiveness of the upper bound and search order optimisation. On the other hand, Alg-2 only needs to explore very small percentage of the possible search space to get the most attribute diversified result, i.e., between 5–20% on average, for different datasets.

5.2   Effectiveness Evaluation

We conducted a case study on IMDb to demonstrate the effectiveness of our proposed attribute diversified community model. We preprocess the IMDb dataset to show an edge between each pair of vertices i.e. persons, who have worked in no less than 5 movies. Each vertex has 3 distinct keyword genres. Each genre is given an abbreviation such as M for Music, D for Drama, B for Biography, Cr for Crime, Wr for War, C for Comedy, R for Romance, etc.

The community found by the attribute diversified community search problem contains 26 film personalities and 44 diversified relationships from the Hollywood film industry that are connected with minimum degree of at least 3. Compared to k-core, attribute diversified community model allows us to find meaningful communities whose relationships exhibit diversity in terms of interested attributes. This community expresses the wide depth across different genres and helps us to find connected personalities who have worked together more often. This diverse community will make the decision making process easy and more refined with proper discussions at a film festival as introduced in the motivating example. In Fig. 9, the result for the attribute diversified community search problem is illustrated for IMDb.

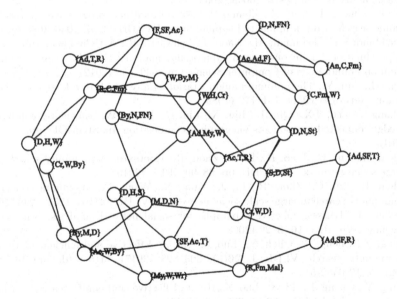

Fig. 9. Case study on IMDb for $k = 3$

6　Conclusion

We studied the problem of finding communities in complex networks by mixing diversity with cohesiveness. We propose a novel community model, attribute diversified community, considering structure cohesiveness and user diversity. We propose an approach that maximizes the average edge diversity. We prove the problem is NP-hard and propose efficient branch and bound algorithms with novel effective pruning bounds. We conduct extensive experiments on real datasets. The results demonstrate the efficiency and effectiveness of the proposed algorithms.

Acknowledgements. This work was supported by the Australian Research Council discovery projects under grant numbers DP170104747, DP180100212.

References

1. Bache, K., Newman, D., Smyth, P.: Text-based measures of document diversity. In: SIGKDD, pp. 23–31. ACM (2013)
2. Batagelj, V., Zaversnik, M.: An O(m) algorithm for cores decomposition of networks. arXiv preprint cs/0310049 (2003)
3. Bi, F., Chang, L., Lin, X., Zhang, W.: An optimal and progressive approach to online search of top-k influential communities. PVLDB **11**(9), 1056–1068 (2018)
4. Buchbinder, N., Feldman, M., Naor, J., Schwartz, R.: A tight linear time (1/2)-approximation for unconstrained submodular maximization. In: Annual Symposium on Foundations of Computer Science, pp. 649–658 (2012)
5. Cai, G., Sun, Y.: The minimum augmentation of any graph to a k edge connected graph. Networks **19**(1), 151–172 (1989)
6. Chang, L., Yu, J.X., Qin, L., Lin, X., Liu, C., Liang, W.: Efficiently computing k-edge connected components via graph decomposition. In: SIGMOD, pp. 205–216 (2013)
7. Chen, L., Liu, C., Liao, K., Li, J., Zhou, R.: Contextual community search over large social networks. In: ICDE, pp. 88–99. IEEE (2019)
8. Chen, L., Liu, C., Zhou, R., Li, J., Yang, X., Wang, B.: Maximum co-located community search in large scale social networks. PVLDB **11**(10), 1233–1246 (2018)
9. Cohen, J.: Trusses: cohesive subgraphs for social network analysis. Nat. Secur. Agency Tech. Rep. **16**, 3–29 (2008)
10. Fang, Y., Cheng, R., Chen, Y., Luo, S., Hu, J.: Effective and efficient attributed community search. VLDB J. **26**(6), 803–828 (2017). https://doi.org/10.1007/s00778-017-0482-5
11. Fang, Y., Cheng, R., Li, X., Luo, S., Hu, J.: Effective community search over large spatial graphs. PVLDB **10**(6), 709–720 (2017)
12. Gallo, G., Grigoriadis, M.D., Tarjan, R.E.: A fast parametric maximum flow algorithm and applications. SIAM J. Comput. **18**(1), 30–55 (1989)
13. Huang, X., Cheng, H., Yu, J.X.: Dense community detection in multi-valued attributed networks. Inf. Sci. **314**(C), 77–99 (2015)
14. Huang, X., Lakshmanan, L.V.: Attribute-driven community search. PVLDB **10**(9), 949–960 (2017)
15. Li, J., Wang, X., Deng, K., Yang, X., Sellis, T., Yu, J.X.: Most influential community search over large social networks. In: ICDE, pp. 871–882. IEEE (2017)

16. Li, R.H., et al.: Skyline community search in multi-valued networks. In: SIGMOD, pp. 457–472. ACM (2018)
17. Li, R.H., Qin, L., Yu, J.X., Mao, R.: Influential community search in large networks. PVLDB 8(5), 509–520 (2015)
18. Nallapati, R.M., Ahmed, A., Xing, E.P., Cohen, W.W.: Joint latent topic models for text and citations. In: SIGKDD, pp. 542–550. ACM (2008)
19. Ruan, Y., Fuhry, D., Parthasarathy, S.: Efficient community detection in large networks using content and links. In: WWW, pp. 1089–1098. ACM (2013)
20. Seidman, S.B.: Network structure and minimum degree. Soc. Netw. 5(3), 269–287 (1983)
21. Wang, K., Cao, X., Lin, X., Zhang, W., Qin, L.: Efficient computing of radius-bounded k-cores. In: ICDE, pp. 233–244. IEEE (2018)
22. Wen, D., Qin, L., Zhang, Y., Chang, L., Chen, L.: Enumerating k-vertex connected components in large graphs. In: ICDE, pp. 52–63. IEEE (2019)
23. Wu, P., Pan, L.: Mining application-aware community organization with expanded feature subspaces from concerned attributes in social networks. Knowl.-Based Syst. 139, 1–12 (2018)
24. Xu, Z., Ke, Y., Wang, Y., Cheng, H., Cheng, J.: A model-based approach to attributed graph clustering. In: SIGMOD, pp. 505–516. ACM (2012)
25. Yang, J., McAuley, J., Leskovec, J.: Community detection in networks with node attributes. In: ICDM, pp. 1151–1156. IEEE (2013)
26. Zhang, F., Zhang, Y., Qin, L., Zhang, W., Lin, X.: When engagement meets similarity: efficient (k, r)-core computation on social networks. PVLDB 10(10), 998–1009 (2017)
27. Zhou, R., Liu, C., Yu, J.X., Liang, W., Chen, B., Li, J.: Finding maximal k-edge-connected subgraphs from a large graph. In: EDBT, pp. 480–491. ACM (2012)
28. Zhou, Y., Cheng, H., Yu, J.X.: Graph clustering based on structural/attribute similarities. PVLDB 2(1), 718–729 (2009)

Business Location Selection Based on Geo-Social Networks

Qian Zeng[1], Ming Zhong[1(✉)], Yuanyuan Zhu[1], and Jianxin Li[2]

[1] School of Computer Science, Wuhan University, Wuhan 430072, China
{wennie,clock,yyzhu}@whu.edu.cn
[2] School of Information Technology, Deakin University, Melbourne, Australia
jianxin.li@deakin.edu.au

Abstract. Location has a great impact on the success of many businesses. The existing works typically utilize the number of customers who are the Reverse Nearest Neighbors (RNN) of a business location to assess its goodness. While, with the prevalence of word-of-mouth marketing in social networks, a business can now exploit the social influence to attract enormous customers to visit it, even though it is not located in the popular but unaffordable business districts with the most RNNs.

In this paper, we propose a novel Business Location Selection (BLS) approach to integrate the factors of both social influence and geographical distance. Firstly, we formally define a BLS model based on relative distance aware influence maximization in geo-social networks, where the goodness of a location is assessed by the maximum number of social network users it can influence via online propagation. To the best of our knowledge, it is the first BLS model that adopts the influence maximization techniques. Then, to speed up the selection, we present two sophisticated candidate location pruning strategies, and extend the Reverse Influence Sampling (RIS) algorithm to select seeds for multiple locations, thereby avoiding redundant computation. Lastly, we demonstrate the effectiveness and efficiency of our approach by conducting the experiments on three real geo-social networks.

Keywords: Business location planning · Influence maximization · Geo-social networks

1 Introduction

Motivation. As indicated by the classic advice "location, location, and location", location can mean the difference between feast or famine for many businesses. Naturally, location selection is the most important one among many issues to consider when looking for a place to establish a new business. Traditionally, the business owners pick their locations mainly based on how many people or potential customers will pass by the place periodically. Therefore, the most ideal locations are always distributed in the popular business districts, like a big shopping mall in a neighborhood with dense population. However,

Y. Nah et al. (Eds.): DASFAA 2020, LNCS 12114, pp. 36–52, 2020.
https://doi.org/10.1007/978-3-030-59419-0_3

the rental of the stores located in such places are very expensive accordingly. It would make the business owners difficult to make satisfied profit.

With the prevalence of online social networks, many businesses start to exploit the word-of-mouth marketing to attract customers via social network platforms. It brings forth the so called "cybercelebrity" (also known widely as "wanghong" in China) shops. With a certain budget, the cybercelebrity shops leverage the users with huge influence in social networks to promote their businesses, and the influence will spread virally among the other users. For example, some famous users with a large number of fans in Tik Tok, the most popular social media platform in China, will release short videos of dining in distinctive but not well-known restaurants to advertise for the restaurants. This kind of advertisements are not costly but could influence a lot of users of social networks.

Interestingly, it is observed that the people influenced by the word-of-mouth effect from social networks are more than willing to experience the products or services (this kind of experience even has a specific name "daka" in China) offered by the cybercelebrity shops no matter where they are. For example, drinking milk tea has become a thing in recent years. Tik Tok has contributed the latest cybercelebrity milk tea franchises like Yidiandian and CoCo in China. Even though some their shops are not located in the most popular business districts, there are always long lines during the peak hours. Obviously, such cybercelebrity shops can gain much larger profit margins since they do not need to pay expensive rentals.

Goal. In this paper, we restudy the long-standing business location selection problem based on geo-social networks. Our goal is to select a location for establishing a new cybercelebrity shop, which enables the shop to attract the maximum number of social network users. To the best of our knowledge, this is the first study to leverage influence maximization in geo-social networks to select a location with the maximum influence, in contrast to many previous works [12–14] that tried to maximize the influence of a given query location.

Related Works. Typically, given a set of customers V, a set of candidate locations C, and a set of locations of existing competitive businesses ES, the business location selection problem is to find the candidate location with the maximum influence. In the traditional approaches (e.g., [2–4]), the candidate location's relative geographical proximity to the customers is considered as a primary metric, and therefore, the problem can be addressed by using Reverse Nearest Neighbor (RNN) [1] query to find a candidate location with the most RNNs, namely, the goodness of a candidate location is the number of its RNNs. Formally, the RNN of a candidate location c is $RNN_c = \{v \in V | \forall es \in ES, dis(v, c) \leq dis(v, es)\}$. However, without considering the influence propagation in social networks, the models used by traditional approaches will be out of time, since the word-of-mouth marketing in social networks has been adopted by almost every business.

A latest research work [5] takes both geographical proximity and social influence into account to address the business location selection problem. It assumes that the influence propagation can only be triggered from the real world.

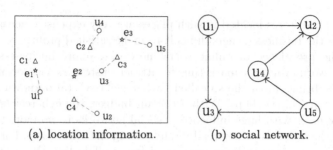

(a) location information.   (b) social network.

Fig. 1. A motivation example.

Specifically, it sets each candidate location's RNNs as seeds and perform the influence propagation from these seeds in social networks. Then they select the candidate location that has the maximum number of social network users residing in a certain region. However, the RNNs of a candidate location may not be the ideal set of seeds for promoting it. There are two reasons. Firstly, the final influence achieved by spreading from the RNNs is not guaranteed to be optimal. Those RNNs may have weak influence in social networks. Secondly, the underlying assumption that all RNNs of a candidate location will promote it is not sound. Just like in the real life, no business will only wait the initial customers to advertise for it. Let us consider the following example.

Example 1. As shown in Fig. 1(a), u_1, u_2, u_3, u_4, u_5 represent customers, e_1, e_2, e_3 represent existing locations, and c_1, c_2, c_3, c_4 represent candidate locations. For the ease of observation, we use the green dotted lines to connect each customer and her nearest location. Figure 1(b) represents the social connections among u_1, u_2, u_3, u_4, u_5. Consider the approach in [5], c_1 has no RNN, c_2's RNN is u_4, c_3's RNN is u_3, and c_4's RNN is u_2. We assume that a user simply follows her outgoing edges to influence her neighbors in the social network iteratively until no user can be influenced, and all RNNs of a candidate location will spread influence for it. Then c_1 has no influence to customers, c_2's influence is 2 (i.e., u_2, u_4), c_3's influence is 3 (i.e., u_1, u_2, u_3), and c_4's influence is 1 (i.e., u_2). The maximum influence to customers achieved by the approach in [5] is 3. However, if we select one seed freely from the social network, the maximum influence we can achieve is 5 and the seed is u_5.

Different from [5], we propose a novel business location selection model based on influence maximization [6] that does not restrict the seeds of influence propagation to be only the RNNs. In contrast, our model leverages the influence maximization techniques to advertise for each candidate location with a fixed budget, so that the influence propagation can be triggered by any fixed-size set of social network users as long as the final influence is maximized. Moreover, our model adopts the relative distance aware influence propagation, which is different from the distance-aware influence propagation proposed in [13].

The distance-aware influence propagation in [13] assumes that the customers who live far away from the specific location will be harder to be influenced. However, in the relative distance aware influence propagation, the customers whose distance to its nearest existing location is closer than their distance to the candidate location will be harder to be influenced by the candidate location. Lastly, the candidate location with the maximum relative distance aware influence will be selected.

Contribution. Our contributions are generalized as follows.

- We formalize a novel business location selection model based on relative distance aware influence maximization in geo-social networks. We use influence maximization techniques to select the most influential location for a cybercelebrity business, which attracts customers through advertising on social networks.
- We propose two rules to prune the candidate locations by bounding their maximum influence, based on two observations. Firstly, the customers too close to an existing location are hard to be influenced. Secondly, for a cluster of candidate locations that are geographically close to a center location, their maximum influence should also be close to the center location's, thus we can prune clusters by the maximum influence of their center locations.
- We develop an influence maximization algorithm for multiple candidate locations based on the RIS framework. Our algorithm can derive the minimum lower bound of sample size for a given set of candidate locations, and then selects seeds for them by using a same set of samples, thereby avoiding the redundant computation exists during performing RIS algorithm on each location respectively.
- We evaluate the performance of the proposed pruning strategies and algorithms on three real geo-social networks. The experiments confirmed the effectiveness and efficiency of the proposed techniques.

Roadmap. The rest of the paper is organized as follows. Section 2 introduces the influence propagation model and gives the definition of the business location selection problem. Section 3 introduces the proposed techniques, including the pruning rules and the seed selection algorithm for multiple candidate locations. Section 4 introduces the whole business location selection algorithm. In Sect. 5, we present our experiment results on three real datasets. Lastly, we introduce the related works in Sect. 6 and conclude the paper in Sect. 7.

2 Preliminary

We present the influence propagation model used in this paper and give the problem definition in this section.

2.1 Influence Propagation Model

We consider a geo-social network as a direct graph $G = (V, E)$, where V represents the set of nodes (the users), and E represents the set of edges (the relationship between users) in G. Each node $v \in V$ has a geographical location (x, y), where x and y denote the latitude and longitude respectively. Given an edge $(u, v) \in E$, u is the incoming neighbor of v, and v is the outgoing neighbor of u. Given $v \in V$, we denote the incoming and outgoing neighbors of v as $N_{in}(v)$ and $N_{out}(v)$ respectively.

In this paper, we exploit the Independent Cascade Model (ICM) [6], which is the classic and well-studied diffusion model. Given an edge $(u, v) \in E$, ICM considers a user v is activated by its incoming neighbors independently by introducing an *activating probability* $P_{u,v}$ to each edge.

To evaluate the influence propagation in geo-social networks, the distance-aware influence spread proposed in [13] has an assumption that users who are close to the location should have higher priority to be influenced, and the influence is weighted by users' distance to a query location. However, such assumption is not valid for the business location selection problem. Given the distance between a node and its nearest existing location (d_1) and the distance between a node and the candidate location (d_2), we need to consider the relative distance between d_1 and d_2 during the influence propagation process. We denote the influence propagation process in the business location selection problem as *relative distance aware influence propagation*, and the formal definition is given as follows. Note that, we refer to relative distance aware influence propagation as influence propagation hereafter if there is no ambiguity.

Definition 1. *Given a geo-social network $G = (V, E)$, a seed set $S \subseteq V$, a set of existing locations ES and a set of candidate locations C, for each $c \in C$, the relative distance aware influence propagation of S is calculated as follows, denoted as $Inf_{rdis}(S, c)$.*

$$Inf_{rdis}(S, c) = \sum_{v \in V} Inf(S, v) rdis(v, c) \tag{1}$$

where $Inf(S, v)$ is the probability that S activates v under ICM [6], and $rdis(v, c) = \lambda e^{-\alpha \frac{dis(v,c)}{min_{es \in ES} dis(v, es)}}$, where $dis(v, c)$ (or $dis(v, es)$) represents the distance between node v and the candidate location c (or the existing location es).

Example 2. As shown in Fig. 1(a), for the candidate location c_1, assumed that there exists only one existing location e_1, if we adopt the distance-aware influence spread in [13], then u_1 is the most likely to be influenced because it is the closest to c_1. However, in the relative distance aware influence propagation process, the probability of u_1 being influenced by c_1 is smaller than the probability of u_3 being influenced because the relative distance $\frac{dis(u_3,c_1)}{dis(u_3,e_1)}$ of u_3 is so smaller that the value of $rdis(u_3, c_1)$ is larger.

2.2 Problem Definition

Based on the definition of *relative distance aware influence propagation*, we present our problem formulation as follows.

Problem 1. Given a geo-social network $G = (V, E)$, a set of existing locations ES, a set of candidate locations C, an integer k, and the seed set S with size k derived from geo-social networks for each candidate location $c \in C$, the business location selection problem is to evaluate the relative distance aware influence propagation of each candidate location c and select the location $c_{max} \in C$ which has the maximum relative distance aware influence propagation, i.e.

$$c_{max} = \arg\max_{c \in C} Inf_{rdis}(S, c) \qquad (2)$$

Problem Hardness. The main challenge of the business location selection problem is to select seeds for multiple candidate locations, which can be considered as the relative distance aware influence maximization problem in geo-social networks. We discuss the problem hardness by simply setting $\lambda = 1$ and $\alpha = 0$, then $Inf_{rdis}(S, c) = \sum_{v \in V} Inf(S, v)$, which is independent with the relative distance, and the relative distance aware influence maximization problem becomes the traditional influence maximization problem. Since the traditional influence maximization problem has been proved NP-hard [6], the relative distance aware influence maximization problem is also NP-hard. Due to the hardness of selecting seeds and evaluating the influence propagation for only one candidate location, the business location selection problem which considers multiple candidate locations is also NP-hard.

Basic Approach. The business location selection problem can be solved by leveraging the existing influence maximization techniques [6–11]. The basic approach is to select seed set S and calculate the influence propagation $Inf_{rdis}(S, c)$ for each candidate location c, and the candidate location with the maximum influence propagation will be selected as the optimal location.

3 Techniques

The main limitation of the basic approach is that it needs to select seeds and calculate influence propagation for every candidate location. Due to the huge number of candidate locations, the computation overhead is unacceptable. Note that, for most candidate locations, their influence propagation may be very small, thus we can avoid performing exact calculation for these candidate locations. Moreover, the seed selection process for each candidate location in the basic approach is independent, if the seed selection process for multiple candidate locations can be performed all at once, many redundant operations can be eliminated. Therefore, we present two pruning rules and the seed selection algorithms for multiple candidate locations in this section.

3.1   Prune Candidate Locations

The first pruning rule is proposed based on an observation that a candidate location which is too close to the existing locations will not be a good choice, because people tend to avoid opening their stores near the similar businesses in reality. The reason is that the influence propagation of two close locations is similar because the potential customers covered by them are almost the same.

Rule 1. Given a set of existing locations ES, a set of candidate locations C and a threshold γ, for any candidate location $c \in C$ and its nearest existing location $es \in ES$, if $dis(c, es) \leq \gamma$, then we can prune c from C. The set of the unpruned candidate locations is denoted as C_1.

Lemma 1. *Given any two locations c_1, c_2 and the threshold γ, if $dis(c_1, c_2) \leq \gamma$, then $Inf_{rdis}(S, c_1) \approx Inf_{rdis}(S, c_2)$, where the seeds of c_1 and c_2 is assumed to be the same because the distance of c_1 and c_2 is very small.*

Proof. For $v \in V$, given v's nearest existing location es, and let λ be simply set as 1, since $dis(c_1, c_2) \leq \gamma$, $rdis(v, c_1) = e^{-\alpha \frac{dis(v,c_1)}{dis(v,es)}}$, and $rdis(v, c_2) = e^{-\alpha \frac{dis(v,c_2)}{dis(v,es)}}$, then $\frac{rdis(v,c_1)}{rdis(v,c_2)} \geq e^{-\alpha \frac{dis(c_1,c_2)}{dis(v,es)}}$. Since $\gamma \geq dis(c_1, c_2)$, when γ is so small that $\gamma \to 0$, then $\frac{rdis(v,c_1)}{rdis(v,c_2)} \to 1$, and $rdis(v, c_1) \approx rdis(v, c_2)$. According to Definition 1, we can know that $Inf_{rdis}(S, c_1) \approx Inf_{rdis}(S, c_2)$.

The number of candidate locations pruned based on Rule 1 is limited, so we propose to do clustering for the candidate locations in C_1 according to the distance. Based on Lemma 1, the influence propagation of the candidate locations belonging to the same cluster should be close to the center of the cluster, so that we can prune clusters by the maximum influence propagation of cluster centers.

Rule 2. Let the candidate locations in C_1 be divided into $|T|$ clusters. Since the best candidate location is derived from cluster t, where the center p_{max} of cluster t has the largest influence to potential customers, then the candidate locations belonging to other clusters can be pruned, and the candidate locations belonging to cluster t compose a new set C_2.

Lemma 2. *Given a set of clusters T, a set of corresponding centers P, a threshold δ, a parameter $\beta \in (\frac{\delta}{4}, \delta)$, and for any candidate locations c_1, c_2 belonging to the same cluster $t \in T$, we have $dis_{c_1,c_2 \in t}(c_1, c_2) \leq \delta$. Let p_{max} be the center with maximum influence propagation, then $Inf_{rdis\{c^* \in t^*|p^* \in t^*, p^* \neq p_{max}\}}(S, c^*) \leq Inf_{rdis\{c \in t|p_{max} \in t\}}(S, c)$.*

Proof. Take two clusters $t_1, t_{max} \in T$, the center of t_1 is p_1 ($p_1 \neq p_{max}$), and $c_1 \in t_1$, while the center of t_{max} is p_{max}, and $c_2 \in t_{max}$. Given $v \in V$ that is influenced by p_{max}, we have $dis(v, c_2) \leq dis(v, p_{max}) + dis(p_{max}, c_2)$, and $dis(v, c_1) \geq dis(v, p_1) - dis(p_1, c_1)$. Since $Inf_{rdis}(S, p_1) \leq Inf_{rdis}(S, p_{max})$ and $dis(v, es)$ is fixed, then $dis(v, p_{max}) \leq dis(v, p_1)$, max $dis(p, c) = \delta/2$, and min $dis(p, c) = 0$.

Based on the triangle inequality, $dis_{\{p_1 \in t_1, p_2 \in t_2 | t_1, t_2 \in T\}}(p_1, p_2) \geq \frac{3\delta}{4}$. Therefore, $dis(v, p_{max}) + dis(p_{max}, c_2) \leq dis(v, p_1) - dis(p_1, c_1) + \beta$, where β is used to "relax" the criteria of the returned candidate locations. Then, we can get $dis(v, c_2) \leq dis(v, c_1)$, i.e. $Inf_{rdis}(S, c_2) \geq Inf_{rdis}(S, c_1)$ is proved.

Since C_2 is composed by the candidate locations in only one cluster, the size of C_2 is relatively small, and we have proved that the candidate locations in C_2 can achieve better influence propagation than other candidate locations. Thus we can select seeds for each candidate location in C_2 and evaluate the corresponding influence propagation to choose the optimal candidate location.

3.2   Seed Selection for Multiple Candidate Locations

The seed selection problem is actually the well-known influence maximization problem, and there exist a large amount of researches devoted to the design of influence maximization algorithms [6–11]. In this paper, we extend the RIS framework [8] to select seeds for multiple candidate locations. Note that, the idea of RIS is that the influence of any seed set S is estimated by selecting random nodes and seeing the portion of the randomly selected nodes which can be reached by S.

Some explanations of related concepts about the RIS framework such as graph distribution and RIS samples can be referred in [14]. Since we apply the relative distance aware influence propagation in the business location selection problem, we need to consider the relative distance $rdis(v, c)$ when calculating the size of the RIS samples, instead of considering the distance weights like [14]. It is crucial for the RIS framework to determine the size of the set of RIS samples R, if the sample size $|R|$ is too large, the seed selection process will run slowly in the worst case. However, to achieve a $1 - 1/e - \epsilon$ approximation result, $|R|$ should not be smaller than θ [9,14] as shown in Eq. (3), note that, the proof of deriving θ can be referred in [9,14], and is omitted here.

$$\theta = \frac{2n \cdot rdis_{max} \cdot ((1 - 1/e) \cdot \varphi_1 + \varphi_2)^2}{OPT \cdot \varepsilon^2} \tag{3}$$

where $rdis_{max}$ denotes the maximum relative distance among candidate locations in C_2, OPT is the influence spread of the optimal seed set, and φ_1, φ_2 can be referred in [9].

Once we apply the basic approach with RIS framework to select seeds for each candidate location in C_2, the process of deriving sample size will perform $|C_2|$ times and the computation overhead will be horrible because the sample size for each candidate location is not small. Thus, we consider to take a uniform sample size for all candidate locations in C_2, and the uniform sample size is the maximum one among all derived sample size for C_2. Since the sample size of each candidate location should not be smaller than θ, the maximum sample size is definitely enough for all candidate locations.

To get the sample size in Eq. (3), we need to get the lower bound of OPT. Since the maximum sample size is needed, so we only need to estimate the minimum lower bound of OPT according to Eq. (3). To avoid calculating the lower bound for every candidate location in C_2, we set p_{max} as an anchor point. We calculate the lower bound of p_{max} and the distance between p_{max} and each candidate location in C_2, then the minimum lower bound can be quickly derived according to Lemma 3.

Lemma 3. *Given the center p_{max} and its seed set $S_{p_{max}}$, a node $v \in V$ and any candidate location $c \in C_2$, L in Eq. (4) is a lower bound of OPT, denoted as $l(c)$.*

$$L = \frac{1 - 1/e - \epsilon_0}{1 - 1/e - \epsilon_0 + \epsilon_2} \cdot e^{-\eta dis(c, p_{max})} \cdot Inf_{rdis}(S_{p_{max}}, p_{max}) \qquad (4)$$

where $\tau_1 = \min_{v \in V} dis(v, es)$, $\eta = \frac{\alpha}{\tau_1}$, and other parameters can be referred in [14].

Proof. The relevant proof is similar to the proof in [14], and is omitted here.

According to Lemma 3, we just need to find the candidate location which is the farthest to center p_{max} and exploit the influence of p_{max} to achieve the minimum lower bound. Note that, the tightness of the lower bound is only decided by the distance between the candidate location $c \in C_2$ and p_{max}.

Minimum Lower Bound Algorithm. Here we propose an algorithm to quickly compute the lower bound of p_{max} [11] and derive the minimum lower bound among C_2. Since the influence decreases as the path length increases, we just consider the two-hop neighbors of the seed set S when evaluating the influence of p_{max}. The detail of the method is presented in Algorithm 1.

As shown in Algorithm 1, for p_{max}, we first sort nodes based on $rdis(u, p_{max})$· $N_{out}(u)$ and select the top-k nodes to make up the seed set S (Line 1–4). Then we set the activated probability $p[u]$ of nodes as 0, except that the activated probability of seeds is set as 1, and the visited status setting is the same with the activated probability setting (Line 5–10). Only the two-hop neighbors of S are considered when calculating the influence of S, and the neighbors are saved in a queue H. The influence is denoted as LB, which is initially set as k. The influence on node v is calculated as $(1 - p[v]) \cdot P_{u,v} \cdot rdis(v, p_{max}) \cdot p[u]$, where $(1 - p[v])$ indicates the probability that v is not activated by other nodes, $P_{u,v} \cdot p[u]$ indicates the probability that u activates v based on u's current activated probability, and $rdis(v, p_{max})$ is v's relative distance based on p_{max} (Line 11–22). After deriving p_{max}'s lower bound, we calculate the distance between any $c \in C_2$ and p_{max} and estimate the minimum lower bound of C_2 based on Lemma 3 (Line 23–27).

Seed Selection Algorithm for Multiple Candidate Locations. The minimum lower bound has been derived based on Algorithm 1, and we can get the maximum sample size based on Eq. (3). Given the maximum sample size, a set of RIS samples R can be constructed, and all candidate locations in C_2 share the same set of RIS samples R. After deriving the set of RIS samples R, for each $c \in C_2$, we follow the steps of the RIS algorithm [14] to select seeds.

Algorithm 1. Minimum Lower Bound Algorithm

Input: a geo-social network G; a set ES of existing locations; a set C_2 of candidate
    locations; the center p_{max}; the size of seed set k.

Output: LB_{min}.

1: precompute the distance between u and its nearest existing location:
    $\min_{es \in ES, u \in V} dis(u, es)$;
2: $LB_{min} \leftarrow 0$;
3: sort nodes based on $rdis(u, p_{max}) \cdot N_{out}(u)$;
4: $S \leftarrow$ top-k nodes;
5: **for** each $u \in V \setminus S$ **do**
6:     $p[u] \leftarrow 0$;
7:     $visited[u] \leftarrow 0$;
8: **for** each $u \in S$ **do**
9:     $p[u] \leftarrow 1$;
10:     $visited[u] \leftarrow 1$;
11: $H \leftarrow \varnothing$; $LB \leftarrow k$;
12: **while** $S \neq \varnothing$ or $H \neq \varnothing$ **do**
13:     **if** $S \neq \varnothing$ **then**
14:         $u \leftarrow S.dequeue$;
15:     **else**
16:         $u \leftarrow H.dequeue$;
17:     **for** each $v \in N_{out}(u)$ **do**
18:         **if** $visited[v] = 0$ and u is dequeued from S **then**
19:             enqueue v into H;
20:             $visited[v] = 1$;
21:         $LB = LB + (1 - p[v]) \cdot P_{u,v} \cdot rdis(v, p_{max}) \cdot p[u]$;
22:         $p[v] = p[v] + (1 - p[v]) \cdot P_{u,v} \cdot rdis(v, p_{max}) \cdot p[u]$;
23: **for** each $c \in C_2$ **do**
24:     calculate the distance $dis(c, p_{max})$;
25: $c_1 \leftarrow$ the farthest candidate location to p_{max};
26: $l(c_1) \leftarrow$ the lower bound of c_1 calculated based on Equation (4);
27: $LB_{min} \leftarrow l(c_1)$;
28: **return** LB_{min};

Note that, when selecting seeds, we exploit the sample location selection method
proposed in [16] rather than the naive sampling methods in [14]. Due to the space
limitation, the detail of the seed selection algorithm is not given. Note that, in
the extended algorithm, the sample generation process is performed only once
rather than $|C_2|$ times, the total computation overhead can be largely decreased.
Moreover, the sample location selection method [16] has been proved to be more
effective than the naive sampling methods in [14], so the extended RIS algorithm
can achieve better results.

4   Business Location Selection Algorithm

In this section, we will present the whole business location selection algorithm combined with Algorithm 1.

As shown in Algorithm 2, we can prune the candidate locations, whose distance to the existing locations is smaller than γ. The unpruned candidate locations compose a new set C_1 (Line 2–6). When clustering, the candidate locations in C_1 are divided into T clusters, and each cluster's center is saved in the center set P (Line 7–8). We exploit the set covering methods [15] to do clustering here, note that, any clustering method is suitable as long as the distance among candidate locations belonging to the same cluster is smaller than δ. Given the set of centers, we evaluate their influence propagation, and derive the center p_{max} with maximum influence propagation. The optimal candidate location is derived from C_2, which is composed by the candidate locations sharing center p_{max} (Line 9–12). Before performing the seed selection for C_2, we should calculate the sample size by utilizing Algorithm 1 and Eq. (3) (Line 13). Then we select seeds for all candidate locations in C_2 and calculate their influence propagation to derive the optimal candidate location (Line 14–19).

Analysis. Compared with the basic approach, we apply two pruning strategies to decrease the computation overhead of the seed selection process. Moreover,

---

Algorithm 2. Improved Business Location Selection Algorithm

Input: a geo-social network G; a set ES of existing locations; a set C of candidate locations; the size of seed set k; the threshold γ.

Output: the optimal business location l.

1: precompute the distance between u and its nearest existing location: $\min_{es \in ES, u \in V} dis(u, es)$;
2: **for** each $c \in C$ **do**
3:     **for** each $es \in ES$ **do**
4:         **if** $dis(c, es) \leq \gamma$ **then**
5:             prune c from the set C of candidate locations;
6: $C_1 \leftarrow$ the set of unpruned candidate locations;
7: divide the candidate locations in C_1 into $|T|$ clusters;
8: $P \leftarrow$ the set of $|T|$ clusters' centers;
9: **for** each $p \in P$ **do**
10:     estimate the influence propagation $Inf_{rdis}(p)$;
11: $p_{max} \leftarrow$ the center with maximum influence propagation;
12: $C_2 \leftarrow$ the set of candidate locations sharing center p_{max};
13: $|R| \leftarrow$ the sample size calculated by calling Algorithm 1;
14: **for** each $c \in C_2$ **do**
15:     $S[c] \leftarrow$ the seed set of c;
16: **for** each $c \in C_2$ **do**
17:     calculate the influence propagation $Inf_{rdis}(c)$;
18: $l \leftarrow$ the candidate location in C_2 that has the largest influence propagation;
19: **return** l;

---

to optimize the calculation of the RIS algorithm [14], we propose the minimum lower bound algorithm, which is applied to make the sample generation process be performed only once and the seeds for each candidate location be selected based on a uniform set of samples. Therefore, much redundant computation can be eliminated and the computation efficiency can be significantly improved.

5 Experiments

Our experiments are conducted on a PC with Intel Core 3.2GHz CPU and 16G memory. The algorithms are implemented in C++ with TDM-GCC 4.9.2.

5.1 Experiment Setup

Algorithms. We compare our algorithms with the method proposed in [5], which addresses the business location selection problem based on RNN query. The algorithms evaluated in this paper are listed as follows.

- RNN-B: the RNN-oriented approach in [5], we set the RNNs as seeds and the process of calculating RNNs' influence propagation is the same as our own algorithms;
- BLS: the business location selection algorithm without any pruning rules;
- BLSI: the business location selection algorithm based on Rule 1;
- BLSII: the business location selection algorithm based on Rule 1 and Rule 2;

Dataset. We use three real-world geo-social networks, and the candidate locations and existing locations are derived from the POI dataset of Foursquare. The concrete information is shown as follows.

- Geo-social networks: 1) Gowalla, consists 100K nodes and 1.4 million edges. 2) Brightkite, consists 58K nodes 428K edges. 3) Foursquare, consists 16K nodes 120K edges.
- Location information: we extract the locations of Chinese restaurants in the POI dataset of Foursquare and get 35489 Chinese restaurants' geo-coordinates. Then we sample some of these locations as the candidate locations, and the rest are served as the existing locations. The default number of candidate locations is 2700.

Parameters. Given an edge (u,v), the activating probability $P_{u,v}$ is set as $1/N_{in}(v)$, the seed set size k is set as 30, and the default value of δ used for clustering is set as 10. To obtain the influence propagation, we run 1000 round simulations for each seed set and take the average of the influence propagation.

5.2   Effectiveness Evaluation

To evaluate the effectiveness of the proposed algorithms, we compare them with the RNN-B approach on three real datasets with varying the value of $|C|$. The results are presented in Fig. 2(a), 2(b), and 2(c) respectively. We can see that the RNN-B algorithm achieves a much smaller influence spread than the proposed algorithms in Gowalla and Foursquare and achieves slightly smaller influence spread than the proposed algorithms in Brightkite. Since the influence propagation of RNN-B is triggered from the real world, that is, the seeds are made up of a candidate location's RNNs. However, a candidate location's RNNs may not be the optimal seeds because the RNNs may have weak influence in social networks. Moreover, we can find that BLS and BLSI have the same influence spread, because Rule 1 only prunes the candidate locations with small influence. The influence spread of BLSII is sometimes slightly smaller than BLS and BLSI. The reason is that the candidate location selected from C_2 may not be the best candidate location among C because of the existence of β, which is used to "relax" the criteria of the returned candidate locations. However, the influence spread of BLSII is at most 5.3% smaller than BLS.

5.3   Efficiency Evaluation

To evaluate the efficiency of the proposed algorithms and pruning strategies, we compare the online time costs of the participant algorithms with varying the value of $|C|$ on three real datasets in Fig. 2(d), 2(e), and 2(f). We can find that the time costs of RNN-B are relatively smaller than BLS and BLSI. With the increase of $|C|$, the time costs of RNN-B increase beyond the time costs of BLSII except in Brightkite. The process of constructing samples and selecting seeds for the candidate locations takes much more time than the process of determining RNNs, so BLS and BLSI incur much more computation overhead than RNN-B. As for BLSII, the small size of C_2 makes the time costs of BLSII less than BLS and BLSI, and BLSII can outperform RNN-B as $|C|$ increases because the size of C has small impact on the time costs of BLSII.

Moreover, Fig. 2(d), 2(e), and 2(f) also show the effectiveness of our pruning strategies. BLSII runs the fastest among the proposed algorithms, and BLSI outperforms BLS. We have pruned many candidate locations with small influence based on Rule 1, and $|C_1|$ is much smaller than $|C|$, so BLSI incurs less time than BLS. As for BLSII, the time costs depend on $|C_2|$. Since δ used for clustering is certain, the size $|C_2|$ of the candidate locations that need to be calculated after clustering is relatively small, thus the computation overhead of BLSII is the smallest among the proposed algorithms.

5.4   Sensitivity Tests

We conduct experiments to demonstrate the impact of threshold γ and $|ES|$ on the time costs of online calculation in three real datasets, and the impact of $|C|$ has been demonstrated in Fig. 2(d), 2(e), and 2(f).

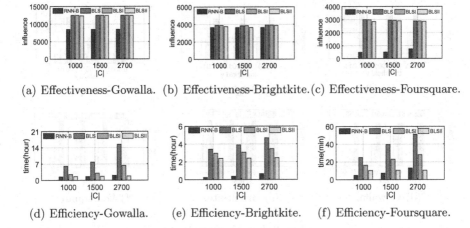

(a) Effectiveness-Gowalla. (b) Effectiveness-Brightkite.(c) Effectiveness-Foursquare.

(d) Efficiency-Gowalla.   (e) Efficiency-Brightkite.   (f) Efficiency-Foursquare.

Fig. 2. Effectiveness and efficiency.

Impact of Threshold γ. As shown in Fig. 3(a), 3(b), and 3(c), we present the results of the proposed algorithms under different values of γ given in Lemma 1, and the values of γ are set as 0.0001, 0.001 and 0.01 respectively. Since γ used in Rule 1 is independent of BLS, the response time of BLS remains the same as the value of γ increases. The response time of BLSI and BLSII decreases as the value of γ increases. Since γ is used to prune the candidate location whose distance to its nearest existing location is less than γ, when γ is larger, the number of candidate locations pruned based on Rule 1 is larger, that is, $|C_1|$ increases as γ decreases. Therefore, we can find that as γ increases, the response time of BLSI decreases greatly, because there are 1494, 1016 and 341 candidate locations that need to be calculated respectively. BLSII incurs less time than BLSI because there are less candidate locations that need to be calculated after clustering based on Rule 2, but the response time of BLSII also decreases with the increase of γ. As γ increases, BLSII consumes 9.07, 6.31, and 2.15 h respectively in Gowalla, while consuming 3.77, 3.45 and 2.95 h respectively in Brightkite and consuming 32.08, 28.03, and 15,80 min respectively in Foursquare.

Impact of $|ES|$. Figure 3(d), 3(e), and 3(f) present the impact of $|ES|$ on the performance of the proposed algorithms, and $|ES|$ is set as 2000, 6000, 10000, and 15000 respectively. With the increase of $|ES|$, the time costs of BLS almost remain the same, while the time costs of BLSI and BLSII decrease. BLS does not adopt any pruning rules, so the increase of $|ES|$ has little impact on the time costs of BLS. As for BLSI, the time costs decrease because there are 2306, 1786, 1411, and 1016 candidate locations that need to be calculated respectively. Since γ is used to prune the candidate location whose distance to its nearest existing location is less than γ, when γ is certain and $|ES|$ is small, the number of candidate locations pruned based on Rule 1 will be small. In other words, the number of candidate locations that need to be calculated is large, thus the time

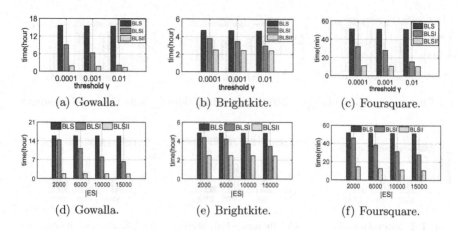

Fig. 3. Sensitivity test.

costs of BLSI increase as $|ES|$ decreases. As $|ES|$ increases, the time costs of BLSII also decrease, such as in Foursquare, the time costs of BLSII are 14.65, 12.72, 11.25, and 10.53 min respectively.

6   Related Works

Business Location Planning. There exist a large amount of researches conduct business location planning based on RNN query [1–5]. The business location planning based on RNN query aims to find a location that has the largest influence, and the influence is defined as the number of a location's RNNs. Huang et al. [2] aim to rank locations based on their influence, and they propose the nearest location circle algorithm and the Voronoi diagram algorithm to process the query. They also study the selection of top-k most influential locations based on RNN in [3,4]. The Estimation Expanding Pruning (EEP) algorithm and the Bounding Influence Pruning (BIP) algorithm are proposed to prune the search space and achieve better performance. Recently, Hung et al. [5] have considered both of RNN and social influence factors for business location planning. They set locations' RNNs as seeds and compute the influence spread of RNNs in social networks, and the location with maximum influence spread will be selected as the optimal location. Unlike the existing works mentioned above, we exploit the influence maximization techniques to derive the seeds of each candidate location from the geo-social networks, and we can ensure that the selected seeds from the geo-social networks can achieve the largest influence propagation for each candidate location, but the RNNs can not.

Location-Aware Product Promotion. The popularity of social media platforms has triggered more and more companies promoting their products in social networks. Influence maximization that leverages the benefit of word-of-mouth effect in social networks, is a key problem in viral marketing and has been widely studied in the literature [6–11]. With the prevalence of location-based social networks, there are many researches focus on the location-aware influence maximization (LAIM) problem [12–14]. Li et al. [12] are the first to study LAIM and aim to find a seed set S with size k that maximize the influence spread over a given geographical region, and the MIA/PMIA model [7] is utilized to calculate the influence spread. Compared with [12], Wang et al. [13,14] focus on the distance-aware influence maximization (DAIM) problem that maximizes the influence spread based on the user's distance to a query location. The MIA/PMIA model [7] is also exploited in [13], and they take some anchor points sampling methods to estimate the influence bound, while in [14], they use the RIS framework [8,9] to select seeds. Different from the existing works, we utilize the influence maximization techniques to select a optimal location for businesses. In other words, we aim to find a new location but not to promote an existing location.

7 Conclusion

Since the "cybercelebrity economy" has become a widely interested phenomenon, we rethink the long-standing business location selection problem with the awareness of word-of-mouth marketing in social networks. Different from the previous works, we define the goodness of a location as the maximum number of users that can be influenced by promoting the location (i.e., influence maximization) in a social network. In addition, the probability that a user can be influenced is also determined by the relative geographical distance between her and the promoted location, with a set of locations of existing competitors. Under the definition, we develop a bunch of techniques to select the optimal location from a probably large candidate set efficiently. The experimental results show that our approach can identify much more influential locations than the previous RNN-based approach, and reduce the time costs of selection significantly. For the businesses that do not want to afford high rental in popular districts, our study can help them to select a location that can gain the most from social network advertising.

In the future, we will investigate how to integrate the user interests into our model, because different businesses have different influence to a same user.

Acknowledgments. This work was supported by the National Natural Science Foundation of China (No. 61202036 and 61972291), Natural Science Foundation of Hubei Province (No. 2018CFB616 and 2018CFB519) and Fundamental Research Funds for the Central Universities (No. 413000078).

References

1. Korn, F., Muthukrishnan, S.: Influence sets based on reverse nearest neighbor queries. ACM SIGMOD Rec. **29**(2), 201–212 (2000). International Conference on Management of Data

2. Huang, J., Wen, Z., Pathan, M., et al.: Ranking locations for facility selection based on potential influences. In: Conference of the Industrial Electronics Society, pp. 2411–2416 (2011)

3. Huang, J., Wen, Z., Qi, J., et al.: Top-k most influential locations selection. In: Conference on Information and Knowledge Management, pp. 2377–2380 (2011)

4. Chen, J., Huang, J., Wen, Z., He, Z., Taylor, K., Zhang, R.: Analysis and evaluation of the top-k most influential location selection query. Knowl. Inf. Syst. **43**(1), 181–217 (2014). https://doi.org/10.1007/s10115-013-0720-0

5. Hung, H., Yang, D., Lee, W., et al.: Social influence-aware reverse nearest neighbor search. ACM Trans. Spatial Algorithms Syst. **2**(3), 1–15 (2016)

6. Kempe, D., Kleinberg, J., Tardos, E., et al.: Maximizing the spread of influence through a social network. In: Knowledge Discovery and Data Mining, pp. 137–146 (2003)

7. Chen, W., Wang, C., Wang, Y., et al.: Scalable influence maximization for prevalent viral marketing in large-scale social networks. In: Knowledge Discovery and Data Mining, pp. 1029–1038 (2010)

8. Borgs, C., Brautbar, M., Chayes, J., et al.: Maximizing social influence in nearly optimal time. In: Symposium on Discrete Algorithms, pp. 946–957 (2014)

9. Tang, Y., Shi, Y., Xiao, X., et al.: Influence maximization in near-linear time: a martingale approach. In: International Conference on Management of Data, pp. 1539–1554 (2015)

10. Tang, Y., Xiao, X., Shi, Y., et al.: Influence maximization: near-optimal time complexity meets practical efficiency. In: International Conference on Management of Data, pp. 75–86 (2014)

11. Wang, X., Zhang, Y., Zhang, W., et al.: Bring order into the samples: a novel scalable method for influence maximization. IEEE Trans. Knowl. Data Eng. **29**(2), 243–256 (2017)

12. Li, G., Chen, S., Feng, J., et al.: Efficient location-aware influence maximization. In: International Conference on Management of Data, pp. 87–98 (2014)

13. Wang, A., Zhang, A., Lin, A., et al.: Distance-aware influence maximization in geo-social network. In: International Conference on Data Engineering, pp. 1–12 (2016)

14. Wang, X., Zhang, Y., Zhang, W., et al.: Efficient distance-aware influence maximization in geo-social networks. IEEE Trans. Knowl. Data Eng. **29**(3), 599–612 (2017)

15. Chvatal, V.: A greedy heuristic for the set-covering problem. Math. Oper. Res. **4**(3), 233–235 (1979)

16. Zhong, M., Zeng, Q., Zhu, Y., Li, J., Qian, T.: Sample location selection for efficient distance-aware influence maximization in geo-social networks. In: Pei, J., Manolopoulos, Y., Sadiq, S., Li, J. (eds.) DASFAA 2018. LNCS, vol. 10827, pp. 355–371. Springer, Cham (2018). https://doi.org/10.1007/978-3-319-91452-7_24

SpEC: Sparse Embedding-Based Community Detection in Attributed Graphs

Huidi Chen[1,2], Yun Xiong[1,2]($\boxtimes$), Changdong Wang[3], Yangyong Zhu[1,2], and Wei Wang[4]

[1] Shanghai Key Laboratory of Data Science, School of Computer Science, Fudan University, Shanghai, China
{hdchen18,yunx,yyzhu}@fudan.edu.cn
[2] Shanghai Institute for Advanced Communication and Data Science, Fudan University, Shanghai, China
[3] School of Data and Computer Science, Sun Yat-sen University, Guangzhou, China
changdongwang@hotmail.com
[4] University of California, Los Angeles, CA, USA
weiwang@cs.ucla.edu

Abstract. Community detection, also known as graph clustering, is a widely studied task to find the subgraphs (communities) of related nodes in a graph. Existing methods based on non-negative matrix factorization can solve the task of both non-overlapping community detection and overlapping community detection, but the probability vector obtained by factorization is too dense and ambiguous, and the difference between these probabilities is too small to judge which community the corresponding node belongs to. This will lead to a lack of interpretability and poor performance in community detection. Besides, there are always many sparse subgraphs in a graph, which will cause unstable iterations. Accordingly, we propose SpEC (Sparse Embedding-based Community detection) for solving the above problems. First, sparse embeddings has stronger interpretability than dense ones. Second, sparse embeddings consume less space. Third, sparse embeddings can be computed more efficiently. For traditional matrix factorization-based models, their iteration update rules do not guarantee the convergence for sparse embeddings. SpEC elaborately designs the update rules to ensure convergence and efficiency for sparse embeddings. Crucially, SpEC takes full advantage of attributed graphs and learns the neighborhood patterns, which imply inherent relationships between node attributes and topological structure information. By coupled recurrent neural networks, SpEC recovers the missing edges and predicts the relationship between pairs of nodes. In addition, SpEC ensures stable convergence and improving performance. Furthermore, the results of the experiments show that our model outperforms other state-of-the-art community detection methods.

Keywords: Community detection · Attributed graph · Non-negative matrix factorization · Graph embedding

© Springer Nature Switzerland AG 2020
Y. Nah et al. (Eds.): DASFAA 2020, LNCS 12114, pp. 53–69, 2020.
https://doi.org/10.1007/978-3-030-59419-0_4

1   Introduction

Graphs can be found everywhere in the real world, such as social networks, Internet networks, and collaboration networks. Community detection, also known as graph clustering, which aims to find several clusters consisting of nodes with high similarity in a whole graph [6]. Community detection is an important application for analyzing graphs. For example, identifying clusters of customers with similar interests in a network of purchase relationships help retailers make better recommendations [6].

There are many models detecting communities on a plain graph. The popular models are spectral clustering [25], Louvain algorithm [2], and NMF (Nonnegative Matrix Factorization) [26]. In many domains, such as natural language processing, computer science, and math, sparse embeddings have been widely used to increase interpretability. Obviously, sparse embeddings cost fewer memory stores and improve calculation efficiency. These advantages of sparse embeddings benefit subsequent community detection. NNSC [8] was proposed for sparse non-negative matrix factorization. SNMF [10] applied sparse non-negative matrix factorization to community detection on plain graphs. However, the real-world graphs are always attributed graphs. Several NMF-based approaches combining both nodes attributes information and topological information have been proposed in [16,27,30]. As the state-of-the-art methods, SCI (Semantic Community Identification) [27], CDE (Community Detection Embedding) [16] improves the way to combine node attributes into embedding. Although these methods perform well so far, we find that the following issues are not considered in the above models:

- **Interpretability.** The soft node-community model [28] usually gets the probability that each node in a graph belongs to each community. Sometimes, there are too many non-zero probabilities that are ambiguous to be determined which communities one node belongs to, especially when the probabilities are close. The key to interpretability is to broadened the gap between each pair of probabilities. To our best knowledge, there is not any model solving this problem on an attributed graph.
- **Efficiency.** To ensure both convergence and non-negative constraints, the common NMF-based community detection models have to project coefficients of multiplication update rules in a positive direction. That embeddings are not updated along the gradients' directions will cause the loss of efficiency. In addition, dense embeddings cost much space and computing resources, which are undoubtedly not beneficial to subsequent community detection.
- **Sparse subgraphs.** A graph consisting of a large number of connected components and the corresponding matrix containing topological information of a graph, such as adjacency matrix and Laplacian matrix, would be a singular matrix. The number of connected components is equivalent to the number of zero eigenvalues of the Laplacian matrix and the corresponding eigenvectors are constant one vectors [25], which results in the poor performance of Laplacian embedding. Besides, a singular matrix cannot guarantee convergence of matrix factorization and affects the embeddings.

For the problem of interpretability, we have applied sparse embeddings to community detection. For the problem of efficiency, besides sparse embeddings, we designed update rules carefully to make sure the coefficients of multiplication positive along the gradients' direction. For the problem of sparse subgraphs, there have been some numerical methods, such as Tikhonov regularization [19], to remain the embeddings of a singular matrix stable. However, these numerical methods do not make full advantage of the topological information. Thus, besides numerical methods, we generate edges between sparse subgraphs to solve the problem. These generated edges are not only a supplement to the missing information but also a prediction of the implicit relationships. Although the existing methods, such as NetGAN [3] and GraphRNN [15] learns the distribution of edges on a graph and recover the edges, these methods work on a plain graph. This means these methods do not utilize the node attributes of a graph. If neighborhood patterns, which imply inherent relationships between node attributes and topological structure information, are fully used, a better solution to the problem will be produced. Thus, we propose a step called AGRNN (Attributed GraphRNN) to learn neighborhood patterns of a graph and generate the edges.

In this paper, we propose SpEC (Sparse Embedding-based Community detection) model to solve the above problems. SpEC model consists of two steps, exploring neighborhood patterns and non-negative matrix factorization with sparse constraints. Our community detection algorithm is more robust than the existing algorithms because our model recovers the edges between sparse subgraphs of a graph and stabilize the matrix factorization. We add sparse constraints for the matrix factorization in order to make our algorithm easier to converge and more interpretable than other models.

Our main contributions are summarized as follows.

- We propose a novel model, SpEC (Sparse Embedding-based Community detection) model and leverage both node attributes and topological information for community detection of attributed graphs.
- We propose an AGRNN step to learn neighborhood patterns and generate real edges on a graph, which ensures the following sparse non-negative matrix factorization on attributed graph convergence. The contrast experiments prove the effectiveness of AGRNN.
- We add sparse constraints to the objective matrix to get sparse embedding and to make our model interpretable and design the corresponding iterative updating rules. We prove the convergence of our model theoretically.
- Extensive experiments results on real-world attributed graphs show that our SpEC model outperforms other state-of-the-art methods.

2 Preliminaries

2.1 Problem Definition

Suppose that an undirected attributed graph $G = (V, E, \mathcal{X})$ is given, where V is the set of nodes, E is the set of edges and $\mathcal{X}$ is the set of node attributes.

The set of nodes V and the set of edges E are represented by the corresponding adjacency matrix A, where $A \in R^{|V| \times |V|}$. Likewise, the set of node attributes are represented by the corresponding node attributes matrix X, where $X \in R^{|V| \times r}$, r is the dimension of node attributes and i-th row of X is the node attributes of the i-th node of the adjacency matrix A. Consequently, an undirected attributed graph can simply be represented by $G = (A, X)$.

Given an attributed graph $G = (A, X)$ and the number of communities c, the community detection is to find c groups of nodes from the graph $G = (A, X)$. In the non-overlapping community detection, the c groups(communities) are not allowed to overlap, namely $\forall \mathcal{O}_i, \mathcal{O}_j \in \{\mathcal{O}_1, \mathcal{O}_2, ..., \mathcal{O}_c\}(i \neq j)$, $\mathcal{O}_i \cap \mathcal{O}_j = \emptyset$, where $\{\mathcal{O}_1, \mathcal{O}_2, ..., \mathcal{O}_c\}$ is the c groups of nodes. In contrast, in the overlapping community detection, the c groups (communities) are allowed to overlap.

2.2 NMF-Based Community Detection

SNMF (Symmetric NMF) [26] factorizes the adjacency matrix A of a graph symmetrically for community detection. SNMF finds a probability matrix U whose element U_{ij} indicates the probability that the i-th node belongs to the j-th community. SNMF can be formulated as Eq. 1.

$$\min_{U \geq 0} \mathcal{L}(U) = \left\| A - UU^T \right\|_F^2, \tag{1}$$

SCI [27] and CDE [16] are based on SNMF and take into account node attributes for community detection. For the convergence of U, they raise the coefficient of their multiplicative update rules to the power of ω, which are formulated as Eq. 2.

$$U_{ij} \leftarrow U_{ij}(I - \mu(U)\nabla_U \mathcal{L})_+^\omega, \tag{2}$$

where I is identity matrix, $\mu(U)$ is a learning rate function of U, $(\cdot)_+$ refers to a non-negative projection. In the implementation of SCI and CDE, they set ω as $\frac{1}{4}$ [7,17]. The numerator of the coefficient $(I - \mu(U)\nabla_U \mathcal{L})$ will be negative, so they project the coefficients to the positive direction, which affects the iteration efficiency.

3 SpEC Model

For non-overlapping community detection and overlapping community detection on attributed graphs, we present SpEC (Sparse Embedding-based Community detection) model. The model includes two phases: i) *exploring neighborhood patterns*, which generates the edges between sparse subgraphs of a graph; ii) *sparse embedding*, which factorizes the new adjacency matrix with new edges and the node attributes matrix sparsely into a probability matrix.

3.1   Exploring Neighborhood Patterns

First, we introduce our algorithm, AGRNN (AttributedGraphRNN), for extracting neighborhood patterns and generates the edges between sparse subgraphs of a graphs. The algorithm is summarized in Fig. 1. We first remove sparse subgraphs $\mathbf{g}_{spa}$ from an original graph G, and divide the remaining part of the graph into a subgraph set $\mathbf{g}$ which consists of the connected components $g_0, g_1, \ldots, g_m$, where m is the number of these connected components. Then we obtain the adjacency vector sequences and node attributes sequences of a graph $g \sim p(\mathbf{g})$ under BFS (breadth-first-search) [32] node orderings π according to two mapping f_A and f_N respectively.

$$A^{\pi(g)} = f_A(g, \pi(g)) = \left(A_1^{\pi(g)}, \ldots, A_n^{\pi(g)}\right)^T, \tag{3}$$

$$X^{\pi(g)} = f_N(g, \pi(g)) = \left(X_1^{\pi(g)}, \ldots, X_n^{\pi(g)}\right)^T, \tag{4}$$

where n is the number of nodes of the graph g, and A_i^{π}, $X_i^{\pi} (1 \leq i \leq n)$ are the corresponding adjacency vector and the node attributes in BFS ordering π of the graph g. There are two main reasons for using BFS orderings. One is to reduce the time complexity when training, the other is that BFS orderings make for refining node attributes of neighbors in order to learn neighborhood patterns of each node.

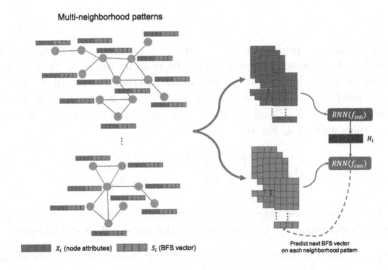

Fig. 1. The AGRNN step: AGRNN learns neighborhood patterns of large connected components. These components are the remaining part of a graph removed sparse subgraphs. The f_{ing} and f_{con} are learnt by two RNN respectively.

Node attributes sequences are refined to neighborhood patterns vector sequences $H^{\pi(g)}$. Since the graph is undirected, adjacency vectors are further compressed through BFS sequence mapping f_S.

$$S_i^{\pi(g)} = f_S(A_i^{\pi(g)}) = \left(S_{i,1}^{\pi(g)}, \ldots, S_{i,s}^{\pi(g)} \right), \tag{5}$$

where s is the sequence length of the i-th compressed adjacency vector, and $s = min\{i, l\}$. l is the maximum number of forward neighbors. Thus, assuming each neighborhood pattern is uncorrelated [32], then the attributed graph generative model is to learn Eq. 6 and Eq. 7.

$$p(S^\pi | H^\pi) = \prod_{i=1}^{n} p(S_i^\pi | H_i^\pi), \tag{6}$$

where

$$p(S_i^\pi | H_i^\pi) = \prod_{j=1}^{s-1} p\left(S_{i,j}^\pi | S_{i,1}^\pi, \ldots, S_{i,j}^\pi, H_i^\pi \right). \tag{7}$$

For learning Eq. 6 and Eq. 7, we use RNN-based neural networks. The first RNN-based neural network learns the inherent neighborhood patterns from node attributes and outputs neighborhood pattern vector, which can be described as Eq. 8. The second RNN-based neural network learns corresponding connection probability with forward nodes in a neighborhood pattern and outputs connection probability vectors, which can be described as Eq. 9.

$$H^\pi = f_{inh}(X^\pi), \tag{8}$$

$$\theta^\pi = f_{con}(H^\pi), \tag{9}$$

where the connection probability $\theta^\pi \in [0,1]^{n \times s}$. n equals to the maximum number of nodes of connected components in the training phase, while n equals to the number of nodes in the entire graph in the inference phase. In the inference phase, we sample from inferred distribution $\theta^{\pi(G)}$ to get each element of new adjacency vectors $S'^{\pi(G)}$. And we update the adjacency vector according to the following rules written as Eq. 10,

$$A'^{\pi(G)}_i = \begin{cases} f_S^{-1}(S'^{\pi(G)}_i), & i \in \mathbf{n}_{spa} \\ A_i^{\pi(G)}, & i \notin \mathbf{n}_{spa} \end{cases}, \tag{10}$$

where f_S^{-1} is an inverse mapping for recovering an adjacency vector from its compressed BFS sequence, the bold $\mathbf{n}_{spa}$ is the nodes set of all the sparse subgraphs $\mathbf{g}_{spa}$, which is equivalent to $\bigcup\limits_{i=1}^{m_{spa}} g_i$, where m_{spa} is the number of sparse subgraphs.

3.2 Sparse Embedding

We factorize node attributes and topological information through ANLS (Alternating Non-negative Least Squares)-based [21] way. Besides AGRNN, we also use the numerical methods to stable iterations. Inspired by Tikhonov regularization [19] and CNMF (Constrained Non-negative Matrix Factorization) [22], we propose an embedding method on attributed graphs with sparse constraints. The objective function in our model are formulated as Eq. 11.

$$\min_{U \geq 0, C \geq 0} \mathcal{L}(U, C) = \frac{1}{2}\|X' - UC\|_F^2 + \alpha\|C\|_F^2 + \alpha\|U\|_1^2 + \beta\|A' - UU^T\|_F^2, \quad (11)$$

where $U \in R^{n \times c}$ and $C \in R^{c \times r}$ represent the probability matrix that nodes belong to each community and the inferred inherent attributes matrix of communities, here n is a number of nodes, c is a number of communities, r is a number of node attributes, $E_{c \times c}$ is a matrix filled with ones.

We update U with fixed C and update C with fixed U iteratively according to the following rules, Eq. 12 and Eq. 13.

$$U_{ij} \leftarrow U_{ij} \frac{\left(X'C^T + 2\beta A'U\right)_{ij}}{\left(UCC^T + 2\alpha U E_{c \times c} + 2\beta UU^T U\right)_{ij}}, \quad (12)$$

$$C_{ij} \leftarrow C_{ij} \frac{U^T X'}{UU^T C + 2\alpha C}, \quad (13)$$

There is a reconstructing trick to reduce the number of computation and obtain another multiplicative update rules.

Proof of Convergence. Since the objective function is not convex, we only get the local optimal solution. Here we prove the convergence of our update rules. This proof is similar to the proof of convergence of Expectation-Minimization algorithm [13].

Definition 1 [13]. $G(U, U')$ is an *auxiliary function* for $\mathcal{L}(U)$ if the conditions

$$G(U, U') \geq \mathcal{L}(U), G(U, U) = \mathcal{L}(U), \quad (14)$$

are satisfied.

Lemma 1 [13]. If G is an auxiliary function, then F is nonincreasing under the update

$$U^{t+1} = \arg\min_U G(U, U^t), \quad (15)$$

Lemma 2 [13]. G is constructed with the following expression

$$G(u, u^t) = \mathcal{L}(u^t) + (u - u^t)^T \nabla \mathcal{L}(u^t) + \frac{1}{2}(u - u^t)^T K(u^t)(u - u^t), \quad (16)$$

where u is any column of U and $K(u^t)$ is a function that needs to be constructed to satisfy that G is an auxiliary function for $\mathcal{L}$. $\mathcal{L}(u)$ is represented by its Taylor expansion

$$\mathcal{L}(u) = L(u^t)^T \nabla \mathcal{L}(u^t) + \frac{1}{2}\mathcal{H}(\mathcal{L})(u - u^t), \quad (17)$$

where $\mathcal{H}(\mathcal{L})$ is Hessian matrix for $\mathcal{L}(u)$.

To prove $G(U, U^t) \geq \mathcal{L}(U)$ is equivalent to prove

$$M(u^t) = (u^t)^T (K(u^t) - \mathcal{H}(\mathcal{L})) u^t, \tag{18}$$

where u^t is any column of U^T, is a semidefinite matrix.

Theorem 1. If U^t is not far away from the KKT stationary point [7], the corresponding $G(u, u^t)$ of the following matrix $K(u^t)$(Eq. 19) is an auxiliary function for $\mathcal{L}(u)$ (Eq. 20).

$$K_{ij}(u^t) = K_{1,ij}(u^t) + K_{2,ij}(u^t) = \frac{\delta_{ij}(\tilde{C}^T \tilde{C} u^t)_i}{u_i^t} + \frac{\delta_{ij} 2\beta (U^t U^{tT} u^t - A' u^t)_i}{u_i^t},$$
$$\tag{19}$$

where u^t is a column of U^{tT}, $\tilde{C} = (C, \sqrt{2\alpha} I_c)^T$, $\delta_{ij} = 1$ when $i = j$ and $\delta_{ij} = 0$ when $i \neq j$.

$$\mathcal{L}(u) = \mathcal{L}_1(u) + \mathcal{L}_2(u)$$
$$= \frac{1}{2} \sum_{i=1}^{n} (\tilde{x}_i - \sum_{k=1}^{r} \tilde{C}_{ik} u_k)^2 + \beta \sum_{i=1}^{n} (a'_i - \sum_{k=1}^{n} U_{ik} u_k)^2, \tag{20}$$

where u is any column of U^T, a' is the corresponding column of A' and $\tilde{x}$ is the corresponding column of $\tilde{X} = (X', O_{c \times n})^T$.

Proof of Theorem 1. Because the sum of two semidefinite matrix is also a semidefinite matrix, this problem is equivalent to prove that $M_1 = (u^t)^T (K_1 - \mathcal{H}(\mathcal{L}_1)) u^t$ and $M_2 = (u^t)^T (K_2 - \mathcal{H}(\mathcal{L}_2)) u^t$ are semidefinite matrices.

$$
\begin{aligned}
\nu^T M_1 \nu &= \sum_{i=1}^{n} \sum_{j=1}^{n} \nu_i M_{1,ij} \nu_j \\
&= \sum_{i=1}^{n} u_i^t (\tilde{C}^T \tilde{C} u^t)_i \nu_i^2 - \sum_{i=1}^{n} \sum_{j=1}^{n} \nu_i u_i^t \left(\tilde{C}^T \tilde{C} \right)_{ij} u_j^t \nu_j \\
&= \sum_{i=1}^{n} \sum_{j=1}^{n} (u_i^t \left(\tilde{C}^T \tilde{C} \right)_{ij} u_j^t \nu_i^2 - \nu_i u_i^t \left(\tilde{C}^T \tilde{C} \right)_{ij} u_j^t \nu_j), \\
&= \sum_{i=1}^{n} \sum_{j=1}^{n} u_i^t \left(\tilde{C}^T \tilde{C} \right)_{ij} u_j^t \left(\frac{1}{2} \nu_i^2 + \frac{1}{2} \nu_j^2 - \nu_i \nu_j \right) \\
&= \frac{1}{2} \sum_{i=1}^{n} \sum_{j=1}^{n} \left(\tilde{C}^T \tilde{C} \right)_{ij} u_i^t u_j^t (\nu_i - \nu_j)^2 \\
&\geq 0
\end{aligned}
\tag{21}
$$

According to Eq. 21, we prove that M_1 is a semidefinite matrix. In addiction, from [7], we know that M_2 is also a semidefinite matrix when u is not far away from KKT stationary point.

4 Experiments

In this section, we evaluate our SpEC model on real-world datasets with attributed graphs by comparing to 10 community detection algorithms. For better comparison with baselines, we conduct our model on each dataset with 10 initializations and report the average of these 10 results as a general rule [16,27]. The codes of our model are available at the website[1]. To demonstrate the effectiveness of our method, we compare our model with these models (summarized in Table 1).

Table 1. Models for community detection. Attr: Attributes, Topo: Topological information, Over: Overlapping community detection, Non-over: Non-overlapping community detection, Emb: Embedding, Spa: Sparse subgraphs.

Model	Attr	Topo	Over	Non-over	Emb	Spa
SpEC (ours)	✓	✓	✓	✓	✓	✓
SNMF [26]		✓	✓	✓	✓	
NC [24]		✓		✓		
CAN [20]	✓			✓		
SMR [9]	✓			✓	✓	
PCL-DC [31]	✓	✓	✓			
SCI [27]	✓	✓	✓	✓	✓	
CDE [16]	✓	✓	✓	✓	✓	
Bigclam [29]		✓	✓		✓	
CESNA [30]	✓	✓	✓		✓	
Circles [14]	✓	✓	✓			

4.1 Datasets

We evaluate the performance of our method on attributed graph datasets with ground-truth communities, whose statistics are shown in Table 2. The memberships in Table 2 represent the average number of communities that each node belongs to. Cora [18], Citeseer [23] are citation networks. In these networks, nodes are documents, edges are citation relationship, node attributes are 0/1-valued word vectors. Cornell, Texas, Washington, Wisconsin are website networks [18]. In these networks, nodes are websites, edges are hyperlinks, node attributes are 0/1-valued word vectors. These citation relationships and hyperlinks are all treated as undirected edges in our method. Facebook ego networks [14] consists of circles of friends from facebook. In these networks, nodes are people, edges are friendship, node attributes are 0/1-valued anonymized personal characteristics. All datasets are available at the website[2,3].

---

[1] https://github.com/wendell1996/SpEC.git.
[2] https://linqs.soe.ucsc.edu/data.
[3] http://snap.stanford.edu/data/index.html.

Table 2. Dataset statistics.

Dataset	Type	Nodes	Edges	Communities	Attributes	Memberships
Cornell	Website network	195	283	5	1703	1
Texas	Website network	187	280	5	1703	1
Washington	Website network	230	366	5	1703	1
Wisconsin	Website network	265	459	5	1703	1
Cora	Citation network	2708	5278	7	1433	1
Citeseer	Citation network	3312	4536	6	3703	1
FB ego 0	Social network	348	2866	24	224	0.93
FB ego 107	Social network	1046	27783	9	576	0.48
FB ego 1684	Social network	793	14810	17	319	0.98
FB ego 1912	Social network	756	30772	46	480	1.41
FB ego 3437	Social network	548	5347	32	262	0.35
FB ego 348	Social network	228	3416	14	161	2.49
FB ego 3980	Social network	60	198	17	42	0.97
FB ego 414	Social network	160	1843	7	105	1.11
FB ego 686	Social network	171	1824	14	63	2.84
FB ego 698	Social network	67	331	13	48	1.27

4.2 Non-overlapping Community Detection

In this experiment, we evaluate the performance of our model on attributed graphs with non-overlapping communities. We compare our model against seven state-of-the-art community detection algorithms. We use the source codes provided by authors.

We choose two evaluation metrics, AC (Accuracy) and NMI (Normalized Mutual Information). The accuracy is calculated after the maximum match of the bipartite graph of predicted communities and ground-truth by Hungarian algorithm [27]. For baseline algorithms, we set their parameters by default as recommended in their papers. We set this parameter k of CDE as 25, the best empirical value of their experiments. In addition, our model set the range of parameters α, β between 0.5 and 2, c as the number of ground-truth communities. We preprocess the node attributes by Gaussian smoothing [11,15], which has experimentally proven effective in many tasks. The smoothing order t is varied from 0 to 2.

The results reported in Table 3 show that SpEC clearly outperforms other baseline methods on these datasets. In addition, these results also prove that our non-negative update rules with sparse constraints are better than CDE and SCI update rules. As the number of nodes increases, the model learns more neighborhood patterns, so there is better performance on Cora and Citeseer. We notice that the values of AC of CDE and SCI are high and the values of NMI of these models are low in Texas dataset. This is because, in the Texas dataset, one type of community has a large proportion and these models tend to divide most

Table 3. Non-overlapping community detection.

Algorithms	Cornell		Texas		Washington		Avg. rank
	AC	NMI	AC	NMI	AC	NMI	
SpEC (ours)	**0.6103**	**0.3103**	0.5508	**0.3107**	**0.6957**	**0.3937**	1.17
CDE	0.5338	0.2280	0.5610	0.1978	0.6400	0.3332	2.42
SCI	0.4769	0.1516	**0.6096**	0.2153	0.5173	0.1304	3.58
PCL-DC	0.3512	0.0873	0.3850	0.0729	0.4608	0.1195	5.17
CAN	0.4154	0.0614	0.4706	0.0908	0.5087	0.1175	5.17
SMR	0.3179	0.0845	0.5401	0.1150	0.4565	0.0381	6.17
SNMF	0.3692	0.0762	0.4019	0.1022	0.3009	0.0321	5.75
NC	0.3538	0.0855	0.4545	0.0706	0.4348	0.0591	6.58
Algorithms	Wisconsin		Cora		Citeseer		Avg. rank
	AC	NMI	AC	NMI	AC	NMI	
SpEC (ours)	**0.7321**	**0.5037**	**0.6795**	**0.4726**	**0.6264**	**0.3726**	1.17
CDE	0.7045	0.4390	0.4145	0.2746	0.5827	0.2985	2.42
SCI	0.5283	0.1823	0.4121	0.2138	0.3260	0.0758	3.58
PCL-DC	0.3773	0.0778	0.5823	0.4071	0.4682	0.2246	5.17
CAN	0.4717	0.0702	0.3021	0.0132	0.2129	0.0079	5.17
SMR	0.4226	0.0777	0.3002	0.0078	0.2111	0.0032	6.17
SNMF	0.3773	0.0842	0.4323	0.2996	0.3079	0.1044	5.75
NC	0.3170	0.0507	0.2622	0.1731	0.4094	0.1998	6.58

of nodes into this community. In contrast, our model is more robust. Overall, our model performs well for non-overlapping community detection.

4.3   Overlapping Community Detection

In this experiment, we evaluate the performance of our model on attributed graphs with overlapping ground-truth communities. We compare our model against five state-of-the-art overlapping community detection algorithms.

We chose the F1-score [30] and Jaccard similarity [16] as the evaluation metrics. We set the parameters of the baseline model by default as recommended in their papers. In order to keep the parameters consistent, our model set the same value as above (Sect. 4.2). From the results in Table 4, our SpEC model outperforms other comparison methods in the task of overlapping community detection on attributed networks. In the ego1684 dataset, our result is not prominent because the number of nodes in each community is small in this dataset and we add edges to the graph, which makes the probability of generating large communities larger. Overall, these experiments prove that our model performs well for overlapping community detection.

Table 4. Overlapping community detection.

Algorithms	FB-ego0		FB-ego107		FB-ego1684		FB-ego1912		FB-ego3437		Avg. rank
	F1	Jaccard	F1	Jaccard	F1	Jaccard	F1	Jaccard	F1	Jaccard	
SpEC (ours)	**0.3335**	**0.2103**	**0.3766**	**0.2616**	0.4173	0.2966	**0.4669**	**0.3311**	**0.3279**	**0.2120**	1.10
CDE	0.2559	0.1559	0.3337	0.2377	**0.5002**	**0.3688**	0.3466	0.2320	0.1529	0.0865	4.80
SCI	0.2104	0.1255	0.1932	0.1203	0.2290	0.1405	0.2787	0.1872	0.1909	0.1130	4.30
CESNA	0.2638	0.1635	0.3526	0.2512	0.3850	0.2656	0.3506	0.2417	0.2125	0.1311	3.55
Circles	0.2845	0.1844	0.2722	0.1755	0.3022	0.1947	0.2694	0.1744	0.1004	0.0545	3.65
BigClaim	0.2632	0.1617	0.3593	0.2589	0.3652	0.2655	0.3542	0.2443	0.2109	0.1226	3.60
Algorithms	FB-ego348		FB-ego3980		FB-ego414		FB-ego686		FB-ego698		Avg. rank
	F1	Jaccard	F1	Jaccard	F1	Jaccard	F1	Jaccard	F1	Jaccard	
SpEC (ours)	**0.5403**	**0.4029**	**0.4772**	**0.3275**	**0.6209**	**0.4930**	**0.5441**	**0.3945**	**0.5543**	**0.4173**	1.10
CDE	0.4223	0.3029	0.3449	0.2398	0.4095	0.2823	0.4191	0.2842	0.3312	0.2120	4.80
SCI	0.4469	0.3068	0.3502	0.2336	0.5617	0.4270	0.4340	0.2915	0.4323	0.2942	4.30
CESNA	0.4943	0.3684	0.4312	0.3041	0.6181	0.4878	0.3638	0.2372	0.5222	0.3811	3.55
Circles	0.5209	0.3937	0.3277	0.2127	0.5090	0.3670	0.5242	0.3828	0.3715	0.2400	3.65
BigClaim	0.4917	0.3649	0.4468	0.3147	0.5703	0.4446	0.3583	0.2272	0.5276	0.3898	3.60

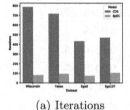
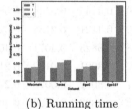
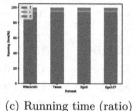

(a) Iterations   (b) Running time   (c) Running time (ratio)

Fig. 2. (a) Comparison of the number of iterations between our model and CDE. (b) Running time of each stage of our model. T represents the training time of each epoch. (c) The proportion of running time of each stage. T: Training stage of AGRNN. I: Inference stage of AGRNN. C: Community detection of sparse embedding.

4.4   Efficiency Analysis

In this section, we evaluate the iterative efficiency of our model. Considering that CDE is the latest model based on NMF and the update rules of SCI are almost the same as the CDE's, we only compare our model with CDE in this section. We report the number of iterations of convergence, as shown in Fig. 2(a). It is seen that the update rules of our model are efficient update rules now that the iterations needed of our model are far less than others.

In addition, we evaluate the running time of each stage of our model. In this experiment, we set the batch size to 3, and train 50 epochs. We report the absolute running time (Fig. 2(c)) and the ratio of each stage (Fig. 2(c)) respectively. It is seen that the major time consuming of our model is the training stage of AGRNN, and the running time of the inference stage of AGRNN and spare embedding is very short. All in all, our model is efficient.

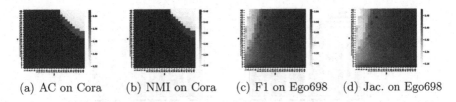

(a) AC on Cora   (b) NMI on Cora   (c) F1 on Ego698   (d) Jac. on Ego698

Fig. 3. Testing the parameter sensitivity of our model for non-overlapping community detection on Cora dataset and testing the parameter sensitivity for overlapping community detection on FB ego698. α and β vary from 1 to 49 respectively.

4.5 Parameter Sensitivity Analysis

In this section, we analyzed the parameter sensitivity of our model. Our model has two major parameters: α is a parameter that controls the sparsity of the probability matrix U and the inherent attributes matrix C, β is a parameter that controls the proportion of topological information and node attributes in the embedding. We vary α and β from 1 to 49 respectively. Figure 3 shows the corresponding results on Cora and FB ego698 dataset. As α is to control sparsity, a too large value of α could cause embedding into a zero vector. Fortunately, our model can run as expected, even when the α is a bit large. Generally, this value is recommended in the range of 0.5 to 1. A community to which a node belongs in an attributed graph is usually dominated by attributes or topological information. However, Fig. 3 shows that our model is well integrated for node attributes and topological information, and is not very dependent on the parameter β, even if it is a parameter that determines the proportion of node attributes and topological information in embeddings.

4.6 Effectiveness of AGRNN

In this section, we compared AGRNN with other generating methods, NetGAN [3] and GraphRNN [32]. The parameters of the sparse embedding step are set as above as well. The parameters of NetGAN [3] and GraphRNN [32] are set as default. NetGAN generates walks of a graph, so the number of generating edges is needed. We set the number as 160. From Fig. 4, we have experimentally proved that AGRNN is effective. There are two major possible reasons why NetGAN performs not well. NetGAN learns how to generate a walk that looks like a true walk. This means NetGAN hardly generates a walk between two connected components, which tends to generate edges of pairs of nodes that are in the same sparse subgraph. In addition, it is difficult to select the number of generating edges. For GraphRNN, the order of nodes of a plain graph generated is random and, in most cases, is not matched as the order of nodes of an input graph.

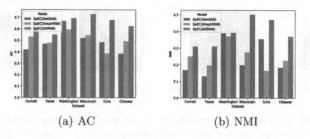

(a) AC   (b) NMI

Fig. 4. Comparing with other generating methods.

With the increase of nodes, this random order would result in much poor performance. Considering the node attributes, whether for learning neighborhood patterns or for overcoming the above shortcomings, are necessary and AGRNN is effective.

4.7 Visualization of Graphs

In this section, our experiments show the visualization of the graph after generating edges by learning neighborhood patterns. We visualized graphs on the Cora dataset. In Fig. 5, the nodes in the center are dense subgraphs and the surrounding nodes are sparse subgraphs. It is seen that most sparse subgraphs are connected with other subgraphs after generating edges from the sparse subgraphs by AGRNN.

5   Related Works

5.1   Generating Graphs

At the earliest, traditional models generate graphs by empirical network features [32], such as Baranasi-Albert model [1], Kronecker graphs [4]. However, these models cannot generate graphs from observed graphs. Models such as GraphRNN [32] and NetGAN [3] solve this problem. NetGAN is a walk-based model, which needs a hyperparameter to control the number of generated edges [3]. GraphRNN is an auto-regressive model that generates a graph without node attributes according to the auto-regressive property of a graph. Nevertheless, these methods do not take advantage of node attributes, so they cannot be used for exploring neighborhood patterns. Some functions based on attribute distance are often used to generate edges on attributed graphs, such as Gaussian function [5,24] and cosine similarity, but such methods based on attribute distance do not consider the property of neighborhood patterns.

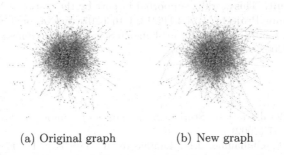

(a) Original graph   (b) New graph

Fig. 5. (a) is the original graph of Cora dataset. (b) is the graph added new edges according to the neighborhood patterns by our model respectively. Nodes in different colors belong to different communities in these figures. A total of 110 edges have been generated, and the ratio of the number of edges to the number of nodes has increased from 1.94 to 1.99. (Color figure online)

5.2   NMF-Based Community Detection

Non-negative matrix factorization is often used for community detection because it not only can detect both non-overlapping communities and overlapping communities but also avoids the problem of *sparse overlaps* [29]. Lee et al. [12] proposed a clustering algorithm based on non-negative matrix factorization. Wang et al. [26] proposed an NMF-based community detection algorithm in graphs without node attributes. They factorize the symmetric adjacency matrix of a graph into a probability matrix whose rows represent the probability that the corresponding nodes belong to every community. SCI [27] treats nodes attributes as the basis to calculate nodes embedding, which could degrade the effectiveness of SCI, CDE tackles the problem by adding a loss item of embedding for node attributes to integrate community structure embedding matrix and node attributes matrix [16].

6   Conclusion

In this paper, we study the algorithm for community detection on attributed graphs. We propose a novel model, namely SpEC, which takes full advantage of both node attributes and topological information and iterates efficiently. Specifically, our model can generate edges according to neighborhood patterns learned by AGRNN. This step can help stabilize the iteration of sparse embedding while recovering or enriching the information of a graph. By adding a sparse constraint of the probability matrix on non-negative matrix factorization and designing multiplication rules guaranteeing that the coefficient of multiplication rules is positive, our model is efficient and robust, and the performance of our model for community detection on attributed graphs is improved. Finally, the experiments on real-world attributed graph datasets show that our model outperforming other state-of-the-art models.

Acknowledgment. This work is supported in part by the National Natural Science Foundation of China Projects No. U1936213, U1636207, the National Science Foundation Projects No. 1565137, 1829071, the Shanghai Science and Technology Development Fund No. 19DZ1200802, 19511121204.

References

1. Albert, R., Barabási, A.L.: Statistical mechanics of complex networks. Rev. Mod. Phys. **74**(1), 47 (2002)
2. Blondel, V.D., Guillaume, J.L., Lambiotte, R., Lefebvre, E.: Fast unfolding of communities in large networks. J. Stat. Mech: Theory Exp. **2008**(10), P10008 (2008)
3. Bojchevski, A., Shchur, O., Zügner, D., Günnemann, S.: NetGAN: generating graphs via random walks. arXiv preprint arXiv:1803.00816 (2018)
4. Cho, K., et al.: Learning phrase representations using RNN encoder-decoder for statistical machine translation. arXiv preprint arXiv:1406.1078 (2014)
5. Defferrard, M., Bresson, X., Vandergheynst, P.: Convolutional neural networks on graphs with fast localized spectral filtering. In: Advances in Neural Information Processing Systems, pp. 3844–3852 (2016)
6. Fortunato, S.: Community detection in graphs. Phys. Rep. **486**(3–5), 75–174 (2010)
7. He, Z., Xie, S., Zdunek, R., Zhou, G., Cichocki, A.: Symmetric nonnegative matrix factorization: algorithms and applications to probabilistic clustering. IEEE Trans. Neural Netw. **22**(12), 2117–2131 (2011)
8. Hoyer, P.O.: Non-negative sparse coding. In: Proceedings of the 12th IEEE Workshop on Neural Networks for Signal Processing, pp. 557–565. IEEE (2002)
9. Hu, H., Lin, Z., Feng, J., Zhou, J.: Smooth representation clustering. In: Proceedings of the IEEE Conference on Computer Vision and Pattern Recognition, pp. 3834–3841 (2014)
10. Kim, J., Park, H.: Sparse nonnegative matrix factorization for clustering. Technical report, Georgia Institute of Technology (2008)
11. Kipf, T.N., Welling, M.: Semi-supervised classification with graph convolutional networks. arXiv preprint arXiv:1609.02907 (2016)
12. Lee, D.D., Seung, H.S.: Learning the parts of objects by non-negative matrix factorization. Nature **401**(6755), 788 (1999)
13. Lee, D.D., Seung, H.S.: Algorithms for non-negative matrix factorization. In: Advances in Neural Information Processing Systems, pp. 556–562 (2001)
14. Leskovec, J., Mcauley, J.J.: Learning to discover social circles in ego networks. In: Advances in Neural Information Processing Systems, pp. 539–547 (2012)
15. Li, Q., Han, Z., Wu, X.M.: Deeper insights into graph convolutional networks for semi-supervised learning. arXiv preprint arXiv:1801.07606 (2018)
16. Li, Y., Sha, C., Huang, X., Zhang, Y.: Community detection in attributed graphs: an embedding approach. In: Thirty-Second AAAI Conference on Artificial Intelligence (2018)
17. Long, B., Zhang, Z.M., Wu, X., Yu, P.S.: Relational clustering by symmetric convex coding. In: Proceedings of the 24th International Conference on Machine Learning, pp. 569–576. ACM (2007)
18. Lu, Q., Getoor, L.: Link-based classification. In: Proceedings of the 20th International Conference on Machine Learning (ICML 2003), pp. 496–503 (2003)
19. Neumaier, A.: Solving ill-conditioned and singular linear systems: a tutorial on regularization. SIAM Rev. **40**(3), 636–666 (1998)

20. Nie, F., Wang, X., Huang, H.: Clustering and projected clustering with adaptive neighbors. In: Proceedings of the 20th ACM International Conference on Knowledge Discovery and Data Mining, pp. 977–986. ACM (2014)
21. Paatero, P., Tapper, U.: Positive matrix factorization: a non-negative factor model with optimal utilization of error estimates of data values. Environmetrics 5(2), 111–126 (2010)
22. Pauca, V.P., Piper, J., Plemmons, R.J.: Nonnegative matrix factorization for spectral data analysis. Linear Algebra Appl. 416(1), 29–47 (2006)
23. Sen, P., Namata, G., Bilgic, M., Getoor, L., Galligher, B., Eliassi-Rad, T.: Collective classification in network data. AI Mag. 29(3), 93 (2008)
24. Shi, J., Malik, J.: Normalized cuts and image segmentation. Departmental Papers (CIS), p. 107 (2000)
25. Von Luxburg, U.: A tutorial on spectral clustering. Stat. Comput. 17(4), 395–416 (2007)
26. Wang, F., Li, T., Wang, X., Zhu, S., Ding, C.: Community discovery using nonnegative matrix factorization. Data Min. Knowl. Disc. 22(3), 493–521 (2011)
27. Wang, X., Jin, D., Cao, X., Yang, L., Zhang, W.: Semantic community identification in large attribute networks. In: Thirtieth AAAI Conference on Artificial Intelligence, pp. 265–271 (2016)
28. Yang, J., Leskovec, J.: Structure and overlaps of communities in networks. arXiv preprint arXiv:1205.6228 (2012)
29. Yang, J., Leskovec, J.: Overlapping community detection at scale: a nonnegative matrix factorization approach. In: Proceedings of the Sixth ACM International Conference on Web Search and Data Mining, pp. 587–596. ACM (2013)
30. Yang, J., McAuley, J., Leskovec, J.: Community detection in networks with node attributes. In: 2013 IEEE 13th International Conference on Data Mining (ICDM), pp. 1151–1156. IEEE (2013)
31. Yang, T., Jin, R., Chi, Y., Zhu, S.: Combining link and content for community detection: a discriminative approach. In: Proceedings of the 15th ACM International Conference on Knowledge Discovery and Data Mining, pp. 927–936. ACM (2009)
32. You, J., Ying, R., Ren, X., Hamilton, W., Leskovec, J.: GraphRNN: generating realistic graphs with deep auto-regressive models. In: International Conference on Machine Learning, pp. 5694–5703 (2018)

MemTimes: Temporal Scoping of Facts with Memory Network

Siyuan Cao[1], Qiang Yang[2], Zhixu Li[1(✉)], Guanfeng Liu[3], Detian Zhang[1], and Jiajie Xu[1]

[1] Institute of Artificial Intelligence, School of Computer Science and Technology,
Soochow University, Suzhou, China
`sycao@stu.suda.edu.cn`, {`zhixuli,detian,xujj`}`@suda.edu.cn`
[2] King Abdullah University of Science and Technology, Jeddah, Saudi Arabia
`qiang.yang@kaust.edu.sa`
[3] Department of Computing, Macquarie University, Sydney, Australia
`guanfeng.liu@mq.edu.au`

Abstract. This paper works on temporal scoping, i.e., adding time interval to facts in Knowledge Bases (KBs). The existing methods for temporal scope inference and extraction still suffer from low accuracy. In this paper, we propose a novel neural model based on Memory Network to do temporal reasoning among sentences for the purpose of temporal scoping. We design proper ways to encode both semantic and temporal information contained in the mention set of each fact, which enables temporal reasoning with Memory Network. We also find ways to remove the effect brought by noisy sentences, which can further improve the robustness of our approach. The experiments show that this solution is highly effective for detecting temporal scope of facts.

Keywords: Temporal Scoping · Memory Network · Iterative Model

1 Introduction

Nowadays, large Knowledge Bases (KBs) of facts play vital roles in many upper-level applications such as Question Answering and Information Retrievel. While some facts hold eternally, many others change over time, such as the president of U.S., the spouse of some people, and the teams of some athlets. However, most KBs including YAGO [6], Wikidata [17] are still sparsely populated in terms of temporal scope, which greatly reduces the usefulness of the data for upper-level applications based on KBs.

The task of adding time interval to facts in KBs is named as Temporal Scoping. It was first well-known as Temporal Slot Filling (TSF) task on Text Analysis Conference (TAC) since 2013. As the example shown in Fig. 1, for a fact such as $(U.S., president, BarackObama)$ waiting for temporal scoping, the TAC-TSF task provides a set of mentions (i.e., sentences that have mentioned the two entities in the fact) with timestamps such as those listed in the figure.

© Springer Nature Switzerland AG 2020
Y. Nah et al. (Eds.): DASFAA 2020, LNCS 12114, pp. 70–86, 2020.
https://doi.org/10.1007/978-3-030-59419-0_5

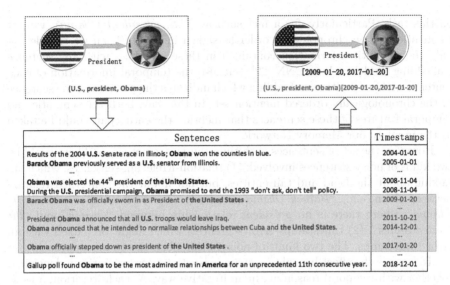

Fig. 1. An example scenario of temporal scoping on TAC-TSF

Based on this mention set, the task requires us to produce temporal scope for the fact, i.e., [*20 Jan. 2009, 20 Jan. 2017*], which indicates that the fact is TRUE for a period beginning at *20 Jan. 2009* and ending at *20 Jan. 2017*.

Since then, plenty of work has been done on temporal scoping. Early methods focus on using syntactic rules or patterns to extract time information contained in Web pages [10,12] or Wikipedia, which suffer from low accuracy. Some later work [20] models temporal scoping into a state change detection problem. They construct temporal profiles for entities with relevant contextual information such as *"elect"* and *"senator"* for *"Barack Obama"*, and then identify change patterns about facts of entities from the contexts. However, they perform poorly on inferring ending time for facts given that contexts relevant to a fact often still mentioned with the entity even after the fact ceases to be valid. Another line of work [2,7,8,15,16] pays attention to infer the temporal information for facts based on temporal KBs of facts, where some facts already have timestamped relations. However, without using external knowledge, these work could hardly reach a satisfied performance.

In this paper, we propose to employ Memory Network [13] in temporal scoping, based on a set of mentioned sentences with timestamps as defined in the TAC-TSF task scenario. Different from previous work which only leverages word-level features of texts and manually supplied constraints (e.g., there can only be one U.S. president at a time) on specified relations, we would like to perform temporal reasoning on the mention set based on both semantic and temporal features of these mentions with a neural model. Given that Memory Network has great advantages in long-term memory and multi-hop reasoning from the past over other models like LSTM and RNN, it becomes a great choice for us on this task. Particularly, for a fact such as $(U.S., president, Obama)$, we put its mention set in chronological order as shown in Fig. 1. Then we encode not only the semantic information contained in these sentences with any sentence embedding model such as CNN [22], but also the temporal information of each sentence with its position embedding which indicates the position of the sentence in the chronologically ordered mention set. In this way, both the semantic and temporal features of these sentences that mention the entity pair could be taken as the input to our Memory Network.

However, the set of sentences with timestamps are not clean. There could be two kinds of noisy sentences involved: (1) relation-irrelevant sentences, which are talking about the head and tail entities, but not about the concerned relation between them, e.g., *"Barack Obama previously served as a U.S. senator from Illinois"*, where there is no *president* relationship between *"Obama"* and *"the United States"*. (2) timestamps-error sentences, which might be parsed with an error timestamps. The two kinds of noises bring obstacles to our approach.

To remove the effect brought by these noisy sentences, we perform denoising, together with temporal reasoning, in an iterative way. At each iteration, a noise weight would be generated for each sentence, reflecting whether the sentence is irrelevant to the target relation according to the temporal reasoning results of this iteration. Then, the noise weights would be integrated into the sentence representations as the input of the next iteration. The iteration process stops when the weights of sentences become stable. After the iteration stops, instead of adopting the temporal reasoning result of the last iteration, we also design a sliding window based way to decide the starting and ending sentence. In this way, we could further improve the robustness of our approach.

In summary, our main contributions are as follows:

- We propose a novel neural model based on Memory Network to do temporal reasoning among sentences for the purpose of temporal scoping.
- We design proper ways to encode both semantic and temporal information contained in the mention set of each fact, which enables temporal reasoning with Memory Network.
- We also propose several ways to remove the effect brought by noisy sentences, which can further improve the robustness of our approach.

Our emperical study on the TAC-TSF evaluation data set shows that our model outperforms several state-of-art approaches on temporal scoping.

Roadmap. The rest of the paper is organized as follows: We define the task of temporal scoping in Sect. 2, and then present our approach in Sect. 3. After reporting our empirical study in Sect. 4, we cover the related work in Sect. 5. We finally conclude in Sect. 6.

2 Problem Definition

This paper works on temporal scoping for facts in a given KB based on a set of sentences with timestamps. More formally, we define the temporal scoping problem as follows.

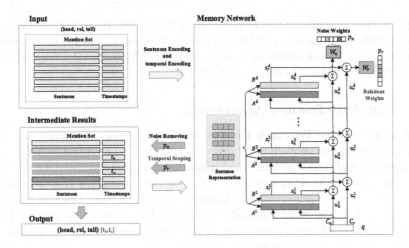

Fig. 2. Architecture of our approach

Definition 1 (Temporal Scoping). *Given a fact (i.e., a knowledge triple)* $f = (head, relation, tail)$, *let* $M(f) = \{(s_1, t_1), (s_2, t_2), ...(s_M, t_M)\}$ *denote its mention set, where* s_i *is the* $i-th$ *sentence that mentions head and tail in* f, *and* t_i *is its associated time spot, the task of* **Temporal Scoping** *aims at finding a time interval* $[t_b, t_e]$, *which indicates that the knowledge triple* f *is TRUE in the time period begining at* t_b *and ending at* t_e.

3   Our Approach

The architecture of our approach is given in Fig. 2: The input is a triple $(head, relation, tail)$ and a set of corresponding mentions, while the output is the time interval $[t_b, t_e]$ for the triple. The entire process of our approach can be roughly divided into three parts: sentence and temporal encoding, iterative memory reasoning and time boundaries determination. More specifically,

- **Sentence and Temporal Encoding.** The pipeline starts with processing the mention set, where sentence encoding and temporal encoding is performed to obtain the sentence representation. We present how we encode both sentences and temporal information for temporal scoping specifically in Sect. 3.1.
- **Iterative Memory Reasoning.** Through iterative and multi-hop memory reasoning, noise weights for sentences are obtained and integrated into the sentence representations and stored in the intermediate result. The intermediate result would be taken as the input for the next iteration. The iteration stops when the noisy weights become stable. More details about this module could be found in Sect. 3.2.
- **Time Boundaries Determination.** When the iteration stops, based on the reasoning result of the last iteration, we finally determine the beginning and

ending time boundaries for the triple. In order to identify the time boundaries accurately, we propose a heuristic method to process the intermediate results, aiming at excluding sentences that are unlikely to be within the time interval. We present this part in details in Sect. 3.3.

3.1 Multi-level Features Acquirement

1) Sentence Level Features Representation. In order to extract sentence level features, we use convolutional neural networks to combine all local features to extract long-distance grammatical information in sentences and finally generate our sentence-level eigenvectors. For a sentence consisting of word tokens, each token is further represented as Word Features (WF) and Position Features (PF), the WF is the word embeddings and the PF is the combination of the relative distances of the current word to the specify entity pairs. By combining the WF and PF, we can convert an instance into a matrix $S \in \mathbb{R}^{s \times d}$, where s is the sentence length and $d = d_w + d_p \times 2$. The matrix S is subsequently fed into the convolution part. Convolution layer extracts local features between convolutional filter and the vector representation of inputs. Let w be the width of a filter, the convolution operation can be expressed as follows:

$$c_{ij} = W_i \cdot S_{j-w+1:j} + b_i \tag{1}$$

where c_{ij} denotes the jth output of filter i, $W_i \in \mathbb{R}^{d \times w}$ is the shared weight parameters of linear layers in the filter and $b_i \in \mathbb{R}^1$ is the bias. Then, we use a single max pooling operation to determine the most significant features:

$$e_i = \max_{1 \leq j \leq s+w-1}(c_{ij}) \tag{2}$$

We then concatenate the output of all convolutional filters and consider it as the sentence level feature:

$$E = [e_1, e_2, e_3, \ldots, e_f] \tag{3}$$

where f denotes the number of filters.

2) Temporal Features Representation. To enable our model to make use of the chronological information of the instances, we must inject positional encoding among sentences. The positional encodings indicate the relative or absolute position of the sentence in the mention set. Because temporal knowledge express differently in sentences, encoding the time spot value is non-trivial for this task. Thus, in this work, we use *sine* and *cosine* functions of different frequencies to learn temporal features:

$$T_{2k} = \sin(o/10000^{2k/d_{pe}}) \tag{4}$$

$$T_{2k+1} = \cos(o/10000^{(2k+1)/d_{pe}}) \tag{5}$$

where o is the order of instances and k is the dimension, d_{pe} is the dimension of position encoding. We chose this function because we hypothesized it would allow the model to easily learn to attend by relative positions.

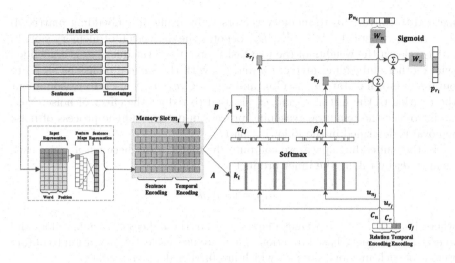

Fig. 3. A single layer version of memory network

3.2   Temporal Reasoning with Memory Network

In this section, we use an overall multi-hop Memory Network to perform temporal reasoning among different instances, and jointly calculate the weighted attention of both noise and relations. The noise weights are given to the input of the next iteration of the Memory Network to reduce the effect of noisy instances, and the relation weights are used to determine the temporal scope.

First, we introduce the input of the Memory Network. We concatenate the encoded sentence and temporal encoding to form the final representation of instance i. Each instance is considered to be a memory slot m_i:

$$m_i = [E_i; T_i] \tag{6}$$

Then, we construct a query by adding the randomly initialized relation embedding which indicates the representation of relations, and we also concatenate the query with the same temporal encoding to obtain the jth query:

$$q_j = [R_j; T_j] \tag{7}$$

where $R_j \in \mathbb{R}^{D_r}$ is the embedding of specific relation r, and D_r denotes the dimension of relation. We utilise both m_i and q_j to iteratively reason and remove noise to determine the final temporal scope.

1) Single Layer Memory Network. In this part, we describe Memory Network in the single layer case, as illustrated in Fig. 3.

First, we explore the importance of all sentences for each relation. We compute the similarity score and importance probability between the query vector q_j and each memory slot m_i. Note that the encoding output m_i and the query vector are not in the same continuous space. So, we adopt linear projection method to convert m_i to memory vectors k_i by using an embedding matrix A: $k_i = A^T m_i$, where $A \in \mathbb{R}^{D_m \times D_r}$, D_m denotes the dimension of memory slot. In order to learn the weights of the two tasks separately, the query q_j is embedded (via another embedding matrix C_r and C_n with the same dimensions as A) to obtain an internal state $u_{r_j} = C_r q_j$ and $u_{n_j} = C_n q_j$, u_{r_j} is used to reason about the relation of the instance, and u_{n_j} is used to reduce the effect of noise. Next we introduce the temporal reasoning process in detail, and the process of noise removal is described in the following part.

In the embedding space, we compute the match score between u_{r_j} and each memory k_i by taking the inner product:

$$a_{i,j} = u_{r_j}^T W_a k_i \tag{8}$$

where $W_a \in \mathbb{R}^{D_r \times D_m}$ is weight matrix, i, j are the indices of memory slots and queries, $a_{i,j}$ is query-based function. Then, we use a selective attention to obtain weight of each memory slots k_i, which are further defined as follows:

$$\alpha_{i,j} = \frac{\exp(a_{i,j})}{\sum_{x=1}^{M} \exp(a_{x,j})} \tag{9}$$

where M is the size of a mention set. Similarly, in order to obtain the memory value v_i, we employ a simple strategy via another embedding matrix B with the same dimensions as A: $v_i = B^T m_i$, which convert memory slot m_i to output memory representation. Therefore, we compute the weighted sum over memory value v_i with the importance probability derived in the previous step:

$$s_{r_j} = \sum_i \alpha_{i,j} v_i \tag{10}$$

So as to generate the final prediction, in the single layer case, we feed the sum of the output vector s_{r_j} and the input embedding u_{r_j} to a binary classifier. Therefore the confidence scores for each relation can be calculated as:

$$p_{r_i} = Sigmoid(W_r(s_{r_j} + u_{r_j})) \tag{11}$$

where $W_r \in \mathbb{R}^{D_r \times D_m}$ is the final weight matrix. All embedding matrices A, B, C_r and C_n, as well as W_a, W_r, and W_n mentioned later are jointly learned during training.

2) Multiple Layer Memory Network. We now extend our model to handle multi-hop operations. Within each hop, we update the query value by adding the output of the previous step, which provides a gradual shift in attention. For instance, the input of $k + 1$ hop is the sum of the output s_r^k and the input u_r^k from layer k: $u_r^{k+1} = s_r^k + u_r^k$. After h hops of temporal reasoning, at the top of the network, the input to W_r also combines the input and the output of the top memory layer and then feed to a binary classifier:

$$p_r = Sigmoid(W_r(s_r^h + u_r^h)) \tag{12}$$

3) Iterative Noise Removal. The process of noise removal is similar to that described above. We use shared memory vector k_i and v_i with temporal reasoning to predict noise weights. The only difference is that we use a different embedding matrix to divide the query into two vectors. Here, we use query vector u_{n_j} to do jointly reasoning in the Memory Network. We obtain the weight of each memory slot $\beta_{i,j}$ and calculate the weighted sum of the memory value s_{n_j} in the same way.

Afterwards, the weights p_n are given to the sentences in the next iteration. We stop the iteration when the denoising results stabilize, and the reasoning result of the final round of Memory Network is our prediction result. A binary label vector $\tilde{p}_{r_i}$ is used to indicate whether the specified relations holds in the instances, where 1 means TRUE, and 0 otherwise. For each relation $r_i, i \in [1, 2, \cdots N]$, we predict its weight as p_{r_i} according to Eq. (12). Following this setting, we use the cross entropy loss function for multi-hop modeling:

$$\mathcal{L}_{TS} = -\sum_{i=1}^{N} \tilde{p}_{r_i} \log(p_{r_i}) + (1 - \tilde{p}_{r_i}) \log(1 - p_{r_i}) \tag{13}$$

where N is the number of relation labels.

We use back propagation to calculate the gradients of all the parameters, and we add small random noise (RN) to the gradients [9] to help regularize the temporal encoding. We randomize other parameters with uniform distribution $U(-0.01, 0.01)$.

3.3 Time Boundaries Determination

In this section, we design a heuristic method to calculate the valid boundary of the relation to further reduce the impact of noisy sentences, which may contain the concerned relation but is not actually within the valid temporal scope. For instance, As shown in Fig. 1 e.g., *"Obama was elected the 44th president of the United States."*, Although the elements in the triple $(U.S., president, Obama)$ are mentioned in the sentence, the timestamp *"2008-11-04"* is not in the Obama presidency. If nothing is done, then the temporal scope of this triple might be falsely determined as *(04 Nov. 2008, 20 Jan. 2017)*.

Let p_n and p_r denote the weight of noisy and the weight of relation of the last iteration, we first take the product of the two weights as follows:

$$p_i = (1 - p_{n_i})p_{r_i} \tag{14}$$

where $i \in \{1, 2, \ldots, M\}$. Here we set a sliding slot of size $j \in \{1, 2, \ldots, k\}$, $k \le \frac{M}{2}$. The diagram of the sliding slot is shown in Fig. 4. Then we calculate the probability of the instance p_{s_i} as follows:

$$p_{s_i} = \omega p_i + (1 - \omega) \sum_{j=1}^{k} 0.5^j p_{i-j:i+j} \tag{15}$$

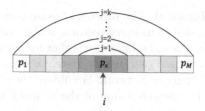

Fig. 4. Slide window diagram

where ω is a learnable parameter, the left side of the formula is the weight of the current position i, and the right side calculates the sum of the weights that are centered on the position i with each j-gram. We try several parameters and finally choose the appropriate penalty factor 0.5^j, so that the closer to the centered i the position j is, the greater the weight it has. This means if position j is far away with the center, it has less influence on i. Then we take the positions of the first and last of the continuous large values and get the corresponding timestamps of the sentence to represent the valid temporal scope of the triple.

4   Experiments

We first introduce our datasets and the metrics for evaluation, and then present the existing state-of-the-art approaches that we would compare with. Finally, we evaluate the performance of our proposed approach by comparing with baselines.

4.1   Datasets and Metrics

We use two datasets to illustrate the performance of our proposed method and comparative baselines for the temporal scoping task. The TAC dataset is from the TAC-TSF task for 2013, which is trained and evaluated for seven relation types, as shown in Table 1. To obtain the training set of TAC, we further extract entity pairs that participate in these seven relations by distant supervision (DS) framework and align the entity pair $(head, tail)$ from a triple $(head, relation, tail)$ extracted from Wikidata with sentences from Wikipedia[1]. We tag each sentence in Wikipedia corpus using either time expression appeared in the sentence or the last appeared time expression. We implement time expression recognition on the basis of TIMEX tool[2]. And then we tag the aligned sentences and order the extracted sentences into a timeline to achieve labels. We employ the TAC evaluation data as other work does, which is publicly available through LDC[3]. For MUL Dataset, we select relations with informative temporal feature in Wikidata such as *educate_at*, *award_received*, etc, and follow the above method to generate dataset. We also filter 6 relations which have very small number of instances. The statistics of the two datasets are given in Table 2.

---

[1] https://zh.wikipedia.org.
[2] https://github.com/nltk/nltk_contrib/blob/master/nltk_contrib/timex.py.
[3] https://catalog.ldc.upenn.edu/.

Table 1. Type of Relations in the TAC-TSF and MUL (Partial)

Index	TAC relations	MUL relations
R1	spouse	title
R2	title	spouse
R3	employee_or_member_of	educate_at
R4	top_employees/members	award_received
R5	cities_of_residence	countries_of_residence
R6	statesorprovinces_of_residence	member_of_sports_team
R7	countries_of_residence	employee_or_member_of

Table 2. Statistics of the Datasets

Datasets		Sentences	Entity pairs	Relations
TAC Dataset	Training	67,843	15,208	7
	Testing	2,061	269	7
MUL Dataset	Training	87,614	18,056	13
	Testing	2,7133	4631	13

As for the evaluation metrics, the performance of comparative experiments are reported by *accuracy* and *StDev*. While *accuracy* reflects the proportion of correctly time interval to entity pairs, *StDev* is standard deviation which reflects the dispersion degree of data relative to mean. Naturally, a higher *accuracy* and a lower *StDev* indicates a better performance. We also evaluate the influence of the iterative time in the iterative model to the prediction results.

4.2 Parameter Setting

Among all experiments, we use 230 convolution kernels with windows size 3. The dropout probability p_d is set to 0.5. We try various max hops and iterations values (from 1 to 3) to test how reasoning works in our model. We train the models with 30 epochs and report the best performance. As for optimization step, we adopt SGD with radient plus Gaussian noise with standard deviation of 0.01, which helps to better generalize. Also, we apply gradient decay of rate ($\rho = 0.5$) over every $\tau = 10$ epochs, and the learning rate is set to 0.01. With regard to inputs, we use 50-d Glove word embeddings pretrained on Wikipedia and Gigaword and 5-d position embedding. The temporal encodings are set directly with T.

4.3 Approaches for Comparison

In this section, we briefly introduce three comparative methods, including our proposed *Baseline Model* and *TAC-TSF systems* on the TAC dataset, as well as

Table 3. Results for TAC-TSF Test Set, Overall and for Individual Slots

	R1(%)	R2(%)	R3(%)	R4(%)	R5(%)	R6(%)	R7(%)	All(%)	StDev
LDC	69.87	60.22	58.26	72.27	81.10	54.07	91.18	68.84	12.32
RPI-Blender	31.19	13.07	14.93	26.71	29.04	17.24	34.68	23.42	7.98
UNED	26.20	6.88	8.16	15.24	14.47	14.41	19.34	14.79	6.07
TSRF	31.94	36.06	32.85	40.12	33.04	31.85	27.35	33.15	3.6
Baseline	42.86	39.30	38.28	41.06	46.02	39.72	42.74	42.23	2.65
MemTimes	**44.43**	**41.04**	**39.25**	**41.93**	**46.82**	**40.59**	**43.69**	**43.75**	**2.59**

the effect of our *Iterative Model* on different number of hops and iterations on the two datasets.

- *Baseline Model*: We only leverage the Memory Network for memory reasoning, and the obtained intermediate results are directly used to judge the valid time boundaries without heuristic calculation.
- *TAC-TSF systems*: We compare MemTimes against three systems: RPI-Blender [1], UNED [4] and TSRF [12]. These systems employ DS strategy to assign temporal labels to relations extracted from text. UNED system uses a document-level representation based on rich graphics to generate novel features. The RPI-Blender uses an ensemble of classifiers to combine surface text-based flat features and dependency paths with tree kernels. And the TSRF learns the relation-specific language model from the Wikipedia infobox tuples and then performs the temporal classification using a model made up of manually selected triggers.
- *Iterative Model*: We consider the effect of different hops and iterations on the *accuracy* of the temporal scoping. In order to demonstrate that our model is adaptive to adding time dimension to multiple relationships, we not only perform experiments on TAC dataset but also on MUL dataset.

4.4   Experimental Results

1) Baseline Model and TAC-TSF systems

In this part, we compare single layer Memory Network and single iteration Mem-Times in conjunction with the results of baseline and the TAC-TSF systems evaluated and the output generated by the LDC human experts. In practice, we also add negative samples to the test set. The experimental results are listed in the Table 3. As can be observed, MemTimes is about one percent higher than the baseline, which indicates that the heuristic computing method we design is very necessary for denoising. We can observe that the most improvement on the *spouse* (R1). We look at the data of R1 and find that there are relatively more noisy sentences. Moreover, Our model achieves approximately 64% of human performance (LDC) and outperforms the other systems in overall score as well as for all individual relations at least 10%. We believe that this performance

Table 4. Different Number of Hops for MemTimes

Dataset	Hop	Accuracy (%)	Hop	Accuracy (%)	Hop	Accuracy (%)
TAC Dataset	1	47.36	2	50.12	3	46.33
MUL Dataset	1	51.23	2	54.15	3	48.55

Table 5. Different Number of Iterations for MemTimes

Dataset	Iteration	Accuracy (%)	Iteration	Accuracy (%)	Iteration	Accuracy (%)
TAC Dataset	1	44.46	2	50.12	3	50.61
MUL Dataset	1	49.33	2	54.15	3	54.69

benefits from the temporal reasoning algorithm employed, Memory Network, which is highly effective in using both semantic and temporal features to decide whether the instance belongs to the relation under consideration or not. Also, MemTimes uses a heuristic method to further clean up the noise introduced by the noisy labeling problem of DS. Apart from that, we can also see that our model achieves a balanced performance on different relations on which it was tested compared with others. As shown in column *StDev* in Table 3, our model achieves the lowest standard deviation in the performance across the relations tested. It is also interesting that MemTimes achieves the best performance on the *cityofresidence* (R5) relations probably in that the noise of the data has less impact on the model. This means our model can be transferred successfully across relations for domain-specific temporal scoping.

2) Iterative Model

In this section, we discuss the influence of different numbers of hops and iterations in MemTimes on both TAC and MUL datasets.

(1) Effect of the number of hops. In order to evaluate how the number of hops affects our proposed model, we fix the number of iterations and vary the value of hops from 1 to 3. We find that when the number of iterations is 2, we achieve best performance from the experiments. The results of the hop number experiment on both datasets are depicted in Table 4. From the results, we can observe that models show better performance with hop number 2. We notice that the performance of the model fluctuates with the increase in the number of hops. The reason might lie in the distribution of the hop distance between origin instance and noisy instance. In order to show that our model can easily adapt to multiple relationships, we also experimented on MUL dataset. We only add a temporal scope to a small number of triples because most of the mention sets are invalid. The effect of hop of the model on this dataset remains roughly consistent. The reason why the model works better on MUL dataset may be that the MUL has a higher semantic match between the training set and the test set. We also experiment with more hops, but there have no positive effect on our model. We consider it possibly due to its influence on hop distance distribution, resulting in the confusion of the original instance and the noise instance.

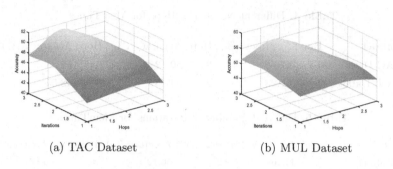

(a) TAC Dataset   (b) MUL Dataset

Fig. 5. The effectiveness of iterations and hops

(2) Effect of the number of iterations. We also evaluate the influence of different number of external iterations for our proposed method. Similarly, in oder to illustrate the effect of the number of iterations, we also fix the other parameter, the number of hops, and change the iteration value from 1 to 3 to evaluate the *accuracy* of our model. We also find when $\#hops = 2$, the best performance is reached. The results of the iteration number experiments on both datasets are shown in Table 5. Note that MemTimes shows better performance with iteration number 3, and the results of the second iteration improves significantly compared with the first iteration, while the results of the third iteration are not much better than those of the second iteration. Hence, the results suggest that MemTimes can benefit from the iteration for getting rid of the noise. But the gains from iteration are limited. Specifically, when the number of iteration is larger than 3, the results first reach a stable state and then decline. Thus, a few iterations can achieve better denoising, otherwise it is easy to mistake correct sentences for noise. On MUL dataset, the overall results are better than the results on the TAC dataset. The possible reason is that the TAC dataset contains meaningless sentences, such as: *"11:42:18 (By Deborah Kuo)"*, it is possible to treat this as a noisy sentence when iterating on denoising. Therefore, it interferes with our iterative model.

(3) Effect of both the number of hops and iterations. We synthesized the results of MemTimes with different hops and iterations, as shown in Fig. 5. We can see that with the increase of the number of hops and iterations, the *accuracy* of our model first goes up and then shows a drop trend slightly. The reasons can be analyzed from two aspects: 1) The iteration process helps to remove noise from hidden states; 2) With the help of multiple hops, our model can reason temporal scope effectively. Here light color represents higher *accuracy* while the dark color stands for lower *accuracy*. And the brighter the color is, the higher the *accuracy* is. Apart from that, we also find that our model achieves best performance when the number of iterations and hops on TAC and MUL datasets both increases to 2. But when they increase to larger values, the performance degrades a lot or even less than the single hop and iteration. We analyze that the main reason for this result is that multiple hops and multiple iterations affect each other and the former has a greater negative effect on the model.

5   Related Work

5.1   Temporal Scoping

Temporal scoping has been studied extensively in the past decade. The existing work could be roughly divided into two categories. While one category extracts temporal information for facts from external resources such as Wikipedia or Web harvested documents, the other category of work pays attention on inferring the temporal information from facts based on temporal KBs of facts.

1) External Resoruces Based Methods. This kinds of methods can be divided into two categories including free text-based extraction methods and statistics-based extraction methods. Early work focuses on the first part, such as YAGO which uses linguistic extraction rules to extract temporal facts from Wikipedia infoboxes [6]. But they are limited by relations where the specific start or end dates in a structured form are not provided contributing to incomplete time information. Recently, Wang et al. [18] propose an iteratively learning unsupervised approach based on Pattern-based IE to extract precise temporal facts from newspaper corpora. It still, however, requires human experts to define some constraints to find the conflicts and doesn't work well for multiple relations. Furthermore, some recent work [5] also leverages temporal expressions annotated in large document corpus to identify time intervals of knowledge graph facts. Apparently, labeling data is costly and rather brittle in that it can not be adapted easily to new relations. The state-of-art systems for TAC-TSF is TSRF [12], which employs a language model consisting of patterns automatically derived from Wikipedia sentences that contain the main entity of a page and temporal slot-fillers extracted from the corresponding infoboxes. The language model can not capture the features of sentences with long-distance dependence. In addition, manually choosing triggers to temporal classification may lead to omissions.

Some statistics based methods are also widely studied, such as CoTS and TISCO [11,14]. The system CoTS relies on document meta-data such as its creation date to assign temporal scopes to facts by using manually edited temporal order constraints. But the main goal of CoTS is to predict temporal ordering of relations rather than to do temporal scoping. TISCO exploits evidence collected from the Web of Data. They devise a three-phase approach which comprises matching, selection and merging for mapping facts to sets of time intervals. As a matter of fact, incorrect information can cause noise and error propagation.

2) Infering Based Methods. Another work pays attention to infer the temporal information for facts based on temporal KBs of facts, where each fact has a timestamped relation. For instance, Jiang et al. [7] considers the time validity of the triples in KGs, and they focus on predicting relation or entity given a time point in which the fact is supposed to be valid. Trivedi et al. [16] propose to learn non-linearly evolving entity representations over time in order to perform temporal reasoning over dynamic KGs. Leblay et al. [8] try to use side information from the atemporal part of the graph for learning temporal embedding to predict time validity for unannotated edges. However, without using external knowledge, these work could hardly reach a satisfied performance.

5.2   Memory Network

Memory Network is first introduced by [19] whose idea is to inference with a long-term memory component, which could be read, written, and jointly learned with the goal for prediction. Feng et al. [3] apply Memory Network to distant supervised relation extraction. The model includes two attention-based memory neural networks: one is based on a word-level Memory Network to weight learning of each context word with specific entity pair while the other is to capture the importance of different sentence instances and the dependencies between relations. Another work [21] is similar to ours, but the difference is that they use Memory Network to perform temporal reasoning to solve relation extraction in different periods while our goal is to predict valid temporal interval of one knowledge triple.

6   Conclusions and Future Work

This paper works on leveraging both temporal reasoning and noise removing to do temporal scoping for triples based on Memory Network. We design a unified framework to encode both semantic and temporal information contained in the mention set of each fact, which enables temporal reasoning with Memory Network. We use the Memory Network to joint learning the relational weights and noise weights, and then we utilize noise weights to remove noise iteratively. Our experiments on real-world datasets demonstrate that our proposed iterative denoising method performs better and our model can add temporal scope for multiple types of relations. In the future work, we would like to design a more complete method to identify the valid boundary of the relationship, which refers to begins or ends.

Acknowledgments. This research is partially supported by Natural Science Foundation of Jiangsu Province (No. BK20191420), National Natural Science Foundation of China (Grant No. 61632016, 61572336, 61572335, 61772356), Natural Science Research Project of Jiangsu Higher Education Institution (No. 17KJA520003, 18KJA520010), and the Open Program of Neusoft Corporation (No. SKLSAOP1801).

References

1. Artiles, J., Li, Q., Cassidy, T., Tamang, S., Ji, H.: CUNY BLENDER TAC-KBP2011 temporal slot filling system description. In: TAC (2011)
2. Bader, B.W., Harshman, R.A., Kolda, T.G.: Temporal analysis of semantic graphs using ASALSAN. In: Seventh IEEE International Conference on Data Mining (ICDM 2007), pp. 33–42. IEEE (2007)
3. Feng, X., Guo, J., Qin, B., Liu, T., Liu, Y.: Effective deep memory networks for distant supervised relation extraction. In: IJCAI, pp. 4002–4008 (2017)

4. Garrido, G., Cabaleiro, B., Penas, A., Rodrigo, A., Spina, D.: A distant supervised learning system for the TAC-KBP slot filling and temporal slot filling tasks. In: TAC (2011)

5. Gupta, D., Berberich, K.: Identifying time intervals for knowledge graph facts. In: Companion Proceedings of the The Web Conference 2018, pp. 37–38. International World Wide Web Conferences Steering Committee (2018)

6. Hoffart, J., Suchanek, F.M., Berberich, K., Lewis-Kelham, E., De Melo, G., Weikum, G.: Yago2: exploring and querying world knowledge in time, space, context, and many languages. In: Proceedings of the 20th International Conference Companion on World Wide Web, pp. 229–232. ACM (2011)

7. Jiang, T., et al.: Towards time-aware knowledge graph completion. In: Proceedings of COLING 2016, the 26th International Conference on Computational Linguistics: Technical Papers, pp. 1715–1724 (2016)

8. Leblay, J., Chekol, M.W.: Deriving validity time in knowledge graph. In: Companion Proceedings of the The Web Conference 2018, pp. 1771–1776. International World Wide Web Conferences Steering Committee (2018)

9. Neelakantan, A., et al.: Adding gradient noise improves learning for very deep networks. arXiv preprint arXiv:1511.06807 (2015)

10. Rula, A., Palmonari, M., Ngonga Ngomo, A.-C., Gerber, D., Lehmann, J., Bühmann, L.: Hybrid acquisition of temporal scopes for RDF data. In: Presutti, V., d'Amato, C., Gandon, F., d'Aquin, M., Staab, S., Tordai, A. (eds.) ESWC 2014. LNCS, vol. 8465, pp. 488–503. Springer, Cham (2014). https://doi.org/10.1007/978-3-319-07443-6_33

11. Rula, A., et al.: TISCO: temporal scoping of facts. J. Web Semant. **54**, 72–86 (2019)

12. Sil, A., Cucerzan, S.P.: Towards temporal scoping of relational facts based on Wikipedia data. In: Proceedings of the Eighteenth Conference on Computational Natural Language Learning, pp. 109–118 (2014)

13. Sukhbaatar, S., Weston, J., Fergus, R., et al.: End-to-end memory networks. In: Advances in Neural Information Processing Systems, pp. 2440–2448 (2015)

14. Talukdar, P.P., Wijaya, D., Mitchell, T.: Coupled temporal scoping of relational facts. In: Proceedings of the Fifth ACM International Conference on Web Search and Data Mining, pp. 73–82. ACM (2012)

15. Tresp, V., Ma, Y., Baier, S., Yang, Y.: Embedding learning for declarative memories. In: Blomqvist, E., Maynard, D., Gangemi, A., Hoekstra, R., Hitzler, P., Hartig, O. (eds.) ESWC 2017. LNCS, vol. 10249, pp. 202–216. Springer, Cham (2017). https://doi.org/10.1007/978-3-319-58068-5_13

16. Trivedi, R., Farajtabar, M., Wang, Y., Dai, H., Zha, H., Song, L.: Know-evolve: deep reasoning in temporal knowledge graphs. arXiv preprint arXiv:1705.05742 (2017)

17. Vrandečić, D., Krötzsch, M.: Wikidata: a free collaborative knowledge base (2014)

18. Wang, X., Zhang, H., Li, Q., Shi, Y., Jiang, M.: A novel unsupervised approach for precise temporal slot filling from incomplete and noisy temporal contexts. In: The World Wide Web Conference, pp. 3328–3334. ACM (2019)

19. Weston, J., Chopra, S., Bordes, A.: Memory networks. arXiv preprint arXiv:1410.3916 (2014)

20. Wijaya, D.T., Nakashole, N., Mitchell, T.M.: CTPs: contextual temporal profiles for time scoping facts using state change detection. In: Proceedings of the 2014 Conference on Empirical Methods in Natural Language Processing (EMNLP), pp. 1930–1936 (2014)

21. Yan, J., He, L., Huang, R., Li, J., Liu, Y.: Relation extraction with temporal reasoning based on memory augmented distant supervision. In: Proceedings of the 2019 Conference of the North American Chapter of the Association for Computational Linguistics: Human Language Technologies, (Long and Short Papers), vol. 1, pp. 1019–1030 (2019)
22. Zeng, D., Liu, K., Lai, S., Zhou, G., Zhao, J., et al.: Relation classification via convolutional deep neural network (2014)

Code2Text: Dual Attention Syntax Annotation Networks for Structure-Aware Code Translation

Yun Xiong[1](✉), Shaofeng Xu[1], Keyao Rong[1], Xinyue Liu[2], Xiangnan Kong[2], Shanshan Li[3], Philip Yu[4], and Yangyong Zhu[1]

[1] Shanghai Key Laboratory of Data Science, School of Computer Science, Fudan University, Shanghai, China
{yunx,sfxu16,18212010024,yyzhu}@fudan.edu.cn
[2] Department of Computer Science, Worcester Polytechnic Institute, Worcester, MA, USA
{xliu4,xkong}@wpi.edu
[3] School of Computer Science, National University of Defense Technology, Changsha, China
shanshanli@nudt.edu.cn
[4] University of Illinois at Chicago, Chicago, IL, USA
psyu@uic.edu

Abstract. Translating source code into natural language text helps people understand the computer program better and faster. Previous code translation methods mainly exploit human specified syntax rules. Since handcrafted syntax rules are expensive to obtain and not always available, a PL-independent automatic code translation method is much more desired. However, existing sequence translation methods generally regard source text as a plain sequence, which is not competent to capture the rich hierarchical characteristics inherently reside in the code. In this work, we exploit the abstract syntax tree (AST) that summarizes the hierarchical information of a code snippet to build a structure-aware code translation method. We propose a syntax annotation network called Code2Text to incorporate both source code and its AST into the translation. Our Code2Text features the dual encoders for the sequential input (code) and the structural input (AST) respectively. We also propose a novel dual-attention mechanism to guide the decoding process by accurately aligning the output words with both the tokens in the source code and the nodes in the AST. Experiments on a public collection of Python code demonstrate that Code2Text achieves better performance compared to several state-of-the-art methods, and the generation of Code2Text is accurate and human-readable.

Keywords: Natural language generation · Tree-LSTM · Abstract syntax tree · Data mining

© Springer Nature Switzerland AG 2020
Y. Nah et al. (Eds.): DASFAA 2020, LNCS 12114, pp. 87–103, 2020.
https://doi.org/10.1007/978-3-030-59419-0_6

1   Introduction

We have witnessed a large amount of source code been released in recent years, manually writing detailed annotations (*e.g.*, comments, pseudocode) for them is a tedious and time-consuming task. However, these annotations play an irreplaceable role in the development of software. For example, it serves as a guideline for the new engineers to quickly understand the functionality of each piece of code, and it helps one grasp the idea of legacy code written in a less popular programming language. Accordingly, an effective automatic source code translation method is desired, where the goal is to translate the code into a corresponding high-quality natural language translation (we also refer the translated text as *annotation* for short throughout this paper).

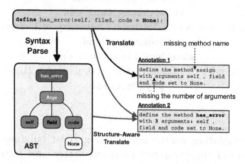

Fig. 1. An illustration of the syntax structure-aware code translation task. We propose to exploit the abstract syntax tree (AST) that reflects the syntax structure of the code snippet for an improved code translation (Annotation 2). While conventional sequence translation methods can only use the sequential input (the code), which may miss key structural information in the code syntax (Annotation 1). (Color figure online)

Given the significance of the code translation problem, most of the existing methods [1,19] only follow the common natural language translation routine by treating the code as a plain sequence. However, programming languages obey much more strict syntax rules than natural languages [8], which should not be ignored in translation. Given the syntax rules of most programming languages are explicitly defined, each piece of legal code could be represented by a structural representation called abstract syntax tree (AST). In Fig. 1, the AST of a simple function declaration (blue box) in Python is illustrated in the underneath yellow box, which depicts the syntax structure of the code. The plain sequence translator who only takes the code as input may have difficulty in capturing the structural information, so the generated annotation may omit crucial details reside in the original code, such as the function name and the number of parameters. We show that this limitation could be fixed by considering the corresponding AST, as it represents the code structure and the hierarchical relations that are hard to learn sequentially.

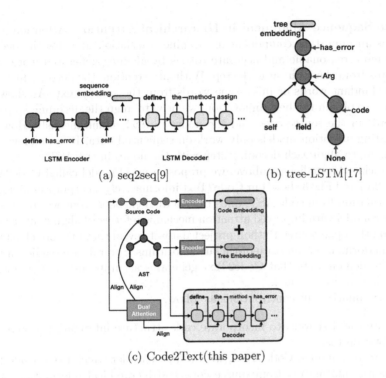

Fig. 2. Comparison of related methods. (a) Seq2seq model consists of an LSTM encoder and an LSTM decoder. (b) Tree structured recurrent neural network which encodes structured text from children to parent nodes. (c) Our Code2Text encodes both sequential and hierarchical information to generate the target sequence. Details on the dual-encoder and dual-attention are illustrated in Fig. 3.

Thus, it is interesting to investigate whether and how one could combine the information in source code and its AST for an improved structure-aware code translation model. Although ASTs are much easier to obtain compared to handcrafted syntax rules, it is not a trivial task to incorporate it into the translation method due to the following reasons.

Sequential and Hierarchical Encoding: The encoder plays a vital role in correctly understanding the semantics of the input text in code translation. Existing methods usually employ RNN/LSTM to learn the dependencies between words according to their sequential orders, as shown in Fig. 2(a). However, it is not applicable to hierarchical inputs such as AST. To additionally consider the hierarchical patterns of the input source code, we need an approach to encode the tree-structured inputs. Besides, we also need the sequential encoder to process the source code since ASTs focus more on structure rather than token-level details. How to effectively combine the sequential encoder and hierarchical encoder is also an open question.

Tree to Sequence Alignment in Hierarchical Attention: Attention mechanism is an important component in machine translation. It helps the decoder choose more reasonable and accurate tokens by aligning generated words with the words from the input at each step. Without attention, the decoder may generate redundant words or miss some words from the source text. As shown in Fig. 2(c), to obtain the best annotation, we need to align the decoding step with the words in the source code and the nodes in AST simultaneously. However, the existing attention models only work on sequential data. It is unknown and challenging to align each decoding step with the nodes in the AST.

To tackle the challenges above, we propose a novel model called Code2Text. As Fig. 2(c) and Fig. 3 show, our Code2Text informatively incorporates hierarchical information from code AST, and trains a dual-encoder sequence to sequence language model with improved attention mechanism for word alignment. Experiments on the open-source Python project dataset reveal that our model achieves better performance than state-of-the-art algorithms. Several case studies are displayed to demonstrate that Code2Text generates accurate and understandable annotations as we pursued.

We summarize our contributions as follow,

1. We are the first work to incorporate code structure information into a code annotation task.
2. We create a model, Code2Text, with a dual-encoder which can encode both semantic information from source code and hierarchical information of the AST. We also propose a dual-attention mechanism to improve the original attention mechanism by extending it to align the structural AST tokens.
3. We perform extensive experiments on a public benchmark Python code dataset. Other than the numerical evaluation, we additionally present case studies to demonstrate the effectiveness of our dual-attention encoder design.

2 Preliminary

2.1 Problem Definition

Let's take a look at the definition of NMT (Neural Machine Translation) first. Suppose we have a dataset $\mathcal{D} = \{(\mathbf{x}^s, \mathbf{y})\}$ and the corresponding annotation. $\mathcal{X}^s$ and $\mathcal{Y}$ are sets of source code and annotation, respectively. $\mathbf{x}^s = (x_1^s, \cdots, x_n^s)$ represents a sequence of source code with n words and $\mathbf{y} = (y_1, \cdots, y_m)$ represents a sequence of annotation with m words. Our task is to translate $\mathbf{x}^s$ to $\mathbf{y}$ for each pair in dataset $\mathcal{D}$, which is the same task as translating a source language to a target language. The overall goal of normal Neural Machine Translation models is to estimate the conditional probability distribution $\Pr(\mathcal{Y}|\mathcal{X}^s)$. Conventional inference approaches usually require i.i.d. assumptions, and ignore dependency between different instances. The inference for each instance is performed independently:

$$\Pr(\mathcal{Y}|\mathcal{X}^s) \propto \prod_{(\mathbf{x}^s, \mathbf{y}) \in \mathcal{D}} \Pr(\mathbf{y}|\mathbf{x}^s) \tag{1}$$

In this work, our model considers not only semantic information but also hierarchical information. Therefore, we derive another symbol $\mathbf{x}^t = (x_1^t, \cdots, x_q^t)$, who contains AST information of relative source code $\mathbf{x}^s$ with q words. $\mathcal{X}^t$ is a set which contains $\mathbf{x}^t$. Accordingly, we create an extended dataset $\mathcal{D}' = \{(\mathbf{x}^s, \mathbf{x}^t, \mathbf{y})\}$. To incorporate hierarchical information, we will modify our probability distribution:

$$\Pr(\mathcal{Y}|\mathcal{X}^s, \mathcal{X}^t) \propto \prod_{(\mathbf{x}^s, \mathbf{x}^t, \mathbf{y}) \in \mathcal{D}'} \Pr(\mathbf{y}|\mathbf{x}^s, \mathbf{x}^t) \tag{2}$$

2.2 Attention Seq2seq Model

Attention seq2seq model is a sophisticated end-to-end neural translation approach, which consists of the encoder process and decoder process with an attention mechanism.

Encoder. In the encoder process, we aim to embed a sequence of source code $\mathbf{x}^s = (x_1^s, \cdots, x_n^s)$ into d-dimension vector space.

We usually replace vanilla RNN [10] unit with LSTM (Long Short Term Memory) [4] unit due to the gradient explosion/vanishing problem. The j-th LSTM unit has three *gates*: an input gate $\mathbf{i}_j^s \in \mathcal{R}^{d \times 1}$, a forget gate $\mathbf{f}_j^s \in \mathcal{R}^{d \times 1}$ and an output gate $\mathbf{o}_j^s \in \mathcal{R}^{d \times 1}$ and two states: a hidden state $\mathbf{h}_j^s \in \mathcal{R}^{d \times 1}$ and a memory cell $\mathbf{c}_j^s \in \mathcal{R}^{d \times 1}$. Update rules for an LSTM unit are below:

$$\mathbf{i}_j^s = \sigma(\mathbf{W}^{(i)} embed(x_j^s) + \mathbf{U}^{(i)} \mathbf{h}_{j-1}^s + \mathbf{b}^{(i)}), \tag{3}$$

$$\mathbf{f}_j^s = \sigma(\mathbf{W}^{(f)} embed(x_j^s) + \mathbf{U}^{(f)} \mathbf{h}_{j-1}^s + \mathbf{b}^{(f)}), \tag{4}$$

$$\mathbf{o}_j^s = \sigma(\mathbf{W}^{(o)} embed(x_j^s) + \mathbf{U}^{(o)} \mathbf{h}_{j-1}^s + \mathbf{b}^{(o)}), \tag{5}$$

$$\tilde{\mathbf{c}}_j^s = \tanh\left(\mathbf{W}^{(\tilde{c})} embed(x_j^s) + \mathbf{U}^{(\tilde{c})} \mathbf{h}_{j-1}^s + \mathbf{b}^{(\tilde{c})}\right), \tag{6}$$

$$\mathbf{c}_j^s = \mathbf{i}_j^s \odot \tilde{\mathbf{c}}_j^s + \mathbf{f}_j^s \odot \mathbf{c}_{j-1}^s, \tag{7}$$

$$\mathbf{h}_j^s = \mathbf{o}_j^s \odot \tanh\left(\mathbf{c}_j^s\right), \tag{8}$$

Here, $\tilde{\mathbf{c}}_j^s \in \mathcal{R}^{d \times 1}$ denotes the state for updating the memory cell $\mathbf{c}_j^s$. Function *embed()* turns a word into a d-dimension embedding vector. It can be assigned with a fixed global word vector or trained by the model itself. $\mathbf{W}^{(\cdot)}, \mathbf{U}^{(\cdot)} \in \mathcal{R}^{d \times d}$ are weight matrix and $\mathbf{b}^{(\cdot)} \in \mathcal{R}^{d \times 1}$ is a bias vector. σ is the logistic function and the operator $\odot$ means element-wise product between two vectors. We initialize $\mathbf{h}_0^s$ as a d-dimension vector of all zeros, and iterate over the sequence and finally obtain $\mathbf{h}_n^s$ at the end of the source sentence, which represents the information of source code.

Decoder. After we obtain source code representation vector $\mathbf{h}_n^s$ from the encoder process, we then predict the annotation sequence with LSTM in a similar way in the decoder process. We define $\mathbf{d}_j$ as the j-th hidden state. Given the input embedding vector $embed(\mathbf{x}^s)$ and previous word sequence $\mathbf{y}_{<j}$, we generate j-th word by estimating the conditional probability:

$$p(y_j|\mathbf{y}_{<j}, embed(\mathbf{x}^s)) = softmax(\mathbf{d}_j), \qquad (9)$$

where $softmax()$ function produces probabilities according to the j-th hidden state $\mathbf{d}_j$, and $\mathbf{d}_j$ is calculated by another non-linear function f_d as follows:

$$\mathbf{d}_j = f_d(\mathbf{d}_{j-1}, embed(y_{j-1})), \qquad (10)$$

We initialize $\mathbf{d}_0 = \mathbf{h}_n^s$ to ensure that our predictor can generate an annotation sequence base on source code sequential information.

Attention Mechanism. Attention mechanism [9] was proposed to align each decoder hidden state with the encoder output states. With the attention process, we can explicitly calculate the contribution each encoder output state made to the word prediction at each step.

Suppose we have hidden state $\mathbf{d}_j$ at time j in the decoder process, and $(\mathbf{h}_1^s, \cdots, \mathbf{h}_n^s)$ are encoder hidden states. According to [9], we first calculate attention weights α_{ij}^s between the i-th hidden state $\mathbf{h}_i^s$ in encoder and the j-th hidden state $\mathbf{d}_j$ in decoder as follows:

$$\alpha_{ij}^s = \frac{\exp(score(\mathbf{h}_i^s, \mathbf{d}_j))}{\sum_{k=1}^n \exp(score(\mathbf{h}_k^s, \mathbf{d}_j))}, \qquad (11)$$

where $score()$ function is used to compare the decoder hidden state $\mathbf{d}_j$ with each of the source hidden states $\mathbf{h}_i^s$, and the result is normalized to produce attention weights (a distribution over source positions). Then based on attention weights we compute j-th context vector $\mathbf{w}_j^s$ as the weighted average of the source encoder hidden states:

$$\mathbf{w}_j^s = \sum_{i=1}^n \alpha_{ij}^s \mathbf{h}_i^s, \qquad (12)$$

Afterward, we apply a non-linear function $tanh$ to the concatenation of the context vector $\mathbf{w}_j^s$ and the current decoder hidden state $\mathbf{d}_j$, and yield the final attention vector $\mathbf{a}_j$:

$$\mathbf{a}_j = \tanh(\mathbf{W}_d \times (\mathbf{d}_j \oplus \mathbf{w}_j^s) + \mathbf{b}_d), \qquad (13)$$

where $\oplus$ means concatenation of $\mathbf{d}_j$ and $\mathbf{w}_j^s$. $\mathbf{W}_d \in \mathcal{R}^{d \times 2d}$ is a weight matrix and $\mathbf{b}_d \in \mathcal{R}^{d \times 1}$ is a bias vector. Once computed, the attention vector $\mathbf{a}_j$ is used to derive the softmax logit:

$$p(y_j|\mathbf{y}_{<j}, \mathbf{x}^s) = softmax(\mathbf{a}_j). \qquad (14)$$

In Sect. 4 our experiments will present performance and cases of this seq2seq model.

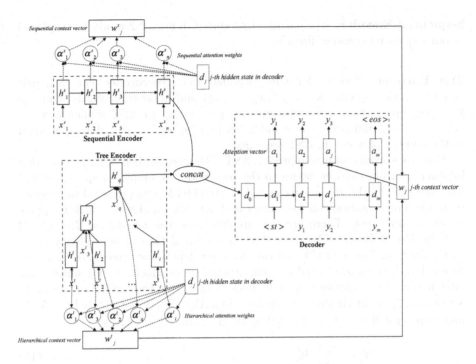

Fig. 3. Architecture of our Code2Text model. $(x_1, x_2, \cdots, x_n)$ are tokens of source code, $(t_1, t_2, \cdots, t_q)$ are tokens of AST and $(y_1, y_2, \cdots, y_m)$ are tokens of natural language annotation. $< st >$ and $< eos >$ are start and end tokens, respectively. Encoder process consists of sequence encoder and tree encoder, and $\mathbf{h}_0$ is initialized with a vector of all zeros. In the decoder process, $\mathbf{d}_0$ is initialized with the bilinear result of $\mathbf{h}_n$ and $\mathbf{h}_q^t$. Context vector $\mathbf{c}$ in attention mechanism is the concatenation of sequence context vector and hierarchical context vector.

3 Proposed Method: Code2Text

In this section, we will formally introduce our model, Code2Text, which is an extension and improvement of original seq2seq models. We first propose a dual-encoder by creating a tree encoder for representing the summary of AST information along with the original sequential encoder in the encoder process, then we explain how our dual-attention mechanism works by incorporating hierarchical outputs from tree encoder. The architecture of our model shows in Fig. 3.

3.1 Dual-Encoder

As Fig. 3 describes, in our encoder process, the sequential encoder produces sequential representation $\mathbf{h}_n^s$, which will be a part of our dual-encoder information. The other part, *tree encoder*, will produce hierarchical representation from AST of source code.

Sequential Encoder. The encoder introduced in Sect. 2.2 would be employed as our sequential encoder directly.

Tree Encoder. Now we formally formulate our tree encoder. For each pair of source code sequence $\mathbf{x}^s = (x_1^s, x_2^s, \cdots, x_n^s)$ and annotation words sequence $\mathbf{y} = (y_1, y_2, \cdots, y_m)$, we preprocess by parsing $\mathbf{x}^s$ to AST sequence $\mathbf{x}^t = (x_1^t, x_2^t, \cdots, x_q^t)$ and AST parent index list $\mathbf{p} = (p_1, p_2, \cdots, p_q)$. Here, q is equal to the number of words in AST sequence.

Our tree encoder aims to represent AST with a vector, hence, for propagating information from children nodes to the root node, we employ a special LSTM unit, *tree-LSTM* [17] to our tree encoder. Tree-LSTM was proposed to improve semantic representations on tree-structured network topologies, which is appropriate for our work. There are two architectures: the *Child-Sum Tree-LSTM* and the *N-ary Tree-LSTM*. Code AST is a natural kind of dependency trees, and Child-Sum Tree-LSTM is a good choice for dependency trees [17]. However, N-ary Tree-LSTMs are suited for constituency trees which are not suitable for AST in our task. Therefore, we will choose Child-Sum Tree-LSTM in our work. We denote $\mathcal{C}_j$ as the children of j-th node in a AST. The hidden state $\mathbf{h}_j^t \in \mathcal{R}^{d \times 1}$ and memory cell $\mathbf{c}_j^t \in \mathcal{R}^{d \times 1}$ for j-th node are updated as follows:

$$\tilde{\mathbf{h}}_j^t = \sum_{k \in \mathcal{C}(j)} \mathbf{h}_k^t, \tag{15}$$

$$\mathbf{i}_j^t = \sigma(\mathbf{W}^{(i^t)} embed(x_j^t) + \mathbf{U}^{(i^t)} \tilde{\mathbf{h}}_j^t + \mathbf{b}^{(i^t)}), \tag{16}$$

$$\mathbf{f}_{jk}^t = \sigma(\mathbf{W}^{(f^t)} embed(x_j^t) + \mathbf{U}^{(f^t)} \mathbf{h}_k^t + \mathbf{b}^{(f^t)}), \tag{17}$$

$$\mathbf{o}_j^t = \sigma(\mathbf{W}^{(o^t)} embed(x_j^t) + \mathbf{U}^{(o^t)} \tilde{\mathbf{h}}_j^t + \mathbf{b}^{(o^t)}), \tag{18}$$

$$\tilde{\mathbf{c}}_j^t = \tanh(\mathbf{W}^{(\tilde{c}^t)} embed(x_j^t) + \mathbf{U}^{(\tilde{c}^t)} \tilde{\mathbf{h}}_j^t + \mathbf{b}^{(\tilde{c}^t)}), \tag{19}$$

$$\mathbf{c}_j^t = \mathbf{i}_j^t \odot \tilde{\mathbf{c}}_j^t + \sum_{k \in \mathcal{C}(j)} \mathbf{f}_{jk}^t \odot \mathbf{c}_k^t, \tag{20}$$

$$\mathbf{h}_j^t = \mathbf{o}_j^t \odot \tanh(\mathbf{c}_j), \tag{21}$$

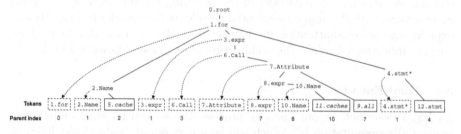

Fig. 4. For each AST, we extract node tokens and put them into a token array, then assign parent index for each token.

where $k \in \mathcal{C}_j$ in Eq. 17, $\tilde{\mathbf{h}}_j^t \in \mathcal{R}^{d \times 1}$ is the sum of children hidden state, $\tilde{\mathbf{c}}_j^t \in \mathcal{R}^{d \times 1}$ denotes the state for updating the memory cell $\mathbf{c}_j^t$. $\mathbf{i}_j^t, \mathbf{o}_j^t, \mathbf{f}_{jk}^t \in \mathcal{R}^{d \times 1}$ are input gate, output gate and forget gate, respectively. $\mathbf{W}^{(\cdot)}, \mathbf{U}^{(\cdot)} \in \mathcal{R}^{d \times d}$ are weight matrix and $\mathbf{b}^{(\cdot)} \in \mathcal{R}^{d \times 1}$ is a bias vector. σ is the logistic function and the operator $\odot$ means element-wise product between two vectors.

3.2 Decoder

From the encoder process, we obtain two embedded vectors, sequential representation vector $\mathbf{h}_n^s$ and tree representation vector $\mathbf{h}_q^t$. Afterward, we initialize decoder hidden state $\mathbf{d}_0$ with the concatenation of $\mathbf{h}_n^s$ and $\mathbf{h}_q^t$ along the sequence length dimension:

$$\mathbf{d}_0 = \mathbf{h}_n^s \oplus \mathbf{h}_q^t, \tag{22}$$

where $\oplus$ means concatenation operation. This decoder initialization considers not only source code sequential summary but also AST structure summary, which could improve predictor performance than the original seq2seq model. The rest decoder process remains the same.

3.3 Dual-Attention Mechanism

After introducing our dual-encoder, we need to improve the attention mechanism to adopt hierarchical hidden outputs from tree encoder. The main difference between our dual-attention and original attention mentioned in Sect. 2.2 is the construction of context vector $\mathbf{w}^t$ for tree encoder. Concretely, as Fig. 3 shows, at j-th step, α_{ij}^s in attention seq2seq still represents sequential attention weights, and hierarchical attention weights α_{ij}^t are calculated by treating them the same as sequential outputs:

$$\alpha_{ij}^t = \frac{\exp(score(\mathbf{h}_i^t, \mathbf{d}_j))}{\sum_{k=1}^{q} \exp(score(\mathbf{h}_k^t, \mathbf{d}_j))}, \tag{23}$$

As in [2], we parameterize the score function $score()$ as a feed-forward neural network which is jointly trained with all the other components of the proposed architecture. Then we compute j-th context vector $\mathbf{w}_j$ as the weighted average of the sequential hidden states and tree hidden states:

$$\mathbf{w}_j = \mathbf{w}_j^s \oplus \mathbf{w}_j^t \tag{24}$$

$$= \sum_{i=1}^{n} \alpha_{ij}^s \mathbf{h}_i^s \oplus \sum_{i=1}^{q} \alpha_{ij}^t \mathbf{h}_i^t, \tag{25}$$

4 Experiments

In this section, we conduct experiments on the task of annotating source code. We first describe our dataset and data preparation steps, then we introduce our training configurations in detail, after that we present experimental results of our model and other baseline algorithms, finally, we show some persuasive generation examples to prove the practicality and readability of our model.

```
class StreamingBuffer(object):    # derive the class StreamingBuffer from the object base class.
    def __init__(self):           #   define the method __init__ with an argument self.
        self.vals = []            #       self.vals is an empty list.
    def write(self, val):         #   define the method write with 2 arguments: self and val.
        self.vals.append(val)     #       append val to self.vals.
    def read(self):               #   define the method read with an argument self.
        ret = b''.join(self.vals) #       join elements of self.vals into a bytes string, substitute the result for ret.
        self.vals = []            #       self.vals is an empty list.
        return ret                #       return ret.
    def flush(self):              #   define the method flush with an argument self.
        return                    #       return nothing.
    def close(self):              #   define the method close with an argument self.
        return                    #       return nothing.
```

Fig. 5. An example of a code snippet with its annotation. The left part is a snippet of Python class, and the right part is its corresponding natural language annotation.

4.1 Dataset

For evaluating our model Code2Text effectively, we choose a high-quality Python-to-English dataset from [12].

Data Description. Python-to-English dataset contains the source code and annotations of Django Project (a Python web application framework). All lines of code are annotated with corresponding annotations by an engineer. The whole corpus contains 18,805 pairs of Python statements and corresponding English annotations, and we split it into a training set and a test set. The training set contains 16,000 statements, and we use it to train our Code2Text model. The rest 2,805 statements in the test set are used to evaluate the model performance. Figure 5 is an example code snippet from the training dataset.

Data Preparation. Since our model exploits hierarchical information of source code, we should first generate code ASTs by applying Python AST Parser[1] to each line of source code in the dataset. Fortunately, Python interpreter itself provides a built-in module called *ast* to help parse source code to its AST. Our model could work on other programming languages such as Java, C++, *etc.* as well if we apply their own open-source libraries for AST parsing[2,3].

For consistency, we need to apply the same preparatory operations to all the source code as follows:

1. For each line of source code, we parse it to an abstract syntax tree with a built-in *ast* module in the Python interpreter.
2. For each AST, we extract node tokens and put them into a tokens array.
3. Then, we index all the nodes and assign the parent's index for every token in the list. We assign index 0 to the root of the tree. The purpose of this step is to reconstruct the tree structure in our training and evaluation phase.

---

[1] https://docs.python.org/3/library/ast.html.
[2] https://github.com/javaparser/javaparser.
[3] https://github.com/foonathan/cppast.

Figure 4 shows an example of how we generate the tokens array and parent index list.

Finally, we extend our dataset by adding a new AST tokens array and an index list for each line of source code. We feed source code sequence into the sequential encoder and feed AST tokens sequence along with the parent index list to the tree encoder.

The goal of our preprocessing steps is decoding an AST into a nodes array(for node embedding) and a parent index array(for reconstruction). The order from the breadth-first search is not important here because the parent index of each node would help us reconstruct the AST tree. During the training process, we would like to accumulate structure information from leaves to root, so we use the parent index list to reconstruct an AST tree and compute root information with the help of parent index array by recursively applying tree-LSTM (As Fig. 2(b) shows).

4.2   Setup

In this part, we introduce our experiment setup, including compared methods and evaluation metric.

Our training objective is the cross-entropy, which maximizes the log probability assigned to the target words in the decoder process.

In the test phase, we have the same inputs in the encoder process for generating $\mathbf{h}_n^s$ and $\mathbf{h}_q^t$. In the decoder process, we predict with the START tag and compute the distribution over the first word y_1^p. We pick the argmax in the distribution and set its embedding vector as the next input y_1, and repeat this process until the END tag is generated. The whole generated sentence $\mathbf{y}^p$ will be our annotation result.

Table 1. Types of models, based on the kinds of features used.

Method	Rules	Seq. Info.	Tree Info.	Atte.
PBMT	$\checkmark$	–	–	–
Seq2seq	–	$\checkmark$	–	–
Seq2seq w/ attention	–	$\checkmark$	–	$\checkmark$
Code2Text w/o seq. info	–	–	$\checkmark$	$\checkmark$
Code2Text w/o attention	–	$\checkmark$	$\checkmark$	–
Code2Text	–	$\checkmark$	$\checkmark$	$\checkmark$

Compared Methods. To validate the improvement of our model, in this paper we compare Code2Text to following state-of-the-art algorithms (summarized in Table 1):

- PBMT [6,7,12]: PBMT is a statistical machine translation framework which uses the phrase-to-phrase relationships between source and target language pairs. Oda et al. apply PBMT to pseudocode generation task [12].
- Seq2seq [16]: Seq2seq model is commonly used in NMT tasks. It consists of encoder process and decoder process, while the encoder process encodes source code sequential information and decoder process learns a language model to predict annotations base on the summary of sequential information.
- Seq2seq w/ attention [9,19]: This version of seq2seq model incorporates attention mechanism which could improve the generation performance.
- Code2Text (w/o sequential encoder): This is one weak version of our method with only tree encoder and hierarchical attention mechanism.
- Code2Text (w/o attention): This weak version of our method combines sequential encoder and tree encoder with no attention mechanism.
- Code2Text: This is our method proposed in Sect. 3, which tries to improve the performance of automatic annotation generation.

Oda et al. [12] also proposed a model with AST information included, however, we have no comparability since we preprocess in different ways.

For all sequential encoders and decoders in both seq2seq models and our model, we use the one-layer LSTM network. The number of epoches is set to 30. All experiments were conducted under a Linux GPU server with a GTX 1080 device.

Metrics. In addition to the direct judgment from real cases, we choose BLEU (Bilingual Evaluation Understudy) score [13] to measure the quality of generated annotations for all the methods. BLEU is widely used in machine translation tasks for evaluating the generated translations. It calculates the similarity of generated translations and human-created reference translations. It is defined as the product of "n-gram precision" and a "brevity penalty" where n-gram precision measures the precision of length n word sequences and the brevity penalty is a penalty for short hypotheses. BLEU outputs a specific real value with range $[0, 1]$ and it becomes 1 when generated hypotheses completely equal to the references. We multiply the BLEU score by 100 in our experiments for display convenience.

4.3   Performance

Our Model vs. Other Models. From Table 2 we can conclude that our model Code2Text outperforms than other compared methods. For PBMT, we only compare BLEU-4 score due to the lack of the other three metrics in [12]. Code2Text has an obvious improvement than PBMT by around 1.6 times and outperforms better than attention seq2seq model since we incorporate hierarchical information in source code.

Table 2. Comparison *w.r.t* BLEU scores. Only BLEU-4 score reported for PBMT due to BLEU-1 to BLEU-3 are not available in source paper.

Models	BLEU-1	BLEU-2	BLEU-3	BLEU-4
PBMT	–	–	–	25.71
Seq2seq w/ atte.	54.11	46.89	42.02	38.11
Code2Text	**65.72**	**55.08**	**48.23**	**42.78**

Table 3. Comparison of BLEU Scores w/o Attention.

Models	BLEU-1	BLEU-2	BLEU-3	BLEU-4
Seq2seq	34.29	28.46	24.61	21.56
+ attention	54.11	46.89	42.02	38.11
Code2Text	36.24	22.96	16.21	11.59
+ attention	**65.72**	**55.08**	**48.23**	**42.78**

Table 4. Effects of tree encoder under BLEU metric.

Models	BLEU-1	BLEU-2	BLEU-3	BLEU-4
Seq2seq + att.	54.11	46.89	42.02	38.11
Code2Text w/o seq.	58.70	47.24	40.14	34.64
Code2Text	**65.72**	**55.08**	**48.23**	**42.78**

Effects of Attention Mechanism. Table 3 tells us how attention mechanism improves model performance. Whether in seq2seq or in our Code2Text, models with attention mechanism both perform better than who without attention mechanism. Meanwhile, the reason that Code2Text without attention mechanism has a lower BLEU score than seq2seq model is dual-encoder compresses more information than the sequential encoder, which makes it harder to capture important information if we do not have an alignment mechanism.

Effects of Tree Encoder. To evaluate the effects of tree encoder in our proposed model, Table 4 reveals that attention seq2seq model and our Code2Text model with only tree encoder has similar performance. Since this weak version of Code2Text neglects sequential encoder, it may not capture order information, which may result in worse performance under BLEU-3 and BLEU-4 than attention seq2seq model. However, our full version of Code2Text achieves the best score.

4.4 Case Study

We present four cases in Fig. 6. Each case has three corresponding annotations apart from the ground truth. For all the four cases, annotations generated by

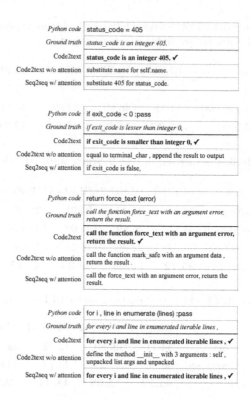

Python code	status_code = 405
Ground truth	*status_code is an integer 405.*
Code2text	**status_code is an integer 405.** ✔
Code2text w/o attention	substitute name for self.name.
Seq2seq w/ attention	substitute 405 for status_code.

Python code	if exit_code < 0 :pass
Ground truth	*if exit_code is lesser than integer 0,*
Code2text	**if exit_code is smaller than integer 0,** ✔
Code2text w/o attention	equal to terminal_char , append the result to output
Seq2seq w/ attention	if exit_code is false,

Python code	return force_text (error)
Ground truth	*call the function force_text with an argument error, return the result.*
Code2text	**call the function force_text with an argument error, return the result.** ✔
Code2text w/o attention	call the function mark_safe with an argument data , return the result .
Seq2seq w/ attention	call the force_text with an argument error, return the result.

Python code	for i , line in enumerate (lines) :pass
Ground truth	*for every i and line in enumerated iterable lines ,*
Code2text	**for every i and line in enumerated iterable lines ,** ✔
Code2text w/o attention	define the method __init__ with 3 arguments : self , unpacked list args and unpacked
Seq2seq w/ attention	**for every i and line in enumerated iterable lines ,** ✔

Fig. 6. Cases of our Code2Text model. Our model could generate readable natural language annotations for various statements.

Code2text without attention mechanism have the least similarity to the ground truth, which meets the BLEU score evaluation results. The reason may be that LSTM performs worse due to the combination of source code tokens and AST tokens. The attention mechanism will help align the annotation words with the source tokens and AST tokens. For the first case and third case, although attention seq2seq model could generate reasonable annotations as well, our Code2Text captures the hidden keyword (*integer, function*) from source code AST and provides more accurate annotations. The fourth case reveals that Code2Text could generate complex expression, which is friendly to beginners.

We visualize our attention matrix α_{ij}^s and α_{ij}^t of the first and third cases in Fig. 7. For the left figure, our model captures the relationship between keyword *integer* and node tokens in AST (*int, n*). Since Python is a Weakly-Typed Language, the base type of a variable (such as integer, string, *etc.*) will be inferenced by the interpreter. Therefore, this type information could only exist in AST, that is why our model can generate keyword *integer*. In the same way, right figure exploits another two relationships. Keyword *function* corresponds with *func* and *argument* corresponds with (*expr*, args*).

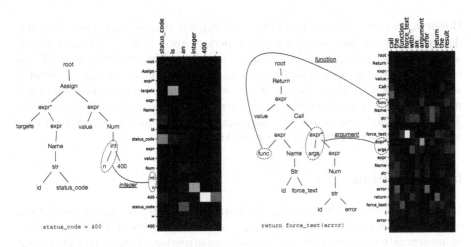

Fig. 7. Two sample alignments above refer to our first and third cases. The x-axis corresponds to the generated annotation, and the y-axis corresponds to the tokens from AST and source code. Each pixel shows the weight α_{ij}^{t} of the annotation of the j-th source word for the i-th target word (see Eq. 23), in grayscale (0: black, 1: white).

5   Related Work

In the early years, various rule-based models were explored by researchers. Sridhara et al. [14,15] focused on automatic comment generation for Java programming language. Their two works both concentrated on designing mapping rules between source code and code comment by hand and generating comments for Java methods by filling out pre-defined sentence templates. Moreno et al. [11] studied comment for Java Classes as well. Their model extracted the class and method stereotypes and used them, in conjunction with heuristics, to select key information to be included in the summaries. Then it generated code snippet summaries using existing lexicalization tools. However, there is a major limitation for the rule-based approach that it lacks portability and flexibility. When new rules that are never seen appear in the source code or we start a new project with another programming language, we have to manually update our rules table and sentence templates.

Later, most researchers tended to data-based approaches in recent years. Wong et al. [18] crawled code-description mappings from online Q&A websites at first, then output code comment by matching similar code segments. Therefore this model does not have a generalization. Haiduc et al. [3] created summaries for source code entities using text retrieval (TR) techniques adapted from automated text summarization [5].

Recently, some deep neural networks are introduced in the annotation generation task. In the task of pseudocode generation, Oda et al. [12] combines the rule-based approach and data-based approach. Their model updated the rules table automatically and generated pseudocode through n-gram language model.

However, its n-gram language model lacks explicit representation of long-range dependency which may affect generation performance. In 2017, Zheng et al. [19] applied attention sequence to sequence neural machine translation model on code summary generation which motivates our work. Allamanis et al. [1] treated source code as natural language texts as well, and learned a convolutional neural network to summarize the words in source code into briefer phrases or sentences. These models for code summary generation task do not consider the hidden hierarchical information inside the source code.

6   Conclusion

In this paper, we studied the problem of structure-aware annotation generation. We proposed a novel model, Code2Text, to translate source code to annotations by incorporating tree encoder and hierarchical attention mechanism. Experiment results showed that our model outperforms among state-of-the-art methods, and example cases prove the practicality and readability. Our model could be also extended to other programming languages easily with the specific parser.

Acknowledgements. This work is supported in part by the National Natural Science Foundation of China Projects No. U1636207, U1936213, the NSF under grants No. III-1526499, III-1763325, III-1909323, and CNS-1930941, the Shanghai Science and Technology Development Fund No. 19DZ1200802, 19511121204.

References

1. Allamanis, M., Peng, H., Sutton, C.: A convolutional attention network for extreme summarization of source code. In: ICML 2016, New York City, NY, pp. 2091–2100 (2016)
2. Bahdanau, D., Cho, K., Bengio, Y.: Neural machine translation by jointly learning to align and translate. arXiv preprint arXiv:1409.0473 (2014)
3. Haiduc, S., Aponte, J., Moreno, L., Marcus, A.: On the use of automated text summarization techniques for summarizing source code. In: WCRE 2010, Beverly, MA, pp. 35–44 (2010)
4. Hochreiter, S., Schmidhuber, J.: Long short-term memory. Neural Comput. **9**(8), 1735–1780 (1997)
5. Jones, K.S.: Automatic summarising: the state of the art. Inf. Process. Manag. **43**(6), 1449–1481 (2007)
6. Karaivanov, S., Raychev, V., Vechev, M.: Phrase-based statistical translation of programming languages. In: SPLASH 2014, Portland, OR, pp. 173–184 (2014)
7. Koehn, P., Och, F.J., Marcu, D.: Statistical phrase-based translation. In: NAACL 2003, Edmonton, Canada, vol. 1, pp. 48–54 (2003)
8. Liu, X., Kong, X., Liu, L., Chiang, K.: TreeGAN: syntax-aware sequence generation with generative adversarial networks. In: ICDM 2018 (2018)
9. Luong, M.T., Pham, H., Manning, C.D.: Effective approaches to attention-based neural machine translation. arXiv preprint arXiv:1508.04025 (2015)
10. Mikolov, T., Karafiát, M., Burget, L., Černocký, J., Khudanpur, S.: Recurrent neural network based language model. In: InterSpeech 2010, Makuhari, Japan (2010)

11. Moreno, L., Aponte, J., Sridhara, G., Marcus, A., Pollock, L., Vijay-Shanker, K.: Automatic generation of natural language summaries for Java classes. In: ICPC 2013, San Francisco, CA, pp. 23–32 (2013)
12. Oda, Y., et al.: Learning to generate pseudo-code from source code using statistical machine translation. In: ASE 2015, Lincoln, NE, pp. 574–584 (2015)
13. Papineni, K., Roukos, S., Ward, T., Zhu, W.J.: Bleu: a method for automatic evaluation of machine translation. In: ACL 2002, Philadelphia, PA, pp. 311–318 (2002)
14. Sridhara, G., Hill, E., Muppaneni, D., Pollock, L., Vijay-Shanker, K.: Towards automatically generating summary comments for Java methods. In: ASE 2010, Lawrence, KS, pp. 43–52 (2010)
15. Sridhara, G., Pollock, L., Vijay-Shanker, K.: Automatically detecting and describing high level actions within methods. In: ICSE 2011, Honolulu, HI, pp. 101–110 (2011)
16. Sutskever, I., Vinyals, O., Le, Q.V.: Sequence to sequence learning with neural networks. In: NIPS 2014, Montreal, Canada, pp. 3104–3112 (2014)
17. Tai, K.S., Socher, R., Manning, C.D.: Improved semantic representations from tree-structured long short-term memory networks. arXiv preprint arXiv:1503.00075 (2015)
18. Wong, E., Yang, J., Tan, L.: AutoComment: mining question and answer sites for automatic comment generation. In: ASE 2013, Palo Alto, CA, pp. 562–567 (2013)
19. Zheng, W., Zhou, H., Li, M., Wu, J.: Code attention: translating code to comments by exploiting domain features. arXiv preprint arXiv:1709.07642 (2017)

Semantic Enhanced Top-k Similarity Search on Heterogeneous Information Networks

Minghe Yu[1,2], Yun Zhang[3(✉)], Tiancheng Zhang[3], and Ge Yu[3]

[1] Software College, Northeastern University, Shenyang, China
yuminghe@mail.neu.edu.cn
[2] Guangdong Province Key Laboratory of Popular High Performance Computers,
Shenzhen University, Shenzhen, China
[3] School of Computer Science and Engineering, Northeastern University, Shenyang, China
1801762@stu.neu.edu.cn, {tczhang,yuge}@mail.neu.edu.cn

Abstract. Similarity search on heterogeneous information networks has attracted widely attention from both industrial and academic areas in recent years, for example, used as friend detection in social networks and collaborator recommendation in coauthor networks. The structure information on the heterogeneous information network can be captured by multiple meta paths and people usually utilized meta paths to design method for similarity search. The rich semantics in the heterogeneous information networks is not only its structure information, the content stored in nodes is also an important element. However, the content similarity of nodes was usually not valued in the existing methods. Although recently some researchers consider both of information in machine learning-based methods for similarity search, they used structure and content information separately. To address this issue by balancing the influence of structure and content information flexibly in the process of searching, we propose a double channel convolutional neural networks model for top-k similarity search, which uses path instances as model inputs, and generates structure and content embeddings for nodes based on different meta paths. Moreover, we utilize two attention mechanisms to enhance the differences of meta path for each node and combine the content and structure information of nodes for comprehensive representation. The experimental results showed our search algorithm can effectively support top-k similarity search in heterogeneous information networks and achieved higher performance than existing approaches.

Keywords: Attention mechanism · Convolutional neural network ·
Heterogeneous information network · Network representation learning · Top-k
similarity search

1 Introduction

In recent years, data mining on heterogeneous information networks has attracted extensive attentions from both industrial and academic areas. For example, in an academic social network, users can search the most similar author or the most relevant paper based on the queries they input [1]. Similarity search is a popular data mining algorithm

© Springer Nature Switzerland AG 2020
Y. Nah et al. (Eds.): DASFAA 2020, LNCS 12114, pp. 104–119, 2020.
https://doi.org/10.1007/978-3-030-59419-0_7

based on some strategies to find the most similar objects to a given object (query statement, node, etc.). Classical similarity search studies including PageRank [2], HITS [3], SimRank [4] focus on Homogeneous Information Network or binary networks, which have only one type of nodes and edges. However, data in the real world is very complicated that using homogeneous information network to describe them cannot completely express the rich semantics, including objects and relations. Therefore, heterogeneous information network (called HIN for short) becomes a great option to describe real data. Recently, searching and recommendation technologies based on heterogeneous information networks are gradually proposed [9–14, 19, 21, 24], which aim to find similar objects in HINs based on given information.

To be specific, top-k similarity search on HINs focuses on obtaining a set of related nodes with the same or different types of target nodes by evaluating how these nodes connected to each other. They rely on predefined meta paths, in which rich structural information can be captured. Unlike homogeneous information networks, different meta paths on heterogeneous networks are specified to represent different semantics. Therefore, it is better to flexibly adjust training different meta paths for different queries. Since nodes have both structure information and content information, and this information have different schemas, how to integrate this information to represent nodes is also a great challenge need to be solved.

There are some unsupervised learning-based methods [10, 25] that consider content and structure information in the node for similarity search on HINs. But these methods utilize fixed mode to combine this information for all nodes. In fact, for nodes with little interaction behaviors, capturing their content information is more important. Meanwhile for active nodes, the main task is to train their structure information on the given heterogeneous information network for similarity search. So, it is necessary to flexibly consider the content information and structure information of nodes.

In the machine learning-based methods, they utilize node embedding on heterogeneous information networks for similarity search, which measure similarity of nodes according to the vectors' similarity. In metapath2vec [16] and other methods [5–7, 9, 17, 18, 20, 23], model training and node embedding are only based on one meta path, which is useful for tasks with specific semantic requirements. In the recommendation task, MCRec [24] uses attention mechanism to combine multiple meta path information to realize similarity measurement between users and items, but it does not consider the content information of nodes. HetGNN [25] also uses attention mechanism to embed nodes. It considers content and structure information for each node. However, the content and structure information are not combined comprehensively. In HetGNN, when the model is trained by the nodes' structure information, the content features would fade or even disappear. In this paper, we will solve this problem.

To overcome this challenge, we propose a model for top-k similarity search based on double channel convolutional neural network on heterogeneous information networks (called **SSDCC**), which uses CNN with two channels to respectively train the content information and structure information at the same time. It firstly generates different structure and content embeddings for each node based on different meta paths. Then, it samples path instances for given meta paths. To train content information and structure information at the same time, Double Channel CNN is used in SSDCC. Our model can distinguish the meta paths for each node by an attention mechanism because the importance of meta paths is not the same for each node. Finally, we propose another attention mechanism in SSDCC to combine node's content and structure information for its comprehensive representation.

In summary, the major contributions of our work are as follows:

- In this paper, we first use the double channel convolution neural network to train the structure information and content information of nodes simultaneously for similarity search on heterogeneous information networks, which avoids the problem that the content information embedded in the previous training of nodes decays with the following structure training.
- Two different attention mechanisms are designed in SSDCC. Firstly, an attention mechanism is used for multiple meta paths when embedding structural information, which flexibly and personally implements comprehensive consideration of structural information under different meta paths for each target node. Secondly, another attention mechanism is used between content representation and structure representation, which individually combines the training results of Double Channel CNN.
- We have implemented our model on the public dataset to evaluate its performance and compared it with the existing methods. Through the analysis of the results, this model can deal with the problem of similarity search on heterogeneous information networks well and achieved higher performance than the existing methods.

2 Related Work

In this section, we will introduce related works of Network Embedding and top-k Similarity Search on heterogeneous information networks.

2.1 Network Embedding

In recent years, with the development of heterogeneous information networks [15], a large number of network embedding methods for HINs emerged. Utilizing node embedding on HIN, the nodes' representations are turned from high-dimensional sparse vectors to high-dimensional vectors while the relationship information among the nodes is preserved. Some researchers use this kind of node vectors to measure the similarity of nodes directly.

One of mainstream node embedding methods is Deep Learning based training method. DeepWalk [5], TriDNR [7] implement the node embedding based on random walk on homogeneous information networks while there are also a lot of researches using GNN [8]. Researches on HIN include metapath2vec [16], HIN2vec [17], HAN [18]. Metapath2vec represents nodes based on random walk, and HIN2vec uses neural network model to learn the representation of nodes and meta path. Both of them only consider a single meta path so that they cannot capture rich path semantic information on HINs. In addition, HAN uses node-level attention mechanism and semantic-level attention mechanism, it distinguishes the meta-paths via semantic-level attention to get semantic information. However, its parameters are shared for all meta-paths so that the attention score of a meta path is the same for different nodes. Therefore, similarity search based on above models cannot flexibly adjust the network status search requirements, and we need a more scalable model.

2.2 Top-k Similarity Search

There have been already many works on HINs to measure the similarity of objects. PathSim [19] utilizes meta paths to resolve similarity search problem, which defines the similarity between two objects of the same type by considering the accessibility and visibility between vertices. HeteSim [20] is an extension of PathSim, which can measure the similarity of different types of objects. W-PathSim [21] proposes another improvement by using the weighted cosine similarity of topics. RoleSim [22] considers both structural features and Jacard similarity of attribute features. Similarity searches are also applied to recommendations, and MCRec [24] uses an attention mechanism to measure the similarities between users and items. However, different users and items affect the attention weights of meta paths, which makes the model inefficiency.

Above approaches only consider a single aspect such as content or structure similarity in heterogeneous information networks, and some only consider one meta path information in the measurement of structural similarity.

In the methods considering both content and structure information [10, 25], nodes are firstly embedded according to the content information. Then they set the content embedding as initial value, and train the model by structure information just based on the initial content embedding. Similar with the optimizing model that seeks the minimum value in the objective function, the initial state often does not have much impact on final result. In other words, as the model training by the nodes' structure information, the content features contained in the initial representation will fade or even disappear.

Therefore, the model proposed in this paper should comprehensively considers the structure information and content information of nodes on heterogeneous networks while overcoming the problems we mentioned above.

3 Preliminary

In this section, we will give definitions that will be used later and formulate the problem.

Definition 3.1 Heterogeneous Information Network (HIN). Given an information network, it can be represented by a directed graph $G = (V, E, A, \mathfrak{R}, W; \phi, \varphi)$, where V is the object set, E is the link set, A is a collection of object types, $\mathfrak{R}$ is the set of link types, and W is the set of attribute values on relations. Besides, $\phi : V \rightarrow A$ is a mapping function from the object set to its type set, where each object ($v \in V$) belongs to a specific object type ($\phi(v) \in A$). And $\varphi : E \rightarrow \mathfrak{R}$ is a mapping function from link set to link type set, where each link ($e \in E$) belongs to a specific link type ($\varphi(e) \in \mathfrak{R}$). When the number of object type $|A| > 1$ (or the number of link type $|\mathfrak{R}| > 1$) and the type of link attribute values $|W| > 1$, this network is Heterogeneous Information Network (HIN).

In the heterogeneous information network, each link type represents a special relationship. And if the attribute values in W are not all equal to 0, the network is weighted heterogeneous information network.

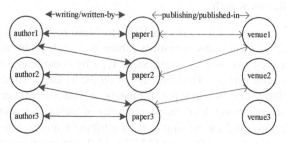

Fig. 1. Example of heterogeneous information network on bibliographic data

Example 3.1 Academic Social Network. Consider heterogeneous information network on bibliographic data in Fig. 1, there are three types of nodes: *authors*, *papers* and *conferences*. And four types of links: *writing* and *written-by* between authors and papers, and *publishing* and *published-in* between papers and the venues.

To express the rich path semantic information in HIN, we use meta path to describe this structure information.

Definition 3.2 Meta Path. A meta path expressed as $A_0 \xrightarrow{\mathfrak{R}_0} A_1 \xrightarrow{\mathfrak{R}_1} \dots \xrightarrow{\mathfrak{R}_{l-1}} A_l$ is a path in an HIN, where $A_0, A_1, \dots, A_l \in A$, and $\mathfrak{R}_0, \mathfrak{R}_1, \dots, \mathfrak{R}_{l-1} \in \mathfrak{R}$. It defines a combination relationship $\mathfrak{R} = \mathfrak{R}_0 \circ \mathfrak{R}_1 \circ \dots \circ \mathfrak{R}_{l-1}$ between object types A_0 and A_l.

For simplicity, we directly represent a meta paths as $P = (A_0, A_1, \dots, A_l)$. And we call a concrete path $p = (a_0, a_1, \dots, a_l)$ of a given meta path P as a path instance.

Definition 3.3 Top-k Similarity Search on HIN. Given a HIN $H = (V, E, A, \mathfrak{R}, W; \phi, \varphi)$, the meta path set $\{P_1, P_2, \dots, P_x\}$, a query node n_q $(n_q \in V)$, and the number of results k. Top-k similarity search in this paper aims to find the most similar nodes of n_q from H based on a similarity measure F. How to define a proper F will be studied in this paper.

4 SSDCC Search Model and Algorithm

In this section, we propose the model for top-k similarity search based on double channel convolutional neural network on HINs (**SSDCC**). Then we give the top-k similarity search algorithm.

4.1 Overview

Our model SSDCC is designed for similarity search on heterogeneous networks, which finds the top-k similar nodes in the HIN for the query node. We first use pre-training methods to generate the content and structure representation of all nodes on HIN dataset. For each given meta path, we generate the model input by sampling path instances starting with target nodes (In the process of model training, every node is traversed. And the node that is trained as the object to be queried is called as target node). Then, we use double channel convolutional neural network to respectively train nodes' content information and structure information at the same time. For different target nodes, the importance of different meta paths is individually obtained through the attention mechanism between them. Finally, we use another attention mechanism on the current node's content embedding and structure embedding to generate its comprehensive representation.

The overall framework of SSDCC is shown in Fig. 2. The model is divided into three parts: Node Representation, Path representation and Combination of Content and Structure. The upper part of Node Representation and Path representation corresponds to structural information training and the lower part corresponds to content information training. The following sections analyze each part in detail.

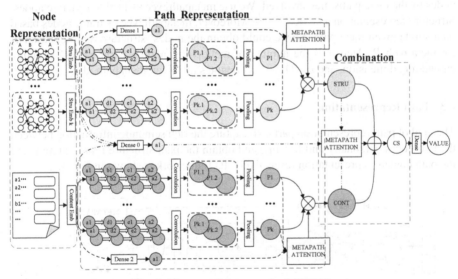

Fig. 2. The overall framework of SSDCC

4.2 Node Representation

Node information in HINs can be divided into two types, including content information and structure information. As shown in Fig. 3, the content information refers to text descriptions, images, labels and other information of nodes, while the structure information refers to the connection relationship between nodes and others on the network.

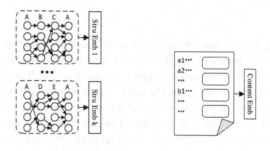

(a) Structure Representation   (b) Content Representation

Fig. 3. Nodes representation

To embed nodes, we use the existing embedding methods to represent nodes themselves. For the content information, we use Doc2vec [23] to generate the nodes' content representation $a^c \in \mathbb{R}^{|A| \times d_1}$. Here $|A|$ is the number of nodes in type A, d_1 is the nodes' content embedding dimension. For the structure information of nodes, because of the meta paths' rich semantics in the HIN, many models have been proposed to embed nodes by the meta paths they involved. We use metapath2vec++ [16] to generate nodes' Structure representation $a^s \in \mathbb{R}^{r \times |A| \times d_2}$. Moreover, metapath2vec++ trains nodes based on a single given meta path. So we could obtain the nodes' structure representation a_i^s for meta path P_i. Here r is the number of meta paths, and d_2 is the nodes' structure embedding dimension, and $i \in \{1, \ldots, k\}$ denotes the given meta path id.

4.3 Path Representation

The goal of path representation part is to capture the rich semantic information between the two nodes personally and flexibly, and obtain the path structure representation and the path content representation respectively. Its framework is shown in Fig. 4.

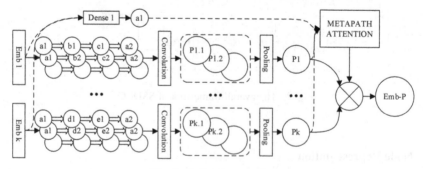

Fig. 4. Path representation

Integrated Embedding of Nodes. In Sect. 4.2, the model generates the structure embeddings a_i^s in r given meta paths, and it also generates one content embedding a^c for each node. We design three Dense layers *Dense 0, Dense 1* and *Dense 2*. We obtain the integrated structure embedding $\boldsymbol{a}^s$ of nodes by inputting $\boldsymbol{a}_i^s$ into *Dense1*, and obtain the integrated embedding $\boldsymbol{a}^{s+c}$ of nodes including both structure and content by inputting $\boldsymbol{a}_i^s$ and $\boldsymbol{a}^c$ into *Dense 0*. We also input $\boldsymbol{a}^c$ into *Dense 2* to maintain the consistency.

Path Instance and Meta Path Representation. Most methods calculate nodes' similarity search on HINs by their embedding vector's similarity directly. Inspired by the recommendation model MCRec [24] which uses path instances as input to capture the nodes' interaction, we represent meta paths in order to capture rich path semantic information between nodes in similarity search tasks. As shown in Fig. 4, for a query node a_1, we take it as a starting point to sample the path on the designated meta path. In the meta path $P_i (i \in \{1, \dots, k\})$, we obtain multiple path instances $P_{i,j} (j \in \{1, \dots, t\}$ is the id of a path instance) from a_1 to another node a_2 on the weighted heterogeneous information network, and calculate the path weight according to the parameters on edge between adjacent nodes. In our model, we use Symmetric Random Walk (SRW) in SSDCC to obtain the weight of the path instance. In addition, in order to reduce the training time, we abandon some path instances with low weight.

During pre-training process, the embedding of nodes on different path instances are generated in the specified meta path respectively. Then for nodes' structure information, we obtain the embedding of path instance $P_{i,j}^s (i \in \{1, \dots, k\}, j \in \{1, \dots, t\})$ through convolutional layers. After that, the model obtains the embedding of meta path P_i^s through a pooling layer. For nodes' content information, we obtain the embedding of path instance $P_{i,j}^c$ and the embedding of meta path P_i^c in the same way.

Path Representation. In order to obtain the comprehensive information including all meta path P_i, we integrate meta path embeddings into one representation by the attention mechanism. We distinguish the meta path attention score to each node. The path attention weight is affected by the target node. We use attention scoring function with Bahdanau's additive style to calculate the importance of each meta path P_i to node $a_m (m \in \{1, \dots, |A|\})$. The importance score $score(a_m, P_i)$ is calculated as follows:

$$score(a_m, P_i) = v_\alpha^T \cdot \tanh(W_1 \cdot a_m + W_2 \cdot \boldsymbol{P}_i), \tag{1}$$

where W_1 and W_2 are two weight matrixes, and v_α^T is a vector to get a score value, $\boldsymbol{P}_i \in \{\boldsymbol{P}_i^s, \boldsymbol{P}_i^c\}$ is the embedding of meta path P_i. After getting the importance score $score(a_m, P_i)$, for node a_m and its similar node a_n, the attention weights of P_i can be calculated through SoftMax function:

$$\alpha_{a_{m,n}-P_i} = \frac{\exp(score(a_m, \boldsymbol{P}_i))}{\sum_{i'=1}^k \exp(score(a_m, \boldsymbol{P}_{i'}))}, \tag{2}$$

which can be interpreted as the contribution of the meta path P_i for node a_m. And n represents another node a_n that measures similarity with node a_m. Based on the attention weights $\alpha_{a_{m,n}-P_i}$ in Function (2), the Path Representation can be computed as the weighted average of the meta path representation $\boldsymbol{P}_i$ shown as follows:

$$Emb_{a_{m,n}} = \sum_i \alpha_{a_{m,n}-P_i} \cdot \boldsymbol{P}_i \tag{3}$$

where $Emb_{a_{m,n}} \in \{S_{a_{m,n}}, C_{a_{m,n}}\}$ includes the path structure representation and the path content representation.

4.4 Double Channel CNN Based Path Representation

The key challenge in SSDCC is how to comprehensively consider the content information and the structure information of the object. The existing methods [26, 27] train the nodes' structure information after content information, which can be considered to be a representation just for structure information. To comprehensively represent the content information and the structure information, we construct a Double Channel Convolutional Neural Network based method. The content representation and structure representation are trained by different channel called *Cont-channel* and *Stru-channel* respectively based on the double channel mechanism.

In *Cont-channel*, nodes on the path instance $P_{i,j}$ are represented with a^c. After Path Representation section, we acquire the content embedding $C_{a_{m,n}}$ for given meta paths. In *Stru-channel*, nodes on the path instance $P_{i,j}$ are represented with a^s. Every node on different meta path P_i has its corresponding embedding a_i^s. We then acquire the structure embedding $S_{a_{m,n}}$ from this channel.

4.5 Combination of Content and Structure

From previous work, we get content embedding C_a, structure embedding S_a and integrated embedding a^{s+c}. In this section we combine them by another attention machine shown in Fig. 5.

The attention score over different representation is:

$$score(a^{s+c}, C_{a_{m,n}}) = v_\beta^T \cdot \tanh(W_3 \cdot a^{s+c} + W_4 \cdot C_{a_{m,n}}) \tag{4}$$

$$score(a^{s+c}, S_{a_{m,n}}) = v_\beta^T \cdot \tanh(W_3 \cdot a^{s+c} + W_4 \cdot S_{a_{m,n}}) \tag{5}$$

where W_3, W_4 are two weight matrixes and v_β^T is the weight vector to get similar score. Based on these two scores, the attention weights of content representation and structure representation are computed as follows:

$$\alpha_C = \frac{\exp(score(a^{s+c}, C_{a_{m,n}}))}{\exp(score(a^{s+c}, C_{a_{m,n}})) + \exp(score(a^{s+c}, S_{a_{m,n}}))}, \tag{6}$$

$$\alpha_S = \frac{\exp(score(a^{s+c}, S_{a_{m,n}}))}{\exp(score(a^{s+c}, C_{a_{m,n}})) + \exp(score(a^{s+c}, S_{a_{m,n}}))}. \tag{7}$$

Then we acquire the combination of content and structure representations with the function shown as follow:

$$CS = \alpha_C \cdot C_{a_{m,n}} \oplus \alpha_S \cdot S_{a_{m,n}}. \tag{8}$$

In this function, we concatenate the results of content and structure representations after multiplying with attention weights. Compared with the methods by summing them together directly like other attention mechanism, utilizing this function can effectively help us to reduce information loss.

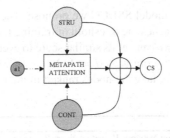

Fig. 5. The attention mechanism for combine content and structure embeddings

4.6 Objective Function and Model Training

From Sect. 4.5, we get the combined representation of content and structure information in the meta paths between target node a_m and its similar node a_n. We need a similar value between these two nodes for top-k similarity search. Therefore, we feed the representation CS into the MLP component to calculate the final output score, that is:

$$y_{m,n} = \text{sigmoid}[f(\mathbf{CS})] \qquad (9)$$

where f is the MLP component that has two Dense layers with the activation function ReLU. Its output will be feed into a sigmoid layer to obtain the result $y_{m,n}$.

We use a logarithmic loss function in the model. Therefore, $y_{m,n}$ of node pair (m, n) in which the node is similar to the other equals to 1 while the dissimilar ones equal to 0. The objective function is shown as follows:

$$o = -\frac{1}{|N|} \sum\nolimits_{i=1}^{|N|} (y_i \log p_i + (1 - y_i) \log(1 - p_i)), \qquad (10)$$

where y_i is the output variable of SSDCC, N is the input sample set, and p_i is the probability that the predicted input instance is similar.

Obviously, a similar node must have more path instances between it and the query node in the given meta path set, and their output score $y_{m,n}$ must be larger than others. So, we use negative sampling to train the model where the negative data set is that including node pairs without path instance to a_n on the given meta path. We set the score of these dissimilar node pairs to zero and then the objective function can be formulated as follows:

$$o = -\frac{1}{|N^+|} \sum\nolimits_{i \in N^+} y_i \log p_i - \frac{1}{|N^-|} \sum\nolimits_{j \in N^-} (1 - y_j) \log(1 - p_j) \qquad (11)$$

where N^+ is the positive sample set and N^- is the negative sample set.

4.7 Top-k Search Algorithm

In Sect. 4.6, we get the output score $y_{m,n}$ between different nodes on the graph. Finally, we propose a Top-k Search Algorithm to obtain the first k most similar nodes of the target node by Algorithm 1. To be specific, we first input the HIN H and the meta path set $\{P_1, P_2, \ldots, P_x\}$ to the model SSDCC we propose. The model is trained until the loss function stabilizes or reaches the threshold of training times we set in advance. For the target node $n_q(n_q \in V)$, we can get its similar score to every node in H. Then we sort other nodes according to the score $y_{n_q,n_i}(i = 1, 2, \ldots, |V|, n_i \in V$, representing each node in graph H). Finally, we take the nodes ranked in the top k position as the output of the search algorithm.

Algorithm 1: Top-k Search

Input: Heterogeneous Information Network H, meta path set P, a query node n_q, k
Output: Result R
1:      Initialize model SSDCC by H and P;
2:      Training SSDCC and generate $y_{n_q,n_i}(i = 1,2,\ldots,|V|, n_i \in V)$;
3:      Sort n_i by $y_{n_q,n_i}(i = 1,2,\ldots,|V|)$;
4:      **For** j **from** 1 **to** k
5:          Add j-th node into R;
6:      **Return** R;

---

5   Experimental Evaluation

In this section, we describe the datasets about our experiment, then evaluate the effectiveness of the SSDCC and compare it with variants and baseline methods.

5.1   Environment and Dataset

All experiments are conducted upon a computer with 2 CPU cores with 2.10 GHz, 1 GB RAM and 2 TB disk space. The operating system is ubuntu 18.04, and our SSDCC based search algorithm is written in python.

We use real-world Academic Social (AS) Network dataset form AMiner [28, 29] as experimental dataset which is publicly available. The AMiner AS dataset contains 2.1M papers including 8M citations, 1.7M authors and 4.3M coauthors. We construct a heterogeneous information network based on three types of nodes: authors, papers, and venues. We also extract different relation edges among three sets of nodes including authors writing papers, papers published on venues, and papers cited by other papers to construct the network. We prune all paper without affiliations, year and publication venue information. Then we extracted the data in 2012, 2013, and 2014, and remained papers and their related information including authors and venues for training the model and comparing the results in effectiveness.

5.2   Experimental Setup – Meta Path and Model Parameters

Meta Paths Selection. Although the similarity measurement methods such as Pathsim [19] cannot synthesize multiple meta paths at the same time to measure node similarity, the meta path used in it to measure the similarity of nodes is reasonable. It presents two meta paths APA and APVPA which represent the scenarios of co-authoring papers and publishing papers in the same respectively. We use the above meta paths to train and test the model SSDCC for similarity measurement of author type objects.

Parameter Setup. We select the text description of the nodes as the content information, and training data by Doc2vec to get content embeddings. Then metapath2vec is used to embed nodes' structure information under different meta paths respectively. We set the dimensionality of both content and structure embedding to 128, that is $d_1 = 128$ and $d_2 = 128$. Besides, all parameters are initialized using a uniform distribution. Here we set the model training epochs to be 5 and the number of negative samples to be 5 in the convolution neural network. The mini-batch Adam optimizer is exerted to optimize these parameters, where the learning rate is set to 0.001, and the batch size is set to 512. Moreover, to make the method scalable, we use multiple processes to train the model. Using 6 processes to train in parallel on the experimental machine results in 6x faster training.

5.3   Experimental Results and Comparison

First, we labeled top-10 results for 10 popular authors (e.g. Jack J. Dongarra, P S Yu, Elisa Bertino, and Jiawei Han) in 2012, 2013 and 2014 respectively according to their research direction, organization and publication of papers to test the quality of the similarity search result. We labeled each result object with relevance score as: 0–not relevant, and 1–relevant.

We compare our SSDCC based search algorithm against the baselines and variants of it to verify its performance:

- SSDCC–: The attention mechanism between content and structure embedding is not considered, while they are directly connected.
- SS1C: Use a single channel model that only considers content representation.
- SS1S: Use a single channel model that only considers structural representation.
- Metapath2vec.apvpa [16]: We use metapath2vec algorithm here and sample the path according to the meta path APVPA to obtain the embedding of the node. According to the Euclidean distance between the vector of different authors, similarity search is carried out.
- Metapath2vec.apa [16]: We use the meta path APA for path sampling to obtain the embedding of nodes. And similarity search is carried out by Euclidean distance.
- Doc2vec [23]: In this method, nodes are embedded according to the content information of the object, and then search the top-k similar objects for each target nodes by using node vectors.

We input the data in 2014 into SSDCC model to obtain the results of top-10 similarity search, and the results of 4 authors are shown in Table 1.

Moreover, we evaluated the precision and normalized discounted cumulative gain (*NDCG*) at rank 10 values of similarity search results and compared them with the baselines and the variants of our model, as shown in Fig. 6 (a). (In Fig. 6, M.apvpa stands for Metapath2vec.apvpa and M.apa stands for Metapath2vec.apa.)

According to the results in Fig. 6 (a), the performance of SSDCC is better than SSDCC–. This is because using the attention mechanism between content and structure

Table 1. Top-10 similarity search results of four popular authors in SSDCC

Rank	Hector Garcia-Molina	Hai Jin	Sushil Jajodia	W Pedrycz
1	Steven Euijong Whang	Heng Chen	Lingyu Wang	Emilio Corchado
2	Ippokratis Pandis	Jun Sun	Hussain M. J. Almohri	Ying Yu
3	Zhifeng Bao	Haoyu Tan	Guohong Cao	Mehmet Sevkli
4	Ryan Johnson	Qiang Sun	Xinyuan Wang	Elankovan Sundararajan
5	Jongwuk Lee	Yuting Chen	Mohamed Shehab	Selim Zaim
6	Kamal Zellag	Yuanjie Si	Dijiang Huang	Amin Jula
7	Bettina Kemme	Yaobin He	Hannes Holm	Erbao Cao
8	Ihab F. Ilyas	Yinliang Zhao	Qinghua Li	Shelton Peiris
9	George Beskales	Xinjun Mao	Nicola Gatti	Peter Wanke
10	S. W. Hwang	Qingjie Zhao	Yang Qin	Heejung Lee

is beneficial to improve the accuracy of the model. The result of SSDCC and SSDCC–
are also better than those of SS1S and SS1C, showing that the model's performance
can be improved by using the Double Channel Convolutional Neural Network to comprehensively consider the content information and structure information. As SSDCC
considers multiple meta paths while the Metapath2vec.apvpa and Metapath2vec.apa are
just trained based on one meta path, the performance of SSDCC and its variants are
all better than the baselines. In addition, Doc2vec has the worst performance mainly
because the authors' structure information in this data set is abundant, while the text
description information is relatively insufficient.

Next, we change the result number k to evaluate its influence for all methods based
on the NDCG values. The results are shown in Fig. 6 (b).

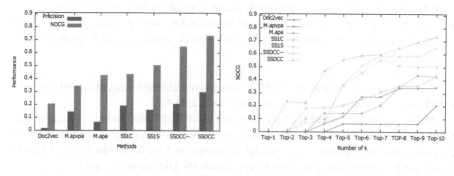

(a) Comparison on precision and NDCG   (b) NDCG from Top-1 to Top-10

Fig. 6. Performance comparison

The experimental results showed our SSDCC model always achieves the best performance in all similarity search tasks. And In most cases, the NDCG accuracy of SS2CC–which considers both content information and structure information is higher than other methods which only consider single node information. In addition, the Doc2vec model has the worst search performance, mainly because the nodes' structure information in academic social network is dense, while the content information is relatively scarce. Therefore, it is very reasonable that the SS1S, which does not consider content information, also has relatively good performance. Because only a single meta path is considered, the search performance of metapath2vec.apvpa and metapath2vec.apa is relatively low, only better than Doc2vec. Moreover, when the number of k is less than 4, the NDCG value is 0. The reason is that many methods cannot search for similar nodes.

6 Conclusion

In this paper, we analyze the problems of similarity search on heterogeneous information networks. In order to overcome the challenge, we propose the model for top-k similarity search based on double channel convolutional neural networks (SSDCC), which uses CNN network with two channels to train the content information and structure information respectively. SSDCC flexibly considers the content information and structure information including multiple meta paths, and uses two kind of attention mechanism for similarity search. The extensive experiment results on public dataset to demonstrate the effectiveness of our proposed model, and our method achieved highest performance than existing methods.

Acknowledgments. This work was supported by the National Natural Science Foundation of China (61902055, U1811261), China Postdoctoral Science Foundation (2019M651134), Guangdong Province Key Laboratory of Popular High Performance Computers (2017B030314073), and Fundamental Research Funds for the Central Universities (N181703006).

References

1. Abbasi, A.: Reviewing academic social network mining applications. In: 2015 International Conference on Information and Communication Technology Convergence (ICTC), pp. 503–508 (2015)
2. Keong, B.V., Anthony, P.: PageRank: a modified random surfer model. In: 2011 7th International Conference on IT in Asia (CITA 2011), p. 6 (2011)
3. Cabrera-Vives, G., Reyes, I., Förster, F., Estévez, P.A., Maureira, J.-C.: Deep-HiTS: rotation invariant convolutional neural network for transient detection. Astrophys. J. **836**(1), 97 (7 pp.) (2017)
4. Zhu, R., Zou, Z., Li, J.: SimRank computation on uncertain graphs, pp. 565–576 (2016)
5. Perozzi, B., Al-Rfou, R., Skiena, S.: DeepWalk: online learning of social representations. In: KDD 2014, New York, NY, USA, 24–27 August 2014
6. Grover, A., Leskovec, J.: node2vec: scalable feature learning for networks. arXiv, p. 10, 3 July 2016

7. Pan, S., Wu, J., Zhu, X., Zhang, C., Wang, Y.: Tri-party deep network representation. In: IJCAI International Joint Conference on Artificial Intelligence, January 2016, pp. 1895–1901 (2016)
8. Scarselli, F., Gori, M., Tsoi, A.C., Hagenbuchner, M., Monfardini, G.: The graph neural network model. IEEE Trans. Neural Netw. **20**(1), 61–80 (2009)
9. Shi, C., Hu, B., Zhao, W.X., Philip, S.Y.: Heterogeneous information network embedding for recommendation. IEEE Trans. Knowl. Data Eng. **31**(2), 357–370 (2019)
10. Ma, X., Wang, R.: Personalized scientific paper recommendation based on heterogeneous graph representation. IEEE Access **7**, 79887–79894 (2019). https://doi.org/10.1109/ACCESS.2019.2923293
11. Xie, F., Chen, L., Lin, D., Zheng, Z., Lin, X.: Personalized service recommendation with mashup group preference in heterogeneous information network. IEEE Access **7**, 16155–16167 (2019)
12. Shi, C., Zhang, Z., Ji, Y., Wang, W., Philip, S.Y., Shi, Z.: SemRec: a personalized semantic recommendation method based on weighted heterogeneous information networks. World Wide Web **22**(1), 153–184 (2019)
13. Xie, F., Chen, L., Ye, Y., Zheng, Z., Lin, X.: Factorization machine based service recommendation on heterogeneous information networks. In: Proceedings of the 2018 IEEE International Conference on Web Services (ICWS), pp 115–122 (2018)
14. Jiang, Z., Liu, H., Fu, B., Wu, Z., Zhang, T.: Recommendation in heterogeneous information networks based on generalized random walk model and Bayesian Personalized Ranking. In: Proceedings of the 11th ACM International Conference on Web Search and Data Mining, WSDM 2018, February 2018, pp. 288–296, 2 February 2018
15. Shi, C., Li, Y., Zhang, J., Sun, Y., Yu, P.S.: A Survey of Heterogeneous Information Network Analysis. arXiv:1511.04854 [cs.SI]
16. Dong, Y., Chawla, N.V., Swami, A.: Metapath2vec: scalable representation learning for heterogeneous networks. In: Proceedings of the ACM SIGKDD International Conference on Knowledge Discovery and Data Mining, KDD 2017, 13 August 2017, Part F129685, pp. 135–144 (2017)
17. Fu, T.-Y., Lee, W.-C., Lei, Z.: HIN2Vec: explore meta-paths in heterogeneous information networks for representation learning. In: Proceedings of the International Conference on Information and Knowledge Management, CIKM 2017, 6 November 2017, Part F131841, p 1797–1806 (2017)
18. Wang, X., et al.: Heterogeneous graph attention network. In: The Web Conference 2019 - Proceedings of the World Wide Web Conference, WWW 2019, pp. 2022–2032, 13 May 2019
19. Sun, Y., Han, J., Yan, X., Yu, P.S., Wu, T.: PathSim: meta path-based top-k similarity search in heterogeneous information networks. Proc. VLDB Endow. **4**(11), 992–1003 (2011)
20. Shi, C., Kong, X., Huang, Y., Yu, P.S., Wu, V.: HeteSim: a general framework for relevance measure in heterogeneous networks. IEEE TKDE **26**(10), 2479–2492 (2014)
21. Pham, P., Do, P., Ta, C.D.C.: W-PathSim: novel approach of weighted similarity measure in content-based heterogeneous information networks by applying LDA topic modeling. In: Nguyen, N.T., Hoang, D.H., Hong, T.-P., Pham, H., Trawiński, B. (eds.) ACIIDS 2018. LNCS (LNAI), vol. 10751, pp. 539–549. Springer, Cham (2018). https://doi.org/10.1007/978-3-319-75417-8_51
22. Jin, R., Lee, V.E., Hong, H.: Axiomatic ranking of network role similarity. In: SIGKDD, pp. 922–930 (2011)
23. Le, Q.V., Mikolov, T.: Distributed representations of sentences and documents. In: Proceedings of the 31st International Conference on Machine Learning, Beijing, China, vol. 32. JMLR: W&CP (2014)
24. Hu, B., Shi, C.: Leveraging meta-path based context for top-n recommendation with a neural co-attention model. In: KDD 2018, 19–23 August 2018

25. Zhang, C., Song, D., Huang, C.: Heterogeneous graph neural network. In: KDD 2019, p. 11 (2019)
26. Severyn, A., Moschitti, A.: Learning to rank short text pairs with convolutional deep neural networks. In: SIGIR 2015, Santiago, Chile, 09–13 August 2015
27. Wang, Z., Zheng, W., Song, C.: Air Quality Measurement Based on Double-Channel Convolutional Neural Network Ensemble Learning. arXiv:1902.06942v2 [cs.CV], 19 February 2019
28. Tang, J., Zhang, J., Yao, L., Li, J., Zhang, L., Su, Z.: ArnetMiner: extraction and mining of academic social networks. In: Proceedings of the Fourteenth ACM SIGKDD International Conference on Knowledge Discovery and Data Mining (SIGKDD 2008), pp. 990–998 (2008)
29. Tang, J., Fong, A.C., Wang, B., Zhang, J.: A unified probabilistic framework for name disambiguation in digital library. IEEE Trans. Knowl. Data Eng. (TKDE) **24**(6), 975–987 (2012)

STIM: Scalable Time-Sensitive Influence Maximization in Large Social Networks

Yuanyuan Zhu[1]($\boxtimes$), Kailin Ding[1], Ming Zhong[1], and Lijia Wei[2]

[1] School of Computer Science, Wuhan University, Wuhan, China
{yyzhu,dk027biu,clock}@whu.edu.cn
[2] Economics and Management School, Wuhan University, Wuhan, China
ljwei.whu@gmail.com

Abstract. *Influence maximization*, aiming to select k seed users to influence the rest of users maximally, is a fundamental problem in social networks. Due to its well-known NP-hardness, great efforts have been devoted to developing scalable algorithms in the literature. However, the scalability issue is still not well solved in the *time-sensitive influence maximization* problem when propagation incurs a certain amount of time delay and only be valid before a deadline constraint, because all possible time delays need to be enumerated along each edge in a path to calculate the influence probability. Existing approaches usually adopt a *path-based search* strategy to enumerate all the possible influence spreading paths for a single path, which are computationally expensive for large social networks. In this paper, we propose a novel scalable time-sensitive influence maximization method, STIM, based on *time-based search* that can avoid a large number of repeated visits of the same subpaths and compute the influence probability more efficiently. Furthermore, based on *time-based search*, we also derive a new upper bound to estimate the marginal influence spread efficiently. Extensive experiments on real-world networks show that STIM is more space and time-efficient compared with existing state-of-the-art methods while still preserving the influence spread quality in real-world large social networks.

Keywords: Influence maximization · Time-sensitive · Social networks

1 Introduction

The *influence maximization* (IM) problem has been attracting much attention from researchers in recent years, along with the increasing popularity of social networks. It aims to select a set of k seed users to influence the largest number of users in a social network, and can be used in a wide range of applications, such as recommendation system [25], information dissemination [3], link prediction [1], incident monitoring [26], advertisement placement [2], etc. Most existing studies on influence maximization follow two classic models: *independent cascade* (IC) model and *linear threshold* (LT) model, under which the influence maximization problem is NP-hard [17]. Thus, many approximate or heuristic methods have been developed in the literature [5,21].

However, the above two classic models cannot fully address the influence maximization problem in real life. It is because information propagation in social networks

Y. Nah et al. (Eds.): DASFAA 2020, LNCS 12114, pp. 120–136, 2020.
https://doi.org/10.1007/978-3-030-59419-0_8

may be *time-sensitive* regarding the following two aspects. On the one hand, the spread of influence may only be valid within a certain *time constraint*. For example, if a company needs to promote a discount on Black Friday, it would not be helpful if this information is propagated to a user after Black Friday. On the other hand, the influence spread is not always instant, and there may be time delays between two adjacent users [11]. Suppose that Bob shared a piece of information on Twitter. His friend Steve who is very active on Twitter may read it in several hours and repost it immediately, while his another friend John who is not so active may read this information several days later. The problem of selecting k seed users to influence the largest number of users with time delays before a given deadline τ is called *time-sensitive influence maximization* (TIM) problem.

To deal with the TIM problem, some approaches have been proposed in recent years, which can be classified into two categories: the *continuous-time* model [9,12,30] where the likelihood of pairwise propagation between nodes can be considered as a continuous distribution of time, and the *discrete-time* model [6,18,23] where a user activates its neighbors with certain probabilities at discrete time steps. The *continuous-time* model was first studied in [12] where a greedy solution based on Monte Carlo (MC) simulation was provided. Later, sampling method based on forward influence sketch [9] and reverse reachable (RR) sketch [15,30] were proposed to accelerate the computation with $1 - 1/e - \epsilon$ approximation guarantee. In these models, a node can always be activated, and its activation time follows a certain continuous distribution. However, in some real-world applications, a node may be activated with a certain probability at a discrete time step. For example, the login event of a user can be considered as a discrete event, and he/she may propagate the information with a certain probability. To deal with such a case, Chen et al. [6] proposed the *independent cascade model with meeting events* (IC-M) by employing the meeting probability between two adjacent nodes to capture the time delay of propagation in a path. They also developed two heuristic solutions MIA-M and MIA-C to find the seed set for a given deadline. Liu et al. [23] proposed a more general model, *latency aware independent cascade model* (LAIC), in which the time delay for each edge follows a certain distribution, and thus IC-M can be considered as a special case. Independently, [18] proposed a different model in which every node has multiple chances to activate its neighbors before a deadline.

Although taking time into account can model the real-world applications more precisely, it is computationally expensive because all possible time delays need to be considered to estimate the influence probability accurately. In the *sampling-based* methods [9,30], a large number of sketches need to be sampled to guarantee the accuracy, which will cost much space and cannot handle large social networks such as twitter on a pc with 128G memory for approximation factor $\epsilon = 0.1$ [15,30]. In the *path-based* approaches [6,23], along with each edge all possible time delays need to be enumerated to estimate the influence probability of a path, and a subpath will be repeatedly visited, which is very time consuming and hinders its scalability on large social networks. In this paper, we re-examine the time-sensitive influence maximization problem and propose a novel space and time-efficient algorithm STIM to select the seed set for large social networks. The contributions of this paper are:

- We formally analyze the scalability issue of the TIM problem by giving the theoretical estimation of the number of influence spreading paths for a single path.
- We propose the *time-based search* strategy to compute the activation probability for the nodes in a path to avoid a large number of repeated visits of the same subpaths in previous *path-based search* methods.
- Based on *time-based search* strategy, we further derive a new upper bound to efficiently and effectively estimate the marginal influence.
- Finally, we conducted extensive experiments to compare our algorithm STIM with the sate-of-the-arts, and validated that STIM can speed up the computation by one to two orders of magnitude while still preserve similar influence spread on large real-world social networks.

Roadmap. Section 2 introduces the preliminaries, including problem statement, state-of-the-art methods, and other related works. Section 3 proposes our new approaches. Section 4 gives the experimental analysis. Section 5 concludes the paper.

2 Preliminaries

We will first give the formulation of the time-sensitive influence maximization problem, then analyze the state-of-the-art solutions and briefly review other related works.

2.1 Problem Statement

A social network is modeled as a directed graph $G = (V, E)$, where V is the node set, and $E \subseteq V \times V$ is the set of directed edges connecting pairs of nodes.

Independent Cascade (IC) Model. Let A_t be the set of nodes activated at time step t. At time $t = 0$, A_0 is initialized as the seed set S. At time $t + 1$, every node $u \in A_t$ has a single chance to activate each of its currently inactive outgoing neighbors v, i.e., $v \in N^{out}(u) \setminus \bigcup_{i=0}^{t} A_i$ where $N^{out}(u)$ is the out-neighbors of u. The success probability of this attempt is $p(u, v)$ and it is independent of all other activation attempts. In the IC model, once a node u is activated, it either activates its currently inactive neighbor v in the immediate next step, or does not activate v at all.

Time-Sensitive Independent Cascade (TIC) Model. The TIC model is extended from the IC model where time delays are encoded into the diffusion process. Here, we consider the discrete TIC model, i.e., besides the activation probability $p(u, v)$, each edge (u, v) is also associated with a time delay probability $\mathcal{P}_u^d(\delta_t)$, where δ_t is the time delay randomly drawn from the delay distribution of node u. Therefore, a node u activated at time t will activate its current inactive neighbor v at time step $t + \delta_t$ with probability $p(u, v)\mathcal{P}_u^d(\delta_t)$. If a node can be influenced by multiple neighbors, it is activated by the node at the earliest activation time while the rest activations will be ignored. The influence propagation terminates when no new nodes can be activated before τ.

Time-Sensitive Influence Maximization (TIM) Problem. Given a graph $G = (V, E)$, an integer k, a deadline τ, and the time delay distribution function $\mathcal{P}_u^d$ for each node $u \in V$, the TIM problem is to find the seed set $S \subseteq V$ of k nodes, such that the expected number of nodes influenced by S under such time delay distribution before the deadline τ, $\sigma_\tau(S)$, is maximized, i.e., finding $S^* = \arg\max_{S \subseteq V, |S| \leq k} \sigma_\tau(S)$.

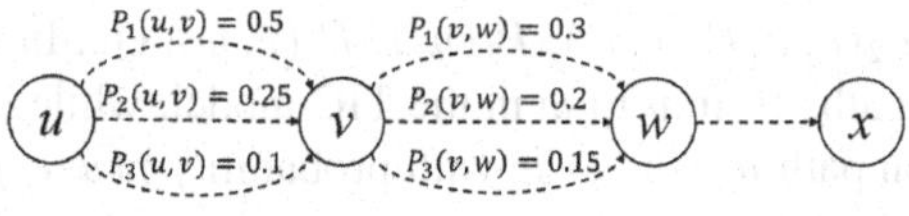

Fig. 1. Example of time-delayed influence spreading paths

2.2 State-of-the-Art Approaches

The TIM problem has been proved to be NP-hard [22]. Fortunately, it still maintains the monotonicity and sub-modularity [22], and thus a greedy algorithm can be applied to achieve the $1 - 1/e$ approximation ratio. As proved in [17], calculating the expected influence spread of a given seed set under the IC model is #P-hard, and this hardness result is also applicable to the TIC model in which the IC model can be considered as a specific case that the time delay is always 1.

Greedy Algorithm Based on MC Simulation [6,22]. In the greedy algorithm, we gradually select the node u with the largest marginal influence to the current seed set S, until the size of S is k. The marginal influence of a node u to a given seed set S is evaluated by $\sigma_\tau(S \cup \{u\}) - \sigma_\tau(S)$, where $\sigma_\tau(S)$ can be computed by MC simulation in $O(|V| + |E|)$ time and $O(|V| + |E|)$ space [23]. Usually, MC simulation need to be repeated for R times to approximate the expected influence spread because of its randomness, and overall time complexity is $O(k|V|R(|V| + |E|))$. Due to the computational curse, the simulation based algorithm is not practical for large social networks. Two improved methods MIA-M/ MIA-C [6] and ISP/MISP [22] were then proposed to estimate $\sigma_\tau(S)$ based on the *influence spreading paths*.

Influence Spreading Path. A social network $G = (V, E)$ can be logically extended to a multigraph $G^\tau = (V, E^\tau)$ to integrate time delay into the network structure. For each $(u, v) \in E$, we put τ influence spreading edges, $e_{uv}^1, e_{uv}^2, ..., e_{uv}^\tau$, from u to v in network G^τ attached with two values, i.e., $len(e_{uv}^t) = t$ and $P_t(u, v) = p(u, v)\mathcal{P}_u^d(t)$ where $len(e_{uv}^t)$ is the augmented length of e_{uv}^t and $P_t(u, v)$ is the probability that u activates v through the influence spreading edge e_{uv}^t. For a simple *influence spreading path* $\overrightarrow{\mathcal{P}} = u_1 \xrightarrow{e_{u_1 u_2}^{t_1}} u_2 \xrightarrow{e_{u_2 u_3}^{t_2}} u_3 \dots \xrightarrow{e_{u_{l-1} u_l}^{t_{l-1}}} u_l$, its augmented length is $\sum_{i=1}^l t_i$ and its propagation probability is $\prod_{i=1}^{l-1} P_{t_i}(u_i, u_{i+1}) = \prod_{i=1}^{l-1} p(u_i, u_{i+1})\mathcal{P}_{u_i}^d(t_i)$. In the IC model, to estimate the probability of node u activating node v along a path, we only need to multiply the propagation probability of each edge on this path, while in the TIC model, we need to consider all possible *influence spreading paths* with different augmented lengths and propagation probabilities. Note that the *influence spreading edge/path* in the multigraph G^τ is different from *edge/path* in the original graph G.

Example 1. Suppose that there is a simple path $u \rightarrow v \rightarrow w$. We can construct the corresponding influence spreading path as shown in Fig. 1. For example, the propagation probability from u to v with time delay 1 is $P_1(u, v) = 0.5$. Similarly, $P_2(u, v) = 0.25$ is with time delay 2 and $P_3(u, v) = 0.1$ is with time delay 3. Suppose u is activated at time step 0, and the deadline τ is 4. To compute the probability of w being activated by u before the deadline 4, we need to consider all the possibilities that w is activated, such as $P_1(u,v)P_1(v,w)$, $P_2(u,v)P_1(v,w)$, $P_1(u,v)P_2(v,w)$, etc. In fact, there are totally 6 influence spreading paths from u to w in the TIC model, while in the IC model there is only one propagation path $u \to v \to w$ with probability $p(u,v)p(v,w)$.

MIA-M/MIA-C [6]. In the MIA-M algorithm, the influence spread is estimated based on *the maximum influence in-arborescence* (MIIA) of each node v, which is the union over the maximum influence paths to v over all $u \in V \setminus \{v\}$. The paths with small propagation probability (less than a pre-defined influence threshold θ) or longer augmented length (more than τ) are further filtered out. Based on the set of remaining paths $\text{MIIA}_\theta^\tau(v)$, they can also derive the node set influenced by v, which is $\text{InfSet}_\theta^\tau(u) = \{u | u \in \text{MIIA}_\theta^\tau(v), v \in V\}$. Let $n_\tau = \max_{v \in V} |\text{MIIA}_\theta^\tau(v)|$ and $m_\tau = \max_{v \in V} |\text{InfSet}_\theta^\tau(v)|$. The time complexity of MIA-M is $O(|V|(t_d + n_\tau \tau^3) + k n_\tau m_\tau (n_\tau \tau + \log |V|))$, where t_d the maximum running time to compute $\text{MIIA}_\theta^\tau(v)$ for any $v \in V$ by Dijkstra's algorithm. In fact, MIA-M is not practical for large social networks especially when τ is large. A faster algorithm MIA-C was further proposed by incorporating meeting probability $m(u,v)$, influence probability $p(u,v)$, and deadline τ by a certain function, which needs $O(|V|t_d + k n_\tau m_\tau (n_\tau \tau + \log |V|))$ time but will loss seed quality.

ISP/MSIP [22,23]. Different from MIA-M/MIA-C [6], ISP utilizes all possible influence spreading paths ending with u to estimate the propagation probability from v to u under the same constraint of θ and τ. Given seed set S, such influence spreading paths starts from seeds in S and ends at u are denoted by $\text{ISP}_\theta^\tau(u, S)$. ISP iteratively calculates the probability that u gets activated by S within time τ through $\text{ISP}_\theta^\tau(u, S)$ based on the *path-based search* strategy to enumerate all the possible influence spreading paths. Let $n_\tau = \max_{|S| \leq k} \sum_{u \in V} |\text{ISP}_{S,\theta}^\tau(u, S)|$. ISP needs $O(k|V|n_\tau)$ time, where n_τ can be as large as $O(|S|d_{max}^\tau)$ for the worst case that $p(u,v)$ is 1 for any edge (u,v) (d_{max} is the largest out-degree among all the nodes). A faster algorithm MISP is also proposed to approximately estimate the marginal influence $\sigma_\tau(S \cup \{v\}) - \sigma_\tau(S)$ by making a discount of $\sigma_\tau(v)$, with time complexity $O(|V|n_\tau + (k-1)d_{max})$. However, the costly n_τ was still not eliminated, which hinders the application of MISP to large social networks.

2.3 Other Related Works

IM Under Classic Models. The influence maximization problem was first studied by Domingo et al. to find a set of nodes to propagate influence so that the spread range can be maximized [8]. Kempe et al. proved that this optimization problem is NP-hard under the IC and LT models, and proposed a greedy algorithm with approximation ratio $1 - 1/e$ based on MC simulation [17]. However, repeatedly computing the influence spread for node sets based on MC simulation is computationally expensive. Thus a number of optimization algorithms were subsequently proposed to reduce the number of influence spread calculations (CELF [19], CELF++ [14], and etc.) or accelerate each influence spread calculation (PMIA [7], IRIE [16], EaSyIm [10], etc.). Recently, some works exploit a number of graph sketches to evaluate influence spread with approximation guarantee, such as RR-Sketch [4], IM [31], IMM [30], SSA/D-SSA [15,24].

Algorithm 1: Greedy Algorithm

---

1 **Input:** A network G, a seed number k, and a deadline τ
2 **Output:** S $S \leftarrow \emptyset$; **for** $i = 1 \rightarrow k$ **do**
3 $\quad \lfloor \quad u \leftarrow \arg\max_{v \in V \setminus S}(\sigma_\tau(S \cup \{v\}) - \sigma_\tau(S)); S \leftarrow S \cup \{u\};$
4 **return** S;

Time-Sensitive Influence Maximization. Existing time-sensitive influence maximization approaches can be classified into two categories: the *continuous-time* model [9,12,30] where the likelihood of pairwise propagation between nodes is considered as a continuous distribution of time, and the *discrete-time* model [6,18,23] where a diffusion process can be considered as a discrete random variable over different time steps. In [12], a greedy algorithm based on MC simulation was provided. Later, [9,13] proposed a sampling method based on the forward influence sketch extracted from the graph so that influence spread can be estimated with a provable guarantee. The IMM algorithm [30] can also be extended to deal with the continuous-time influence maximization model. In these models, a node can always be activated, and its activation time follows a certain distribution. However, the number of sampled sketches can be huge for large networks to guarantee the accuracy in the time-sensitive influence maximization problem. Note that the methods tailored for the continuous-time model cannot be directly applied to the discrete-time, because besides the difference of the delay distribution, each edge is also attached with an activation probability. Thus, [6,23] were proposed as introduced above. Besides, [18] proposed a different model independently where every active node has multiple chances to activate its neighbors before τ.

Besides, different variants of IM problem were also investigated, including location information [32], budget restrictions [20], adaptive influence maximization [33], target influence maximization [27], multi-round influence maximization [28], online influence maximization [29], etc. More works can be found in [5,21].

3 Our Approach

Our approach also follows the greedy framework, as shown in Algorithm 1. The first step is to choose a node with the maximum influence spread. Then, we gradually select the rest of the nodes by evaluating their marginal influence spreads. Thus in the following, we mainly focus on how to estimate the influence spread efficiently.

3.1 Time-Based Search for Node Influence

The influence spread of node u is the sum of the probabilities of each node $v \in V \setminus \{u\}$ being activated by u before τ, i.e., $\sigma_\tau(u) = \sum_{v \in V \setminus \{u\}} AP^\tau(v, u)$, where $AP^\tau(v, u)$ is the probability of v being activated by u before τ. Recall that in the TIC model each node can only be activated once, which means events that v gets activated at different time steps are mutually exclusive. Thus, we can further derive $AP^\tau(v, u) = \sum_{t=0}^{\tau} AP_t(v, u)$ where $AP_t(v, u)$ is the probability of v being activated by u at time t.

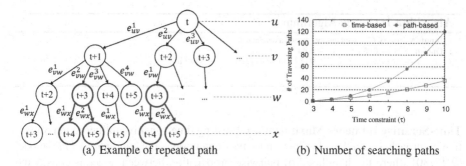

(a) Example of repeated path   (b) Number of searching paths

Fig. 2. Example and statistics of searching paths

Obviously, with the constraint of τ, we only need to consider the nodes that can be reached by u with path length no larger than τ and propagation probability no less than θ as in [6,23]. To further simplify the computation, we only consider the path with the maximum propagation probability from u to each node v as in [6] instead of using all the paths that satisfy the probability and length constraints in [23]. Specifically, we can construct an out-arborescence for each node u, where root u can reach each leaf within τ steps with propagation probability no less than θ. We call such a tree rooted at u *maximum-influence propagation tree* (MP tree), and denote it as $MP_\theta^\tau(u)$. Thus, the influence spread of u can be simplified as $\sigma_\tau(u) = \sum_{v \in MP_\theta^\tau(u)} AP^\tau(v, u)$.

Since there is only one path from root u to each node v in $MP_\theta^\tau(u)$, i.e., v only has one ingoing neighbor w on the path from u to v, the probability of v being activated by u at t can be the sum of the probabilities that w is activated by u at any time $t' < t$ and then v is activated by w with time delay $t-t'$. Therefore, $AP^\tau(v, u)$ can be further derived as $AP^\tau(v, u) = \sum_{t=0}^{\tau} \sum_{t'=0}^{t-1} AP_{t'}(w, u)p(w, v)\mathcal{P}_w^d(t - t')$. However, this will involve a large number of influence spreading path, as shown in the following lemma.

Lemma 1. *Given a path composed of n nodes and a deadline $\tau \geq n - 1$, if the start node is the only seed activated at time step 0, then there are $\sum_{t=n-1}^{\tau} C_{t-1}^{n-2}$ possible influence spreading paths from the start node to the end node.*

To enumerate all the possible influence spreading paths in above equation, existing approach such as ISP [23] adopts the *path-based search* strategy, where the number of influence spreading paths can be as large as $\sum_{t=n-1}^{\tau} C_{t-1}^{n-2}$ for a path composed of n nodes and a deadline $\tau \geq n - 1$, and thus is not applicable for large social networks. Fortunately, after a careful examination of all the influence spreading paths, we discover that there are a large number of repeated subpaths among these influence spreading paths.

Example 2. Take the path $u \rightarrow v \rightarrow w \rightarrow x$ in Fig. 1 as an example. All the possible influence spreading paths from u to x are organized as a tree shown in Fig. 2(a). Suppose that u is activated at time t and the deadline is $\tau = t + 5$. Node v at the first level may be activated at time steps $t + 1, \ldots, t + 5$ by propagation edges $e_{uv}^1, \ldots, e_{uv}^5$, respectively. The number of leaf nodes in this tree is exactly the number of possible

---

Algorithm 2: Computing $\sigma_\tau(u)$

Input: Maximum propagation tree $MP_\theta^\tau(u)$ and deadline τ
Output: $\sigma_\tau(u)$

1   $depth \leftarrow 0; cnt \leftarrow 1; Q \leftarrow \{u\};$
2   **while** $Q \neq \varnothing$ **do**
3      $v \leftarrow Q.pop(); cnt \leftarrow cnt - 1;$
4   **for** $w \in N^{out}(v)$ **do**
5          $Q \leftarrow Q \cup \{w\};$
6   **for** $t' = depth \rightarrow \tau$ **do**
7   **for** $t = t' + 1 \rightarrow \tau$ **do**
8                  $AP_t(w,u) \leftarrow AP_t(w,u) + AP_{t'}(v,u)p(v,w)\mathcal{P}_v^d(t - t');$
9   **for** $t = depth + 1 \rightarrow \tau$ **do**
10             $\sigma_\tau(u) \leftarrow \sigma_\tau(u) + AP_t(w,u);$
11   **if** $cnt = 0$ **then**
12          $depth \leftarrow depth + 1; cnt \leftarrow Q.size();$
13   **return** $\sigma_\tau(u);$

---

influence spreading paths from u to x, which is $\sum_{t=n-1}^{\tau} C_{t-1}^{n-2} = \sum_{t=4-1}^{5} C_{t-1}^{4-2} = 10$. As shown in the bold red line in Fig. 2(a), to compute the probability of w being activated at $t + 3$ and then x being activated at time step $t + 4$, we need to traverse subpath e_{wx}^1 and e_{wx}^2 twice. In fact, we can compute the probability of w being activated at time step $t + 3$ first, i.e., $AP_{t+3}(w,u) = P_1(u,v)P_2(v,w) + P_2(u,v)P_1(v,w)$. In other words, we can merge the two red $t + 3$ nodes at level w first, and then only visit subpaths e_{wx}^1 and e_{wx}^2 at level x once to save the computation cost.

Time-Based Search Strategy. Based on above observation, we now give our *time-based search* strategy to compute $AP^\tau(v,u)$ over the propagation tree. Suppose that there are n nodes $x_0 = u, x_1, \ldots, x_{n-1} = v$ along the path from u to v, where u is activated at t and the deadline is τ. Starting from $i = 1$, we gradually compute the probabilities that x_i is activated by u at time steps $t + i, \ldots, \tau$ respectively, i.e., $AP_{t+i}(x_i,u), \ldots, AP_\tau(x_i,u)$. By doing this, we can obtain the probability of being activated by u for each node along this path. Moreover, to compute $\sigma_\tau(u)$, we can search the whole MP tree $MP_\theta^\tau(u)$, and obtain $AP_t(v,u)$ for $v \in MP_\theta^\tau(u)$ at each tim step t by traversing the tree only once. The detailed process is shown in Algorithm 2.

Based on above steps, we can easily derive that the overall time complexity of Algorithm 2 is $O(m_\tau \tau^2)$, where $m_\tau = \max_{u \in V} |MP_\theta^\tau(u)|$. The *time-based search* strategy can substantially reduce the number of paths visited compared with *path-based search* method as shown in the following theorem.

Theorem 1. *Given a path composed of n nodes ($n > 2$) and a deadline $\tau \geq n - 1$, if the start node is the only seed activated at time step 0, then the number of paths visited by time-based search strategy from the start node to the end node is $\frac{(\tau-n+3)(\tau-n+2)}{2}$.*

Proof. Suppose that the pre-node before the end node is w, which can be activated from time $n - 2$ to $\tau - 1$. When w is activated at time $n - 2$, there are $\tau - n + 2$ possible time-delayed influence propagation edges from w to the end node. When w is activated at $\tau - 1$, there is 1 time-delayed influence propagation edge from w to the end node

---

Algorithm 3: Computing $MG^\tau(v, u)$ (by Theorem 3)

Input: A node v, a seed set S and a node u, a deadline τ
Output: $MG^\tau(v, u)$

1 $S' \leftarrow S \cup \{u\}; N \leftarrow \emptyset;$
2 $AP_{t,N}(v, S') \leftarrow 0,$ for $0 \le t \le \tau;$
3 **for** $w \in N^{in}(v) \setminus N^{out}(v)$ **do**
4 $\quad AP_w^\tau(v, S') \leftarrow 0; AP_N^\tau(v, S') \leftarrow 0;$
5 $\quad$ **for** $t = 1 \rightarrow \tau$ **do**
6 $\quad\quad$ **for** $t' = 0 \rightarrow t - 1$ **do**
7 $\quad\quad\quad \lfloor\ AP_{t,w}(v, S') + = AP_{t'}(w, S')p(u, v)\mathcal{P}_u^d(t - t');$
8 $\quad\quad AP_w^\tau(v, S') \leftarrow AP_w^\tau(v, S') + AP_{t,w}(v, S');$
9 $\quad\quad AP_{t,N \cup \{w\}}^\tau(v, S') \leftarrow AP_{t,N}(v, S')(1 - AP_w^\tau(v, S')) + AP_{t,w}(v, S')(1 - AP_N^\tau(v, S'));$
10 $\quad\quad AP_N^\tau(v, S') \leftarrow AP_N^\tau(v, S') + AP_{t,N}(v, S');$
11 $\quad \lfloor\ N \leftarrow N \cup \{w\};$
12 **for** $t = 1 \rightarrow \tau$ **do**
13 $\quad \lfloor\ MG^\tau(v, u) \leftarrow MG^\tau(v, u) + AP_N^t(v, S') - AP_N^t(v, S);$
14 **return** $MG^\tau(v, u);$

---

with time delay 1. Thus, there will be $(\tau - n + 2) + (\tau - n + 1) \ldots + 1 = \frac{(\tau - n + 3)(\tau - n + 2)}{2}$ paths to be visited in time-based search. $\qquad\qquad\qquad\qquad\qquad\qquad\Box$

Clearly, the value of $\frac{(\tau - n + 3)(\tau - n + 2)}{2}$ is much smaller than that of $\sum_{t=n-1}^\tau C_{t-1}^{n-2}$, especially for large n and τ. As τ and n increase, the gap between the *path-based search* and *time-based search* becomes larger. Figure 2(b) shows the numbers of paths visited to compute the propagation probability from u to x in Figure 1 for different τ.

3.2 Upper Bounds for Marginal Influence

After selecting the node with the maximum influence into the seed set S, now we move on to the selection of the following seeds by evaluating the marginal influence $MG^\tau(u) = \sigma_\tau(S \cup \{u\}) - \sigma_\tau(S)$ of node $u \in V \setminus S$. Since node u can only influence the nodes in tree $MP_\theta^\tau(u)$, the marginal influence of u can be rewritten as $MG^\tau(u) = \sum_{v \in MP_\theta^\tau(u)}(AP^\tau(v, S \cup \{u\}) - AP^\tau(v, S))$ where $AP^\tau(v, S)$ is the probability of v being activated by seed set S before a deadline τ. Since events that v gets activated at different time steps are mutually exclusive, we have $AP^\tau(v, S) = \sum_{t=0}^\tau AP_t(v, S)$ where $AP_t(v, S)$ is the probability of v being activated by set S at time t. Thus, we have $MG^\tau(u) = \sum_{v \in MP_\theta^\tau(u)} \sum_{t=0}^\tau (AP_t(v, S \cup \{u\}) - AP_t(v, S))$.

Now we discuss how to compute $AP_t(v, S)$ for $v \in V \setminus S$. Let $N^{in}(v)$ be the set of ingoing neighbors of v. For a node $w \in N^{in}(v)$, we use $AP_{t,w}(v, S)$ to denote the probability that v is activated by S through w at time step t. Then it can be computed as $AP_{t,w}(v, S) = \sum_{t'=0}^{t-1} AP_{t'}(w, S)p(w, v)\mathcal{P}_w^d(t - t')$, which is the sum of the probabilities that w is activated by S at any time $t' < t$ and then w activates v by influence propagation edge $e_{wv}^{t-t'}$.

Theorem 2. *Given a seed set S, the probability of $v \in V \setminus S$ being activated by S at time t can be computed as $AP_t(v, S) = \prod_{w_i \in N^{in}(v)} \left(1 - \sum_{t'=0}^{t-1} AP_{t',w_i}(v, S)\right) -$ $\prod_{w_i \in N^{in}(v)} \left(1 - \sum_{t'=0}^{t} AP_{t',w_i}(v, S)\right)$.*

Proof. It can be easily derived from the fact that events that v gets activated at different time t are mutually exclusive. Thus, $\prod_{w_i \in N^{in}(v)} (1 - \sum_{t'=0}^{t-1} AP_{t',w_i}(v, S))$ is the probability that v has not been activated by S before or at $t - 1$. Similarly, $\prod_{w_i \in N^{in}(v)} (1 - \sum_{t'=0}^{t} AP_{t',w_i}(v, S))$ is the probability that v has not been activated by S before or at t. Thus, their difference is the probability that v is activated at t.   □

Theorem 3. *Given a seed set S, for each node $v \in V \setminus S$, we use $AP_{t,N'}(v, S)$ to denote the probability of v being activated through a subset of its neighbors $N' \subset N^{in}(v)$. Then for any node $w \in N^{in}(v) \setminus N'$, we have $AP_{t,N' \cup \{w\}}(v, S) = AP_{t,N'}(v, S)(1 - \sum_{t'=0}^{t} AP_{t',w}(v, S)) + AP_{t,w}(v, S)(1 - \sum_{t'=0}^{t-1} AP_{t',N'}(v, S))$.*

Proof. From Theorem 2, we have

$$AP_{t,N' \cup \{w\}}(v, S)$$

$$= (1 - \sum_{t'=0}^{t} AP_{t',w}(v, S) + AP_{t,w}(v, S)) \prod_{w_i \in N'} (1 - \sum_{t'=0}^{t-1} AP_{t',w_i}(v, S))$$

$$- (1 - \sum_{t'=0}^{t} AP_{t',w}(v, S)) \prod_{w_i \in N'} (1 - \sum_{t'=0}^{t} AP_{t',w_i}(v, S))$$

$$= (1 - \sum_{t'=0}^{t} AP_{t',w}(v, S)) \times (\prod_{w_i \in N'} (1 - \sum_{t'=0}^{t-1} AP_{t',w_i}(v, S)) - \prod_{w_i \in N'} (1 - \sum_{t'=0}^{t} AP_{t',w_i}(v, S)))$$

$$+ AP_{t,w}(v, S) \prod_{w_i \in N'} (1 - \sum_{t'=0}^{t-1} AP_{t',w_i}(v, S))$$

$$= AP_{t,N'}(v, S)(1 - \sum_{t'=0}^{t} AP_{t',w}(v, S)) + AP_{t,w}(v, S) \prod_{w_i \in N'} (1 - \sum_{t'=0}^{t-1} AP_{t',w_i}(v, S))$$

Since $\prod_{w_i \in N'} (1 - \sum_{t'=0}^{t-1} AP_{t',w_i}(v, S))$ is the probability that S cannot activate v via N' before or at t, we rewrite it as $1 - \sum_{t'=0}^{t} AP_{t',N'}(v, S)$ and complete the proof.   □

Based on above theorem, we can compute the marginal influence of u given a seed set S as follows. Suppose that for each node v activated before τ, we have recorded the probabilities of v being activated by S at each time step $t \leq \tau$, $AP_t(v, S)$. The computation of $MG^\tau(v, u)$ based on Theorem (3) is shown in Algorithm 3. We can compute $AP_t(v, S \cup \{u\})$ for any node $v \in MP_\theta^\tau(u)$ at any time step $t \leq \tau$ (lines 4–11) by traversing the tree from the root to leaves and obtain the marginal influence of u to v by $MG^\tau(v, u)$ in lines 12–13. The time complexity of computing $MG^\tau(u)$ is $O(|MP_\theta^\tau(u)|d_{max}\tau^2)$, which is still expensive. Thus, next we will focus on how to approximately estimate the marginal influence of a node.

Theorem 4. *Given a seed set S and a node u. If the pre-nodes of v on the path from u to v cannot be influenced by S, the probability of v being activated by set $S \cup \{u\}$ at time step t can be simplified as $AP_t(v, S \cup \{u\}) = AP_t(v, S)(1 - \sum_{t'=0}^{t} AP_{t'}(v, u)) + AP_t(v, u)(1 - \sum_{t'=0}^{t-1} AP_{t'}(v, S))$.*

Algorithm 4: STIM Algorithm

Input: A network G, the seed number k, and a deadline τ
Output: S
1 Build MP tree $MP_\theta^\tau(u)$ for each node $u \in V$;
2 **for** $u \in V$ **do**
3   compute $\sigma_\tau(u)$ by Algorithm 1;
4      $MG'^\tau(u) \leftarrow \sigma_\tau(u)$; $status(u) \leftarrow 0$;
5 **while** $|S| < k$ **do**
6      $u \leftarrow \arg\max_{v \in V \setminus S}(MG'^\tau(v))$;
7   **if** $status(u) = |S|$ **then**
8   **for** $v \in MP_\theta^\tau(u)$ **do**
9   **for** $t = 1 \rightarrow \tau$ **do**
10   Compute $AP_t(v, S \cup \{u\})$ by Theorem 4;
11         $S \leftarrow S \cup \{u\}$;
12   **else**
13         $status(u) \leftarrow |S|$; $MG'^\tau(u) \leftarrow 0$;
14   **for** $v \in MP_\theta^\tau(u)$ **do**
15   **for** $t = 1 \rightarrow \tau$ **do**
16   Compute $MG'_t(u, v)$ by Equation (1);
17                 $MG'^\tau(u) \leftarrow MG'^\tau(u) + MG'_t(u, v)$;

18 return S;

Proof. Let $w \in N^{in}(v)$ be the node through which u activates v and $N' \subset N^{in}(v)$ be the set of nodes through which S activates v. Based on Theorem 3, we have $AP_t(v, S \cup \{u\}) = AP_{t,N'}(v, S \cup \{u\})(1 - \sum_{t'=0}^{t} AP_{t',w}(v, S \cup \{u\}) + AP_{t,w}(v, S \cup \{u\})(1 - \sum_{t'=0}^{t-1} AP_{t',N'}(v, S \cup \{u\}))$ As S cannot influence w and u cannot influence any other in-neighbors of v except w, we have $AP_{t,w}(v, S \cup \{u\}) = AP_t(v, u)$ and $AP_{t,N'}(v, S \cup \{u\}) = AP_t(v, S)$. Thus, we simplify the equation in Theorem 3 as $AP_t(v, S \cup \{u\}) = AP_t(v, S)(1 - \sum_{t'=0}^{t} AP_{t'}(v, u)) + AP_t(v, u)(1 - \sum_{t'=0}^{t-1} AP_{t'}(v, S))$. □

Note that Theorem 4 cannot hold if the pre-nodes of v on the path from u to v can be influenced by S. In fact the influence obtained by Theorem 4 is an upper bound of $AP_t(v, S \cup \{u\})$ in this situation. Consider the case that u and S influence w together, and w influences v. If u has activated w at time t', then S cannot activate w before t'. Nevertheless, the probability that S activates w before t' is included in the equation of Theorem 4. Therefore, it is an upper bound for $AP_t(v, S \cup \{u\})$, and can be used to obtain an approximation value of $MG_t(u, v)$, which is

$$MG'_t(u, v) = AP_t(v, u)(1 - \sum_{t'=0}^{t-1} AP_{t'}(v, S)) - AP_t(v, S) \sum_{t'=0}^{t} AP_{t'}(v, u) \quad (1)$$

Finally, the marginal influence of u can be estimated as $MG'^\tau(u) = \sum_{v \in MP_\theta^\tau(u)} \sum_{t=1}^{\tau} MG'_t(u, v)$. Since for each node v, we have record $AP_t(v, u)$ during the computation of $\delta_\tau(u)$, and maintain $AP_t(v, S)$ after each update of S, we only need $O(t)$ time to compute $MG'_t(u, v)$. Moreover, since $MG'_t(u, v)$ needs to be computed for $0 \leq t \leq \tau$, we can simplify the complexity to $O(1)$ by recording the sum in Eq. (1). Thus, the overall complexity of computing $MG'^\tau(u)$ is only $O(|MP_\theta^\tau(u)| \times \tau)$.

Table 1. Statistics of real-world networks

Datasets	DBLP	LiveJournal	Twitter
# of nodes	317K	3,017K	41.7M
# of edges	2,099K	87,037K	1.47G
Average degree	6.62	28.85	35.97

3.3 The Overall STIM Algorithm

Now we wrap up above techniques and show our *scalable time-sensitive influence maximization* (STIM) method in Algorithm 3. After building $MP_\theta^\tau(u)$, we compute $\sigma_\tau(u)$ for each $u \in V$ by time-based search (line 3). Then we iteratively select the node u with the maximal marginal influence. We use $status(u)$ to record the size of the seed set S when $MG'(u)$ is updated. If the marginal influence of u is updated by current S, we select it as the new seed and insert it into S. Then, for all v that can be influenced by u, we update the probability of v being activated by $S \cup \{u\}$ (lines 7–11). If $MG'(u)$ is not up-to-date, we update the marginal influence of u by Eq. (1) (lines 13–17).

The computational complexity of loop in lines 2–3 is $O(|V|m_\tau\tau^2)$ because the computation complexity for a single node is $O(m_\tau\tau^2)$, where $m_\tau = \max_{u \in V} |MP_\theta^\tau(u)|$. In the loop of selecting k seeds, lines 7–11 need $O(km_\tau\tau)$ time, and lines 13–17 need $O(n'km_\tau\tau)$ time, where n' is the maximum number of updates when selection each seed. In practice, $n' << |V|$. The overall time complexity of STIM is $O(|V|t_d + |V|m_\tau\tau^2 + n'km_\tau\tau)$, where t_d is the maximum running time to build $MP_\theta^\tau(u)$ for any $v \in V$ by Dijkstra's algorithm.

4 Experiments

In this section, we evaluate the performance of our algorithm STIM and the state-of-the-art methods experimentally to show the efficiency and effectiveness of our methods.

4.1 Experimental Setup

First of all, we evaluate our algorithm STIM with the following up-to-date approaches under the same TIC model used in this paper: MIA-M/MIA-C [6], MISP [23], and the greedy baseline algorithm MC based on Monte Carlo simulation. Note that we did not compare with ISP in [23] as it is much slower than MISP with similar influence spread. Moreover, we also compared our algorithm with IMM [15,30] based on RR-set. Note that IMM [30] is tailored for the continuous-time model, which cannot be directly applied to the discrete-time model studied in this paper. It is because besides the difference in the delay distribution, each edge is also attached with an activation probability in the discrete-time model. Thus the discrete-time model cannot be considered as a special case of the continuos-time model. To deal with the discrete TIC model studied in this paper, we revised the IMM code obtained from the authors as follows. To sampling the RR set $R(v)$ for node v, we invoke the Dijkstra's algorithm to traverse

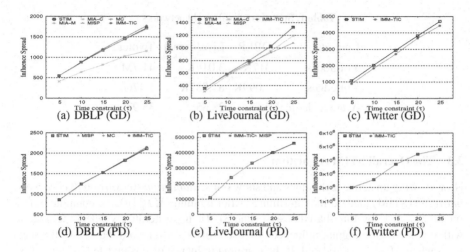

Fig. 3. Influence spread with different τ ($k = 50$)

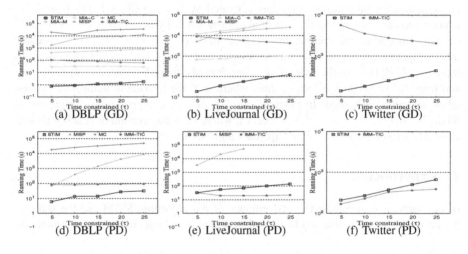

Fig. 4. Running time with different τ ($k = 50$)

G starting from v, following only the incoming edge. Each time we encounter an edge (u, v), we first sample (u, v) with probability $p(u, v)$, and then sample its length from its time-delay distribution. We denote the revised algorithm as IMM-TIC. For MC, we run 10,000 simulations to estimate the influence spread. We obtained the code of MISP from the authors and implemented other algorithms by C++. All the testings run on a Linux Server with intel Xeon 2.60 GHz CPU and 128GB memory.

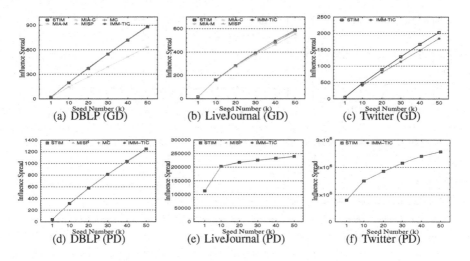

Fig. 5. Influence spread with different k ($\tau = 10$)

Datasets. We use three real-world datasets that are publicly available from SNAP[1] and COSNET[2] as in [6,15,23,30]. DBLP is a large co-authorship network, and we double the undirected edges to construct directed graphs. LiveJournal is a much larger social network where users can keep a blog, journal, or diary. Twitter shows the relationship between Twitter users. Table 1 shows the details of our datasets.

Parameters. Following [6,23], we set $p(u,v)$ to $1/|N^{in}(v)|$, and consider two types of time delay distributions, Geometric distribution (GD) and Poisson distribution (PD). For GD, we set the parameter to $5/(|N^{out}(u)|+5)$ [6]. For PD, we use a random integer between 1 and 20 as the parameter [23]. The threshold θ in MIA-M is set to 1/320 [6]. The threshold μ is set to 10^{-5} in MISP [23]. Among the two settings of ϵ in IMM (0.5 in [30] and 0.1 in [15]), we choose $\epsilon = 0.1$ since it will not loose the seed quality too much in the TIC model (Note that we set $\epsilon = 0.2$ for Twitter as IMM will run out of memory due to the large number of RR sets when $\epsilon = 0.1$.). For the fairness of comparison, we fix the rest of parameters before we run any algorithm so that all the algorithms run with the same parameter value. Each selected seed set is evaluated by 10,000 MC simulations based on TIC model.

4.2 Experimental Results

We will evaluate the efficiency (influence spread) and effectiveness (running time) of the algorithms by varying the deadline τ and the number of seeds k. We will not report the result for the cases that the algorithms cannot run to complete in 10 h.

Varying τ. Figure 3 shows how τ affects the influence spread of different algorithms. Note that MIA-M and MIA-C are not depicted in Fig. 3(e)–(h) as they can only sup-

---
[1] http://snap.stanford.edu/data.
[2] https://www.aminer.cn/cosnet.

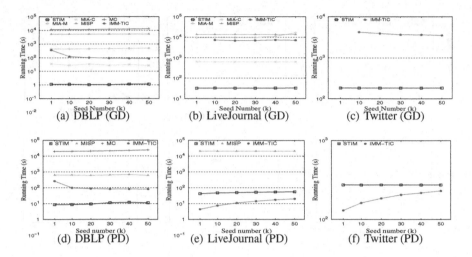

Fig. 6. Running time with different k ($\tau = 10$)

port Geometric distribution. In both distributions, the influence spread increases as τ increases, which validates that more nodes can be activated if there is sufficient time. In the Poisson distribution, STIM, MISP, IMM-TIC, and MC achieve similar influence spread on all the datasets. In the Geometric distribution, STIM, MISP, MIA-M, IMM-TIC, and MC achieve similar influence spread while MIA-C achieves less influence spread on all the datasets except LiveJournal. Specifically, MIA-C has 8% and 10% loss for $\tau = 20$ and $\tau = 25$ on DBLP. On LiveJournal, both MISP and MIA-C obtain less influence spread compared with STIM and MIA-M. On Twitter, the influence spread of IMM-TIC is slightly less than that of STIM in Geometric distribution. Figure 4 shows the running time, which increases as τ increases except for IMM-TIC, as more RR-sets need to be sampled for smaller k when k is small. STIM and MIA-C are nearly 100x faster than MIA-M in Geometric distribution. The running time of MIA-C keeps stable as τ increases because its complexity is irrelevant to τ. MISP takes a much longer time than MIA-M on DBLP because it is derived from the undirected graph that will result in a large number of propagation paths enumerated by MISP while MIA-M only considers the maximum influence path. Overall, STIM is 10–100 faster than MIA-M, MISP, IMM-TIC, and MC, and achieves better scalability and efficiency for the large time constraint. Although the running times of MIA-C and STIM are similar, the influence spread of MIA-C is not as good as that of STIM. The running time of STIM is close to that of IMM-TIC in Poisson distribution, but it is much less than that of IMM-TIC in Geometric distribution.

Varying k. Figure 5 shows the influence spread when we vary k, which increases as k increases. All the algorithms achieve similar influence spread except MIA-C. The gap between MIA-C and other algorithms also becomes larger when more seeds are selected. Figures 6 shows the running time of the evaluated algorithms when varying the number of seeds k, which increases when k becomes larger except MISP. As k increases, the running time of MISP almost stays stable. It is because MISP spends much time on the selection of the first seed by considering all the possible influence spreading paths and use a fast method to estimate the marginal influence. Our method STIM is about 1000x faster than MC, 100x faster than MISP for most cases, and 50x faster than MIA-M, and 20x-10x faster than MIA-C. The running time of STIM is close to that of IMM-TIC in Poisson distribution, but it is much less than that of IMM-TIC in Geometric distribution.

5 Conclusion

In this paper, we study the time-sensitive influence maximization problem under the TIC model, which has been proved to be NP-hard. Although the greedy search framework can bee utilized due to its monotonicity and sub-modularity properties, existing algorithms still need to repeatedly visit the subpaths in the graphs to compute the influence probability. We propose a novel space and time-efficient algorithm STIM equipped with a new time-based search strategy to estimate node influence and marginal influence. We evaluated our algorithms and the state-of-arts on publicly available real-world datasets and validated the efficiency and effectiveness of our approach.

Acknowledgment. The work was supported in part by grants of Natural Science Foundation 61972291, Natural Science Foundation of Hubei Province 2018CFB519 and 2018CFB616, Fundamental Research Funds for the Central Universities 413000078, The Key program of New Generation Information Technology Innovation of the Ministry of Education 2019J01011.

References

1. Backstrom, L., Huttenlocher, D., Kleinberg, J., Lan, X.: Group formation in large social networks: membership, growth, and evolution. In: KDD, pp. 44–54 (2006)
2. Bakshy, E., Eckles, D., Yan, R., Rosenn, I.: Social influence in social advertising: evidence from field experiments. In: CEC, pp. 146–161 (2012)
3. Bakshy, E., Hofman, J.M., Mason, W.A., Watts, D.J.: Everyone's an influencer: quantifying influence on Twitter. In: WSDM, pp. 65–74 (2011)
4. Borgs, C., Brautbar, M., Chayes, J., Lucier, B.: Maximizing social influence in nearly optimal time. In: SODA, pp. 946–957 (2014)
5. Chen, W., Lakshmanan, L.V.S., Castillo, C.: Information and Influence Propagation in Social Networks. Synthesis Lectures on Data Management (2013)
6. Chen, W., Lu, W., Zhang, N.: Time-critical influence maximization in social networks with time-delayed diffusion process. In: AAAI (2012)
7. Chen, W., Wang, C., Wang, Y.: Scalable influence maximization for prevalent viral marketing in large-scale social networks. In: KDD, pp. 1029–1038 (2010)
8. Domingos, P., Richardson, M.: Mining the network value of customers. In: KDD, pp. 57–66 (2001)
9. Du, N., Song, L., Gomez-Rodriguez, M., Zha, H.: Scalable influence estimation in continuous-time diffusion networks. In: NIPS, pp. 3147–3155 (2013)
10. Galhotra, S., Arora, A., Roy, S.: Holistic influence maximization: combining scalability and efficiency with opinion-aware models. In: SIGMOD, pp. 743–758 (2016)
11. Gomez-Rodriguez, M., Balduzzi, D., Schölkopf, B.: Uncovering the temporal dynamics of diffusion networks. In: ICML, pp. 561–568 (2011)

12. Gomez-Rodriguez, M., Schölkopf, B.: Influence maximization in continuous time diffusion networks. In: ICML (2012)
13. Gomez-Rodriguez, M., Song, L., Du, N., Zha, H., Schölkopf, B.: Influence estimation and maximization in continuous-time diffusion networks. TIS **34**(2), 9:1–9:33 (2016)
14. Goyal, A., Lu, W., Lakshmanan, L.V.S.: CELF++: optimizing the greedy algorithm for influence maximization in social networks. In: WWW, pp. 47–48 (2011)
15. Huang, K., Wang, S., Bevilacqua, G.S., Xiao, X., Lakshmanan, L.V.S.: Revisiting the stop-and-stare algorithms for influence maximization. PVLDB **10**(9), 913–924 (2017)
16. Jung, K., Heo, W., Chen, W.: IRIE: scalable and robust influence maximization in social networks. In: ICDM, pp. 918–923 (2012)
17. Kempe, D., Kleinberg, J., Tardos, E.: Maximizing the spread of influence through a social network. In: KDD, pp. 137–146 (2003)
18. Lee, W., Kim, J., Yu, H.: CT-IC: continuously activated and time-restricted independent cascade model for viral marketing. In: ICDM, pp. 960–965 (2012)
19. Leskovec, J., Krause, A., Guestrin, C., Faloutsos, C., VanBriesen, J., Glance, N.: Cost-effective outbreak detection in networks. In: KDD, pp. 420–429 (2007)
20. Li, J., Cai, Z., Yan, M., Li, Y.: Using crowdsourced data in location-based social networks to explore influence maximization. In: INFOCOM, pp. 1–9 (2016)
21. Li, Y., Fan, J., Wang, Y., Tan, K.: Influence maximization on social graphs: a survey. TKDE **30**(10), 1852–1872 (2018)
22. Liu, B., Cong, G., Xu, D., Zeng, Y.: Time constrained influence maximization in social networks. In: ICDM, pp. 439–448 (2012)
23. Liu, B., Cong, G., Zeng, Y., Xu, D., Chee, Y.M.: Influence spreading path and its application to the time constrained social influence maximization problem and beyond. TKDE **26**(8), 1904–1917 (2014)
24. Nguyen, H.T., Thai, M.T., Dinh, T.N.: Stop-and-stare: optimal sampling algorithms for viral marketing in billion-scale networks. In: SIGMOD, pp. 695–710 (2016)
25. Rashid, A.M., Karypis, G., Riedl, J.: Influence in ratings-based recommender systems: an algorithm-independent approach. In: SDM, pp. 556–560 (2005)
26. Sakaki, T., Okazaki, M., Matsuo, Y.: Earthquake shakes Twitter users: teal-time event detection by social sensors. In: WWW, pp. 851–860 (2010)
27. Song, C., Hsu, W., Lee, M.: Targeted influence maximization in social networks. In: CIKM, pp. 1683–1692 (2016)
28. Sun, L., Huang, W., Yu, P.S., Chen, W.: Multi-round influence maximization. In: KDD, pp. 2249–2258 (2018)
29. Tang, J., Tang, X., Xiao, X., Yuan, J.: Online processing algorithms for influence maximization. In: SIGMOD, pp. 991–1005 (2018)
30. Tang, Y., Shi, Y., Xiao, X.: Influence maximization in near-linear time: a martingale approach. In: SIGMOD, pp. 1539–1554 (2015)
31. Tang, Y., Xiao, X., Shi, Y.: Influence maximization: near-optimal time complexity meets practical efficiency. In: SIGMOD, pp. 75–86 (2014)
32. Wang, X., Zhang, Y., Zhang, W., Lin, X.: Efficient distance-aware influence maximization in geo-social networks. TKDE **29**(3), 599–612 (2017)
33. Yuan, J., Tang, S.: No time to observe: adaptive influence maximization with partial feedback. In: IJCAI, pp. 3908–3914 (2017)

Unsupervised Hierarchical Feature Selection on Networked Data

Yuzhe Zhang[1], Chen Chen[2], Minnan Luo[1(✉)], Jundong Li[3,4], Caixia Yan[1], and Qinghua Zheng[1,5]

[1] School of Computer Science and Technology, Xi'an Jiaotong University, Xi'an, China
{zhangyuzhe,yancaixia}@stu.xjtu.edu.cn, {minnluo,qhzheng}@xjtu.edu.cn
[2] Google Inc., Menlo Park, USA
chenannie45@gmail.com
[3] Department of Electrical and Computer Engineering, University of Virginia, Charlottesville, USA
jundong@virginia.edu
[4] Department of Computer Science and School of Data Science, University of Virginia, Charlottesville, USA
[5] National Engineering Lab for Big Data Analytics, Xi'an Jiaotong University, Xi'an, China

Abstract. Networked data is commonly observed in many high-impact domains, ranging from social networks, collaboration platforms to biological systems. In such systems, the nodes are often associated with high dimensional features while remain connected to each other through pairwise interactions. Recently, various unsupervised feature selection methods have been developed to distill actionable insights from such data by finding a subset of relevant features that are highly correlated with the observed node connections. Although practically useful, those methods predominantly assume that the nodes on the network are organized in a flat structure, which is rarely the case in reality. In fact, the nodes in most, if not all, of the networks can be organized into a hierarchical structure. For example, in a collaboration network, researchers can be clustered into different research areas at the coarsest level and are further specified into different sub-areas at a finer level. Recent studies have shown that such hierarchical structure can help advance various learning problems including clustering and matrix completion. Motivated by the success, in this paper, we propose a novel unsupervised feature selection framework (HNFS) on networked data. HNFS can simultaneously learn the implicit hierarchical structure among the nodes and embed the hierarchical structure into the feature selection process. Empirical evaluations on various real-world datasets validate the superiority of our proposed framework.

Keywords: Feature selection · Attributed networks · Hierarchical structure · Pseudo labels

© Springer Nature Switzerland AG 2020
Y. Nah et al. (Eds.): DASFAA 2020, LNCS 12114, pp. 137–153, 2020.
https://doi.org/10.1007/978-3-030-59419-0_9

1   Introduction

Networked data is ubiquitous in many application domains. Typical examples include social networks, collaboration platforms, biological systems, and transportation networks. Normally, nodes in the above-mentioned systems are not only structurally connected, but also are associated with high-dimensional features/attributes. For example, in the biology networks, genes are connected by mutual interactions, while each of them contains numerous fragments which bring in high-dimensional features. Another representative instance is the social networks in which users are connected with each other and a diverse of user activities (e.g., posting, retweet) brings high-dimensional features. In fact, the high-dimensional data is often notoriously to tackle due to the curse of dimensionality [14]. Meanwhile, high-dimensional data not only increases the requirement of memory storage and the cost of computation, but also deteriorate the effectiveness of the algorithm due to the redundant and noisy information. To alleviate these problems, various dimensionality reduction techniques have been explored, among which feature selection has shown its effectiveness for various data mining and machine learning tasks. In particular, a feature selection algorithm can be seen as the combination of a search technique for selecting a subset of high-quality features, along with an evaluation measure which scores different subsets. The selected features would be efficient and effective to the subsequent learning tasks as the storage and computational cost is greatly reduced while the redundant and noisy information is significantly eliminated.

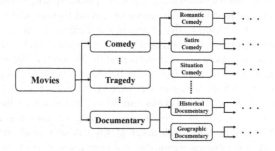

Fig. 1. Hierarchical structure of Douban movies.

Varying by the availability of labels, feature selection methods can be categorized into supervised methods and unsupervised methods. Supervised methods such as [8,25] usually gain better performance as label information is involved in the selection process. However, due to the expensive cost of amassing substantial labeled data, unsupervised feature selection has received more attention in recent years. A family of unsupervised feature selection methods employ the pseudo clustering labels of data to guide the selection phase, typical algorithms along this line include Nonnegative Discriminative Feature Selection (NDFS) [16], Robust Unsupervised Feature Selection (RUFS) [24], and Consensus Guided

Unsupervised Feature Selection (CGUFS) [18]. Although empirically effective, these cluster labels are still generated by all features which may lead to suboptimal results. Thus, some works like [15,26] chose to generate pseudo labels from external resources like connections among different data samples and has shown to be very effective. Nonetheless, these algorithms assume that the nodes on the network are organized in one-layer flat structure, which is often not the case in reality. Take the douban[1] movie rating network as an example, the movies in the platform can be classified into different genres, such as comedy, tragedy, action, *etc.* For each genre of the movies, we can further divide it into several sub-categories, which can be further divided again and again in a hierarchical manner as illustrated in Fig. 1. The data hierarchical structure has been proved to be effective in many other tasks such as representation learning [27] and recommendation [29]. Thus, it motivates us to investigate whether the success can be shifted to guide the selection of more relevant features when the label information is not available.

To address the aforementioned issues, in this paper, we propose a novel unsupervised feature selection algorithm, *i.e.*, HNFS to exploit the implicit hierarchical structure embedded on the network. Specifically, we propose to learn the implicit hierarchical structure from the network structure and measure its correlation with the node attribute information for unsupervised feature selection. The major contributions of this paper are as follow:

- Providing a principled way to learn implicit hierarchical structures of network data.
- Proposing a novel unsupervised feature selection framework which embeds the hierarchical structure learning into feature selection.
- Providing an effective alternating algorithm for the proposed algorithm.
- Demonstrating the effectiveness of the proposed framework on four commonly used real-world datasets.

2   The Proposed Framework

We first summarize the notations used throughout the paper. For a given matrix $\mathbf{A}$, $\mathbf{A}(i,j)$ denotes the (i,j)-th entry of $\mathbf{A}$. $Tr(\mathbf{A})$ denotes the trace of $\mathbf{A}$ if $\mathbf{A}$ is a square matrix. $\langle \mathbf{A}, \mathbf{B} \rangle$ equals $Tr(\mathbf{A}^T \mathbf{B})$, which means the standard inner product between two matrices. $\mathbf{I}$ is the identity matrix and $\mathbf{1}$ is a vector whose elements are all 1. For any matrix $\mathbf{A} \in \mathbb{R}^{n \times d}$, its Frobenius norm and $l_{2,1}$-norm are respectively defined as $\|\mathbf{A}\|_F = \sqrt{\sum_{i=1}^{n} \sum_{j=1}^{d} \mathbf{A}(i,j)^2}$ and $\|\mathbf{A}\|_{2,1} = \sum_{i=1}^{n} \sqrt{\sum_{j=1}^{d} \mathbf{A}(i,j)^2}$.

2.1   Unsupervised Feature Selection

Sparse learning has been regarded as a potent tool for feature selection [5,16]. In particular, one popular method is to embed feature selection into a clustering

---
[1] https://movie.douban.com/.

algorithm by selecting latent features with sparse learning [28]. Following this approach, we choose to embed feature selection into a low-rank matrix construction algorithm and apply $\ell_{2,1}$-norm on the latent representation of the original data. Let $\mathbf{X} \in \mathbb{R}^{n \times d}$ be the feature matrix which collects the feature vector of all the n nodes. Our basic model decomposes the feature matrix $\mathbf{X}$ into two matrices, $i.e.$, $\mathbf{V} \in \mathbb{R}^{n \times k}$ and $\mathbf{W} \in \mathbb{R}^{d \times k}$, and perform $\ell_{2,1}$-norm on $\mathbf{W}$ as follows:

$$\min_{\mathbf{V},\mathbf{W}} \left\| \mathbf{X} - \mathbf{V}\mathbf{W}^T \right\|_F^2 + \alpha \left\| \mathbf{W} \right\|_{2,1}, \quad s.t. \mathbf{V}^T\mathbf{V} = \mathbf{I}, \mathbf{V} \geq 0, \tag{1}$$

where $\mathbf{V}$ is the clustering indicator matrix, $\mathbf{W}$ is the latent feature matrix, and k is the number of predefined clusters. In supervised feature selection, we can regard the label information as the clustering indicator $\mathbf{V}$ to steer the selection process. But when it comes to the unsupervised situation, there is no such ground truth information, thus we choose to generate the pseudo labels by resorting to side information such as the structure information among data instances.

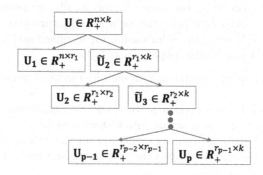

Fig. 2. Hierarchical structure of nodes via deeply factorizing the latent feature matrix.

2.2 Latent Representation of Network Structure

Given a network G with adjacency matrix $\mathbf{A}$, we can model the latent representations of its nodes with nonnegative matrix factorization [17] as

$$\min_{\mathbf{U},\mathbf{V}} \left\| \mathbf{A} - \mathbf{U}\mathbf{V}^T \right\|_F^2, \quad s.t. \ \mathbf{U} \geq 0, \mathbf{V} \geq 0, \tag{2}$$

where $\mathbf{U} \in \mathbb{R}_+^{n \times k}$ and $\mathbf{V} \in \mathbb{R}_+^{n \times k}$. Optimizing Eq. (2) can be viewed as a clustering process over the network. Specifically, each column of $\mathbf{U}$ represents the potential definition of a community, and each row of $\mathbf{V}$ denotes the membership of a node to all k communities. Naturally, $\mathbf{U}(i,l)\mathbf{V}(l,j)$ can be regarded as the contribution of the l-th community to the edge $\mathbf{A}(i,j)$. Thus, $\widetilde{\mathbf{A}}(i,j) = \sum_{l=1}^{k} \mathbf{U}(i,l)\mathbf{V}(l,j)$ should be the result of the relationship between node i and j. Moreover, the membership of nodes obtained in $\mathbf{V}$ can act as the role of pseudo labels for unsupervised feature selection.

Through Eq. (2), we learn a one-layer representation of clustering (i.e., communities in network) $\mathbf{U}$ and a community membership matrix $\mathbf{V}$). However, it assumes that the nodes on the network are organized in a one-layer flat structure, which omits the diversified and complicated organizational patterns in real-world networks as described in [34]. To learn more accurate representation of the communities on the network, we decide to further factorize the one-layer latent representation $\mathbf{U}$ to capture the implicit hierarchical structure among nodes embedded in the network. Specifically, we factorize the adjacency matrix $\mathbf{A}$ into $p+1$ nonnegative factor matrices, as follows:

$$\mathbf{A} \approx \mathbf{U}_1\mathbf{U}_2 \ldots \mathbf{U}_p\mathbf{V}^T, \tag{3}$$

where $\mathbf{V} \in \mathbb{R}_+^{n \times k}$, $\mathbf{U}_i \in \mathbb{R}_+^{r_{i-1} \times r_i}(1 \leq i \leq p)$, and $n = r_0 \geq r_1 \geq \ldots \geq r_{p-1} \geq r_p = k$.

Additionally, the widely used Frobenius norm for reconstruction error measuring is often very sensitive to the anomaly nodes in the network, while the $\ell_{2,1}$-norm error is often more preferred as it can enhance the robustness of the model. Hence, to collectively capture the hierarchical structures of the communities on the networks and ensure the robustness of the model, we propose to learn the latent representations of the nodes and the community assignment through the following optimization problem:

$$\min_{\mathbf{U}_i,\mathbf{V}} \left\| \mathbf{A} - \mathbf{U}_1\mathbf{U}_2 \ldots \mathbf{U}_p\mathbf{V}^T \right\|_{2,1} \tag{4}$$
$$s.t. \ \ \mathbf{V} \geq 0, \mathbf{U}_i \geq 0, i \in 1, 2, ..., p,$$

where the original flat-structured community matrix $\mathbf{U}$ is firstly decomposed into two nonnegative matrices $\mathbf{U_1} \in \mathbb{R}_+^{n \times r_1}$ and $\widetilde{\mathbf{U_2}} \in \mathbb{R}_+^{r_1 \times k}$. Following the same procedure, the latent feature matrix $\mathbf{U}$ can be further factorized into p nonnegative matrices as illustrated in Fig. 2. This formulation will lead to more accurate community membership results, i.e., a better community assignment matrix $\mathbf{V}$.

2.3   The Proposed Framework – HNFS

With a hierarchy of p layers latent representations of the network, we combine Eq. (1) and Eq. (4) into a unified framework—HNFS by solving the following optimization problem:

$$\min_{\mathbf{U}_i,\mathbf{V},\mathbf{W}} \left\| \mathbf{A} - \mathbf{U}_1\mathbf{U}_2 ... \mathbf{U}_p\mathbf{V}^T \right\|_{2,1} + \alpha \left\| \mathbf{X} - \mathbf{V}\mathbf{W}^T \right\|_F^2 + \beta \left\| \mathbf{W} \right\|_{2,1} \tag{5}$$
$$s.t. \ \ \mathbf{V}^T\mathbf{V} = \mathbf{I}, \mathbf{V} \geq 0, \mathbf{U}_i \geq 0, i \in 1, 2, ..., p,$$

where α controls the balance between the network structure and feature information for community assignment learning; while β is a parameter to decide the sparsity of the model. With $\mathbf{W}$ fixed, the latent representation $\mathbf{V}$ is associated with both the network structure (*i.e.*, $\mathbf{A}$) and the features (*i.e.*, $\mathbf{X}$). When fixing the latent representations $\mathbf{V}$, the nodes membership learned in $\mathbf{V}$ can be viewed as the pseudo labels to guide the feature selection. As a result, the feature selection part and latent representation learning part could compliment each other and leads to a better model.

3   Optimization Algorithm

3.1   Solution

The objective function is not jointly convex *w.r.t.* all the variables, but it is convex *w.r.t.* each variable individually. Therefore, we can optimize the variables in an alternative update manner. Following [13], we propose to solve the problem with Alternating Direction Method of Multiplier (ADMM) [11]. First, we introduce two auxiliary variables $\mathbf{Z}$ and $\mathbf{E}$, and rewrite the optimization problem as:

$$\min_{\mathbf{U}_i,\mathbf{V},\mathbf{W},\mathbf{E},\mathbf{Z}} \|\mathbf{E}\|_{2,1} + \alpha \left\|\mathbf{X} - \mathbf{V}\mathbf{W}^T\right\|_F^2 + \beta \|\mathbf{W}\|_{2,1}$$
$$s.t. \quad \mathbf{Z} = \mathbf{U}_1\mathbf{U}_2...\mathbf{U}_p, \mathbf{E} = \mathbf{A} - \mathbf{Z}\mathbf{V}^T, \mathbf{V}^T\mathbf{V} = \mathbf{I} \tag{6}$$
$$\mathbf{Z} \geq 0, \mathbf{V} \geq 0, \mathbf{U}_i \geq 0, i \in 1, 2, ..., p.$$

The problem in Eq. (6) can be formulated as the following ADMM problem:

$$\min_{\mathbf{U}_i,\mathbf{V},\mathbf{W},\mathbf{E},\mathbf{Z}} \|\mathbf{E}\|_{2,1} + \alpha \left\|\mathbf{X} - \mathbf{V}\mathbf{W}^T\right\|_F^2 + \beta \|\mathbf{W}\|_{2,1}$$
$$+ \langle \mathbf{Y}_1, \mathbf{Z} - \mathbf{U}_1\mathbf{U}_2...\mathbf{U}_p \rangle + \langle \mathbf{Y}_2, \mathbf{A} - \mathbf{Z}\mathbf{V}^T - \mathbf{E} \rangle$$
$$+ \frac{\mu}{2}(\|\mathbf{Z} - \mathbf{U}_1\mathbf{U}_2...\mathbf{U}_p\|_F^2 + \left\|\mathbf{A} - \mathbf{Z}\mathbf{V}^T - \mathbf{E}\right\|_F^2) \tag{7}$$
$$s.t.\mathbf{V}^T\mathbf{V} = \mathbf{I}, \mathbf{Z} \geq 0, \mathbf{V} \geq 0, \mathbf{U}_i \geq 0, i \in 1, 2, ..., p,$$

where $\mathbf{Y}_1, \mathbf{Y}_2$ are two Lagrangian multipliers, and μ is a scalar to control the penalty for the violation of equality constraints (*i.e.*, $\mathbf{Z} = \mathbf{U}_1\mathbf{U}_2...\mathbf{U}_p$ and $\mathbf{E} = \mathbf{A} - \mathbf{Z}\mathbf{V}^T$).

Update E. Fixing all other variables except $\mathbf{E}$, the objective function can be reformulated as:

$$\min_{\mathbf{E}} \frac{1}{2} \left\|\mathbf{E} - (\mathbf{A} - \mathbf{Z}\mathbf{V}^T + \frac{1}{\mu}\mathbf{Y}_2)\right\|_F^2 + \frac{1}{\mu} \|\mathbf{E}\|_{2,1}. \tag{8}$$

The equation has a closed-form solution by using the following Lemma [19].

Lemma 1. Let $\mathbf{Q} = [\mathbf{q}_1; \mathbf{q}_2; ...; \mathbf{q}_m]$ be a given matrix and λ be a positive scalar. If the optimal solution of

$$\min_{\mathbf{W}} \frac{1}{2} \|\mathbf{W} - \mathbf{Q}\|_F^2 + \lambda \|\mathbf{W}\|_{2,1} \tag{9}$$

is $\mathbf{W}^*$, then the ith row of $\mathbf{W}^*$ is

$$\mathbf{w}_i^* = \begin{cases} (1 - \frac{\lambda}{\|\mathbf{q}_i\|})\mathbf{q}_i \ if \ \|\mathbf{q}_i\| > \lambda \\ 0 \qquad\qquad\qquad otherwise. \end{cases} \tag{10}$$

Suppose $\mathbf{Q} = \mathbf{A} - \mathbf{Z}\mathbf{V}^T + \frac{1}{\mu}\mathbf{Y}_2$, $\mathbf{E}$ can be updated as follow by using Lemma 1:

$$\mathbf{e}_i = \begin{cases} (1 - \frac{1}{\mu\|\mathbf{q}_i\|})\mathbf{q}_i \ if \ \|\mathbf{q}_i\| > \frac{1}{\mu} \\ 0 \qquad\qquad\qquad otherwise. \end{cases} \tag{11}$$

Update V. We follow the same strategy in [23] to update $\mathbf{V}$. Note that the constraints of $\mathbf{V}$ are the same in [23] and ours. Removing irrelevant terms to $\mathbf{V}$ from Eq. (7), the optimization problem can be rewritten as:

$$\min_{\mathbf{V}^T\mathbf{V}=\mathbf{I}} \frac{\mu}{2} \left\| \mathbf{A} - \mathbf{Z}\mathbf{V}^T - \mathbf{E} + \frac{1}{\mu}\mathbf{Y}_2 \right\|_F^2 + \alpha \left\| \mathbf{X} - \mathbf{V}\mathbf{W}^T \right\|_F^2. \tag{12}$$

After expanding the objective function and dropping terms that are independent of $\mathbf{V}$, we get

$$\min_{\mathbf{V}^T\mathbf{V}=\mathbf{I}} \frac{\mu}{2} \|\mathbf{V}\|_F^2 - \mu\langle\mathbf{N}, \mathbf{V}\rangle, \tag{13}$$

where $\mathbf{N} = (\mathbf{A}^T - \mathbf{E}^T + \frac{1}{\mu}\mathbf{Y}_2^T)\mathbf{Z} - \frac{2\alpha}{\mu}\mathbf{X}\mathbf{W}$. The above equation can be further simplified to a more compact form as $\min_{\mathbf{V}^T\mathbf{V}=\mathbf{I}} \|\mathbf{V} - \mathbf{N}\|_F^2$. According to [13], $\mathbf{V}$ can be updated by the following equation in which $\mathbf{P}$ and $\mathbf{Q}$ are left and right singular values of the SVD decomposition of $\mathbf{N}$:

$$\mathbf{V} = \mathbf{P}\mathbf{Q}^T. \tag{14}$$

Update W. The update rule for $\mathbf{W}$ is similar as $\mathbf{E}$. When other variables except $\mathbf{W}$ are fixed and terms that are irrelevant to $\mathbf{W}$ are removed, the optimization problem for $\mathbf{W}$ can be rewritten as:

$$\min_{\mathbf{W}} \alpha \left\| \mathbf{X} - \mathbf{V}\mathbf{W}^T \right\|_F^2 + \beta \|\mathbf{W}\|_{2,1}. \tag{15}$$

Using the fact that $\mathbf{V}^T\mathbf{V} = \mathbf{I}$, it can be reformulated as

$$\min_{\mathbf{W}} \frac{1}{2} \left\| \mathbf{W} - \mathbf{V}\mathbf{X}^T \right\|_F^2 + \frac{\beta}{2\alpha} \|\mathbf{W}\|_{2,1}. \tag{16}$$

Again, the above equation has a closed-form solution according to Lemma 1. Let $\mathbf{K} = \mathbf{V}\mathbf{X}^T$, then

$$\mathbf{w}_i = \begin{cases} (1 - \frac{\beta}{2\alpha\|\mathbf{k}_i\|})\mathbf{k}_i \ if \ \|\mathbf{k}_i\| > \frac{\beta}{2\alpha} \\ 0 \qquad\qquad\qquad otherwise. \end{cases} \tag{17}$$

Algorithm 1. Algorithm 1 The Proposed HNFS algorithm

Input: The data matrix $\mathbf{X}$ and the adjacency matrix $\mathbf{A}$

        The layer size of each layer $\mathbf{r}_i$

        The regularization parameter α, β

        The number of selected features m

Output: The most m relevant features.

1: Initialize $\mu = 10^{-3}$, $\rho = 1.1$, $\mathbf{U}_i = 0$, $\mathbf{V} = 0$ (or initialized using K-means)

2: **while** not convergence **do**

3:      Calculate $\mathbf{Q} = \mathbf{A} - \mathbf{Z}\mathbf{V}^T + \frac{1}{\mu}\mathbf{Y}_2$

4:      Update $\mathbf{E}$ by Eq. (11)

5:      Calculate $\mathbf{K} = \mathbf{V}\mathbf{X}^T$

6:      Update $\mathbf{W}$ by Eq. (17)

7:      Calculate $\mathbf{T} = \frac{1}{2}[(\mathbf{A} - \mathbf{E} + \frac{1}{\mu}\mathbf{Y}_2)\mathbf{V} + \mathbf{U} + \frac{1}{\mu}\mathbf{Y}_1]$

8:      Update $\mathbf{Z}$ by Eq. (19)

9:      Calculate $\mathbf{S}_i = (\mathbf{H}_i^T\mathbf{H}_i)^{-1}\mathbf{H}_i^T(\mathbf{Z} - \frac{\mathbf{Y}_1}{\mu})\mathbf{B}_i^T(\mathbf{B}_i\mathbf{B}_i^T)^{-1}$

10:     Update $\mathbf{U}_i$ by Eq. (24)

11:     Calculate $\mathbf{N} = (\mathbf{A}^T - \mathbf{E}^T + \frac{1}{\mu}\mathbf{Y}_2^T)\mathbf{Z} - \frac{2\alpha}{\mu}\mathbf{X}\mathbf{W}$

12:     Update $\mathbf{V}$ by Eq. (14)

13:     Update $\mathbf{Y}_1, \mathbf{Y}_2$ and μ by Eq. (25), Eq. (26) and Eq. (27)

14: **end while**

15: Sort each feature of $\mathbf{X}$ according to $\|\mathbf{w}_i\|_2$ in descending order and select the top-m features

---

Update Z. By removing other irrelevant parts to $\mathbf{Z}$, the objective function can be rewritten as:

$$\min_{\mathbf{Z} \geq 0} \frac{\mu}{2}\left\|\mathbf{A} - \mathbf{Z}\mathbf{V}^T - \mathbf{E} + \frac{1}{\mu}\mathbf{Y}_2\right\|_F^2 + \frac{\mu}{2}\left\|\mathbf{U}_1\mathbf{U}_2 \ldots \mathbf{U}_p - \mathbf{Z} + \frac{1}{\mu}\mathbf{Y}_1\right\|_F^2. \tag{18}$$

By setting the derivative of Eq. (18) *w.r.t.* $\mathbf{Z}$ to zero, we get $2\mathbf{Z} = (\mathbf{A} - \mathbf{E} + \frac{1}{\mu}\mathbf{Y}_2)\mathbf{V} + \mathbf{U} + \frac{1}{\mu}\mathbf{Y}_1$. Let $\mathbf{T} = \frac{1}{2}[(\mathbf{A} - \mathbf{E} + \frac{1}{\mu}\mathbf{Y}_2)\mathbf{V} + \mathbf{U} + \frac{1}{\mu}\mathbf{Y}_1]$. Then $\mathbf{Z}$ can be updated as:

$$\mathbf{Z}_{i,j} = max(\mathbf{T}_{i,j}, 0). \tag{19}$$

Update $\mathbf{U}_i$. By fixing all the variables except $\mathbf{U}_i$, the objective function in Eq. (7) is reduced to:

$$\min_{\mathbf{U}_i \geq 0} \frac{\mu}{2}\left\|\mathbf{H}_i\mathbf{U}_i\mathbf{B}_i - \mathbf{Z} + \frac{1}{\mu}\mathbf{Y}_1\right\|_F^2, \tag{20}$$

where $\mathbf{H}_i$ and $\mathbf{B}_i$, $1 \leq i \leq p$, are defined as:

$$\mathbf{H}_i = \begin{cases} \mathbf{U}_1\mathbf{U}_2 \ldots \mathbf{U}_{i-1} & if \quad i \neq 1 \\ \mathbf{I} & if \quad i = 1, \end{cases} \tag{21}$$

and

$$\mathbf{B}_i = \begin{cases} \mathbf{U}_{i+1}\mathbf{U}_{i+2} \ldots \mathbf{U}_p & if \quad i \neq p \\ \mathbf{I} & if \quad i = p. \end{cases} \tag{22}$$

By setting the derivative of Eq. (20) *w.r.t.* $\mathbf{U}_i$ to zero, we get

$$\mathbf{H}_i^T \mathbf{H}_i \mathbf{U}_i \mathbf{B}_i \mathbf{B}_i^T - \mathbf{H}_i^T (\mathbf{Z} - \frac{\mathbf{Y}_1}{\mu}) \mathbf{B}_i^T = 0, \tag{23}$$

where $\mathbf{H}_i^T \mathbf{H}_i$ is a positive semi-definite matrix, the same is true of $\mathbf{B}_i \mathbf{B}_i^T$. Let $\mathbf{S}_i = (\mathbf{H}_i^T \mathbf{H}_i)^{-1} \mathbf{H}_i^T (\mathbf{Z} - \frac{\mathbf{Y}_1}{\mu}) \mathbf{B}_i^T (\mathbf{B}_i \mathbf{B}_i^T)^{-1}$, then $\mathbf{U}_i$ has a closed-form solution:

$$\mathbf{U}_i(j, k) = max(\mathbf{S}_i(j, k), 0). \tag{24}$$

Update $\mathbf{Y}_1, \mathbf{Y}_2$ **and** μ. After updating the variables, the ADMM parameters should be updated. According to [4], they can be updated as follows:

$$\mathbf{Y}_1 = \mathbf{Y}_1 + \mu(\mathbf{Z} - \mathbf{U}_1 \mathbf{U}_2 \dots \mathbf{U}_p), \tag{25}$$

$$\mathbf{Y}_2 = \mathbf{Y}_2 + \mu(\mathbf{A} - \mathbf{Z}\mathbf{V}^T - \mathbf{E}), \tag{26}$$

$$\mu = \rho\mu. \tag{27}$$

Here, $\rho > 1$ is a parameter to control the convergence speed. The larger ρ is, the fewer iterations we require to get the convergence, while the precision of the final objective function value may be sacrificed. In Algorithm 1, we summarize the procedures for optimizing Eq. (6).

Table 1. Detailed information of the datasets.

	Wiki	BlogCatalog	Flickr	DBLP
#Users	2,405	5,196	7,575	18,448
#Features	4,937	8,189	12,047	2,476
#Links	17,981	171,743	239,738	45,611
#Classes	19	6	9	4

4 Experiments

4.1 Experimental Settings

Datasets. The experiments are conducted on four commonly used real-world networks datasets, including Wiki[2], BlogCatalog[3], Flickr (See footnote 3) and DBLP[4]. The detailed statistics of these datasets are listed in Table 1.

---

[2] https://github.com/thunlp/OpenNE/tree/master/data/wiki.
[3] http://dmml.asu.edu/users/xufei/datasets.html.
[4] https://www.aminer.cn/citation.

- **Wiki:** Wiki is a document network which is composed of hyperlinks between wikipedia documents. Each document is displayed by a high-dimensional vector which indicates the word frequency count of itself. These documents are classified into dozens of predefined classes.
- **BlogCatalog:** BlogCatalog is a social blog directory in which users can register their blogs under different predefined categories [31]. Names, ids, blogs, the associated tags and blog categories form the content information while the class label is selected from a predefined list of categories, indicating the interests of each user.
- **Flickr:** Flickr is a content sharing platforms, with a focus on photos, where users can share their contents, upload tags and subscribe to different interest groups [32]. Besides, users interact with others forming link information while groups that users joined can be treated as class labels.
- **DBLP:** DBLP is a part of the DBLP bibliographic network dataset. It contains papers from four research areas: Database, Data Mining, Artificial Intelligence and Computer Vision. Each paper's binary feature vectors indicate the presence/absence of the corresponding word in its title.

Baseline Methods. We compare our proposed framework HNFS with the following seven unsupervised feature selection algorithms, which can be divided into two groups. The first five algorithms only consider the attribute information while the latter two take both attribute information and structure information into consideration. Following are the comparing methods used in our experiment.

- **LapScore:** Laplacian Score is a filter method for feature selection which is independent to any learning algorithm [2]. The importance of a feature is evaluated by its power of locality preserving, or, Laplacian Score [12].
- **RUFS:** RUFS is a robust unsupervised feature selection approach where robust label learning and robust feature learning are simultaneously performed via orthogonal nonnegative matrix factorization and joint $\ell_{2,1}$-norm minimization [24].
- **UDFS:** UDFS incorporates discriminative analysis and $\ell_{2,1}$-norm minimization into a joint framework for unsupervised feature selection under the assumption that the class label of input data can be predicted by a linear classifier. [33].
- **GreedyFS**: GreedyFS is an effective filter method for unsupervised feature selection which first defines a novel criterion that measures the reconstruction error of the selected data and then selects features in a greedy manner based on the proposed criterion [10].
- **MCFS:** By using spectral regression [6] with $\ell_{2,1}$-norm regularization, MCFS suggests a principled way to measure the correlations between different features without label information. Thus, MCFS can well handle the data with multiple cluster structure [5].
- **NetFS:** NetFS is an unsupervised feature selection framework for networked data, which embeds the latent representation learning into feature selection [15].

- **HNFS-flat:** HNFS-flat is a variant of our proposed framework which only considers the flat structure of networks by setting p = 1 in our model.

Metrics and Settings. Following the standard ways to assess unsupervised feature selection, we evaluate different feature selection algorithms by evaluating the clustering performance with the selected features. Two commonly adopted clustering performance metrics [14] are used: (1) *normalized mutual information* (NMI) and (2) *accuracy* (ACC). The parameter settings of the baseline methods all follow the suggestions by the original papers [5,12,33]. For our proposed method, we tune the model parameters by a "grid-search" strategy from {0.001,0.01,0.1,1,10,100,1000} and the best clustering results are reported. We implement HNFS with the number of layers $p = 2$. Although different layers $p \in \{2, 3, 4, 5, 6\}$ are tried, the performance improvement is not significant while more running time is required. Meanwhile, we specify the size of layer $r_1 = 256$ and we will explain the reason later. In the experiments, each feature selection algorithm is first used to select a certain number of features, then we use K-means to cluster nodes into different clusters based on the selected features. Since K-means may converge to local optima, we repeat the experiments 20 times and report the average results.

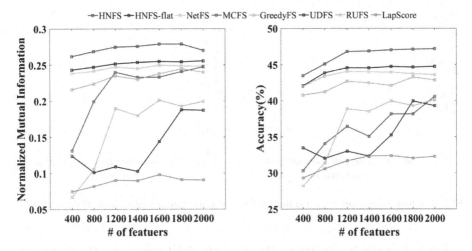

Fig. 3. Clustering results with different feature selection algorithms on Blogcatalog dataset.

4.2   Quality of Selected Features

In this subsection, we compare the quality of the selected features by our model and other baseline methods on all the datasets. The number of selected features varies from {400, 800, 1200, 1400, 1600, 1800, 2000}. The results are shown

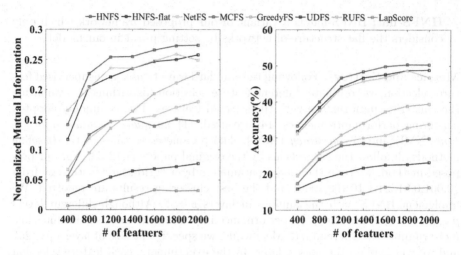

Fig. 4. Clustering results with different feature selection algorithms on Flickr dataset.

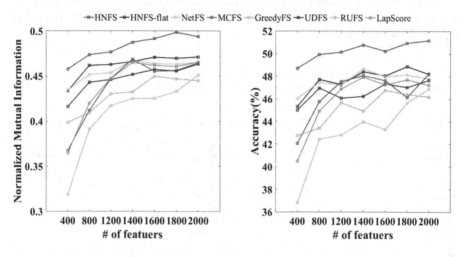

Fig. 5. Clustering results with different feature selection algorithms on Wiki dataset.

in Fig. 3, 4, 5 and 6. The higher the ACC and NMI values are, the better the feature selection performance is. We have the following observations based on the experimental results:

- The methods that consider both attribute information and structure information obtain much better results than the ones that only exploit feature information especially in Blogcatalog and Flickr. It is because that these two datasets contain abundant structure information while Wiki and DBLP have sparse adjacency matrix. Despite of this, methods that consider structure information can still benefit from it in latter two datasets in most situations.

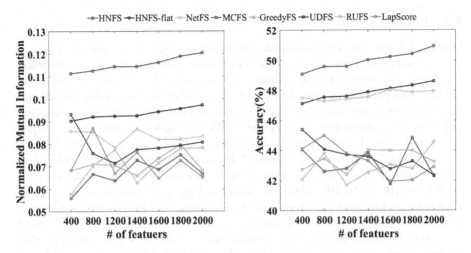

Fig. 6. Clustering results with different feature selection algorithms on DBLP dataset.

It implies that when the label information is not explicitly given, network structure indeed can help us select more relevant features.

– HNFS and NetFS consider the network structure differently. NetFS regard it as a flat-structure while real-world networks usually exhibit hierarchical structures which should be fully considered. The results between our model and NetFS prove that implicit hierarchical structures of networks can improve the performance of feature selection. The observations are further confirmed by the improvement of HNFS over its flat-structure variant HNFS-flat.

– In Wiki and DBLP datasets, HNFS performs well with only a few hundred of features. BlogCatalog and Flickr have more features than the first two, but HNFS still obtains good clustering performance with only around 1/10 and 1/20 of total features, respectively.

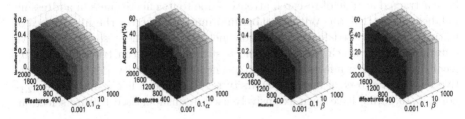

The effect of parameter α.   The effect of parameter β.

Fig. 7. Parameter analysis on Wiki.

4.3   Parameter Analysis

Our model has two regularization parameters α and β. α controls the balance between the network structure and feature information for feature selection, while β determines the sparsity of the model. To discuss the influences of these two parameters, we choose to fix one parameter each time and change the other one to see how the clustering results change. Due to space limit, we only report the results on Wiki in Fig. 7. We first make the parameter β equal to 10 and vary the parameter α as $\{0.001, 0.01, 0.1, 1, 10, 100, 1000\}$. We can see from Fig. 7 that when α is around 10 we can get a relatively better clustering performance. Then we make α equal to 1 and vary the parameter β as $\{0.001, 0.01, 0.1, 1, 10, 100, 1000\}$. As shown in Fig. 7, with the increase of β, the clustering performance first increases then becomes stable. The reason is that a small α will reduce the feature sparsity of the model, which is not of great benefit to feature selection. Besides, the experimental results show that the clustering performance is more sensitive to the number of selected features compared with α and β. However, picking the proper number of features is still an open problem that requires deep investigation.

5   Related Work

5.1   Traditional Feature Selection

Depending on the existence of the label information, feature selection algorithms can be broadly divided into supervised and unsupervised methods. Supervised feature selection algorithms assess feature relevance via its correlation with the class labels [22,35]. According to the adopted strategies, we can further divide supervised feature selection into filter methods and wrapper methods [14]. Filter methods pay attention to feature selection part which means they are independent of any learning algorithms. On the contrary, wrapper methods have a close relationship with the learning algorithm. They use the learning performance to access the quality of selected features iteratively, which is often computationally expensive. Unsupervised feature selection algorithms, on the other hand, have attracted a surge of research attention due to its effectiveness in addressing unlabeled data [1,3,33]. Without label information to access the importance of features, unsupervised feature selection methods [7,30] need some alternative criteria to decide which features to select, such as data reconstruction error [9], local discriminative information [16,33], and data similarity [12,36]. To effectively select a subset of features, sparsity regularizations like l_1-norm and $\ell_{2,1}$-norm [5,15,16,33] have been extensively used in unsupervised feature selection.

5.2   Unsupervised Feature Selection with Pseudo Labels

Furthermore, to compensate the shortage of labels, many unsupervised feature selection methods tend to explore some other information among data instances to guide the feature selection procedure, namely pseudo labels. The result of clustering has been commonly used as pseudo labels in many unsupervised feature selection works. For example, NDFS [16] combines the result of spectral clustering with the traditional feature selection and obtain better performance. EUFS [28] and RUFS [24] utilize the result of spectral clustering in the same way and only change the part of feature selection to make it more robust. Since the spectral clustering can help to select feature, so do other clustering methods. CGUFS [18] proposes to learn a consensus clustering results from multitudinous clustering algorithms, which leads to better clustering accuracy with high robustness. But, the clustering result obtained by high-dimensional feature matrix may contain numerous noise. Thus, some other methods attempt to utilize structure information as the pseudo labels, namely the adjacency matrix. LUFS [26] first extract social dimensions and then utilize them for selecting discriminative features on the attributed networks while NetFS [15] embeds the latent representation obtained from structure information into feature selection. However, these works are substantially different from our proposed framework HNFS as they omit the hierarchical structure among data instances. The hierarchical information has demonstrated its importance in supervised feature selection [20,21], which facilitates the investigation of HNFS in this paper. Besides, HNFS provides an iterative way to learn the implicit hierarchical structures and feature importance measures simultaneously and the feature selection part becomes more robust compared with other unsupervised feature selection algorithms.

6   Conclusion and Future Work

In this paper, we propose an unsupervised feature selection framework HNFS for networked data. Specifically, the proposed method can effectively capture the implicit hierarchical structure of the network while measuring its correlation with node attributes for feature selection. Methodologically, we perform Alternating Direction Method of Multiplier (ADMM) to optimize the objective function. Extensive experimental results on four real-world network datasets have validated the effectiveness of our model.

There are several directions worth further investigation. First, it would be meaningful to study the effectiveness of other hierarchical network representation methods in contrast to the nonnegative matrix factorization method used in this work. Second, real-world networks are evolving over time, which means both the network structure and the features are changing timely. Thus how to generalize the proposed method in a dynamic setting would be another interesting research direction.

Acknowledgement. This work was supported by National Nature Science Foundation of China (No. 61872287, No. 61532015, and No. 61872446), Innovative Research Group of the National Natural Science Foundation of China (No. 61721002), Innovation Research Team of Ministry of Education (IRT_17R86), and Project of China Knowledge Center for Engineering Science and Technology. Besides, this research was funded by National Science and Technology Major Project of the Ministry of Science and Technology of China (No. 2018AAA0102900).

References

1. Alelyani, S., Tang, J., Liu, H.: Feature selection for clustering: a review. Data Cluster.: Algorithms Appl. **21**, 110–121 (2013)
2. Belkin, M., Niyogi, P.: Laplacian eigenmaps and spectral techniques for embedding and clustering. In: NIPS (2001)
3. Boutsidis, C., Mahoney, M.W., Drineas, P.: Unsupervised feature selection for the k-means clustering problem. In: NIPS (2009)
4. Boyd, S., Parikh, N., Chu, E., Peleato, B., Eckstein, J.: Distributed optimization and statistical learning via the alternating direction method of multipliers. Foundations & Trends in Machine Learning **3**, 1–122 (2011)
5. Cai, D., He, X.: Unsupervised feature selection for multi-cluster data. In: KDD (2010)
6. Cai, D., He, X., Han, J.: Spectral regression for efficient regularized subspace learning. In: ICCV (2007)
7. Dong, X., Zhu, L., Song, X., Li, J., Cheng, Z.: Adaptive collaborative similarity learning for unsupervised multi-view feature selection. In: IJCAI (2018)
8. Fan, M., Chang, X., Zhang, X., Wang, D., Du, L.: Top-k supervise feature selection via ADMM for integer programming. In: IJCAI (2017)
9. Farahat, A.K., Ghodsi, A., Kamel, M.S.: An efficient greedy method for unsupervised feature selection. In: ICDM (2011)
10. Farahat, A.K., Ghodsi, A., Kamel, M.S.: Efficient greedy feature selection for unsupervised learning. Knowl. Inf. Syst. **2**, 285–310 (2013). https://doi.org/10.1007/s10115-012-0538-1
11. Gabay, D.: A dual algorithm for the solution of nonlinear variational problems via finite element approximation. Comput. Math. Appl. **2**, 17–40 (1976)
12. He, X., Cai, D., Niyogi, P.: Laplacian score for feature selection. In: NIPS (2005)
13. Huang, J., Nie, F., Ding, C.: Robust manifold nonnegative matrix factorization. ACM Trans. Knowl. Discov. Data **8**, 11:1–11:21 (2014)
14. Li, J., et al.: Feature selection: a data perspective. ACM Comput. Surv. **50**, 94:1–94:45 (2017)
15. Li, J., Hu, X., Wu, L., Liu, H.: Robust unsupervised feature selection on networked data. In: SDM (2016)
16. Li, Z., Yang, Y., Zhou, X., Lu, H.: Unsupervised feature selection using nonnegative spectral analysis. In: AAAI (2012)
17. Lin, C.: Projected gradient methods for nonnegative matrix factorization. Neural Comput. **19**, 2756–2779 (2007)
18. Liu, H., Shao, M., Fu, Y.: Consensus guided unsupervised feature selection. In: AAAI (2016)
19. Liu, J., Ji, S., Ye, J.: Multi-task feature learning via efficient $\ell_{2,1}$-norm minimization. In: UAI (2009)
20. Liu, J., Ye, J.: Moreau-Yosida regularization for grouped tree structure learning. In: NIPS (2010)
21. Liu, Y., Wang, J., Ye, J.: An efficient algorithm for weak hierarchical lasso. ACM Trans. Knowl. Discov. Data **10**, 32:1–32:24 (2014)
22. Nie, F., Huang, H., Cai, X., Ding, C.H.: Efficient and robust feature selection via joint $ell_{2,1}$-norms minimization. In: NIPS (2010)

23. Pan, W., Yang, Q.: Transfer learning in heterogeneous collaborative filtering domains. Artif. Intell. **197**, 39–55 (2013)
24. Qian, M., Zhai, C.: Robust unsupervised feature selection. In: IJCAI (2013)
25. Tang, J.: Feature selection with linked data in social media. In: SDM (2012)
26. Tang, J., Li, H.: Unsupervised feature selection for linked social media data. In: KDD (2012)
27. Trigeorgis, G., Bousmalis, K., Zafeiriou, S.P., Schuller, B.W.: A deep semi-NMF model for learning hidden representations. In: ICML (2014)
28. Wang, S., Liu, H.: Embedded unsupervised feature selection. In: AAAI (2015)
29. Wang, S., Tang, J., Wang, Y., Liu, H.: Exploring implicit hierarchical structures for recommender systems. In: IJCAI (2015)
30. Wang, S., Wang, Y., Tang, J., Aggarwal, C., Ranganath, S., Liu, H.: Exploiting hierarchical structures for unsupervised feature selection. In: SDM (2017)
31. Wang, X., Tang, L., Gao, H., Liu, H.: Discovering overlapping groups in social media. In: ICDM (2010)
32. Wang, X., Tang, L., Liu, H., Wang, L.: Learning with multi-resolution overlapping communities. Knowl. Inf. Syst. **36**, 517–535 (2013). https://doi.org/10.1007/s10115-012-0555-0
33. Yang, Y., Shen, H., Ma, Z., Huang, Z., Zhou, X.: $\ell_{2,1}$-norm regularized discriminative feature selection for unsupervised learning. In: IJCAI (2011)
34. Ye, F., Chen, C., Zheng, Z.: Deep autoencoder-like nonnegative matrix factorization for community detection. In: CIKM (2018)
35. Yu, L., Liu, H.: Feature selection for high-dimensional data: a fast correlation-based filter solution. In: ICML (2003)
36. Zhao, Z., Liu, H.: Spectral feature selection for supervised and unsupervised learning. In: ICML (2007)

Aspect Category Sentiment Analysis with Self-Attention Fusion Networks

Zelin Huang[1], Hui Zhao[1,2(✉)], Feng Peng[1], Qinhui Chen[1], and Gang Zhao[3]

[1] Software Engineering Institute, East China Normal University, Shanghai, China
alan_huang96@163.com, hzhao@sei.ecnu.edu.cn, leodpen@gmail.com,
qinhui_chen97@163.com
[2] Shanghai Key Laboratory of Trustworthy Computing, Shanghai, China
[3] Microsoft, Beijing, China
gang.zhao@microsoft.com

Abstract. Aspect category sentiment analysis (ACSA) is a subtask of aspect based sentiment analysis (ABSA). It aims to identify sentiment polarities of predefined aspect categories in a sentence. ACSA has received significant attention in recent years for the vast amount of online reviews toward the target. Existing methods mainly make use of the emerging architecture like LSTM, CNN and the attention mechanism to focus on the informative sentence spans towards the aspect category. However, they do not pay much attention to the fusion of the aspect category and the corresponding sentence, which is important for the ACSA task. In this paper, we focus on the deep fusion of the aspect category and the corresponding sentence to improve the performance of sentiment classification. A novel model, named Self-Attention Fusion Networks (SAFN) is proposed. First, the multi-head self-attention mechanism is utilized to obtain the sentence and the aspect category attention feature representation separately. Then, the multi-head attention mechanism is used again to fuse these two attention feature representations deeply. Finally, a convolutional layer is applied to extract informative features. We conduct experiments on a dataset in Chinese which is collected from an online automotive product forum, and a public dataset in English, Laptop-2015 from SemEval 2015 Task 12. The experimental results demonstrate that our model achieves higher effectiveness and efficiency with substantial improvement.

Keywords: Aspect category sentiment analysis · Self-attention fusion networks · Multi-head attention mechanism

1 Introduction

With the development of Internet technology, the manner of interaction between customers and retailers/producers has been dramatically changed. Nowadays, more and more firms tend to collect customer's feedback, in terms of reviews, online forum discussions, etc. to improve customer experience, product design,

© Springer Nature Switzerland AG 2020
Y. Nah et al. (Eds.): DASFAA 2020, LNCS 12114, pp. 154–168, 2020.
https://doi.org/10.1007/978-3-030-59419-0_10

and more. However, one main challenge we are facing is how to perform the distillation from the numerously overloaded information. Sentiment analysis is a crucial stage to address the above issue. Sentiment analysis can provide much valuable information for firms and customers. Some firms are interested not only in the overall sentiment on a given product, but also in the sentiment on fine-grained aspects of a product. Aspect category sentiment analysis (ACSA) is a research task catering to these needs. It is a subtask of aspect based sentiment analysis (ABSA).

In rational scenarios, customers would not merely express their opinions on one aspect category for a given product in one comment. Customers may review the product on design, price and customer service, and have different attitudes subject to these aspect categories. For example, there is one comment from an online automotive product forum, '这车价位上还可以，主要在于装配及调校 还需要提高，开起来到处异响真不省心。' (The price of this car is nice, however, the assembly and adjustment need to be improved. In addition, there is too much strange noise everywhere when I am driving, which bothers me a lot.). The user mentions three aspect categories, '价格' (price), '质量' (quality) and '异响' (strange noise), and expresses a positive sentiment over '价格' and negative sentiments over '质量' and '异响'. Another example comes from SemEval 2015, 'Battery could be better but it has robust processor and plenty of RAM so that is a trade off I suppose'. The sentiment polarity for 'battery#general' is negative while for 'CPU#operation_performance' and 'memory#design_features' is positive. These reviews lead to two tasks to be solved: aspect category detection and aspect category sentiment polarity detection. In this paper, we focus on the latter.

Some existing methods have been proposed to solve the ACSA task. Wang et al. [1] concatenate the aspect category embedding and its sentence embedding as inputs. The attention mechanism [2] is employed to focus on the text spans related to the given aspect category. Ma et al. [3] focus on modeling aspects separately, especially with the aid of contexts and propose interactive attention networks (IAN). Tay et al. [4] challenge the necessity of the naive concatenation of aspect and words at both the LSTM layer and attention layer and propose Aspect Fusion LSTM (AF-LSTM). Xue et al. [5] notice that the convolutional layers with multiple filters can efficiently extract features and propose the Gated Convolutional network with Aspect Embedding (GCAE).

All the above methods mainly take advantage of the emerging architecture like LSTM, CNN, and the attention mechanism to focus on the informative sentence spans towards the aspect category. However, they do not pay much attention to the deep fusion of the aspect category and the corresponding sentence, which is important for the ACSA task. For example, 'But the touchpad is soooooooooo big that whenever I am typing, my palm touches some part of it and then the cursor goes somewhere else and it's really annoying', the aspect categories of this sentence are 'mouse#design_features' and 'mouse#usability'. These two aspect categories do not occur explicitly in the sentence. We argue that fusing information on the aspect category and the corresponding sentence can enhance the performance of the ACSA task.

In this paper, we emphasize the deep fusion of the aspect category and the corresponding sentence to improve the performance of sentiment classification. We propose a novel model, named Self-Attention Fusion Networks (SAFN). The core idea is to utilize the multi-head attention mechanism to fuse information on the aspect category and the corresponding sentence deeply. Specifically, the multi-head self-attention mechanism is utilized to obtain the sentence and the aspect category attention feature representation separately. Then, we apply the multi-head attention mechanism to fuse the sentence and the aspect category attention feature representation deeply. Finally, we use a convolutional layer to extract informative features. For the convenience of narration, we use 'category' to briefly express 'aspect category'.

Experiments are conducted on a dataset in Chinese, which is collected from an online automotive product forum and a public dataset in English, Laptop-2015 from SemEval 2015 Task 12 to evaluate our proposed model. SAFN outperforms ATAE-LSTM [1] in Accuracy by 2.2%, in Macro-Averaged F1 by 3.16% on the dataset in Chinese, and in Accuracy by 0.69%, in Macro-Averaged F1 by 3.28% on Laptop-2015. In addition, SAFN has the shortest training time compared with the baseline models. The experimental results prove that our model achieves higher effectiveness and efficiency with substantial improvement.

The rest of this paper is organized as follows: Sect. 2 discusses related research. Section 3 elaborates on the details of our proposed model. Section 4 describes the experimental details and Sect. 5 concludes our work.

2   Related Work

Aspect category sentiment analysis (ACSA) is a subtask of aspect based sentiment analysis (ABSA), which is a key task of sentiment analysis. In recent years, thanks to the increasing progress of the deep neural network research and the continuous increase of online review data, many novel methods have been proposed to tackle this task.

Some methods are developed based on Recurrent Neural Network (RNN) architecture. Wang et al. [1] use the Attention-based Long Short-Term Memory Network (AT-LSTM) and incorporate the aspect category embedding with each word embedding as inputs, which is called aspect embedding (AE) to generate aspect-specific text representations. Ruder et al. [6] propose a hierarchical bidirectional LSTM (H-LSTM) to model the interdependencies of sentences in a review. Ma et al. [3] argue that existing methods all ignore modeling targets separately, especially with the aid of contexts and propose interactive attention networks (IAN). Tay et al. [4] challenge the necessity of the naive concatenation of aspect and words at both the LSTM layer and attention layer, and propose Aspect Fusion LSTM (AF-LSTM) to model word-aspect relationships. Cheng et al. [7] propose a novel model, namely hierarchical attention (HEAT) network, which jointly models the aspect and the aspect-specific sentiment information from the context through a hierarchical attention module. Zhu et al. [8] argue that the predefined aspect information can be used to supervise the learning procedure besides the sentiment polarity and propose a novel aspect aware learning (AAL) framework.

Some researchers also use Convolutional Neural Network (CNN) to solve the ACSA problem. Xue and Li [5] use two separate convolutional layers and novel gate units to build a CNN-based model, which is called Gated Convolutional network with Aspect Embedding (GCAE).

Some other researchers propose joint models to tackle this task. They jointly model the detection of aspect categories and the classification of their sentiment polarity. Schmitt et al. [9] propose two joint models: End-to-end LSTM and End-to-end CNN, which produce all the aspect categories and their corresponding sentiment polarities. Li et al. [10] propose a novel joint model, which contains a contextualized aspect embedding layer and a shared sentiment prediction layer.

ACSA is still an underexploited task [8]. Many proposed methods in other subtasks of sentiment analysis are also valuable for the ACSA task. Such methods include Recurrent Attention on Memory (RAM) [11], BILSTM-ATT-G [12] and Transformation Networks (TNet) [13] for ATSA task, and a siamese bidirectional LSTM with context-aware attention model [14] for topic-based sentiment analysis.

Overall, current studies do not pay much attention to deep fusion of the aspect category and the corresponding sentence to improve the performance of the ACSA task, which is important for the task.

3   Proposed Self-Attention Fusion Networks

In this section, we elaborate on our proposed model, SAFN (Self-Attention Fusion Networks). We first formulate the problem. There are K predefined aspect categories $\mathbf{C} = \{\mathbf{C}_1, \mathbf{C}_2, ..., \mathbf{C}_K\}$ in the dataset. Given a category-sentence pair $(\mathbf{w}^c, \mathbf{w}^s)$, where $\mathbf{w}^c = \{w_1^c, w_2^c, ..., w_m^c\} \in \mathbf{C}$ is the category of $\mathbf{w}^s = \{w_1^s, w_2^s, ..., w_n^s\}$, and the corresponding word embeddings $\mathbf{x}^c = \{x_1^c, x_2^c, ..., x_m^c\}$ and $\mathbf{x}^s = \{x_1^s, x_2^s, ..., x_n^s\}$, the task is to predict the sentiment polarity $y \in \{P, N, O\}$ of the sentence $\mathbf{w}^s$ over the category $\mathbf{w}^c$, where P, N and O denote 'positive', 'negative' and 'neutral' sentiments respectively.

The architecture of the proposed SAFN is shown in Fig. 1. SAFN contains four modules: a category representation encoder, a sentence representation encoder, a sentence-category fusion layer and a convolutional feature extractor. Firstly, a multi-head self-attention component and a feed forward layer transform the category's word embedding and position embedding to the category attention feature representation. Secondly, another multi-head self-attention component transforms the sentence's word embedding and position embedding to the sentence attention feature representation. Thirdly, the third multi-head attention component takes the category attention feature representation as the inputs of key and value, and the sentence attention feature representation as the input of query. And then, a feed forward layer uses the output of the multi-head attention component to obtain the sentence-category fusion feature representation.

Fourthly, a convolutional and max-pooling layer extracts informative features for classification from the sentence-category fusion feature representation. Finally, a fully-connected layer with *softmax* function uses the output of the convolutional and max-pooling layer to predict the sentiment polarity.

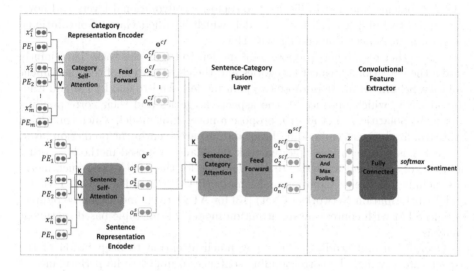

Fig. 1. The architecture of SAFN

3.1   Category Representation Encoder

Self-attention is the most important component in google's transformer model, which is a special attention mechanism to obtain the sequence's representation. It has been successfully applied to many tasks, such as machine translation [15], reading comprehension [16], and multi-turn dialogue generation [17]. One important advantage of self-attention is that it has the ability to well capture the long distant dependency information [15]. Another critical advantage of it is that it allows for significantly more parallelization and requires significantly less time to train than the RNN architecture [15]. That's why we use this attention mechanism in our model.

In SAFN, we adopt the multi-head attention mechanism [15]. The input of the self-attention mechanism includes a matrix of n query vectors $Q \in \mathbb{R}^{n \times d}$, key vectors $K \in \mathbb{R}^{n \times d}$ and value vectors $V \in \mathbb{R}^{n \times d}$, where d is the dimension of the hidden units in our networks and n is the length of the sentence. These three vectors are used to calculated the scaled dot-product attention, which is described formally as the following equation:

$$Attention(Q, K, V) = softmax(\frac{QK^T}{\sqrt{d}})V \qquad (1)$$

To enhance the efficiency and effectiveness, we use H parallel heads. We linearly project the Q, K and V matrices H times to d_H dimension, where $d_H = d/H$. Each of these projected queries, keys and values are used to calculated the scaled dot-product attention in parallel to obtain H heads. And then, the H heads are concatenated together and linearly projected to obtain o.

$$o = Concat(head_1, ..., head_H)W^o$$
$$head_i = Attention(QW_i^Q, KW_i^K, VW_i^V)$$
(2)

where parameter weight matrices $W_i^Q \in \mathbb{R}^{d \times d_H}$, $W_i^K \in \mathbb{R}^{d \times d_H}$, $W_i^V \in \mathbb{R}^{d \times d_H}$ and $W^o \in \mathbb{R}^{d \times d}$ are the projections.

It is widely accepted that self-attention mechanism cannot utilize position information [15]. Thus, it is important to encode each position information. In our model, we parameterize position embedding $PE_i \in \mathbb{R}^d$, $i = 1, ..., N$. Given the category's word embedding $\mathbf{x}^c = \{x_1^c, x_2^c, ..., x_m^c\}$ as the input, the position embedding is simply added to the category's word embedding $\mathbf{x}^c$ to obtain the category position union embedding $\mathbf{x}^{cp} = \{x_1^{cp}, x_2^{cp}, ..., x_m^{cp}\}$. $\mathbf{x}^{cp}$ is fed as queries, keys, and values matrices to the multi-head attention component by using different linear projections. Then the multi-head attention $\mathbf{o}^c = \{o_1^c, o_2^c, ..., o_m^c\}$ is computed as Eq. 1 and Eq. 2. Finally, a feed forward layer [15] is employed to generate the category attention feature representation $\mathbf{o}^{cf} = \{o_1^{cf}, o_2^{cf}, ..., o_m^{cf}\}$.

3.2 Sentence Representation Encoder

In this module, the self-attention mechanism is applied to obtain the sentence attention feature representation. Given the sentence's word embedding $\mathbf{x}^s = \{x_1^s, x_2^s, ..., x_n^s\}$ as the input, the position embedding is added to $\mathbf{x}^s$ to generate the sentence position union embedding $\mathbf{x}^{sp} = \{x_1^{sp}, x_2^{sp}, ..., x_n^{sp}\}$ as mentioned in the Sect. 3.1. Then $\mathbf{x}^{sp}$ is fed as queries, keys, and values matrices to another multi-head attention component by using different linear projections to obtain the sentence attention feature representation $\mathbf{o}^s = \{o_1^s, o_2^s, ..., o_n^s\}$ as Eq. 1 and Eq. 2 in the Sect. 3.1.

3.3 Sentence-Category Fusion Layer

In this module, the information of the sentence and category is fused deeply for classification. The category attention feature representation $\mathbf{o}^{cf}$ is fed as keys and values matrices, and the sentence attention feature representation $\mathbf{o}^s$ is fed as queries matrices to the third multi-head attention component. The output is denoted as $\mathbf{o}^{sc}$. Then a new feed forward layer [15] is used to obtain the sentence-category fusion feature representation $\mathbf{o}^{scf} = \{o_1^{scf}, o_2^{scf}, ..., o_n^{scf}\}$.

3.4 Convolutional Feature Extractor

After fusing the sentence and category attention feature representation, we have obtained the sentence-category fusion feature representation $\mathbf{o}^{scf}$. A convolutional and max-pooling layer is employed to extract informative features for classification in this module. $\mathbf{o}^{scf}$ is fed to a convolutional layer to obtain the feature map $\mathbf{c} \in \mathbb{R}^{n-e+1}$ as follows:

$$c_i = ReLU(\mathbf{o}^{scf}_{i:i+e-1} * \mathbf{w}_{conv} + b_{conv}) \tag{3}$$

where $\mathbf{o}^{scf}_{i:i+e-1} \in \mathbb{R}^{e \times d}$ is the concatenated vector of $o^{scf}_i, ..., o^{scf}_{i+e-1}$ and e is the kernel size. $\mathbf{w}_{conv} \in \mathbb{R}^{e \times d}$ and $b_{conv} \in \mathbb{R}$ are trainable weights. $*$ denotes convolution operation. The max-pooling is employed to capture the most informative features. Then we use n_k convolutional kernels to generate the final feature representation $z \in \mathbb{R}^{n_k}$:

$$z = [max(\mathbf{c}_1), ..., max(\mathbf{c}_{n_k})]^\top \tag{4}$$

Finally, z is fed to a fully connected layer with *softmax* to obtain the sentiment polarity probability $\hat{y}$:

$$\hat{y} = Softmax(W_z z + b_z) \tag{5}$$

where $W_z \in \mathbb{R}^{3 \times n_k}$ and $b_z \in \mathbb{R}^3$ are trainable parameters.

The model is trained to minimize the cross-entropy loss between the one-hot ground-truth y and the predicted probability $\hat{y}$ for all training examples:

$$\mathcal{L} = -\sum_i y_i^\top log\hat{y}_i \tag{6}$$

where i is the index of a training example.

4   Experiments

4.1   Datasets

We evaluate the proposed model SAFN on two datasets. One is a dataset in Chinese, which is collected from an online automotive product forum. Another is a public dataset in English, Laptop-2015 from SemEval 2015 Task 12. The statistics of the two datasets are illustrated in Table 1 and Table 2. In laptop-2015, the number of categories is more than 9 as shown in Table 2. The format of categories in Laptop-2015 is like 'laptop#general', 'battery#operation_performance' and 'os#general'. We merge the categories for statistics as the reference [8], while we use the complete categories in our experiments.

4.2   Experimental Setup

It is necessary to do some preprocess work before evaluating our model. For the dataset in Chinese, we split the Chinese sentence by word. For example, given a Chinese sentence '我觉得这家门店的销售人员很不错', we split it into '我', '觉', '得', '这', '家', '门', '店', '的', '销', '售', '人', '员', '很', '不' and '错'. Each word corresponds to w_i^s or w_i^c denoted in the Sect. 3. For Laptop-2015, the special symbols '#' and '_' in categories are replaced by a blank space. Thus, the format of categories in our experiments is like 'laptop general', 'battery operation performance' and 'os general'. All tokens in Laptop-2015 is lowercased. For both two datasets, the sentences are zero-padded to the length of the longest sentence.

In our experiments, we use 300-dimension GloVe vectors which are pretrained on unlabeled data of 6 billion tokens [18] to initialize the word embedings for Laptop-2015. For the dataset in Chinese, the word embedding vectors are initialized with 256-dimension GloVe vectors which are pretrained on our unlabeled data.

Table 1. Statistics of the dataset in Chinese.

Category	Positive		Negative		Neutral	
	Train	Test	Train	Test	Train	Test
版本 (Version)	18	5	35	3	260	31
保险 (Insurance)	17	2	28	7	213	38
车身颜色 (The color of car body)	175	24	59	10	376	39
贷款 (Loan)	100	11	112	18	623	73
动力系统 (Dynamic system)	93	14	143	23	278	41
价格 (Price)	294	40	287	30	1112	134
配置功能/性能 (Configure function / performance)	163	24	151	14	326	34
提车周期 (Pickup cycle)	44	3	121	25	125	18
外观细节特征 (Appearance detail)	143	10	119	12	178	18
外观整体风格 (Overall appearance style)	408	53	135	16	132	19
网点覆盖 (Network coverage)	32	1	95	16	162	19
维修保养政策 (Maintenance and repair policy)	206	27	88	20	423	56
销售人员态度及专业性 (Attitude and specialization of salesman)	40	6	164	31	61	8
异响 (Strange noise)	31	7	141	17	87	9
优惠 (Discount)	392	32	212	32	897	107
油耗 (Oil consumption)	239	35	330	29	758	86
丰富的配置 (Rich configuration)	157	16	103	13	249	30
赠品 (Gift)	56	8	31	5	172	24
整体动力评价 (Overall dynamic evaluation)	127	21	71	6	88	13
质量 (Quality)	272	37	540	82	297	32
Total	3007	376	2965	409	6817	829

To alleviate overfitting, the dropout is used on the category and sentence position union embedding (p_{posi}), the output of the multi-head attention component (p_{att}) and the output of the convolutional and max-pooling layer (p_{conv}).

Table 2. Statistics of Laptop-2015.

Category	Positive		Negative		Neutral	
	Train	Test	Train	Test	Train	Test
connectivity	17	6	15	15	0	3
design_features	150	71	67	39	33	16
general	401	197	168	79	10	15
miscellaneous	71	43	35	21	12	5
operation_performance	164	88	114	77	9	6
portability	36	5	8	2	0	1
price	41	38	25	5	22	17
quality	115	61	289	65	10	5
usability	108	32	44	26	10	11
total	1103	541	765	329	106	79

We set $p_{posi} = 0.2$, $p_{att} = 0.2$ and $p_{conv} = 0.1$ for both two datasets. The maximal epoch is set to 40. We apply Adam [19] as the optimizer using the decay rates in the original paper. All the models are run on a NVIDIA GTX 1080Ti GPU card with Tensorflow. Other hyper-parameters of SAFN for the dataset in Chinese are listed in Table 3. Other hyper-parameters of SAFN for Laptop-2015 are nearly the same as the dataset in Chinese, except that d is set to 300 and H is set to 6.

We adopt Accuracy and Macro-Averaged F1 to evaluate the models. The Macro-Averaged F1 is adopted because of the unbalanced classes of the dataset. In addition to these two metrics, we also adopt the training time of per epoch to evaluate the speed of training.

Table 3. The hyper-parameters of SAFN.

Hyper-params	Learning rate	Batch size	d	H	e	n_k
Settings	0.0001	64	256	8	2,3,4	50

*There are three kernel sizes and there are 50 kernels for each kernel size.

4.3 Comparison Methods

We compare our method against the following baseline methods:

- **ATAE-LSTM** [1] ATAE-LSTM is a LSTM model with the attention mechanism and it incorporates the aspect category embedding with each word embedding as the input.
- **SI-BILSTM-ATT** [14] This is a siamese bidirectional LSTM with context-aware attention model. The authors do not name their model and we name

it SI-BILSTM-ATT. It is proposed for topic-based sentiment analysis. The category in our datasets is considered as the topic in SI-BILSTM-ATT.

4.4 Experimental Results

In experimental results, Accuracy, Macro-Averaged F1 and training time of per epoch of all competing models over the test datasets are reported. Every experiment is repeated three times. The mean results are reported in Table 4 and Table 5. As Table 4 and Table 5 show, SAFN achieves the best performance in all three metrics on both two datasets, which proves the effectiveness and efficiency of our model.

Table 4. Experimental results on the dataset in Chinese.

Models	Accuracy (%)	Macro-F1 (%)	Training time/epoch (s)
ATAE-LSTM	67.34	62.55	41
SI-BILSTM-ATT	67.86	63.51	90
SAFN (without CFE.*)	68.36	64.00	–
SAFN	**69.54**	**65.71**	**20**

*CFE. refers to Convolutional Feature Extractor.
*Best Accuracy, Macro-F1 and training time/epoch is marked with bold font.

Table 5. Experimental results on Laptop-2015.

Models	Accuracy (%)	Macro-F1 (%)	Training time/epoch (s)
ATAE-LSTM	74.58	55.64	6
SI-BILSTM-ATT	74.08	55.68	10
SAFN	**75.27**	**58.92**	**4**

* Best Accuracy, Macro-F1 and training time/epoch is marked with bold font.
* Some other experimental results on Laptop-2015 can be seen in the reference [8].

ATAE-LSTM has the worst performance of all reported models in Accuracy and Macro-Averaged F1 on the dataset in Chinese, while it has nearly the same performance as SI-BILSTM-ATT on Laptop-2015. It also spends longer time to train than SAFN because of the inherent properties of RNN architecture. This model uses LSTM to obtain the features for the classification. It simply concatenates the given aspect category embedding with each word embedding as the input and applies the attention mechanism to enforce the model to attend to the important part of a sentence.

SI-BILSTM-ATT has a little improvement compared with ATAE-LSTM in Accuracy and Macro-Averaged F1 on the dataset in Chinese, while it has nearly the same performance as ATAE-LSTM on Laptop-2015. The training time of SI-BILSTM-ATT is the worst of all because of the inherent properties of RNN architecture and the complexity of the model. This model applies a siamese bidirectional LSTM to obtain the sentence and category features. After the siamese bidirectional LSTM obtaining the sentence and category feature representation, this model simply concatenates these two feature representations and applies the context-aware attention mechanism.

SAFN increases Accuracy by 1.68%, Macro-Averaged F1 by 2.2% compared with SI-BILSTM-ATT on the dataset in Chinese, and Accuracy by 0.69%, Macro-Averaged F1 by 3.28% compared with ATAE-LSTM on Laptop-2015. In addition, the training time is significantly shorter than the others because of the better parallelization. SAFN applies the multi-head self-attention mechanism to obtain the sentence and category attention feature representation separately. These two attention feature representations are deeply fused by the multi-head attention mechanism. The fusion makes the category and the corresponding sentence interact with each other, which benefits the prediction of the sentiment a lot. We also employ a convolutional and max-pooling layer to extract the informative features for classification.

To prove the effectiveness of the convolutional feature extractor, we run SAFN without this component on the dataset in Chinese. This component helps SAFN extract informative features from the results which are computed by previous modules. The experimental results indicate that the convolutional feature extractor component improves the model. In addition, the performance of SAFN without the convolutional feature extractor component is also better than the performance of ATAE-LSTM and SI-BILSTM-ATT, which indicates the effectiveness of the other components except for the convolutional feature extractor.

4.5   Impact of the Number of Layer

In this section, we examine the impact of the number of layer L. The category multi-head self-attention component, the sentence multi-head self-attention component and the sentence-category multi-head attention component all involve multiple layers. In SAFN, the number of these layers is same. The model is changed by increasing L from 1 to 6 stepped by 1. The experimental results in terms of Accuracy and Macro-Averaged F1 on two datasets by changing the layer number are shown in Fig. 2 and Fig. 3 respectively. It can be observed that the model on two datasets both obtain the best results when $L = 2$. However, when L is set 3 or 3 above, the performance is down and the model cannot be trained totally. We investigate the gradient during the training and find that the gradients become 'NAN'. This implies the gradient exploding.

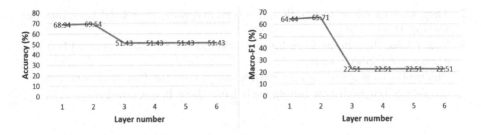

Fig. 2. Impact of the number of layer on the dataset in Chinese

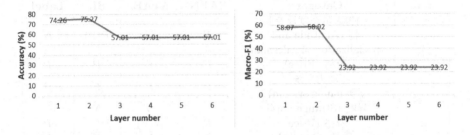

Fig. 3. Impact of the number of layer on Laptop-2015

4.6   Case Study

In this section, we show some sample cases from the two datasets. The prediction results of ATAE-LSTM, SI-BILSTM-ATT, and our model are shown in Table 6. The notations P, N and O in the table represent positive, negative and neutral respectively. The sentences in the table are listed below:

1. 从发动机角度来说确实还挺不错的, 售价也公道, 但营销真的很不受人欢迎啊。 (For the engine, it's really good. The price is fair, but the sales service is not satisfying.)
2. 这车价位上还可以, 主要在于装配及调校还需要提高, 开起来到处异响 真不省心。 (The price of this car is nice, however, the assembly and adjustment need to be improved. Also, there is too much strange noise everywhere when I am driving, which bothers me a lot.)
3. 这车除了动力系统差一点, 性价比超高。(Except for the poor power system, this car has high cost performance.)
4. 准确的说, 全系车型现在的质量都挺到位, 当然小问题肯定会有, 但是大事没有。 (To be exact, the quality of all the Series is satisfying now. Of course, there will be some small problems, but no big ones.)
5. The only reason I didn't give it a 5 star is, I feel I have to reach so far to type on the key board.
6. But the touchpad is soooooooooo big that whenever I am typing, my palm touches some part of it and then the cursor goes somewhere else and it's really annoying.

7. They were excellent to handle our return and post our refund back into our account the next day.
8. Now I have dual booted my laptop to Windows 8.1 and Ubuntu 14.04 and it's not got too slow either.
9. The keyboard is a little wonky with having to use the Function button to get the F-keys but with how infrequently i use those it is a non-issue.

Table 6. Comparision results of some cases.

Sen.	Category	SAFN	ATAE.	SI.	Label
1.	销售人员态度及专业性 (Attitude and specialization of salesman)	N	N	$P(F)$	N
	价格 (Price)	P	P	P	P
	动力系统 (Dynamic system)	P	P	P	P
2.	异响 (Strange noise)	N	N	N	N
	价格 (Price)	P	$N(F)$	$N(F)$	P
	质量 (Quality)	N	N	N	N
3.	价格 (Price)	$N(F)$	$N(F)$	$N(F)$	P
	动力系统 (Dynamic system)	N	N	N	N
4.	质量 (Quality)	P	$N(F)$	$N(F)$	P
5.	laptop#general	P	P	$N(F)$	P
	keyboard#usability	N	N	N	N
6.	mouse#design_features	N	$P(F)$	$P(F)$	N
	mouse#usability	N	$P(F)$	$P(F)$	N
7.	support#quality	P	$N(F)$	$N(F)$	P
8.	laptop#operation_performance	P	$N(F)$	$N(F)$	P
9.	keyboard#design_features	N	N	N	N
	keyboard#usability	$N(F)$	$N(F)$	$N(F)$	O

* **Sen.**, **ATAE.** and **SI.** refer to **Sentence**, **ATAE-LSTM** and **SI-BILSTM-ATT**.
* (F) indicates the wrong prediction.

As Table 6 shows, SAFN can predict the sentiment polarity over the category more accurately than ATAE-LSTM and SI-BILSTM-ATT. For example, for the category '价格' in the 2nd sentence. Only our model can distinguish the opposite positive sentiment for '价格', while the sentiments for the other two categories in this sentence is both negative. For the category '质量 ' in the 4th sentence, our model focuses on the meaning of the whole sentence and predict the positive sentiment by deep fusion of the category and the corresponding sentence. However, the other models may be interfered by the sentence spans, '小问题肯定会有' and give a wrong prediction. For the categories 'mouse#design_features' and 'mouse#usability' in the $6th$ sentence, these two categories do not occur explicitly in the sentence. SAFN predicts the sentiments of the $6th$ sentence over these two categories correctly, while the other two models do not.

We also find that all these models cannot give correct predictions for the category '价格' in the $3th$ sentence and for the category 'keyboard#usability' in the last sentence. For the category '价格' in the $3th$ sentence, all these models may not understand the Chinese phrase '性价比', which means the performance of a product compared with its price. The last sentence has a twist in the second half of the sentence and the models may not understand the word 'non-issue'. These two elements may cause the wrong prediction.

5 Conclusion

In this paper, we conduct research on aspect category sentiment analysis on online reviews. We contribute a new aspect category sentiment analysis model, namely SAFN, to deeply fuse the information on the aspect category and the corresponding sentence utilizing the multi-head attention mechanism. We run experiments on a dataset in Chinese constructed from an online automotive product forum and a public dataset in English, Laptop-2015 from SemEval 2015 task 12. The experimental results confirm the effectiveness and efficiency of our model. SAFN can be applied in the scenarios when firms are interested in the sentiment on fine-grained aspects of a given product. In future work, we would like to solve the gradient exploding problem when the number of layers is more than two.

References

1. Wang, Y., Huang, M., Zhu, X., Zhao, L.: Attention-based LSTM for spect-level sentiment classification. In: Proceedings of EMNLP, pp. 606–615 (2016)
2. Bahdanau, D., Cho, K., Bengio, Y.: Neural machine translation by jointly learning to align and translate. In: ICLR, CoRR abs-1409.0473 (2014)
3. Ma, D., Li, S., Zhang, X., Wang, H.: Interactive attention networks for aspect-Level sentiment classification. In: International Joint Conference on Artificial Intelligence, pp. 4068–4074 (2017)
4. Tay, Y., Tuan, L.A., Hui, S.C.: Learning to attend via word-aspect associative fusion for aspect-based sentiment analysis. In: The Thirty-Second AAAI Conference on Artificial Intelligence, pp. 5956–5963 (2018)
5. Xue, W., Li, T.: Aspect based sentiment analysis with gated convolutional networks. In: Proceedings of the 56th Annual Meeting of the Association for Computational Linguistics (Long Papers), pp. 2514–2523 (2018)
6. Ruder, S., Ghaffari, P., Breslin, J.G.: A hierarchical model of reviews for aspect-based sentiment analysis. In: Proceedings of the 2016 Conference on Empirical Methods in Natural Language Processing, pp. 999–1005 (2016)
7. Cheng, J., Zhao, S., Zhang, J., King, I., Zhang, X., Wang, H.: Aspect-level sentiment classification with heat (hierarchical attention) network. In: Proceedings of the 2017 ACM on Conference on Information and Knowledge Management, pp. 97–106 (2017)

168   Z. Huang et al.

8. Zhu, P., Chen, Z., Zheng, H., Qian, T.: Aspect aware learning for aspect category sentiment analysis. ACM Trans. Knowl. Discov. Data (TKDD), **13**(6), Article No. 55 (2019)
9. Schmitt, M., Steinheber, S., Schreiber, K., Roth, B.: Joint aspect and polarity classification for aspect-based sentiment analysis with end-to-end neural networks. In: Proceedings of the 2018 Conference on Empirical Methods in Natural Language Processing, pp. 1109–1114 (2018)
10. Li, Y., et al.: A joint model for aspect-category sentiment analysis with contextualized aspect embedding. arXiv preprint arXiv:1908.11017 (2019)
11. Chen, P., Sun, Z., Bing, L., Yang, W.: Recurrent attention network on memory for aspect sentiment analysis. In: Proceedings of the 2017 Conference on Empirical Methods in Natural Language Processing, pp. 452–461 (2017)
12. Liu, J., Zhang, Y.: Attention modeling for targeted sentiment. In: Proceedings of the 15th Conference of the European Chapter of the Association for Computational Linguistics, vol. 2, Short Papers, pp. 572–577 (2017)
13. Li, X., Bing, L., Lam, W., Shi, B.: Transformation networks for target-oriented sentiment classification. In: Proceedings of the 56th Annual Meeting of the Association for Computational Linguistics, pp. 946–956 (2018)
14. Baziotis, C., Pelekis, N., Doulkeridis, C.: Datastories at semEval-2017 task 4: deep LSTM with attention for message-level and topic-based sentiment analysis. In: Proceedings of the 11th International Workshop on Semantic Evaluations (SemEval-2017), pp. 747–754 (2017)
15. Vaswani, A., et al.: Attention is all you need. In: Advances in Neural Information Processing Systems 30: Annual Conference on Neural Information Processing Systems, pp. 6000–6010 (2017)
16. Yu, A.W., et al.: Qanet: combining local convolution with global self-attention for reading comprehension. In: International Conference on Learning Representations (2018)
17. Zhang, H., Lan, Y., Pang, L., Guo, J., Cheng, X.: ReCoSa: detecting the relevant contexts with self-attention for multi-turn dialogue generation. In: Proceedings of the 57th Annual Meeting of the Association for Computational Linguistics, pp. 3721–3730 (2019)
18. Pennington, J., Socher, R., Manning, C.: Glove: global vectors for word representation. In: Proceedings of EMNLP, pp. 1532–1543 (2014)
19. Kingma, D., Ba, J.: Adam: a method for stochastic optimization. In: Proceedings of ICLR (2015)

Query Processing

A Partial Materialization-Based Approach to Scalable Query Answering in OWL 2 DL

Xiaoyu Qin[1], Xiaowang Zhang[1(✉)], Muhammad Qasim Yasin[1], Shujun Wang[1],
Zhiyong Feng[1,2], and Guohui Xiao[3]

[1] College of Intelligence and Computing, Tianjin University, Tianjin 300350, China
`xiaowangzhang@tju.edu.cn`
[2] College of Intelligence and Computing,
Shenzhen Research Institute of Tianjin University, Tianjin University, Tianjin, China
[3] Faculty of Computer Science, Free University of Bozen-Bolzano, Bolzano, Italy

Abstract. This paper focuses on the efficient ontology-mediated querying (OMQ) problem. Compared with query answering in plain databases, which deals with fixed finite database instances, a key challenge in OMQ is to deal with the possibly infinite large set of consequences entailed by the ontology, i.e., the so-called chase. Existing techniques mostly avoid materializing the chase by query rewriting to address this issue, which, however, comes at the cost of query rewriting and query evaluation at runtime, and the possibility of missing optimization opportunity at the data level. Instead, pure materialization technology is adopted in this paper. The query-rewriting is unnecessary at materialization. A query analysis algorithm (QAA) is proposed for ensuring the completeness and soundness of OMQ over partial materialization for rooted queries in $DL\text{-}Lite_{horn}^{\mathcal{N}}$. We also soundly and incompletely expand our method to deal with OWL 2 DL. Finally, we implement our approach as a prototype system SUMA by integrating off-the-shelf efficient SPARQL query engines. The experiments show that SUMA is complete on each test ontology and each test query, which is the same as Pellet and outperforms PAGOdA. In addition, SUMA is highly scalable on large data sets.

Keywords: Ontology reasoning · Query answering · Materialization

1 Introduction

Ontology-mediated querying (OMQ) is a core reasoning task in many ontology query answering applications [2]. Queries against plain databases can only query the records explicitly declared in tables. However, OMQ makes implicit information a query object through ontology. It includes not only queries but also ontology reasoning. Ontology can significantly impact OMQ efficiency. One of the biggest impacts is that the consequences entailed by ontologies (also known as the *universal model* or *chase*) can be infinite [14]. Solving the problem of infinite materialization is a big challenge for OMQ.

© Springer Nature Switzerland AG 2020
Y. Nah et al. (Eds.): DASFAA 2020, LNCS 12114, pp. 171–187, 2020.
https://doi.org/10.1007/978-3-030-59419-0_11

The first way to avoid infinite materialization is to rewrite the query [4,5]. The query rewriting, however, significantly increases the cost of query because rewriting is performed at runtime. Secondly, the materialization-based program adopts a tableau algorithm with a roll-up technique [9] to solve infinite materialization, e.g., Pellet [24]. Pellet is not scalable for large datasets. It can only be applied to small and medium-sized datasets due to the high complexity of the tableau algorithm. PAGOdA [25] is scalable by delegating a large amount of the computational load to a datalog reasoner [19,20]. It uses the hypertableau algorithm [18] only when necessary. PAGOdA is incomplete in terms of infinite materialization (we proved it in Sect. 5). gOWL [13] proposes a partial materialization-based approach that deals with acyclic queries. The materialization algorithm of gOWL has a high time and space complexity due to its poor indexes for the storage and rules. Besides, gOWL cannot handle cyclic queries, and its approximation rules lose most of the semantics of the OWL 2 DL. Thirdly, there is a hybrid approach that computes the canonical model and rewrites the queries [11] or uses a filter mechanism to filter spurious answers [14]. They are limited to lightweight ontology languages (such as logics DL-Lite$_\mathcal{R}$ [1]).

Generally speaking, the users are mainly interested in the first few levels of the anonymous part of the universal models [8]. We develop a query analysis algorithm to ensure the soundness and completeness of query answering over partial materialization. Considering the following query,

Q : select $?Z$ where $\{?Z$ type *Student*. $?Z$ advisor $?Y_0$. $?Y_0$ teaches $?Y_1$.$\}$.

If we execute the query Q on the infinite universal model, as shown in Fig. 1. In this paper, we select only one part from the entire model (the sRDF) for answering the query as ans(sRDF, Q) = ans(RDF, Q).

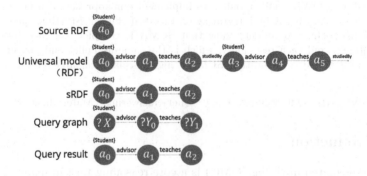

Fig. 1. A running example of partial materialization

The main contributions of our work are stated as below:

– We propose a complete partial materialization approach in $DL\text{-}Lite_{horn}^{\mathcal{N}}$ [1] for a given rooted query [15] with n quantified variables.

- We soundly and incompletely generalize our approach to deal with OWL 2 DL to make our approach unlimited to lightweight ontology languages.
- We implement our approach as a prototype system SUMA. From a system perspective, SUMA allows us to design an off-line modular architecture to integrate off-the-shelf efficient SPARQL [7] query engines. In this way, it makes online queries more efficient.

Finally, we evaluate SUMA on the realistic and benchmark datasets and manually extend the LUBM [6] and UOBM [12] ontologies to test SUMA on the infinite universal model. Experiments show that SUMA is highly efficient, only taking 202 s to materialize LUBM(1000) and 515 s to materialize UOBM(500).

2   Preliminaries

In this section, we briefly introduce the syntax and semantics of $DL\text{-}Lite_{horn}^{\mathcal{N}}$, OWL 2 DL, conjunctive query, and universal model.

$DL\text{-}Lite_{horn}^{\mathcal{N}}$. Let $\mathsf{N_I}$ be the individual set. In this paper, by default, we use a, b, c, d, e (with subscripts) represent individual names, A, B, Z denote concept names, C (with subscripts) are concepts, P, S are role names, and R (with subscripts) are roles. Roles and concepts in $DL\text{-}Lite_{horn}^{\mathcal{N}}$ are defined as follows:

$$R := P \mid P^-, \qquad C := \bot \mid A \mid \geq m\ R.$$

$DL\text{-}Lite_{horn}^{\mathcal{N}}$ presents $\exists R$ as $\geq 1R$ and defines $(P^-)^-$ with P. Let $\mathsf{N_R^-}$ denote the set of roles.

Formally, a KB $\mathcal{K}$ is of the form $(\mathcal{T}, \mathcal{A})$ where $\mathcal{T}$ is a TBox and $\mathcal{A}$ is an ABox. A $DL\text{-}Lite_{horn}^{\mathcal{N}}$ $\mathcal{T}$ is a finite collection, including concept inclusions (CIs) axioms, that are in the form of $C_1 \sqcap \ldots \sqcap C_n \sqsubseteq C$. A $DL\text{-}Lite_{horn}^{\mathcal{N}}$ ABox consists of concept assertions $A(a)$ and role assertions $P(a, b)$. Let $\mathrm{Ind}(\mathcal{A})$ be a set of individual names occurring in $\mathcal{A}$.

The semantics of $DL\text{-}Lite_{horn}^{\mathcal{N}}$ are defined by the interpretation $\mathcal{I} = (\Delta^{\mathcal{I}}, \cdot^{\mathcal{I}})$, where $\Delta^{\mathcal{I}}$ is a non-empty domain. The function is denoted by $\cdot^{\mathcal{I}}$, which can map each A into the set $A^{\mathcal{I}}$, each P into the relation $P^{\mathcal{I}}$ and each a to an element $a^{\mathcal{I}}$. $A^{\mathcal{I}}$ and $P^{\mathcal{I}}$ are subsets of $\Delta^{\mathcal{I}}$ and $\Delta^{\mathcal{I}} \times \Delta^{\mathcal{I}}$, respectively. The $a^{\mathcal{I}}$ is an element of $\Delta^{\mathcal{I}}$. Besides, $DL\text{-}Lite_{horn}^{\mathcal{N}}$ implements the *unique name assumption* (UNA), that is, if v and w are distinct, then $v^{\mathcal{I}}$ is different from $w^{\mathcal{I}}$.

$\cdot^{\mathcal{I}}$ interprets each complex concept or role in the following ways: (1) $\bot^{\mathcal{I}} = \emptyset$; (2) $(S^-)^{\mathcal{I}} = \{(v, w) \mid (w, v) \in S^{\mathcal{I}}\}$; and (3) $(\geq mS)^{\mathcal{I}} = \{w \mid \sharp\{v \mid (w, v) \in S^{\mathcal{I}}\} \geq m\}$. Here $\sharp$ denotes the cardinality.

The $\mathcal{I}$ satisfies a CIs axiom $\mathcal{T}_1$ in the form of $C_1 \sqcap \ldots \sqcap C_n \sqsubseteq C$ if only if $\bigcap_{i=1}^n C_i^{\mathcal{I}} \subseteq C^{\mathcal{I}}$, denoted as $\mathcal{I} \models \mathcal{T}_1$. If $a^{\mathcal{I}} \in A^{\mathcal{I}}$ then $\mathcal{I} \models A(a)$ holds. If $(a^{\mathcal{I}}, b^{\mathcal{I}}) \in P^{\mathcal{I}}$ then $\mathcal{I} \models P(a, b)$ holds. If $\mathcal{I}$ satisfies all TBox and ABox axioms of $\mathcal{K}$ then $\mathcal{I}$ is a model of $\mathcal{K}$.

$\mathcal{SROIQ}$. The underlying logic of OWL 2 DL is $\mathcal{SROIQ}$. A $\mathcal{SROIQ}$ $\mathcal{K}$ consists of RBox $\mathcal{R}$, TBox $\mathcal{T}$ and ABox $\mathcal{A}$.

The RBox is a limited collection of either role inclusion axioms like $R_1 \sqsubseteq R_2$ or $R_1 \circ R_2 \sqsubseteq R_3$, or disjointness axioms in the form of $\mathrm{Dis}(R_1, R_2)$. The inverse role is denoted as $\mathrm{Inv}(R)$ with $\mathrm{Inv}(R) = R^-$, the symmetric role is denoted as $\mathrm{Sym}(R)$ (defined as $\mathrm{Inv}(R) \equiv R$), and the transitive role is denoted as $\mathrm{Trans}(R)$ (defined as $R \circ R \sqsubseteq R$). $\mathrm{Fun}(R)$ represents functional role.

The concept of $\mathcal{SROIQ}$ is defined as: $C := \bot \mid \top \mid A \mid \neg A \mid \{a\} \mid \geq mR.A \mid \exists R.A$. Besides, $A \sqcup B$, $\forall R.A$, and $\leq nR.A$ are used to abbreviate $\neg(\neg A \sqcap \neg B)$, $\neg \exists R.\neg A$, and $\neg \geq (n+1)R.A$, respectively. Enumeration $\{a_1, a_2, \ldots, a_n\}$ is equal to $\{a_1\} \sqcup \{a_2\} \sqcup \ldots \sqcup \{a_n\}$.

A TBox $\mathcal{T}$ is a finite set of general concept inclusions (GCIs). Disjointness axioms and Equivalent concepts are in the form of $\mathrm{Dis}(C_1, C_2)$ and $C_1 \equiv C_2$, respectively. A $\mathcal{SROIQ}$ ABox $\mathcal{A}$ without UNA includes individual equality $a \doteq b$ ($\doteq$ is called sameAs in OWL) and individual inequality $a \neq b$.

Given an interpretation $\mathcal{I}$, we write $\mathcal{I} \models \mathrm{Dis}(C_1, C_2)$ if $C_1^{\mathcal{I}} \cap C_2^{\mathcal{I}} = \emptyset$, $\mathcal{I} \models \mathrm{Dis}(R_1, R_2)$ if $R_1^{\mathcal{I}} \cap R_2^{\mathcal{I}} = \emptyset$. If $\mathcal{I} \models a \doteq b$ then $a^{\mathcal{I}} = b^{\mathcal{I}}$. If $\mathcal{I} \models a \neq b$ then $a^{\mathcal{I}} \neq b^{\mathcal{I}}$. If the role R is functional, then $\mathcal{I} \models (\geq 2R \sqsubseteq \bot)$. Besides, if both $(a, b) \in R^{\mathcal{I}}$, $(a, d) \in R^{\mathcal{I}}$ and $b \neq d \notin \mathcal{A}$, then $b \doteq d$.

Conjunctive Query. We use $\mathsf{N_V}$ to denote a collection of variables, $A(t)$ the *concept* atomic form, and $P(t, t')$ *role* atomic form with $t, t' \in \mathsf{N_I} \cup \mathsf{N_V}$, respectively. A *conjunctive query* (CQ) $q = \exists \boldsymbol{u} \psi(\boldsymbol{u}, \boldsymbol{v})$. It is making up of concept and role atoms. It connects these atoms by conjunction. The vector $\boldsymbol{v}$ consists of free variables. If $|\boldsymbol{v}| = 0$, we call q *Boolean*. The vector $\boldsymbol{u}$ comprises a collection of variables that are quantified. Since disconnected queries can be divided into connected subqueries for processing, this article only considers connected conjunctive queries. If a CQ is connected and not Boolean, it is a rooted CQ.

The notions of *answers* and *certain answers* of CQ are introduced as follows [11]. Let $q(\boldsymbol{v})$ be a CQ with $|\boldsymbol{v}| = k$, and $\mathcal{I}$ be an interpretation. The $\mathsf{N_T}$ is used for indicating the collection of all terms in q, that is $\mathsf{N_T} = \mathsf{N_I} \cup \mathsf{N_V}$. Let π be a *mapping* which maps each term of q to $\Delta^{\mathcal{I}}$ and each constant a to $a^{\mathcal{I}}$, we call $\mathcal{I}$ satisfies q under π if only if for every $A(t) \in q$, $\pi(t) \in A^{\mathcal{I}}$ and for every $P(t, t') \in q$, $(\pi(t), \pi(t')) \in P^{\mathcal{I}}$. The π is called a *match* for CQ in $\mathcal{I}$. The vector $\boldsymbol{a} = a_1 \ldots a_k$ is an *answer* of q, when given a mapping π with $\pi(v_i) = a_i^{\mathcal{I}}$ ($i \leq k$) and $\mathcal{I} \models^{\pi} q$. The $\mathrm{ans}(q(\boldsymbol{v}), \mathcal{K})$ represents the collection of all answers of $q(\boldsymbol{v})$. Let's call $\boldsymbol{a}$ a certain answer when $\boldsymbol{a}$ is a subset of $\mathrm{Ind}(\mathcal{A})$ and each model of $\mathcal{K}$ satisfies $q(\boldsymbol{a})$. The certain answer collection is denoted as $\mathrm{cert}(q(\boldsymbol{v}), \mathcal{K})$.

Universal Model. A role R is called *generating in* $\mathcal{K}$ if there exists $a \in \mathrm{Ind}(\mathcal{A})$ and $R_1, \ldots, R_n = R$ such that the followings hold: (agen) $\mathcal{K} \models \exists R_1(a)$ but $R_1(a, b) \notin \mathcal{A}$, for all $b \in \mathrm{Ind}(\mathcal{A})$ (written $a \leadsto c_{R_1}$); (rgen) for $i < n$, $\mathcal{T} \models \exists R_i^- \sqsubseteq \exists R_{i+1}$ and $R_i^- \neq R_{i+1}$ (written $c_{R_i} \leadsto c_{R_{i+1}}$) [11]. If R is generating in $\mathcal{K}$, then c_R is called an *anonymous* individual. And, the anonymous individual collection is denoted as $\mathsf{N_I^{\mathcal{T}}}$, which is disjoint from $\mathrm{Ind}(\mathcal{A})$.

The last element of $\sigma = a c_{R_1} \cdots c_{R_n}$ is denoted by $\mathrm{tail}(\sigma)$. The σ in $\mathcal{I}_{\mathcal{K}}$ contains limited individuals and $R = R_1, \ldots, R_n$ is generating in $\mathcal{I}_{\mathcal{K}}$.

We compare the definitions of *canonical interpretation* $\mathcal{I}_{\mathcal{K}}$ and *universal model* $\mathcal{U}_{\mathcal{K}}$ given in [11].

canonical interpretation
$\Delta^{\mathcal{I}_{\mathcal{K}}} = \mathrm{Ind}(\mathcal{A}) \cup \{c_R \mid R \in \mathsf{N_R^-}, R \text{ is generating in } \mathcal{K}\};$
$a^{\mathcal{I}_{\mathcal{K}}} = a$, for all $a \in \mathrm{Ind}(\mathcal{A});$
$A^{\mathcal{I}_{\mathcal{K}}} = \{a \in \mathrm{Ind}(\mathcal{A}) \mid \mathcal{K} \models A(a)\} \cup \{c_R \in \Delta^{\mathcal{I}_{\mathcal{K}}} \mid \mathcal{T} \models \exists R^- \sqsubseteq A\};$
$P^{\mathcal{I}_{\mathcal{K}}} = \{(a,b) \in \mathrm{Ind}(\mathcal{A}) \times \mathrm{Ind}(\mathcal{A}) \mid P(a,b) \in \mathcal{A}\} \cup \{(d, c_P) \in \Delta^{\mathcal{I}_{\mathcal{K}}} \times \mathsf{N_I^{\mathcal{T}}} \mid d \rightsquigarrow c_P\}$ $\cup \{(c_{P^-}, d) \in \mathsf{N_I^{\mathcal{T}}} \times \Delta^{\mathcal{I}_{\mathcal{K}}} \mid d \rightsquigarrow c_{P^-}\}.$
universal model
$\Delta^{\mathcal{U}_{\mathcal{K}}} = \{a \cdot c_{R_1} \cdots c_{R_n} \mid a \in \mathrm{Ind}(\mathcal{A}), n \geq 0, a \rightsquigarrow c_{R_1} \rightsquigarrow \ldots \rightsquigarrow c_{R_n}\};$
$a^{\mathcal{U}_{\mathcal{K}}} = a$, for all $a \in \mathrm{Ind}(\mathcal{A});$
$A^{\mathcal{U}_{\mathcal{K}}} = \{\sigma \in \Delta^{\mathcal{U}_{\mathcal{K}}} \mid \mathrm{tail}(\sigma) \in A^{\mathcal{I}_{\mathcal{K}}}\};$
$P^{\mathcal{U}_{\mathcal{K}}} = \{(a,b) \in \mathrm{Ind}(\mathcal{A}) \times \mathrm{Ind}(\mathcal{A}) \mid P(a,b) \in \mathcal{A}\} \cup \{(\sigma, \sigma \cdot c_P) \in \Delta^{\mathcal{U}_{\mathcal{K}}} \times \Delta^{\mathcal{U}_{\mathcal{K}}} \mid \mathrm{tail}(\sigma) \rightsquigarrow c_P\} \cup \{(\sigma \cdot c_{P^-}, \sigma) \in \Delta^{\mathcal{U}_{\mathcal{K}}} \times \Delta^{\mathcal{U}_{\mathcal{K}}} \mid \mathrm{tail}(\sigma) \rightsquigarrow c_{P^-}\}.$

Example 1 and Example 2 are used to explain $\mathcal{I}_{\mathcal{K}}$ and $\mathcal{U}_{\mathcal{K}}$.

Example 1. Let $\mathcal{K}$ consist of $\mathcal{T} = \{B \sqsubseteq \exists S, \exists S^- \sqsubseteq \exists S\}$ and $\mathcal{A} = \{B(d_0)\}$.
    Then $\Delta^{\mathcal{I}_{\mathcal{K}}} = \{d_0, d_1\}$, $B^{\mathcal{I}_{\mathcal{K}}} = \{d_0\}$, and $S^{\mathcal{I}_{\mathcal{K}}} = \{(d_0, d_1), (d_1, d_1)\}$.
    $\Delta^{\mathcal{U}_{\mathcal{K}}} = \{d_0, d_1, d_2, d_3, \ldots\}$, $B^{\mathcal{U}_{\mathcal{K}}} = \{d_0\}$, and $S^{\mathcal{U}_{\mathcal{K}}} = \{(d_0, d_1), (d_1, d_2), \ldots\}$.

Example 2. Let $\mathcal{K}$ consist of $\mathcal{T} = \{\text{Student} \sqsubseteq \exists\text{advisor}, \exists\text{advisor}^- \sqsubseteq \exists\text{teaches}, \exists\text{teaches}^- \sqsubseteq \exists\text{studiedBy}, \exists\text{studiedBy}^- \sqsubseteq \text{Student}\}$, and $\mathcal{A} = \{\text{Student}(a_0)\}$.
    The universal model of this example is shown in Fig. 1.

Theorem 4. [11] For every consistent $DL\text{-}Lite_{horn}^{\mathcal{N}}$ KB $\mathcal{K}$ and every CQ q, we have $\mathrm{cert}(q, \mathcal{K}) = \mathrm{ans}(q, \mathcal{U}_{\mathcal{K}})$.

3 n-step Universal Model

3.1 n-step Universal Model in $DL\text{-}Lite_{horn}^{\mathcal{N}}$

n-step Universal Model. Query answering over $\mathcal{A}$ with $\mathcal{T}$ is equal to query answering over the universal model. However, the universal model can be infinite.
    Considering such axioms, it satisfies three characteristics. Firstly, it belongs to the concept inclusions. Secondly, its head and body contain both existential quantifiers. Thirdly, the roles included in this axiom are inverse to each other. We refer to this kind of axiom as cyclic existential quantifiers axioms (CEQ, for short) in this paper. The simple form of CEQ axioms is $\exists R^- \sqsubseteq \exists R$ or $\exists R^- \sqsubseteq A$, $A \sqsubseteq \exists R$. When ontology contains CEQ axioms, the universal model is infinite [14], as shown in Example 1 and Example 2.
    We propose an n-step universal model to replace the possible infinite universal model. Intuitively, the process of materialization is to extend ABox to a universal model. The process can be thought of as a sequence $U = \{\text{ABox}, \mathcal{U}_{\mathcal{K}}^1, \mathcal{U}_{\mathcal{K}}^2, \ldots, \mathcal{U}_{\mathcal{K}}^n, \ldots\}$. And, $\mathcal{U}_{\mathcal{K}}^i \subseteq \mathcal{U}_{\mathcal{K}}^{i+1}$, $i < |U|$. The (agen) and (rgen) (see Sect. 2)

are the fundamental reasons for expanding $\mathcal{U}_{\mathcal{K}}^i$ to $\mathcal{U}_{\mathcal{K}}^{i+1}$. The element $\mathcal{U}_{\mathcal{K}}^n$ of U is called the n-step universal model in our method. $\mathcal{U}_{\mathcal{K}}^n$ is always finite. The core technology of our approach is to select $\mathcal{U}_{\mathcal{K}}^n$ from the $\mathcal{U}_{\mathcal{K}}$ for query answering.

To formalize $\mathcal{U}_{\mathcal{K}}^n$, we require some preliminary definitions. We label R as n-step generating in $\mathcal{K}$, when a is an individual of $\mathrm{Ind}(\mathcal{A})$, and the $R = R_1 \dots R_n$ satisfies (agen) and (rgen). The N_R^n denotes the set of roles that are $\leq n$-step generating in $\mathcal{K}$. Based on $\Delta^{\mathcal{I}_{\mathcal{K}}^n} = \mathrm{Ind}(\mathcal{A}) \cup \{c_R \mid R \in \mathsf{N}_\mathsf{R}^n\}$ and $A^{\mathcal{I}_{\mathcal{K}}^n} = \{a \in \mathrm{Ind}(\mathcal{A}) \mid \mathcal{K} \models A(a)\} \cup \{c_R \in \Delta^{\mathcal{I}_{\mathcal{K}}^n} \mid \mathcal{T} \models \exists R^- \sqsubseteq A\}$, the n-step universal model ($\mathcal{U}_{\mathcal{K}}^n$) is defined as follows:

- $\Delta^{\mathcal{U}_{\mathcal{K}}^n} = \{a \cdot c_{R_1} \cdots c_{R_l} \mid a \in \mathrm{Ind}(\mathcal{A}), R_l \in \mathsf{N}_\mathsf{R}^n, a \rightsquigarrow c_{R_1} \rightsquigarrow \dots \rightsquigarrow c_{R_l}\}$,
- $a^{\mathcal{U}_{\mathcal{K}}^n} = a$, for all $a \in \mathrm{Ind}(\mathcal{A})$,
- $A^{\mathcal{U}_{\mathcal{K}}^n} = \{\sigma \in \Delta^{\mathcal{U}_{\mathcal{K}}^n} \mid tail(\sigma) \in A^{\mathcal{I}_{\mathcal{K}}^n}\}$,
- $P^{\mathcal{U}_{\mathcal{K}}^n} = \{(a,b) \in \mathrm{Ind}(\mathcal{A}) \times \mathrm{Ind}(\mathcal{A}) \mid P(a,b) \in \mathcal{A}\} \cup \{(\sigma, \sigma \cdot c_P) \in \Delta^{\mathcal{U}_{\mathcal{K}}^n} \times \Delta^{\mathcal{U}_{\mathcal{K}}^n} \mid tail(\sigma) \rightsquigarrow c_P\} \cup \{(\sigma \cdot c_{P-}, \sigma) \in \Delta^{\mathcal{U}_{\mathcal{K}}^n} \times \Delta^{\mathcal{U}_{\mathcal{K}}^n} \mid tail(\sigma) \rightsquigarrow c_{P-}\}$.

Example 3. We use Example 1 to illustrate $\mathcal{U}_{\mathcal{K}}^n$. $\Delta^{\mathcal{U}_{\mathcal{K}}^n} = \{d_0, d_1, \dots, d_n\}, d_0^{\mathcal{U}_{\mathcal{K}}^n} = d_0, B^{\mathcal{U}_{\mathcal{K}}^n} = \{d_0\}, S^{\mathcal{U}_{\mathcal{K}}^n} = \{(d_0, d_1), (d_1, d_2), \dots, (d_{n-1}, d_n)\}$.

Example 4. The 2-step universal model (sRDF) of Example 2 is shown in Fig. 1.

Query Analysis Algorithm. We design a query analysis algorithm (QAA) to ensure that the n-step universal model can always compute the same answer as the universal model. QAA takes a rooted query as input and calculates the number of quantified variables n. Obviously, n can be calculated in $\mathcal{O}(|\mathsf{N}_\mathsf{T}|)$. The number of quantified variables denotes the step of the universal model. If the step size is n, then the n-step universal model, i.e., the $\mathcal{U}_{\mathcal{K}}^n$ in U, can produce the same answers for q as the universal model. This method is proved in Theorem 2. We define a new triple relation on q. σ denotes a path that consists of terms. δ represents a path that includes roles.

Definition 1. *Let $q = \exists u \varphi(u, v)$ be a CQ, $\mathcal{T}$ be a TBox and π be a mapping, we define a triple relation $f_\rho = \cup_{i \geq 0} f_\rho^i \subseteq \mathsf{N}_\mathsf{T} \times \mathsf{N}_\mathsf{T}^* \times \mathsf{N}_\mathsf{R}^*$ with $\rho = \{t \mid t \in \mathsf{N}_\mathsf{T}, \pi(t) \in \mathrm{Ind}(\mathcal{A})\}$, where*

- $f_\rho^0 = \{(t, t, \varepsilon) \mid t \in \rho\}$;
- $f_\rho^{i+1} = f_\rho^i \cup \{(t, \sigma st, \delta SR) \mid (s, \sigma s, \delta S) \in f_\rho^i, R(s,t) \in q, tail(\pi(s)) \rightsquigarrow tail(\pi(t))\} \cup \{(t, \sigma, \delta) \mid (s, \sigma s, \delta R^-) \in f_\rho^i, R(s,t) \in q, tail(\pi(s)) \rightsquigarrow tail(\pi(t))\}$.

A path $\sigma = t_0 \cdot t_1 \cdots t_{n-1} \cdot t_n$ is a *certain path* of q, if (t_0) is mapped to $\mathrm{Ind}(\mathcal{A})$ and all other terms are mapped to $\mathsf{N}_\mathsf{I}^{\mathcal{T}}$. A certain path collection is represented as $\mathrm{CertPath}(q, \pi)$. The *max certain path* is defined as $\mathrm{MaxCertPath}(q, \pi) := \{\sigma \mid |\sigma_i| \leq |\sigma|, \text{ for all } \sigma_i \in \mathrm{CertPath}(q, \pi)\}$. Let π be a mapping. We set the *depth* of q as $\mathrm{dep}(q, \pi) := |\sigma| - 1$, with $\sigma \in \mathrm{MaxCertPath}(q, \pi)$.

The anonymous part of the universal model is a forest-shaped structure, as shown in Example 3 and 4. Thus, the part of q wants to be mapped in an anonymous part must have a forest-shaped structure. If a term t in q wants to be mapped to an anonymous part of $\mathcal{U}_{\mathcal{K}}$, it can only be mapped in this way, $\pi(t) = \pi(t_0) \cdot c_{R_1} \cdots c_{R_n}$ with $(t, \sigma, \delta) \in f_\rho$, where $\sigma = t_0 t_1 \cdots t_n$ is a certain path, and $\delta = R_1 R_2 \cdots R_n$.

Theorem 1. *For every consistent DL-Lite$_{horn}^{\mathcal{N}}$ KB $\mathcal{K}$, every rooted conjunctive query $q = \exists u \varphi(\boldsymbol{u}, \boldsymbol{v})$, and every mapping π, with $\mathrm{dep}(q, \pi) = n$, we have $\mathcal{U}_{\mathcal{K}} \models^\pi q$ if only if $\mathcal{U}_{\mathcal{K}}^n \models^\pi q$.*

Proof. ($\Rightarrow$) For every π with $\mathrm{dep}(q, \pi) = n$, then there exists a max certain path $\sigma = t_0 t_1 \cdots t_{n-1} t_n$ and $R(t_{n-1}, t_n) \in q$. Thus, R is n-step generating, and all other roles are $\leq n$-step generating. By definition of $\mathcal{U}_{\mathcal{K}}^n$, we have that for every $R \in q$, if $(a, b) \in R^{\mathcal{U}_{\mathcal{K}}}$ then $a \in \Delta^{\mathcal{U}_{\mathcal{K}}^n}$ and $b \in \Delta^{\mathcal{U}_{\mathcal{K}}^n}$. Thus, $(a, b) \in R^{\mathcal{U}_{\mathcal{K}}^n}$. Since q is connected, for every $A \in q$, suppose $a \in A^{\mathcal{U}_{\mathcal{K}}}$, then $(a, *) \in R^{\mathcal{U}_{\mathcal{K}}}$ or $(*, a) \in R^{\mathcal{U}_{\mathcal{K}}}$. Thus, $a \in \Delta^{\mathcal{U}_{\mathcal{K}}^n}$ and $a \in A^{\mathcal{U}_{\mathcal{K}}^n}$. In conclusion, $\mathcal{U}_{\mathcal{K}}^n \models q$.

($\Leftarrow$) For every $R \in q$, if $(a, b) \in R^{\mathcal{U}_{\mathcal{K}}^n}$ then $a \in \Delta^{\mathcal{U}_{\mathcal{K}}^n}$ and $b \in \Delta^{\mathcal{U}_{\mathcal{K}}^n}$. Because $\Delta^{\mathcal{U}_{\mathcal{K}}^n}$ is a subset of $\Delta^{\mathcal{U}_{\mathcal{K}}}$, $a \in \Delta^{\mathcal{U}_{\mathcal{K}}}$, $b \in \Delta^{\mathcal{U}_{\mathcal{K}}}$ and $(a, b) \in R^{\mathcal{U}_{\mathcal{K}}}$. For every $A \in q$, if $a \in A^{\mathcal{U}_{\mathcal{K}}^n}$, then $a \in \Delta^{\mathcal{U}_{\mathcal{K}}^n}$. Thus $a \in \Delta^{\mathcal{U}_{\mathcal{K}}}$ and $a \in A^{\mathcal{U}_{\mathcal{K}}}$.

Therefore, $\mathcal{U}_{\mathcal{K}}^n \models q(\boldsymbol{a}, \boldsymbol{b})$ if only if $\mathcal{U}_{\mathcal{K}} \models q(\boldsymbol{a}, \boldsymbol{b})$.

Let f_ρ is a function if for every term t, $f_\rho(t)$ is singleton set or if $(t, \sigma, \delta) \in f(t)$ and $(t, \sigma', \delta') \in f(t)$, then $\delta = \delta'$.

Lemma 1. *For every mapping, if $\mathcal{U}_{\mathcal{K}} \models^\pi q$, then f_ρ is a function and every $\sigma \in \mathrm{MaxCertPath}(q, \pi)$ is finite and $\mathrm{dep}(q, \pi) \leq |\boldsymbol{u}|$.*

Proof. Suppose f_ρ is not a function, then there exists $t \in \mathsf{N}_\mathsf{T}$, with $(t, \sigma_i, \delta_i) \in f_\rho(t)$ and $(t, \sigma_j, \delta_j) \in f_\rho(t)$, where $\delta_i \neq \delta_j$. We labeled δ_i and δ_j as $\delta_i = R_i^0 \cdot R_i^1 \cdots R_i^{n_i}$ and $\delta_j = R_j^0 \cdot R_j^1 \cdots R_j^{n_j}$, respectively. Thus, $\mathcal{U}_{\mathcal{K}} \not\models^\pi q$ due to $c_{R_i^0} \cdots c_{R_i^{n_i}} \neq c_{R_j^0} \cdots c_{R_j^{n_j}}$. This creates a contradiction.

Because N_T is finite, if σ is not finite, then there exists t', with $\sigma = \sigma_1 t' \sigma_2 t' \sigma_3$. Thus, $(t', \sigma_1 t', \delta_1) \in f_\rho(t')$ and $(t', \sigma_1 t' \sigma_2 t', \delta_2) \in f_\rho(t')$. By the definition of f_ρ, we have that δ_1 is a subsequence of δ_2 because $\sigma_1 t'$ is a subsequence of $\sigma_1 t' \sigma_2 t'$. Thus, $f_\rho(t)$ is not a function. However, we have proved that if $\mathcal{U}_{\mathcal{K}} \models^\pi q$, then f_ρ is a function. This creates a contradiction.

Suppose $\mathrm{dep}(q, \pi) > |\boldsymbol{u}|$, then there exists a σ with $|\sigma| > |\boldsymbol{u}| + 1$. Because free variables can only be mapped into $\mathrm{Ind}(\mathcal{A})$, a quantified variable repeatedly appears in the path σ exists. Thus, σ is infinite. We have proved that if $\mathcal{U}_{\mathcal{K}} \models^\pi q$, then every $\sigma \in \mathrm{MaxCertPath}(q, \pi)$ is finite. This creates a contradiction.

Based on Lemma 1 and Theorem 1, we can conclude that, for every rooted query, we extend the model at most $|\boldsymbol{u}|$ steps. The core of the QAA:

Theorem 2. *For each consistent DL-Lite$_{horn}^{\mathcal{N}}$ KB $\mathcal{K}$ and each rooted conjunctive query $q = \exists u \varphi(\boldsymbol{u}, \boldsymbol{v})$, with $|\boldsymbol{u}| = n$, we have $\mathrm{cert}(q, \mathcal{K}) = \mathrm{ans}(q, \mathcal{U}_{\mathcal{K}}^n)$.*

Proof. Theorem 4 (see Sect. 2) states that $\mathrm{cert}(q, \mathcal{K}) = \mathrm{ans}(q, \mathcal{U}_\mathcal{K})$. Thus, we only need to proof $\mathrm{ans}(q, \mathcal{U}_\mathcal{K}) = \mathrm{ans}(q, \mathcal{U}_\mathcal{K}^n)$.

($\Rightarrow$) Lemma 1 shows that for every mapping π, if $\mathcal{U}_\mathcal{K} \models^\pi q$, then $n^* = \mathrm{dep}(q, \pi) \leq |\boldsymbol{u}|$. Based on Theorem 1, we can conclude that $\mathcal{U}_\mathcal{K}^{n^*} \models^\pi q$. Because $n^* \leq |\boldsymbol{u}|$, $\Delta^{\mathcal{U}_\mathcal{K}^{n^*}} \subseteq \Delta^{\mathcal{U}_\mathcal{K}^n} \subseteq \Delta^{\mathcal{U}_\mathcal{K}}$. We can conclude that $\mathcal{U}_\mathcal{K}^n \models^\pi q$.

($\Leftarrow$) Because $\Delta^{\mathcal{U}_\mathcal{K}^n} \subseteq \Delta^{\mathcal{U}_\mathcal{K}}$, if $\mathcal{U}_\mathcal{K}^n \models^\pi q$ then $\mathcal{U}_\mathcal{K} \models^\pi q$.

In conclusion, $\mathrm{ans}(q, \mathcal{U}_\mathcal{K}) = \mathrm{ans}(q, \mathcal{U}_\mathcal{K}^n)$, that is, $\mathrm{cert}(q, \mathcal{K}) = \mathrm{ans}(q, \mathcal{U}_\mathcal{K})$.

The example in Sect. 1 proves our idea. QAA(Q) = 2, thus, the sRDF(2- step universal model) makes ans(sRDF, Q) = ans(RDF, Q).

3.2 *n*-step Universal Model in OWL 2 DL

OWL 2 DL has complex restrictions, such as property restrictions and arbitrary cardinality. Besides, UNA is not adopted by OWL 2 DL. Thus, complex information about the described domain can be captured by OWL 2 DL. We extend our approach to support OWL 2 DL to enjoy the high expressiveness.

We implement the support for OWL 2 DL through approximation and rewriting mechanisms. Given an OWL 2 DL ontology, we first attempt to rewrite it as an equivalent $DL\text{-}Lite_{horn}^\mathcal{N}$ TBox axiom, if possible. Otherwise, we have three choices, approximate processing or adding additional data structures or other ABox transformation rules. The last two methods can preserve the semantics that approximate processing would lose.

Table 1. OWL 2 DL rewriting rules

No.	TBox axiom	Rewriting TBox axiom
1	$\prod_{i=1}^{n} C_i \equiv C$	$\prod_{i=1}^{n} C_i \sqsubseteq C, C \sqsubseteq \prod_{i=1}^{n} C_i$
2	$C \sqsubseteq \prod_{i=1}^{n} C_i$	$C \sqsubseteq C_i, 1 \leq i \leq n$
3	$\exists R.\{a_1, a_2, a_3\} \equiv C$	$\exists R.\{a_1, a_2, a_3\} \sqsubseteq C, C \sqsubseteq \exists R.\{a_1, a_2, a_3\}$
4	$\bigsqcup_{i=1}^{n} C_i \equiv C$	$\bigsqcup_{i=1}^{n} C_i \sqsubseteq C, C \sqsubseteq \bigsqcup_{i=1}^{n} C_i$
5	$\bigsqcup_{i=1}^{n} C_i \sqsubseteq C$	$C_i \sqsubseteq C, 1 \leq i \leq n$
6	$\mathrm{Dis}(A, B)$	$A \sqsubseteq \neg B, B \sqsubseteq \neg A$
7	$\mathrm{Dis}(r, s)$	$r \sqsubseteq \neg s, s \sqsubseteq \neg r$
8	$r \equiv s$	$r \sqsubseteq s, s \sqsubseteq r$

TBox transformation is presented in Table 1 and Table 2 (r and s denote role names). Given a $\mathcal{SROIQ}$ TBox axiom, we first rewrite it to an equivalent one according to the rewriting rules in Table 1, also known as normalization in [22]. We add concept and its complement to a complement table (CT), which is designed to record complement semantic. Because of the normalized TBox axioms $\mathcal{T}$ beyond the expressiveness of $DL\text{-}Lite_{horn}^\mathcal{N}$, and an axiom in the form of

$C \sqsubseteq \overset{n}{\underset{i=1}{\sqcup}} C_i$ or $C \sqsubseteq \exists R.\{a_1, a_2, a_3\}$ will lead to non-determinism, we syntactically approximate partial axioms by their complement, as shown in Table 2. Specially, we construct a new concept for nominals at No. 10 approximation rule. All TBox transformation rules are sound, as shown in [22].

Table 2. Approximation rules

No.	TBox axiom	Approximation axiom
9	$C \sqsubseteq \overset{n}{\underset{i=1}{\sqcup}} C_i$	$\overset{n}{\underset{i=1}{\sqcap}} \neg C_i \sqsubseteq \neg C$
10	$C \sqsubseteq \exists R.\{a_1, a_2, a_3\}$	$C_1 \equiv \{a_1, a_2, a_3\}, \forall R.\neg C_1 \sqsubseteq \neg C$
11	$C \sqsubseteq \le mR.C_1$	$\ge (m+1)R.C_1 \sqsubseteq \neg C$

The semantics of $DL\text{-}Lite_{horn}^{\mathcal{N}}$ do not cover the axiom shown in Table 3. Thus, we design tractable ABox transformation rules, i.e., ABox reasoning rules for them based on the semantics of $\mathcal{SROIQ}$.

Table 3. ABox transformation

No.	TBox axiom	ABox reasoning rules
12	$\{C \equiv \{a_1, a_2, \cdots, a_n\}\}$	$\mathcal{A} = \mathcal{A} \cup \{C(a_i)\}, 1 \le i \le n$
13	$\text{Fun}(r)$	$r(a,b) \in \mathcal{A} \wedge r(a,c) \in \mathcal{A} \wedge b \ne c \notin \mathcal{A} \rightarrow \mathcal{A} = \mathcal{A} \cup \{b \doteq c\}$
14	$\text{Trans}(r)$	$r(a,b) \in \mathcal{A} \wedge r(b,c) \in \mathcal{A} \wedge r(a,c) \notin \mathcal{A} \rightarrow \mathcal{A} = \mathcal{A} \cup r(a,c)$
15	$\text{Sym}(r)$	$r(a,b) \in \mathcal{A} \wedge r(b,a) \notin \mathcal{A} \rightarrow \mathcal{A} = \mathcal{A} \cup r(b,a)$

An *extended TBox* $\mathcal{T}^*$ is a set of axioms obtained from $\mathcal{T}$ by applying TBox transformation rules and adding ABox transformation rules. Let $\mathcal{K}^* := (\mathcal{T}^*, \mathcal{A})$. Let n be a natural number. The n-step universal model of $(\mathcal{T}^*, \mathcal{A})$ is called an *extended n-step universal model*, denoted by $\mathcal{U}_{\mathcal{K}^*}^n$, of $\mathcal{K}$.

By Theorem 2 and the definition of the transformation rules, we can conclude:

Proposition 1 (Approximation). *Let* $\mathcal{K} = (\mathcal{T}, \mathcal{A})$ *be a consistent KB and* $q = \exists \boldsymbol{u} \varphi(\boldsymbol{u}, \boldsymbol{v})$ *be a rooted CQ with* $|\boldsymbol{u}| = n$. *For every* $\boldsymbol{a} \subseteq \text{Ind}(\mathcal{A})$ *with* $|\boldsymbol{a}| = |\boldsymbol{v}|$, *if* $\boldsymbol{a} \in \text{ans}(q, \mathcal{U}_{\mathcal{K}^*}^n)$ *then* $\boldsymbol{a} \in \text{cert}(q, \mathcal{K})$.

Example 5. Let $\mathcal{K} = (\{\alpha_1, \alpha_2, \alpha_3\}, \{\beta_1, \beta_2, \beta_3, \beta_4, \beta_5\})$ be a KB shown in the following table. GraduateStudent was abbreviated as GS.

Axiom	Expression	Axiom	Expression
α_1	GS $\equiv$ Person $\sqcap$ $\geq$ 3takeCourse	β_2	takeCouse(b, c_1)
α_2	Person $\equiv$ Woman $\sqcup$ Man	β_3	takeCouse(b, c_2)
α_3	Dis(Woman, Man)	β_4	takeCouse(b, c_3)
β_1	GS(a)	β_5	Man(b)

First step: we get new TBox axioms:

α_1^1 GS $\sqsubseteq$ Person (α_1, No1, No2) α_2^2 Man $\sqsubseteq$ Person (α_2, No4, No5)

α_1^2 GS $\sqsubseteq$ $\geq$ 3takeCourse.Thing (α_1, No1, No2) α_3^1 Woman $\sqsubseteq$ $\neg$Man (α_3, No6)

α_1^3 Person $\sqcap$ $\geq$ 3takeCourse $\sqsubseteq$ GS (α_1, No1) α_3^2 Man $\sqsubseteq$ $\neg$Woman (α_3, No6)

α_2^1 Woman $\sqsubseteq$ Person (α_2, No4, No5)

Second step: from the above axioms, we can get the following new facts:

β_1^1 Person(a) (β_1, α_1^1) β_1^2 takeCourse(a, a_1) (β_1, α_1^2)

β_5^1 Person(b) (β_5, α_2^2) β_1^3 takeCourse(a, a_2) (β_1, α_1^2)

β_6 GS(b) ($\beta_2 - \beta_4, \beta_5^1, \alpha_1^3$) β_1^4 takeCourse(a, a_3) (β_1, α_1^2)

β_5^2 $\neg$Woman(b) (β_5, α_3^2) β_8 ($a_1, \neq, a_2$) ($\beta_1^2, \beta_1^3, \alpha_1^2$)

β_7 ($a_1, \neq, a_3$) ($\beta_1^2, \beta_1^4, \alpha_1^2$) β_9 ($a_2, \neq, a_3$) ($\beta_1^3, \beta_1^4, \alpha_1^2$)

4   The System and Implementation of SUMA

4.1   An Overview of SUMA

SUMA computes model off-line and executes queries online. The off-line stage consists of three modules: ontology processor, storage, and materialization (Fig. 2).

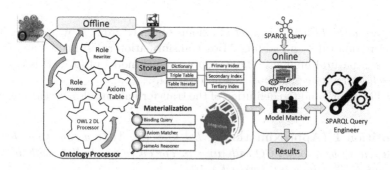

Fig. 2. The architecture of SUMA

The ontology processor module has three submodules. OWL 2 DL processor parses the ontology through the OWL API [10]. The rule index is generated at the role processor. Role rewriter rewrites the axioms to $DL - Lite$ axioms according to technologies shown in Sect. 3.2.

Algorithm 1. Materialization Algorithm

Input: $\mathcal{I}$: a collection of facts, $\mathcal{T}$: a collection of axioms
Output: $\mathcal{I}$: a collection of expanded facts

1: **while** $F = \mathcal{I}$.next and $F \neq \varepsilon$ **do**
2:     $G = \Omega$.getSameAsMapping(F);
3:     **if** G.equals(F) **then**
4:         **if** F is the form of $(d, \doteq, e)$ **then**
5:             sameAsReasoning(F);
6:         **else**
7:             **if** F is the form of $(d, P, e) \wedge P$.isFunctionalProperty **then**
8:                 **for** $e^* \in \mathcal{I}$.getIndividual(d, P) **do**
9:                     **if** $\mathcal{I}$.contains($(e, \neq, e^*)$) **then**
10:                         result in contradiction;
11:                     **else** $\mathcal{I}$.add($(e, \doteq, e^*)$);
12:             **else**
13:                 **for** each $(type, r, F) \in$ matchAxiom($F, \mathcal{T}$) **do**
14:                     **for** $F' \in \mathcal{I}$.evaluate($type, r, F$) **do**
15:                         $\mathcal{I}$.add(F');

The storage module uses the Jena API to load RDF data and generates a dictionary by encoding each RDF resource in integer ID. The RDF data is stored as a triple table with three types of indexes, e.g., a primary index, a secondary index, and a tertiary index. TableIterator can traverse the triple table efficiently. It maintains an index array that records the n-step model corresponding index ranges in the triple table.

The materialization module has three submodules: binding query, axiom matcher, and sameAs reasoner. The detailed materialization algorithm is shown in Algorithm 1. It iteratively reads a new triple F from the triple table through TableIterator. If F has an equivalent triple G (For instance, $F = (d, R, e)$, $G = (d', R, e')$, $d \doteq d'$, $e \doteq e'$), the program does not process F to improve reasoning efficiency. The reasoning of F can be divided into three situations.

Firstly, if F is the form of $(d, \doteq, e)$, it is processed by Algorithm 2. The sameAs reasoning function puts the individual d and e into an equivalent pool and selects the individual with the smallest ID as the identifier. We set the sameAs mapping of d as c if there exists $(d, \doteq, c)$ and c is the smallest ID of the equivalent pool. Secondly, if the role in F is a functional or an inverse functional role. For instance, $F = (d, P, e)$, and P is a functional role. Then, all triples like $(d, P, *)$ are returned by $\mathcal{I}$.getIndividual(d, P). A new fact $(e, \doteq, *)$ is added to $\mathcal{I}$ if $\mathcal{I}$ does not contain a fact $(e, \neq, *)$. Thirdly, axiom matcher returns all axioms that can match the triple F through matchAxiom($F, \mathcal{T}$). Binding query function converts these partially matched axioms into partial binding queries. The function $\mathcal{I}$.evaluate executes these queries over $\mathcal{I}$ and returns a new fact.

The online part includes a SPARQL processor and a model matcher. The SPARQL processor applies QAA technologies to compute the step size (n) of the model. The model matcher takes n as input and passes the n-step universal model to the SPARQL query engine. The SPARQL query engine executes SPARQL queries and returns query results.

Algorithm 2. sameAs Reasoning Algorithm

Input: $\mathcal{I} : d$, e : individual name, $pool$: a list of equivalent pool
1: idx = mergeEquivalentPool(d, e);
2: c = selectNewIdentifier($pool$[idx]);
3: **for** $i \in pool$[idx] **do**
4:     i.setSameAsMapping(c);

---

5   Experiments and Evaluations

SUMA delegates SPARQL queries to RDF-3X [21] at this experiment. The experimental environment is a 24 core machine that is equipped with 180GB RAM and Ubuntu 18.04. We compare SUMA with two well-known approaches, Pellet and PAGOdA. We test two aspects, (i) the soundness and completeness of the answer, (ii) the scalability of the query answering system. The first aspect is testing the number of queries that the system can answer correctly under the certain answer semantic. The evaluation of the scalability of the query answering system is to test the pre-processing time, consists of data load time and materialization time, and the query processing time on the increasing datasets.

- Data load time: This time includes all the data pre-processing steps before materialization, such as constructing a dictionary, generating an index, etc.
- Materialization time: The time taken by reasoner to compute consequences.
- Query processing time: The time taken by the system to execute a query on the extended data and return the query results.

Pellet is sound and complete in OWL 2 DL. It is adopted as the criterion for soundness and completeness evaluation. PAGOdA employs RDFox for highly scalable reasoning. Therefore, we mainly test the scalability of SUMA with PAGOdA. We perform two types of experiments, query answering over the finite universal model and query answering over the infinite universal model.

Table 4 gives a summary of all datasets and queries. Besides the 14 standard queries of LUBM, we also test ten queries from PAGOdA. The DBPedia [3] axiom is simple. It could be captured by OWL 2 RL [16]. Therefore, we adopt the DBPedia+ axiom and 1024 DBPedia+ queries provided by PAGOdA. It includes additional tourism ontologies. The LUBM, UOBM, DBPedia+ all have a finite universal model. They are not suitable for the second experiment. We add some manual CEQ axioms to the LUBM and UOBM ontologies, respectively. We also customize some additional queries to test LUBM+ and UOBM+.

5.1   Query Answering over Finite Universal Model

The Soundness and Completeness Evaluation. Because Pellet cannot give query results on LUBM(100), UOBM(100), and DBPedia+ in two hours, we did not display the results of it. As shown in Table 5, SUMA can correctly answer all queries on each test dataset.

Table 4. The information of datasets   **Table 5.** The quality of the answers

Data	Expressivity	Axioms	Facts	Queries
LUBM(n)	EL++	243	$n * 10^5$	24
LUBM+(n)	EL++	245	$n * 10^5$	33
UOBM(n)	$\mathcal{SHION}(\mathcal{D})$	502	$2.6n * 10^5$	15
UOBM+(n)	$\mathcal{SHION}(\mathcal{D})$	504	$2.6n * 10^5$	20
DBPedia+	$\mathcal{SHION}(\mathcal{D})$	3000	$2.6 * 10^7$	1024

Solved	SUMA	Pellet	PAGOdA
LUBM(1)	24	24	24
LUBM(100)	24	*	24
UOBM(1)	15	15	15
UOBM(100)	15	*	15
DBPedia+	1024	*	1024

The Scalability Test. We set the growth step size of the LUBM dataset and UOBM dataset as n = 200 and n = 100, respectively. For each dataset and ontology, we test the pre-processing time (*pre-time*), data load time, materialization time (*mat-time*), and average query processing time (*avg-time*).

Pre-processing Time Evaluation. As shown in Fig. 3, SUMA significantly reduces pre-processing time. For instance, it only takes 202 s to materialize LUBM(1000). The pre-processing time of SUMA on LUBM(1000) is 627 s, which is faster than PAGOdA's 1692 s. The time taken by SUMA to materialize UOBM(500) is 515 s. The total pre-processing time is 966 s. Compared with the 5937 s pre-processing time of PAGOdA, SUMA is much faster. SUMA takes 20 s to materialize DBPedia+. The pre-processing time of SUMA on DBPedia+ is 73 s, which is still faster than PAGOdA's 309 s.

Average Query Processing Time Evaluation. The average query processing time of SUMA on LUBM (1) and UOBM (1) is four and five orders of magnitude faster than Pellet, respectively.

We test the average query processing time of 24 LUBM queries on six LUBM datasets, 15 UOBM queries on five UOBM datasets, and 1024 DBPedia+ queries on one DBPedia+ dataset. As shown in Fig. 4(a), SUMA has a faster average query processing time than PAGOdA on all LUBM datasets except LUBM(100). (Time(SUMA, LUBM(100)) = 0.62 s, Time(PAGOdA, LUBM(100)) = 0.57 s). The significant decrease in the query processing time of SUMA on LUBM (500) is related to RDF-3X. RDF-3X can provide shorter query time on larger data by building different efficient indexes.

Figure 4(b) shows the average query processing time of SUMA is an order of magnitude faster than that of PAGOdA on all UOBM datasets. The average query processing time of SUMA on DBPedia+ is 24.337 ms, which is faster than PAGOdA's 33.235 ms.

5.2 Query Answering over Infinite Universal Model

Besides, the queries included in the first experiment, we add nine queries for LUBM+ and five queries for UOBM+, respectively. Two queries of LUBM+ contain a cyclic structure, and nine queries of LUBM+ include more than two

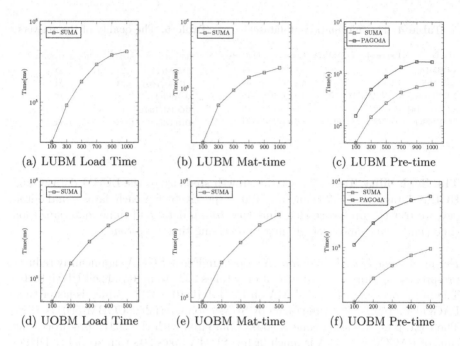

Fig. 3. Pre-processing experimental results

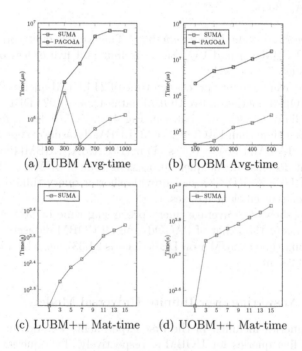

Fig. 4. Experimental results

quantified variables. The number of UOBM+ queries that consist of a cyclic structure is two. The number of UOBM+ queries that contain more than two quantified variables is three.

The Soundness and Completeness Evaluation. The experiment shows that SUMA can calculate all the correct answers for all queries of LUBM+ and UOBM+. Whereas PAGOdA is incomplete on five queries (Q2, Q4, Q5, Q6, Q7)[1] in LUBM+ and three queries (Q1, Q2, Q3) in UOBM+.

The Scalability Test. According to statistical analysis of the actual SPARQL queries, more than 96% of the queries include up to 7 triple patterns [8]. Therefore, in most cases, we only need to consider n not being greater than 7. Besides, we find that SUMA is also efficient when n is more than 7.

SUMA shows high scalability on LUBM+ and UOBM+. The average query processing time of SUMA on LUBM+(1) and UOBM+(1) is 1.99 ms and 6.48 ms, respectively. It is faster than the PAGOdA's 11.78 ms and 10.40 ms and five orders of magnitude faster than Pellet.

We focus on testing the materialization time of the infinite universal model. To make our test more challenging, we manually add 100 CEQ axioms to LUBM+ and UOBM+ ontologies, named as LUBM++ and UOBM++.

The materialization time of the 15-step universal model of LUBM++(1000) and UOBM++(500) is 351.144s and 698.536 s, respectively. When n = 7, the materialization time of LUBM++(1000) and UOBM++(500) is 253.583 s and 590.440 s, respectively. SUMA is highly scalable on the infinite universal model.

6 Discussion

This paper proposes a partial materialization approach for efficient ontology-mediated querying over OWL 2 DL. SUMA significantly reduces off-line materialization costs by building efficient indexes for facts and rules. The low complexity materialization algorithm makes SUMA can support real-time reasoning of large-scale datasets. In future works, we are interested in extending this proposal to support Boolean conjunctive queries. And, materializing complex roles in SUMA remains a significant cause of reducing materialization performance. Inspired by the role rewriting algorithm [23], we will further optimize materialization time and memory consumption caused by complex roles.

Acknowledgments. We thank Guilin Qi for valuable comments. This work is supported by the National Key Research and Development Program of China (2017YFC0908401), the National Natural Science Foundation of China (61972455, 61672377), and Shenzhen Science and Technology Foundation (JCYJ2017081609 3943197). Xiaowang Zhang is supported by the Peiyang Young Scholars in Tianjin University (2019XRX-0032).

---

[1] https://github.com/SUMA-2019/SUMA.

References

1. Artale, A., Calvanese, D., Kontchakov, R., Zakharyaschev, M.: The DL-Lite family and relations. J. Artif. Intell. Res. **36**, 1–69 (2009). https://doi.org/10.1613/jair. 2820
2. Bienvenu, M.: Ontology-mediated query answering: harnessing knowledge to get more from data. In: Proceedings of IJCAI 2016, pp. 4058–4061 (2016)
3. Bizer, C., Lehmann, J., Kobilarov, G., Auer, S., Becker, C., Cyganiak, R., et al.: DBpedia: a crystallization point for the web of data. J. Web Semant. **7**(3), 154–165 (2009). https://doi.org/10.1016/j.websem.2009.07.002
4. Botoeva, E., Calvanese, D., Santarelli, V., Fabio Savo, D., Solimando, A., Xiao, G.: Beyond OWL 2 QL in OBDA: rewritings and approximations. In: Proceedings of AAAI 2016, pp. 921–928 (2016)
5. Calvanese, D., Cogrel, B., Komla-Ebri, S., Kontchakov, R., Lanti, D., Rezk, M., et al.: Ontop: answering SPARQL queries over relational databases. Semant. Web **8**(3), 471–487 (2017). https://doi.org/10.3233/SW-160217
6. Guo, Y., Pan, Z., Heflin, J.: LUBM: a benchmark for OWL knowledge base systems. J. Web Semant. **3**(2–3), 158–182 (2005). https://doi.org/10.1016/j.websem.2005. 06.005
7. Harris, S., Seaborne, A., Apache, T., Foundation, S., Prud, E.: SPARQL 1. 1 query language. W3C Recommendation (2013)
8. Han, X., Feng, Z., Zhang, X., Wang, X., Rao, G., Jiang, S.: On the statistical analysis of practical SPARQL queries. In: Proceedings of WebDB 2011, p. 2 (2016). https://doi.org/10.1145/2932194.2932196
9. Horrocks, I., Tessaris, S.: Querying the semantic web: a formal approach. In: Horrocks, I., Hendler, J. (eds.) ISWC 2002. LNCS, vol. 2342, pp. 177–191. Springer, Heidelberg (2002). https://doi.org/10.1007/3-540-48005-6_15
10. Horridge, M., Bechhofer, S.: The owl API: a java API for OWL ontologies. Semant. Web **2**(1), 11–21 (2011)
11. Kontchakov, R., Lutz, C., Toman, D., Wolter, F., Zakharyaschev, M.: The combined approach to query answering in DL-Lite. In: Proceedings of KR 2010 (2010)
12. Ma, L., Yang, Y., Qiu, Z., Xie, G., Pan, Y., Liu, S.: Towards a complete OWL ontology benchmark. In: Sure, Y., Domingue, J. (eds.) ESWC 2006. LNCS, vol. 4011, pp. 125–139. Springer, Heidelberg (2006). https://doi.org/10.1007/11762256_12
13. Meng, C., Zhang, X., Xiao, G., Feng, Z., Qi, G.: gOWL: a fast ontology-mediated query answering. In: Proceedings of ISWC 2016 (P&D) (2018)
14. Lutz, C., Seylan, İ., Toman, D., Wolter, F.: The combined approach to OBDA: taming role hierarchies using filters. In: Alani, H., et al. (eds.) ISWC 2013. LNCS, vol. 8218, pp. 314–330. Springer, Heidelberg (2013). https://doi.org/10.1007/978-3-642-41335-3_20
15. Lutz, C.: The complexity of conjunctive query answering in expressive description logics. In: Armando, A., Baumgartner, P., Dowek, G. (eds.) IJCAR 2008. LNCS (LNAI), vol. 5195, pp. 179–193. Springer, Heidelberg (2008). https://doi.org/10. 1007/978-3-540-71070-7_16
16. Motik, B., Cuenca Grau, B., Horrocks, I., Wu, Z., Fokoue, A., Lutz, C.: OWL 2 web ontology language profiles (second edition). W3C Recommendation, World Wide Web Consortium (2012)
17. Motik, B., Patel-Schneider, P.F., Parsia, B.: OWL 2 web ontology language structural specification and functional-style syntax (Second Edition). W3C Recommendation, World Wide Web Consortium (2012)

18. Motik, B., Shearer, R., Horrocks, I.: Hypertableau reasoning for description logics. J. Artif. Intell. Res. **36**, 165–228 (2009)

19. Motik, B., Nenov, Y., Piro, R., Horrocks, I., Olteanu, D.: Parallel materialisation of datalog programs in centralised, main-memory RDF systems. In: Proceedings of AAAI 2014, pp. 129–137 (2014)

20. Nenov, Y., Piro, R., Motik, B., Horrocks, I., Wu, Z., Banerjee, J.: RDFox: a highly-scalable RDF store. In: Arenas, M., et al. (eds.) ISWC 2015. LNCS, vol. 9367, pp. 3–20. Springer, Cham (2015). https://doi.org/10.1007/978-3-319-25010-6_1

21. Neumann, T., Weikum, G.: RDF-3X: a RISC-style engine for RDF. PVLDB **1**(1), 647–659 (2008). https://doi.org/10.14778/1453856.1453927

22. Pan, J., Ren, Y., Zhao, Y.: Tractable approximate deduction for OWL. Artif. Intell. **235**, 95–155 (2016). https://doi.org/10.1016/j.artint.2015.10.004

23. Qin, X., Zhang, X., Feng, Z.: Optimizing ontology materialization with equivalent role and inverse role rewriting. In: Proceedings of WWW 2020 (Poster) (2020, accepted)

24. Sirin, E., Parsia, B., Grau, B., Katz, Y.: Pellet: a practical OWL-DL reasoner. J. Web Semant. **5**(2), 51–53 (2007). https://doi.org/10.1016/j.websem.2007.03.004

25. Zhou, Y., Grau, B., Nenov, Y., Kaminski, M., Horrocks, I.: PAGOdA: pay-as-you-go ontology query answering using a datalog reasoner. J. Artif. Intell. Res. **54**, 309–367 (2015). https://doi.org/10.1613/jair.4757

DeepQT : Learning Sequential Context for Query Execution Time Prediction

Jingxiong Ni[1], Yan Zhao[2], Kai Zeng[3], Han Su[1], and Kai Zheng[1(✉)]

[1] University of Electronic Science and Technology of China, Chengdu, China
nijingxiong@std.uestc.edu.cn, {hansu,zhengkai}@uestc.edu.cn
[2] School of Computer Science and Technology, Soochow University, Suzhou, China
zhaoyan@suda.edu.cn
[3] Alibaba Group, HangZhou, China
zengkai.zk@alibaba-inc.com

Abstract. Query Execution Time Prediction is an important and challenging problem in the database management system. It is even more critical for a distributed database system to effectively schedule the query jobs in order to maximize the resource utilization and minimize the waiting time of users based on the query execution time prediction. While a number of works have explored this problem, they mostly ignore the sequential context of query jobs, which may affect the performance of prediction significantly. In this work, we propose a novel <u>Deep</u> learning framework for <u>Q</u>uery execution <u>T</u>ime prediction, called <u>DeepQT</u>, in which the sequential context of a query job and other features at the same time are learned to improve the performance of prediction through jointly training a recurrent neural network and a deep feed-forward network. The results of the experiments conducted on two datasets of a commercial distributed computing platform demonstrate the superiority of our proposed approach.

Keywords: Deep learning · Query-time prediction · Distributed database · Jointly training

1 Introduction

Modern database management can greatly benefit from the prediction of query execution time, which aims to predict the time between the start of a SQL query job in database management and the end of the job. The prediction of query time is an important research topic and can be used in many scenarios, such as admission control decisions [23], query scheduling decisions [6], query monitoring [19], system sizing [25], and so forth. Within the generation of distributed computing platforms such as MaxCompute developed by Alibaba, we can obtain high volumes of information about users' query jobs which contains different SQL statement, query execution plans, and job configurations. Based on these historical data, it is possible for us to make an accurate prediction about query execution time using deep neural networks.

© Springer Nature Switzerland AG 2020
Y. Nah et al. (Eds.): DASFAA 2020, LNCS 12114, pp. 188–203, 2020.
https://doi.org/10.1007/978-3-030-59419-0_12

In existing works, query time prediction problem is conventionally tackled by cost-based analytical modeling approach [27] or traditional machine learning techniques, such as Multiple Linear Regression (MLR) [2] and Support Vector Regression (SVR) [24]. While the cost-based analytical modeling approaches are good at comparing the costs of alternative query execution plans, they are poor predictors of query execution time, especially in the commercial distributed database management where there are more sophisticated factors to consider and less information about the query execution cost than centralized database management. As for the previous works using traditional machine learning techniques, there are two limitations. First, due to the lack of real-world data, the training data used in previous work is generated by a small amount of benchmark query statement like TPC-H, ignoring the complexity and diversity of users' real queries. Second, traditional machine learning techniques lack the ability to model complex patterns in a large amount of real-world data. Moreover, most of the previous works take the number of operators in the execution plan as the input feature without considering the order and dependency between the operators in query execution plan. However, such information in query jobs can greatly affect the query execution time.

In this paper, we take the order and strong dependency between operators in a query's execution plan into consideration. Figure 1 presents an example of query execution plan, which can be expressed as sequences after topologically sorting. Thus, we model the query execution plan as a sequential context to learn the information in the order and dependency between operators. In addition, since the number of operators in a query execution plan is not known, the length of the execution plan topology is not available in advance as well. Therefore we adopt a Recurrent Neural Network (RNN) in modeling the sequential context of a user's query job, because RNN has the powerful ability to handle sequential data whose length is not known beforehand and has superiority in encoding dependencies [21].

In addition to the query execution plan, we can extract many critical features from the job configuration generated by the query optimizer, such as PlanMem, PlanCpu (see Table 1) and so on. We leverage the Deep Neural Networks (DNN) to learn the information provided by these features. DNN has been successfully applied to various difficult problems and has achieved excellent performance, due to its ability to perform arbitrary parallel computation for a modest number of steps.

In summary, RNN is good at learning the sequential features, and DNN has the superior performance in learning information from non-sequential features. Therefore we aim to combine them together to improve the performance of query execution time prediction.

Our main contributions are summarized as follows:

- We propose a learning framework based on deep neural network, namely DeepQT, to solve the problem of query execution time prediction. By leveraging

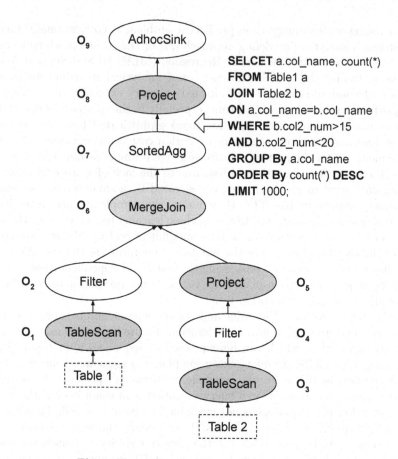

Fig. 1. Illustration for the query execution plan

the large volume of historical user query job data stored in MaxCompute[1], DeepQT has the ability of learning sequential context to make an accurate query time prediction by jointly training a recurrent model component and a deep model component.

- Taking the sequential context of query jobs into consideration, we devise a novel neural network architecture to learn the pattern across different job configurations (e.g., PlanMem, PlanCpu and RunningMode (see Table 1)) and the sequential contextual information (i.e., the order and dependency between the operators in query execution plan) simultaneously. Such sequential contextual information can help to significantly improve the accuracy of the prediction.

---

[1] MaxCompute (previously known as ODPS) is a general purpose, fully managed, multi-tenancy data processing platform for large-scale data warehousing. https://www.alibabacloud.com/product/maxcompute.

Table 1. The features of query job configuration

Features	Comment
PlanMem	the memory allocated to the job
PlanCpu	the cpu allocated to the job
RunningMode	the running mode of the job
ExecutorNum	the number of executor allocated to the job
InputRecord	the number of records the query need to scan
TableNum	the number of table relevant to the job
taskNum	the number of task in execution plan
instNum	the number of instance in execution plan
RunningCluster	the running cluster of the job

- We evaluate our model on the data about real-world users' query jobs, which is more important and challenging than evaluating on the data about benchmark query jobs. The experimental results demonstrate the advantage of our model over existing methods.

2   Problem Definition

When users submit query jobs to the system, query optimizer of the system would generate an execution plan of the query job and the corresponding job configuration. The job configuration includes general descriptions about the execution environment of query jobs (as shown in Table 1). The query execution plan consists of many operators in tree-based structures (see Fig. 1), which is also named Physical Operators Tree (POT). Thus, the topology of a query execution plan is used to represent the query job.

Let $U = \{u_1, u_2, ..., u_N\}$ denote a set of users and $P_n = \{q_1, q_2, ..., q_{c_n}\}$ denote all query execution plans (POTs) for each user u_n, where $q_{i=1,2,...,c_n}$ is the representation of query execution plan and c_n is the number of plans. There should be a corresponding time spent by execution plan, which is defined as $T_n = \{t_1, t_2, ..., t_{c_n}\}$. Moreover, we use $J_n = \{j_1, j_2, ..., j_{c_n}\}$ to denote the corresponding job configuration of execution plan.

Problem Statement: Given the historical data P and J of users in U, as well as their corresponding execution time T, our goal is to learn a predictor to estimate the execution time of the newly query job of users based on the configuration and execution plan of the job.

3   DeepQT Model

3.1   Model Overview

In our study, we divide the input features into two categories: 1) sequential feature (i.e., POT) and 2) non-sequential features (e.g., InputRecord and PlanMem).

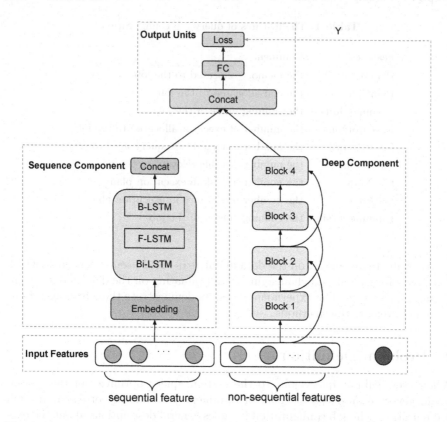

Fig. 2. Overview of the DeepQT architecture for query execution time prediction (F-LSTM: Forward Long Short-Term Memory; B-LSTM: Backward Long Short-Term Memory; Bi-LSTM: Bidirectional Long Short-Term Memory; Block: the combination of a series of consecutive operations whose details are shown in Fig. 4)

Inspired by the DeepFM model [11] that can well utilize the advantages of Factorization Machine (FM) and feed-forward neural network to learn both low-order and high-order feature interactions, we propose a new neural network model, DeepQT, to handle the above two kinds of features.

As shown in Fig. 2, our model consists of two main components, a sequence component and a deep component. Taking the topology of POT as input in sequence component and other non-sequential features as input in deep component, the two components are jointly trained for learning sequential contextual information and learning non-sequential features. The structure of the joint training framework we devised has the ability to integrate the knowledge learned by the above two components.

Specifically, through a lookup table operation, the POT is first mapped to the embedding vectors with a fixed dimension, then fed into RNN layers to learn the sequential pattern of POT. The deep component is a feed-forward neural network with skip connection, which aims to learn the pattern of the other

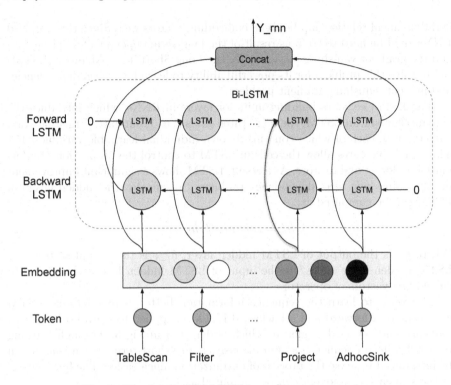

Fig. 3. The detail of Bi-LSTM in architecture overview

non-sequential features. In the output units, the outputs of the sequence component and the deep component are concatenated first, then fed into a fully connected layer to get the final output, which can be defined as follows:

$$\widehat{y} = W^{out}[y_{deep} : y_{sequence}] + b^{out}, \tag{1}$$

where $\widehat{y}$ is the predicted execution time, y_{deep} is the output of deep component, $y_{sequence}$ is the output of sequence component, $[y_{deep} : y_{sequence}]$ is the concatenation of the outputs, W^{out} and b^{out} are the learnable parameters. Then, the $\widehat{y}$ will be fed into a loss function for joint training.

3.2 Sequence Component

The sequence component is used to learn the dependency between operators in the query execution plan, which is illustrated in Fig. 3. Through tokenization, we represent the original operators in a query plan q_n as $O^n = [o_1^n, ..., o_{c_{q_n}}^n]$, and apply the commonly used embedding layer to map each operator into an embedding vector with fixed-sized dimension, denoted by $E^n = [e_1^n, ..., e_{c_{q_n}}^n]$. The embedding vectors can be learned during training to obtain more accurate representations of each operator. Then, we feed the embedding vectors into RNN layers which are sensitive to operator order and can learn the complex sequential and dynamical relationship between embedding vectors well. Since the standard RNN would be hard to train as result of the long-term dependencies in sequence and the gradient vanishing [3], we apply the Long Short Term Memory (LSTM) cell to solve this problem for its powerful ability to learn long-term dependencies and prevent vanishing gradient problem.

The LSTM cells are building units for layers in RNNs, which introduce the gate mechanism. In LSTM, there are three gates (i.e., input gate, forget gate and output gate), each of which contains its own individual learnable variables [15]. These multiple gates allow the cells in LSTM to control the proportion of information to forget and to store. As a result, LSTM shows a significant improvement in addressing long-term dependency problem. Specifically, the hidden layers of LSTM can be computed as:

$$h_l^n = LSTM(e_l^n, h_{l-1}^n), \tag{2}$$

Where h_l^n is the output of LSTM hidden layer, h_{l-1}^n is the output of previous LSTM hidden layer and e_l^n is the input of LSTM hidden layer, which refer to embedding vectors of operators.

Moreover, to learn the sequence information better, a bidirectional LSTM consisting of a forward and a backward LSTM is applied to learn both forward sequence and reversed sequence, which is depicted in Fig. 3. Through learning reversed order of sequence simultaneously, many short-term dependencies can be introduced to make the process of optimization much easier. The final output of the sequence component can be computed as:

$$y_{sequence} = [\overleftarrow{h_{c_{q_n}}^n} : \overrightarrow{h_{c_{q_n}}^n}], \tag{3}$$

where $\overleftarrow{h_{c_{q_n}}^n}$ is the last hidden state of backward LSTM, $\overrightarrow{h_{c_{q_n}}^n}$ is the last hidden state of forward LSTM, and the concatenation of these two states is the final output of sequence component.

3.3   Deep Component

The deep component is a feed-forward neural network using the skip connection, which aims to create short paths from previous layers to the subsequent layers. In our implementation, we combine some consecutive operations into a block, which is illustrated in Fig. 4. In a block, we adopt the Batch Normalization (BN) [14] right after Fully Connected (FC) layer and before Rectified Linear Unit (ReLU) [10] activation function to increase the speed of model training. The Batch Normalization (BN) layer is a novel mechanism for reducing the internal covariate shift, which refers to the change in the distribution of network activation caused by the change in network parameters during training. Inside layers, the normalization is performed for each input mini-batch. Besides, two additional learnable parameters are introduced to ensure the representation ability of the network [14]. In previous researches, it has been proved that the BN is an effective and promising way for improving the gradient propagation and the training speed of the network.

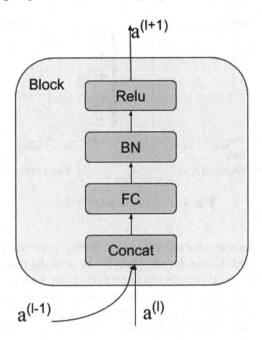

Fig. 4. The detail of the block (FC: Fully-connected; BN: Batch Normalization)

In addition, compared to traditional feed-forward neural network that is composed of fully connection, the deep component in DeepQT adds residual skip connections (called RES) between all blocks (see Fig. 2). For each block, the input is the concatenation of the outputs of the precedent block, and its output is used as input in the later layers. Due to the success of residual skip connection in alleviating the vanishing gradient problem and enhancing the feature propagation [12,13], we adopt it as our connection mode, which is beneficial for both convergence rate of the deep component and the performance of prediction. Specifically, the forward process can be denoted as:

$$a^{(l)} = H^{(l)}(W^{(l)}[a^{(l-2)}, a^{(l-1)}] + b^{(l)}), \tag{4}$$

where $a^{(l)}$ refers to the output of the l-th block. $a^{(0)}$ denotes the input vector, and $[a^{(l-1)}, a^{(l-1)}]$ is the concatenation of the outputs produced in $(l-1)$-th block and $(l-2)$-th block. We define a function $H^{(l)}$, which begins with the batch normalization, followed by a ReLU activation function.

3.4 Periodicity Analysis

In this section, we extract periodicity information to get a more accurate prediction result. According to our observations, the execution time of query jobs submitted by the same user usually changes periodically, which means that the execution time of query jobs at a certain time interval is similar to the same time

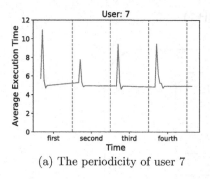
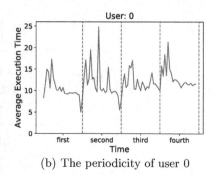

(a) The periodicity of user 7   (b) The periodicity of user 0

Fig. 5. Periodicity information

interval of the previous day for the same user. In Fig. 5, we collect the query jobs of two users submitted in four days to analyze the periodic characteristics of the execution time. Figure 5(a) depicts the daily periodicity with four obvious peaks. Although the execution time of query jobs is more turbulent in Fig. 5(b), we can still find the daily periodicity, especially at the early part of the day where there are two to three peaks. Moreover, it is observed that the execution time in the late part of the day is obviously less than that in the early part of the day, which implies that users tend to submit large query jobs in the morning and small query jobs late in the day. The periodicity information can be regarded as the supplementary of the learnt features. Therefore, we add the periodicity data as a non-sequential feature into the deep component.

For extracting the periodicity information, the granularity of a timestamp is first set as 1 h (i.e., 10:00 am–11:00 am) and then represented as a one-hot vector. Finally, we feed all the one-hot vectors into the deep component as an input feature to capture the pattern of daily periodicity for execution time.

4   Experiment

In this section, we first describe our training datasets and training details. Moreover, we conduct extensive experiments to evaluate our proposed model on these datasets.

4.1   Training Details

The model is trained to minimize the L2 loss between the training data and the predictions. Formally, the loss function can be computed as $L = \sum_{q \in Q} \left(\widehat{y}^q - y^q \right)^2$, where $\widehat{y}^q$ is the predicted value, y^q is the real value and Q is the training data. In addition, for the implementation of our model, we utilize TensorFlow. We use the ReLU [20] activation function after every fully connected layer. The weights in DeepQT are initialized with Xavier initialization [9]. The model is trained with the batch size of 1024 and a mini-batch stochastic gradient

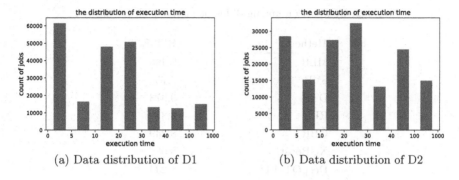

(a) Data distribution of D1   (b) Data distribution of D2

Fig. 6. Distribution of execution time in datasets

descent using Adaptive Moment Estimation (Adam) optimizer [16] with a learning rate 0.001. Moreover, the hyperparameters of our best-performing model in experiments are shown as follows. The dimension of the word embedding is 64; the dropout is applied before all fully connected layers with the ration of 0.3; the size of each hidden recurrent layer in the sequence component is 128; sizes of fully connected layers are 258, 126, 64, and 32 respectively. Besides, the early stopping strategy is adopted to terminate the training process when the model achieves the best performance on validation datasets. For the hardware environment, one NVIDIA GTX 1080ti GPU with 8 GB of RAM is used to train the model.

4.2 Datasets

MaxCompute is a commercial distributed computing platform for large-scale data warehousing, in which users submit millions of queries every day. When users' queries are submitted to the platform, the MaxCompute will generate a corresponding job. Such query job contains an execution plan of the query and a job configuration file which describes the execution environment of the job. We generate two training datasets by extracting query logs from MaxCompute in a period of time. In particular, the first dataset, **D1**, contains more than 217,000 query jobs of users, allocated from 13th November to 13th December in 2019. The second dataset, **D2**, contains about 156,000 query jobs of users, allocated from 13th October to 9th December in 2019. In order to highlight the generality of the model, the dataset **D2** we obtain has different data distribution from the **D1**, as shown in Fig. 6. We can find that the execution time of most query jobs on the **D1** is less than 30 s, and the distribution of execution time on the **D2** is more homogeneous. Two datasets have the same fields of features and these 29 fields of features in the datasets are divided into three categories: 1) sequential context (i.e., execution plan (POT)), 2) job configuration (such as PlanMem, PlanCpu and InputRecord (see Fig. 1)) and 3) the submitting time of query jobs. All fields of the features are used in the experiment. Furthermore, 30% of each dataset are randomly selected as test data and the rest as training data.

Table 2. Comparison among different methods on D1 dataset

Method	RMSE
MLR	9.498
SVR	9.756
DTR	9.973
RFR	8.310
GBR	8.249
XGBoost	8.087
DeepQT-4-DNN	8.234
DeepQT-4-DNN-RES	8.118
DeepQT-4-DNN-RNN	8.005
DeepQT-4-DNN-RNN-RES	**7.998**
DeepQT-5-DNN-RNN-RES	8.016
DeepQT-3-DNN-RNN-RES	8.116

4.3 Baselines

The method proposed in this paper is compared with the following baseline methods:

- **SVR** [24]: As a traditional machine learning technology, Support Vector Regression (SVR) has been proved to be able to achieve good performance in the regression problem.
- **MLR** [2]: As the most common form of linear regression analysis, Multiple Linear Regression (MLR) is used to predict the query execution time.
- **DTR**: By using Decision Tree Regression (DTR), a regression model is built in the form of a tree structure to perform the prediction.
- **RFR**: Random Forest Regression (RFR) is an ensemble technique, which can perform the regression task using multiple decision trees and a technique called Bootstrap Aggregation. The RFR implementation from the scikit-learn library is used in this paper.
- **GBR**: Using Gradient Boosting Regression (GBR), a prediction model is built in the form of an ensemble of weak prediction models. It has been proved that it has the powerful ability in regression tasks.
- **XGBoost**: As a popular and efficient ensemble method, XGBoost [4] has been widely applied in many regression problems because of its remarkable performance.

Except for the job execution plan (i.e., POT), the input features for baseline methods are the same as those for the method proposed in this paper for the purpose of fair comparisons. The number of different operations in the POT is used as input features in baseline methods. In contrast, the whole topology of POT is used as an input feature in the proposed method.

4.4 Effectiveness Comparisons

Performances of our method and other baseline methods are evaluated by Root Mean Square Error (RMSE). Formally, $\widehat{y}^q$ is used to denote the predicted value of y^q, and Q is used to denote the test data. Then, the RMSE can be defined as follows:

$$RMSE = \sqrt{\frac{1}{|Q|} \sum_{q \in Q} \left(\widehat{y}^q - y^q\right)^2} \tag{5}$$

We first compare DeepQT with the baselines on the **D1** dataset, which is depicted in Table 2. Meanwhile, we compare 6 variants of DeepQT with different components and layers to evaluate the effect of each component and the number of hidden layers. Taking DeepQT-4-DNN-RNN-RES as an example, it adds RES connections between 4 block layers in the deep component, where each block contains a BN layer, an FC layer, a concatenation layer and an activation function ReLU. Besides, it has a sequential component for learning sequential context. On the contrary, DeepQT-4-DNN means the model that only has a deep component.

It can been seen from Table 2 that the results of some tradition machine learning methods, such as MLR, SVR, and DTR, are obviously larger than other methods. The advanced methods based on ensemble learning (i.e., RFR, GBR and XGBoost) all provide a result of well-performance prediction, among which XGBoost achieves the best prediction accuracy among the baseline methods in terms of RMSE. The method proposed in this paper outperforms other existing baseline methods. It can be seen from the experiments that the RMSE of DeepQT-4-DNN- RNN-RES is 7.998, which significantly improves the prediction accuracy. The results of DeepQT-4-DNN and DeepQT-4-DNN-RES show that the deep feed-forward neural network with skip connections can achieve a well-performance result compared with other traditional methods; that is, skip connections indeed can improve the performance of DeepQT. The reason may be that the skip connections can enhance the gradient propagation and avoid losing some shallow information during propagation. Moreover, taking sequential context into consideration, the topology of POT is fed into the sequence component (BiLSTM in our framework) in DeepQT-4-DNN-RNN. The results show that DeepQT-4-DNN-RNN is further promoted, which indicate that the sequence component can exactly capture the pattern of order and dependency in sequential context. Meanwhile, the joint training framework in this paper has the ability to supplement the knowledge learned in the deep component with the knowledge learned in the sequential context. In addition, from the results of the experiments, we can find that DeepQT with 4 blocks can obtain a better result than DeeQT with 3 or 5 blocks. Although the effectiveness of DeepQT-4-DNN-RNN-RES is slightly improved compared with DeepQT-4-DNN-RNN, its efficiency of convergence is greatly improved, which we will show in Sect. 4.5.

Table 3 shows the results of experiments on the **D2**. From the results, we can see that our proposed approach achieves the best performance on this dataset. We can find that the RMSE of the DeepQT-4-DNN-RNN-RES has relatively

Table 3. Comparison among different methods on D2 dataset

Method	RMSE
MLR	12.339
SVR	12.867
DTR	13.583
RFR	12.567
GBR	11.174
XGBoost	10.858
DeepQT-4-DNN	11.412
DeepQT-4-DNN-RES	10.887
DeepQT-4-DNN-RNN	10.475
DeepQT-4-DNN-RNN-RES	**10.452**
DeepQT-5-DNN-RNN-RES	10.566
DeepQT-3-DNN-RNN-RES	10.626

from 3.8% up to 29.9% lower than these baselines on the **D2**, which demonstrates the excellent generalization and superiority of our proposed approach on the datasets with different data distribution.

Since the execution time of most queries in datasets is between 0 and 100 s (shown in Fig. 6), and the number of samples is large enough, the minor improvement of RMSE, 0.1 or 0.4, would represent a significant improvement in efficiency. Specifically, for a small number of long-term tasks in datasets, it would be an error difference of one minute or two minutes. Hence, it is essential for downstream tasks (e.g., Task Scheduling [6]).

4.5 Efficiency Analysis

In this section, we conduct experiments on two datasets to evaluate the efficiency of all the DeepQT variants by comparing their convergence rates. To ensure a fair comparison, we maintain the same hyperparameters for different variants and train the variants using Adam method [16] with batch size of 1024 and learning rate 0.001. The weights in our framework are initialized with Xavier initialization [9]. Learning curves are presented in Fig. 7.

It is noted that the variant with the sequence component (marked as DNN+RNN) and the variant with the RES connection (marked as DNN+RES) converge much faster than the variants with a single component (marked as DNN), which means the proposed model can be trained by using less epochs. Furthermore, the sequence component and the RES connection not only increase the prediction accuracy but also improve the convergence rate. Not surprisingly, the combination of both sequence component and RES connection (marked as DNN+RNN+RES) can achieve the best performance in convergence rate.

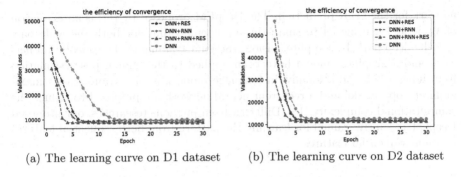

(a) The learning curve on D1 dataset   (b) The learning curve on D2 dataset

Fig. 7. Efficiency of convergence on datasets

To sum up, it can be found from the experiments that the proposed framework can well learn the sequential context to increase the prediction accuracy and reduce the iterative time for convergence under different data distributions.

5   Related Work

5.1   Query Execution Time Prediction

As a significant problem in the database management research, the query execution time prediction has received great attention over the last decades. [8] proposes an approach to address this problem based on a Kernel Canonical Correlation Analysis (KCCA) modeling technique. After then many machine learning techniques are applied to query time prediction, such as Support Vector Regression (SVR) [24], Multiple Additive Regression-Tree (MART) [17] and Multiple Linear Regression (MLR) [2]. Unlike these work based on machine learning technique, [27] proposes a method based on calibrating cost models of the query optimizer. However, the above works always assume the workload of the database is static, which is unrealistic. To relieve this problem, some works are proposed by considering the workload is concurrent and dynamic for generality, such as [1,7,26]. However, these works are not general enough. The query data they use for the experiments is still baseline queries rather than real-world queries. Moreover, the database management they use for the experiments is still centralized rather than distributed.

5.2   Deep Learning

Deep learning techniques have been applied to many difficult problems in various domains and achieved excellent success [5,12,21]. In this paper, a recurrent neural network and a deep feed-forward network are combined into a model for joint training to solve the difficulties in predicting query execution time. The idea of jointly training is inspired by previous researches such as Wide & Deep [5], which explores the joint training of a linear model and a feed-forward neural network for the CTR prediction. The DeepFM proposed by [11] is an extension of Wide & Deep model to share the input embedding for both the wide part and the deep part. In computer vision, the joint training of the convolution network and a graphical model have been applied to the human pose estimation from images [22]. Additionally, in language models, a joint training of a maximum entropy model and a recurrent neural network are proposed to reduce the computational complexity [18]. Different from previous researches, our proposed DeepQT is jointly trained for learning the pattern from both sequential features and non-sequential features.

6   Conclusion

In this paper, we study the problem of query execution time prediction. We propose a learning framework named DeepQT, which jointly trains a recurrent neural network and a deep feed-forward neural network. Our approach has the ability to learn sequential context to improve prediction accuracy and reduce the number of convergent iterations. Extensive experiments are conducted on real-world datasets, whose results show that our model outperforms the existing methods. Moreover, our framework has excellent generality and can achieve great performance on datasets with different data distributions. One of our future work is to incorporate our framework into the database management system so that it can be used in more applications.

Acknowledgement. This work is partially supported by Natural Science Foundation of China (No. 61972069, 61836007, 61832017, 61532018, 61802054) and Alibaba Innovation Research (AIR).

References

1. Ahmad, M., Duan, S., Aboulnaga, A., Babu, S.: Predicting completion times of batch query workloads using interaction-aware models and simulation. In: EDBT, pp. 449–460 (2011)
2. Akdere, M., Çetintemel, U., Riondato, M., Upfal, E., Zdonik, S.B.: Learning-based query performance modeling and prediction. In: ICDE, pp. 390–401 (2012)
3. Bengio, Y., Simard, P., Frasconi, P.: Learning long-term dependencies with gradient descent is difficult. IEEE Trans. Neural Netw. 5(2), 157–166 (1994)
4. Chen, T., Guestrin, C.: Xgboost: a scalable tree boosting system. In: SIGKDD, pp. 785–794 (2016)
5. Cheng, H.T., et al.: Wide & deep learning for recommender systems. In: DLRS, pp. 7–10 (2016)
6. Chi, Y., Moon, H.J., Hacigümüş, H.: icbs: incremental cost-based scheduling under piecewise linear SLAs. PVLDB 4(9), 563–574 (2011)
7. Duggan, J., Cetintemel, U., Papaemmanouil, O., Upfal, E.: Performance prediction for concurrent database workloads. In: SIGMOD, pp. 337–348 (2011)
8. Ganapathi, A., et al.: Predicting multiple metrics for queries: Better decisions enabled by machine learning. In: ICDE, pp. 592–603 (2009)

9. Glorot, X., Bengio, Y.: Understanding the difficulty of training deep feedforward neural networks. In: AISTATS, pp. 249–256 (2010)
10. Glorot, X., Bordes, A., Bengio, Y.: Deep sparse rectifier neural networks. In: AIStats, pp. 315–323 (2011)
11. Guo, H., Tang, R., Ye, Y., Li, Z., He, X.: Deepfm: a factorization-machine based neural network for CTR prediction. In: IJCAI, pp. 1725–1731 (2017)
12. He, K., Zhang, X., Ren, S., Sun, J.: Deep residual learning for image recognition. In: CVPR, pp. 770–778 (2016)
13. Huang, G., Liu, Z., Van Der Maaten, L., Weinberger, K.Q.: Densely connected convolutional networks. In: CVPR, pp. 2261–2269 (2017)
14. Ioffe, S., Szegedy, C.: Batch normalization: accelerating deep network training by reducing internal covariate shift. In: ICML, pp. 448–456 (2015)
15. Karpathy, A., Johnson, J., Fei-Fei, L.: Visualizing and understanding recurrent networks. arXiv preprint arXiv:1506.02078 (2015)
16. Kingma, D.P., Ba, J.: Adam: a method for stochastic optimization. arXiv preprint arXiv:1412.6980 (2014)
17. Li, J., König, A.C., Narasayya, V., Chaudhuri, S.: Robust estimation of resource consumption for SQL queries using statistical techniques. PVLDB 5(11), 1555–1566 (2012)
18. Mikolov, T., Deoras, A., Povey, D., Burget, L., Černocký, J.: Strategies for training large scale neural network language models. In: ASRU Workshop, pp. 196–201 (2011)
19. Mishra, C., Koudas, N.: The design of a query monitoring system. TODS 34(1), 1 (2009)
20. Nair, V., Hinton, G.E.: Rectified linear units improve restricted Boltzmann machines. In: ICML, pp. 807–814 (2010)
21. Sutskever, I., Vinyals, O., Le, Q.V.: Sequence to sequence learning with neural networks. In: NIPS, pp. 3104–3112 (2014)
22. Tompson, J.J., Jain, A., LeCun, Y., Bregler, C.: Joint training of a convolutional network and a graphical model for human pose estimation. In: NIPS, pp. 1799–1807 (2014)
23. Tozer, S., Brecht, T., Aboulnaga, A.: Q-cop: avoiding bad query mixes to minimize client timeouts under heavy loads. In: ICDE, pp. 397–408 (2010)
24. Van Wouw, S.: Performance evaluation of distributed SQL query engines and query time predictors (2014)
25. Wasserman, T.J., Martin, P., Skillicorn, D.B., Rizvi, H.: Developing a characterization of business intelligence workloads for sizing new database systems. In: DOLAP, pp. 7–13 (2004)
26. Wu, W., Chi, Y., Hacígümüş, H., Naughton, J.F.: Towards predicting query execution time for concurrent and dynamic database workloads. PVLDB 6(10), 925–936 (2013)
27. Wu, W., Chi, Y., Zhu, S., Tatemura, J., Hacigümüs, H., Naughton, J.F.: Predicting query execution time: Are optimizer cost models really unusable? In: ICDE, pp. 1081–1092 (2013)

DARS: Diversity and Distribution-Aware Region Search

Siyu Liu[1], Qizhi Liu[1(✉)], and Zhifeng Bao[2]

[1] State Key Laboratory for Novel Software Technology, Nanjing University,
Nanjing, China
mg1833049@smail.nju.edu.cn, lqz@nju.edu.cn
[2] RMIT University, Melbourne, Australia
zhifeng.bao@rmit.edu.au

Abstract. Recent years have seen the rapid development of Location Based Services (LBSs). Many users of these services are making use of them to, for example, plan trips, find houses or explore their surroundings. In this paper we introduce a novel problem called the *diversity and distribution-aware region search* (DARS) problem. In particular, DARS aims to find regions of size $a \times b$ where the number of different categories is maximized such that objects of different categories are not too scattered from each other and objects of the same category are within reasonable distance (which is a tunable parameter to cater for different users' needs). We propose several methods to tackle the problem. We first design a sweepline based method, and then design various techniques to further improve the efficiency. We have conducted extensive experiments over real datasets and demonstrate both the usefulness and the efficiency of our methods.

Keywords: Spatial region search · Spatial databases · Points of interest

1 Introduction

With the quick growth of Location Based Services (LBSs) in recent years, many online businesses (e.g., *Yelp*[1], *Dianping*[2]) are now incorporating geographic data such as point of interest (POI) in their services. Apart from geographic locations, POIs are also often tagged with category information. Those data can be utilized to help explore a large area more efficiently, such as finding a best location for facility deployment [4–6,18–20]. Apart from finding a single best location, the study [7] by Feng et al. aimed to identify a "best region" that contains the largest number of POI categories. In our work we argue that interrelationship among the POIs within a region is also important, as illustrated in the following example.

---

[1] www.yelp.com.
[2] www.dianping.com.

© Springer Nature Switzerland AG 2020
Y. Nah et al. (Eds.): DASFAA 2020, LNCS 12114, pp. 204–220, 2020.
https://doi.org/10.1007/978-3-030-59419-0_13

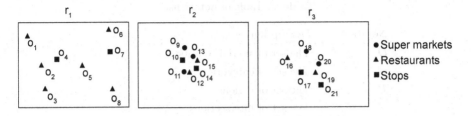

Fig. 1. A motivating example.

Example 1. John Doe plans to buy a new property in Anytown, and he wants to first find a region that has many kinds of facilities (e.g., POIs such as super markets, restaurants, metro or bus stops, parking lots etc.). If the locations of such POIs in a region are not too scattered, then John can make use of them quite efficiently without having to waste time on commuting. John might also prefer that there are at least some candidates for each category so he would have backup options if some POIs of certain category cannot satisfy his needs (e.g., different operation hours). How can John select such a "diversified" and "well-distributed" region in Anytown so that he can choose the location of his new property regarding the region?

Figure 1 shows three possible candidate regions for John. Although r_1 contains eight POIs, which is more than that in r_2 and r_3, r_1 contains POIs of two categories while r_2 and r_3 contain three, thus on diversity r_1 loses. Furthermore, POIs of different categories in r_1 are much scattered than those in r_2 and r_3, hence in r_2 or r_3 John can spend less time and energy traveling to different POIs. In this case r_2 and r_3 shall be considered as answers prior to r_1. We can even say r_2 is better than r_3 because POIs of different categories are even closer in r_2. For POIs of one category (e.g., restaurants), it is better to have some POIs of the same category nearby so John can have more options. Regarding this we can see that r_2 is also better than r_3 because POIs of same categories are also closer in r_2. These distribution of POIs within a region is the interrelationship of POIs we want to consider. At the same time, there might be other regions just as good as r_2. Therefore, returning multiple regions helps when there are multiple relatively good regions. As a result, with less time spent on commuting between different POIs, John can quickly find public transportations during his work in the daytime, have dinners of different styles to explore different dining options and shop at different markets to find the general goods he need in r_2. Note that it is critical that John offers the size of such a region that is of interest to him. In fact, other users might also want to search for a region satisfying their needs the most, but of different size.

From the example we can see that, given the size of a query rectangle, users wish to find regions that are **diversified** and the POIs within the regions are **well-distributed**, such that the number of different categories is maximized

Table 1. Table of notations.

Notation	Description
O	The set of spatial objects
C	The set of categories
R	The set of regions
T	The set of intersections
o, r, t	An object, a region, an intersection
$a \times b$	Size of a region
O_r	The set of spatial objects covered by r
C_r	The set of categories region r possesses
O_{r_m}	All POIs of category m in region r
$o^i_{r_m}$	The ith object of category m in r
$d'(O_{r_m}, O_{r_n})$	Minimum distance between two categories

while POIs of different categories are not too scattered from each other. As for POIs of the same category, different users might have different requirements for their distribution. This kind of interrelationship between POIs in a region is never considered in region search to the best of our knowledge. We refer to the problem in our paper as *diversity and distribution-aware region search* (DARS) problem. Briefly, in a spatial object database O, given a query rectangle of size $a \times b$, DARS aims to find regions that have the most diversified collection of POIs, minimize the average distance among POIs of different categories, and POIs of the same categories hold a distance no greater than a threshold between each other. In summary, we make the following contributions (Table 1):

- We propose and study the DARS problem for the first time, to capture diversity and distribution of POIs in a region simultaneously. Specifically, we use distance to capture the interrelationship among POIs in a region to measure the distribution at a region level (Sect. 2).
- We first design a baseline algorithm for scanning and finding regions, and then propose two optimizations to further improve the efficiency (Sect. 3).
- We conduct extensive experiments over two real-world and one synthetic datasets to demonstrate the efficiency and effectiveness of our solutions (Sect. 4).

2   Problem Formulation

In this section we formally introduce the *diversity and distribution-aware region search* (DARS) problem, starting from the following two preliminaries to define the diversity.

Definition 1 (Average Minimum Distance). *The average minimum distance among all POIs of different categories is:*

$$g(O_r) = \frac{1}{|C|} \sum_{1 \leq m < n \leq |C|} d'(O_{r_m}, O_{r_n}).$$

When considering the diversity of a region, we aim to minimize the average distance among POIs of different categories. For a region r, we can get the average minimum distance with $g(O_r)$. In $g(O_r)$, we use the minimum distance between POIs of different categories. The reason is that if the average minimum distance of POIs of different categories in a region is small, not only that POIs of different categories are quite near to each other, but also that we can avoid impacts from stray points. As a result users can spend less time in a region finding POIs that satisfy their different needs. We then need to find a way to define a "score" for the diversity of a region. We have two intuitions to follow. First, regions with more categories shall be considered prior to regions with less categories. Second, when two regions have the same number of categories, the distribution of POIs will most likely differ. Intuitively, POIs of different categories should be close to each other so that users can utilize different services efficiently. At the same time, for POIs of each category there shoule better be some POIs of the same category nearby for users to have more candidates to choose from.

Definition 2 (In-Region Diversity). *Given a region r, its in-region diversity of r is defined as*

$$f(r) = e^{-(\frac{|C_r| \times (|C_r|-1)}{2} + e^{-g(O_r)})}.$$

The definition of in-region diversity is driven by the two intuitions above. $\frac{|C_r| \times (|C_r|-1)}{2}$ is always monotone to $|C_r|$, so the difference with different $|C_r|$ is always greater than one. For example, consider $|C_r| = 3$, then $\frac{|C_r| \times (|C_r|-1)}{2} = 3$, $e^{-g(O_r)}$ is a value in range $(0, 1]$, so $\frac{|C_r| \times (|C_r|-1)}{2} + e^{-g(O_r)}$ is a value in range $(3, 4]$. But when $|C_r| = 2$, $\frac{|C_r| \times (|C_r|-1)}{2} + e^{-g(O_r)}$ is a value in range $(1, 2]$. These two ranges never intersect, so regions with more categories is always superior. From Fig. 1 in Example 1 we can see that r_2 and r_3 both possess three categories, and POIs of different categories are closer to each other in r_2 than those in r_3, then $g(O_{r_2}) < g(O_{r_3})$, so $f(r_2) < f(r_3)$. Furthermore, POIs of same categories are also closer to each other in r_2. In fact, this is just one possible form of the function f. Any other forms should also work as long as they consider regions with more categories prior to other regions, or as long as they consider regions that have POIs of different categories closer to each other prior to other regions. To this end, we proceed to present our problem formulation as below.

Definition 3 (Diversity and distribution-aware Region Search (DARS)). *In a spatial object dataset O, given a query rectangle size $a \times b$, a diversity function f, and a distance threshold δ, DARS finds a region r of size $a \times b$ that minimizes $f(r)$, while the distance between at least one pair*

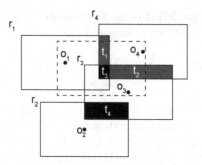

Fig. 2. Rectangle intersections.

of POIs of the same category should be no greater than a threshold δ: Formally, DARS finds a region r i.e., $r = \arg\min_{r \in R} f(r)$, subject to the constraint $\exists i, j \ \ d(o_{r_m}^i, o_{r_m}^j) \leq \delta, 1 \leq m \leq |C|$.

Note that the constraint requires a distance threshold δ. It means that there should be at least a pair of POIs of the same category that hold a distance of at most δ from each other. This is a tunable parameter to cater for needs in different preferences and circumstances – some users might think that the more the better, while others might just want a small but enough amount of POIs around. When δ becomes the diagonal's length of the query rectangle, there is no constraint for POIs of the same category. We use as $\delta - constraint$ to denote this constraint set for POIs of the same category in a region. Moreover, the definition can be generalized so that we can even consider the constraint for POIs of different categories and try to minimize the minimum average distance among POIs of the same category. This is how the DARS problem can be generalized. In this paper we follow the case in Definition 3.

3   Our Solutions

In this section we explain the details of our solutions. We first show that the DARS problem can be reduced to the rectangle intersection problem. Then we propose a baseline method called DPOF to solve the rectangle intersection problem in order to find the answers to the DARS problem. However, DPOF lacks in efficiency as we will see in experiments, so we present an optimization strategy. Last, we further optimize the method via a space partitioning to enhance the efficiency.

3.1   Finding Regions

Before we score a region, we must find one. Suppose a user issues a query request with a rectangle query range of size $a \times b$, the goal is to find some candidate regions in the whole search space (e.g., search in a city or a province) as results.

Algorithm 1: FindMaxIntersects

Input: A set R of rectangles
Output: All maximal intersections T

1 $T \leftarrow \emptyset, G \leftarrow \emptyset$;
2 **while** Sweep a vertical line l_v from left to right **do**
3   **if** l_v meets the left of a rectangle r **then**
4          $G \leftarrow G \cup \{r\}$;
5   **else if** l_v meets the right of a rectangle r **then**
6   **if** A left edge and a right edge has been scanned continuously **then**
7   Get the maximal intersection t of rectangles in G;
8   **if** t is not fully covered by any intersection in T **then**
9   Delete intersections that are covered by t from T;
10                 $T \leftarrow T \cup \{t\}$;
11         $G \leftarrow G \backslash \{r\}$;
12 **return** T;

Algorithm 2: DARS Post-Filter (DPOF)

Input: A set R of rectangles of size $a \times b$
Output: Candidate regions R_c

1 $R_c \leftarrow \emptyset$;
2 $T \leftarrow \emptyset, A \leftarrow \emptyset$;
3 **while** Sweep a horizontal line l_h from bottom to top **do**
4   **if** l_h meets the bottom of a rectangle r **then**
5          $A \leftarrow A \cup \{r\}$;
6   **else if** l_h meets the top of a rectangle r **then**
7   **if** A bottom edge and a top edge has been scanned continuously **then**
8              $T' \leftarrow$ FindMaxIntersects(A);
9              $T \leftarrow T \cup T'$;
10         $A \leftarrow A \backslash \{r\}$;
11 **foreach** t in T **do**
12     $O_t \leftarrow$ Set of objects at the centers of rectangles that form t;
13   **if** O_t satisfies the $\delta - constraint$ **then**
14         $r_c \leftarrow$ a new region of size $a \times b$, centered at a the center of t;
           `// r_c contains all objects in o_t`
15         $R_c \leftarrow R_c \cup \{r_c\}$;
16 **return** R_c

Unfortunately, it is prohibitively expensive to search the whole space because there is an infinite number of rectangles of size $a \times b$.

Reduction of the DARS Problem. We first reduce the step of finding regions to the problem of *max-enclosure* in [14]. The goal of *max-enclosure* is to find the position of a rectangle that encloses a maximum number of points. This enclosure problem can be transformed to the *rectangle intersection* problem according to [14]. Hence our first goal becomes finding where the most rectangles intersect in a given set of rectangles.

Consider the example in Fig. 2. There are four objects, and we draw rectangles with the same size $a \times b$ centered at each object. The highlighted areas are possible intersections. r_2 and r_3 form an intersection t_4. r_1 and r_4 form an intersection t_1, r_4 and r_3 form an intersection t_2. Here, t_1 and t_2 again intersect and form an intersection t_3. We can see that t_3 is formed by a set of three rectangles r_1, r_3 and r_4, which is a superset of the rectangles that form t_1 and t_2. As a result, for these four rectangles we find two "most important" intersections t_3 and t_4. For an intersection we find, we can simply choose an arbitrary point inside the intersection and draw a rectangle of size $a \times b$ centered at the point. The newly drawn rectangle will cover the centers of all rectangles that form the intersection. For instance, the dashed-line rectangle can cover o_1, o_3 and o_4 because it is centered at an point inside t_3. Hence we can find all such intersections and rate the corresponding regions to get the answers to the DARS problem.

Finding Intersections. From the example in Fig. 2 we know that we only need to find and check those "most important intersections" because they are formed by a maximal number of rectangles among all intersections. We refer to these most important intersections as *maximal intersections*. In general, we use the sweep-line method to find the maximal intersections. We summarize the steps in Algorithm 1. For a set of rectangles, we can scan from left to right (Line 1.2–1.11) and during the scanning we can check if a possible group of rectangles is a superset of other groups that have already been found. If they do include some other groups, we discard those groups and preserve this newly found group (Line 1.8–1.10)

With the help of Algorithm 1, we can start scanning all rectangles to find regions. We summarize the method in Algorithm 2. We use a horizontal line to scan from the bottom to the top. If the line meets the bottom of a rectangle, we add this rectangle to an *Active rectangles* set. If the line meets the top of a rectangle, we remove this rectangle from the *Active rectangles* set (Line 2.3–2.10). If the line meets a bottom edge and a top edge consecutively, it means that we can find some possible groups of intersecting rectangles from *Active rectangles* set now (Line 2.7–2.9). The whole scanning process terminates until the topmost horizontal edge is met. We can get a series of regions by drawing rectangles centered at a point in intersections attained from Algorithm 1. Finally we can filter out the regions that does not satisfy the $\delta - constraint$, then we have the remaining regions as results (Line 2.11–2.15).

With Algorithm 1 and Algorithm 2, we now have a basic workflow of solving the DARS problem. In the following sections we will focus on the optimization of this workflow.

3.2   Checking the $\delta - Constraint$ During Sweeping

Algorithm 1 finds all maximal intersections. When the size of O grows or the query rectangle size grows, it is time-consuming to find all maximal intersections.

Algorithm 3: Stripe Pre-Filter (SPRF)

Input: A set R of rectangles of size $a \times b$, threshold δ
Output: Candidate regions R_c

1 $T \leftarrow \emptyset$;
2 Cut the wholse space in horizontal direction into a set of stripes S;
3 **foreach** s in S **do**
4      $R' \leftarrow$ Rectangles in R that intersect with s;
5      $R'_c \leftarrow$ DARS Pre-Filter(R', δ);
6      $R_c \leftarrow R_c \cup R'_c$;
7 **return** R_c;

Is it possible that we can reduce some computations so we do not need to rate *every* possible region?

The answer is positive. In Algorithm 1, if we find one maximal intersection, we need to check if it is already covered by other intersections or if it covers other intersections, and this is time-consuming. Therefore by first filtering with the $\delta - constraint$ we can reduce a lot of computation. We implement this filter idea as follows. In Algorithm 1, when sweeping (Line 1.2–1.11), we can first use the $\delta - constraint$ to filter out some groups of rectangles. This is because if a group of rectangles intersect, and their centers do not satisfy the $\delta - constraint$, we can simply ignore this group of rectangles. We can simply add a checking process before Line 1.8. If centers of rectangles in G does not satisfy the $\delta - constraint$, then we can terminate the current loop and start the next loop. In the end we will get some groups of rectangles and their corresponding maximal intersections, and the objects that each group contains will satisfy the $\delta - constraint$. With these groups that satisfy the $\delta - constraint$ found, we then do not need to check $\delta - constraint$ when processing each maximal intersections (Line 2.11–2.15). Therefore we can remove the if check at Line 2.13.

With these modifications done to Algorithm 1 and Algorithm 2, we then have an optimized workflow of solving the DARS problem. We name this modified algorithm DPRF (DARS Pre-Filter).

3.3 Further Optimization with Stripes

As a matter of fact, for every possible group of intersecting rectangles, if we want to know if one such group satisfies $\delta - constraint$, we need to scan this group, so we cannot just ignore it like BRS [7] does. Thus, our further optimization strategy needs to focus on optimizing the execution of the algorithms.

By dividing the space range in one dimension into a series of intervals, we can further reduce the computations needed to be done. Here we cut the space with vertical lines, so in the horizontal direction there will be a series of intervals. These intervals and the splitting vertical lines will form a series of rectangle spaces. We will denote as "stripes" these rectangle spaces. If there is a group of intersecting rectangles, the furthest distance between the leftmost vertical edge and the rightmost vertical edge will always be less than $2 \times a$, otherwise they will not be intersecting each other, so we will set the width of each stripe to $2 \times a$. We illustrate this in Fig. 3. By doing so, a group of rectangles will only intersect with at most two stripes. Therefore, we can issue a new instance of DPRF for each stripe. Although the total number of loops will increase, things done in each loop will be reduced drastically compared to the increase in number of loops. We will see this effect in experiment section.

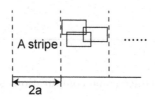

Fig. 3. Cutting space into stripes

We summarize the process in Algorithm 3. We issue a new DPRF instance for each stripe (Line 3.2–3.6, input to DPRF will be the rectangles that intersect with the stripe and the distance constraint δ, and these rectangles can be found by using the interval tree), then combine the results from each stripe and we will get the final results in the whole space. We call Algorithm 3 SPRF (Stripe Pre-Filter).

Table 2. Summary of datasets.

Property	BrightKite	Yelp	Synthetic
# of objects	772,966	157,776	60,000
Width	40,075 km	3,933 km	84 km
Height	20,015 km	2,012 km	152 km

Time Complexity Analysis. Suppose there are n rectangles, and denote the average size of active rectangles (A used in DPOF (Algorithm 2) and DPRF) as $|A|$, the average number of POIs in a region as $|r|$. It takes $O(|A||r|)$ time to check for maximal intersections, so the total complexity of DPOF is $O(n|A||r|)$. For DPRF, in the worst case it has the same complexity as DPOF (every possible group satisfies $\delta - constraint$). As for SPRF, we construct an interval tree in $O(nlogn)$ time to get rectangles (Denoted as R' in Algorithm 3) that intersect with a stripe. It takes $O(logn + |R'|)$ to find them with the interval tree ($|R'| \ll n$). There are at most $\frac{n}{2}$ stripe, and each DPRF instance has a complexity of $O(|R'||r|^2)$. Thus, the toal complexity of SPRF is $O(nlogn + \frac{n}{2}(logn + |R'| + |R'||r|^2)) = O(nlogn + n|R'||r|^2)$.

4   Experiments

We first explain our setups for the experiments. Then we evaluate our methods on both quality and efficiency. Last, we run the scalability test.

4.1   Experiment Setup

Datasets. We use two real-life datasets BrightKite[3] and Yelp[4] to evaluate our methods. We will also use synthetic datasets for evaluation and scalability test. Summary of different datasets is in Table 2. POIs in BrightKite dataset are scattered over the whole globe. POIs in Yelp dataset are mainly in North America. POIs in synthetic dataset are generated in the scope of a normal city.

Query Rectangle Size. Different datasets come with different cardinality, and this will have impact on efficiency of the algorithms. Thus, we set different unit query rectangle size for different datasets. Given a dataset with cardinality $|O|$, we can get its minimum bounding rectangle of size $w \times h$, and the size of a unit query rectangle is $a \times b$, where $a = \frac{w}{|O|}$ and $b = \frac{h}{|O|}$. We can also apply a multiplier k to a and b so the query rectangle size becomes $ka \times kb$. Sometimes a or b might actually become a very small value (e.g., $2\,\text{m} \times 1\,\text{m}$) and this is irrational. Under such circumstances we will expand a or b until they become rational values, and we will use a and b after the expansion as the query unit size $q = a \times b$. We will indicate this expansion with different k in the figures. Different dataset has different k, hence leading to different sizes kq. In Fig. 4 we demonstrate how we choose a k value for Yelp dataset. For following experiments we choose $k = 8$ for Yelp dataset so that the region size is about $200\,\text{m} \times 100\,\text{m}$ which is proper. For other datasets, we omit reporting the figures due to space limit ($k = 4$ for BrightKite dataset and $k = 100$ for Synthetic dataset).

Performance Measures. We mainly focus on these measures: (a) Efficiency, which is the runtime of each algorithm. For SPRF, we set the width of each stripe to $2 \times a$ as discussed in Sect. 3.3. (b) The quality of returned regions that satisfy the $\delta - constraint$. Quality is the value calculated according to f in Definition 1. Since DPOF, DPRF and SPRF achieve the same quality, we will use a united name DARS to compare with BRS. For BRS, we use f from Definition 1 for its pruning strategy. For evaluations involving no number of categories, we use $|C| = 3$. For evaluations involving no changing of $\delta - constraint$, we use $\delta = \frac{a}{10}$ by default. Note that according to Definition 2 and 3, the smaller the quality is, the better the corresponding region is.

Since BRS [7] ignores a lot of possible groups of intersecting rectangles, it cannot guarantee that the result it finds satisfies $\delta - constraint$ (the result it finds might not be an answer to the DARS problem), so when compared with other methods, runtime is the main factor. But for the completeness we also compare BRS with other methods on quality.

---

[3] http://snap.stanford.edu/data/index.html.
[4] https://www.yelp.com.sg/dataset/.

Implementation. All algorithms are implemented in Java with JDK 11. We run all experiments on a PC with Intel i7-8700K 3.70 GHz CPU and 32 GB RAM.

4.2   Evaluation Results

For all the runtime evaluations we did not show the breakdown when reporting all efficiency results because the time taken for scoring a region was only 1%–2% compared to finding regions. Each experiment is repeated ten times, and the average result is reported.

Runtime (Different Datasets). First we evaluate the runtime of different algorithms on different datasets with proper k values. From Fig. 5 we can see that as $|O|$ grows, the runtime of DPOF increases fairly fast. It always scans and checks intersections without filtering with $\delta - constraint$. But it is also slightly faster than BRS and DPRF when the size of data is small (60,000 POIs in synthetic dataset), because DPOF does nothing but simply scanning and finding intersections, while BRS needs to first estimate an upper bound for each vertical interval, and DPRF needs to check $\delta - constraint$ whenever it runs into a possible group of rectangles. When the size of data is small, such as 60,000 POIs in synthetic dataset, these extra efforts in BRS and DPRF make them only a bit slower. But when the size of data grows, the advantages of BRS and DPRF stand out because they do not need to check intersections as often as DPOF does. The improved SPRF algorithm runs one magnitude faster than others on average.

Runtime (Different Query Rectangle Sizes). Figure 6 illustrates the runtime for different algorithms on different datasets with different query rectangle sizes. We find as the size of query rectangle grows, the runtime of each algorithm also grows because more intersections among rectangles are likely to appear. The runtime of DPOF grows still quite fast, while the runtime of BRS and DPRF grows slower. On the synthetic dataset with a relatively small size, DPOF still runs only a little faster because it does nothing during the scanning process except for finding the maximal intersections. SPRF is still the fastest here (1–2 magnitude faster).

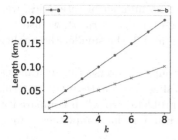

Fig. 4. a and b with different k.

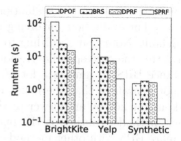

Fig. 5. Runtime on different dataset with proper k.

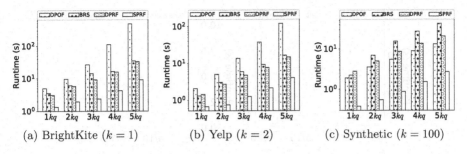

(a) BrightKite ($k = 1$)   (b) Yelp ($k = 2$)   (c) Synthetic ($k = 100$)

Fig. 6. Runtime with different query rectangle sizes.

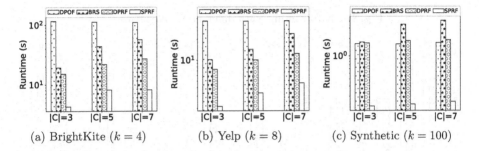

(a) BrightKite ($k = 4$)   (b) Yelp ($k = 8$)   (c) Synthetic ($k = 100$)

Fig. 7. Runtime with different number of categories.

Runtime (Different Number of Categories). We set different number of categories to evaluate its impact on the runtime of different algorithms in Fig. 7. As the number of categories grows, the runtime of DPOF stays almost the same and it is the slowest. As for the remaining algorithms, the runtime of each algorithm grows because more calculation needs to be done in $f(r)$. The growth on the runtime of DPRF and SPRF is the slowest because when there are more categories, there are fewer POIs of the same category needed to be checked against $\delta - constraint$. Again SPRF outperforms the rest (1–2 magnitude faster).

Runtime (Different Constraints for DARS). In Fig. 8 we set different $\delta - constraints$ to evaluate the runtime of DPRF and SPRF. Since DPOF does not use $\delta - constraint$ during the process of scanning, the impact of different δ is tested only on DPRF and SPRF here. We set the constraint based on the width of a query rectangle by dividing the width with different values. We can see that as δ becomes smaller, the runtime of both algorithms decreases because when δ is small, fewer regions would satisfy the $\delta - constraint$, so that fewer operations would be required to check new intersections and delete old intersections in Algorithm 1. SPRF is still faster than DPRF (3–10 times faster).

Quality (Different Query Rectangle Sizes). Because DPOF always finds the same answers as DPRF and SPRF do, we will use a united name DARS to compare with BRS. As DARS finds multiple top regions, we use the best one in order to compare with BRS because BRS always finds a single region as

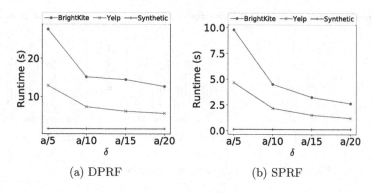

(a) DPRF   (b) SPRF

Fig. 8. DARS runtime with different $\delta - constraint$.

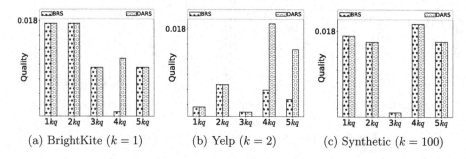

(a) BrightKite ($k = 1$)   (b) Yelp ($k = 2$)   (c) Synthetic ($k = 100$)

Fig. 9. Quality with different query rectangle sizes.

result. One thing about BRS is that it always finds the region that minimizes $f(r)$ without considering $\delta - constraint$. So for $f(r)$, if BRS gets a value v_1 and DARS gets a value v_2, $v_1 \leq v_2$ always holds. Now we set different query rectangle size for quality evaluations in Fig. 9. The difference between some values seems huge, but in fact the quality values differ at around 6–7 digits after the decimal point. For example, in Fig. 9a at $4kq$ the quality value from BRS is 0.018315693 while DARS returns a quality value of 0.018315712, and the quality value at $1kq$ is 0.018315724. This is the same for difference of quality values in Fig. 9b and Fig. 9c.

Quality (Different Number of Categories). In Fig. 10 we check the quality with different number of categories. As we can see difference brought by different $|C|$ is very obvious (due to how $f(O_r)$ works). When $|C| = 7$, quality returned from synthetic dataset does not reach a value as big as that from BrightKite and Yelp. Because in a relatively small dataset we might find regions with less than $|C|$ categories of POIs.

Quality (Different Constraints for DARS). We check the impact of different δ on the quality of a returned region with $|C| = 3$ by default. From Fig. 11 we can see that when the size of data is fairly small, the change in quality is obvious (the polyline of Synthetic dataset with stroke markers). When the size

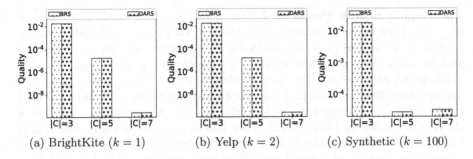

(a) BrightKite ($k = 1$)   (b) Yelp ($k = 2$)   (c) Synthetic ($k = 100$)

Fig. 10. Quality with different number of categories.

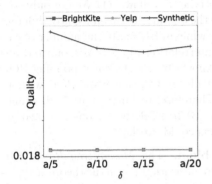

Fig. 11. DARS quality with different $\delta - constraint$.

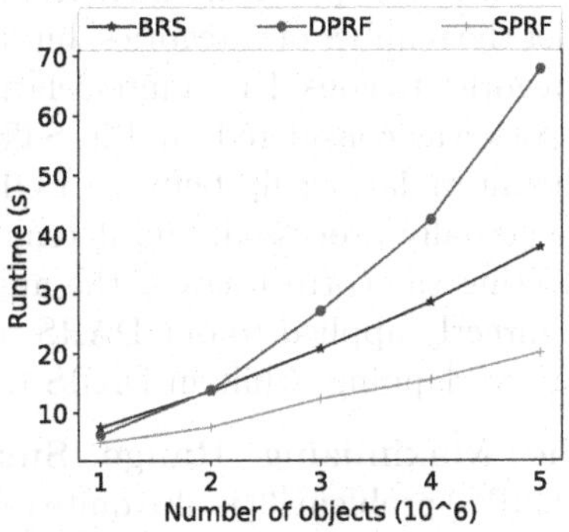

Fig. 12. Scalability test.

of data grows big, there is almost no change in quality because we are always likely to find all $|C|$ categories of POIs when there are more POIs no matter what the constraint value is.

Scalability Test. Now we generate synthetic datasets of different size to test the scalability of DPRF and SPRF because DPOF finds the same results and does not perform well on big dataset. We also test BRS here. We generate synthetic datasets in the same scope as in BrightKite dataset. The runtime is illustrated in Fig. 12. Neither DPRF nor BRS scales that well. SPRF scales a bit better, although when there are more than 3 million objects it still takes more than 10 s to terminate, it can already handle real-life scenarios where there are not that many objects.

5 Related Work

The DARS problem is closely related to the *region search* problem. There are also other related studies such as the *maximizing range sum* problem [3]. Table 3 summarizes existing methods.

The Region Search Problem. Region search has been studied in the past few years. Feng et al. [7] studied the problem of *best region search* (BRS). Given a set O of all spatial objects, a rectangle r of a given size and a submodular and monotone function which extracts a property for a region (e.g., Number of categories in a region), BRS aims to identify a single "best region" that can maximize this function, and proposed algorithms and pruning techniques around the function. Another problem called *subjected-oriented top-k hot region query* (STR) was studied by Liu et al. [12]. They aim to find the top-k non-overlapping square regions that have the highest scores computed by the number of feature objects (e.g., objects of cultural features include museums, libraries, exhibition halls, etc.) and their weights.

Our DARS problem differs from these studies as follows. (1) We not only consider the number of categories, but also use the distance between POIs of different categories to consider an interrelationship, while in BRS only the number of categories was considered, and in STR this interrelationship was also omitted. (2) The interrelationship between POIs is unknown before the corresponding POIs are actually processed, and it remains a challenge to prune some POIs without affecting the correctness of the answers. Therefore, techniques in [7,12] cannot be directly applied to our DARS problem. (3) In STR the regions returned are non-overlapping, while in DARS the regions could overlap.

The Maximizing Range Sum Problem. The *Maximizing Range Sum* (MaxRS) problem [3] is also quite relevant to our work. First in the theoretical perspective for MaxRS Imai et al. [10] and Nandy et al. [14] aimed to find a rectangle to enclose the maximum number of points based on a classical *distribution-sweep* paradigm [9]. Given a set O of weighted points and a rectangle r of a given size, MaxRS aimed to find a rectangle range of the same size that maximizes the sum of weights of all the points covered by the rectangle (e.g., sum of influence of all points). A naive solution is to issue an infinite number of *range aggregate* (RA) operations [1,11,15–17] but this is prohibitively expensive. Choi et al. [3] solved the MaxRS problem with a scalable method in spatial databases.

Our DARS problem differs from MaxRS in these aspects: (1) we take into account the interrelationship between points in a region; (2) our objective function completely differs from a simple aggregate function in MaxRS (e.g., SUM, COUNT). As a result, DARS does not consider a weight for each POI while MaxRS does.

There are also other studies less relevant to this work. To name a few, some studies tried to capture the features of regions and find similar regions to a given query region [8,13]. Region-wise deployment of a set of facilities or advertisements are studied in [21,22], aiming to maximize the number of users influenced. Choi et al. [2] proposed *nearest neighbourhood search* to find circular nearest regions that contains some number of POIs through pure geometric computations.

Table 3. Summary of methods

Methods	Diversity	Interrelationship	Objective function
BRS	#Category	None	Submodular and monotone
MaxRS	#Category	None	Aggregate
DARS	#Category	Distance between POIs	Not submodular, not aggregate

6   Conclusions

In this paper we introduced a novel problem called the *diversity and distribution-aware region search* (DARS). The goal of DARS is to find regions that are diversified (contain many categories of POIs) and the POIs within these regions are well-distributed so that users can enjoy several kinds of services they want without spending much time on commuting. We proposed several methods including the baseline method DPOF, an improved method DPRF and a further optimized method SPRF. SPRF can handle many kinds of scenarios quite efficiently in extensive experiments over real-world datasets. In the future we would like to discuss how to solve the DARS problem in a road network environment, or assigning different weights to different categories.

Acknowlegements.. Special thanks to Kunkui Yang at BaiZhi Data Technology Co., Ltd., Nanjing, China for providing essential inspirations. This research is supported in part by ARC DP200102611, DP180102050, NSFC 91646204.

References

1. Cho, H., Chung, C.: Indexing range sum queries in spatio-temporal databases. Inf. Softw. Technol. **49**(4), 324–331 (2007)
2. Choi, D., Chung, C.: Nearest neighborhood search in spatial databases. In: ICDE, pp. 699–710. IEEE Computer Society (2015)
3. Choi, D., Chung, C., Tao, Y.: A scalable algorithm for maximizing range sum in spatial databases. PVLDB **5**(11), 1088–1099 (2012)
4. Choudhury, F.M., Culpepper, J.S., Bao, Z., Sellis, T.: Finding the optimal location and keywords in obstructed and unobstructed space. VLDB J. **27**(4), 445–470 (2018). https://doi.org/10.1007/s00778-018-0504-y
5. Drezner, Z., Hamacher, H.W.: Facility Location - Applications and Theory. Springer, Heidelberg (2002)
6. Du, Y., Zhang, D., Xia, T.: The optimal-location query. In: Bauzer Medeiros, C., Egenhofer, M.J., Bertino, E. (eds.) SSTD 2005. LNCS, vol. 3633, pp. 163–180. Springer, Heidelberg (2005). https://doi.org/10.1007/11535331_10
7. Feng, K., Cong, G., Bhowmick, S.S., Peng, W., Miao, C.: Towards best region search for data exploration. In: SIGMOD Conference, pp. 1055–1070. ACM (2016)
8. Feng, K., Cong, G., Jensen, C.S., Guo, T.: Finding attribute-aware similar region for data analysis. PVLDB **12**(11), 1414–1426 (2019)
9. Goodrich, M.T., Tsay, J., Vengroff, D.E., Vitter, J.S.: External-memory computational geometry (preliminary version). In: FOCS, pp. 714–723. IEEE Computer Society (1993)

10. Imai, H., Asano, T.: Finding the connected components and a maximum clique of an intersection graph of rectangles in the plane. J. Algorithms **4**(4), 310–323 (1983)
11. Lazaridis, I., Mehrotra, S.: Progressive approximate aggregate queries with a multi-resolution tree structure. In: SIGMOD Conference, pp. 401–412. ACM (2001)
12. Liu, J., Yu, G., Sun, H.: Subject-oriented top-k hot region queries in spatial dataset. In: CIKM, pp. 2409–2412. ACM (2011)
13. Liu, Y., Zhao, K., Cong, G.: Efficient similar region search with deep metric learning. In: KDD, pp. 1850–1859. ACM (2018)
14. Nandy, S., Bhattacharya, B.: A unified algorithm for finding maximum and minimum object enclosing rectangles and cuboids. Comput. Math. Appl. **29**(8), 45–61 (1995)
15. Papadias, D., Kalnis, P., Zhang, J., Tao, Y.: Efficient OLAP operations in spatial data warehouses. In: Jensen, C.S., Schneider, M., Seeger, B., Tsotras, V.J. (eds.) SSTD 2001. LNCS, vol. 2121, pp. 443–459. Springer, Heidelberg (2001). https://doi.org/10.1007/3-540-47724-1_23
16. Sheng, C., Tao, Y.: New results on two-dimensional orthogonal range aggregation in external memory. In: PODS, pp. 129–139. ACM (2011)
17. Tao, Y., Papadias, D.: Range aggregate processing in spatial databases. IEEE Trans. Knowl. Data Eng. **16**(12), 1555–1570 (2004)
18. Wong, R.C., Özsu, M.T., Yu, P.S., Fu, A.W., Liu, L.: Efficient method for maximizing bichromatic reverse nearest neighbor. PVLDB **2**(1), 1126–1137 (2009)
19. Xiao, X., Yao, B., Li, F.: Optimal location queries in road network databases. In: ICDE, pp. 804–815. IEEE Computer Society (2011)
20. Zhang, D., Du, Y., Xia, T., Tao, Y.: Progressive computation of the min-dist optimal-location query. In: VLDB, pp. 643–654. ACM (2006)
21. Zhang, P., Bao, Z., Li, Y., Li, G., Zhang, Y., Peng, Z.: Trajectory-driven influential billboard placement. In: KDD, pp. 2748–2757. ACM (2018)
22. Zhang, Y., Li, Y., Bao, Z., Mo, S., Zhang, P.: Optimizing impression counts for outdoor advertising. In: KDD, pp. 1205–1215. ACM (2019)

I/O Efficient Algorithm
for c-Approximate Furthest Neighbor
Search in High-Dimensional Space

Wanqi Liu[1]([✉])[iD], Hanchen Wang[1], Ying Zhang[1], Lu Qin[1], and Wenjie Zhang[2]

[1] University of Technology Sydney, Sydney, Australia
{wanqi.liu,hanchen-1.wang}@student.uts.edu.au,
{ying.zhang,lu.qin}@uts.edu.au
[2] University of New South Wales, Sydney, Australia

Abstract. Furthest Neighbor search in high-dimensional space has been widely used in many applications such as recommendation systems. Because of the "curse of dimensionality" problem, c-approximate furthest neighbor (C-AFN) is a substitute as a trade-off between result accuracy and efficiency. However, most of the current techniques for external memory are only suitable for low-dimensional space.

In this paper, we propose a novel algorithm called reverse incremental LSH based on Indyk's LSH scheme to solve the problem with theoretical guarantee. Unlike the previous methods using hashing scheme, reverse incremental LSH (RI-LSH) is designed for external memory and can achieve a good performance on I/O cost. We provide rigorous theoretical analysis to prove that RI-LSH can return a c-AFN result with a constant possibility. Our comprehensive experiment results show that, compared with other c-AFN methods with theoretical guarantee, our algorithm can achieve better I/O efficiency.

Keywords: Locality-sensitive hashing · Furthest neighbour search · Similarity search

1 Introduction

Furthest Neighbor (FN) search is a logical opposite of the well-known problem Nearest Neighbor (NN) search. Given a set of d-dimensional objects and a query object, similar to Nearest Neighbor search, the Furthest Neighbor search aims to find an object which has the longest distance to the query point. It has been widely applied in many domains such as recommendation systems to increase the diversity of the recommendation [12,13]. For example, if a user is interested in a product q, the recommendation algorithm would recommend some similar products o to the user as well. Suppose we have a set of "similar" objects, which means all the objects in the set are close to q, we might hope to recommend the furthest point o in the set to q since it won't increase the possibility of a sale very significantly if two products are too similar. Besides, FN is also a key

Y. Nah et al. (Eds.): DASFAA 2020, LNCS 12114, pp. 221–236, 2020.
https://doi.org/10.1007/978-3-030-59419-0_14

component in some fundamental problems such as maximum spanning tree [1] and non-linear dimensionality reduction problem [14]. There are several existing solutions for furthest neighbor search in low-dimensional space. But when it comes to the high-dimensional space, FN faces the similar problem with NN search, which is called "curse of dimensionality". It becomes very expensive to find the exact furthest neighbor for a given query object. To solve this problem, an approximate FN is an acceptable solution as a trade-off between efficiency and accuracy. In this paper, we focus on solving approximate furthest neighbor problem with theoretical guarantee, which is also called c-AFN problem. Given an approximation ratio c ($c > 1$) and a success possibility δ, a c-AFN query returns a c-approximate furthest neighbor with confidence at least δ. In many applications, the dataset is huge and hard to be fitted in the main memory. So we aim to develop an external memory algorithm which is I/O efficient.

1.1 Motivation

Locality Sensitive hashing (LSH) [10] is a well known solution for c-approximate nearest neighbor search in the high-dimensional space. It has a good property that the close objects in the original space are likely to be close as well after projection. Using such property, we can also design an algorithm for c-AFN search using LSH.

The projection distance on a LSH function is the most important information to reflect the distance between two points in the original space. In most LSH schemes, there is a "bucket" with a certain width. If two points o_1 and o_2 fall in the same bucket, we say o_1 and o_2 collide over this function. In the NN problem, the number of collisions of a point o and query point reflects the possibility of o to be a NN points. As an opposite problem of Nearest Neighbor, it is easy to get a conclusion: if o_1 and o_2 don't collide in a LSH function, the Euclidean distance between o_1 and o_2 is likely to be a large value in the original space.

Most of the LSH based algorithms use a virtual rehashing to expand the searching radius. The state-of-the-art algorithm for c-AFN with theoretical guarantee is RQALSH [9], which proposed a solution to set the bucket width R at a large value and then decrease it to $\frac{R}{c^k}$ at each round, where k is a constant number, and c is the approximation ratio. It adopts the bucket exponential reducing strategy which means the bucket widths always be a power of c and will encounter too many objects. For example, in Fig. 1(a), the objects which are in bucket R_1 but not in bucket R_2 are dense. If the radius reduces from R_1 to R_2, most of the points will be involved. Another drawback of RQALSH is that the approximation ratio is c^2 but not c, which leads to larger index size and more I/O cost.

Based on the observations, we follow the framework of RQALSH to enjoy the high efficient sequential I/O brought by B+ tree and propose a more I/O efficient algorithm by optimizing the search strategy to decrease the I/O cost and index size. Besides, we provide rigorous analysis to show that the continuous decreasing search strategy shown in Fig. 1(b) can hold the theoretical guarantee. The algorithm, named RI-LSH, improves the I/O performance significantly by

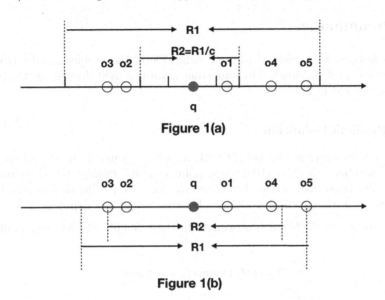

Figure 1(a)

Figure 1(b)

Fig. 1. Motivation for RI-LSH

tightening the approximation ratio from c^2 to c. For example, if the approximation ratio is 4, our algorithm can use $c = 4$ directly in the algorithm, while RQALSH has to use $c = 2$ to get a 4-approximate solution.

Example 1. In Fig. 1, RQALSH's searching radius will decrease from R to R/c and then all the objects falling in the bucket (o_2, o_3, o_4, o_5) will be loaded. While RI-LSH decreases the radius continuously, and only one point will be visited. If the data points aggregate at one bucket after projection, using RQALSH's searching strategy will lead to a result that a large amount of points have to be loaded into memory and to be visited, while our algorithm can avoid such situation.

1.2 Contribution

The principal contributions of our work are summarized as follows:

- We propose a novel c-AFN algorithm called RI-LSH for high-dimensional data. It uses continuous searching strategy on each projection dimension.
- We prove that our algorithm has theoretical guarantee and has a more strict approximation bound.
- We conduct extensive performance evaluation against two c-AFN algorithms regarding I/O cost, running time and result accuracy. The results show that our algorithm can achieve a better performance on both accuracy and efficiency.

2   Preliminaries

In this section, we present the formal definition of the problem, and introduce some existing related work. The important notations used throughout the paper are summarized in Table 1.

2.1   Problem Definition

Given a d-dimensional dataset D with n points, donated by R^d, where d is a large number (e.g., $d \geq 100$), each point o in the dataset has d coordinate values. The coordinate value of o on the i_{th} dimension is denoted as $o[i]$. For a query point q, the Euclidean distance between o and q is denoted by $\|o, q\| = \sqrt{\sum_{i=1}^{d}(o[i] - q[i])^2}$. The c-approximate furthest neighbor is defined as follows:

Table 1. Summary of notations

Notation	Definition
n	The number of point objects in the dataset
d	The dimensionality of the dataset
m	The number of hash functions (projected dimensions)
q	The query object
$\|o_1, o_2\|$	The Euclidean distance between o_1 and o_2
o^*, R^*	The FN object of q, with distance R^*
o_{max}, R_{max}	c-AFN object returned, with distance R_{max}
c and δ	Approximate ratio and success probability
w	The initial bucket width in LSH functions
R and r	$R = \frac{2r}{w}$, radius of ball $B(q, R)$ in $\mathcal{R}^d$ space regarding search radius r

Definition 1. *c-approximate furthest neighbor (c-AFN). Given a query object q and a d-dimensional dataset D and a furthest neighbor o^* of q whose distance to q is R^*, a c-approximate furthest neighbor of q is a point $o \in D$ which satisfied $\|o, q\| \geq \frac{R^*}{c}$, where c is the approximation ratio $(c > 1)$.*

Problem Statement. In this paper, we focus on proposing an efficient algorithm to solve the c-AFN problem in a high-dimensional Euclidean space with theoretical guarantee, which means given the approximation ratio c, a query point q and the probabilistic threshold δ, the algorithm will return a c-AFN of q with probability at least δ.

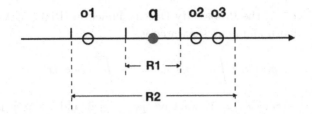

Fig. 2. Example for separation

2.2 LSH Family for Furthest Neighbor

Just like we discussed in Sect. 1, our algorithm is designed on the LSH scheme, so first we will introduce the traditional LSH for NN search. Locality-Sensitive Hashing was actually designed for (R, c)-NN search in high-dimensional space. It means that the radius R is given, and the LSH returns an object o satisfied $\|o, q\| \leq cR$. A LSH family consists of a set of LSH functions, and each one is defined as (r_1, r_2, p_1, p_2)-sensitive if and only if the hash function satisfies that for any two points x and y and a given radius r_1 and r_2 $(r_1 < r_2)$, $\begin{cases} \Pr_{H \in \mathcal{H}}[H(x) = H(y)] \geq p_1, & \text{if } f(x, y) \leq r_1 \\ \Pr_{H \in \mathcal{H}}[H(x) = H(y)] \leq p_2, & \text{if } f(x, y) \geq r_2 \end{cases}$. It means that the possibility of mapping two points into a same hash bucket decreases as the Euclidean distance between the two points grows.

To solve the c-AFN problem, the inequation should be modified:
$\begin{cases} \Pr_{H \in \mathcal{H}}[H(x) \text{ separates with } H(y)] \geq p_1, & \text{if } f(x, y) \geq r_1 \\ \Pr_{H \in \mathcal{H}}[H(x) \text{ separates with } H(y)] \leq p_2, & \text{if } f(x, y) \leq r_2 \end{cases}$. The definition of "separate" is given in Definition 2.

Definition 2. Separate. *Given a constant value w and a query point o, if the projection distance of o and q is larger than $\frac{w}{2}$ ($\|H(o), H(q)\| \geq \frac{w}{2}$), then we say o and q are separated on $H(\cdot)$.*

In Fig. 2, q is the query points and there are two bucket widths R_1 and R_2. For bucket R_1, all the objects are separated with q, while for bucket R_2, all the points are collided with q.

2.3 $(c, 1, p_1, p_2)$-Sensitive Reverse LSH

Similar as RQALSH, a reverse LSH function can be formally represented by $H_a(o) = a \cdot o$ where o is the d-dimensional object and a is a d-dimensional vector for the random projection, whose elements are following the p-stable distribution. To solve the Euclidean space problem, p is set to 2, e.g., the normal distribution.

Let $s = \|o, q\|$, $r_1 = c$ and $r_2 = 1$. We can compute p_1 and p_2. According to the property of p-stable distribution, $(a \cdot o - a \cdot q)$ has the same distribution with sX where X is a random variable chosen from the normal distribution $\mathcal{N}(0, 1)$.

Let $\phi(x) = \frac{1}{\sqrt{2\pi}} e^{-\frac{x^2}{2}}$, the Probability Density Function (PDF) of $\mathcal{N}(0,1)$, then the possibility that o and q are separated is:

$$p(s) = 2 \int_{-\infty}^{-\frac{w}{2s}} \phi(x)dx = 1 - \int_{-\frac{w}{2s}}^{\frac{w}{2s}} \phi(x)dx$$

When $r_1 = c$ and $r_2 = 1$, we have $p_1 = p(c)$ and $p_2 = p(1)$. Because $p(s) = 2norm(-\frac{w}{2s})$ where $norm(x)$ is the CDF of $\mathcal{N}(0,1)$ and is a monotonic increasing function of x, and the value of $-\frac{w}{2s}$ increases when s increases. Thus, $p(s)$ increases monotonically as s increases. So the reverse LSH function is $(c, 1, p_1, p_2)$-sensitive.

2.4   Related Works

To solve the c-AFN problem, there are basically two classifications: (1) using hash functions or (2) tree-based solutions. Qiang Huang proposed two efficient c-AFN algorithms in TKDE 2017 [9] called RQALSH and RQALSH*. Both of the two algorithms are based on the LSH scheme proposed by Indyk [7], and the difference is that RQALSH holds a theoretical guarantee, while RQALSH* utilized machine learning skill to heuristically pre-process the data. RQALSH used a query-aware reverse hashing function to project the d-dimensional dataset to a m-dimensional dataset and use B+ trees to store the hash values. The advantage of RQALSH is that it uses a query aware hash function so the query point is always located at the center of each bucket, and it adopts B+ tree to build the index and can have a good I/O performance since searching on B+ tree cost a sequential I/O. DrusillaSelect [6] used a novel hashing strategy for approximate furthest neighbor search that selects projection bases using the data distribution. Instead of using random projections, DrusillaSelect chooses a small number of data objects as candidates based on data distribution and also has a variant which gives an absolute approximation ratio. Another method with theoretical guarantee is QDAFN [11] which has been adapted for external memory. Compared with Indyk's implementation, QDAFN is simpler to implement and also keeps the theoretical guarantee. It also has a heuristic version called QDAFN*. For the tree-based algorithms, two of the typical algorithms are [15] proposed by B Yao and [5] proposed by RR Curtin. Both of the two algorithm build a tree using the data objects and then adopt the branch-and-bound pruning strategy to prune the nodes which is not possible to be a FN of the query point. Besides, most of the tree-based methods for NN problem can also be used for c-AFN problem as well such as R+ tree [2], kd-tree [3]and etc.

3   Approach

In this section, we present our RI-LSH algorithm in detail. First, in Sect. 1.1, we briefly introduce our motivation and then introduce our algorithm in Sect. 3.2. Section 3.3 shows that our algorithm can also be easily extended to solve top-k c-approximate furthest neighbor problem (c-k-AFN).

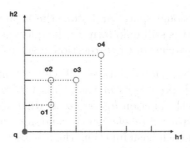

Fig. 3. Motivation for RI-LSH

3.1 Motivation

From Sect. 2 we can get that for two points o_1, o_2 and a reverse LSH function h_i, the projection distance $\|h_i(o_1), h_i(o_2)\|$ reflects the Euclidean distance between o_1 and o_2. If o has a far projection distance on most hash functions, o is likely to be a c-AFN of q. Most LSH based algorithms used a "bucket" to determine the original distance relationship between two points. To expand the radius, the algorithm using buckets has to involve all the points falling into the bucket at one time. While in our method, the key idea is to continuously reduce the radius R and only one point on a single hash function is involved in each iteration.

For example, in Fig. 3, there are two hash functions h_1 and h_2, and q is located at the origin, i.g. the hash value of q on h_1 and h_2 are both 0. The point who has the longest projection distance to q is o_4 on h_1, and $\|h_1(o_4), h_1(q)\| = 3$. So point o_4 will be accessed first, and then the algorithm visits the next furthest point. o_4 on h_2 is the next point to visit, because $\|h_2(o_4), h_2(q)\| = 3$ and all the other points have a smaller projection distance. The searching area is a square with a decreasing size when there are two hash functions. The points will be accessed one by one according to their projection distances on the functions.

To boost the success possibility, we need a set of reverse LSH functions. Suppose there are m hash functions, and a point o separated with q on all of the functions, o is a c-AFN of q with a high probability. But if we set the separation threshold as m, it is too strict to have enough candidates. In our scheme, we use a parameter α $(p_2 < \alpha < p_1)$. When a point o is separated with q for at least αm times, it could be far from o with a high enough probability. Following the traditional LSH scheme, o will be added to the candidate set S. Only the candidates in S will be calculated its real distance to q. We keep set S as a limited size to guarantee that the I/O performance is much better than linear search.

3.2 Approach

In this subsection, we describe our algorithm in details and give the pseudo-code. The setting of parameters will be given in Sect. 4. There are two steps of RI-LSH: indexing and querying. In the indexing step, we project the whole d-dimensional dataset into a m-dimensional space, and store the hash values in B+ trees. In this step, the query point is still unknown. In the querying step, the query point comes. The algorithm searches in the projection space to find the c-AFN result.

Indexing. Given a d-dimensional data set D with n points, we use m 2-stable functions to project each point o to m hash values. $h_i(o)$ denotes the i_{th} hash value of o. A 2-stable hash function $h_i(o) = \boldsymbol{a} \cdot \boldsymbol{o}$ where $\boldsymbol{o}$ is the data point and $\boldsymbol{a}$ is a d-dimensional vector whose element is randomly chosen from the normal distribution $\mathcal{N}(0,1)$. For each hash function, the n pairs $(ID(o), h_i(o))$ are stored in a B+ tree, where $ID(o)$ is the ID of object o and $h_i(o)$ is the hash value. To set up the initial radius for query processing, we store min_i and max_i for the minimum hash value and maximum hash value for each hash function $h_i(\cdot)$.

Query Processing. When the query point q comes, we first project q to m hash functions and insert the hash values of q into B+ trees, then the points in the dataset and their hash values will be accessed according to their projection distance to q on the m projected dimensions. Since we are looking for c-AFN of q, the algorithm starts from the furthest point over the m projection dimensions and decreases the projection distance continuously. For example, in Fig. 3, $m = 2$, and the furthest projection distance to q is $\|h_1(o_1), h_1(q)\|$. So o_1 will be accessed first, and then the second furthest projection distance to q is $\|h_2(o_2), h_2(q)\|$. o_2 is the second object to be visited. A point will be considered as a candidate if it has been accessed at αm times, where α is a parameter related to c and the initial bucket width. There are two conditions to stop the algorithm: (1) if there are βn candidates in the candidate set S, RI-LSH will terminate and return the furthest point in S as the c-AFN fo q, or (2) if there is a point o satisfied that $\|o, q\| \geq R/c$ then return o as the result, where R is corresponding l_2 distance of the current searching radius r.

Algorithm 1: Reverse incremental LSH search($\mathcal{B}$, q)

    Input   : $\mathcal{B}$: m B+ tree indices for object IDs and hash values;
                q: the query object;
                w: the initial bucket width;
    Output : o: the c-AFN object

1   $n_{can} := 0$;
2   Apply m hash functions on q;
3   **while** $n_{can} < \beta n$ **do**
4 | $o \leftarrow$ next object with largest projected distance;
5 | $i \leftarrow$ the projection dimension o comes from;
6 | $r \leftarrow \| h_i(o), h_i(q) \|$;
7 | $R \leftarrow \frac{2r}{w}$;
8 | $cn(o) := cn(o) + 1$;
9 | **if** $cn(o) == \alpha m$ **then**
10 | | compute $\| o, q \|$ and update o_{max} and R_{max};
11 | └ $n_{can} := n_{can} + 1$;
12 | **if** $R_{max} \geq R/c$ **then**
13 | └ break;

14 **return** o_{max}

Algorithm 1 shows the details of the query processing step. The input consists of m B+ trees indices, the query point q and an initial bucket width w. Let n_{can} denotes the number of candidates found in the searching procedure. Firstly the query point q will be projected to the m hash functions (line 2) and then the searching start at the furthest point to q over all of the m B+ trees (line 3 to line 13). At each iteration, the next object with largest projected distance to q will be accessed (line 4). We use o and i to denote the object and the projection dimension separately. $r = \|h_i(o), h_i(q)\|$ is the projection distance between o and q on i. Using r, the corresponding radius R can be calculated (line 7). $cn(o)$ demonstrates the visited times of o. If $cn(o) == \alpha m$ then o will be added to the candidate set S. The real distance between o and q will be computed (line 10). If $\|o, q\| > R_{max}$ then update o as the current c-AFN result o_{max} and set $R_{max} = \|o, q\|$ (line 10). The candidate set size n_{can} will be added to 1 (line 11). If the current largest distance R_{max} satisfies the termination condition 2 (12), the algorithm will terminate (line 12) otherwise it continues until the first termination condition is satisfied (line 3). Finally, the point o_{max} will be returned as the c-AFN.

I/O Costs. The majority of I/O cost comes from reading data from the B+ trees for reverse incremental search and the computation of real distance between candidates and the query point. We use the number of leaf-node visited to evaluate the I/O cost because all the hash values and data IDs are in the leaf-nodes. To compute the real distance between two points, one random I/O will be invoked to read the original d-dimensional data from the disk. For the reverse incremental search, each time we load l leaves together and cost 1 random I/O and l sequence I/O. Thus the total I/O cost is $O(\frac{s}{n_e} \cdot I_{seq} + \frac{s}{n_{el}} \cdot I_{ran} + n_{can} \cdot I_{ran})$, where s is the total iteration times, n_e is the number of entries per page, n_{can} is the number of candidate objects accessed and l_{seq}, $l_r an$ denote the unit cost of sequence I/O and random I/O respectively.

3.3   c-k-AFN

Algorithm 1 can be easily extended to solve the c-k-AFN search problem by the following changes: (1) instead of at most βn candidates in the set, we require $\beta n + k - 1$ candidates as the termination condition; (2) instead of o_{max}, we maintain k furthest candidate points o_k, and their distances to q will be used for the termination condition test, and (3) there will be k points returned by the algorithm as the result.

4   Analysis

In this section, we provide the theoretical guarantee of RI-LSH: given a query point q, approximation ratio c ($c > 1$) and success possibility δ ($0 \leq \delta \leq 1$), our algorithm can return a c-AFN of q with probability at least δ.

Correctness of (R, c)-FN. Given a radius R, a reverse LSH family can return a point o satisfies that $\|o, q\| \geq R/c$. First, we prove the correctness when the radius R is given. Suppose $R = c$, and q is the query point, the reverse LSH family is $(c, 1, p_1, p_2)$-sensitive. There are two termination conditions.

$\mathcal{E}1$: There is a point o which satisfies $\|o, q\| \geq R/c$.
$\mathcal{E}2$: There are βn candidates have been found in the searching procedure.

We use $P_1 = Pr[\mathcal{E}1]$ and $P_2 = Pr[\mathcal{E}2]$ to denote the possibility that our algorithm is terminated by $\mathcal{E}1$ or $\mathcal{E}2$ and returns a correct result, separately.

Proof. A point o will be considered as a candidate if $\#collision(o) \geq \alpha m$. For $\forall o \notin B(q, R)$,

$$Pr[\#collision(o) \geq \alpha m] = \sum_{i=\alpha m}^{m} C_m^i p^i (1-p)^{m-i},$$

where $p = Pr[o \text{ separates with } q]$. According to C2LSH [8], $Pr[\#collision(o) \geq \alpha m] \geq 1 - exp(-2(p_1 - \alpha)^2 m)$. This is the value of P_1. To compute P_2, we consider for any data object $o \in B(q, R/c)$,

$$Pr[\#collision(o) \geq \alpha m] = \sum_{i=\alpha m}^{m} C_m^i p^i (1-p)^{m-i},$$

where $p = Pr[o \text{ separates with } q] \leq p_2 < \alpha$, $j = 1, 2, ..., m$. Similarly, based on Hoeffding's Inequality, the upper bound of $Pr[\#collision(o) \geq \alpha m]$ is $exp(-2(\alpha - p_2)^2 m)$. And based on Markov's Inequality, we have

$$P_2 > 1 - \frac{1}{\beta} \cdot exp(-2(\alpha - p_2)^2 m).$$

When $m = \lceil max(\frac{1}{2(p_1 - \alpha)} ln \frac{1}{\delta}, \frac{1}{2(\alpha - p_2)^2} ln \frac{2}{\beta}) \rceil$, we have $P_1 \geq 1 - \delta$ and $P_2 > \frac{1}{2}$. So the total success possibility is $P_1 + P_2 - 1 > \frac{1}{2} - \delta$.

Reverse Incremental Search. The correctness of RI-LSH is depending on the (r_1, r_2, p_1, p_2)-sensitive property of the reverse LSH function. Lemma 1 indicates that a $(1, c, p_1, p_2)$-sensitive hash function with bucket width w is (k, ck, p_1, p_2)-sensitive if the bucket width is set to kw.

Lemma 1. *Given a $(c, 1, p_1, p_2)$-sensitive hash function $h(\cdot)$, it is (ck, k, p_1, p_2)-sensitive if the bucket width is set to kw for any real value $k > 0$.*

Proof. According to the definition of (r_1, r_2, p_1, p_2)-sensitive hash function, for the bucket width w, we have

$$p_2 = \eta(1, w) = 1 - \int_{-\frac{w}{2}}^{\frac{w}{2}} \phi(x) dx$$

and

$$p_1 = \eta(c, w) = 1 - \int_{-\frac{w}{2c}}^{\frac{w}{2c}} \phi(x) dx.$$

Then for the bucket width kw, we have

$$\eta(k, kw) = 1 - \int_{-\frac{kw}{2k}}^{\frac{kw}{2k}} \phi(x)dx = 1 - \int_{-\frac{w}{2}}^{\frac{w}{2}} \phi(x)dx = p_2$$

and

$$\eta(ck, kw) = 1 - \int_{-\frac{kw}{2ck}}^{\frac{kw}{2ck}} \phi(x)dx = 1 - \int_{-\frac{w}{2c}}^{\frac{w}{2c}} \phi(x)dx = p_1.$$

Therefore, the reverse hash function $h(\cdot)$ is (ck, k, p_1, p_2)-sensitive with bucket width kw. In other words, the possibility p_1 and p_2 won't change when the radius r_1, r_2 changes if the bucket width is changed as well.

According to Lemma 1, each bucket width kw has a corresponding $B(q, R)$ in $\mathcal{R}^d$ centered at q with radius $R = k$, which means for any point $o \notin B(q, R)$, o separates with q (i.e., $\|h(o), h(q)\| > \frac{kw}{2}$) with possibility at least p_1, and for any point $o \in B(q, R/c)$, o separates with q with possibility at most p_2.

Since k can be any real value greater than 0, kw and ck can be any nonnegative real value as well. Let r be the current searching radius, we can always find a k satisfied $ck = r$. So the hash function is always (cr, r, p_1, p_2)-sensitive.

Correctness of Algorithm. We have proved the correctness of (R, c)-AN situation, and proved the correctness won't be broken when w changes with the bucket width R. So RI-LSH can return a c-AFN with at least $\frac{1}{2} - \delta$ possibility.

Approximation Ratio. The existing c-AFN algorithm with theoretical guarantee RQALSH achieves a c^2 approximation ratio. Since the number of B-trees required to keep the theoretical guarantee m is related to the approximation ratio c, a c^2-approximate furthest neighbor algorithm will require a larger size of index and a larger I/O cost compared with c-approximate algorithm under the same condition. Our algorithm reduce the approximation ratio to c and can use a much smaller size of index to keep the theoretical guarantee.

Let P_1 denote the condition that if there is a point o falling out of $B(q, R)$, $\|h(o), h(q)\| \geq r$. Let P_2 denote the condition that the total false positive number is less than βn, which means when the algorithm finds βn candidates, at least one candidate is a c-AFN of q.

Lemma 2. *When both of P_1 and P_2 hold, given a query point q, suppose the furthest neighbor of q is o^* and $\|o^*, q\| = R^*$, rilsh will stop at a radius R which satisfies $R \geq R^*$.*

Proof. Since P_1 is satisfied, the furthest neighbor of q must have a projection distance larger than r^*, where r is the corresponding projection radius of $B(q, R^*)$, which means $\|h(o^*), h(q)\| \geq r$. So when all the points falling out of $B(q, R^*)$ have been checked, there are two possible conditions: (1) There are more than βn points visited, then the algorithm will be terminated by C_1 since P_2 holds. Then we have $R \geq R^*$; (2) Before RI-LSH visits βn candidates, it finds o^* and then return it as the result since o^* falls out of r according to P_1. We still have $R \geq R^*$.

According to Lemma 2, we have that under a constant possibility $1/2 - \delta$, the algorithm can always return a c-approximate furthest neighbor of q. So the approximation ratio is c.

5   Experiment

In this section, we present results of comprehensive experiments to evaluate the I/O efficiency and accuracy of the proposed technique in the paper compared with the existing c-AFN algorithms. We choose the state-of-the-art algorithm RQALSH and use the same setting to conduct experiments.

5.1   Experiment Setup

In this section, we introduce the experiment settings of our performance evaluation including the chosen benchmark methods, the datasets, the evaluation metrics we use to compare the algorithm fairly and the initial parameter setting.

Benchmark Methods. RQALSH is the state-of-the-art c-AFN algorithm with theoretical guarantee.

- RQALSH was proposed by Qiang Huang [9] in 2017. The source code is from the author's website https://github.com/HuangQiang/RQALSH.
- RI-LSH is the algorithm proposed in this paper.

Datasets. We conduct experiments on several million-scale real-world high-dimensional datasets.

- Sift contains 1 million 128-dimensional SIFT vectors.
- Tiny contains 5 million GIST feature vectors in a 384-dimensional space.
- MillionSong contains 1 million 420-dimensional data points.

Evaluation Metrics. We evaluate the algorithms using three evaluation metics: I/O cost, running time and I/O/ratio. We also compare the index size between RI-LSH and RQALSH under the same theoretical guarantee. Since the algorithms are built for external memory, we use I/O cost as the primary evaluation metric to evaluate the algorithms. Each random I/O contributes one I/O cost to the I/O costs, and each sequential I/O contributes 0.1 I/O cost.

Parameter Setting. To fairly compare the algorithms, we set the theoretical guarantee to be $1/2 - 1/e$, which means the algorithm will return a c-AFN with at least δ possibility. And the default ratio c is 4. Since RI-LSH is a c-approximate algorithm but RQALSH is c^2-approximate algorithm, we set c to be 4 for our algorithm and 2 for RQALSH. All the other settings follow the default setting in RQALSH. The success possibility is $\delta = 1/2 - 1/e$ and the required hash function number m is calculated using c, β and δ. In our algorithm, we set $\beta = 0.001$ and when $c = 4$, m is a constant value 16 for all the datasets. The page size B is set to 8192 bytes for all the algorithms and all the datasets.

All the experiments were executed on a PC with intel(R) Xeon(R) CPU E3-1231 v3 with 3.04 GHz, 8 cores and 32G memory. The program was implemented in C++ 11. We select 100 query points for each dataset and use the average result to evaluate the algorithms.

5.2 Index Size

Using the default approximation ratio $c = 4$, the index sizes of our algorithm and RQALSH are shown in Table 2. Since we set m to be a constant value and changes β to achieve the theoretical guarantee, our algorithm always requires 16 B+ trees when $c = 4$. RQALSH sets β to be a constant and changes m to hold the guarantee, so the index size varies dramatically.

Table 2. Index size

Dataset (d)	Sift (128)	Tiny (384)	Song (420)
m	77	86	77
RQALSH (MB)	372	2000	391
m	16	16	16
RI-LSH (MB)	151	760	155

5.3 I/O Costs

For a c-AFN algorithm with theoretical guarantee, the best solution to evaluate its I/O efficiency is to calculate the minimum I/O cost required to hold the theoretical guarantee. For the given approximation ratio $c = 4$ and success possibility $\delta = 1/2 - 1/e$, we conduct experiments for the two algorithms over the real-world datasets. The value of k is varying from 1 to 100 and the default value is 30. The experiment results are given in Fig. 4. From the figures, we have the following observations:

- The I/O consumption of RQALSH is larger than RI-LSH over all the four datasets under the save theoretical guarantee and success possibility. This is because RQALSH is c^2-approximate algorithm while RI-LSH is c-approximate algorithm. When the approximation ratio is same, RQALSH requires much larger hash functions to project the dataset, and causes a larger I/O cost.
- When $c = 2$, the difference between our algorithm and RQALSH becomes larger. RQALSH costs more than 1500 I/O per query on the sift dataset.
- The I/O cost of both RQALSH and RI-LSH increase steadily when k grows. Because to solve the c-k-AFN problem, the algorithm has to visit more points to find more candidates.

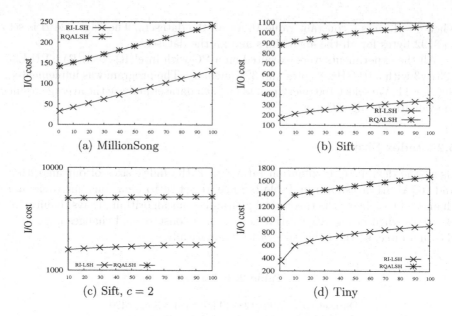

Fig. 4. I/O costs varying k

5.4 Running Time

Another evaluation metric is to compare the running time of the two algorithms. Since both methods use external memory, the I/O time is also included in the total running time. To get a fair and accurate result, we conduct experiments on the same PC and repeat for several times. The experiment results are shown in Fig. 5. From the experiment result, we get the following conclusion:

- The running time is related to I/O cost. Since RQALSH always costs more I/O than our algorithm, it also has a larger time consumption over all the datasets. The running time of RQALSH is at least 2 times of the RI-LSH's running time.
- On some easy dataset, like millionSong, the running time of RI-LSH basically doesn't increase when k becomes larger, but QALSH's running cost grows dramatically. The main reason is on easy dataset, the number of candidates finally found won't be much greater than k. Our algorithm requires less B+ trees so that the value of k doesn't affect running time a lot.

5.5 I/O and Ratio

To evaluate the accuracy of the two algorithms fairly, we conduct experiment to compare the ratio-I/O of RI-LSH and RQALSH instead of ratio-k. A I/O efficient algorithm means that it can achieve a same ratio using less I/O for the same theoretical guarantee. We set $k = 20$ and modify the termination condition:

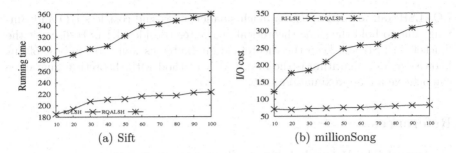

Fig. 5. Running time varying k

for a given upper bound of the candidate set size, the algorithm will terminate if the point number in the candidate set has achieved the upper bound. The experimental results are given in Fig. 6. From the results we have the following observations:

- For all the datasets, RI-LSH can get a higher accuracy result when using the same I/O cost compared with RQALSH. Since the two algorithms use a similar scheme, the ratio difference is small. Our algorithm uses a continuous searching strategy on each B+ tree, so it can find a better solution earlier than RQALSH.
- Another reason that RI-LSH can get a better I/O-ratio result is our algorithm requires a smaller number of B+ trees to build index. Then in the searching step, it is easier to find a candidate. So using similar I/O cost, our algorithm can visit and verify more points, which leads to a better ratio.

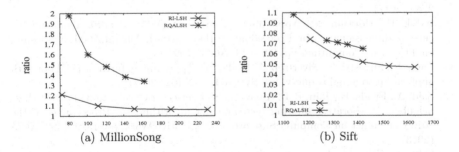

Fig. 6. I/O-ratio

6 Conclusion

In this paper, we proposed a novel I/O efficiency c-AFN method based on the classical LSH scheme. It can return a c-AFN result with a constant possibility regardless of the data distribution. Compared with existing work such as

RQALSH, our algorithm uses a much smaller index and cost less I/O and running time to hold the same theoretical guarantee. Besides, RI-LSH discards the 'bucket' in LSH and keep the theoretical guarantee as well. To the best of our knowledge, our algorithm is the first c-AFN method with theoretical guarantee can achieve a c approximation ratio.

References

1. Agarwal, P.K., Matoušek, J., Suri, S.: Farthest neighbors, maximum spanning trees and related problems in higher dimensions. Comput. Geom. **1**(4), 189–201 (1992)
2. Beckmann, N., Kriegel, H.P., Schneider, R., Seeger, B.: The R*-tree: an efficient and robust access method for points and rectangles. In: ACM SIGMOD Record, vol. 19, pp. 322–331. ACM (1990)
3. Bentley, J.L.: Multidimensional binary search trees in database applications. IEEE Trans. Softw. Eng. **4**, 333–340 (1979)
4. Bespamyatnikh, S.: Dynamic algorithms for approximate neighbor searching. In: CCCG, pp. 252–257 (1996)
5. Curtin, R.R., et al.: MLPACK: a scalable C++ machine learning library. J. Mach. Learn. Res. **14**, 801–805 (2013)
6. Curtin, R.R., Gardner, A.B.: Fast approximate furthest neighbors with data-dependent hashing. arXiv preprint arXiv:1605.09784 (2016)
7. Datar, M., Immorlica, N., Indyk, P., Mirrokni, V.S.: Locality-sensitive hashing scheme based on p-stable distributions. In: Proceedings of the Twentieth Annual Symposium on Computational Geometry, pp. 253–262. ACM (2004)
8. Gan, J., Feng, J., Fang, Q., Ng, W.: Locality-sensitive hashing scheme based on dynamic collision counting. In: Proceedings of the 2012 ACM SIGMOD International Conference on Management of Data, pp. 541–552. ACM (2012)
9. Huang, Q., Feng, J., Fang, Q., Ng, W.: Two efficient hashing schemes for high-dimensional furthest neighbor search. IEEE Trans. Knowl. Data Eng. **29**(12), 2772–2785 (2017)
10. Indyk, P., Motwani, R.: Approximate nearest neighbors: towards removing the curse of dimensionality. In: Proceedings of the Thirtieth Annual ACM Symposium on Theory of Computing, pp. 604–613. ACM (1998)
11. Pagh, R., Silvestri, F., Sivertsen, J., Skala, M.: Approximate furthest neighbor with application to annulus query. Inf. Syst. **64**, 152–162 (2017)
12. Said, A., Fields, B., Jain, B.J., Albayrak, S.: User-centric evaluation of a k-furthest neighbor collaborative filtering recommender algorithm. In: Proceedings of the 2013 Conference on Computer Supported Cooperative Work, pp. 1399–1408. ACM (2013)
13. Said, A., Kille, B., Jain, B.J., Albayrak, S.: Increasing diversity through furthest neighbor-based recommendation. In: Proceedings of the WSDM 2012 (2012)
14. Vasiloglou, N., Gray, A.G., Anderson, D.V.: Scalable semidefinite manifold learning. In: 2008 IEEE Workshop on Machine Learning for Signal Processing, pp. 368–373. IEEE (2008)
15. Yao, B., Li, F., Kumar, P.: Reverse furthest neighbors in spatial databases. In: 2009 IEEE 25th International Conference on Data Engineering, pp. 664–675. IEEE (2009)

An Efficient Approximate Algorithm for Single-Source Discounted Hitting Time Query

Kaixin Liu, Yong Zhang$^{(\boxtimes)}$, and Chunxiao Xing

BNRist, Department of Computer Science and Technology, RIIT,
Institute of Internet Industry, Tsinghua University, Beijing, China
lkx17@mails.tsinghua.edu.cn, {zhangyong05,xingcx}@tsinghua.edu.cn

Abstract. Given a graph G, a source node s and a target node t, the *discounted hitting time (DHT)* of t with respect to s is the expected steps that a random walk starting from s visits t for the first time. For a query node s, the *single-source DHT (SSDHT)* query returns the top-k nodes with the highest DHT values from all nodes in G. SSDHT is widely adopted in many applications such as query suggestion, link prediction, local community detection, graph clustering and so on. However, existing methods for SSDHT suffer from high computational costs or no guaranty of the results. In this paper, we propose $FBRW$, an effective SSDHT algorithm to compute the value of DHT with guaranteed results. We convert DHT to the ratio of personalized PageRank values. By combining Forward Push, Backward propagation and Random Walk, FBRW first evaluates personalized PageRank values then returns DHT values with low time complexity. To our knowledge, this is the first time to compute SSDHT with personalized PageRank. Extensive experiments demonstrate that FBRW is significantly ahead of the existing methods with promising effectiveness at the same time.

Keywords: Discounted hitting time · Personalized PageRank · Forward push · Backward propagation · Random walk

1 Introduction

Given a graph G, a source node s and a target node t, the *hitting time (HT)* is expected number of steps in a random walk starting from s before node t is visited for the first time [2]. As an empirically effective proximity measure, HT can capture the holistic feature of the graphic topology and has robustness toward noise [18]. Recently, as a variant of HT, the *discounted hitting time (DHT)* is proposed [10,19] to avoid the drawbacks of *HT* including sensitive to long distances [12] and preference for high degree nodes [6]. Although there are different versions of DHT, the core idea is the same, which is to add an decay factor to the path to reduce the impact of long paths. Intuitively, DHT is more reasonable

© Springer Nature Switzerland AG 2020
Y. Nah et al. (Eds.): DASFAA 2020, LNCS 12114, pp. 237–254, 2020.
https://doi.org/10.1007/978-3-030-59419-0_15

than HT and other variants while having good theoretical properties. Thus DHT has received more and more attentions.

The *single source DHT* query is a basic and useful query that takes the source node s and parameter k as input, and returns the top-k nodes in G that have the highest DHT values with respect to s. The SSDHT query is useful for many applications, such as query suggestion [16], link prediction [13], local community detection [27], graph clustering [7], etc.

However, DHT computation can be immensely expensive because it requires extracting the eigenvalues of a Laplace matrix ($n \times n$ matrix), where n is the number of nodes in the graph. Note that some large graphs may contain billions of nodes. DHT can be calculated as the target node of a recursive definition. Based on the recursive definition, the DHT computation is transformed into the computation of the solution of the linear system, but the computation cost is still very expensive. Since the DHT value varies with different pairs of source and destination nodes, storing DHT requires $O(n^2)$ complexity, which is also not feasible. For the above reasons, dynamic programming algorithms [19,28,29] are developed to compute the DHT values from other nodes to the query node and the *Monte-Carlo(MC)* method is designed to compute the DHT values from the query node to other nodes. Note that DHT is an asymmetrical proximity measurement, that is, the DHT from s to t is different from t to s. Both methods focus on approximate DHT computation and provide absolute or relative error bounds. This paper also studies approximate DHT computation, which is approximate SSDHT query defined in Sect. 2.1.

According to the definition of DHT, it is difficult to use dynamic programming to compute DHT values from the query node because it is difficult to define smaller subproblems. The simple solution is to run the dynamic programming algorithm at all other points. In this way, the running time has a linear relationship with the node size of the graph, and the complexity of the dynamic programming solution is relatively high, so it is not feasible. As for the Monte-Carlo method, its complexity is also very high, which will be proved in this article.

Unlike DHT, there exist many very efficient algorithms for *personalized PageRank (PPR)* computation, which is another similarity measure based on random walks. FORA [24] is designed for single source PPR query whose main idea is combine random walk and forward push (a local update algorithm). BiPPR [14] is the state of the art for answering pairwise PPR queries by combining random walk and backward propagation (a local update algorithm). Define the PPR value from s to t as $\pi_s(t)$, we notice that DHT value $h_s(t)$ can be computed by the ratio of $\pi_s(t)$ and $\pi_t(t)$, which means SSDHT can be computed by the result of SSPPR and O(n) pairwise PPR query from node to itself. Noted that although there are only O (n) $\pi_t(t)$ in graph, it is not feasible to pre-compute $\pi_t(t)$ for all nodes and then execute the SSPPR query. The reason is that many graphs are highly dynamic in nature, such as transaction networks. Thus frequent maintenance of the PPR from the node to itself will bring high costs. However without dynamic maintenance, there will be no guarantee about the accuracy of results.

In this paper, we propose FBRW (short for Forward push, Backward propagation and Random Walk), an efficient approximate algorithm for SSDHT query. The basic idea of FBRW is to use forward push [4] and random walk [9] to get the PPR values from the source node to other nodes. Then FBRW use backward propagation [3] and random walk to get the PPR from the node to itself. FBRW can accurately estimate the time complexity of the two parts to improve efficiency by optimizing parameters. The major contributions of this paper are as follows.

- First, we design an efficient algorithm denoted as FBRW which can answer SSDHT query providing rigorous guarantees on result quality. FBRW does not require any index or precomputation. To the best of our knowledge, it is the first time to compute SSDHT with personalized PageRank.
- Second, compared with previous algorithms, we use an iterative method to estimate the complexity of the algorithm, optimize the setting of intermediate parameters. In addition, we cache random walk to further improve efficiency.
- We validate the effectiveness and efficiency of our algorithm on several real-world datasets. The experimental results show FBRW considerably outperforms other state-of-art methods.

The rest of this paper is organized as follows. Section 2 formally defines SSDHT query and shows some related algorithms. Section 3 introduces our method. Section 4 discuss related Work about DHT and PPR. Section 5 contains experimental results. Finally, we conclude in Sect. 6.

2   Preliminaries

2.1   Discounted Hitting Time and Personalized PageRank

Let $G(V, E)$ be a directed graph with the node set V of size n and an edge set of size m. Consider an α-discounted random walk on G as follows: after we set the start node s and a decay factor α, the random walker which starts from s randomly moves to its out-neighbors or stops at the current node with α probability in each step.

Definition 1. *The discounted hitting time (DHT) from node s to node t, denoted by $h_s(t)$, is the expected number of steps in an α-discounted random walk starting from s before node t is visited for **the first time**.*

There are several different forms of DHT [10,19]. Sarkar [30] gives the general form of DHT, defined as follows:

Definition 2. *Given a source node s and a target node t, The general form of DHT, denoted by $h_s(t)$, is:*

$$h_s(t) = \beta \sum_{i=1}^{\infty} (1 - \alpha)^i \mathcal{P}_i^h(s, t) + \gamma$$

where $\alpha \in (0,1)$ is the decay factor mentioned before, $\beta (\neq 0)$ and γ are coefficients varies with different versions of the definition. $\mathcal{P}_i^h(s,t)$ is the hitting probability that a random walker starting from node s first hits node t after i steps.

From the above definition, we can know that $\sum_{i=1}^{\infty}(1-\alpha)^i \mathcal{P}_i^h(s,t)$ is the key part of DHT and the most complicated in calculation. So in this paper, we focus on computing this item by set $\beta = 1$ and $\gamma = 0$. The definition of PPR is very similar to DHT:

Definition 3. *The* personalized PageRank *(PPR) from node s to node t $\pi_s(t)$ is the probability that a random walk starting from s stops at t.*

Intuitively, the hitting time and personalized PageRank can be interpreted as Weighted sum of the ensemble of paths which meets the definition respectively. Paths contributing to PPR value $\pi_s(t)$ can pass through node t multiple times. When computing DHT $h_s(t)$, if paths reach the destination point t is allowed to jump to the neighbor, PPR $\pi_s(t)$ can be obtained. The following lemma gives the exact relationship between PPR and DHT.

Lemma 1. *The discounted hitting time can be expressed as a ratio of personalized PageRank:*

$$h_s(t) = \pi_s(t)/\pi_t(t) \tag{1}$$

Proof. Define $\mathcal{P}_i(s,t)$ is the probability that a random walker starting from node s reach t in the i-th step, $\pi_s(t)$ can be expressed as: $\pi_s(t) = \alpha \sum_{i=1}^{\infty}(1-\alpha)^i \mathcal{P}_i(s,t)$. As mentioned before, when computing PPR, random walker can meet target node t multiple times, which lead to: $\mathcal{P}_i(s,t) = \sum_{k=1}^{i-1} \mathcal{P}_k^h(s,t) \cdot \mathcal{P}_{i-k}(s,t)$. Thus we have:

$$\pi_s(t) = \alpha \sum_{i=1}^{\infty}(1-\alpha)^i \mathcal{P}_i(s,t)$$

$$= \alpha \sum_{i=1}^{\infty}(1-\alpha)^k \sum_{k=1}^{i-1} \mathcal{P}_k^h(s,t) \cdot (1-\alpha)^{i-k}\mathcal{P}_{i-k}(s,t)$$

$$= \sum_{k=1}^{\infty}(1-\alpha)^k \mathcal{P}_k^h(s,t) \cdot \alpha \sum_{j=1}^{\infty}(1-\alpha)^j \mathcal{P}_j(s,t)$$

$$= h_s(t) \cdot \pi_t(t)$$

2.2 Problem Definition

As a simple version, an approximate whole-graph SSDHT query that does not need to calculate the top-k nodes is defined. Note that this type of query has been studied by many existing works [14,15,22,24].

Definition 4 *(Approximate Whole-Graph SSDHT). Given a source node s, a threshold δ, an error bound ϵ, and a failure probability p_{fail}, an approximate* *whole-graph SSPPR query returns an estimated DHT $\hat{h}_s(v)$ for each node $v \in V$, such that for any $h_s(v) > \delta$,*

$$|h_s(v) - \hat{h}_s(v)| \leq \epsilon \cdot h_s(v) \tag{2}$$

holds with at least $1 - p_f$ probability.

Following previous work [22,24], the approximate Top-k SSDHT query is defined in Definition 5.

Definition 5 *(Approximate Whole-Graph SSDHT). Given a source node s, a threshold δ, an error bound ϵ, and a failure probability p_{fail}, an approximate top-k SSPPR query returns a sequences of k nodes, $v_1, v_2, ..., v_k$, such that with at least $1 - p_f$ probability, for any $i \in [i, k]$ with $h_s(v^*) > \delta$,*

$$|h_s(v_i) - \hat{h}_s(v_i)| \leq \epsilon \cdot h_s(v_i) \tag{3}$$

$$h_s(v_i) \geq (1 - \epsilon) \cdot h_s(v_i^*) \tag{4}$$

holds with at least $1 - p_f$ probability, where v_i^* is the node whose actual DHT with respect to s is the i-th largest.

Note that we guarantee the result quality for the nodes with above-average PPR values by setting $\delta = O(1/n)$, which is common in existing work [14,15,22, 24]. The frequently-used notations in this paper are summarized in Table 1.

2.3 Basic Techniques

In what follows, main methods designed for computing DHT and PPR are discussed. Noted that it is recognized that dynamic programming algorithms, which are difficult to be applied in calculating SSDHT [1,20], so dynamic algorithms are not discussed here.

Monte-Carlo. The Monte-Carlo (MC) method [9,10,20] is effective for both DHT and PPR computations. Given a source node s, MC samples ω independent random walks from s and each random walk stops when termination conditions are met. If there are c_t random walks stops at t, $\frac{c_t}{\omega}$ is the estimate of $h_s(t)$ or $\pi_s(t)$. The number of random walks is determined by Theorem 1. As Lemma 2 indicates, although MC is simple and effective, its complexity is intolerable when the required error bound is relatively small.

Lemma 2. *For any node v with $h(s,v) > \delta$, in order to obtain $P(\exists v \in V, |\hat{h}_s(v) - h_s(v)| \geq \epsilon h_s(v)) \leq p_{fail})$, numbers of random walk samples ω should at least be $\frac{(2\epsilon/3+2)\cdot log(2/p_f)}{\epsilon^2 \delta}$.*

Proof. Due to $a_i = 1 \forall i \in [1, \omega]$, $v = h_s(v) < 1$. Replacing λ with $\epsilon h_s(v)$ Lemma 2 is proved.

Table 1. Frequently used notations.

Notation	Description
$G = (V, E)$	The input graph G with node set V and edge set E
n, m	The number of nodes and edges of G, respectively
$N^{out}(v), N^{in}(v)$	The set of out-neighbors and in-neighbors of node v
$d_{out}(v), d_{in}(v)$	The out-degree and in-degree of node v
$\pi_s(t)$	The exact PPR value of t with respect to s
α	The decay factor
$\epsilon, \delta, p_{fail}$	The error bound, threshold and failure probability of approximate DHT computed by FBRW
$\epsilon^f, \delta^f, p_{fail}^f$	The error bound, threshold and failure probability of approximate PPR computed by forward push and random walk
$\epsilon^b, \delta^b, p_{fail}^b$	The error bound, threshold and failure probability of approximate PPR computed by backward propagation and random walk
$r_v^f(t), \pi_v^f(t)$	The residue vector and the reserve vector from s to all nodes in the forward push
$r_v^b(t), \pi_v^b(t)$	The residue vector and the reserve vector from s to all nodes in the backward propagation
$r_v^f\ r_{sum}^f$	The sum of all nodes' residues during in the forward
$\mathbf{e}_v$	the row unit vector whose v^{th} is equal to 1
M	The random walk transition matrix of size $n * n$, $M_{ij} = 1/d_{out}$ if j is a out neighbor of i, otherwise 0

Theorem 1 [8]. *Let $X_1, ... X_\omega$ be independent random variables with $Pr[X_i = 1] = p_i$ and $Pr[X_i = 0] = 1 - p_i$. Let $X = \frac{1}{\omega} \cdot \sum_{i=1}^{\omega} a_i X_i$ with $a_i > 0$, and $v = \frac{1}{\omega} \cdot \sum_{i=1}^{\omega} a_i^2 X_i$. Then,*

$$Pr[|X - E[X]| \geq \lambda] \leq 2 \cdot \exp(-\frac{\lambda^2 \cdot \omega}{2v + 2a\lambda/3}),$$

where $a = \max a_1, ..., a_\omega$.

Forward Push. Forward push [4] is a local update algorithm for PPR computation. According to the definition, $\pi_s(t) = \alpha \sum_{t=0}^{\infty}(1 - \alpha)^t \cdot \mathbf{e}_s M^t \mathbf{e}_t^T$. The high level idea of the local algorithm forward push showed in Algorithm 1 can be expressed to compute $\alpha \sum_{t=0}^{\infty}(1 - \alpha)^t \cdot (\mathbf{e}_s M^t)$ for the source node s in a decentralized way. As shown in Algorithm 1, $r_s^f(v)$ is set to $\mathbf{e}_s(v)$ and $\pi_s^f(v)$ is set to 0 for all nodes in graph during initialization (Lines 1–2). Subsequently, it iteratively propagates residuals to *out-neighbors* until the ratio of the residual to out-degree is less than the threshold r_{max}^f for all nodes (Lines 3–7).

Algorithm 1: Forward Push.

Input: Graph G, source node s, decay factor α, threshold r_{max}^f
Output: Forward residue $r_s^f(v)$ and reserve $\pi_s^f(v)$ for all $v \in V$
1   $r_s^f(s) \leftarrow 1$ and $r_s^f(v) \leftarrow 0$ for all $v \neq s$;
2   $\pi_s^f(v) \leftarrow 0$ for all v;
3   **while** $\exists u \in V$ *such that* $r_s^f(u)/d_{out}(u) \geq r_{max}^f$ **do**
4   **for** *each* $v \in N^{out}(u)$ **do**
5          $\left\lfloor \; r_s^f(v) \leftarrow r_s^f(v) + (1 - \alpha) \cdot \frac{r_s^f(u)}{d_{out}(u)} \right.$;
6      $\pi_s^f(u) \leftarrow \pi_s^f(u) + \alpha \cdot r_s^f(u)$;
7      $r_s^f(v) \leftarrow 0$;

Algorithm 2: Backward Propagation.

Input: Graph G, target node t, decay factor α, threshold r_{max}^b
Output: Backward residue $r_s^b(v)$ and reserve $\pi_s^b(v)$ for all $v \in V$
1   $r_t^b(t) \leftarrow 1$ and $r_v^b(t) \leftarrow 0$ for all $v \neq t$;
2   $\pi_v^b(t) \leftarrow 0$ for all v;
3   **while** $\exists v \in V$ *such that* $r_v^b(t) \geq r_{max}^b$ **do**
4   **for** *each* $u \in N^{in}(v)$ **do**
5          $\left\lfloor \; r^b(u,t) \leftarrow r^b(u,t) + (1 - \alpha) \cdot \frac{r_v^b(t)}{d_{in}(u)} \right.$;
6      $\pi_v^b(t) \leftarrow \pi_v^b(t) + \alpha \cdot r_v^b(t)$;
7      $r_v^b(t) \leftarrow 0$;

The time complexity of Algorithm 1 is $O(1/r_{max})$. Obviously, $\pi_s^f(t)$ approaches $\pi_s(t)$ and the cost increases as r_{max}^f decreases. However, no explicit error guarantee is provided.

Backward Propagation. The backward propagation algorithm [3] is the reverse version of forward push whose pseudo-code is shown in Algorithm 2. Different from forward push, it is designed to compute $\alpha \sum_{t=0}^{\infty} (1 - \alpha)^t \cdot (M^t \mathbf{e}_t^T)$. In the initial part of the algorithm, $r_v^b(t) = \mathbf{e}_t(v)$ and $\pi_v^b(t) = 0$ for all nodes in graph. Then similar with forward push, if the residual of any node is greater than the threshold r_{max}^b, the residual is propagated to *in-neighbors* and finally PPR values from other nodes to target node t are returned. Backward propagation guarantees that the estimation error of $\pi_s(t)$ is not higher than r_{max}^b, and the time complexity of the algorithm is $O(\frac{m}{n\alpha r_{max}^b})$.

3   The FBRW Framework

In this section, we first present FBRW for whole graph SSDHT in Sect. 3.1. Section 3.2 describes FBRW with top-k selection.

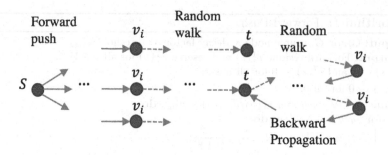

Fig. 1. The framework of FBRW

3.1 SSDHT on a Whole Graph

As mentioned in Sect. 1, it is difficult to adopt dynamic programming algorithms to answer SSDHT query while the Monte-Carlo method is quite inefficient. Since the relationship between DHT and PPR has been proven in Lemma 1, the high-level idea of FBRW is to transform the SSDHT query into two parts. The first part calculates SSPPR using forward push and random walk, which returns $\pi_s(t)$ $\forall t \in V$. The second part calculates multi-pair PPR using backward propagation and random walk, which returns $\pi_t(t)$ $\forall t \in V$. The framework of FBRW is shown in Fig. 1.

Lemma 3 shows the relationship between the error bound ϵ_f of the first part and the error bound ϵ_b of the second part. The error boundary is closely related to the time complexity of the two parts, which will be explained later. The total time cost can be reduced by adjusting the value of error bounds, if the time complexity of the two parts is known. But we cannot get accurate estimates of the complexity of SSPPR and multi-pair PPR due to the following reasons. (i) In the multi-pair PPR part, the number of nodes that need to calculate cannot be predicted in advance. (ii) When calculating SSPPR, the number of random walks required cannot be accurately estimated even with given parameters [24]. (iii) The actual running cost of the forward push and backward propagation is much lower than the complexity given by [3,4]. Rather than setting parameters at the beginning, FBRW continuously modifies parameters to improve efficiency in an iterative manner.

The details of FBRW is shown in Algorithm 3. Given a graph G, a source node s and a decay factor α, FBRW returns the estimated DHT values for all nodes. In the initialization phase (Line 1–3), set ϵ^f/ϵ^b to n assuming $Cost_f$ is much less than $Cost_b$. Due to $\alpha < \pi_t(t) < 1$, we set $\delta_b = 1/\alpha$ and $\delta^f = \alpha/n$. The values of failure probability p_{fail}^f, p_{fail}^b are set by Boole's inequality. When computing SSPPR (Line 5–9), we keep updating ϵ^f and ϵ_b and computing r_{max}^f. Then Algorithm 1 is executed until the total cost is minimized. Then random walks are sampled to get the estimate result of SSPPR. When computing multi-pair PPR, backward propagation and random walk are used. Line 16 gives the number of random walks needed based on the r_{max}^b calculated in the previous phase. For

---

Algorithm 3: FBRW for Whole Graph SSPPR.

Input: Graph G, source node s, decay factor α

Output: Estimated DHT $\hat{h}_s(v)$ for all $v \in V$

1   Let λ be the ratio of ϵ^f and ϵ^b, init $\lambda = n$;

2   Let $Cost_f, Cost_b$ be the cost of computing SSPPR and multi-pair PPR;

3   $\delta^f \leftarrow \alpha/n, \delta^b \leftarrow 1/\alpha, p_{fail}^f \leftarrow p_{fail}/2, p_{fail}^b \leftarrow p_{fail}/2, \epsilon^f + (1+\epsilon) \cdot \epsilon^b \leq \epsilon$;

4   //Computing SSPPR

5   **do**

6   Get r_{max}^f, r_{max}^b according to $\epsilon^f/\epsilon^b = \lambda$ and Lemma 3;

7      $[r^f, \pi^f] \leftarrow$ ForwardPush(r_{max}^f);

8   Compute $Cost_f, Cost_b$ and set $\lambda \leftarrow Cost_f/Cost_b$;

9   **while** $Cost_f \approx Cost_b$;

10   Let $r_{sum}^f \leftarrow \sum_{v_i \in V} r_s(v_i)$ and $\omega^f \leftarrow r_{sum} \cdot \frac{(2\epsilon/3+2) \cdot log(2/p_{fail}^f)}{\epsilon^2 \cdot \delta}$;

11   **for** $v_i \in V$ *with* $r_s(v_i) > 0$ **do**

12   Let $\omega_i^f = \lceil r_s(v_i) \cdot \omega^f / r_{sum}^f \rceil$;

13   **for** $i = 1$ *to* ω_i^f **do**

14   Sample a random walk from v_i whose end node denoted as t;

15          $\pi_s^f(t) += r_s(v_i)/\omega_i^f$;

16   //Computing Multi-pair PPR

17   Let $\omega^b \leftarrow r_{max}^b \cdot \frac{(2\epsilon/3+2) \cdot log(2/p_{fail}^f)}{\epsilon^2 \cdot \delta}$;

18   **for** $v_i \in V$ *with* $\pi_s^f(v_i) > \delta^f$ **do**

19      $[r^b, \pi^b] \leftarrow$ BackwardPropagation(r_{max}^b);

20   **for** $i = 1$ *to* ω^b **do**

21   Sample a random walk from v_i whose end node denoted as t;

22          $\pi_{v_i}^b(v_i) += r_{v_i}^b(t)/\omega^b$;

23   **return** $\hat{h}_s(v_1) = \frac{\pi_s^f(v_1)}{\pi_{v_1}^b(v_1)}, ..., \hat{h}_s(v_n) = \frac{\pi_s^f(v_n)}{\pi_{v_n}^b(v_n)}$;

---

each node v that $h_s(v_i)$ may exceed $1/n$, we call Algorithm 2 and sample random walks to calculate $\pi_{v_i}^b(v_i)$ (Line 17–21). After all v_i are processed, FBRW returns the approximate DHT value $\hat{h}_s(v_i)$ for each node v_i. Note that in order to avoid wasting calculations, random walks are cached during the calculation.

Lemma 3. *The error bounds of* $\pi_s(t), \pi_t(t)$ *and* $h_s(t)$ *satisfy the following inequality:* $\epsilon^f + (1+\epsilon) \cdot \epsilon^b \leq \epsilon$.

Proof. Note that by Lemma 1, the above inequality can be obtained by substituting the error bound.

Correctness. The residuals and reserves returned by forward push and backward propagation satisfy the following equation:

$$\pi_s(t) = \pi_s^f(t) + \sum_{v_i \in V} r_s^f(v_i) \cdot \pi_{v_i}(t) \tag{5}$$

$$\pi_t(t) = \pi_t^b(t) + \sum_{v_i \in V} \pi_t^b(v_i) \cdot r_{v_i}(t) \tag{6}$$

First we discuss the calculation of SSPPR. For j-th random walk W_j, if it starts from node v_i, let $X_j(v_i, t)$ be a Bernoulli variable that takes value 1 if W_j terminates at t, and value 0 otherwise. Plainly, $E[X_j(v_i, t)] = \pi_{v_i}(t)$. Let $X_j(t)$ denotes W_j terminates at t and $\omega^{f'} = \sum_{i=1}^n$ denote the total number of random walks. As the probability W_i starts from node v_i is $\omega_i^f / \omega^{f'}$, we have

$$\mathbb{E}\left[\sum_{j=1}^{\omega^{f'}} X_j\right] = \mathbb{E}\left[\sum_{j=1}^{\omega^{f'}} \left[\frac{\omega_i^f}{\omega^{f'}} \cdot \sum_{i=1}^n \frac{r_s^f(v_i)}{\omega_i^f} \cdot X_j(v_i, t)\right]\right] = \mathbb{E}\left[\sum_{i=1}^n r_s^f(v_i) \cdot \pi_{v_i}(t)\right]$$

Combining above equation and Eq. 5, we can see the expectation of $\pi_s^f(v)$ computed by FBRW equals $\pi_s(v)$ for each node v. By setting a_i to $\frac{r_s^f(v_i)}{r_{sum}^f}$ and ω to ω^f, we can apply Theorem 1 to guarantee that $|\pi_s^f(v) - \pi_s(v)| < \epsilon \cdot \pi_s(v)$ with at least $1 - p_{fail}$ probability. It is easy to get the same conclusion for the estimated value $\pi_b(v, v)$. All in all, we have the following result.

Lemma 4. *For any node v with $h(s, v) > \delta$, Algorithm 3 returns an approximated DHT $\hat{h}(s, v)$ that satisfies Eq. 2 with at least $1 - p_{fail}$ probability.*

Complexity Analysis. The Forward Push runs in $O(1/r_{max}^f)$ and its corresponding random walk part's time complexity is $O\left(\frac{m \cdot r_{max}^f \cdot \log(2/p_{fail}^f)}{(\epsilon^f)^2 \delta^f}\right)$, due to $r_{sum}^f < m \cdot r_{max}^f$. When $r_{max}^f = \epsilon \cdot \sqrt{\frac{\delta}{m \cdot \log(2/p_{fail}^f)}}$, the expected time complexity of both reaches the minimum of $\sqrt{m \cdot n \cdot \log(n)}/\epsilon^f$. Similarly, the complexity of Backward Propagation and random walk for all nodes is $O(\sqrt{m \cdot n \cdot \log(n)})/\epsilon^b$. Therefore, FBRW can improve efficiency by adjusting λ.

3.2 Top-k SSDHT

Intuitively answering top-k queries can directly apply Algorithm 3 to get the DHT of all nodes and then return the top-k nodes. However, as the exact k-th DHT value $h_s(v_k^*)$ is unknown, we have to set $\delta = 1/n$, which lead unnecessary computation. In addition, too accurate computation of $\pi(v, v)$ for v that is not top-k is also meaningless.

We propose a trial-and-error approach to answer top-k SSDHT query. The details are shown in Algorithm 4. In each round, we adjust the parameters to obtain more accurate estimate of PPR from source node to other nodes (Line 4–8) or PPR from node to itself (Line 9–13), until we get the results that meets the definition. Note that the number of iteration for computing $\pi(s, t)$ or $\pi(t, t)$ is at most $\log(n)$. We assume that in each iteration the bounds are correct. As the failure probability is set to $p_{fail}/(n \cdot \log(n))$ in Line 1 of Algorithm 4, the result returned by Algorithm 4 satisfies Definition 5 with at least $1 - p_{fail}$ probability. Given the number of random walks sampled in each iteration, UB^f, LB^f, UB^b, LB^b can be computed by Theorem 1. In detail, let $\lambda_j^f = \frac{\omega_j^{f'} \cdot r_{sum}^f}{\omega_j^f}$ in j-th iteration. As it is proved that $\mathbb{E}\left[\sum_{j=1}^{\omega^{f'}} X_j\right] = \mathbb{E}\left[\sum_{i=1}^{n} r_s^f(v_i) \cdot \pi_{v_i}(t)\right]$, we have $\Pr[|\pi_s(t) - \hat{\pi}_s(t)| > \lambda_j^f] \leq 2 \cdot \exp\left(-\frac{(\lambda_j^f)^2 \cdot \omega_j^{f'}}{r_{sum}^f \cdot (2\pi_s(t) + 2\lambda_j^f/3)}\right) \leq p'_{fail}$. Similarly, we can get absolute error bound λ_j^b and relative error bound $\epsilon_j^f, \epsilon_j^b$. UB^f, LB^f, UB^b, LB^b are updated based on these parameters.

The stop condition and the approximation guarantee for the results of Algorithm 4 are demonstrated in Theorem 2 [24].

Theorem 2. *Let $v_1', ..., v_k'$ be the nodes with the largest lower bounds in j-th iteration of Algorithm 4. Let $S = \{v_1', ..., v_k'\}$ and U be the set of nodes $u \in V \backslash S$ such that $UB_j(u) > (1 + \epsilon) \cdot LB_j(v_k')$, if $UB(, v_i') < (1 + \epsilon) \cdot LB_j(v_i')$ for $i \in [1, k], LB_j(v_k') \geq \delta$, and there exists no $u \in U$ such that $UB_j(u) < (1 + \epsilon) \cdot LB_j(u)/(1 - \epsilon)$, then returning $v_1', ..., v_k'$ and their estimated DHT values would satisfy the requirements in Definition 5 with at least $1 - p_{fail}$ probability.*

4   Other Related Work

This section first discusses the related work of DHT. Since we use the algorithm to calculate PPR, the related work of PPR is also involved. Although the ways to calculate DHT and PPR are similar, the details are different.

4.1   Discounted Hitting Time

The methods of computing DHT can be roughly divided into three categories, exact solution, dynamic programming and the Monte-Carlo method. Guan et al. [10] derive a linear system by the recursive definition of DHT which can give the exact results. However, computing the solution of a linear system is too expensive. More efforts have been made in studying efficient dynamic programming algorithms. Zhang et al. [29] first do a best-first search get candidate nodes, then apply iteration algorithm on candidate nodes. Wu et al. [28] execute an iterative algorithm to calculate the upper and lower bounds of the nodes in the candidate set after a step of best first search, which guarantees the quality of the results. However, as Sarkar et al. [1] mentioned, dynamic programming algorithms are not suitable for single-source queries. Sarkar et al. [20] propose a simple but useful Monte-Carlo algorithm for computing truncated hitting time which is another variant of hitting time. Guan et al. [10] also exploit an effective Monte-Carlo algorithm for DHT from a source node to a target set. Zhang et al. [30] calculate multi-way joins over DHT using the Monte-Carlo method.

Algorithm 4: Top-k FBRW

Input: Graph G, source node s, decay factor α
Output: k nodes with the highest approximate DHT scores
1 Set failure probability as $p'_{fail} = p_{fail}/(n \cdot \log(n))$;
2 Init parameters as Line1-3 in Algorithm 3;
3 **while** $\delta^f < 1/n$ *and* $\delta^b < 1/n$ **do**
4   **if** $Cost_f <= Cost_b$ **then**
5        $\delta^f \leftarrow \delta^f/2$;
6        $[r^f, \pi^f] \leftarrow$ ForwardPush(r^f_{max});
7   Sample random walks as Line 10-14 in Algorithm 3;
8   Update $LB^f(u)$ and $UB^f(u)$ for all node u;
9   **else**
10       $\delta^b \leftarrow \delta^b/2$;
11   **for** $v_i \in C$ **do**
12   Compute $\pi^b_{v_i}(v_i)$ by Line 16-21 in Algorithm 3;
13   Update $LB^b(u)$ and $UB^b(u)$ for all node $u \in C$;
14   Update $LB(u)$ and $UB(u)$ for all nodes;
15   Let C be the set that contains the nodes with upper bound greater than the top-k largest top-k lower bounds;
16   **if** *Got top-k nodes* **then**
17   **return** The nodes with the top-k largest DHT;
18   **else**
19   Compute $Cost_f$ and $Cost_b$;

4.2   Personalized PageRank

For personalized PageRank, we introduce research about pairwise and single source queries which are relevant with our work. Fast-PPR [15], BiPPR [14], HubPPR [22] are all designed for pair-wise PPR query. FastPPR is not as efficient as BiPPR and HubPPR can be seen the indexed version of BiPPR. As for single source top-k PPR query, Gupta et al. [11] and Avrachenkov et al. [5] generate the top-k nodes using forward and random walk respectively. However, they cannot guarantee giving error bounds and results estimates at the same time. TopPPR [26] is the state-of-art algorithm for the top-k query. It is faster than other algorithms but does not guarantee the order of top-k nodes. FORA [23, 24] can quickly calculate the top-k nodes and give estimated values with an error guarantee. A parallel algorithm for SSPPR is presented in [21]. Similar to FBRW, PRSim [25] answers SimRank queries based on the connection of PPR and SimRank.

5   Experiments

In this section, we experimentally evaluate the effectiveness and efficiency of our proposed FBRW algorithm against the states of art. All programs are written in

C++. All experiments are conducted on a Linux machine with an Intel 2.5GHz CPU and 110GB memory. We open the source code on Github[1].

5.1 Experimental Setting

Datasets and Query Sets. We use 5 real graphs: *DBLP, Web-St, Pokec, Lj and Orkut* which are widely used in recent work [14,24]. Table 2 summarizes the statistics of the data. For each dataset, 50 nodes are sampled uniformly at random as query nodes. For top-k query, we set $k = 100, 200, ...500$. Following previous work [14,22,24], we set $\delta = 1/n, p_{fail} = 1/n$, and $\epsilon = 0.5$.

Methods. We compare FBRW against three methods. The first is the Monte-Carlo approach, dubbed as MC. The number of random walks ω is determined by Lemma 2. For each random walk W_i, the DHT value of each node that appears on W_i should be updated. The second method directly uses FORA [24] to calculate the SSPPR and then uses BiPPR [14] to calculate the PPR from the node to itself. For convenience, we represent this method as FORA+BiPPR. According to the estimation of complexity, we set the error bounds of FORA and BiPPR to be the same and set other parameters according to FBRW. The third method is Dynamic Neighborhood Expansion (DNE) [29] which is a dynamic programming algorithm. DNE uses the best-first expansion strategy to calculate the top-k nodes without guaranteeing the quality of the results. For fair comparison we set the error bound for DNE the same as FBRW and the maximum number of visited nodes is 30,000. Although dynamic programming algorithms are difficult to apply to SSDHT, we also implement DNE to compare efficiency.

Table 2. Datasets ($K = 10^2, M = 10^6$).

Name	n	m	Type	Linking site
DBLP	613.6K	2.0M	Undirected	www.dblp.com
Web-St	281.9K	2.3M	Directed	www.stanford.edu
Pokec	1.6M	30.6M	Directed	pokec.azet.sk
LJ	4.8M	69.0M	Directed	www.livejournal.com
Orkut	3.1M	117.2M	Undirected	www.orkut.com

---

[1] https://github.com/thu-west/FBRW.

Table 3. Whole-graph SSDHT performance (s) ($K = 10^3$).

	MC	DNE	FORA+BiPPR	FBRW
DBLP	22.62	48.02	3.97	3.42
Web-St	21.02	4.46	0.85	0.82
Pokec	366.72	396.45	411.50	345.21
LJ	1.17K	537.14	667.21	539.71

5.2   Whole-Graph SSDHT Queries

In our first set of experiments, we evaluate the efficiency of each method for whole-graph SSDHT queries. The experimental results are shown in Table 3. As we can observe, our FBRW and FORA+BiPPR outperform MC. This verifies the feasibility of using PPR to calculate SSDHT in terms of efficiency. Due to the optimized set of parameters and repeated use of the random walk, FBRW is also superior to FORA+BiPPR. We set the maximum number of visit nodes for DNE, which makes DNE stops early when some query points do not reach the error limit. Setting the upper limit of the node to 30,000, DNE often return results within the error bound on small graph, but this is not the case for large graph. Therefore, the time comparison of DNE on the small graph is more meaningful.

5.3   Top-k SSDHT Queries

In our second set of experiments, we evaluate the efficiency and accuracy of each method for top-k SSDHT. Figure 2 reports the query efficiency of all methods on four representative datasets: *DBLP, Pokec, Lj, Orkut*. Note that the y-axis is in log-scale. As we can see, FBRW considerably outperforms other competitors. In particular, FBRW is at least an order of magnitude faster than MC. The running time of the DNE algorithm increases sharply with the graph, because best-first expansion strategy it uses requires access to a particularly large number of nodes when the error margin is small. Although FORA+BiPPR also has an early-termination mechanism, it does not adjust the parameters according to the actual cost, which results in an imbalance in the two parts of time. In addition, FBRW caches and reuses random walks generated during calculation, which avoids a lot of calculation waste and significantly improves efficiency.

To compare the accuracy of the top-k results returned by each method, we first use the *Power Iteration* [17] method with 100 iterations to get the PPR from the source node to other nodes. Then we apply backward propagation algorithm with r^b_{max} of 10^{-10} for each node that may be in the top-k nodes. Combining the results of these two parts, we get DHT values that can be used as ground truth. Note that the precision and recall are the same for the top-k SSDHT queries. Figure 3 reports the accuracy of the top-k query algorithms on four datasets: *Web-St, DBLP,Pokec, Lj*. As we can see, all methods provide high

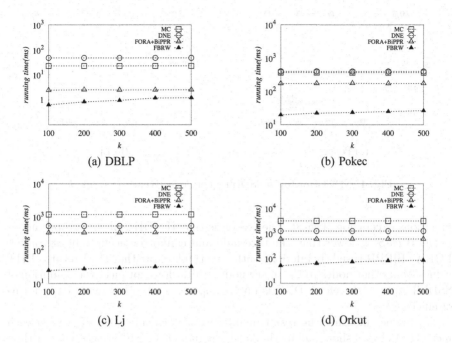

Fig. 2. Top-k SSDHT query efficiency: varying k

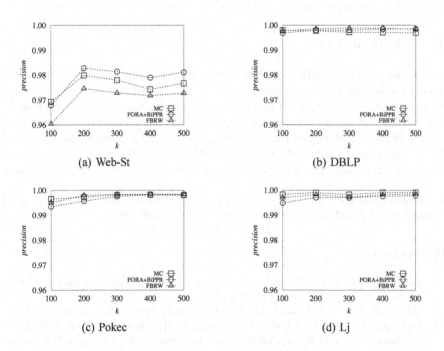

Fig. 3. Top-k SSDHT query accuracy: varying k

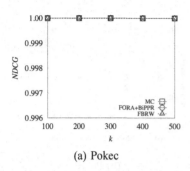
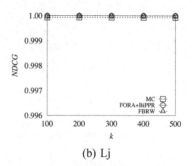

(a) Pokec   (b) Lj

Fig. 4. NDCG for top-k SSDHT query effectiveness: varying k

precisions except for *Web-St*. The reason is that the DHT of some top-k nodes in the *Web-St* is lower than the threshold and cannot be accurately estimated. FORA+BiPPR is sightly inferior to other methods. Note that DNE returns DHT values from other nodes to the query node, which does not meet the definition of SSDHT query. Therefore, DNE is only compared with other methods in terms of efficiency.

To further measure the result quality, we also calculate the NDCG of each method. As Fig. 4 shows, all methods offer high NDCG. MC is sightly lower than other methods. Note that we perform experiments on all datasets, the results on the other datasets are similar, and are omitted due to the space constraint.

6   Conclusion

This paper presents FBRW, an efficient algorithm for approximate single-source discounted hitting time computation. To our best knowledge, it is the first time answering single-source discounted hitting time query with personalized PageRank. FBRW can adjust parameters to improve efficiency based on accurate estimation of the complexity of Forward Push, Backward Propagation and Random Walk. Moreover, FBRW caches random walks to avoid the waste of computing. Extensive experiments demonstrate that FBRW outperforms existing solutions.

Acknowledgments. This work was supported by National Key R&D Program of China (2018YFB1404401, 2018YFB1402701), NSFC (91646202).

References

1. Aggarwal, C.C. (ed.): Social Network Data Analytics. Springer, Boston (2011). https://doi.org/10.1007/978-1-4419-8462-3
2. Aldous, D., Fill, J.: Reversible Markov chains and random walks on graphs (1995)
3. Andersen, R., Borgs, C., Chayes, J., Hopcraft, J., Mirrokni, V.S., Teng, S.-H.: Local computation of PageRank contributions. In: Bonato, A., Chung, F.R.K. (eds.) WAW 2007. LNCS, vol. 4863, pp. 150–165. Springer, Heidelberg (2007). https://doi.org/10.1007/978-3-540-77004-6_12

4. Andersen, R., Chung, F.R.K., Lang, K.J.: Local graph partitioning using PageRank vectors. In: FOCS, pp. 475–486 (2006)

5. Avrachenkov, K., Litvak, N., Nemirovsky, D., Smirnova, E., Sokol, M.: Quick detection of top-k personalized PageRank lists. In: Frieze, A., Horn, P., Prałat, P. (eds.) WAW 2011. LNCS, vol. 6732, pp. 50–61. Springer, Heidelberg (2011). https://doi.org/10.1007/978-3-642-21286-4_5

6. Brand, M.: A random walks perspective on maximizing satisfaction and profit. In: SDM, pp. 12–19 (2005)

7. Chen, M., Liu, J., Tang, X.: Clustering via random walk hitting time on directed graphs. In: AAAI, vol. 8, pp. 616–621 (2008)

8. Chung, F.R.K., Lu, L.: Concentration inequalities and martingale inequalities: a survey. Internet Math. $3(1)$, 79–127 (2006)

9. Fogaras, D., Rácz, B., Csalogány, K., Sarlós, T.: Towards scaling fully personalized PageRank: algorithms, lower bounds, and experiments. Internet Math. $2(3)$, 333–358 (2005)

10. Guan, Z., Wu, J., Zhang, Q., Singh, A.K., Yan, X.: Assessing and ranking structural correlations in graphs. In: SIGMOD, pp. 937–948 (2011)

11. Gupta, M.S., Pathak, A., Chakrabarti, S.: Fast algorithms for topk personalized PageRank queries. In: WWW, pp. 1225–1226 (2008)

12. Katz, L.: A new status index derived from sociometric analysis. Psychometrika $18(1)$, 39–43 (1953)

13. Liben-Nowell, D., Kleinberg, J.: The link-prediction problem for social networks. J. Am. Soc. Inform. Sci. Technol. $58(7)$, 1019–1031 (2007)

14. Lofgren, P., Banerjee, S., Goel, A.: Personalized PageRank estimation and search: a bidirectional approach. In: WSDM, pp. 163–172 (2016)

15. Lofgren, P., Banerjee, S., Goel, A., Comandur, S.: FAST-PPR: scaling personalized PageRank estimation for large graphs. In: ACM SIGKDD, pp. 1436–1445 (2014)

16. Mei, Q., Zhou, D., Church, K.: Query suggestion using hitting time. In: CIKM, pp. 469–478 (2008)

17. Page, L., Brin, S., Motwani, R., Winograd, T.: The PageRank citation ranking: bringing order to the web. Technical report, Stanford InfoLab (1999)

18. Sarkar, P., Moore, A.W.: A tractable approach to finding closest truncated-commute-time neighbors in large graphs. In: UAI, pp. 335–343 (2007)

19. Sarkar, P., Moore, A.W.: Fast nearest-neighbor search in disk-resident graphs. In: ACM SIGKDD, pp. 513–522 (2010)

20. Sarkar, P., Moore, A.W., Prakash, A.: Fast incremental proximity search in large graphs. In: ICML, pp. 896–903 (2008)

21. Wang, R., Wang, S., Zhou, X.: Parallelizing approximate single-source personalized PageRank queries on shared memory. VLDB J. $28(6)$, 923–940 (2019). https://doi.org/10.1007/s00778-019-00576-7

22. Wang, S., Tang, Y., Xiao, X., Yang, Y., Li, Z.: HubPPR: effective indexing for approximate personalized PageRank. PVLDB $10(3)$, 205–216 (2016)

23. Wang, S., et al.: Efficient algorithms for approximate single-source personalized PageRank queries. ACM Trans. Database Syst. $44(4)$, 18:1–18:37 (2019)

24. Wang, S., Yang, R., Xiao, X., Wei, Z., Yang, Y.: FORA: simple and effective approximate single-source personalized PageRank. In: ACM SIGKDD, pp. 505–514 (2017)

25. Wei, Z., et al.: PRSim: sublinear time SimRank computation on large power-law graphs. In: SIGMOD, pp. 1042–1059. ACM (2019)

26. Wei, Z., He, X., Xiao, X., Wang, S., Shang, S., Wen, J.: TopPPR: top-k personalized PageRank queries with precision guarantees on large graphs. In: SIGMOD, pp. 441–456 (2018)
27. Wu, Y., Jin, R., Li, J., Zhang, X.: Robust local community detection: on free rider effect and its elimination. PVLDB **8**(7), 798–809 (2015)
28. Wu, Y., Jin, R., Zhang, X.: Fast and unified local search for random walk based k-nearest-neighbor query in large graphs. In: SIGMOD, pp. 1139–1150 (2014)
29. Zhang, C., Shou, L., Chen, K., Chen, G., Bei, Y.: Evaluating geo-social influence in location-based social networks. In: CIKM, pp. 1442–1451 (2012)
30. Zhang, W., Cheng, R., Kao, B.: Evaluating multi-way joins over discounted hitting time. In: ICDE, pp. 724–735 (2014)

Path Query Processing Using Typical Snapshots in Dynamic Road Networks

Mengxuan Zhang$^{(\boxtimes)}$, Lei Li, Pingfu Chao, Wen Hua, and Xiaofang Zhou

School of Information Technology and Electrical Engineering,
The University of Queensland, Brisbane, Australia
{mengxuan.zhang,l.li3,p.chao,w.hua}@uq.edu.au, zxf@itee.uq.edu.au

Abstract. The shortest path query in road network is a fundamental operation in navigation and location-based services. The existing shortest path algorithms aim at improving efficiency in the static/time-dependent environment. However, the real-life road networks are dynamic, so they can hardly meet the requirement in practice. In this paper, we aim to support the path query in dynamic road networks by identifying the typical snapshots from the snapshot sequences, building the path indexes on them, and finally processing the query with the most suitable typical snapshot. Specifically, we first use the typical OD pairs to capture the dynamic information and represent the snapshots. Then the snapshot similarity is measured by considering the shortest path error and the shortest path similarity of these OD pairs. Because the OD pair number is huge and they have different power in capturing the traffic condition, we further propose a hot region-based OD selection method that could select a small but powerful OD set. Lastly, we use the distance-based χ-quantile error for the query accuracy evaluation and conduct experiments in a large real-world dynamic road network to verify the effectiveness of our method compared with the state-of-the-art.

Keywords: Shortest path · Dynamic road network · Typical snapshot

1 Introduction

Finding the shortest path between the origin and destination (called OD) is the fundamental operation in road network routing and navigation. A road network can be represented as a graph $G(V, E, W)$, where V is the intersection set, $E \subseteq V \times V$ is the road set, and W is the weight associated with each road $e \in E$. Here we focus on the scenario where both V and E are static while W is dynamic. In other words, the road network structure keeps stable and the edge weights change over time. There are two network models depending on whether the edge weight changing is predicable or not: the *time-dependent network* which uses a function to tell the edge weight at different time, and the *dynamic network* whose edge weights change in ad-hoc. In this paper, we attempt to solve the shortest path query in the dynamic road network from a novel perspective.

© Springer Nature Switzerland AG 2020
Y. Nah et al. (Eds.): DASFAA 2020, LNCS 12114, pp. 255–271, 2020.
https://doi.org/10.1007/978-3-030-59419-0_16

In the past decades, various techniques for the shortest path computation are proposed. The fundamental algorithms like *Dijkstra's* [4] and A^* [8] can process the queries on both static and not-static graph since they find the path only with the current graph as the input information. Another large number of well-designed algorithms are quite efficient in the static network, such as the *2-hop labeling* [3], the *hierarchical techniques* [5] and the *path oracle* [16]. These algorithms all highly rely on the pre-computed indexes, which are big in size, slow to construct, and hard to update when the weight changes. Therefore, they are not practical for real-life networks because the traffic condition keeps changing. To cope with this problem, one straightforward approach is building an index for each network snapshot. Obviously, it is space-consuming and contains lots of redundant information. Another way is using the time-dependent network model [11,12] and build a large index for the entire time domain like *TCH* [1] and *T2Hop* [9]. But the premise of these methods is that the time-dependent functions do not change, which essentially requires the network being static from the perspective of "change". To improve the query efficiency in the dynamic environment, the existing works [10,18,19] extend A^* by sharing the computation among the queries. However, their performance is not comparable to the index-based approach. If we summarize all the discussion above, one natural question arises: can we introduce the index back to the dynamic environment for fast query answering, even at the cost of some accuracy? As we know, although the traffic condition changes almost all the time, some snapshots are quite similar to each other such as the traffic condition in every Monday morning, or during most midnights. Hence, we can identify some typical traffic conditions and build indexes on them. When the traffic changes, we choose the most similar typical snapshot and use its index to answer the query efficiently. Therefore, facing the existing non-static algorithms, we try to seek a more flexible solution between them: given a sequence of snapshots in a dynamic road network, we first identify several typical snapshots, assign every snapshot to one of the typical snapshots and process the queries on the dynamic road network by using the indexes of these typical snapshots.

However, it is unclear how to select the typical snapshots. One straightforward idea is clustering these snapshots and using the central ones as the typical. But it is hard to find a suitable similarity measurement for the clustering task. Because each snapshot is essentially a graph, *graph similarity* seems promising to help. The first kind of similarity is *graph edit distance* [7,13,21], which is the minimum edit operation number to transfer one graph to another. But this type of distance is related to the structural or attributive evolvement, while the road network studied here is structurally static and without attribute information, so it is not suitable. Another one is the *feature-based distance* [2], which measures the similarity on the graph abstraction and not related to the weight changes. The most related work [20] in this category analyzes the weight variance and converts the graph to an edge-based vector or a vertex-based vector, then it computes the similarity using these vectors. Nevertheless, it ignores the connectivity of the network and fails to capture the influence of the weight change over the paths.

In fact, it is the OD query that we aim to support, and the accuracy is evaluated over the shortest paths between OD, so we use the OD pair to capture the snapshot features directly in this paper. Specifically, because the OD paths in different snapshots are different, the connectivity influence is incorporated naturally and the traffic condition changes along all these paths are all considered. However, it is challenging to utilize OD pair to select the typical snapshots. The first challenge is how to measure the error of the queries in different snapshots. One common way to measure the approximate shortest path error [6,15] is the average distance error. However, based on our observation, many query errors are close to zero since the weights of the road segments only change during peak hours and most weights keep unchanged most of the time, which further hides the difference between the snapshots. In fact, even when the average error between two snapshots is close to 0, it does not mean putting them into the cluster is a good choice because there also exists a small part of queries whose errors can reach up to 1. What's more, we have different error expectations of different ODs. For example, an error of 0.1 is acceptable for the OD of 5 min (with 0.5 min deviation), but it is horrible for the OD of 90 min (with 9 min deviation). In order to capture the true difference between snapshots and reduce the error for those ODs with large errors, we propose to use the distance-based χ-quantile error instead of the average error to evaluate the accuracy.

The second challenge is how to represent the snapshot and measure the similarity with the OD pairs. One direct idea is representing the snapshot as a vector where each dimension denoting the shortest path distance of the corresponding OD. But the shortest distance alone is not enough to reflect the dynamics as the actual path can be totally different while the distance only changes a little because the shortest distance is always the minimum accumulation value of the edges connecting the OD pair. In other words, the actual path difference is also important to capture the snapshot features from a global perspective. Therefore, we propose a similarity measurement that takes both path error and path similarity into consideration to achieve higher query accuracy.

The third challenge is how to choose the OD pairs. The basic choice is to use the all-pair OD, but the computation cost is extremely high. Besides, it is not always better to select more OD pairs, because the shortest paths on the snapshot sequences are almost the same for many OD pairs which help little for the snapshot differentiation. Based on these observations, we first identify the hot regions in the road network that can capture the network changes. After that, we propose a greedy algorithm to choose the top-k hot regions with the $(1-1/e)$ approximation of the optimal. Finally, OD pairs within and between the hot regions are regarded as candidate pairs, and we further provide an entropy-based refinement approach to determine the final typical OD set. With this small set of typical OD pairs, we are able to cluster the snapshots with much lower errors, which makes the indexing the dynamic road network applicable.

The contributions of this paper can be summarized as follows:

- We formally study the problem of path query in the dynamic road network and propose a typical snapshot solution to provide support for path index.
- We propose the *distance-based χ-quantile error* to measure the query accuracy more reasonably, and propose a new snapshot similarity measurement that considers both path error and path similarity.
- We design a hot-region based typical OD selection method that can capture the weight changes in the snapshots more accurately, cover enough areas and OD types, while remains a small OD set size.
- We conduct extensive evaluations using a large real-world road network and traffic condition. The results verify the effectiveness of our approaches.

The remaining of this paper is organized as follows: we discuss the network similarity measurement briefly in Sect. 2. In Sect. 3, we define our problem and demonstrate the framework. Our solution of snapshot similarity measurement and typical OD selection are described in Sect. 4 and Sect. 5. Section 6 presents the experimental results, followed by a conclusion in Sect. 7.

2   Related Work

There are two categories for the graph similarity measurement: *graph edit distance* [7,13,21] and *feature-based distance* [2]. The *graph edit distance* is calculated as the minimal edit operations that can transfer one graph to another, including vertex insertion/deletion/alteration and edge insertion/deletion/alteration. Obviously, all these operations are related to the structural or attributive evolvement, while the network studied in this work is structurally static without attribute information. Therefore, it can hardly apply to our problem. In the *feature-based distance*, the similarity is not measured on the graph directly but on its abstraction. For example, some labeled edges are selected as the features and one graph is represented as a feature vector where each dimension indicates the existence or the frequency of the corresponding edge [17]. The structural closeness and attribute similarity are combined by the proposed neighborhood random walk model for the graph clustering [22]. These works consider the attributive or structural similarity, which cannot be used here likewise. There is one *feature-based* work [20] focusing on the edge weights change. It represents the snapshot using edges or vertices and then calculates the snapshot distance based on them. Specifically, the edges whose weight varies frequently are selected and one graph is expressed as a vector with each dimension denoting the edge weight. And for the vertex-based representation, each vertex corresponds to a vertex set in each snapshot that can be reached within the same time interval, and the vertices with large vertex set fluctuation are chosen as the typical vertices. Then the graph is represented as a vector by these typical vertices where each dimension is a reachable vertex set. Nevertheless, both these two types of representation ignore the connectivity of road network, since the network is not formed by separated edges or vertices, and the shortest path can pass through the entire network rather than the simple combination of edges or vertices. Therefore, we propose to use the OD pairs for the graph representation and further for the graph similarity measurement.

3 Preliminary

3.1 Problem Definition

The dynamic road network can be viewed as a sequence of its snapshots that are taken every small time interval during which the edge weights are supposed to be still. We define it formally as follows:

Definition 1. Road Network Snapshot. *The road network snapshot is denoted as a weighted directed graph $G_{t_i}(V, E, W)$ with time interval t_i representing the snapshot time. Specifically, each weight $w_{t_i}(u, v) \in W$ denotes the cost passing through (u, v) during t_i. V and E in every snapshot G_{t_i} are the same such that the road network structure does not change.*

A path from o to d in road network G is a sequence of vertices $p = <o = v_0, v_1, \ldots, v_k = d>$ with path length $d_G(p) = \sum_{l=0}^{k-1} w(v_l, v_{l+1})$. For a shortest path query $q(o, d)$ on G, the returned path $p(o, d)$ should be the one with the minimum length from o to d. Similarly, the shortest path query $q_{t_i}(o, d)$ asks for the path $p_{t_i}(o, d)$ of the minimum cost from o to d on snapshot G_{t_i}.

For a sequence of snapshots, a naive way to process the shortest path queries on them is building a shortest path index on each of them. Suppose each index is of size I and the snapshots number is n, then the total index size is $n * I$. Since the time can be unbounded, the value of n can be very large leading to expensive index overhead. To reduce the index construction and storage space consumption, we select some of the snapshots to build the indexes. When there comes a query in one snapshot without index, we assign it to an existing indexed snapshot to process it. Errors could exist as these two snapshots might be different and we define the error below:

Definition 2. Shortest Path Query Error. *Given a shortest path query $q_{t_i}(o, d)$, suppose its shortest path is $p = <v_0, v_1, v_2, \ldots, v_k>$ with path length $d_{G_{t_i}}(p) = \sum_{l=0}^{k-1} w_{t_i}(v_l, v_{l+1})$. If it is processed on the index of snapshot G_{t_j} with the shortest path $p' = <v'_0, v'_1, \ldots, v'_m>$ returned, then the shortest path query error is $e_{t_j}(q_{t_i}) = \frac{|d_{G_{t_i}}(p) - d_{G_{t_i}}(p')|}{d_{G_{t_i}}(p)}$ with $d_{G_{t_i}}(p') = \sum_{l=0}^{m-1} w_{t_i}(v'_l, v'_{l+1})$.*

The average error is the common evaluation of query accuracy. In our application, the error $e_{t_j}(q_{t_i})$ of processing $q_{t_i}(o, d)$ on G_{t_j} is close to zero for many queries. In this way, the average error is usually small and the maximum value is less than 0.1 according to our experimental observation. However, some of query errors can also be more than 1, but their existence are hidden by the massive amount of small error queries. This phenomenon creates an illusion that two snapshots are similar and provides poor query accuracy for some queries. Therefore, the average error cannot fully reflect the query performance. In addition, the same error value for queries of different running time would result in different time deviation. Then we have different tolerant error for query of different running time. For example, the 0.1 error is more acceptable for a query of 5 min (with half minute deviation) than that of 90 min (with 9 min deviation). As a

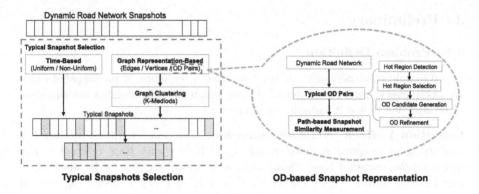

Fig. 1. The framework

result, we test the χ-quantile error for queries of different time ranges (called *distance-based χ-quantile error*), which is defined as follows.

Definition 3. χ-***quantile error.*** *Given a set of queries $\{q_{t_i}(o,d)\}$ with size Ψ on the snapshot set $\{G_{t_i}\}$ and $\{d_{G_{t_i}}(p)\}$ have similar length. If these queries are processed in another snapshot G_{t_j}, then each query would have an error and there would be Ψ errors in total. Suppose the χ-quantile error is ξ, it indicates that there are $\chi \times \Psi$ errors whose value is smaller then ξ.*

Now we are ready to define our problem formally as follows:

Problem Definition. Given a snapshot sequence $G_S = \{G_{t_1}, G_{t_2}, \ldots, G_{t_n}\}$ of the road network G and the OD pair set $P = \{p_1, p_2, \ldots, p_k\}$ with $p_j = (o_j, d_j)$ on every snapshot G_{t_i} corresponding to a shortest path query $q_{t_i}(o_j, d_j)$. Then there are $n \times k$ shortest path queries in total. Set the typical snapshots number m and the snapshot assignment function $f(\cdot)$ such that the χ-quantile error of $\{e_{f(t_i)}(q_{t_i}(o_j, d_j))\}(1 \leq i \leq n, 1 \leq j \leq k)$ is minimized, that is

$$\arg\min_{m, f(\cdot)} \chi - quantile\ error(\{e_{f(t_i)}(q_{t_i}(o_j, d_j))\}) \tag{1}$$

To minimize the χ-quantile errors, two aspects should be focused on: *snapshots selection* and *assignment function*. To select the typical snapshots, the first step is clustering the similar snapshots together. The second step is to choose one representative snapshot in each cluster as one of the typical snapshots and we choose the most central snapshot here. For the snapshot assignment, we just assign one snapshot to its most similar typical snapshots. Therefore, the snapshot similarity measurement is the key point in both *snapshots selection* and *assignment function* and is elaborated on in Sect. 4. In addition, the typical OD pair selection has great impact on the snapshot similarity measurement and we design the OD selection method in Sect. 5.

3.2   Framework

A framework of the system is illustrated in Fig. 1. The *typical snapshot selection* takes the original snapshots as input and uses two ways to identify the typical ones. The first approach is *time-based*, and it can produce typical snapshots directly. The second approach first converts the snapshot into three different representations, and then the graph clustering is applied to select the typical ones of each cluster. After that, the indexes are built on each of these typical snapshots to boost the query efficiency. In this paper, we focus on the OD-based typical snapshots selection, as magnified on the right. We use the typical OD to represent the snapshot and select them based on the hot regions (the weight changes frequently). Given a dynamic road network, the hot regions are firstly identified and selected, then the candidate OD pairs are generated within and between these hot regions, and finally the OD candidates with low typicality are pruned by the OD refinement. Having the typical OD pairs, we represent the snapshot as the shortest paths between these OD pairs. And the path-based similarity is used to measure the similarity between snapshots, which can then be used as the input of the *K-medoids* method for snapshot clustering. At last, the snapshot in the center of each cluster is selected as the typical snapshot and all other snapshots in the same cluster is assigned as the typical snapshot.

4   Snapshot Similarity

When we use r OD pairs to represent snapshots, their similarity is calculated depending on how to represent a snapshot specifically. The aim of snapshot representation is to recognize similar and different snapshots by assigning the similarity value in proportion to the snapshot similarity.

The most straightforward way to utilize OD is using the shortest distance between them. In this way, each snapshot can be denoted as a r-dimensional vector with the lth dimension being the shortest path length of the lth OD pair. The *cosine distance* or the *L1 norm distance* of the r-dimensional vectors can be used to calculate the snapshot similarity. However, the shortest path itself is the result of avoiding the traffic heavy areas and choose the most "fluent" road segments (the segment with less traffic and less weight) to pass through, so as to make the path as short as possible. This property could result in a phenomenon that even when two snapshots are very different and the shortest paths of the OD pair on them are also very different, the shortest distances of them are still close to each other. Such a misleading result has no impact on the current OD, but could have dramatic impact on the others. Therefore, the shortest distance representation can not fully reflect the variance of the road network traffic condition.

To capture the difference between the actual shortest paths, we use the shortest path to represent the snapshot. The snapshot G_{t_i} is still denoted as a r-dimensional vector but with the lth dimension signifying the shortest path $p_{t_i}(o_l, d_l)$ (p_{il} for short). In this way, G_{t_i} is represented as $[p_{i1}, p_{i2}, \ldots, p_{ir}]$. However, the *cosine distance* and *L1 norm distance* can only deal with the numeric

values so they are not applicable here. Besides, the path error is also an important factor, so we propose the *snapshot similarity* that considers both the path
error and path similarity of two snapshots G_{t_i}, G_{t_j} as below:

$$Simi(G_{t_i}, G_{t_j}) = \frac{1}{r} \sum_{l=1}^{r} (1 - \alpha)(1 - e_{t_j}(q_{t_i}(o_l, d_l))) + \alpha LCSSsimi(p_{il}, p_{jl}) \quad (2)$$

where $e_{t_j}(q_{t_i}(o_l, d_l))$ is the shortest path query error caused by answering query
$q_{t_i}(o_l, d_l)$ on G_{t_j} and $LCSSsimi(p_{il}, p_{jl})$ denotes the $LCSS$ similarity of paths
p_{il}, p_{jl}. It should be noted that this measurement is non-symmetric, that is,
$Simi(G_{t_i}, G_{t_j}) \neq Simi(G_{t_j}, G_{t_i})$.

Suppose we use one OD pair (o_l, d_l) to represent the snapshot, if we assign G_{t_j}
as the typical snapshot of G_{t_i}, then the error of query $q_{t_i}(o_l, d_l)$ caused by this
assignment is $e_{t_j}(q_{t_i}(o_l, d_l))$ according to the Definition 2. Larger error means
smaller similarity between these two snapshots. And we use the average error
sum as the snapshot distance if r pairs of typical OD are used. However, this type
of measurement is from the aspect of the OD pairs. Can it reflect the snapshot
similarity completely? In the same idea of *shortest distance representation*, this
error-based representation can hardly show the snapshot difference when the
shortest paths are totally different (which shows the variance of snapshots) while
the path lengths are almost the same.

For each pair of path, we use the *Longest Common SubSequence (LCSS)* for
the similarity measurement. $LCSS$ is widely used in string similarity calculation.
We apply it here because it is more robust to noise and can tell the difference
between the paths. For all the point pairs along the paths, only the ones with
distance shorter than the given threshold ρ would be chosen as correlative points
and the $LCSS$ between two paths $LCSS(p_1, p_2)$ is the number of the correlative
point pairs, which is defined as below:

$$LCSS(p_1, p_2) = \begin{cases} 0, & if \quad n = 0 \quad or \quad m = 0 \\ LCSS(W(p_1), W(p_2)) + 1, if & dist(H(p_1), H(p_2)) < \rho \\ \max\{LCSS(p_1, W(p_2)), LCSS(W(p_1), p_2)\}, otherwise \end{cases} \quad (3)$$

where n, m are the vertex number along the path, $H(p_1), H(p_2)$ represent the
head of path and $W(p_1), W(p_2)$ represent the rest of path. We use the Euclidean
distance to compute the $dist$ here. Since the $LCSS$ is only the number of correlative pair of points, the similarity between two paths is defined as follow:

$$LCSSsimi(p_1, p_2) = \frac{LCSS(p_1, p_2)}{\min\{n, m\}} \quad (4)$$

Based on the snapshot similarity measurement, the snapshots are clustered
by the *k-medoids clustering* method.

5   Typical OD Selection

As discussed in the previous sections, the OD pairs are the foundation for the
snapshot similarity. One basic idea for the OD-based snapshot representation is

to use the all-pair OD with the size of $|V| \times |V|$. As the preprocessing cost for snapshot similarity is $r \times n \times c$ (c denotes the shortest path query cost), fewer typical OD pairs can speed up the similarity computation. In addition, more OD pairs do not necessarily contribute to the better result of typical snapshot selection and assignment. For some OD pairs, the shortest paths on the snapshot sequences are almost the same. So these ODs can narrow down the snapshot distance and a large amount of these ODs can even offset the effectiveness of the ODs which help to differentiate the snapshot. What is more, many OD pairs are similar as they will pass through the overlapping road segments according to the path coherence property [18,19]. These similar OD pairs are redundant for the snapshot representation. Therefore, it is necessary to select the *typical* OD pairs, which can differentiate the snapshots for the typical snapshot selection and snapshot assignment from the all-pair ODs. Therefore, in this section, we propose a typical OD selection method to generate a small but powerful set of OD pairs that can improve the snapshot selection accuracy.

Because the traffic condition is the origin of the network evolvement, we can tell whether two snapshots are similar or not by comparing the paths passing through the "frequently-traffic-changing areas" (called hot region). Hence, the first step is detecting the hot region detection. Then we propose a greedy algorithm to select a set of hot regions as the influential area. After that, the candidate OD pairs are generated with and between the areas. Finally, the candidate ODs are refined by the *OD typicality* to improve the OD quality. The details in each step are explained as follows.

5.1 Hot Region Detection

Inspired by the vertex-based [20] snapshot representation, we detect the hot regions by finding the "hot vertex" around which the traffic condition changes frequently. First of all, we use the following block coefficient of a vertex v_a in g_i to capture how the region evolves as the weight changes:

$$b(v_{ia}) = max\{|S_{1a}|, |S_{2a}|, \ldots, |S_{na}|\}/|S_{ia}| \tag{5}$$

where S_{ia} is the reachable vertex set from one vertex v_a in the snapshot g_i within a constant time t_0 and $|S_{1k}|, |S_{2k}|, \ldots, |S_{nk}|$ is the reach vertex set size in each snapshot sequence, respectively. The block coefficient indicates the traffic congestion level around v_a in snapshot g_i. We define the block fluctuation of vertex as

$$f(v_a) = \frac{\sigma\{b(v_{1a}), b(v_{2a}), \ldots, b(v_{na})\}}{\mu\{b(v_{1a}), b(v_{2a}), \ldots, b(v_{na})\}} \tag{6}$$

where σ and μ denotes the standard deviation and the mean of $\{b(v_{1a}), b(v_{2a}), \ldots, b(v_{na})\}$. If $f(v_a)$ is larger than a threshold β, then v_a is regarded as a hot vertex. The union of the reachable vertex set around v_a is detected as the hot region R_a, which is denoted as

$$R_a = \bigcup\{S_{ia}\}(i = 1, 2, \ldots, n) \tag{7}$$

5.2   Hot Region Selection

Now we have a set of hot vertices $\{v_a\}$ and each of them covers a region R_a that the traffic condition changes frequently. Obviously, these vertices capture the traffic fluctuation information of the whole graph, and their corresponding hot regions have overlapping. However, how to utilize them is not trivial. Firstly, it is inefficient to use all of them for OD selection because it could produce a huge amount of OD pairs, which further prolongs the similarity computation time and does not guarantee a better result. Therefore, we limit the selected hot vertices/regions number to k. Secondly, the overlapping of these hot regions introduce lots of redundant information. If we choose the top-k hot regions depending on the fluctuation directly, only a small part of the graph would be covered and the selected regions are highly redundant. In this section, we propose a greedy selection algorithm that could cover the hot regions evenly and has an $(1 - 1/e)$ approximate ratio of the optimal result. The hot region selection problem can be formally defined as below:

Definition 4 *(Top-k Hot Region Selection (kHRS)). Given a set of hot vertices $\{v_a\}$ and their hot regions $R = \{R_a\}$, select a subset $R^* = \{R_1^*, \ldots, R_k^*\}$ of k regions such that the vertex number it covers $\bigcup_{i=1}^{k} |R_a^*|$ is maximum.*

Theorem 1. *Top-k Hop Region Selection Problem is NPH.*

Proof. Consider an instance of the NPH *Maximum Set Cover* problem: Given a set of items $U = \{u_i\}$ and a collection of subsets $S_i \subseteq U$, select at most k subsets such that $\bigcup^{k} |S_i|$ is maximum. If we let $U = V$ and each hot region R_i to a S_i, then the maximum set cover can be solved by *kHRS*. Therefore, *kHRS* is as hard as the maximum set cover problem.

Because returning the sizes of the hot region and region union satisfies the submodular property [14] naturally, we use the following greedy algorithm to achieve the $(1 - 1/e)$ approximation of the optimal result. The time complexity of it is $O(k \times |R| \times max|R_i|)$.

Algorithm 1: Top-k Hot Region Selection

Input: Hot regions $R = \{R_i\}$, k
Output: $R^* = \bigcup R_i^*, |R^*| = k$
1 $R^* \leftarrow \phi$;
2 **for** $i \leftarrow 1$ *to* k **do**
3 $\quad$ $R_i^* \leftarrow argmax_{R_i \in R}|R_i|$, $R^* \leftarrow R^* \cup \{R_i^*\}$, $R \leftarrow R \setminus \{R_i^*\}$;
4 $\quad$ **foreach** $R_i \in R$ **do**
5 $\quad\quad$ $\lfloor$ $R_i = R_i \setminus R_i^*$;
6 **return** R^*

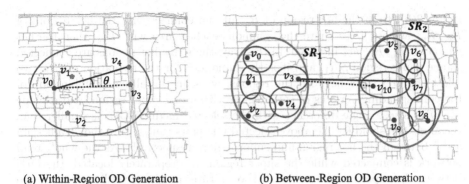

(a) Within-Region OD Generation   (b) Between-Region OD Generation

Fig. 2. OD selection (Color figure online)

5.3 OD Candidate Generation

After determining the top-k hot regions, we generate the candidate OD pairs based on them. Meanwhile, the query of different length can all contribute to the query accuracy from different aspects. Therefore, we generate the OD pairs both within and between these hot regions.

Within-Region OD Generation. Given a hot region R_a and its corresponding hot vertex v_a, we use $S_{v_a}^\cap$ to denote the set of vertex that can be reached from v_a in every snapshot, which can be computed as the intersection of set $S_{1v_a}, S_{2v_a}, \ldots, S_{nv_a}$. As shown in Fig. 2(a), the region within the yellow circle is $S_{v_0}^\cap$ and within the blue circle is R_{v_0}. These two vertex sets indicates that we can always arrive $v_1 \in S_{v_0}^\cap$ from v_0 within t_0, but sometimes we cannot reach $v_2, v_3, v_4 \in R_{v_0} - S_{v_0}^\cap$ from v_0 within the same time interval. The reason behind this phenomenon is that there must exist the "traffic-jam area" between the OD pairs $(v_0, v_2), (v_0, v_3), (v_0, v_4)$ such that v_2, v_3, v_4 are not reachable in every snapshot. Therefore, it is reasonable to use these ODs because they contain the traffic changing information.

For each OD pair with $O = v_a$ and $D \in R_{v_a} - S_a^\cap$, we first sort them in decreasing order of the fluctuation of D. Then we select the destination one by one with the condition that the angle difference of newly selected OD and the selected OD is large than an angle threshold θ, to avoid the similar OD pair. For example, (v_0, v_4) and (v_0, v_3) cannot be both selected as the angle difference between them is smaller than the angle threshold.

Between-Region OD Generation. In the road network, the hot regions are not usually "alone". For example, there can be multiple hot regions in the Central Business District or around the airport and train station. Based on this observation, we connect the hot regions with the overlapping reached area as the one super region SR until no other hot regions can be included. As shown in Fig. 2(b), from the hot region of R_{v_2}, it can connect with the hot regions of R_{v_1}, R_{v_4} as they both share the common reached area with R_{v_2}, then the hot region of R_{v_0}, R_{v_3} are included. In this way, these five hot regions are connected as one super region SR_1, and it is the same idea for the formation of SR_2. To select the OD pairs between the regions, we scan the all-pair combination of the corresponding vertices of each region within one super region, and exclude the ODs whose angle difference with the selected medium ODs is smaller than the angle threshold θ. The above selected OD pairs can capture the "traffic-jam area" between the hot region and super region, respectively. As the whole road network is connected while the super regions are separately located, the ODs between the super regions are needed to capture the global traffic condition. To select the long ODs, we scan the all-pairs combination among vertices in two super regions and also prune the ODs with small angle difference. For example, if (v_3, v_7) is selected, (v_3, v_{10}) would be excluded.

5.4   Typicality-Based OD Refinement

After obtaining the candidate OD pairs, we refine the OD pairs according to the their typicality. In other words, we would like to select the OD pairs which are typical enough to differentiate snapshots. Suppose we use one OD pair (o, d) to represent the snapshot and its shortest path on snapshot set $\{G_{t_1}, G_{t_2}, \ldots, G_{t_n}\}$ are $p_1, p_2, \ldots, p_n$. On one extreme, it is possible that all these paths are the same, and we can select any one snapshot as the typical snapshot and the shortest path query error for this (o, d) is always zero. But it does not make sense since we cannot differentiate the snapshots according to this OD pair. Another totally opposite case is that every paths are different, then this OD pair is typical enough to distinguish the snapshots. Besides the above two extreme cases, the common case can be that some paths can be the same or of high similarity. Therefore, we evaluate the OD typicality as the entropy of the number of different paths in each group and it is defined as

Definition 5. *OD typicality.* *For one OD pair (o, d), its shortest paths on the snapshot set $\{G_{t_1}, G_{t_2}, \ldots, G_{t_n}\}$ are $p_1, p_2, \ldots, p_n$. After gathering the similar paths (with similarity larger then δ) in one group, there are h path groups with path number $n'_1, n'_2, \ldots, n'_h$ $(\sum_{i=1}^{h} n'_i = n)$ respectively, and the typicality of this OD pair is calculated as $\sum_{i=1}^{h} -\frac{n'_i}{n} log(\frac{n'_i}{n})$.*

Therefore, in the above two cases, their *OD typicality* values are zero and reach to the maximum value, respectively. Since the OD pair with higher typicality can differentiate the snapshot more, we prune the OD pairs whose typicality value is smaller than the threshold γ in the OD refine stage. And for the shortest path set of one OD pair, two paths can hardly be exactly the same since any slight weight variance of the road segments can cause the path difference in a small area. Therefore, we put the paths into the same group if their similarity is larger than δ.

6 Experiments

In this section, we experimentally evaluate the proposed snapshot similarity measurements and OD selection method in the real-life road network and traffic condition against the current state-of-the-art methods.

6.1 Experimental Setup

All the algorithms are implemented in C++, compiled with -O3 optimization, and tested on a Dell R730 PowerEdge Rack Mount Server which has two Xeon E5-2630 2.2 GHz (each has 10 cores and 20 threads) and 378G memory. The data are stored on a 12 × 4 TB Raid-50 disk.

Dataset. We obtain our road network of Beijing form *NavInfo*. It consists of 296,710 intersections and 387,588 roads, which covers a 184 km × 185 km spatial range. The snapshot data is extracted from Beijing taxi trajectory collected from 1st April 2016 to 6th April 2016. We take one snapshot every 5 min and there are 1728 snapshots in total. For the test set, we take 1000 randomly selected OD as the test data in each time range: 10, 20, 30, 40, 50, 60, 70, 80, and 90 min. For the typical snapshot number, we suppose the snapshots are uniformly sampled every 10 min, 15 min, 20 min, 30 min, 1 h, 2 h, 3 h, 4 h, half day, and 1 day. So we set different typical snapshot numbers to 864, 576, 432, 288, 144, 72, 48, 36, 24, 12, and 6. Totally, there are 9000 test queries in each snapshot. After clustering these snapshots by *K-medoids clustering* method, we test the χ-quantile error under different time ranges and typical snapshot numbers.

Comparison Methods. For the snapshot similarity measurements, we use *PathBased* to denote the proposed *path-based snapshot similarity* and compare it with the *Uniform Sampling (Sampling)*, *Shortest distance-based similarity (ShortDis)*, *Edge-based similarity (Edge-Based)* and *Vertex-based similarity (Vertex-Based)* measurements. For the *OD selection* method, we denote the proposed method as *Refined OD* and compare it with ODs without refinement *(Select OD)* and randomly selected OD *(Random OD)*. As for the influence of different OD numbers, we use *R5000*, *R10000*, and *R20000* to denote the 5000, 10000, and 20000 randomly selected OD pairs. For each time range of the test query, we show the 0.95- and 0.85-quantile errors under different typical snapshot numbers. And for each typical snapshot number, we show the 0.95- and 0.85-quantile errors under different time ranges of the test set. The parameters in our experiments are: $\alpha = 0.5, \rho = 50, \theta = 30, \delta = 0.9, \gamma = 1.0$.

6.2 Snapshot Similarity Measurement

Figure 3(a)–3(d) show the χ-quantile errors under different time ranges with 432 and 48 typical snapshots, respectively. The *uniform sampling* causes the largest

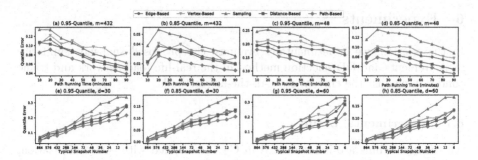

Fig. 3. Distance measurement evaluation

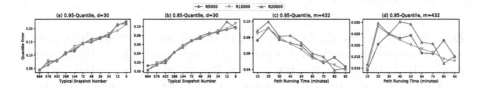

Fig. 4. OD number evaluation

χ-quantile error most times, and the *path-based similarity* always performs better than the others. Specifically, when the typical snapshot number is set as 48, about 15 min (90 min × 0.17) deviation would cause for 5% queries using the *uniform sampling*, and it decreases to 10 min using the *shortest distance-based* measurement and decreases further under 9 min using the *path-based* measurement. Figure 3(e)–3(h) show the χ-quantile errors under different typical snapshot number with the time range of 30 and 60 min. Firstly, the fewer typical snapshot we select, the larger χ-quantile error we end up with. Secondly, the proposed *path-based* similarity measurement always performs the best. For the query with 60 min time range, when only 6 typical snapshots are selected, about 12 min deviation would cause for 15% queries using the *uniform sampling*, and the deviation drop to 6 min using the *path-based* similarity measurement. This group of experimental results show the superiority of our proposed snapshot similarity measurement under different time ranges and typical snapshot numbers.

6.3   Typical OD Selection

We first show the influence of different typical OD numbers. Figure 4 shows the χ-quantile errors when we set the randomly OD number as 5000, 10000, and 20000. The χ-quantile errors are almost the same and it seems hard to tell which one is better. This group of results experimentally shows that the error is not always smaller when selecting more typical OD. In fact, it is the quality (typicality) of the OD rather than the quantity that helps to reduce the query error.

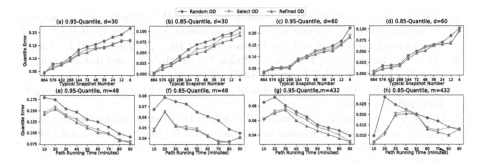

Fig. 5. OD selection performance

Then we test χ-quantile errors of different OD selection methods under different time range and typical snapshot number. We set $t_0 = 6\,\text{min}$ and select 500 hot regions. And 5596 OD pairs are generated and there are left 3002 OD pairs after refine. As shown in Fig. 5, the *Refined OD Selection* performs better than the method without refinement. Moreover, the *Random OD Selection* performs worse than both of the other two methods. These results indicate that the proposed OD selection can help to reduce the query error with the same typical snapshot number.

7 Conclusion

In this paper, we study the problem of selecting typical snapshots for supporting the shortest path index in the dynamic road network. Although there exist indexes for the time-dependent environment, they are essentially static and cannot satisfy the practical application in the real dynamic environment. Therefore, we aim to solve this problem by selecting the typical snapshots and building indexes on them, assigning the other snapshots to one typical snapshot and sharing its index at the cost of query accuracy. We first propose to use the χ-quantile to better evaluate the clustering quality. After that, we propose a snapshot similarity measurement that considers both the path error and path similarity. Finally, we propose a hot-region based typical OD selection that can capture the snapshot features with only a small set of ODs. Our extensive experiments show the effectiveness of the proposed snapshot similarity measurement and typical OD selection method.

Acknowledgment. This research is partially supported by the Australian Research Council (DP200103650 and LP180100018).

References

1. Batz, G.V., Delling, D., Sanders, P., Vetter, C.: Time-dependent contraction hierarchies. In: Proceedings of the Meeting on Algorithm Engineering & Experiments, pp. 97–105. Society for Industrial and Applied Mathematics (2009)

2. Chen, L., Gao, Y., Zhang, Y., Jensen, C.S., Zheng, B.: Efficient and incremental clustering algorithms on star-schema heterogeneous graphs. In: 2019 IEEE 35th International Conference on Data Engineering (ICDE), pp. 256–267. IEEE (2019)
3. Cohen, E., Halperin, E., Kaplan, H., Zwick, U.: Reachability and distance queries via 2-hop labels. SIAM J. Comput. **32**(5), 1338–1355 (2003)
4. Dijkstra, E.W.: A note on two problems in connexion with graphs. Numer. Math. **1**(1), 269–271 (1959)
5. Geisberger, R., Sanders, P., Schultes, D., Delling, D.: Contraction hierarchies: faster and simpler hierarchical routing in road networks. In: McGeoch, C.C. (ed.) WEA 2008. LNCS, vol. 5038, pp. 319–333. Springer, Heidelberg (2008). https://doi.org/10.1007/978-3-540-68552-4_24
6. Geisberger, R., Schieferdecker, D.: Heuristic contraction hierarchies with approximation guarantee. In: Third Annual Symposium on Combinatorial Search (2010)
7. Gouda, K., Hassaan, M.: CSI_GED: an efficient approach for graph edit similarity computation. In: 2016 IEEE 32nd International Conference on Data Engineering (ICDE), pp. 265–276. IEEE (2016)
8. Hart, P.E., Nilsson, N.J., Raphael, B.: A formal basis for the heuristic determination of minimum cost paths. IEEE Trans. Syst. Sci. Cybern. **4**(2), 100–107 (1968)
9. Li, L., Wang, S., Zhou, X.: Time-dependent hop labeling on road network. In: 2019 IEEE 35th International Conference on Data Engineering (ICDE), pp. 902–913, April 2019. https://doi.org/10.1109/ICDE.2019.00085
10. Li, L., Zhang, M., Hua, W., Zhou, X.: Fast query decomposition for batch shortest path processing in road networks. In: 2020 IEEE 36th International Conference on Data Engineering (ICDE) (2020)
11. Li, L., Hua, W., Du, X., Zhou, X.: Minimal on-road time route scheduling on time-dependent graphs. Proc. VLDB Endow. **10**(11), 1274–1285 (2017)
12. Li, L., Zheng, K., Wang, S., Hua, W., Zhou, X.: Go slow to go fast: minimal on-road time route scheduling with parking facilities using historical trajectory. VLDB J. **27**(3), 321–345 (2018). https://doi.org/10.1007/s00778-018-0499-4
13. Li, Z., Jian, X., Lian, X., Chen, L.: An efficient probabilistic approach for graph similarity search. In: 2018 IEEE 34th International Conference on Data Engineering (ICDE), pp. 533–544. IEEE (2018)
14. Nemhauser, G.L., Wolsey, L.A., Fisher, M.L.: An analysis of approximations for maximizing submodular set functions–I. Math. Program. **14**(1), 265–294 (1978). https://doi.org/10.1007/BF01588971
15. Potamias, M., Bonchi, F., Castillo, C., Gionis, A.: Fast shortest path distance estimation in large networks. In: Proceedings of the 18th ACM Conference on Information and Knowledge Management, pp. 867–876. ACM (2009)
16. Sankaranarayanan, J., Samet, H., Alborzi, H.: Path oracles for spatial networks. Proc. VLDB Endow. **2**(1), 1210–1221 (2009)
17. Yan, X., Yu, P.S., Han, J.: Substructure similarity search in graph databases. In: Proceedings of the 2005 ACM SIGMOD International Conference on Management of Data, pp. 766–777. ACM (2005)
18. Zhang, M., Li, L., Hua, W., Zhou, X.: Batch processing of shortest path queries in road networks. In: Chang, L., Gan, J., Cao, X. (eds.) ADC 2019. LNCS, vol. 11393, pp. 3–16. Springer, Cham (2019). https://doi.org/10.1007/978-3-030-12079-5_1
19. Zhang, M., Li, L., Hua, W., Zhou, X.: Efficient batch processing of shortest path queries in road networks. In: 2019 20th IEEE International Conference on Mobile Data Management (MDM), pp. 100–105. IEEE (2019)

20. Zhang, M., Li, L., Hua, W., Zhou, X.: Typical snapshots selection for shortest path query in dynamic road networks. In: Borovica-Gajic, R., Qi, J., Wang, W. (eds.) ADC 2020. LNCS, vol. 12008, pp. 105–120. Springer, Cham (2020). https://doi. org/10.1007/978-3-030-39469-1_9
21. Zhao, X., Xiao, C., Lin, X., Wang, W.: Efficient graph similarity joins with edit distance constraints. In: 2012 IEEE 28th International Conference on Data Engineering, pp. 834–845. IEEE (2012)
22. Zhou, Y., Cheng, H., Yu, J.X.: Graph clustering based on structural/attribute similarities. Proc. VLDB Endow. **2**(1), 718–729 (2009)

Dynamic Dimension Indexing for Efficient Skyline Maintenance on Data Streams

Rui Liu[1] and Dominique Li[1,2]([⊠])

[1] Computer Science Department, University of Tours, 41000 Blois, France
{liurui,dominique.li}@univ-tours.fr
[2] LIFAT Laboratory, 37000 Tours, France

Abstract. Skyline computation receives much attention in research and application domains, for which many algorithms have been developed during decades. However, maintaining the skyline in data streams is much challenging because of the continuous updates of skyline with respect to non stop adding of incoming tuples and removing of expired tuples. In this paper, we present a dynamic dimension indexing based approach RSS to skyline computation on high dimensional data streams, which is efficient at both count-based and time-based sliding windows regardless the dimensionality of data. Our analysis shows that the time complexity of RSS is bounded by a subset of the instant skyline, and our evaluation shows the efficiency of RSS on both of low and high dimensional data streams.

Keywords: Skyline · Dynamic dimension indexing · Date stream

1 Introduction

Skyline computation receives continuous attention since the introduction of the *skyline query* [2] that aims at returning the exact set of *dominating tuples* in multidimensional data, for which many efficient algorithms have been developed during the past two decades. However, the *skyline maintenance* problem in the context of streaming data is much challenging because it requires instantaneous updates of the skyline regarding the arrival of new data and the expiration of too early data that are probably worthless and shall be discarded due to the time sensibility of data streams.

As the most useful technique in processing streaming data, a *sliding window* is usually considered while formulating stream based queries [12]. The model of window is generally *count-based* that covers a number of the most recent tuples at every time instant, or *time-based* that is bounded by a number of time units coinciding with timestamps of the stream. Figure 1 shows a paradigm of sliding window based skyline maintenance on the price-distance relation, where we consider the prices (Y axis) of hotels with respect to their distances (X axis) to some place, such as the city center, the beach, or the railway station. Let us consider a count-based window $W = 5$ and the arrival order of hotel tuples from a to f, then initially there are 4 skyline tuples $\{a, c, d, e\}$ as shown in Fig. 1(a).

© Springer Nature Switzerland AG 2020
Y. Nah et al. (Eds.): DASFAA 2020, LNCS 12114, pp. 272–287, 2020.
https://doi.org/10.1007/978-3-030-59419-0_17

At the instant while f arrives, the earliest tuple a must be discarded in order to keep the size of window. As a consequence, since a is the only tuple that dominates b, b becomes a skyline tuple when a is discarded; further, the incoming tuple f dominates current skyline tuples d and c, so d and c must be removed from current skyline, and finally the updated skyline is therefore $\{b, c, f\}$, as shown in Fig. 1(b).

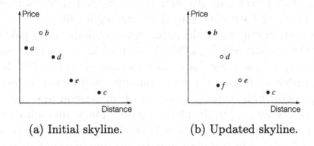

(a) Initial skyline.   (b) Updated skyline.

Fig. 1. Skyline maintenance example.

In fact, such dynamic updates of skyline make skyline maintenance difficult over data streams: (1) *an expired skyline tuple may release dominated tuples so that they become skyline tuples*; (2) *an incoming tuple dominates existing skyline tuples so that the latter shall be removed from the skyline*. Most of the existing skyline algorithms designed for static data do not fit such requirements imposed by streaming data. For instance, regarding each update of skyline invoked by incoming/expired tuples with respect to a sliding window, the straightforward algorithm BNL [2] requires a full scan of the window, sorting-based algorithms [1,3,4,14] require re-sorting all tuples in the window, and partitioning-based algorithms [5–7,11,13,17] require updating the partitioning structure that is costly. Hence, the dynamic update of skyline and depended data structures are two important issues of skyline maintenance on data streams [8,10,15].

In this paper, we present a dynamic dimension indexing algorithm RSS (**R**ange **S**earch for **S**tream) extended from our recent work SDI [9], which outperforms the state-of-the-art skyline algorithms, for *sliding window skyline* maintenance on data streams. Our main contributions include:

- We construct a linked sorting structure for maintaining a dynamic dimension index that serve the sliding window.
- We show that with the proposed dynamic dimension indexing method, the dominance tests involved by the above operations (1) and (2) are limited to optimal subsets of current skyline tuples in the sliding window.
- We evaluate our proposed algorithm with both count-based and time-based sliding window, on both synthetic and real datasets, and show its efficiency on both low and high dimensional data streams.

The remainder of this paper is organized as follows. Section 2 reviews related skyline algorithms. Section 3 presents the basis of our approach. We detail the algorithm RSS in Sect. 4 with analysis of its efficiency. In Sect. 5, we report the performance evaluation of RSS and finally we conclude in Sect. 6.

2 Related Work

Many algorithms have been developed for skyline computation and maintenance during the past two decades. In this section, we review mainstream sorting-based, partitioning-based, and stream-oriented skyline algorithms.

BNL [2] is a straightforward algorithm that compares each pair of tuples in order to find the skyline. SFS [3], LESS [4], and SaLSa [1] are three sorting-based algorithms that share the same dominance test mechanism of BNL by presorting all tuples in order to reduce the total number of dominance tests. Different from BNL-like algorithms, Index [14] sorts each dimension in order to find local skylines that can be merged to the global skyline. The main limitation of sorting-based algorithms on streaming data is that all tuples must be re-sorted while the window sliding, and in comparison with our proposed algorithm RSS, which can be considered as sorting-based, the above algorithms cannot benefit from existing sliding window skyline for efficient dominance tests.

Partitioning-based algorithms focus on grouping tuples into regions so the skyline can be determined by performing region-level dominance tests. D&C [2] uses a divide-and-conquer fashion to partition data, and from which derives the best known time complexity in the worst case of skyline computation [13]. NN [5] and BBS [11] partition data by using the nearest-neighbor search with respect to R-tree that allows to prune non skyline tuples at region level, which shows the efficiency on low dimensional data. In the framework Z-SKY [7], the algorithm ZSearch performs region-level tuple pruning with Z-ordering on ZB-tree. OSP [17] and the state-of-the-arts algorithm BSkyTree [6] partition data into dominance regions and incomparable regions that take the incomparability of tuples into account. Note that BBS, Z-SKY, OSP, and BSkyTree can incrementally update the skyline while inserting or deleting tuples. Our proposed algorithm can also be regarded as partitioning-based since each dimension index can be considered as a region of tuples and dominance tests are performed at region level.

On data streams, the n-of-N model [8,16] focuses on retrieving the skyline from n tuples of the most recent N tuples. The incremental skyline update makes possible adapting partitioning-based algorithms to data streams, for instance, R-tree based Lazy/Eager [15] and quadtree based LookOut [10] are derived from BBS (which are however inefficient on high dimensional data streams). To the best of our knowledge, there are no reported adaptions of Z-SKY, OSP, and BSkyTree in streaming environments. Note that our base algorithm SDI [9] is much efficient than BSkyTree on high dimensional data.

Although the partial ordered domains (POD) is not covered by this paper, we incidentally found that the column-store of tuples designed in SCL [18] for POD skyline computation is similar to our dimension indexing schema [9]. However, the dominance tests in SCL is based on intersections over tuple IDs, which are inefficient on high dimensional data with high cardinality.

3   Dynamic Dimension Indexing

Let t be a d-dimensional tuple, we denote $t[i]$, for $1 \leq i \leq d$, the *dimension value* of the tuple t on the dimension i. We define a *preference order* $\prec$ as a total order on all dimensions such that given two tuples t and u, $t[i] \prec u[i]$ denotes that $t[i]$ is *better than* $u[i]$ and $t[i] \preceq u[i]$, that is, $(t[i] \prec u[i]) \vee (t[i] = u[i])$, denotes that $u[i]$ is *not worse* than $t[i]$. We have that $t[i] \prec u[i] \Rightarrow t[i] \preceq u[i]$.

A tuple t *dominates* a tuple u, denoted by $t \prec u$, if and only if for each dimension $1 \leq i \leq d$, we have $t[i] \preceq u[i]$, and for at least one dimension $1 \leq k \leq d$, we have $t[k] \prec u[k]$. Given two tuples t and u, we denote $t \not\prec u$ that t does not dominate u, and we denote $t \sim u$ that t and u are *incomparable*, that is, $(t \not\prec u) \wedge (u \not\prec t)$. In order to simplify our formal description, we extend $\{\prec, \not\prec, \preceq, \sim\}$ to sets of tuples such that, for instance, $t \prec X$ denotes that the tuple t dominates each tuple in X, $t \sim X$ denotes that the tuple t is incomparable with each tuple in X.

Let $\mathcal{D}$ be a d-dimensional database, a tuple $t \in \mathcal{D}$ is a *skyline tuple* if and only if $\nexists u \in \mathcal{D}$ such that $u \prec t$. The *skyline* $\mathcal{S}$ of $\mathcal{D}$ is the complete set of all skyline tuples, that is, $\mathcal{S} = \{t \in \mathcal{D} \mid \nexists u \in \mathcal{D} \text{ such that } u \prec t\}$.

Definition 1 (Dimension Index). *Let $\mathcal{D}$ be a database of d dimensions. The dimension index $\mathcal{I}$ of $\mathcal{D}$ is the ensemble of d ordered lists of index entries, where each index entry corresponds to an individual tuple, which contains the dimension value, a dimension link that points to the entry of the same tuple in the next dimension, and a header link that points to the tuple header. Each list $I_i \in \mathcal{I}$ is a sub-index of the dimension i, $1 \leq i \leq d$, in which the index entries are sorted by the dimension value on the preference order $\prec$.*

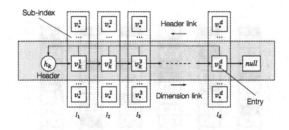

Fig. 2. Dimension index data structure.

Figure 2 illustrates the dimension index data structure by following the links on all index entries of a tuple t_k, where the node h_k depicts the tuple header of t_k and the nodes $v_k^1, v_k^2, \ldots, v_k^d$ depict the index entries of t on dimensions $1, 2, \ldots, k$. Since all entries in a sub-index is sorted by dimension values, we note other tuple IDs as $*$ in our illustration, for instance, v_*^3 denotes the dimension value of a tuple $*$ on the dimension 3. This design of linked entries allows fast tuple access from any dimension.

Let us consider 64-bit operating systems and 64-bit double-precision floating-point values, a d-dimensional tuple requires $24 \times d$ Bytes to store all index entries, including values, header links, and dimension links; a header stores a tuple ID as a 64-bit integer number, a 64-bit dominance list pointer (see Sect. 4, Fig. 5), and a skyline flag, that requires 20 Bytes. Therefore, to store a tuple in our dimension index, totally $24 \times d + 20$ Bytes are required. Note that the space complexity of B-tree is $\mathcal{O}(N)$, where N is the number of nodes. If 2 64-bit pointers per node are required to maintain the tree, then totally $16 \times N \times (24 \times d + 20)$ Bytes are required by our proposed dynamic dimension index, that is, 100 MB of memory space is enough to handle a 10 K window of 20-dimensional streaming data.

Theorem 1. *Let $\mathcal{I}$ be the dimension index of a d-dimensional database $\mathcal{D}$, $I_i \in \mathcal{I}$ be an arbitrary sub-index of $\mathcal{I}$, and t be a tuple. Then, t is a skyline tuple if and only if $\nexists r \in \mathcal{D}$ such that $(r[i] = t[i]) \wedge (r \prec t)$ and one of the following conditions is satisfied: (1) $\nexists u \in \mathcal{D}$ such that $u[i] \prec t[i]$, or (2) $\forall s \in S$ such that $s[i] \prec t[i] \Rightarrow s \nprec t$.*

Proof. We consider an arbitrary dimension i, $1 \leq i \leq d$. First, let us consider the conditions (1) and (2) without repeated dimension values, that is, $\nexists r \in \mathcal{D}$ such that $r[i] = t[i]$. For (1), t appears on the top of I_i, so t is a skyline tuple because no tuple is better than t. For (2), let s be a skyline tuple such that $t[i] \prec s[i]$, we have $s \nprec t$; if for any skyline tuple s such that $s[i] \prec t[i] \Rightarrow s \nprec t$, we further have that t is incomparable to any skyline tuple, thus, t is a skyline tuple. Then, let us consider that $\exists r \in \mathcal{D}$ such that $r[i] = t[i]$. In this case, the above conditions (1) and (2) cannot exclude $r \prec t$: (1) does not cover $(r[i] = t[i]) \wedge (r \prec t)$; (2) does not cover $r \in \mathcal{S}$. Thus, in order to establish (1) and (2), for $\forall r \in \mathcal{D}$ such that $r[i] = t[i]$, $r \nprec t$ must be established. $\square$

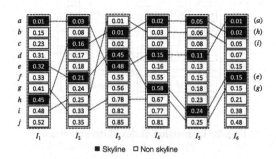

Fig. 3. A example of dimension index.

Let us regard the example shown in Fig. 3 that consists of 6 sub-indexes $I_1, I_2, \ldots, I_6$ and 10 tuples in total. In order to make our example simple, we concern with only 5 tuples $\{a, e, g, h, i\}$, by assuming others are not skyline tuples, and ignore tuple headers that lead linked entries so the dashed lines depict dimension links. According to Theorem 1, a is a skyline tuple, which can be independently concluded from I_1, I_2, and I_6. If we focus only on I_3, we have both $h[3] = i[3]$ and no tuple is better than h and i in this dimension, for which both $h \prec i$ and $i \prec h$ must be tested to determine whether h and i are skyline tuples (indeed we have $h \prec i$ and $i \not\prec h$). If we apply Theorem 1 only to I_4, h is immediately a skyline tuple, and we have $(h[4] < i[4]) \wedge (h \prec i)$, so i is not a skyline tuple; if we go ahead, the next one is not skyline tuple neither till to a, where $(h[4] < a[4]) \wedge (h \not\prec a)$, so a is a skyline tuple. Finally, with I_6 only, 3 consecutive entries contain the dimension value 0.15 so these 3 tuples must be first locally compared in order to filter skyline tuples locally present, in our example, e, to which Theorem 1 can be therefore applied. We call such skyline tuples locally present in I_3 and I_6 the *local skyline*. Obviously, each index entry contains a distinct dimension value is itself a local skyline.

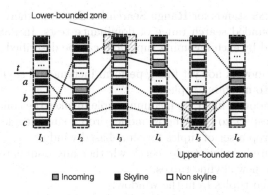

Fig. 4. A example of dimension index.

Based on Theorem 1, we now present a dimension index update method that allows efficient skyline maintenance. We illustrate our method with the example shown in Fig. 4, where we assume that all values on any dimension are unique. Without loosing the generality, the symbol $\cdots$ signifies other index entries.

Let t be an incoming tuple, we first locate the index entries of t on each sub-index with respect to the preference order $\prec$ (as the gray entries). Then we find a dimension *low* that contains the minimum number of skyline tuples s such that $s[lower] \prec t[lower]$, which we call the *lower-bounded* dimension of t where a *lower-bounded zone* can be detected by bounding the lower side from the top of the sub-index. We call the set of skyline tuples contained in the lower-bounded zone the *lower-bounded skyline*, to which we apply Theorem 1 to test whether t is a skyline tuple. For instance, in our example, the lower-bounded skyline is located in I_3, comparing with which t and c is incomparable so t is a skyline tuple. The incoming tuple t may dominate existing skyline tuples, Theorem 1 can also applied to filter such existing skyline tuples. We find a dimension *up* that contains the minimum number of skyline tuples u such that $t[upper] \prec u[upper]$, and we call such a dimension *up* the *upper-bounded* dimension of t where a *upper-bounded zone* can be detected as well as the lower-bounded zone. For each tuple u contained in the *upper-bounded skyline*, if $t \prec u$, then u can be removed from the skyline. In our example, current skyline tuples a and b are dominated by t so they will be no longer in the skyline. Otherwise, if t is not a skyline tuple, no upper-bounded skyline detection is necessary.

While deleting a skyline tuple t, then for each non skyline tuple x contained in the lower-bounded zone of t such that $t \prec x$, the lower-bounded zone must be detected in order to test whether x is dominated by its lower-bounded skyline: if not, x shall be a new skyline tuple. No dominance test is required while deleting a non skyline tuple.

4 Skyline Maintenance for Streaming Data

The algorithm RSS stands for **R**ange **S**earch for **S**tream, where *Range Search* is based on the bounded search range of dominance tests. In RSS, the update of skyline is invoked by each incoming tuple and can be described as the following:

1. When the window is not yet filled, perform incremental skyline computation with respect to incoming tuples.
2. When the window is filled, an incoming tuple invokes the collection of expired tuples with respect to the window mode: drop a single tuple for count-based window, or drop a set of tuples for time-based window.
3. For each expired skyline tuple, check whether any dominated tuples can be released to be new skyline tuples.
4. Discard expired tuples from the window.
5. Add the incoming tuple to the window by performing incremental skyline computation.

Algorithm 1 outlines RSS, which takes a data stream $\mathcal{D}$ and a window W as input, and updates the skyline index $\mathcal{S}$ while sliding the window W over $\mathcal{D}$ by accepting each incoming tuple t from the data stream.

First, RSS deletes expired tuples from the dimension index $\mathcal{I}$ (lines 5–17). RSS fetches the set U of all expired tuples with respect to the count-based or time-based window W (line 4). For each expired skyline tuple $u \in U$, RSS finds by *DominatedTuples* the set X of all tuples directly dominated by u (line 7) then disables u in $\mathcal{I}$ for not affecting following dominance tests. For each tuple $x \in X$, RSS calls *RangeSearch* (Algorithm 2) to determine new skyline tuples (lines 9–13). To finish tuple deletion, RSS removes u from the skyline index $\mathcal{S}$ if $u \in \mathcal{S}$ (line 14), and definitively removes u from $\mathcal{I}$ (line 16). Then, RSS inserts the incoming tuple into $\mathcal{I}$ (lines 18–25). RSS calls *RangeSearch* to determine whether t is a skyline tuple (line 18). If t is a skyline tuple, RSS calls *DominatedTuples* that returns the set R of the tuples directly dominated by t and removes each tuple $r \in R$ (lines 19–22). Finally, RSS adds t to $\mathcal{S}$ (line 23), and commits the insertion of t to $\mathcal{I}$ (line 25). Notice that if no expired tuple can be fetched, RSS directly proceeds tuple insertion.

Algorithm 1: RSS (Range Search for Stream)

    Input: Data stream $\mathcal{D}$, Window W

    Output: Instant update of skyline $\mathcal{S}$

1   Dimension Index $\mathcal{I} = \emptyset$

2   Skyline Index $\mathcal{S} = \emptyset$

3   **while** $t \leftarrow \mathcal{D}$ **do**

4      $U \leftarrow ExpiredTuples(\mathcal{I}, W)$

5   **foreach** $u \in U$ **do**

6   **if** $u \in \mathcal{S}$ **then**

7            $X \leftarrow DominatedTuples(\mathcal{I}, u)$

8   Disable u in $\mathcal{I}$

9   **foreach** $x \in X$ **do**

10   **if** $RangeSearch(\mathcal{I}, x)$ **then**

11                $\mathcal{S} \leftarrow \mathcal{S} \cup \{x\}$

12   **end**

13   **end**

14            $\mathcal{S} \leftarrow \mathcal{S} \setminus \{u\}$

15   **end**

16   Remove u from $\mathcal{I}$

17   **end**

18   **if** $RangeSearch(\mathcal{I}, t)$ **then**

19         $R \leftarrow DominatedTuples(\mathcal{I}, t)$

20   **foreach** $r \in R$ **do**

21   Remove r from $\mathcal{I}$

22   **end**

23         $\mathcal{S} \leftarrow \mathcal{S} \cup \{t\}$

24   **end**

25   Insert t to $\mathcal{I}$

26 **end**

In our design of dimension index (Fig. 2), each sub-index is a B-tree of index entries with respect to the preference order $\prec$ on dimension values. Three additional data structures, including a *skyline index* that contains pointers to skyline tuple headers, a *header list* that contains all headers, and a *dominance index* that contains pointers to all directly dominated tuples, are used to maintain the dimension index. While inserting tuple to the dimension index, a header is first created to identify the tuple and appended to the header list with respect to the incoming order; then, a linked list of per-dimension index entries are created from the tuple with respect to dimension links, and each index entry corresponding to a dimension i will be inserted into the sub-index I_i.

The *RangeSearch* method is listed in Algorithm 2, which follows the dynamic dimension indexing described in Sect. 3 by determining lower-bounded and upper-bounded skylines for the tuple t within the dimension index $\mathcal{I}$. Given a tuple t, *RangeSearch* first calculates the lower-bounded dimension for t in order to perform dominance tests (line 2). According to Theorem 1, *RangeSearch* tests

---

Algorithm 2: RangeSearch

Input: Dimension index $\mathcal{I}$, tuple t
Output: **true** if t is a skyline tuple

1 Skyline Index S
2 $lower \leftarrow LowerBoundedDimension(\mathcal{I}, t)$
3 $B \leftarrow GetBlock(t, I_{lower})$
4 **if not** $BNL(B, t)$ **then**
5 $\quad$ | $\quad$ **return false**
6 **end**
7 $S \leftarrow LowerBoundedSkyline(I_{lower}, t)$
8 **foreach** $s \in S$ **do**
9 $\quad$ | $\quad$ **if** $s \prec t$ **then**
10 $\quad$ | $\quad$ | $\quad$ **return false**
11 $\quad$ | $\quad$ **end**
12 **end**
13 $upper \leftarrow UpperBoundedDimension(\mathcal{I}, t)$
14 $B \leftarrow GetBlock(t, I_{upper})$
15 **foreach** $b \in B$ **do**
16 $\quad$ | $\quad$ **if** $b \in S$ **and** $t \prec b$ **then**
17 $\quad$ | $\quad$ | $\quad$ $UpdateDominanceList(t, b)$
18 $\quad$ | $\quad$ | $\quad$ $S = S \setminus \{b\}$
19 $\quad$ | $\quad$ **end**
20 **end**
21 $S \leftarrow UpperBoundedSkyline(I_{upper}, t)$
22 **foreach** $s \in S$ **do**
23 $\quad$ | $\quad$ **if** $t \prec s$ **then**
24 $\quad$ | $\quad$ | $\quad$ $UpdateDominanceList(t, s)$
25 $\quad$ | $\quad$ | $\quad$ $S = S \setminus \{s\}$
26 $\quad$ | $\quad$ **end**
27 **end**
28 **return true**

---

whether t is a local skyline in the set B of all tuples having the same dimension value (line 3) by the embedded BNL algorithm[1], which returns *true* if t is a local skyline; otherwise *RangeSearch* stops (lines 4–6). If t is a local skyline, *RangeSearch* further retrieves the lower-bounded skyline S for t (line 7), then tests t is a *true* skyline against S: if not, *RangeSearch* stops by returning *false* (lines 8–12). If t is found being a skyline tuple, *RangeSearch* calculates the upper-bounded dimension for t in order to eliminate skyline tuples eventually dominated by t (line 13). At this step, repeated dimension values shall also be taken into account: if t dominates any skyline tuple b having the same dimension values, then b must be removed from the skyline index (lines 14–20). *RangeSearch* retrieves the upper-bounded skyline S for t (line 21) removes all

---

[1] Any skyline algorithm can be applied to compute local skylines, BNL is the simplest one with respect to a small number of repeated dimension values.

skyline tuples dominated by t (lines 22–27). *RangeSearch* returns *true* if t is a skyline tuple.

The update of dominance index is performed by the *UpdateDominanceList* method in *RangeSearch* while the skyline status of any tuple is modified, to make *DominatedTuples* efficient. A call like *UpdateDominanceList*(t, b) joins the dominance index of b to which of t. Figure 5 shows how dominance indexes are updated while an incoming tuple dominates existing skyline tuples, where a count-based window $W = 15$ is concerned. The update of dominance indexes can be illustrated by the following steps (let the gray band depict the skyline index and assume that the incoming tuple 33 is a skyline tuple): (a) the incoming tuple 33 makes the current earliest tuple 18 expired; (b) while removing the expired tuple 18, assume that tuple 25 becomes skyline tuple, and assume that the incoming tuple 33 dominates the skyline tuple 21; (c) tuple 21 is appended to the tuple 33 and removed from the skyline index and its dominance entries are moved to tuple 33.

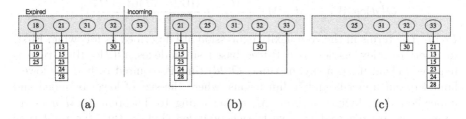

(a)   (b)   (c)

Fig. 5. Dynamic update of dominance indexes.

With the above description, it is easy to see that *DominatedTuples* returns all non-expired headers of the tuples directly dominated by a given tuple t, but not the complete set of all tuples that can be dominated by t. Indeed, if a tuple x is dominated by a skyline tuple s, then for any incoming tuple t such that $t \nsim s$, no dominance test between t and x will be performed though we may have $t \prec x$, since *RangeSearch* takes only skyline tuples into account. Therefore, even if $t \prec x$ in fact, x will not appear in the dominance index of t thus will not be returned by *DominatedTuples*. Effectively, according to the arrival order of incoming tuples in stream data, such incompleteness of dominance indexes does not affect the update of non skyline tuples dominated by expired tuples.

Theorem 2. *The update of non skyline tuples dominated by expired tuples listed in Algorithm 1 (lines 7–14) is correct and complete.*

Proof. If a tuple x is dominated by a tuple s and s is incomparable with the incoming tuple t, then s and x must be expired before t. Thus, x will not be taken into account while removing t. $\square$

In order to determine the lower-bounded and upper-bounded dimensions for a tuple t, we use the following estimation formula instead of counting the exact sizes of lower-bounded and upper-bounded skylines.

$$I_{lower} = \arg\min_{I_i}(|\frac{v_t^i - \min(I_i)}{\max(I_i) - \min(I_i)}|) \tag{1}$$

$$I_{upper} = \arg\max_{I_i}(|\frac{v_t^i - \min(I_i)}{\max(I_i) - \min(I_i)}|) \tag{2}$$

The above approximate estimation of two bounded dimensions requires $\mathcal{O}(d)$ time, which is a trade-off between selecting the best dimension with the smallest bounded skyline and calculating in the least time. Within a sliding window W over d-dimensional streaming data, let M be the size of skyline and assume that dominance test requires $\mathcal{O}(d)$ time, we have the following analysis.

Theorem 3. *In the worst case, RSS inserts an incoming skyline tuple t in time*

$$\mathcal{O}\left(d\log|W| + 2(d+M) + \frac{M}{d}(d + \frac{(|W|-M)(d+M)}{d})\right).$$

Proof. The terms in the complexity expression are described as following. First, to insert d index entries to d B-tree based sub-indexes, $\mathcal{O}(d\log|W|)$ time is required. Then, if t is a skyline tuple, $\mathcal{O}(2d)$ time is required to find the lower-bounded and upper-bounded dimensions, where the size of lower-bounded and upper-bounded skylines is both M/d. According to Theorem 1, M/d dominance tests are required by t with each bounded skyline, thus the total time is bounded in $\mathcal{O}(2d + 2M)$. In the worst case, t dominates all M/d tuples in the upper-bounded skyline, then for each upper-bounded skyline tuple, $\mathcal{O}(d)$ time is required to find the upper-bounded dimension, where $(|W| - M)/d$ non skyline tuples are dominated by each upper-bounded skyline tuple. Finally, for each of such non skyline tuples, $\mathcal{O}(d)$ time is required to find the lower-bounded dimension and the size of lower-bounded is M/d in the worst case, that is, $\mathcal{O}(d + M)$ time is required to test whether it will be a skyline tuple.   □

Theorem 4. *In the worst case, RSS deletes an expired skyline tuple t in time*

$$\mathcal{O}\left(d + \frac{(|W|-M)(d+M)}{d} + d\log|W|\right).$$

Proof. To delete a skyline tuple, $\mathcal{O}(d)$ time is required to find the upper-bounded dimension. Then, as shown in the proof of Theorem 3, $(|W| - M)/d$ non skyline tuples in the upper-bounded zone shall be tested in the worst case, each test requires $\mathcal{O}(d + M)$ time. Finally, $\mathcal{O}(d\log|W|)$ time is require to remove d index entries from the B-trees.   □

With the above analysis, it is clear that RSS requires $\mathcal{O}(d\log|W| + d + M)$ time to insert a non skyline incoming tuple and $\mathcal{O}(d\log|W|)$ time to delete a non skyline expired tuple. Furthermore, the higher the dimensionality d of data, the smaller the factor $(|W| - M)$. Indeed, our performance evaluation shown in Sect. 5 confirms the efficiency of RSS in high dimensional data streams.

5   Performance Evaluation

This section presents our performance evaluation of RSS on both count-based window and time-based window over streams of both synthetic and real datasets. We implemented RSS in C++, and all executables are compiled by LLVM Clang with -03 option[2]. All experiments were conducted on an Intel Core i5 3.1 GHz processor with 16 GB DDR3 RAM running macOS 10.14.5 operating system. All results reported are the average performance over 5 iterations except per incoming tuple processing time. Note that the efficiency of dimension indexing in skyline computation in comparison with the state-of-the-art algorithm can be found in our recent work [9].

We consider *anti-correlated* (AC), *correlated* (CO), and *uniform independent* (UI) synthetic datasets generated by Skyline Benchmark Data Generator[3], and 3 real-world datasets HOUSE ($d = 6$), NBA ($d = 8$), and WEATHER ($d = 15$)[4]. The dimensionality of synthetic datasets varies from 2 to 24 dimensions. Different window sizes are considered, $W \in \{1K, 2K, 4K, 8K, 16K, 32K\}$ tuples for count-based window and $W \in \{10, 20, 40, 80, 160, 320\}$ seconds for time-based window. Particularly for time-based window, the arrival speed of data stream is uniformly randomized between 100 and 1000 tuples per second. As a reference, Table 1 lists the mean per window skyline ratio (%) with respect to the dimensionality of synthetic data. All results are reported after the sliding window is filled, and subsequently, fixed window size and dimensionality is chosen to analyze the detail process of one incoming tuple.

Table 1. Mean per window skyline ratio (%) w.r.t. dimensionality.

d	02	04	06	08	10	12	14	16	18	20	22	24
AC	1.10	23.46	61.00	83.35	93.45	97.21	98.90	99.48	99.73	99.89	99.95	99.98
CO	0.09	0.20	0.90	1.83	2.68	7.63	11.60	18.47	27.15	31.92	40.29	48.73
UI	0.27	2.84	13.37	33.78	56.57	75.43	89.10	95.07	98.40	99.51	99.85	99.94

(a) Count-based window.

d	02	04	06	08	10	12	14	16	18	20	22	24
AC	0.82	19.97	56.87	80.65	92.00	96.46	98.56	99.32	99.64	99.85	99.93	99.97
CO	0.07	0.15	0.69	1.46	2.12	6.54	10.09	15.60	24.40	28.15	37.31	44.89
UI	0.19	2.33	11.46	30.45	52.72	71.95	87.42	94.20	97.94	99.35	99.79	99.92

(b) Time-based window.

Figure 6 shows the mean processing time for synthetic data by varying the dimensionality from $d = 2$ to 24 stepped by 2. In Fig. 6(c), there are irregular points when $d = 2$ and $W \geq 8K$ that the skyline updates spend more time but size is quite small. We investigate the detailed process when $d = 2$ and

---

[2] https://github.com/skyline-sdi/sdi-rss.
[3] http://pgfoundry.org/projects/randdataset.
[4] https://github.com/sean-chester/SkyBench.

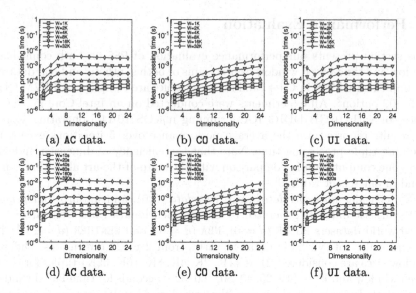

Fig. 6. Mean skyline update time for count-based window (a, b, c) and time-based window (d, e, f) w.r.t dimensionality on synthetic data.

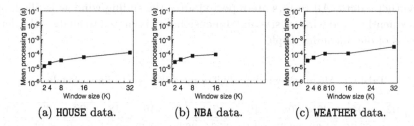

Fig. 7. Mean skyline update time for count-based window on real-world data.

find out that the first expired tuple is a skyline in all cases. Theoretically, the dataset is uniformly distributed and independent on each dimensions, in addition the skyline size is very small in comparing with the window size, so the first expired tuple is likely to dominate approximately half-window (W/d) tuples. For $W = 32K$, the experiment shows that the first expired tuple dominates more than $15\,K$ tuples in the window and it spends more than 1s to check all its dominated non skyline tuples. Same situation happens to $d = 2$ and $W = 16K$ as well as $d = 2$ and $W = 8K$, of which the first expired tuple dominates respectively about 8K and 4K tuples, and the processing time is respectively about 0.2 s and 0.04 s. Figure 6(f) also shows exceptional points as which is resulted from the same reason. Therefore, the mean processing time is affected by deleting the first expired tuple. Notice that except the first expired tuple, the processing time per new tuple is 10^{-5} on average.

A common observation is that the performance reaches stable or keeps in one magnitude regarding the incremental of the dimensionality. For AC data, the performance is stable while $d \geq 8$ with both window mode, and $d \geq 12$ for UI data; however, the processing time keeps increasing for CO data. This observation confirms our theoretical analysis in Sect. 4.

Figure 7 shows the mean per window skyline update time on real-world data (where the NBA dataset is too small), where RSS shows the same performance as well as on synthetic data. Due to the space limitation, we do not include the mean skyline update time for time-based window on real-world data, however the curves are very similar to which shown in Fig. 7.

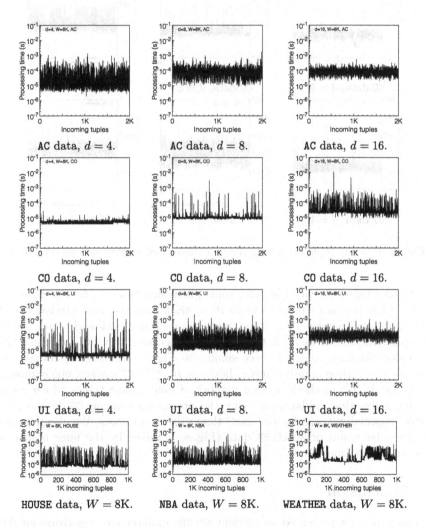

Fig. 8. Incoming tuple processing time for count-based window on synthetic and real-world data.

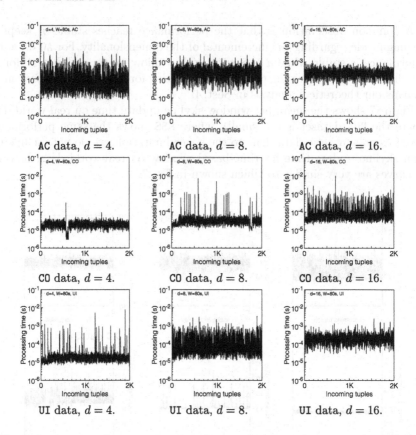

Fig. 9. Incoming tuple processing time for time-based window on synthetic data.

To investigate the per incoming tuple processing time, we fix the dimension to $d \in \{4, 8, 16\}$ and window size to $W = 8K$ for count-based window (Fig. 8) and time-based window (Fig. 9) on synthetic data and real-world data (due to the space limitation, we do not include the time-based window processing time on real-world data, which is very similar to Fig. 8). Although it reveals that the processing time mainly varying between 10^{-4} ~10^{-5}s, when one expiration is skyline, it has to check a large number of non skyline tuples which results in high spike, that explains also why the processing time on CO data is much stable (especially $d = 4$, for instance) than other data types. We also note that the irregular result on WEATHER data (Fig. 8) is raised by the huge number of repeated dimension values.

6   Conclusion

In this paper, we presented an efficient skyline maintenance algorithm for data streams. We proposed a dynamic dimension indexing method, with which the dominance tests are limited to optimal subsets of current skyline tuples in the sliding window. We developed the RSS algorithm and our theoretical analysis shows that the time complexity on sliding window skyline update per incoming tuple is bounded. We evaluated our proposed algorithm with both count-based and time-based sliding window, on both synthetic and real datasets, and show its efficiency on both low and high dimensional data streams. As a possible future work, we are interested in the adaption of RSS to distributed systems over data streams.

References

1. Bartolini, I., Ciaccia, P., Patella, M.: Efficient sort-based skyline evaluation. ACM Trans. Database Syst. **33**(4), 31 (2008)
2. Borzsony, S., Kossmann, D., Stocker, K.: The Skyline operator. In: ICDE, pp. 421–430 (2001)
3. Chomicki, J., Godfrey, P., Gryz, J., Liang, D.: Skyline with presorting. In: ICDE, vol. 3, pp. 717–719 (2003)
4. Godfrey, P., Shipley, R., Gryz, J.: Maximal vector computation in large data sets. In: VLDB, pp. 229–240 (2005)
5. Kossmann, D., Ramsak, F., Rost, S.: Shooting stars in the sky: an online algorithm for skyline queries. In: VLDB, pp. 275–286 (2002)
6. Lee, J., Hwang, S.-W.: Scalable skyline computation using a balanced pivot selection technique. Inf. Syst. **39**, 1–21 (2014)
7. Lee, K.C.K., Lee, W.-C., Zheng, B., Li, H., Tian, Y.: Z-SKY: an efficient skyline query processing framework based on Z-order. VLDB J. **19**(3), 333–362 (2010)
8. Lin, X., Yuan, Y., Wang, W., Lu, H.: Stabbing the sky: efficient skyline computation over sliding windows. In: ICDE, pp. 502–513 (2005)
9. Liu, R., Li, D.: Efficient skyline computation in high-dimensionality domains. In: EDBT (2020)
10. Morse, M., Patel, J.M., Grosky, W.I.: Efficient continuous skyline computation. Inf. Sci. **177**(17), 3411–3437 (2007)
11. Papadias, D., Tao, Y., Greg, F., Seeger, B.: Progressive skyline computation in database systems. ACM Trans. Database Syst. **30**(1), 41–82 (2005)
12. Patroumpas, K., Sellis, T.: Window specification over data streams. In: EDBT, pp. 445–464 (2006)
13. Sheng, C., Tao, Y.: Worst-case I/O-efficient skyline algorithms. ACM Trans. Database Syst. **37**(4), 26 (2012)
14. Tan, K.-L., Eng, P.-K., Ooi, B.C.: Efficient progressive skyline computation. In: VLDB, pp. 301–310 (2001)
15. Tao, Y., Papadias, D.: Maintaining sliding window skylines on data streams. IEEE Trans. Knowl. Data Eng. **18**(03), 377–391 (2006)
16. Lin, X., Lu, H., Xu, J., Yu, J.X.: Continuously maintaining quantile summaries of the most recent n elements over a data stream. In: ICDE, pp. 362–373 (2004)
17. Zhang, S., Mamoulis, N., Cheung, D.W.: Scalable skyline computation using object-based space partitioning. In: SIGMOD, pp. 483–494 (2009)
18. Zhang, S., Mamoulis, N., Cheung, D.W., Kao, B.: Efficient skyline evaluation over partially ordered domains. PVLDB **3**(1–2), 1255–1266 (2010)

SCALE: An Efficient Framework for Secure Dynamic Skyline Query Processing in the Cloud

Weiguo Wang[1], Hui Li[1(✉)] (iD), Yanguo Peng[2], Sourav S. Bhowmick[3], Peng Chen[1], Xiaofeng Chen[1], and Jiangtao Cui[2]

[1] School of Cyber Engineering, Xidian University, Xi'an, China
{wgwang,pchen97}@stu.xidian.edu.cn, {hli,xfchen}@xidian.edu.cn
[2] School of Computer Science and Technology, Xidian University, Xi'an, China
{ygpeng,cuijt}@xidian.edu.cn
[3] School of Computer Science and Engineering, Nanyang Technological University, Singapore, Singapore
assourav@ntu.edu.sg

Abstract. It is now cost-effective to outsource large dataset and perform query over the cloud. However, in this scenario, there exist serious security and privacy issues that sensitive information contained in the dataset can be leaked. The most effective way to address that is to encrypt the data before outsourcing. Nevertheless, it remains a grand challenge to process queries in ciphertext efficiently. In this work, we shall focus on solving one representative query task, namely *dynamic skyline query*, in a secure manner over the cloud. However, it is difficult to be performed on encrypted data as its dynamic domination criteria require both subtraction and comparison, which cannot be directly supported by a single encryption scheme efficiently. To this end, we present a novel framework called SCALE. It works by transforming traditional dynamic skyline domination into pure comparisons. The whole process can be completed in single-round interaction between user and the cloud. We theoretically prove that the outsourced database, query requests, and returned results are all kept secret under our model. Empirical study over a series of datasets demonstrates that our framework improves the efficiency of query processing by nearly **three orders of magnitude** compared to the state-of-the-art.

Keywords: Skyline · Secure · Cloud · Query

1 Introduction

With the rapid expansion in data volumes, many individuals and organizations are increasingly inclined to outsource their data to public cloud services since they provide a cost-effective way to support large-scale data storage and query processing. As a major type of query and fundamental building block for various applications, *skyline query* [6] has become an important issue in database

© Springer Nature Switzerland AG 2020
Y. Nah et al. (Eds.): DASFAA 2020, LNCS 12114, pp. 288–305, 2020.
https://doi.org/10.1007/978-3-030-59419-0_18

research for extracting interesting objects from multi-dimensional datasets. The skyline query processing is widely adopted in many applications that require multi-criteria decision making such as market research [12], location based systems [14], web services study [2], etc. The *skyline operator* filters out a set of interesting points based on a group of evaluation criteria from a large set of points. A point is considered as interesting, if there does not exist a point that is at least as good in all criteria and better in at least one criteria. However, similar to other types of query, outsourcing skyline query workload to a public cloud will inevitably raise privacy issues. Since a real-world database may often contain sensitive information such as personal electronic mails, health records, financial transactions, etc., a cloud service provider may illegally spy on the data and invade the privacy of the data owner and users.

In this paper, we focus on the problem of *secure* skyline querying on the cloud aiming to protect the security of outsourced data, query request and results. Secure query processing on encrypted data has been extensively studied during recent years [17,24]. For instance, fully homomorphic encryption schemes [13] ensure strong security while enabling arbitrary computations on the encrypted data. Modular Order-preserving encryption [4,8] provides an intuitive security model which supports comparison over the ciphertext without decryption. Despite the promising achievements in the area of secure query processing, it remains a grand challenge for processing *dynamic* skyline queries over ciphertext, where the *skyline operator* is executed with respect to some query point [19] and is adopted in many applications [26]. The main reason for the problem is as follows. Given a query request, a dynamic skyline query requires performing *both* comparison and distance evaluation online simultaneously. Unfortunately, accomplishing this task over ciphertext cannot be realized efficiently via existing encryption schemes.

For instance, suppose that a medical institution wishes to outsource its electronic diabetes records to some public cloud service. Naturally, the medical institution would like to prevent any leak of the contents of the records to the cloud server. An electronic diabetes record consists of a series of attributes, including *ID, age, FBGL (fasting blood glucose level)*, etc. Let $P = \{p_1, \ldots, p_n\}$ denote a set of electronic diabetes records. When the medical institution receives a new record (i.e., patient) q, it expects the cloud server to retrieve a similar record to enhance and personalize the treatment for the new patient q. However, it is usually difficult or even impossible to uniformly assign weights to all the attributes to return the nearest neighbor (*e.g.*, p_1 is the nearest if only *age* is involved while p_2 is the nearest if only *FBGL* is taken into account). In light of that, dynamic skyline query provides all possible Pareto records that are not dominated by any other ones. Given a query q, we can compute the difference between each attribute for p_i and q. Let t_i be the difference tuple between p_i and q, and $t_i[j] = |p_i[j] - q[j]|$ for each dimension j. An object t_i dominates t_j if it is better than t_j in at least one dimension and not worse in every other dimensions. If an object cannot be dominated by any other object, this object is one of the skyline points that needs to be returned. As shown in Fig. 1, there are five patient

Records	Age	FBGL	Projected Age (w.r.t. q)	Projected FBGL (w.r.t. q)
p_1	29	7.2	35	11.4
p_2	35	8.7	35	9.9
p_3	21	8.1	3	10.5
p_4	22	11.2	42	11.2
p_5	30	12.7	34	12.7
q	32	9.3		

(a) Original and projected samples   (b) Dynamic skyline

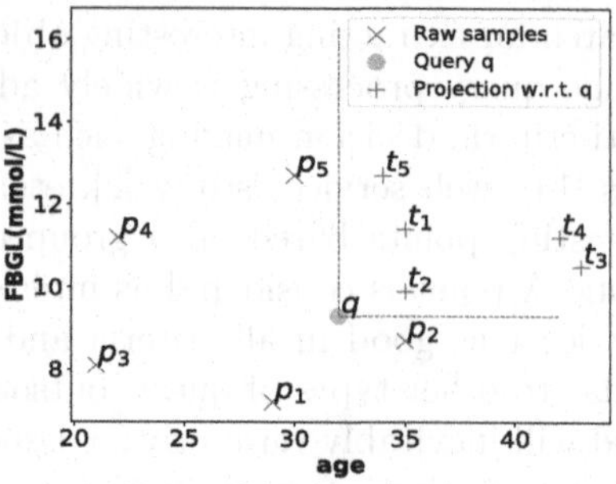

Fig. 1. Dynamic skyline query example.

records $p_1, \ldots, p_5$. Given a query record q, we calculate $t_1, \ldots, t_5$ and can easily identify the skyline points as t_5 and t_2. Therefore, p_5 and p_2 are the results for the dynamic skyline query w.r.t q.

Notably, in the above example, a dynamic skyline query requires performing both subtraction and comparison online. As there is no practical encryption scheme supporting both operators over ciphertext, existing model employs secure multiparty computation over at least two third-party non-collusion clouds and processes the query with multiple rounds of interactions. In this work, we present a novel framework called SCALE (SeCure dynAmic skyLine quErying) by transforming traditional skyline domination criteria, which requires both subtraction and comparison, into comparison only. In this way, we are able to present a new scheme that can support dynamic skyline query over ciphertext without any help from a second cloud and can be completed in a single-round interaction between a user and the cloud. We theoretically prove that the outsourced database, query requests, and returned results are all kept secret under our model. Empirical study over four datasets including both synthetic and real-world ones demonstrate that our framework outperforms the state-of-the-art method by nearly **three orders of magnitude**. Notably, as a special case of dynamic skyline query, skyline computation can also be processed securely and efficiently under our model (with trivial modifications). In summary, this work makes the following contributions.

- We propose a new scheme to encrypt the outsourced database and query request. Based on the scheme, dynamic skyline query can be answered without decrypting the database or the query. Within the scheme, the cloud server and data user need only one interaction during the query.
- We theoretically prove that our model is secure if the cloud is curious-but-honest.
- Empirical study over both synthetic and real-world datasets justify that our model is superior to the state-of-the-art w.r.t the query response time.

The rest of this paper is organized as follows. In Sect. 2 we review related works. In Sect. 3 we formally present the problem definition and system model for this work. The detailed designs of encryption scheme and query framework are discussed in Sect. 4. Empirical study and corresponding results are shown in Sect. 5. In Sect. 6, we conclude this work.

2   Related Work

The skyline query is particularly important for several applications involving multi-criteria decision making. The computation of the skyline is equivalent to determining the maximal vector problem in computational geometry [15,23], or equivalently the Pareto optimal set [23] problem. *static* skyline query has been extensively studied in the database field [6,11,21,27]. A *dynamic* skyline query is a variation of skyline computation that was first introduced in [19,20]. Instead of computing the skyline points purely from the given dataset, dynamic skyline query returns series of points that are not dominated by any others with respect to q. In another word, skyline computation can be viewed as a special case of dynamic skyline query where q is fixed as origin point and only the comparison (without distance evaluation) is required.

Table 1. Summary of notations

Notation	Definition
$\mathsf{Enc}(q)$ ($\mathsf{Enc}(2q)$)	Ciphertext of the query (doubled query) tuple
$P = \{p_1, \ldots, p_n\}$	A database with n tuples
$E(P)$	Ciphertexts of tuples for P
$E(\Phi)$	Ciphertexts of the pairwise sums for tuples in P
$p_i[j]$	The $j - th$ attribute of p_i
$keys[\cdot]$	The set of private keys

With the development of cryptography, Encryption technology is gradually applied in the database field. Bothe *et al.* [7] presented an approach for skyline computation over Encrypted Data. It provided efficiency analysis and empirical study for computing skyline points and decrypting the results. However, it failed to provide any formal security guarantee. Another work [9] proposed three novel schemes that enable efficient verification of skyline query results returned by an unauthentic cloud server. This work focuses on the verification but not privacy issues, and does not work on ciphertext. It is orthogonal to the scope of this paper. Liu *et al.* [17] proposed the first semantically secure protocol for dynamic skyline query over the cloud platform. The scheme adopts both Paillier cryptosystem [18] and Secure Multi-party Computation (SMP) as building blocks. Although it is proved to be semantically secure, the protocol suffers from huge computation cost and strict system model. In fact, as a query framework, the response time is the most important issue for the success of the application, but the performance of [17] is far from satisfactory in this aspect.

Order-Preserving Encryption (OPE) scheme [1], whose ciphertext preserve the original ordering of the plaintexts, has been extensively applied in range query over encrypted databases. The ideal security goal for an order-preserving scheme, IND-OCPA [3], is to reveal no additional information about the plaintext values besides their order. Following that, a series of schemes have been proposed in literature [5,22]. Chenette *et al.* [10] built efficiently implementable order-revealing encryption based on pseudorandom functions. Lewi *et al.* [16] improved the above scheme. The ORE scheme in [16] is adopted for this work. We will discuss it further in Sect. 4.

3 Problem Definition

In this section, we shall first introduce a group of key concepts for skyline query, then describe the system and security models utilized in this paper. For ease of discussion, the key notations used throughout this paper are summarized in Table 1.

3.1 Skyline Query Definition

In this part, we shall introduce a series of key concepts for skyline problem that is important for our subsequent discussions.

Definition 1 (Dynamic Domination). *Given two points p_α, p_β and a query point q in d-dimensional space, we say p_α dynamically dominates p_β with respect to q (denoted by $p_\alpha \prec_q p_\beta$), if $\forall i \in \{1, \ldots, d\}$, $|p_\alpha[i] - q[i]| \leq |p_\beta[i] - q[i]|$, and $\exists i \in \{1, \ldots, d\}$, $|p_\alpha[i] - q[i]| < |p_\beta[i] - q[i]|$.*

Definition 2 (Dynamic Skyline Query). *Given a dataset $P = \{p_1, \ldots, p_n\}$ and a query q in d-dimensional space, dynamic skyline query returns the set $S \subseteq P$, such that $\forall p \in S$, $\nexists p' \in P$ such that $p' \prec_q p$ (i.e., $\forall p \in S, p' \in P$, p' cannot dynamically dominate p with respect to q).*

A common algorithm (*i.e.*, BNL [6]) for dynamic skyline query is shown in Algorithm 1. It first calculates the differences (*i.e.*, t_i) between each tuple (*i.e.*, p_i) and the query request (*i.e.*, q) in every dimension (Lines 1–3). When a tuple p_i is read from P, it is added to S if S is empty (Lines 5–6). Otherwise, we shall compare p_i's corresponding difference tuple with respect to q, namely t_i, with that of each tuple in S. In case $t_i \prec t_j$, where $p_j \in S$, we shall delete p_j from S. If there is no $p_j \in S$ such that $t_j \prec t_i$, we shall add p_i to S (Lines 10–11, 16–18). The algorithm repeats this process for the remaining tuples in P, and finally returns S (Line 21). We shall use this as the basis for our secure skyline model. Notably, this is not the most efficient algorithm for plaintext skyline query. We select this method as our building block for the following reasons. Firstly, the state-of-the-art solution for secure dynamic skyline is [17], which adopts BNL [6] as the basic building block. In line with [17] and to make a fair comparison, our solution is constructed according to the same query framework. Secondly, BNL is a common and popular iterative algorithm for answering dynamic skyline query in plaintext. Thirdly, as discussed in Sect. 1, the key challenge in secure dynamic skyline query lies in the solution for performing both subtraction and comparison over ciphertext. A secure model building on any other (plaintext) dynamic skyline query algorithm inevitably has to address that. In other words, although our solution in this work adopts Algorithm 1 as the foundation, it can be easily adapted to other (plaintext) dynamic skyline query algorithms.

3.2 System Model and Design Goals

Our system model involves three types of participants: a data owner, a cloud server and a group of query users. The cloud server is assumed to have large storage and computation ability, and provide outsourcing storage and computation services. As Fig. 2 shows, the data owner employs the cloud service to store his private database. To preserve data privacy, the data owner will encrypt his dataset, and only outsource the encrypted dataset to the cloud. Every query user may submit a query point (*i.e.*, q) to the system. The query request may be locally encrypted before sending to the cloud server. Then, the cloud server will perform dynamic skyline query over encrypted database and query request without decryption. Afterwards, it returns the encrypted results to the user. Finally, the user decrypts these results using his own private keys.

Security Model. We parameterize the security model by a collection of leakage functions $\mathcal{L} = (\mathcal{L}_{Encrypt}, \mathcal{L}_{Query}, \mathcal{L}_{Insert}, \mathcal{L}_{Delete})$. These functions describe what information the protocol leaks to the adversary $\mathcal{A}$. Our model ensures that the scheme does not reveal any information beyond what can be inferred from the leakage functions.

Algorithm 1. Basic Skyline Query Algorithm

Require: The dataset P and a query tuple q
Ensure: The result set of skyline points S
 1: **for** i in $1, \ldots, n$ and j in $1, \ldots, d$ **do**
 2:     let $t_i[j] = |p_i[j] - q[j]|$
 3: **end for**
 4: **for** i in $1, \ldots, n$ **do**
 5:     **if** S is empty **then**
 6:         add p_i to S
 7:     **else**
 8:         $flag \leftarrow True$
 9:         **for** each $p_j \in S$ **do**
10:             **if** $t_j \prec t_i$ **then**
11:                 $flag \leftarrow False$
12:             **else if** $t_i \prec t_j$ **then**
13:                 delete p_j from S
14:             **end if**
15:         **end for**
16:         **if** $flag == True$ **then**
17:             add p_i to S
18:         **end if**
19:     **end if**
20: **end for**
21: **return** S

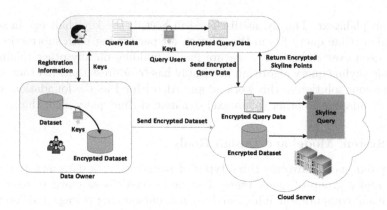

Fig. 2. The system model of secure skyline query

We define two games $\text{Game}_{\mathcal{R},\mathcal{A}}$ and $\text{Game}_{\mathcal{S},\mathcal{A}}$ as follows. The adversary repeatedly encrypts data and queries skyline points, and receives the transcripts generated from $Encrypt()$ and $Query()$ algorithms in the real game $\text{Game}_{\mathcal{R},\mathcal{A}}$ or receives the transcripts generated by the simulator $\mathcal{S}(\mathcal{L}_{Encrypt})$ and $\mathcal{S}(\mathcal{L}_{Query})$ in the ideal game $\text{Game}_{\mathcal{S},\mathcal{A}}$. Eventually, $\mathcal{A}$ outputs a bit 0 ($\text{Game}_{\mathcal{R},\mathcal{A}}$) or 1 ($\text{Game}_{\mathcal{S},\mathcal{A}}$).

Definition 3 (Adaptively secure). *A scheme is $\mathcal{L}$-adaptively-secure if for all probabilistic polynomial-time algorithm $\mathcal{A}$, there exists an efficient simulator $\mathcal{S}$ such that:* $|Pr[\text{Game}_{\mathcal{R},\mathcal{A}}(\lambda) = 1] - Pr[\text{Game}_{\mathcal{S},\mathcal{A}}(\lambda) = 1]| \leq negl(\lambda)$.

Design Goals. Our design goals contain both efficiency and privacy, including database privacy, query privacy, and result privacy. The details are as follows.

- Data owners need to encrypt the database before it is sent to the cloud server. Meanwhile, the content in the database is not leaked to the cloud server.
- Query request, as well as the results, should not be revealed to the cloud server throughout query processing.
- As a query processing framework, efficiency should be considered as one of the most important issue for measuring its success. Although the entire query processing is performed in ciphertext here, it should minimize the additional cost associated with it.

4   The SCALE Framework

As discussed above, processing dynamic skyline query given a query point q requires performing both subtraction and comparison. Addressing both tasks in ciphertext form is challenging as there is no practical encryption scheme that supports both operations simultaneously. To address this challenge, we re-investigate the entire dynamic skyline query workflow described in Definition 3.1 and Algorithm 1. Our investigation revealed an important fact that may lead to an effective solution. Notably, to answer a dynamic skyline query given a request q, quantifying the differences between each point p_i and q through all dimensions is not mandatory. Instead, what we need is the relative order of such differences for a group of different p_i.

Observation 1. *In order to evaluate whether p_α dynamically dominates p_β with respect to q, we do not need to know the exact values for the difference vectors T_α and T_β, where $T_i[j] = |p_i[j] - q[j]|$ for $j \in [1, \ldots, d]$. In fact, what we really need to know is whether $T_\alpha[j] \leq T_\beta[j]$ or $T_\alpha[j] < T_\beta[j]$ for $j \in [1, \ldots, d]$. For simplicity, for an arbitrary dimension j, we need to know whether $p_\alpha[j]$ or $p_\beta[j]$ is close to $q[j]$. To answer that, we have to consider two possible cases depending on whether $q[j]$ falls in the interval between $p_\alpha[j]$ and $p_\beta[j]$. Figure 3a and Fig. 3b depict the cases. In Fig. 3a, the order between $T_\alpha[j]$ and $T_\beta[j]$ can be interpreted as the relationship between $p_\alpha[j]$ and $p_\beta[j]$. In the case of Fig. 3b, the order between $T_\alpha[j]$ and $T_\beta[j]$ can be interpreted as the relationship between $p_\alpha[j] + p_\beta[j]$ and $q[j] + q[j]$.*

In the aforementioned study, we notice that the multi-type-operation requirement (*i.e.*, with both subtraction and comparison) in dynamic skyline query can be transformed to **uni-type-operation involving only comparison**. Inspired by this critical point, current encryption schemes that support comparison over ciphertext can be adopted in our framework to realize our design goals.

4.1 Database Encryption

In our scheme, we adopt a state-of-the-art encryption scheme that supports comparison, namely *order-revealing encryption* [16]. We first present the formal definition of order-revealing encryption.

Definition 4. (Order-Revealing Encryption). *An order-revealing encryption (ORE) scheme [16] is a tuple of three algorithms including* Setup, Encrypt *and* Compare *defined over a well-ordered domain D with the following properties:*

- Setup(1^λ) $\rightarrow$ *sk: On input a security parameter λ, the setup algorithm outputs a secret key sk.*
- Encrypt(sk, m) $\rightarrow$ *ct: On input a secret key sk and a message $m \in D$, the encryption algorithm outputs a ciphertext ct.*
- Compare(ct_1, ct_2) $\rightarrow$ *b: On input two ciphertexts ct_1, ct_2, the compare algorithm outputs a bit $b \in \{-1, 0, 1\}$.*

With the help of the ORE scheme, evaluating the dynamic domination relation between p_α and p_β can be carried out securely in ciphertext form as outlined in Algorithm 2. For ease of subsequent discussion, we shall denote Enc(x) as the ORE ciphertext for the original message x.

Minimizing the Number of Keys. Following Observation 1, a data owner needs to encrypt database P and the sum of any two tuples in P in each dimension, namely $p_\alpha[j] + p_\beta[j]$, where $\alpha \neq \beta, \alpha, \beta \in [1, n], j \in [1, d]$. The above two

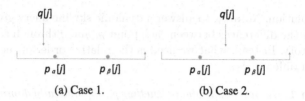

(a) Case 1.   (b) Case 2.

Fig. 3. Cases for the relationship between q and (p_α, p_β)

Algorithm 2. SecureCompare Algorithm

Require: The ORE ciphertext for $\mathsf{Enc}(p_\alpha[j])$, $\mathsf{Enc}(p_\beta[j])$, $\mathsf{Enc}(q[j])$, as well as $\mathsf{Enc}(p_\alpha[j] + p_\beta[j])$, $\mathsf{Enc}(2q[j])$.
Ensure: The comparison result as $-1, 0, 1$ denoting that $p_\alpha[j]$ is closer to (*resp.*, equivalent with, farther from) $q[j]$ than $p_\beta[j]$.
1: **if** ORE.Compare($\mathsf{Enc}(p_\alpha[j])$, $\mathsf{Enc}(p_\beta[j])$) $== 0$ **then**
2:    return 0
3: **else if** ORE.Compare($\mathsf{Enc}(p_\alpha[j])$, $\mathsf{Enc}(p_\beta[j])$) $== -1$ **then**
4:    **if** $\mathsf{Enc}(q[j])$ falls outside the interval **then**
5:       **return** ORE.Compare($\mathsf{Enc}(q[j])$, $\mathsf{Enc}(p_\alpha[j])$)
6:    **else**
7:       **return** ORE.Compare($\mathsf{Enc}(2q[j])$, $\mathsf{Enc}(p_\alpha[j] + p_\beta[j])$)
8:    **end if**
9: **else**
10:    **if** $Enc(q[j])$ falls outside the interval **then**
11:       **return** ORE.Compare($\mathsf{Enc}(q[j])$, $\mathsf{Enc}(p_\beta[j])$)
12:    **else**
13:       **return** ORE.Compare($\mathsf{Enc}(p_\alpha[j] + p_\beta[j])$, $\mathsf{Enc}(2q[j])$)
14:    **end if**
15: **end if**

---

ciphertexts are denoted as $E(P)$ and $E(\Phi)$, respectively. In this step, if we use the same private key on both $E(P)$ and $E(\Phi)$, the sum of paired tuples in $E(\Phi)$, although encrypted, will leak more message about plaintext beyond the order.

For example, assume that P contains five tuples, whose values in a particular dimension are a, b, c, d, e, respectively. Suppose that after sorting the values in ascending order, we get b, c, a, e, d. Then their sums can be listed as $b + c, b + a, b + e, b + d, c + a, c + e, c + d, a + e, a + d, e + d$. For ease of discussion, in the following we shall refer to these values as *pairs of sums*. If we encrypt the results for these pairs of sums using the same key as $E(P)$, an attacker can get the ordering of plaintexts. Therefore, he may possibly know $b + e \leq c + a$, and then infer that $e - a \leq c - b$. In this way, besides the order, the distribution of values in plaintext tuples is also leaked.

However, according to the security model in this work, except the order of tuples in some dimensions, the cloud should not be able to infer the content of the tuples. Therefore, we have to avoid leaking the distribution of data by adopting different keys in ORE. Intuitively, an ideal method is to encrypt each pair of sums using a different key, as it is not required to perform comparison among any pair of $p_\alpha[j], p_\beta[j]$ according to Algorithm 2. However, the increased number of keys will further introduce key management and storage problems. We propose a novel method to address this problem. As shown in Fig. 4, b, c, a, e, d are the sorted values for five tuples in P on a particular dimension. According to

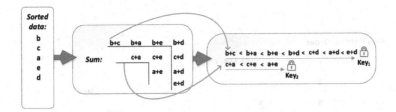

Fig. 4. A novel encryption scheme for pairs of tuples

Algorithm 2, these values should be encrypted using the same key as comparisons over their ciphertext are required. As a result, given that $\mathsf{Enc}(b), \ldots, \mathsf{Enc}(d)$ are encrypted using the same key under ORE, any adversary can easily infer that $b + c < b + a < b + e < b + d$ regardless that $b + c, \ldots, b + d$ are encrypted with different keys or not. Therefore, it is not beneficial to use multiple keys for such a group of sums.

Definition 5. (Order-Obvious Class). *Given the order of n elements, whose exact values are unknown, if the order of two summations over paired elements can be inferred, we call them Order-Obvious. All the $n(n-1)/2$ paired summations can be divided into several disjoint subsets accordingly, such that all the summations in each subset are Order-Obvious. We refer to each subset as an Order-Obvious Class (abbrev. OOC).*

Generally, we can find all OOCs, which is classified using the solid lines in Fig. 4. The relations for sums in the same OOC (*e.g.*, line) can be inferred easily purely from $E(P)$. In light of that, we can use the same key to encrypt the sums in the same OOC, and adopt different keys across OOCs. In this way, any adversary cannot get additional information over the ciphertexts besides the order, and we can effectively minimize the number of keys. In particular, the minimum number of keys, denoted as κ, (*e.g.*, the number of lines in Fig. 4) must satisfy the following theorem.

Theorem 1. *In order to satisfy the predefined security model, the minimum number of encryption keys in a dimension should be* $\kappa = \lceil \frac{2*n-3}{4} \rceil$.[1]

Remark. Through the above strategy, we have minimized the required number of encryption keys. In spite of that, κ is still linear to n, which may introduce key management burden if n is very large. To address this, we suggest the following implementations. For each row in Fig. 4, we assign it a random Id_i. The data owner only needs to store one master key mk and a series of random Id_i. Then, key_i for encrypting each row is generated by $mk \oplus Id_i$. In this way, we can effectively generate κ different keys based on mk.

Accessing the Pairs of Sums. As required by Algorithm 2, in order to compare $t_\alpha[j]$ and $t_\beta[j]$, it is always required to retrieve the ciphertext of $p_\alpha[j] + p_\beta[j]$.

---

[1] The proof for all theories can be found in our Technical Report [25].

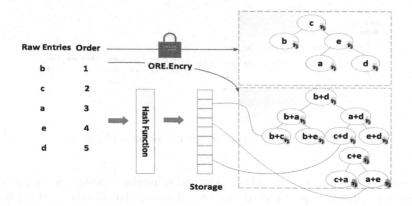

Fig. 5. The complete ciphertext storage structure

Therefore, it is necessary to build a map between the elements of $E(P)$ with the corresponding sums in $E(\Phi)$. That is, we need to build an index that maps $\mathsf{Enc}(p_\alpha[j])$ and $\mathsf{Enc}(p_\beta[j])$ to $\mathsf{Enc}(p_\alpha[j] + p_\beta[j])$. To this end, we present an index based on hash function. Formally, we define a hash function as $h : \mathbb{N}^2 \to \mathbb{N}$, where $\mathbb{N}$ denote the set of natural numbers. The hash function h should satisfy the following property, $\forall x_1, y_1, x_2, y_2 \in \mathbb{N}$, $h(x_1, y_1) = h(x_2, y_2)$ if and only if $x_1 = x_2$ and $y_1 = y_2$.

Assume the indices for $\mathsf{Enc}(p_\alpha[j])$ and $\mathsf{Enc}(p_\beta[j])$ in $E(P)$ are denoted as α and β, respectively. Then the index of $\mathsf{Enc}(p_\alpha[j] + p_\beta[j])$ in $E(\Phi)$ can be easily acquired as $h(\alpha, \beta)$. Figure 5 presents an example for the hash function. There are five encrypted values in $E(P)$, namely $a, \ldots, e$. The hash function in this example is simply designed as a regular traversal order for the corresponding sums. In fact, any hash function that satisfies the aforementioned property can be adopted here.

Indexing the Pairs of Sums. Additionally, as all the pairs of sums within a particular OOC are encrypted by ORE using the same key, we need to exploit additional index structures for efficient retrieval of corresponding entries for these *pairs of sums*. Therefore, we also design an index scheme for management of these ORE encrypted *pairs of sums*. In SCALE, we adopt AVL-Tree based structure to construct the indexing structure, as it provides excellent efficiency when querying for a particular range. Specifically, it is possible for us to build an AVL-Tree to index all these encrypted sums in the same OOC. Notably, each AVL-Tree is rooted at the median of each OOC and all the nodes in an AVL-Tree are the corresponding ciphertexts for *pairs of sums* in the same OOC.

For instance, given the records in Fig. 4, there are two OOCs. We shall build two different AVL-Trees for indexing the corresponding ciphertexts for each OOC, respectively. That is, the first OOC centered at $b + d$ corresponds to an AVL-Tree rooted at $\mathsf{Enc}(b + d)$; another OOC centered at $c + e$ corresponds to another AVL-Tree rooted at $\mathsf{Enc}(c + e)$ (as shown in Fig. 5).

Algorithm 3. Dataset Encryption

Require: The dataset P
Ensure: The ciphertexts sets $E(P)$, $E(\Phi)$
1: generate $d + \lceil \frac{2*n-3}{4} \rceil * d$ keys with ORE.Setup as $keys[]$
2: **for** $p \in P$ and j in $1, \ldots, d$ **do**
3:    $\mathsf{Enc}(p[j]) \leftarrow$ ORE.Encrypt$(keys[j], p[j])$
4:    let $\mathsf{Enc}(p) = \{\mathsf{Enc}(p[1]), \ldots, \mathsf{Enc}(p[d])\}$ and add $\mathsf{Enc}(p)$ to $E(P)$
5: **end for**
6: let $m = 1$
7: **for** j in $1, \ldots, d$ **do**
8:    $\Lambda = (p^{(1)}[j], \ldots, p^{(n)}[j]) \leftarrow$ sort $p_1[j], \ldots, p_n[j]$ in ascending order
9:    **while** Λ is not empty **do**
10:       **for** i in $2, \ldots, len(\Lambda)$ **do**
11:          add ORE.Encrypt$(keys[d + m], p^{(1)}[j] + p^{(i)}[j])$ to $E(\Phi)$
12:       **end for**
13:       **for** i in $2, \ldots, len(\Lambda) - 1$ **do**
14:          add ORE.Encrypt$(keys[d + m], p^{(n)}[j] + p^{(i)}[j])$ to $E(\Phi)$
15:       **end for**
16:       remove the first and last elements in Λ, let $m = m + 1$
17:    **end while**
18: **end for**
19: **return** $E(P)$, $E(\Phi)$

Algorithm 4. Query Request Encryption

Require: The query data q, keys from data owner $keys[]$
Ensure: The ciphertexts $\mathsf{Enc}(q)$, $\mathsf{Enc}(2q)$
1: **for** j in $1, \ldots, d$ **do**
2:    $\mathsf{Enc}(q[j]) \leftarrow$ ORE.Encrypt$(keys[j], q[j])$
3:    **for** m in $1, \ldots, \lceil \frac{2*n-3}{4} \rceil$ **do**
4:       let $key\_num = d + (j - 1) * \lceil \frac{2*n-3}{4} \rceil + m$
5:       $\mathsf{Enc}(2q_m[j]) \leftarrow$ ORE.Encrypt$(keys[key\_num], 2q[j])$
6:    **end for**
7: **end for**
8: **return** $\mathsf{Enc}(q)$, $\mathsf{Enc}(2q)$

In fact, data structures other than AVL-Tree can also be adopted to index the ORE ciphertexts for each OOC. We select AVL-Tree as the default setting in SCALE as it provides the best query response time among alternative choices.

Database Encryption. We have now all the ammunitions in place to demonstrate the entire process of encrypting the database (Algorithm 3). First, the data owner generates $d + \lceil \frac{2*n-3}{4} \rceil * d$ keys (Line 1), and for each column (*i.e.*, attribute) in P we encrypt the entries using the same key (Lines 3–5), resulting in $E(P)$. Then, the data owner sorts the entries (Line 8) in each column (*i.e.*, attribute) and computes the sums for pairs of entries in each dimension. Afterwards, the sums are then encrypted using the corresponding keys as shown in Fig. 4 (Lines 9–17), resulting in $E(\Phi)$. Finally, the data owner sends $E(P)$, $E(\Phi)$ to the cloud server.

Besides, the data owner also creates a hash function h that maps each pair of elements in $E(P)$ and the corresponding sums in $E(\Phi)$, and sends h to the cloud server. It is now possible for the cloud to quickly locate the ciphertext of the corresponding sums for each pair $(p_\alpha[j], p_\beta[j])$.

Algorithm 5. Secure Skyline Query Algorithm

Require: The ciphertext for dataset $E(P)$ and $E(\Phi)$, query request $\mathsf{Enc}(q)$ and $\mathsf{Enc}(2q)$
Ensure: The encrypted result set of skyline points S
1: **for** i in $1, \ldots, n$ **do**
2:    **if** S is empty **then**
3:        add $\mathsf{Enc}(p_i)$ to S
4:    **else**
5:        $flag\_cur \leftarrow True$
6:        **for** each $\mathsf{Enc}(p_j) \in S$ **do**
7:            **for** m in $1, \ldots, d$ **do**
8:                $flag[m] \leftarrow SecureCompare(\mathsf{Enc}(p_i[m]), \mathsf{Enc}(p_j[m]), \mathsf{Enc}(q[m]), \mathsf{Enc}(p_i[m] + p_j[m]),$
                $\mathsf{Enc}(2q[m]))$
9:            **end for**
10:           **if** $\forall m, flag[m] \geq 0$, and $\exists k$ such that $flag[k] > 0$ **then**
11:               $flag\_cur \leftarrow False$
12:           **else if** $\forall m, flag[m] \leq 0$, and $\exists k$ such that $flag[k] < 0$ **then**
13:               delete $\mathsf{Enc}(p_j)$ from S
14:           **end if**
15:        **end for**
16:        **if** $flag\_cur$ is $True$ **then**
17:            add $\mathsf{Enc}(p_i)$ to S
18:        **end if**
19:    **end if**
20: **end for**
21: **return** S

4.2 Query Processing

A data user needs to register their information to the Data owner and securely get the keys. Then the query user encrypts the request according to Algorithm 4 before sending it to the cloud server.

As shown in Algorithm 4, user encrypts each dimension of the query tuple using corresponding keys (Line 2) and encrypts the doubled entries for the query tuple using other keys (Lines 4–5). Finally, the user sends $\mathsf{Enc}(q)$, $\mathsf{Enc}(2q)$ to the cloud server. As mentioned in Algorithm 1, given an encrypted query q as shown Algorithm 4, the cloud server needs to perform comparisons and computations over encrypted data. According to the approach shown in Fig. 3, the cloud server can perform skyline query via the comparison relationship with encrypted tuples, encrypted query request, encrypted sums, and encrypted doubled request. As a result, the process described in Algorithm 1 can be now performed in ciphertext without decryption, which is shown in Algorithm 5. To illustrate the entire protocol, we provide a running example in the following.

Example 1. Assuming that P contains five tuples, whose entries in dimension 1 are sorted as $7, 13, 21, 32, 53$. For simplicity, hereby we show only one dimension. According to Algorithm 3, we shall first compute the sums for all pairs of values, e.g., $7+13 = 20$, $7+21 = 28$, $7+32 = 39$, $7+53 = 60$, $13+21 = 34$, $13+32 = 45$, $13+53 = 66$, $21+32 = 53$, $21+53 = 74$, $32+53 = 85$. As shown in Theorem 1, the number of encryption keys required for these sums can be calculated as $\lceil \frac{2*5-3}{4} \rceil = 2$. Therefore, we use two keys to encrypt the above sums, resulting in $\mathsf{Enc}_1(20)$, $\mathsf{Enc}_1(28)$, $\ldots$, $\mathsf{Enc}_1(85)$ and $\mathsf{Enc}_2(34)$, $\mathsf{Enc}_2(45)$, $\mathsf{Enc}_2(53)$.

Besides, we also need to use another key to encrypt the original tuples, e.g., $\mathsf{Enc}_3(7), \mathsf{Enc}_3(13), \ldots, \mathsf{Enc}_3(53)$. Suppose that a user submits a query with

$q[1] = 23$. Then q and $2q$ need to be encrypted according to our scheme, resulting in $\mathsf{Enc}_1(46), \mathsf{Enc}_2(46), \mathsf{Enc}_3(23)$. These ciphertexts are then sent to the cloud server. The cloud server compares ciphertexts one by one according to the protocol. Through ORE.Compare (Definition 4.1) and Algorithm 2, the cloud server can easily determine that $\mathsf{Enc}_3(32)$ dominates $\mathsf{Enc}_3(53)$ following the case shown in Fig. 3a. Similarly, $\mathsf{Enc}_3(21)$ dominates $\mathsf{Enc}_3(7)$ and $\mathsf{Enc}_3(13)$. In the case shown in Fig. 3b, $\mathsf{Enc}_3(21)$ dominates $\mathsf{Enc}_3(32)$ because ORE.Compare$(\mathsf{Enc}_2(53), \mathsf{Enc}_2(46)) = 1$. Algorithm 5 will iteratively repeat this process for all dimensions and remaining tuples. ∎

4.3 Security Analysis

The presented SCALE framework is constructed based on ORE scheme proposed in [16], which is secure with leakage function $\mathcal{L}_{BLK}$.

Lemma 1. *The ORE scheme is secure with leakage function $\mathcal{L}_{BLK}$ assuming that the adopted pseudo random function (PRF) is secure and the adopted hash functions are modeled as random oracles. Here, $\mathcal{L}_{BLK}(m_1, \ldots, m_t) = \{(i, j, BLK(m_i, m_j)) | 1 \le i < j \le t\}$ and $BLK(m_i, m_j) = (ORE.Compare (m_i, m_j), \mathsf{ind}_{\mathsf{diff}}(m_i, m_j))$, in which $\mathsf{ind}_{\mathsf{diff}}$ is the first differing block function that is the first index $i \in [n]$ such that $x_i = x_j$ for all $j < i$ and $x_i \ne x_j$. (The proof of this lemma is in Appendix 4.1 in [16])*

In order to formally prove the security of SCALE, we extend $\mathcal{L}$-adaptively-secure model for keyword searching scheme as shown in Definition 3.

Theorem 2. *Let the adopted PRF in ORE is secure. The presented SCALE framework is $\mathcal{L}$-adaptively-secure in the (programmable) random oracle model, where the leakage function collection $\mathcal{L} = (\mathcal{L}_{Encrypt}, \mathcal{L}_{Query}, \mathcal{L}_{Insert}, \mathcal{L}_{Delete})$ is defined as follows,*

$$\mathcal{L}_{Encrypt} = \mathcal{L}_{BLK}(\cup_{k=1}^{d} X_k^{(n)}), \mathcal{L}_{Query} = \mathcal{L}_{BLK}(\cup_{k=1}^{d} X_k^{(n)'}),$$

$$\mathcal{L}_{Insert} = \mathcal{L}_{BLK}(\cup_{k=1}^{d} X_k^{(n+1)}), \mathcal{L}_{Delete} = \mathcal{L}_{BLK}(\cup_{k=1}^{d} X_k^{(n)})$$

where $X_k^{(n)} = \cup_{t=1}^{(\lceil \frac{2*n-3}{4} \rceil)}(Y_t^k)$, $X_k^{(n)'} = \cup_{t=1}^{(\lceil \frac{2*n-3}{4} \rceil)}(Y_t^k \cup \mathsf{Enc}_t(q) \cup \mathsf{Enc}_t(2q))$ and $Y_t^k = \{\mathsf{Enc}(p_t[k] + p_j[k]) | t < j < n - t + 1\} \cup \{\mathsf{Enc}(p_j[k] + p_{(n-t+1)}[k]) | t < j < n - t + 1\}$.

The advantage for any probabilistic polynomial-time adversary is,

$$|\Pr[\mathrm{Game}_{\mathcal{R}, \mathcal{A}(\lambda)} = 1] - \Pr[\mathrm{Game}_{\mathcal{S}, \mathcal{A}(\lambda)} = 1]|$$

$$\le \mathrm{negl}(\lambda) = d \cdot (\mathrm{negl}^{ORE}(\lambda) + (2n - 1)\mathrm{poly}(\lambda)/2^\lambda).$$

4.4 Complexity Analysis

In the encryption phase, the plaintext data from the data owner can be sorted and encrypted in advance. We need $O(d + \lceil \frac{2*n-3}{4} \rceil)$ encryption operations every time when a user submits a query following Algorithm 4.

During the querying phase, our scheme replaces the original plaintext subtraction and comparison operations with a limited number of comparisons over ciphertext. The time taken for encryption and ciphertext comparisons is only affected by the block size in ORE, key length in AES, and plaintext length. Therefore, the main logic for dynamic skyline query processing is unchanged. Hence, the complexity for query processing in our scheme is consistent with that of [6], *i.e.*, $O(n^2)$ for the worst case.

5   Experimental Study

In this section, we evaluate the performance and scalability of SCALE under different parameter settings over four datasets, including both real-world and synthetic ones. We also compare our model with another baseline, namely BSSP [17], which is the only solution for secure dynamic skyline query.

5.1   Experiment Settings

All algorithms are implemented in C, and tested on the platform with a 2.7GHz CPU, 8GB memory running MacOS. We use both synthetic and real-world datasets in our experiments. In particular, we generated independent (INDE), correlated (CORR), and anti-correlated (ANTI) datasets following the seminal work in [6]. In line with [17], we also adopt a dataset that contains 2500 NBA players who are league leaders of playoffs[2]. Each player is associated with six attributes that measure the player's performance: Points, Offensive Rebounds, Defensive Rebounds, Assists, Steals, and Blocks.

5.2   Performance Results

In this subsection, we evaluate our protocols by varying the number of tuples (n), the number of dimensions (d), the ORE block setting, and the length of key (K).

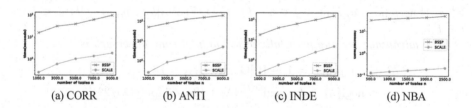

(a) CORR   (b) ANTI   (c) INDE   (d) NBA

Fig. 6. Response time by varying n (with $d = 3, block = 16, K = 256$)

---

[2] https://stats.nba.com/alltime-leaders/?SeasonType=Playoffs.

Varying the Number of Tuples. Figure 6 shows the time cost by varying the number of tuples, namely n, on the four datasets. In this group of experiments, we fix the number of dimensions, ORE block size and key length as 3, 16 and 256, respectively. We observe that for all datasets, the time cost increases almost linearly with respect to n. This phenomenon is consistent with our complexity study in Sect. 4.4. Notably, for the real-world dataset (*i.e.*, NBA), the query response time is less than 0.2 s, which is efficient enough in practice. Compared to the state-of-the-art [17], SCALE is more than 3 orders of magnitude faster. In the following, we shall fix $n = 2500$ and focus on evaluating the effects of the three parameters in our scheme.

Impact of d. Figure 7a shows the time cost for different d on the four datasets, where we fix the ORE block size and key length as 16 and 256, respectively. For all datasets, as d increases from 2 to 6, the response time in all four datasets increases almost exponentially as well. This fact is consistent with the ordinary dynamic skyline querying in plaintext. This is because an increase in d leads to more comparison operations for the decision of dynamic dominance criteria.

Impact of ORE Block. Encrypting plaintext based on block cipher, different block sizes may take different time. Figure 7b plots the time cost by varying the block sizes used in the ORE scheme, where d and K are fixed as 3 and 256, respectively. As mentioned in [16], this ORE scheme leaks the first block of δ-bits that differs, therefore, increasing the block size brings higher security. Observe that the response time increases slightly with respect to the size of ORE block. That is, higher security level in ORE has to sacrifice some response time.

Impact of K. Figure 7c shows the time cost by varying the lengths for the keys in the ORE scheme. This ORE scheme uses AES as the building block, therefore, increasing the encryption key size brings in higher security. Similar to that of block size, the response time also increases linearly with respect to the size of encryption keys. Comparing Fig. 7c against 7d, we find the following phenomenons. First, increasing the security level for ORE will definitely sacrifice some efficiency. Second, the key length in AES exhibits more significant impact on the efficiency comparing to that of ORE block size.

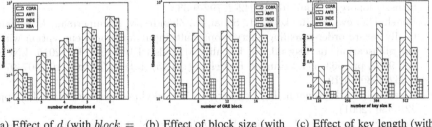

(a) Effect of d (with $block = $ 16, $K = 256$)   (b) Effect of block size (with $d = 3, K = 256$)   (c) Effect of key length (with $block = 16, d = 3$)

Fig. 7. The effects of different parameters ($n = 2500$)

6   Conclusions

In this paper, we have presented a new framework called SCALE to address the secure dynamic skyline query problem in the cloud platform. A distinguishing feature of our framework is the conversion of the requirement of both subtraction and comparison operations to only comparisons. As a result, we are able to use ORE to realize dynamic domination protocol over ciphertext. Based on this feature, we built SCALE on top of BNL. In fact, our framework can be easily adapted to other plaintext dynamic skyline query models. We theoretically show that the proposed scheme is secure under our system model, and is efficient enough for practical applications. Moreover, there is only one interaction between a user and the cloud, which minimizes the communication cost and corresponding threats. Experimental study over both synthetic and real-world datasets demonstrates that SCALE improves the efficiency by at least three orders of magnitude compared to the state-of-the-art method. As part of our future work, we plan to further enhance the security of our scheme and explore how the scheme can be adapted to support other variations of skyline query.

Acknowledgments. This work is supported by National Natural Science Foundation of China (No. 61672408, 61972309, 61702403, 61976168) and National Engineering Laboratory (China) for Public Safety Risk Perception and Control by Big Data (PSRPC). Sourav S Bhowmick is partially funded by Huawei grant M4062170.

References

1. Agrawal, R., Kiernan, J., Srikant, R., Xu, Y.: Order preserving encryption for numeric data. In: SIGMOD, pp. 563–574. ACM (2004)
2. Alrifai, M., Skoutas, D., Risse, T.: Selecting skyline services for QoS-based web service composition. In: WWW, pp. 11–20 (2010)
3. Boldyreva, A., Chenette, N., Lee, Y., O'Neill, A.: Order-preserving symmetric encryption. In: Joux, A. (ed.) EUROCRYPT 2009. LNCS, vol. 5479, pp. 224–241. Springer, Heidelberg (2009). https://doi.org/10.1007/978-3-642-01001-9_13
4. Boldyreva, A., Chenette, N., O'Neill, A.: Order-preserving encryption revisited: improved security analysis and alternative solutions. In: Rogaway, P. (ed.) CRYPTO 2011. LNCS, vol. 6841, pp. 578–595. Springer, Heidelberg (2011). https://doi.org/10.1007/978-3-642-22792-9_33
5. Boneh, D., Lewi, K., Raykova, M., Sahai, A., Zhandry, M., Zimmerman, J.: Semantically secure order-revealing encryption: multi-input functional encryption without obfuscation. In: Oswald, E., Fischlin, M. (eds.) EUROCRYPT 2015. LNCS, vol. 9057, pp. 563–594. Springer, Heidelberg (2015). https://doi.org/10.1007/978-3-662-46803-6_19
6. Börzsönyi, S., Kossmann, D., Stocker, K.: The skyline operator. In: ICDE, pp. 421–430 (2001)
7. Bothe, S., Cuzzocrea, A., Karras, P., Vlachou, A.: Skyline query processing over encrypted data: an attribute-order-preserving-free approach. In: PSBD@CIKM, pp. 37–43 (2014)

8. Chatterjee, S., Das, M.P.L.: Property preserving symmetric encryption revisited. In: Iwata, T., Cheon, J.H. (eds.) ASIACRYPT 2015. LNCS, vol. 9453, pp. 658–682. Springer, Heidelberg (2015). https://doi.org/10.1007/978-3-662-48800-3_27

9. Chen, W., Liu, M., Zhang, R., Zhang, Y., Liu, S.: Secure outsourced skyline query processing via untrusted cloud service providers. In: INFOCOM, pp. 1–9 (2016)

10. Chenette, N., Lewi, K., Weis, S.A., Wu, D.J.: Practical order-revealing encryption with limited leakage. In: Peyrin, T. (ed.) FSE 2016. LNCS, vol. 9783, pp. 474–493. Springer, Heidelberg (2016). https://doi.org/10.1007/978-3-662-52993-5_24

11. Chomicki, J., Godfrey, P., Gryz, J., Liang, D.: Skyline with presorting. In: ICDE, pp. 717–719 (2003)

12. Dellis, E., Seeger, B.: Efficient computation of reverse skyline queries. In: VLDB, pp. 291–302 (2007)

13. Gentry, C.: A fully homomorphic encryption scheme. Stanford University (2009)

14. Kriegel, H., Renz, M., Schubert, M.: Route skyline queries: a multi-preference path planning approach. In: ICDE, pp. 261–272 (2010)

15. Kung, H.T., Luccio, F., Preparata, F.P.: On finding the maxima of a set of vectors. J. ACM 22(4), 469–476 (1975)

16. Lewi, K., Wu, D.J.: Order-revealing encryption: new constructions, applications, and lower bounds. In: CCS, pp. 1167–1178 (2016)

17. Liu, J., Yang, J., Xiong, L., Pei, J.: Secure skyline queries on cloud platform. In: ICDE, pp. 633–644 (2017)

18. Paillier, P.: Public-key cryptosystems based on composite degree residuosity classes. In: Stern, J. (ed.) EUROCRYPT 1999. LNCS, pp. 223–238. Springer, Heidelberg (1999). https://doi.org/10.1007/3-540-48910-X_16

19. Papadias, D., Tao, Y., Fu, G., Seeger, B.: An optimal and progressive algorithm for skyline queries. In: SIGMOD, pp. 467–478 (2003)

20. Papadias, D., Tao, Y., Fu, G., Seeger, B.: Progressive skyline computation in database systems. ACM Trans. Database Syst. 30(1), 41–82 (2005)

21. Park, Y., Min, J., Shim, K.: Efficient processing of skyline queries using mapreduce. IEEE Trans. Knowl. Data Eng. 29(5), 1031–1044 (2017)

22. Popa, R.A., Li, F.H., Zeldovich, N.: An ideal-security protocol for order-preserving encoding. In: SP, pp. 463–477 (2013)

23. Preparata, F.P., Shamos, M.I.: Computational Geometry - An Introduction. Springer, Heidelberg (1985). https://doi.org/10.1007/978-1-4612-1098-6

24. Sun, W., Zhang, N., Lou, W., Hou, Y.T.: When gene meets cloud: enabling scalable and efficient range query on encrypted genomic data. In: INFOCOM, pp. 1–9 (2017)

25. Wang, W., et al.: An efficient secure dynamic skyline query model. arXiv:2002.07511 (2020)

26. Wang, W.-C., Wang, E.T., Chen, A.L.P.: Dynamic skylines considering range queries. In: Yu, J.X., Kim, M.H., Unland, R. (eds.) DASFAA 2011. LNCS, vol. 6588, pp. 235–250. Springer, Heidelberg (2011). https://doi.org/10.1007/978-3-642-20152-3_18

27. Zhou, X., Li, K., Zhou, Y., Li, K.: Adaptive processing for distributed skyline queries over uncertain data. IEEE Trans. Knowl. Data Eng. 28(2), 371–384 (2016)

Authenticated Range Query Using SGX for Blockchain Light Clients

Qifeng Shao[1,2], Shuaifeng Pang[1], Zhao Zhang[1(✉)], and Cheqing Jing[1]

[1] School of Data Science and Engineering, East China Normal University,
Shanghai, China
{shao,sfpang}@stu.ecnu.edu.cn, {zhzhang,cqjin}@dase.ecnu.edu.cn
[2] School of Software, Zhongyuan University of Technology, Zhengzhou, China

Abstract. Due to limited computing and storage resources, light clients and full nodes coexist in a typical blockchain system. Any query from light clients must be forwarded to full nodes for execution, and light clients verify the integrity of query results returned. Since existing authenticated query based on Authenticated Data Structure (ADS) suffers from significant network, storage and computing overheads by virtue of Verification Objects (VO), an alternative way turns to Trust Execution Environment (TEE), with which light clients have no need to receive or verify any VO. However, state-of-the-art TEE cannot deal with large-scale application conveniently due to limited secure memory space (i.e, the size of enclave in Intel SGX is only 128MB). Hence, we organize data hierarchically in both trusted (enclave) and untrusted memory and only buffer hot data in enclave to reduce page swapping overhead between two kinds of memory. Security analysis and empirical study validate the effectiveness of our proposed solutions.

Keywords: Blockchain · Authenticated query · MB-tree · Intel SGX

1 Introduction

Blockchain, the core technology of Bitcoin [9], is a decentralized, trustless, tamper-proof and traceable distributed ledger managed by multiple participants. Specifically, by integrating P2P protocol, asymmetric cryptography, consensus algorithm, hash chain structure and so forth, blockchain can achieve trusted data sharing among untrusted parties without the coordination of any central authority.

All nodes of blockchain are usually classified as full node and light node. Full node holds a complete copy of block data, while the light node (also called light client) only stores block header or other verification information due to limited storage resource. Queries from a light node are forwarded to full nodes for execution, but the integrity of query results returned from full node needs to be authenticated by light client itself.

Current blockchain systems have limited ability to support authenticated queries of light clients. Most of them are merely suitable for digital currency field,

Y. Nah et al. (Eds.): DASFAA 2020, LNCS 12114, pp. 306–321, 2020.
https://doi.org/10.1007/978-3-030-59419-0_19

e.g., Simple Payment Verification (SPV) in Bitcoin can only answer queries of transaction existence. To the best of our knowledge, no present system is able to handle range query and verification accordingly, like selecting transactions satisfying "2019-11 $\leq Timestamp \leq$ 2019-12". With the popularization of blockchain technology among traditional industries, the desire to support various authenticated queries becomes stronger.

In this study, we focus on authenticated range query, a representative task. The authenticated range query can be tracked back to outsourcing databases, where clients delegate data to remote database servers and initiate database query. Although the signature chaining [10] and Authenticated Data Structure (ADS), e.g., Merkle Hash Tree (MHT) [8] and Merkle B-tree (MB-tree)[3], are widely adopted to guarantee the correctness and completeness of query results, neither fits for blockchain since the former triggers tremendous signature computing cost and the latter raises the cost of VO, e.g., Merkle cryptographic proofs. Since computational cost of signature chain is determined by hardware and may not diminish in a short time, we propose a solution of blockchain based on ADS. Nevertheless, applying existing ADS to blockchain is quite challenging.

- The full node returns query results along with cryptographic proofs, known as VO. On the client side, the splicing and authentication of these VO require significant network and computing resources.
- As updates lead to hash computing and signature costs, traditional ADS solutions assume that the databases they're serving have fixed or fewer updates, which is not applicable to the case of blockchain since blocks are appended to the blockchain periodically.

Consequently, a new solution that assures the integrity of query results and further effectively reduces the costs of verification and maintenance needs to be devised.

Recently, the emergance of trusted hardware (Trusted Execution Environment, TEE) that supports secure data accessing offers a promising direction of designing range query authentication schemes. For example, Intel Software Guard Extensions (SGX) [7] can protect code and data from disclosure or modification, and enforce the security level of application. SGX allows to create one or more isolated contexts, named enclaves, which contain segments of trusted memory. Hence, to guarantee integrity and confidentiality, sensitive codes are installed in enclave and run on untrusted machines.

The special region of isolated memory reserved for enclave is called Enclave Page Cache (EPC). Currently, EPC has a maximal size of 128 MB, of which only 93 MB are utilizable for applications. EPC page faults occur when the code accesses the pages outside of enclave. Page swapping is expensive, because enclave memory is fully encrypted and its integrity also needs to be protected. Two built-in wrapper codes, *ecall* and *ocall*, respectively invoke enter and exit instructions to switch the execution context. These two codes add overhead of approximately 8,000 CPU cycles, compared to 150 cycles of a regular OS system call [12]. Though the emergence of SGX solves the secure remote computing problem of sensitive data on untrusted servers, the performance implications of SGX remain an open question. When applying Intel SGX to blockchain, effective optimization strategies must be considered.

In summary, this paper proposes an efficient query authentication scheme for blockchain by combining ADS-based MB-tree with Intel SGX. To the best of our knowledge, it is the first step toward investigating the problem of query authentication with SGX over blockchain. Our main contributions are as follows.

- An efficient SGX-based query authentication scheme for blockchain is proposed, with which light clients have no need to receive or verify any VO.
- A solution of integrating MB-tree with SGX is devised in view of the space limitation of enclave memory. Only frequently-used MB-tree nodes are cached in enclave.
- To reduce the cascading hash computing cost brought by item-by-item updates on MB-tree, a hybrid index consisting of an MB-tree and a skip list in enclave is provided.
- We conduct empirical study to evaluate the proposed techniques. Experimental results show that the efficacy of the proposed methods.

The rest of the paper is organized as follows. Section 2 reviews existing works. Section 3 introduces the problem formulation. Section 4 presents our solution of authenticated range query with SGX. The batch update is discussed in Sect. 5. The security analysis is presented in Sect. 6. The experimental results are reported in Sect. 7. Section 8 concludes this paper.

2   Related Work

To the best of our knowledge, no existing work explores authenticated range queries using SGX for blockchain. In the following, we briefly review related studies and discuss relevant techniques.

Query Authentication over Traditional Database. Query authentication has been extensively studied to guarantee the results' integrity against untrusted service providers. There are two basic solutions to ensure correctness and completeness, signature chaining [10] and ADS. For signature chaining, requiring every tuple being signed, the servers using aggregated signatures always return only one signature regardless of the result set size, and the client can process aggregate verification. Signature chaining features small VO size and communication cost, but it cannot scale up to large data sets because of high cost of signing all tuples. For ADS, MHT [8] and MB-tree [3] are widely adopted. MHT solves the authentication problem for point queries. MB-tree combines MHT with B^+-tree to support authenticated range queries. As MB-tree enables efficient search as B^+-tree and query authentication as MHT, our proposed solution uses this approach. MB-tree has also been studied to support authenticated join [14] and aggregation queries [4]. These works are more about outsource databases, insufficient for the case of blockchain.

Query Authentication over Blockchain. SPV, introduced by Satoshi Nakamoto [9], can only verify if a transaction exists in the blockchain or not. Hu et al. [2] leverage smart contract for trusted query processing over blockchain, focusing on file-level keyword searching without investigating the indexing issue. To support verifiable query over blockchain, Xu et al. [13] propose an accumulator-based ADS scheme for dynamic data aggregation over arbitrary query attribute, but blockchain clients need to receive and verify VO. Zhang et al. [15] present a gas-efficient scheme to support authenticated range query by utilizing multiple MB-trees.

Blockchain with Intel SGX. Present blockchain systems mainly perform software-based cryptographic algorithms to ensure the trusty of data. The appearance of trusted hardware, Intel SGX, opens up new possibility to enhance integrity and confidentiality of blockchain. Town Crier [16], an authenticated data feed system between existing web sites and smart contracts, employs SGX to furnish data to Ethereum. Ekiden [1] enables efficient SGX-backed confidentiality-preserving smart contracts and high scalability. BITE [6] leverages SGX on full nodes, and serves privacy-preserving requests from light clients of Bitcoin. Although these existing studies harmonize blockchain and SGX, none of them explore query authentication with SGX.

3   System Overview

Architecture. Figure 1 elucidates our system model that consists of a full node and a light client. The queries from the light client are forwarded to the full node for processing. The full node must prove it executes queries faithfully and returns all valid results, since query results may be maliciously tampered with. Traditional solutions organize data with MB-tree, and provide light clients with both query results and cryptographic proofs (VO) for further authentication. In our case, however, a big VO, especially when processing range queries, may be beyond the processing capacity of light clients like mobile devices. Consequently, our system is equipped with Intel SGX, providing integrity and confidentiality guarantees on untrusted full nodes. The query results are returned to clients through a secure channel. Clients can trust these query results without receiving or verifying any VO. Considering limited enclave memory, we organize data hierarchically in trusted memory (enclave) and untrusted memory, where skip list and MB-tree are adopted in trusted memory and untrusted memory respectively. Skip list in enclave buffers appended blocks. It merges block data into MB-tree periodically once exceeding the predefined threshold. It is worthy to note that a hot cache, residing in enclave, caches the frequently-used MB-tree nodes. These nodes will no longer be verified in future queries. More details are discussed in Sects. 4 and 5.

Adversary Model. In this study, we assume that there is no specific affiliation between light clients and full nodes. The full node is treated as a potential adversary since no participant in the blockchain network trust others. To address such

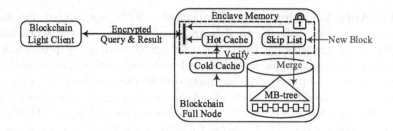

Fig. 1. System architecture.

a threat and free the clients from trivial verification process, we apply TEE-based Intel SGX to process query with integrity assurance. Since enclave memory is limited, we employ an authenticated index structure, MB-tree, outside of enclave to guarantee data integrity. Even though an adversary may compromise OS and other privileged softwares on a full node, it cannot break the hardware security enforcement of Intel SGX. With our hardware-based model, clients can trust the correctness and the completeness of query results under the following criteria.

– **Correctness.** All results that satisfy the query conditions haven't been tampered with.
– **Completeness.** No valid result is omitted regarding the range query.

4   Authenticated Range Query with SGX

SGX can protect code and data from disclosure or modification. Hence, an ideal solution to guarantee query results' integrity is to install the entire storage engine and execute all queries in enclave, which eliminates computing and network overheads induced by VO in traditional solutions. However, the memory limitation of enclave makes it inadequate to handle large scale applications. In this study, we design a mechanism to organize data hierarchically in untrusted and trusted memory. Meanwhile, the data in untrusted memory is organized as MB-tree and the frequently-accessed internal nodes are cached in enclave as trusted checkpoints. A skip list, maintained in trusted memory, buffers newly appended block data. Once the capacity of the skip list reaches a threshold, a merge operation is launched from skip list to MB-tree.

4.1   MB-tree in SGX

In this solution, the root node of MB-tree is always resident in enclave, and the rest are loaded into enclave according to the query request during runtime. After verifying the Merkle proofs of a node, it is believed to be trusted and used for searching. The frequently-used nodes are cached in enclave to implement authenticated query cheaply, and the rest nodes are outside enclave to alleviate the size limitation of enclave.

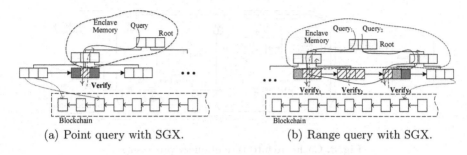

(a) Point query with SGX.   (b) Range query with SGX.

Fig. 2. Query and verify on MB-tree with SGX.

In MB-tree, each node contains $f - 1$ index keys and f pointers to the child nodes, where f is the fanout. Like MHT, each pointer is augmented with a corresponding digest. In the leaf node, each digest is a hash value $h = H(r)$, where r is the record pointed by pointer entry. In the internal node, each digest is a hash value $h = H(h_1||...||h_f)$, where $h_1, ..., h_f$ are the hash values of the child nodes. Then, recursively up to the root node, the contents of all nodes in the entire tree are reflected to the root node by hash. Since the root node involves digest about every node, the entire tree data can be verified based on the root node. Therefore, attackers cannot modify or replay any value in the tree. MB-tree can be created either from scratch or based on existing data. The enclave on an untrusted full node is firstly authenticated through remote attestation of Intel SGX. Once passing the remote attestation, we provision the root node into the enclave through a secure channel. When the MB-tree maintaining thread in enclave receives a new block, transactions are extracted and verified based on the verification rules of blockchain before updating the corresponding index items.

We now illustrate the query and authentication process on an MB-tree with SGX. Figure 2(a) demonstrates authenticated point queries on MB-tree in enclave. The query process is the same as the traditional one, during which, accessing nodes from root to leaf, appending hashes of sibling nodes to VO and returning the query results. SGX can perform all the authentication works, so light clients don't have to receive or verify any VO. Since enclave may cache previously verified nodes, when computing Merkle proofs from bottom to up, i.e., verification path, the authenticating process can be early terminated once encountering a node located in enclave, as Fig. 2(a) illustrates (the red dashed arrow).

Different from point queries, authenticated range queries ensure the correctness and completeness of results at the same time. Thus, the left and right boundaries of results should also be included in VO for further completeness authentication. As demonstrated by red dash arrows in Fig. 2(b), the results of range query involve multiple consecutive leaf nodes, i.e., a number of verification paths. The authenticating process goes like this: the leftmost and rightmost leaf nodes are responsible for computing the node digest by considering query results,

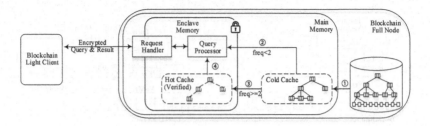

Fig. 3. Cache architecture of query processing.

sibling hashes and boundaries synthetically, while the leaf nodes located between them only compute the digest based on all results in the node. Same as point query, a verified node in enclave can cut short the verification path and further reduce the computational cost. Specifically, when all the leaf nodes covered by the results are in enclave, there is no need to perform any verification. Thus, SGX simplifies and improves the query authentication.

4.2 Cache Architecture of Query Processing

To improve accessing efficiency of MB-tree, we design three-level storage architecture, including disk storage, cold cache and hot cache. Figure 3 details the cache architecture. Disk storage at the lowest level persists the entire MB-tree. Cold cache, on untrusted memory, caches the MB-tree nodes to reduce I/O cost. Hot cache, on trusted enclave memory, only caches frequently-used and verified MB-tree nodes to alleviate verifying cost. We integrate these two types of caches and design an efficient cache replacement strategy.

When applying simple and efficient LRU cache replacement algorithm to MB-tree in enclave, burst accesses and sequential scans will make enclave read in nodes only accessed once. These nodes will not be swapped out of enclave in a short time, which lowers the utilization of the enclave memory. Motivated by the LRU-2 cache replacement algorithm [11] that keeps around the last 2 reference times for each page to estimate evicted page, we propose a replacement algorithm, Hierarchical Least Recently Used (H-LRU), for the two-level cache architecture composed of hot cache and cold cache. H-LRU considers more of the reference history besides the recent access for each node and addresses the problems of correlated references.

As shown in Algorithm 1, when a node of MB-tree is accessed for the first time, it's read out from the disk and buffered in the cold cache, thus avoiding the I/O cost in the subsequent accesses. When such node is accessed again, if that's been quite a while since the last access, i.e., uncorrelated reference, it is promoted to the hot cache, thus eliminating the verifying cost in the future queries. Algorithm 1 uses the following data structure.

- **HIST(n,t)** denotes the history of reference times of node n, discounting correlated references. $HIST(n, 1)$ denotes the last reference, $HIST(n, 2)$ the second to the last reference.

Algorithm 1: H-LRU

Input: Node n, Time t /* n is referenced at time t */
1 **if** $n \in HotCache$ **then**
2   **if** $isUncorrelated(n, t)$ **then**
3   ⌊ move n to the head of HotCache;

4 **else if** $n \in ColdCache$ **then**
5   **if** $isUncorrelated(n, t)$ **then**
6   **if** $HotCache.isFull()$ **then**
7   ⌊ move the tail out of HotCache;
8   ⌊ add n to the head of HotCache;

9 **else**   /* n is not in memory */
10   **if** $ColdCache.isFull()$ **then**   /* select replacement victim */
11         $min \leftarrow t$;
12   **foreach** $Node\ i \in ColdCache$ **do**
13   **if** $t - LAST(i) > CR\_Period \&\& HIST(i, 2) < min$ **then**
               /* CR_Period: Correlated Reference Period */
14                 $victim \leftarrow i$;   /* eligible for replacement */
15   ⌊ $min \leftarrow HIST(i, 2)$;
16   ⌊ move $victim$ out of ColdCache;
17   add n to ColdCache;
18     $HIST(n, 2) \leftarrow HIST(n, 1)$;
19     $HIST(n, 1) \leftarrow t$;
20   ⌊ $LAST(n) \leftarrow t$;

21 **function** $IsUncorrelated(n, t)$
22     $flag \leftarrow FALSE$;
23   **if** $t - LAST(n) > CR\_Period$ **then**   /* an uncorrelated reference */
24         $HIST(n, 2) \leftarrow LAST(n)$;
25         $HIST(n, 1) \leftarrow t$;
26         $LAST(n) \leftarrow t$;
27         $flag \leftarrow TRUE$;
28   **else**   /* a correlated reference */
29   ⌊ $LAST(n) \leftarrow t$;
30   ⌊ **return** $flag$;

- **LAST(n)** denotes the time of the last reference to node n, which may be a correlated reference or not.

Simple LRU may replace frequently referenced pages with pages unlikely to be referenced again. H-LRU moves hot nodes to the enclave memory and permits less referenced nodes to stay in normal memory only. When hot cache is full, it evicts the least recently used node. When cold cache is full, it evicts the node whose second-most recent reference is furthest in the past. Our algorithm takes both the recentness and frequency into consideration and avoids the interference of related references, so as to improve the utilization of the enclave memory and achieve better performance.

5 Batch Updates

For MB-tree, whichever leaf node is updated, its digest will be propagated up to the root node, which incurs significant computational cost. If the entire subtree to be updated is cached in enclave, and after a certain period, the digest changes caused by multiple updates are merged and written back to the root node for one time, the update cost of MB-tree will be significantly reduced. In addition, different from traditional databases that are randomly updated at any time, the characteristic of blockchain that periodically submits transactions by block is very suitable for the scenario of batch updates undoubtedly.

Since only the signed root node is trusted in traditional MB-tree, its digest changes must be propagated to the root node immediately when any leaf node is updated. When dealing with frequent updates, such a pattern will surely downgrade system performance. With SGX, all nodes cached by enclave are verified and trusted as mentioned before. The propagation of digest changes can end at an internal node located in enclave memory.

As shown in Fig. 4(a), $Update_1$, $Update_2$, and $Update_3$ respectively represent three update operations on different leaf nodes. When combined with SGX, the parent node of three updates is verified and trusted, so they just need to propagate the digest to the parent node. When the parent node is swapped out by the replace algorithm, or its structure changes due to splitting or merging, the updated digest reflecting three changes will be propagated to the root node.

5.1 Batch Updates with Hybrid Index

In addition to the cost of propagating digest, if the updated MB-tree occurs node splitting and merging, it will further amplify the update cost. The lock operation for updating node will block the query and limit concurrency. To alleviate the cost of MB-tree update, previous works generally adopt batch update that defers the installation of a single update and waits for a batch of updates to process at specific time intervals. For blockchain, it accumulates multiple update transactions to a block according to the time interval or the number of transactions and submits them by blocks, so the blockchain is more suitable for batch update.

This paper presents a dual-stage hybrid index architecture. As shown in Fig. 4(b), it allocates additional trusted space in enclave to create a skip list buffering the incoming new blocks. Compared to B^+-tree, red-black tree and other balanced trees, skip list is more suitable for memory index and has no additional rebalancing cost. Our hybrid index is composed of skip list and MB-tree, where the skip list located in enclave, is used to index newly appended block, and the MB-tree, located in disk, is used to index historical block. Query processor searches the skip list and MB-tree at the same time, and then merges the results returned by skip list and MB-tree to get the full results. A bloom filter atop of the skip list is added to improve query efficiency.

5.2 Merge

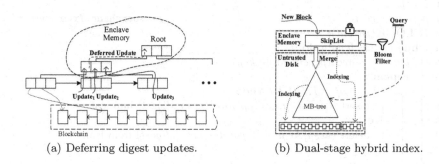

(a) Deferring digest updates.   (b) Dual-stage hybrid index.

Fig. 4. Batch update and merge.

The main purpose of applying merge mechanism in this paper is to utilize batch updates to alleviate the cost of the digest propagation in MB-tree. Blockchain updates data by block periodically, which is different from existing databases, so its merging algorithm and merging strategy are different from previous designs.

There are two solutions for batch updates in MB-tree: full rebuild and delta update. Full rebuild merges and reorders existing leaf nodes of MB-tree with new data and rebuilds the entire MB-tree. Delta update adds new sorted data to the MB-tree in batch. When full rebuild is applied to MB-tree, it will incur considerable cost to recompute the digest of the entire MB-tree and block queries for a long time. Therefore, delta update is selected as our merging algorithm, as shown in Algorithm 2.

The batch update algorithm is efficient because it performs searching and propagates digest updates only once for all keys belonging to the same leaf. When advancing down the tree, the algorithm applies the lock-coupling strategy of nodes, which means only the leaf node and its parent are locked exclusively. The parent is kept locked until all child nodes have been updated and the digest changes from them have been applied.

To find the leaf for a search key, the procedure *Search* starts from the root and advances down along a path by using search key. If an internal node is full or half full, the split or merge operation is triggered accordingly.

Since our system has hot and cold caches for query processing, and updates are processed in batch to reduce MB-tree update cost, we treat the skip list as a write buffer that continuously accumulates new blocks from the blockchain network and move the entire data out of skip list at one time.

Further, it is necessary to determine the specific threshold about how many blocks are buffered for one merge. If the number of buffered blocks is too small to form a considerable sequence length, the MB-tree update cost will not be effectively reduced. Considering the query time in skip list, too many buffered blocks will take longer to query and process the merge.

Algorithm 2: Batch Updates

    Input: Node *root*, Transaction[] *txs*   /* *txs* is sorted in skip list */
1 $i \leftarrow 1$;
2 *parent* ← *root*;   /* searching from root */
3 **X-LOCK**(*parent*);
4 **while** $i \leq txs.length$ **do**
5   *leaf* ← *Search*(*parent*, *txs*[*i*].*key*);
6   **repeat**
7   **if** *txs*[*i*].*op* = *INSERT* **then**
8 | insert (*txs*[*i*].*key*, *txs*[*i*].*poniter*, *txs*[*i*].*digest*) into leaf;
9   **else**
10   ⌊ delete (*txs*[*i*].*key*, *txs*[*i*].*pointer*, *txs*[*i*].*digest*) from leaf;
11          $i \leftarrow i + 1$;
12   **until** *all txs belonging to leaf have been inserted/deleted* **OR** *parent has either n − 1 children when deleting or 2n children when inserting*;
13   **if** *parent is not in enclave* **then**
14   ⌊ verify *parent* and move it into enclave;
15   *updateDigest*(*leaf*, *parent*); /* propagating digests to parent only */
16   **UNLOCK**(*leaf*);
17   **if** *txs*[*i*].*key is in the range of parent and parent has either n − 1 or 2n children* **OR** *txs*[*i*].*key is not in the range of parent* **then**
18   *updateDigest*(*parent*, *root*);   /* propagating digests to root */
19   **UNLOCK**(*parent*);
20   *parent* ← *root*;   /* re-searching from root */
21   ⌊ **X-LOCK**(*parent*);
22 **UNLOCK**(*parent*);

23 **function** *Search*(*parent, k*)
24   *node* ← *getChildNode*(*parent*, *k*);   /* *node* is the child of *parent* */
25   **X-LOCK**(*node*);
26   **while** *node is not a leaf node* **do**
27   **if** *node contains 2n keys* **then**
28 | *split*(*parent*, *node*);
29   **else if** *node contains n − 1 keys* **then**
30   ⌊ *merge*(*parent*, *node*);
31   **UNLOCK**(*parent*);
32   *parent* ← *node*;   /* making *node* as new *parent* */
33   *node* ← *getChildNode*(*parent*, *k*);
34   ⌊ **X-LOCK**(*node*);
35   **return** *node*;

---

6 Security Analysis

In this section, we perform security analysis. Our basic security model of range query authentication is secure, if the underlying hash function is collision-resistant and security enforcement of SGX cannot be broken.

Tampering Attack. As the frequently-accessed nodes of MB-tree and the entire skip list are resident in enclave, attackers cannot tamper with them. Although, the other nodes of MB-tree located in normal memory can be tampered by adversaries, the integrity of query results returned can been authenticated by the verified nodes in enclave. Using the query results and VO, query processor reconstructs the digests up to the root, and compares the root digest against that in enclave. Considering a node being successfully tampered with, there exist two MB-trees with different nodes but the same root digest. This implies a successful collision of the underlying hash function, which leads to contradiction.

Rollback Attack. In rollback attack, the untrusted node can replace the MB-tree with an old version, which makes the clients read stale results. A trusted monotonic counter can protect the latest version of MB-tree. To detect and defend the rollback attack, we use SGX monotonic counter service or rollback-protection system such as ROTE [5] to guarantee the freshness of query results.

Untrusted Blockchain Data. In our solution, the SGX on an untrusted full node performs all verification for clients, yet the untrusted full node can deliver incorrect or incomplete blocks, even not send the latest block to the enclave. To protect against such compromise, a client needs to acquire the latest block hash from other sources, compares the block hash to that from the SGX, and deduces if the results are integrity or not.

7 Implementation and Evaluation

In this section, we evaluate the performance of our proposed scheme that integrates MB-tree and Intel SGX, including authenticated query, cache architecture and batch updates.

7.1 Experimental Setup

We use BChainBench [17], a mini benchmark for blockchain database, to generate synthetic blockchain dataset that consists of 1 million transactions, of which each key has 8 bytes and value has 500 bytes. We implement and construct an MB-tree with 2 KByte page size. For each node of MB-tree, both the key and pointer occupy 8 bytes and the digest uses 20 bytes, which makes each node index 56 entries ($\lfloor (2048 - (8 + 20))/(8 + 8 + 20) \rfloor$=56). Initially, our MB-tree is stored on disk, except that a copy of the root node is located in enclave. All experiments were conducted on a server, which is equipped with a 32 GB RAM and an Intel Core i7-8700k CPU @2.70Hz, and runs Ubuntu 16.04 OS with Intel SGX Linux SDK and SGXSSL library.

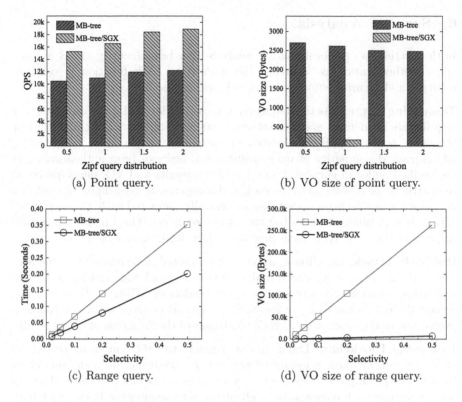

Fig. 5. Query performance and VO size.

7.2 Query Performance

Figure 5(a) manifests the performance of point query in Zipfian distribution. With the increment of skew parameter, the throughput of MB-tree in SGX is about 1.6 times more than traditional MB-tree, because the frequently-used MB-tree nodes in enclave cut short the verification path. In Fig. 5(b), the VO size of MB-tree in SGX decreases by one or two order of magnitude. For traditional MB-tree, the query authentication is evaluated on the server, so the performance of light clients on mobile devices becomes worse when the verification is performed locally. For MB-tree in SGX, the verification is accomplished by SGX on full node, so that light clients avoid receiving and processing VO.

Figure 5(c) demonstrates the performance of range query in Zipfian distributions. The executing time of MB-tree in SGX is merely 60% of the traditional MB-tree when the selectivity is set to 50%. In Fig. 5(d), the reduction of VO size is more remarkable, since range query owns much more verification information than point query. Tens of kilobytes VO exhibit significant network overhead for lights clients, especially for mobile devices.

7.3   Cache Performance

Figure 6(a) reports the performance of H-LRU and LRU. We run 100,000 point queries, and report cache hit rate, i.e., the number of accessing MB-tree nodes located in hot cache to the number of accessing all nodes. We change the cache size from 5% to 40% of the cache size of the highest hit rate. H-LRU provides about 10% improvement over traditional LRU. The performance boost is higher with smaller cache size.

In Fig. 6(b), we randomly mix some range queries in point queries, which will start scan operations occasionally. We set the probability of starting a range query to 0.1, i.e., one-tenth of the generated queries are range queries. We vary the selectivity based on the cache size of the highest hit rate. The experiments confirm that H-LRU is more adaptable than LRU.

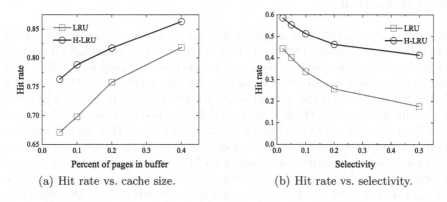

(a) Hit rate vs. cache size.   (b) Hit rate vs. selectivity.

Fig. 6. The effect of the H-LRU.

7.4   Update Performance

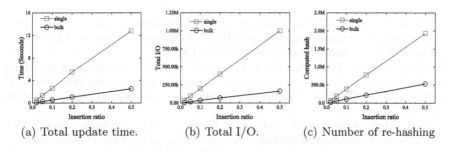

(a) Total update time.   (b) Total I/O.   (c) Number of re-hashing

Fig. 7. Update performance.

Figure 7 presents the performance of batch update. Due to insufficient space, batch update only consists of a number of uniform insertions, ranging from 1% to 50% of the total blockchain data size. When the insertion ratio reaches 50%, the update time and the number of re-hashing are diminished by about 4 times, and I/O cost is reduced by about 6 times. It is because MB-tree is bulk-loaded with 70% utilization, and batch insertions quickly lead to many split operations, which creates a lot of new nodes. Although most of improvements are contributed by reducing of I/O cost, our batch update algorithm avoids hash computing being propagated to the root node and reduces the lock operations. Traditional MB-tree requires one expensive signature re-computation for every update. In order to show only the update performance of the tree, we did not consider that factor.

8 Conclusion

We explore the problem of authenticated range queries using Intel SGX for blockchain light clients. The main challenge lies in how to design an authenticated query scheme with memory-limited enclave. We propose a solution by integrating MB-tree with SGX, which caches frequently-used MB-tree nodes in trusted enclave and the rest nodes in untrusted memory. An efficient cache replacement algorithm, H-LRU, is devised for the two-level cache architecture, which considers more of the reference history to improve enclave memory utilization. To reduce the cascading hash computing brought by updates on MB-tree, we provide a hybrid index consisting of an MB-tree and a skip list in enclave, which buffers multiple new blocks in skip list and regularly merges them to MB-tree in batch. Security analysis and empirical results substantiate the robustness and efficiency of our proposed solution. In future, we plan to extend our idea to process other authenticated queries, such as join and aggregation.

Acknowledgment. This research is supported in part by National Science Foundation of China under grant number U1811264, U1911203, 61972152 and 61532021.

References

1. Cheng, R., et al.: Ekiden: a platform for confidentiality-preserving, trustworthy, and performant smart contracts. In: 2019 IEEE European Symposium on Security and Privacy (EuroS&P), pp. 185–200. IEEE (2019)
2. Hu, S., Cai, C., Wang, Q., Wang, C., Luo, X., Ren, K.: Searching an encrypted cloud meets blockchain: A decentralized, reliable and fair realization. In: IEEE INFOCOM 2018-IEEE Conference on Computer Communications, pp. 792–800. IEEE (2018)
3. Li, F., Hadjieleftheriou, M., Kollios, G., Reyzin, L.: Dynamic authenticated index structures for outsourced databases. In: Proceedings of the 2006 International Conference on Management of Data, pp. 121–132. ACM (2006)
4. Li, F., Hadjieleftheriou, M., Kollios, G., Reyzin, L.: Authenticated index structures for aggregation queries. ACM Trans. Inf. Syst. Secur. (TISSEC) 13(4), 32 (2010)

5. Matetic, S., et al.: ROTE: rollback protection for trusted execution. In: 26th USENIX Security Symposium (USENIX Security 17), pp. 1289–1306 (2017)
6. Matetic, S., Wüst, K., Schneider, M., Kostiainen, K., Karame, G., Capkun, S.: BITE: bitcoin lightweight client privacy using trusted execution. In: 28th USENIX Security Symposium (USENIX Security 19), pp. 783–800 (2019)
7. McKeen, F., et al.: Innovative instructions and software model for isolated execution. HASP@ISCA **10**(1) (2013)
8. Merkle, R.C.: A certified digital signature. In: Brassard, G. (ed.) CRYPTO 1989. LNCS, vol. 435, pp. 218–238. Springer, New York (1990). https://doi.org/10.1007/0-387-34805-0_21
9. Nakamoto, S., et al.: Bitcoin: a peer-to-peer electronic cash system (2008)
10. Pang, H., Tan, K.L.: Authenticating query results in edge computing. In: Proceedings of the 20th International Conference on Data Engineering, pp. 560–571. IEEE (2004)
11. Robinson, J.T., Devarakonda, M.V.: Data cache management using frequency-based replacement, vol. 18. ACM (1990)
12. Weisse, O., Bertacco, V., Austin, T.: Regaining lost cycles with hotcalls: a fast interface for SGX secure enclaves. ACM SIGARCH Comput. Archit. News **45**(2), 81–93 (2017)
13. Xu, C., Zhang, C., Xu, J.: vChain: enabling verifiable Boolean range queries over blockchain databases. In: Proceedings of the 2019 International Conference on Management of Data, pp. 141–158. ACM (2019)
14. Yang, Y., Papadias, D., Papadopoulos, S., Kalnis, P.: Authenticated join processing in outsourced databases. In: Proceedings of the 2009 ACM SIGMOD International Conference on Management of Data, pp. 5–18. ACM (2009)
15. Zhang, C., Xu, C., Xu, J., Tang, Y., Choi, B.: Gem^2-tree: a gas-efficient structure for authenticated range queries in blockchain. In: 2019 IEEE 35th International Conference on Data Engineering (ICDE), pp. 842–853. IEEE (2019)
16. Zhang, F., Cecchetti, E., Croman, K., Juels, A., Shi, E.: Town crier: an authenticated data feed for smart contracts. In: Proceedings of the 2016 ACM SIGSAC Conference on Computer and Communications Security, pp. 270–282. ACM (2016)
17. Zhu, Y., Zhang, Z., Jin, C., Zhou, A., Yan, Y.: SEBDB: semantics empowered blockchain database. In: 2019 IEEE 35th International Conference on Data Engineering (ICDE), pp. 1820–1831. IEEE (2019)

Stargazing in the Dark: Secure Skyline Queries with SGX

Jiafan Wang, Minxin Du, and Sherman S. M. Chow$^{(\boxtimes)}$ (ID)

Department of Information Engineering, The Chinese University of Hong Kong,
Shatin, N.T., Hong Kong
{wj016,dm018,sherman}@ie.cuhk.edu.hk

Abstract. Skylining for multi-criteria decision making is widely applicable and often involves sensitive data that should be encrypted, especially when the database and query engine are outsourced to an untrusted cloud platform. The state-of-the-art designs (ICDE'17) of skylining over encrypted data, while relying on two *non-colluding* servers, are still slow – taking around three hours to get the skyline for 9000 2-D points.

This paper proposes a very efficient solution with a trusted processor such as SGX. A challenge is to support dynamic queries while keeping the memory footprint small and simultaneously preventing unintended leakage with only lightweight cryptographic primitives. Our proposed approach iteratively loads data to the memory-limited SGX on-demand and builds a binary-tree-like index for logarithmic query time. For millions of points, we gain $6000 - 28000\times$ improvement in query time (ICDE'17).

1 Introduction

Skyline, also known as the *Pareto optimality* in Economics, has received considerable attention since its introduction by Börzsönyi et al. [3] as part of the SQL syntax due to its various applications, including multi-criteria decision making. For a set of multi-dimensional points, the skyline operator [3] retrieves all points that are not dominated by others. A point *dominates* another if it is as good or better in all dimensions and strictly better in at least one dimension. Without loss of generality, we use the min function to evaluate the goodness. A querier can also issue a dynamic query point [18], of which the relative coordinates for all data points are used for the determination of domination. More vividly, there is a *space mapping* stage – the queried point is regarded as the origin for a new data space, in which every data point is mapped to by taking the absolute dimension-wise distances to the query point as its new coordinates. Compared to the well-known k nearest neighbor (kNN) query, which is parameterized with

---

The first two authors contributed equally and share the "co-first author" status. Sherman S. M. Chow is supported by General Research Funds (CUHK 14209918 and 14210217) of the Research Grants Council, UGC, Hong Kong. The authors would like to thank Shuaike Dong and Di Tang for their advice on the experiments, and the anonymous reviewers for their suggestions and comments.

© Springer Nature Switzerland AG 2020
Y. Nah et al. (Eds.): DASFAA 2020, LNCS 12114, pp. 322–338, 2020.
https://doi.org/10.1007/978-3-030-59419-0_20

ID	Age	Trestbps	ID_{map}	Age	Trestbps
P_1	38	124	P'_1	52	136
P_2	60	152	P'_2	60	152
P_3	54	115	P'_3	54	145
P_4	28	118	P'_4	62	142
P_5	70	115	P'_5	70	145
P_6	27	165	P'_6	63	165
P_7	58	102	P'_7	58	162
P_8	46	157	P'_8	46	157

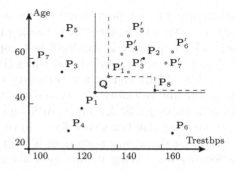

Fig. 1. An example of skyline (query point $Q(130, 45)$) over EHR

predefined dimensional weights, skylining is more flexible as it does not need this hard-to-specify knowledge and essentially considers all relative weights.

Consider electronic health records (EHRs) in Fig. 1, where a physician issues a skyline query $Q(130, 45)$ for a 45-year old patient whose resting blood pressure (trestbps) is 130. All the raw records are then mapped to $P'_1, \ldots, P'_8$ in a new space where the absolute dimension-wise distances to Q become the new coordinates. The query retrieves (P_1, P_8) as Q's skyline. The physician can then make a better diagnosis focusing on these two patients with similar trestbps/age.

Skylining is especially useful when the database is large. In such situations, the data owner may resort to an external cloud server for alleviating local costs. Yet, the data can be sensitive, but existing cloud facilities are often not fully trusted. One may employ a standard encryption scheme. Yet, it hinders any basic operations (such as comparisons) by the cloud, not to say skyline query.

Secure skyline query was first studied in eSkyline [4], where points (viewed as high-dimensional vectors) are encrypted by invertible random matrices, and the dominations among points are determined by computing scalar products of the sub-vectors. It traverses the entire dataset and uses "domination-preserving" encryption, which already leaks point-wise dominations before querying.

Recently, order-revealing encryption (ORE) is proposed, which features a special operation that reveals the relative order of two pieces of encrypted data. However, using ORE to support dynamic skylining probably requires a number of comparisons *quadratic* in the database size for each dimension. Moreover, practical ORE schemes often have a somewhat "non-intuitive" leakage profile, *e.g.*, equality pattern of most significant differing-bit [5]. Whether a smart attacker can exploit this for inferring more than the skyline is not entirely clear [1].

1.1 Challenges in Supporting Skyline Queries over Encrypted Data

Processing encrypted data can be supported by using fully-homomorphic encryption (FHE), which allows any number of addition/multiplication (and hence any circuit) over the encrypted plaintexts. Space mapping and determination of domination can then be replaced by homomorphic operations over encrypted points. Yet, such an approach is still inefficient, especially for a large database.

Two recent secure skylining schemes [16] resort to additively-homomorphic encryption (AHE) and secure two-party computation (2PC) protocols [8,23] among *two non-colluding servers* to compute, *e.g.*, comparisons, multiplications, and minima needed in skylining[1]. Although AHE is more efficient than FHE, the space mapping still takes time linear in the database size[2]. Also, invoking 2PC protocols for the atomic operations is both time- and bandwidth-consuming.

Security-wise, being in the non-colluding setting, their schemes would be totally broken when a server turns malicious and successfully compromised the other one. Also, they gave no precise definition of the allowed "harmless" leakage. Such leakage-based definition is a well-established approach to define security in the related literature of symmetric searchable encryption (SSE) [6,10,14].

1.2 Technical Overview and Contributions

To eliminate the need for two non-colluding servers, we thus turn our attention to using trusted hardware such as Intel SGX (see Sect. 2.1), which can create secure isolation for sensitive data and operations in hostile environments. With SGX, we do not invoke costly 2PC protocols (hence avoiding interactions among two servers) or rely on expensive AHE/FHE as in (purely-) cryptographic designs. We restrict to only (more efficient) symmetric-key cryptographic primitives and shift some "hard-to-compute" operations (over encrypted data) to SGX.

We start with the idea of using nearest neighbor searches as the classical skyline computation on plaintext data [11,18]. We aim to find skyline points iteratively in the original space, thus avoiding the standard space-mapping preparation that incurs a linear overhead. In each iteration, we find the nearest neighbor (NN) to the query point over a "to-do list" as a skyline point, and then prune points dominated by it from the to-do list. One naïve method is to load the entire encrypted database into SGX for decryption, but this can easily exceed the current memory limit (*i.e.*, 128 MB) of SGX. Our strategy is to adapt a secure multi-dimensional range query scheme [21] for these two steps.

In more detail, to find NN, we heuristically issue range queries on a gradually expanding area with the query point as its center and get NN (as a skyline point) from the query results in SGX. This returns initial skyline points quickly without scanning the entire database (*i.e.*, progressiveness [18]). To refine the to-do list, we identify the area containing points not dominated by the NN, based on which another range query is issued. The query result becomes the new to-do list. This allows us to directly work on the original space. Due to a tree index of the range

---

[1] We remark that the assumption of two non-colluding servers and the usage of AHE enables both addition and multiplication over encrypted data [19].

[2] Two other works [17,22] also use AHE over distributed encrypted datasets (*cf.* computing on data encrypted under multiple keys [19]), but only for static skyline.

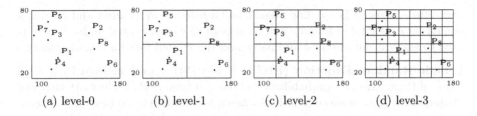

Fig. 2. Hierarchical hyper-rectangle encoding for $d = 2$ and $\ell = 3$

query scheme, we achieve a logarithmic query time. It is important to note that we use SGX for only simple arithmetics, and the range-query results are loaded on-demand, which keeps a small memory consumption.

On the theory side, we formalize a security model under the simulation-based framework [6] and achieve a tradeoff for practical efficiency by only allowing the difficult-to-avoid leakage, *e.g.*, the database size and the access pattern.

Practically, via experiments on both real-world and some synthetic datasets, we show that our approach is practical for a million-scale database and can achieve up to $28\,000\times$ query time improvement compared to the prior art [16].

2 Preliminaries

We consider multi-dimensional data points. Let $\Delta = (T_1, \ldots, T_d)$, where T_i is the upper bound of the i-th dimension. $\mathbb{L}_\Delta = [T_1] \times \cdots \times [T_d]$ defines a d-dimensional data space (or lattice), where $[T_i]$ denotes the set of integers $\{1, \ldots, T_i\}$. A data point $\mathbf{P}$ in $\mathbb{L}_\Delta$ is defined by $\mathbf{P} = (p_1, \ldots, p_d)$, where $p_i \in [T_i]$ ($\forall i \in [d]$). We also refer its i-th dimension by $\mathbf{P}[i]$. A hyper-rectangle $\mathbf{R}$ in $\mathbb{L}_\Delta$ is defined by $\mathbf{R} = (\mathbf{r}_1, \ldots, \mathbf{r}_d)$, where $\mathbf{r}_i$ is a range $\mathbf{r}_i \subseteq [1, T_i]$, $\forall i \in [d]$.

Definition 1 (Skyline). *Given a dataset $D = \{\mathbf{P}_1, \ldots, \mathbf{P}_n\}$ with n points in $\mathbb{L}_\Delta$, for $a \neq b$, we say $\mathbf{P}_a$ dominates $\mathbf{P}_b$, denoted by $\mathbf{P}_a \prec \mathbf{P}_b$, if $\forall j \in [d]$, $\mathbf{P}_a[j] \leq \mathbf{P}_b[j]$, and for at least one j^*, $\mathbf{P}_a[j^*] < \mathbf{P}_b[j^*]$. The skyline is a set of points that are not dominated by any other point.*

Definition 2 (Dynamic Skyline Query [18]). *Given a query $\mathbf{Q}$ and a dataset $D = \{\mathbf{P}_1, \ldots, \mathbf{P}_n\}$ in $\mathbb{L}_\Delta$, for $a \neq b$, we say $\mathbf{P}_a$ dynamically dominates $\mathbf{P}_b$ with respect to $\mathbf{Q}$, also denoted by $\mathbf{P}_a \prec \mathbf{P}_b$, if $\forall j \in [d]$, $|\mathbf{P}_a[j] - \mathbf{Q}[j]| \leq |\mathbf{P}_b[j] - \mathbf{Q}[j]|$, and for at least one j^*, $|\mathbf{P}_a[j^*] - \mathbf{Q}[j^*]| < |\mathbf{P}_b[j^*] - \mathbf{Q}[j^*]|$. The skyline is a set of points that are not dynamically dominated by others with respect to $\mathbf{Q}$.*

2.1 Cryptographic Tools and Software Guard Extensions (SGX)

Symmetric-Key Encryption (SKE). SKE consists of three polynomial-time algorithms $\mathsf{SKE} = (\mathsf{Gen}, \mathsf{Enc}, \mathsf{Dec})$. Gen takes a security parameter λ as input and returns a secret key k. Enc takes k and a message m as input and returns a ciphertext c. Dec takes k and c and returns m iff c is encrypted under k. SKE is used to encrypt data-point coordinates and the related files for confidentiality.

Pseudo-Random Function (PRF). F is a PRF family indexed by a λ-bit key k, if there exists no probabilistic polynomial-time (PPT) adversary that can distinguish it from random functions when k is kept secret and randomly chosen, even given access to an oracle computing F using k. A bijective PRF is pseudorandom permutation (PRP). We use it to derive a pseudonym for sensitive data.

Instantiation. Our experiments use AES-128 and HMAC-SHA256 to instantiate SKE and PRF/PRP – 1K runs of Enc and Dec for a 128-bit message take about 23ms and 21ms, and 1K runs of PRF/PRP only take about 3.5 ms.

SGX. With SGX, applications can create protected memory regions called *enclaves*. Sensitive data/code is isolated from any other (privileged) programs, and the isolation is enforced by hardware access controls. SGX provides a sealing mechanism that allows an enclave to store its data in non-volatile and untrusted memory securely, and unsealing to restore the sealed data. Given a remote SGX platform, we can use remote attestation to establish a secure channel with an enclave on it and authenticate the target program is executed benignly.

2.2   Hierarchical Hyper-rectangle Encoding

We run a multi-dimensional range query scheme [21] as a subroutine. Its core is to use hierarchical cube encoding to assign discrete codes for data points (and query ranges) and then organize them in a binary tree. We slightly generalize it to hierarchical hyper-rectangle encoding (HRE) and review its key idea below. In a nutshell, it produces pseudorandom codes that uniquely identify each hyper-rectangle of a particular size at a certain location (*cf.* geospatial query [9]).

To encode a d-dimensional dataset with n points via HRE, we first define a level-0 hyper-rectangle where the length of each dimension depends on the minimal and maximum values of that dimension, *i.e.*, $[T_1], \ldots, [T_d]$. All data points are then included in the level-0 hyper-rectangle. Each hyper-rectangle at level-i is then recursively divided into 2^d sub-hyper-rectangles of equal size at level-$(i + 1)$ until the length of the shortest side is smaller than a specified *threshold* ρ. The number of levels in the hierarchy is denoted by ℓ. We also need to ensure that the number of points in every level-ℓ hyper-rectangle is small. This prevents performance degradation when data is skewed. Figure 2 gives an example of hierarchical hyper-rectangles for data points in Fig. 1.

We assign each hyper-rectangle a discrete pseudorandom code derived from its center and the level it belongs to. For a level-i hyper-rectangle whose center is $(x_1, \ldots, x_d)$, its code is computed by a PRF F under a key k as $F(k, x_1|| \cdots ||x_d||i)$, where $||$ denotes concatenation. For example, the code of the top-left rectangle in Fig. 2b, whose level is 1 and center is $(120, 65)$, is $F(k, 120||65||1)$. Any data point $\mathbf{P}$ belongs to ℓ hyper-rectangles under an ℓ-level hierarchy. The codes for these ℓ hyper-rectangles are the *data code set* for

$\mathbf{P}$, denoted by $DCSet(\mathbf{P})$. No one can learn the coordinates of $\mathbf{P}$ without k as F outputs pseudorandom strings.

For a range query $\mathbf{Q}_r$, we also assign a *query code set* $QCSet(\mathbf{Q}_r)$ to it. $QCSet(\mathbf{Q}_r)$ consists of the codes for the fewest hyper-rectangles that fully cover the area specified by $\mathbf{Q}_r$. Concretely, it is obtained in a top-down manner: if hyper-rectangles at level-i are fully covered by $\mathbf{Q}_r$, add their HRE codes into $QCSet(\mathbf{Q}_r)$; otherwise, check hyper-rectangles at level-$(i+1)$. These two steps repeat until reaching level-ℓ: those codes for hyper-rectangles which only have overlaps with (but not fully covered by) $\mathbf{Q}_r$ are also added in $QCSet(\mathbf{Q}_r)$. As a result, *false positives exist* since the aggregation of all these hyper-rectangles is larger than the queried range $\mathbf{Q}_r$ (*but can be removed* by SGX in our scheme). It is easy to see that a smaller threshold ρ results in a lower false-positive rate.

Finally, the query is handled by performing set membership testing between $QCSet(\mathbf{Q}_r)$ and $DCSet(\mathbf{P}_i)$ (instead of making dimension-wise comparisons among all data points). We still take Fig. 2 as an example, suppose the rectangle at level-0 has one code c_0, rectangles from top-left to bottom-right at level-1 have 4 codes $c_1, \ldots, c_4$, and so forth. For a query $\mathbf{Q}_r = [100, 140] \cap [50, 80]$ which is fully covered by the top-left rectangle at level-1, its $QCSet(\mathbf{Q}_r)$ is $\{c_1\}$. Three points $\mathbf{P}_3, \mathbf{P}_5, \mathbf{P}_7$ with code c_1 in their $DCSet$'s fall in the queried range.

2.3 Indistinguishable Bloom Filter (IBF)

Bloom filter (BF) [2] is a space-efficient probabilistic data structure for a set of elements such that anyone can test if a certain element is in the set. Indistinguishable BF (IBF) [15] only allows those with a secret key to test membership.

IBF consists of an array B of g twins (or entries), L *keyed* hash functions $h_1, h_2, \ldots, h_L$, and a hash function H. Every twin has two cells, each of which stores one bit opposite to the other. For each twin, H outputs a bit to determine the cell to be chosen. The chosen one is initialized to be 0; 1 otherwise. Given an element e, we hash it to L twin locations (via L keyed hash functions). For each of these location, the cell chosen by the output bit of H is set to 1; 0 otherwise. Thus, every IBF always has g 1's and 0's with cells chosen randomly from H. Any PPT adversary cannot distinguish between IBFs of different sets.

The array B can be instantiated by a two-dimensional array where the bit in the i-th row and j-th column is denoted by $B[i][j]$. We introduce an extra keyed hash function h_{L+1} to prevent an adversary from learning which cell is chosen in advance. To insert an HRE code e, we set $B[H(h_{L+1}(h_l(e)) \oplus r_B)][h_l(e)]$ as 1 and $B[1 - H(h_{L+1}(h_l(e)) \oplus r_B)][h_l(e)]$ as 0 for $l \in [L]$, where r_B is a randomness peculiar to this IBF to eliminate linkability among different IBFs. To perform membership testing of e, we calculate *the set of such hashes* over e, denoted as position set $PosSet(e)$. We check whether those cells indicated by $PosSet(e)$ are all 1. If so, e is in the set with a high probability (due to false positives).

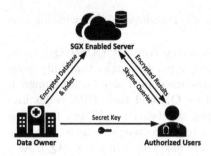

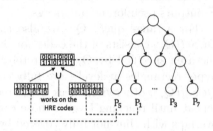

Fig. 3. System model   **Fig. 4.** An example of tree index

3 Problem Formulation

3.1 System Model

Figure 3 provides the system model, consisting of a data owner, authorized users, and an untrusted SGX-enabled cloud server (and a trusted SGX enclave within).

The data owner (*e.g.*, a hospital) runs the setup, which derives a secret key. The data owner pre-processes its data (*e.g.*, a gigantic EHR database) to build a secure index using the secret key. The secure index, along with the associated data (which is also encrypted under the secret key), is then outsourced to the cloud. The data owner uses the remote attestation for authenticating a benign enclave running at the cloud and then provisions the key to the enclave securely.

In our setting, the data owner will delegate the secret key to authorized users (*e.g.*, physicians). With the secret key, a user can then issue an encrypted query (from an EHR of a patient) to request the skyline with respect to this query (for making diagnoses). Given the secure index and queries, the cloud server returns encrypted results (including skyline points and the associated data) to the users.

Below, we formally define the syntax of a secure skyline query scheme.

Definition 3 (Secure Skyline Query Scheme). *A secure skyline query system* $\Pi = (\mathsf{Setup}, \mathsf{Index}, \mathsf{Search})$ *is a tuple of two algorithms and a protocol.*

$(K, \mathbb{L}_\Delta) \leftarrow \mathsf{Setup}(1^\lambda)$ *is a probabilistic algorithm that takes as input a security parameter* $\lambda \in \mathbb{N}$. *It outputs a secret key* K *and a d-dimensional data space* $\mathbb{L}_\Delta$. $\mathbb{L}_\Delta$ *is a public parameter, which is implicitly taken as input by* Index *and* Search.

$(\mathsf{EDB}, \Gamma) \leftarrow \mathsf{Index}(K, \mathsf{DB})$ *is a probabilistic algorithm that takes as input* K *and a database* DB *in* $\mathbb{L}_\Delta$. *It outputs an encrypted database* EDB *and a secure index* Γ.

$\mathcal{R} \leftarrow \mathsf{Search}((K, \mathbf{Q}); (\mathsf{EDB}, \Gamma))$ *is a protocol between an authorized user with the secret key* K *and a dynamic skyline query* $\mathbf{Q}$; *and the server with the encrypted database* EDB *and secure index* Γ. *The protocol outputs an encrypted result* $\mathcal{R}$.

Correctness. Π *is correct if, for all* $\lambda \in \mathbb{N}$, *all* $(K, \mathbb{L}_\Delta) \in \mathsf{Setup}(1^\lambda)$, *all* $(\mathsf{EDB}, \Gamma) \in \mathsf{Index}(K, \mathsf{DB})$, *and all sequences of* Search, *it returns complete and correct skyline points and the associated data except with negligible probability in* λ.

3.2 Threat Model and Security Definitions

We assume the cloud server supports any trusted execution environment with capabilities similar to SGX, which ensures the integrity and confidentiality of data and operations inside. We consider the server as a semi-honest adversary, which provides reliable storage and query services exactly as specified by the scheme, but tries to learn any extra information from what it receives.

Ideally, any adversary learns no extra information other than those allowed by the system. But it is known to be hard to prevent all leakages without heavyweight cryptographic tools. Typical SSE schemes (*e.g.*, [7,12,13,20]) thus relax the ideal definition by allowing *well-defined* leakages. We adopt the simulation-based framework [6] parameterized with stateful leakage functions, which captures what is leaked by (EDB, $\varGamma$) and executions of Search.

Our scheme also involves interactions between the server and SGX that can be seen by the adversary, beyond the interactions between the data owner and the server covered by the security definitions of SSE [6]. Thus, our leakage functions include any possible leakage from the inputs and outputs of SGX. However, we do not consider hardware attacks (*e.g.*, denial of service) or side-channel attacks on SGX (*e.g.*, cache-timing attacks and page-fault attacks).

Definition 4 (Adaptive Security). *Let Π be a secure skyline query scheme. We say Π is $\mathcal{L}$-adaptively-secure, where $\mathcal{L} = \{\mathcal{L}^{\mathsf{Idx}}, \mathcal{L}^{\mathsf{Srch}}\}$ is a set of two leakage functions (to be defined in Sect. 4.3 for our scheme), if for any PPT adversary $\mathcal{A}$, there exists a PPT simulator $\mathcal{S}$ such that:*

$$| \Pr[\mathbf{Real}_{\mathcal{A}}(1^\lambda) = 1] - \Pr[\mathbf{Ideal}_{\mathcal{A},\mathcal{S}}(1^\lambda) = 1]| \leq \mathsf{negl}(\lambda),$$

*with the following probabilistic experiments **Real** and **Ideal**.*

$\mathbf{Real}_{\mathcal{A}}(1^\lambda)$: *The challenger runs* Setup(1^λ) *to obtain a secret key K and the specification of $\mathbb{L}_\Delta$ (available to $\mathcal{A}$). $\mathcal{A}$ outputs a database* DB. *The challenger runs* Index(K, DB) *and sends the pair* (EDB, $\varGamma$) *to $\mathcal{A}$. $\mathcal{A}$ adaptively makes a polynomial number of skyline queries $\mathbf{Q}$. The challenger returns the output of* Search *protocol given $\mathbf{Q}$. Finally, $\mathcal{A}$ returns a bit b that is output by the experiment.*

$\mathbf{Ideal}_{\mathcal{A},\mathcal{S}}(1^\lambda)$: *$\mathcal{A}$ outputs a database* DB. *Given $\mathcal{L}^{\mathsf{Idx}}(\mathsf{DB})$, $\mathcal{S}$ generates the pair* (EDB, $\varGamma$) *and sends it to $\mathcal{A}$. $\mathcal{A}$ adaptively makes a polynomial number of skyline queries $\mathbf{Q}$. For each $\mathbf{Q}$, $\mathcal{S}$ is given $\mathcal{L}^{\mathsf{Srch}}(\mathbf{Q})$ and returns the simulated output of* Search *protocol to $\mathcal{A}$. Finally, $\mathcal{A}$ returns a bit b that is output by the experiment.*

4 Our Construction

4.1 General Ideas

We employ a progressive algorithm [18] that iteratively finds the NN to $\mathbf{Q}$ within a to-do list and then refines the list. To realize it securely, we incorporate a tree-index-based multi-dimensional range query scheme [21]. Such an index structure is built using only lightweight symmetric-key primitives, and avoids linear computation for the space mapping – both boost the query efficiency.

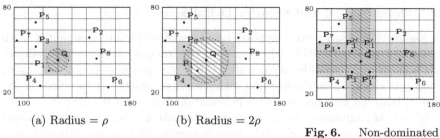

(a) Radius = ρ (b) Radius = 2ρ

Fig. 5. Nearest neighbors search of **Q**

Fig. 6.   Non-dominated area of $\mathbf{P}_1$ regarding **Q**

To construct the index, we utilize HRE to encode the data points and organize them in a binary tree[3], where each leaf node points to a randomly chosen data point and stores an IBF constructed from its HRE codes, while each internal node stores an IBF constructed from the union of its left and right children HRE codes. Figure 4 shows an example of an index from data points in Fig. 1.

Given an encrypted **Q**, we cannot accurately decide in advance which area NN of **Q** locates in. Our solution is to issue a multi-dimensional range query for shortlisting a candidate set and let SGX find NN within it. The key issue is how to determine the queried range while ensuring there indeed exist candidate points. We thus consider a hyper-sphere centered at **Q** whose radius is increased by a fixed amount for each trial query. This hyper-sphere serves as the query range $\mathbf{Q}_r$ from which $QCSet(\mathbf{Q}_r)$ is derived. Figure 5 gives an example where the shadowed circle is $\mathbf{Q}_r$, and the gray rectangles represent $QCSet(\mathbf{Q}_r)$. The query result is then loaded in SGX for finding NN as a skyline point.

To remove data points dominated by the skyline point, we directly find points that are not dominated using the range query scheme. Since we avoid the data space mapping, we need to "probe" possible areas as follows. Given the skyline point, all its symmetric points against axes (with **Q** as the origin) can be obtained. Based on them, we can learn the areas which contain non-dominated points to serve as the query range $\mathbf{Q}_r$. As shown in Fig. 6, assume $\mathbf{P}_1$ is the skyline point found in the previous step, $\mathbf{P}_1', \mathbf{P}_1'', \mathbf{P}_1'''$ are its symmetric points, the shadowed cross is $\mathbf{Q}_r$, which is fully covered by the grey cross for deriving $QCSet(\mathbf{Q}_r)$. This may incur false positives (due to a larger grey cross), which can be removed inside SGX. The results update the to-do list for future iterations.

4.2   Detailed Construction

Setup. Given a security parameter λ, the data owner sets up the environment (*e.g.*, a d-dimensional data space $\mathbb{L}_\Delta$) and generates a secret key set $K = \{k_1, k_2, \mathsf{hk}_1, \ldots, \mathsf{hk}_{L+1}\}$, where k_1 is used by a PRF F to derive HRE

---

[3] We do not exploit multi-dimensional tree structures, such as R-tree, kd-tree, or their variants, since they will leak the distribution of the entire database before querying.

codes, k_2 is used by an SKE scheme to encrypt the coordinates of query points[4], and $hk_1, \ldots, hk_{L+1}$ are used by keyed hash functions to insert codes into IBFs.

Using the remote attestation, the data owner authenticates a benign enclave (on the server) and provisions K to it through an established secure channel.

Database Encryption and Index Construction. The data owner encrypts the concatenated coordinates of each data point by SKE. These encrypted data points associated with their encrypted files form the encrypted database EDB. The data owner derives $DCSet$ – the HRE codes for every data point using k_1, and indexes them by a (balanced) binary tree. Each leaf node stores a pointer to a data point and an IBF constructed from $DCSet$ of the point using keyed hashes. The data point is randomly chosen, say, via a PRP taking as input its identifier, thwarting any possible linkage in the original space, *e.g.*, near data points will not be stored in near leaves. Each internal node stores an IBF constructed from the union of its left and right children's HRE codes. Thus, the IBF of an internal node contains all codes in its descendant leaf nodes, and the root covers $DCSets$ of the whole dataset. The binary tree storing IBFs forms the secure index Γ.

Dynamic Skyline Query. The Search protocol iteratively performs multi-dimensional range queries for finding NN and non-dominated points with respect to the query. Its pseudocode is listed in Algorithm 1, which we explain below.

Given a query point $\mathbf{Q}$, an authorized user encrypts the concatenation of its coordinates and sends it to the server. After receiving the encrypted $\mathbf{Q}$, the server initializes a to-do list $\mathcal{T}$ (for storing non-skyline and non-dominated points) initialized as the entire database, a candidate set $\mathcal{C}$ (for storing possible nearest neighbors), and a result set $\mathcal{R}$. Sets $\mathcal{C}$ and $\mathcal{R}$ are initialized to be empty.

Finding NN. The enclave loads the encrypted $\mathbf{Q}$ for decryption and initializes a monotonically increasing counter i to be 0. It issues range queries on a gradually expanding area for shortlisting possible nearest neighbors (lines 6–12). Concretely, the area denoted by $\odot_i$ is the *hyper-sphere* whose center is $\mathbf{Q}$ and radius $i \cdot \rho$, where ρ is the threshold length. We denote the *nearby area* of $\mathbf{Q}$ by $NearArea_i$, which is $\odot_1$ for $i = 1$, or $\odot_i/\odot_{i-1}$ for $i > 1$, which represents the *non-overlapping area* between $\odot_i$ and $\odot_{i-1}$.

We derive $QCSet$ from $NearArea_i$ instead of $\odot_i$ to avoid duplicates of codes. Given $QCSet$, the enclave computes its position set $PosSet$ (with $L + 1$ keyed hash functions). Starting from the root, the server performs set-membership tests against IBFs in the index Γ using $PosSet$. The retrieved leaf nodes are added to an intermediate result set R_C, which is then added into $\mathcal{C}$, *i.e.*, $\mathcal{C} \cup R_C$. Also, the server prunes the points that are not in $\mathcal{T}$ by computing the intersection.

The enclave decrypts and refines $\mathcal{C}$ to $\mathcal{C}'$ by removing false positives not in $\odot_i$ (lines 13–15). The removed points are retained in $\mathcal{C}$ since they may be NN in the future. If $\mathcal{C}'$ is not empty, the enclave finds $\mathbf{P}^*$ nearest to $\mathbf{Q}$ as a skyline

---

[4] Associated files are encrypted under another SKE key, generated by the data owner and shared with authorized users. The server (or SGX) cannot decrypt these files.

Algorithm 1. Search Protocol

Input: An authorized user contributes the key K and a skyline query $\mathbf{Q}$.
The server contributes an encrypted database EDB and a secure index Γ.
Output: A result set $\mathcal{R}$ (with associated files).

User:
1: Encrypt $\mathbf{Q}$ under k_2 and send it to server;
Server:
2: Initialize to-do list $\mathcal{T}$ as the entire database, candidate set $\mathcal{C}$ as $\emptyset$, and result set $\mathcal{R}$ as $\emptyset$;
Enclave:
3: $i := 0$;   ▷*an increasing counter*
4: Load and decrypt encrypted $\mathbf{Q}$ using k_2;
Enclave:
5: **while** $\mathcal{T} \neq \emptyset$ **do**
6:     Identify $\odot_i$ and $NearArea_i$ of $\mathbf{Q}$;
7:     Set $\mathbf{Q}_{r_{NN}} := NearArea_i$;
8:     Derive $QCSet(\mathbf{Q}_{r_{NN}})$ using k_1;
9:     Derive $PosSet(\mathbf{Q}_{r_{NN}})$ using $\{hk_i\}_{i=1}^{L+1}$;
10:     Send $PosSet(\mathbf{Q}_{r_{NN}})$ to server;
Server:
11:     Search on Γ using $PosSet(\mathbf{Q}_{r_{NN}})$;
        Let the search result be $R_{\mathcal{C}}$.
12:     Update $\mathcal{C} := (\mathcal{C} \cup R_{\mathcal{C}}) \cap \mathcal{T}$;
Enclave:
13:     Load and decrypt $\mathcal{C}$ using k_2;
14:     Set $\mathcal{C}' := \{\mathbf{P}|(\mathbf{P} \in \mathcal{C}) \wedge (\mathbf{P} \in \mathbf{Q}_{r_{NN}})\}$;
            ▷*remove false positives in $\mathcal{C}$*
15:     Update $\mathcal{C} := \mathcal{C} \setminus \mathcal{C}'$ and send $\mathcal{C}$ to server;

16:     **while** $\mathcal{C}' \neq \emptyset$ **do**
17:         Select the NN $\mathbf{P}^*$ of $\mathbf{Q}$ from $\mathcal{C}'$;
18:         Set $\mathbf{Q}_{r_{ND}} := NonDominArea_{\mathbf{Q}}(\mathbf{P}^*)$;
19:         Derive $QCSet(\mathbf{Q}_{r_{ND}})$ using k_1;
20:         Derive $PosSet(\mathbf{Q}_{r_{ND}})$ using $\{hk_i\}_{i=1}^{L+1}$;
21:         Set $\mathcal{C}' := \mathcal{C}' \setminus \{\mathbf{P}^*\}$;
22:         Send $PosSet(\mathbf{Q}_{r_{ND}})$, $\mathbf{P}^*$ to server;
Server:
23:         Set $\mathcal{R} := \mathcal{R} \cup \{\mathbf{P}^*\}$ and $\mathcal{T} := \mathcal{T} \setminus \{\mathbf{P}^*\}$;
24:         Search on Γ using $PosSet(\mathbf{Q}_{r_{ND}})$;
            Let the search result be $R_{\mathcal{T}}$.
25:         Update $\mathcal{T} := \mathcal{T} \cap R_{\mathcal{T}}$;
Enclave:
26:         Load and decrypt $\mathcal{T}$ using k_2;
27:         Set $\mathcal{T} := \{\mathbf{P}|(\mathbf{P} \in \mathcal{T}) \wedge (\mathbf{P} \in \mathbf{Q}_{r_{ND}})\}$;
                ▷*remove false positives in $\mathcal{T}$*
28:         Update $\mathcal{C}' := \mathcal{C}' \cap \mathcal{T}$;
29:         Send $\mathcal{T}$ to server as its new $\mathcal{T}$;
30:     **end while**
31:     Compute $i := i + 1$;
32: **end while**
Server:
33: Return $\mathcal{R}$ with the encrypted files to user;

point (lines 16–17) and forwards the ciphertext of $\mathbf{P}^*$ to the server, which is added into $\mathcal{R}$ and removed from $\mathcal{T}$ (line 23). Otherwise, the enclave increases the counter by 1, resulting in a larger hyper-sphere for the next round (line 31).

Finding Non-dominated Points. We exploit the range query scheme to find points that are not dominated by $\mathbf{P}^*$. The result is then used to update $\mathcal{T}$ (lines 18–29).

Without the space mapping, we note that the area containing non-dominated points can be represented in the original space. For easy presentation, we assume 2-D points in the following. For the query $\mathbf{Q} = (\mathbf{Q}[1], \mathbf{Q}[2])$ and the skyline point $\mathbf{P}^* = (\mathbf{P}^*[1], \mathbf{P}^*[2])$, we denote the area containing non-dominated points by $NonDominArea_{\mathbf{Q}}(\mathbf{P}^*)$. The symmetric points of $\mathbf{P}^*$ against axes (with $\mathbf{Q}$ as the origin) are computed as $(2\mathbf{Q}[1] - \mathbf{P}^*[1], \mathbf{P}^*[2])$, $(\mathbf{P}^*[1], 2\mathbf{Q}[2] - \mathbf{P}^*[2])$, and $(2\mathbf{Q}[1] - \mathbf{P}^*[1], 2\mathbf{Q}[2] - \mathbf{P}^*[2])$. For the example in Fig. 6, the shadowed cross determined by the four points is indeed $NonDominArea_{\mathbf{Q}}(\mathbf{P}_1)$. The 2-D case can be easily generalized to the d-dimensional case by first computing all $(2^d - 1)$ symmetric points via exhausting all possibilities for each coordinate being $\mathbf{P}^*[i]$ or $2\mathbf{Q}[i] - \mathbf{P}^*[i]$ for $i \in [d]$. $NonDominArea_{\mathbf{Q}}(\mathbf{P}^*)$ is decided by these 2^d points.

Based on $NonDominArea_{\mathbf{Q}}(\mathbf{P}^*)$, another multi-dimensional range query is issued; namely, the enclave adds HRE codes of hyper-rectangles fully covering $NonDominArea_{\mathbf{Q}}(\mathbf{P}^*)$ to $QCSet$ and derives its position set $PosSet$ accordingly (line 20). Given $PosSet$, the server performs set-membership tests over the encrypted index Γ, yielding a result set $R_{\mathcal{T}}$ of retrieved leaves (which contains false positives). The server also updates $\mathcal{T}$ to its intersection with $R_\mathcal{T}$. To eliminate false positives in $\mathcal{T}$, the enclave loads it for decryption and removes points that are not in $NonDominArea_\mathbf{Q}(\mathbf{P}^*)$. It also refines $\mathcal{C}'$ by removing points not in the updated $\mathcal{T}$ for finding NN as the next skyline point. The updated $\mathcal{T}$ is output to the server, and the iterative search terminates until it becomes empty.

Optimizations. Observing that the size of $\mathcal{T}$ reduces with the increasing number of iterations, we can set an extra threshold (depending on the memory limit of SGX) for better performance. When $\mathcal{T}$ gets smaller than the threshold, the enclave performs the remaining iterations without interacting with the server (via range queries). In most cases, only a few iterations can result in a small enough $\mathcal{T}$ to be processed inside the enclave (as also validated in our experiments). With this, our scheme shows good scalability to a large-scale database. Another positive implication is less information (*e.g.*, access pattern on tree nodes) leaked due to fewer range queries executed at the server. It is worth noting that our scheme is also highly parallelizable. We can optimize it with multi-thread techniques where each thread performs a traversal from the root to a leaf node separately.

An Illustrative Example. With reference to Fig. 1, to find NN of $\mathbf{Q}$, the enclave issues a range query for $NearArea_1$, *i.e.*, the shadowed circle with radius ρ in Fig. 5a. Its $QCSet$ is then derived from the shaded rectangles fully covering the circle. After this query, the server gets $R_\mathcal{C} = \{\mathbf{P}_1\}$ and adds it to $\mathcal{C}$. Since $\mathbf{P}_1$ does not fall in sphere $\odot_1$, we keep it in $\mathcal{C}$, and the enclave obtains an empty $\mathcal{C}'$.

The enclave then continues to issue another query for $NearArea_2$ with its $QCSet$ derived from rectangles fully covering $NearArea_2$ (in Fig. 5b). The server gets a new $R_\mathcal{C} = \{\mathbf{P}_3, \mathbf{P}_4\}$ and updates the old $\mathcal{C}$ to $\{\mathbf{P}_1, \mathbf{P}_3, \mathbf{P}_4\}$. After removing points not in $\odot_2$, the enclave selects $\mathbf{P}_1$ (from $\mathcal{C}' = \{\mathbf{P}_1\}$), which is closest to $\mathbf{Q}$, as a skyline point. $\mathbf{P}_1$ is added into $\mathcal{R}$ and removed from $\mathcal{T}$.

Given $\mathbf{P}_1$, the points not dominated by it will be found via a range query. The enclave derives $QCSet$ from shaded rectangles fully covering the shadowed cross of $NonDominArea_\mathbf{Q}(\mathbf{P}_1)$ determined by $\{\mathbf{P}_1, \mathbf{P}_1', \mathbf{P}_1'', \mathbf{P}_1'''\}$ in Fig. 6. The server uses the query result $R_\mathcal{T} = \{\mathbf{P}_3, \mathbf{P}_8\}$ to update $\mathcal{T}$ by computing their intersection. The enclave loads the updated $\mathcal{T}$ for pruning the false-positive point $\mathbf{P}_3$ (with only $\mathbf{P}_8$ left in $\mathcal{T}$). The enclave updates $\mathcal{C}'$ and outputs $\mathcal{T} = \{\mathbf{P}_8\}$ to the server. These procedures repeat. $\mathbf{P}_8$ is then added into $\mathcal{R}$. Since no point exists in $\mathcal{T}$, the algorithm terminates and returns $\mathcal{R} = \{\mathbf{P}_1, \mathbf{P}_8\}$ as the skyline of $\mathbf{Q}$.

Remark. Our scheme can support insertion or removal of data points. We need to maintain an extra list at the server to manage leaves not assigned with data points. To add a point, a leaf is chosen from the list, and the path from the root to the chosen leaf is updated as in building the index. To delete a point, the server just marks the corresponding leaf as "free" and then inserts it to the list.

4.3 Security Analysis

EDB (of encrypted data points and files) and query points accessed by the server are all encrypted by an SKE scheme that is indistinguishable against adaptive chosen-ciphertext attack (IND-CCA secure). The integrity and the confidentiality of data and codes are protected inside SGX. The security thus boils down to the secure index Γ and the range query scheme executed at the server. For efficiency, our scheme tolerates reasonable leakages defined as below.

– $\mathcal{L}^{\mathsf{Idx}}(\mathsf{DB})$ for setup: Given the database DB as input, this function outputs the number of data points n, the number of dimensions d, the bit length of IBF g, and the tree structure. We assume that for two databases with the same n, the tree structure is fixed and hence achieves indistinguishability.
– $\mathcal{L}^{\mathsf{Srch}}(\mathsf{DB}, \mathbf{Q})$ for skyline query: Given the database DB and a skyline query $\mathbf{Q}$ as input, this function outputs *search pattern* $\mathsf{SP}(\mathbf{Q})$ and *access pattern* $\mathsf{AP}(\mathbf{Q})$, both of which are defined below.

The search-pattern $\mathsf{SP}(\mathbf{Q})$ reveals whether a query has appeared before. Since the point coordinates are encrypted by an IND-CCA secure scheme, even the same plaintext yields a different ciphertext. Therefore, $\mathsf{SP}(\mathbf{Q})$ only contains the search pattern of a series of range queries derived from $\mathbf{Q}$, *i.e.*, the repetition of positions computed by L hash functions on *NearArea* or *NonDominArea* of $\mathbf{Q}$.

Compared to existing SSE schemes [6], our access pattern additionally covers the interactions between the server and the enclave. Concretely, $\mathsf{AP}(\mathbf{Q})$ has two parts: (1) a set of identifiers of skyline points (as the result set leaked in typical SSE); (2) the extra interactions which contain multiple sets of identifiers of leaf nodes (*i.e.*, data points) touched by the range queries. The accessed paths on the secure index Γ are also uniquely determined given the second part of $\mathsf{AP}(\mathbf{Q})$.

These two patterns suffice for simulating the ideal experiment[5].

Theorem 1. *Assuming tree nodes (on the index) are indistinguishable (due to IBFs) and H is a hash function modeled as a random oracle outputting a single bit, our scheme is $(\mathcal{L}^{\mathsf{Idx}}, \mathcal{L}^{\mathsf{Srch}})$-adaptively-secure.*

5 Experiments

We implemented our scheme to evaluate its efficiency in practice. To our best knowledge, our implementation is the first one to handle secure skyline queries over datasets of up to one million data points. Another highlight from the results shows that our scheme is orders of magnitude faster than the prior art [16].

Our experiments operate on an SGX-enabled machine equipped with an Intel Core i5-6500 3.20 GHz CPU and 8 GB RAM. We implement the algorithms in C/C++ with the latest version of Intel SGX SDK. We instantiate all the cryptographic primitives with SGX-SSL library (adapted from OpenSSL): AES-128 for SKE, and HMAC-SHA256 for PRF and keyed hash functions.

---

[5] Due to the page limit, we defer proof of Theorem 1 to the full version of this paper.

Table 1. Characteristics of datasets

Name	Type	Dimension	# of points
Indep/Corr/Anti	Synthetic	2 – 4	32.77K–1.05M
CLBP	Real-world	3	19.57K

Table 2. Performance of database encryption and indexing

Dataset	Dimension	# of points	Time (s)		Storage (GB)	
			Γ	EDB	Γ	EDB (est.)
Indep	2	1.05M	372.95	1395.92	3.74	15.36
	3		464.23			
	4		563.45			
CLBP	3	19.57K	8.67	33.06	0.12	0.29

Our experiments are evaluated over both real-world and synthetic datasets. Their main characteristics are summarized in Table 1. We synthesize three classical categories of datasets: Independent (Indep), Correlated (Corr), and Anti-correlated (Anti), where each dimension ranges from 0 to 1K. A description of these datasets can be found in the seminal work [3]. We also use the Cuff-Less Blood Pressure (CLBP) estimation dataset, where values on each dimension are scaled up appropriately for compatibility. For both synthetic and real-world datasets[6], we generate artificial text of size around 15.36 KB on average as the electronic medical record (EMR) and randomly assign it to each data point.

Table 2 reports the time and storage consumptions for constructing the secure index Γ and encrypting the database to EDB. Since the performance of this phase is independent of the data point distribution, we only take Indep and CLBP datasets for evaluation. For Indep with over one million EMRs, it takes about 23.27 min for obtaining EDB (of AES-encrypted EMRs), dominating the one-time initialization phase. The time costs of index construction range from 6.22 to 9.39 min, which are affected by the number of dimensions when generating HRE codes of each data point. The storage costs for Γ and EDB are 3.74 GB and 15.36 GB, respectively. For the CLBP dataset, both time and storage costs are extremely efficient due to the smaller dataset size. Besides, building the tree-based index and encrypting EMRs by AES are highly-parallelizable, we can use a cluster of machines to reduce the initialization time considerably.

For each skyline query, we sample the query point uniformly at random from the data space. The (total) query time counts from the start of issuing a query to the end of decrypting all the skyline points (at the client). Figure 7 shows the query performance evaluated on various datasets.

---

[6] CLBP: https://archive.ics.uci.edu/ml/datasets.php, EMR: http://www.emrbots. org.

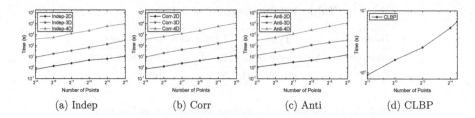

(a) Indep	(b) Corr	(c) Anti	(d) CLBP

Fig. 7. Query time of dynamic skyline

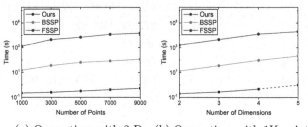

(a) Query time with 2-D   (b) Query time with 1K points

Fig. 8. Comparisons to the state-of-the-art

For query time, our experiment over synthetic datasets (Figs. 7a, 7b, and 7c) validated that it increases with the number of dimensions and data points. Given a common use case of 2-D data (*e.g.*, spatial data), the averaged query times for all three synthetic datasets with one million points are about 10s, which is extremely fast. It may take some hours to handle a query for 4-D data, yet, we can optimize it via multi-threading as discussed. Also, the distribution of datasets affects the query performance, *i.e.*, the query times for Anti datasets are slightly slower than those of other datasets. Figure 7d shows the query times of the CLBP dataset (with different scales), which are all within 10 s.

Comparison to the State-of-the-Art. To highlight the query time improvement, we compare ours to the best solutions [16] to the best of our knowledge: Basic Secure Skyline Protocol (BSSP) and Fully Secure Skyline Protocol (FSSP)[7], We test over Indep with the same configurations as the existing experiments [16]: the number of points varies from 1K to 9K with the dimension of the points fixed to 2-D, and the number of dimensions varies from 2 to 5 for 1K data points.

Figure 8 shows the query time gains compared to the prior work. For searching over 2-D data points (Fig. 8a), our savings are up to 200× (against BSSP) and up to 27 000× (against FSSP). For searching over 1K data points, ours achieves a speedup of up to 200× (against BSSP) and 28 000× (against FSSP).

---

[7] The extended version of [16] studies optimizations of BSSP and FSSP, which discusses the influence of parameters separately rather than providing a complete optimized scheme. We thus focus on the comparisons with regular BSSP and FSSP.

The main reason for such a performance gap is due to the dependency on heavy cryptographic sub-protocols in BSSP and FSSP. The cost of these sub-protocols repeats for every single query. In contrast, our solution builds an index in advance, this one-time set-up cost is still significantly less than that of the setup phases for BSSP and FSSP, which involve costly public-key primitives.

6 Conclusion

In this paper, we propose an efficient SGX-aided approach for secure skyline queries using the arguably most basic cryptographic primitives. The major challenge is that SGX only supports at most 128 MB of memory, which obviously fails to handle a large-scale (encrypted) database due to the I/O bottleneck (and the high cost for page swap). Our idea is to use a lightweight, yet secure multi-dimensional range query scheme, to shortlist candidates on-the-fly. The security of our solution is proven under precisely-defined leakages. We implement and evaluate our scheme on both synthetic and real-world datasets, confirming its remarkable efficiency gain compared to the state-of-the-art. Our scheme is the first-of-its-kind to be practical even for a database with millions of data points.

References

1. Bindschaedler, V., Grubbs, P., Cash, D., Ristenpart, T., Shmatikov, V.: The Tao of inference in privacy-protected databases. PVLDB **11**(11), 1715–1728 (2018)
2. Bloom, B.H.: Space/time trade-offs in hash coding with allowable errors. Commun. ACM **13**(7), 422–426 (1970)
3. Börzsönyi, S., Kossmann, D., Stocker, K.: The skyline operator. In: ICDE, pp. 421–430 (2001)
4. Bothe, S., Karras, P., Vlachou, A.: eSkyline: processing skyline queries over encrypted data. PVLDB **6**(12), 1338–1341 (2013)
5. Cash, D., Liu, F.-H., O'Neill, A., Zhandry, M., Zhang, C.: Parameter-hiding order revealing encryption. In: Peyrin, T., Galbraith, S. (eds.) ASIACRYPT 2018. LNCS, vol. 11272, pp. 181–210. Springer, Cham (2018). https://doi.org/10.1007/978-3-030-03326-2_7
6. Curtmola, R., Garay, J., Kamara, S., Ostrovsky, R.: Searchable symmetric encryption: improved definitions and efficient constructions. In: CCS, pp. 79–88. ACM (2006)
7. Du, M., Wang, Q., He, M., Weng, J.: Privacy-preserving indexing and query processing for secure dynamic cloud storage. IEEE TIFS **13**(9), 2320–2332 (2018)
8. Elmehdwi, Y., Samanthula, B.K., Jiang, W.: Secure k-nearest neighbor query over encrypted data in outsourced environments. In: ICDE, pp. 664–675 (2014)
9. Hu, P., Chow, S.S.M., Aloufi, A.: Geosocial query with user-controlled privacy. In: WiSec, pp. 163–172. ACM (2017)
10. Kamara, S., Papamanthou, C., Roeder, T.: Dynamic searchable symmetric encryption. In: CCS, pp. 965–976. ACM (2012)
11. Kossmann, D., Ramsak, F., Rost, S.: Shooting stars in the sky: an online algorithm for skyline queries. In: VLDB, pp. 275–286 (2002)

12. Lai, R.W.F., Chow, S.S.M.: Structured encryption with non-interactive updates and parallel traversal. In: ICDCS, pp. 776–777 (2015)
13. Lai, R.W.F., Chow, S.S.M.: Parallel and dynamic structured encryption. In: Deng, R., Weng, J., Ren, K., Yegneswaran, V. (eds.) SecureComm 2016. LNICST, vol. 198, pp. 219–238. Springer, Cham (2017). https://doi.org/10.1007/978-3-319-59608-2_12
14. Lai, R.W.F., Chow, S.S.M.: Forward-secure searchable encryption on labeled bipartite graphs. In: Gollmann, D., Miyaji, A., Kikuchi, H. (eds.) ACNS 2017. LNCS, vol. 10355, pp. 478–497. Springer, Cham (2017). https://doi.org/10.1007/978-3-319-61204-1_24
15. Li, R., Liu, A.X.: Adaptively secure conjunctive query processing over encrypted data for cloud computing. In: ICDE, pp. 697–708 (2017)
16. Liu, J., Yang, J., Xiong, L., Pei, J.: Secure skyline queries on cloud platform. In: ICDE, pp. 633–644 (2017)
17. Liu, X., Choo, K.R., Deng, R.H., Yang, Y., Zhang, Y.: PUSC: privacy-preserving user-centric skyline computation over multiple encrypted domains. In: TrustCom, pp. 958–963 (2018)
18. Papadias, D., Tao, Y., Fu, G., Seeger, B.: An optimal and progressive algorithm for skyline queries. In: SIGMOD, pp. 467–478 (2003)
19. Wang, B., Li, M., Chow, S.S.M., Li, H.: A tale of two clouds: computing on data encrypted under multiple keys. In: CNS, pp. 337–345. IEEE (2014)
20. Wang, J., Chow, S.S.M.: Forward and backward-secure range-searchable symmetric encryption. IACR ePrint Archive: 2019/497 (2019)
21. Wu, S., Li, Q., Li, G., Yuan, D., Yuan, X., Wang, C.: ServeDB: secure, verifiable, and efficient range queries on outsourced database. In: ICDE, pp. 626–637 (2019)
22. Zheng, Y., Lu, R., Li, B., Shao, J., Yang, H., Choo, K.R.: Efficient privacy-preserving data merging and skyline computation over multi-source encrypted data. Inf. Sci. **498**, 91–105 (2019)
23. Zhu, H., Meng, X., Kollios, G.: Privacy preserving similarity evaluation of time series data. In: EDBT, pp. 499–510 (2014)

Increasing the Efficiency of GPU Bitmap Index Query Processing

Brandon Tran[1], Brennan Schaffner[1], Jason Sawin[1(✉)], Joseph M. Myre[1],
and David Chiu[2]

[1] Department of Computer and Information Sciences,
University of St. Thomas, St. Paul, MN, USA
jason.sawin@stthomas.edu
[2] Department of Mathematics and Computer Science,
University of Puget Sound, Tacoma, WA, USA

Abstract. Once exotic, computational accelerators are now commonly available in many computing systems. Graphics processing units (GPUs) are perhaps the most frequently encountered computational accelerators. Recent work has shown that GPUs are beneficial when analyzing massive data sets. Specifically related to this study, it has been demonstrated that GPUs can significantly reduce the query processing time of database bitmap index queries. Bitmap indices are typically used for large, read-only data sets and are often compressed using some form of hybrid run-length compression.

In this paper, we present three GPU algorithm enhancement strategies for executing queries of bitmap indices compressed using Word Aligned Hybrid compression: 1) data structure reuse 2) metadata creation with various type alignment and 3) a preallocated memory pool. The data structure reuse greatly reduces the number of costly memory system calls. The use of metadata exploits the immutable nature of bitmaps to pre-calculate and store necessary intermediate processing results. This metadata reduces the number of required query-time processing steps. Preallocating a memory pool can reduce or entirely remove the overhead of memory operations during query processing. Our empirical study showed that performing a combination of these strategies can achieve 33× to 113× speedup over the unenhanced implementation.

Keywords: Bitmap indices · Big data · Query processing · GPU

1 Introduction

Modern companies rely on big data to drive their business decisions [9,11,18]. A prime example of the new corporate reliance on data is Starbucks, which uses big data to determine where to open stores, target customer recommendations, and menu updates [17]. The coffee company even uses weather data to adjust its digital advertisement copy [5]. To meet this need, companies are collecting astounding amounts of data. The shipping company UPS stores over 16 petabytes of data to meet their business needs [9]. Of course, large repositories

© Springer Nature Switzerland AG 2020
Y. Nah et al. (Eds.): DASFAA 2020, LNCS 12114, pp. 339–355, 2020.
https://doi.org/10.1007/978-3-030-59419-0_21

of data are only useful if they can be analyzed in a timely and efficient manner. In this paper we present techniques that take advantage of synergies between hardware and software to speed up the analysis of data.

Indexing is one of the commonly-used software techniques to aid in the efficient retrieval of data. A bitmap index is a binary matrix that approximates the underlying data. They are regularly used to increase query-processing efficiency in data warehouses and scientific data. It has been shown that bitmap indices are efficient for some of the most common query types: point, range, joins, and aggregate queries. They can also perform better than other indexing schemes like B-trees [26]. One of the main advantages of bitmap indices is that they can be queried using hardware-enabled bitwise operators. Additionally, there is a significant body of work that explores methods of compressing sparse bitmap indices [6,8,10,12,22,24]. The focus of most compression work is on various forms of hybrid run-length encoding schemes. These schemes not only achieve substantial compression, but the compressed indices they generate can be queried directly, bypassing the overhead of decompression. One such commonly used compression scheme is Word Aligned Hybrid (WAH) [24]. To improve query processing the WAH scheme compresses data to align with CPU word size.

One of the oft-cited shortcomings of bitmap indices is their static nature. Once a bitmap is compressed, there is no easy method to update or delete tuples in the index. For this reason, bitmap indices are most commonly used for read-only data sets. However, the immutable nature of bitmaps can be exploited to increase the efficiency of query algorithms. Specifically, as bitmap indices are not often updated, it is relatively cheap to build and maintain metadata that can be used to aid in query processing. Additionally, static data structures can be preallocated to reduce query processing overhead.

Meanwhile, recent work has shown how graphics processing units (GPUs) can exploit data-level parallelism inherent in bitmap indices to significantly reduce query processing time. GPUs are massively-parallel computational accelerators that are now standard augmentations to many computing systems. Previously, Andrezejewski and Wrembel [1] proposed GPU-WAH, a system that processes WAH compressed bitmap indices on the GPU. To fully realize the data parallel potential inherent in bitmaps, GPU-WAH must first decompress the bitmap. Nelson *et al.* extended GPU-WAH so that it could process range queries [19]. Nelson *et al.* demonstrated that tailoring the range query algorithm to the unique GPU memory architecture can produce significant improvements (an average speedup of $1.48\times$ over the naive GPU approach and $30.22\times$ over a parallel CPU algorithm).

In this paper, we explore techniques that use metadata, data structure reuse and preallocation tailored to speed up processing WAH range queries on GPUs.

The specific contributions of this paper are:

- We describe how reusing data structures in GPU-WAH decompression algorithm can reduce the number of synchronized memory calls by over 50%.

– We present a tiered investigation of ways to incorporate precompiled meta-data into the processing of WAH queries on the GPU. Each successive tier reduces the amount of work performed by the GPU-WAH decompression algorithm but increases the memory overhead. Additionally, we explore how data type selection can align our algorithms to the architecture of the GPU.
– We present a technique that exploits the static nature of bitmap indices to create a fixed size memory pool. The pool is used to avoid all synchronous dynamic memory allocation at query time.
– We present an empirical study of our proposed enhancements to the GPU-WAH decompression algorithm applied to both real and synthetic data sets. Our experimental results show that an implementation using both metadata and a static memory pool can achieve an average speedup of 75.43× over an unenhanced version of GPU-WAH.

The remainder of the paper is organized as follows. In Sect. 2, we provide an overview of bitmap indices and WAH compression. Section 3 describes a procedure for executing WAH range queries on the GPU. Section 4 describes our enhancement strategies. We present our methodology in Sect. 5, our results in Sect. 6 and discuss the results in Sect. 7. We briefly describe related works a in Sect. 8. We conclude and present future work in Sect. 9.

2   Bitmap Indices and WAH Compression

In this section, we describe the creation of bitmap indices and the WAH compression algorithm. A bitmap index is created by discretizing a relation's attribute values into bins that represent distinct values or value-ranges. Table 1 shows a relation and a corresponding bitmap index. The table above shows a possible bitmap for the **Stocks** relation to its left. The s_i columns in the bitmap are the bins used to represent the **Symbol** attribute. As stock symbols are distinct values, each value is assigned a bin (*e.g.*, s_0 represents the value GE, s_1 represents WFC, and so on). The p_j bins represent ranges of values into which **Price** values can fall. p_0 represents the range $[0, 50)$, p_1 denotes $[50, 100)$, p_2 is $[100, 150)$, and p_3 represents $[150, \infty)$.

Table 1. Example relation (left) and a corresponding bitmap (right).

Stocks		Symbol Bins						Price Bins			
Symbol	**Price**	s_0	s_1	s_2	s_3	s_4	s_5	p_0	p_1	p_2	p_3
GE	11.27	1	0	0	0	0	0	1	0	0	0
WFC	54.46	0	1	0	0	0	0	0	1	0	0
M	15.32	0	0	1	0	0	0	1	0	0	0
DIS	151.58	0	0	0	1	0	0	0	0	0	1
V	184.51	0	0	0	0	1	0	0	0	0	1
CVX	117.13	0	0	0	0	0	1	0	0	1	0

Consider the first tuple in the **Stocks** relation (Table 1). This tuple's **Symbol** value is GE, and thus in the bitmap a 1 is placed in s_0 and all other s bins are set to 0. The **Price** value is 11.27. This value falls into the $[0, 50)$ range, so a 1 is assigned to the p_0 bin, and all other p bins get 0.

The binary representation of a bitmap index means that hardware primitive bitwise operations can be used to process queries. For example, consider the following query: SELECT * FROM Stocks WHERE Price>60;. This query can be processed by solving $p_2 \vee p_2 \vee p_3 = res$. Only the rows in res that contain a 1 corresponds to a tuple that should be retrieved from disk for further processing. WAH compression operates on stand-alone bitmap bins (which are also referred to as *bit vectors*). An example WAH compression of 252-bits is shown in Figures 1(a) and (b). Assuming a 64-bit architecture, WAH clusters a bit vector into consecutive *(system word length)* − 1 (or 63) bit "chunks." In Fig. 1(a) the first chunk is heterogeneous and the remaining 3 chunks are homogeneous.

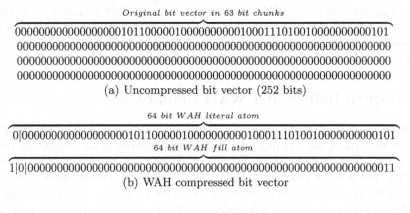

Original bit vector in 63 bit chunks

```
000000000000000000101100000100000000000100011101001000000000101
000000000000000000000000000000000000000000000000000000000000000
000000000000000000000000000000000000000000000000000000000000000
000000000000000000000000000000000000000000000000000000000000000
```
(a) Uncompressed bit vector (252 bits)

64 bit WAH literal atom

```
0|000000000000000000101100000100000000000100011101001000000000101
```
64 bit WAH fill atom

```
1|0|00000000000000000000000000000000000000000000000000000000000011
```
(b) WAH compressed bit vector

Fig. 1. An example of WAH compression.

WAH then encodes each chunk into system word sized (64-bit) atoms. Heterogeneous chunks are encoded into *literal atoms* of the form *(flag, lit)*, where the most-significant-bit (MSB), or *flag*, is set to zero to indicate a *literal*. The remaining 63-bits record the original heterogeneous chunk from the bit vector. The first chunk in Fig. 1(a) is heterogeneous and encoded into a *literal* atom.

Homogeneous chunks are encoded as *fill* atoms of the form *(flag, val, len)*, where the MSB *(flag)* is set to 1 to indicate a *fill* and the second-MSB *(val)* records the value of the homogeneous sequence of bits. The remaining 62-bits *(len)* record the run length of identical chunks in the original bit vector. The last three chunks in Fig. 1(a) are homogeneous and are encoded into a *fill* atom, where the *val* bit is set to 0 and the *len* field is set to 3 (as there are three consecutive repetitions of the homogeneous chunk).

3   GPU Processing of WAH Range Queries

WAH compressed bitmaps can be queried directly without the need for decompression. It has been shown that the system word alignment used by WAH can lead to faster querying than other compression schemes [23]. However, this approach is tailored to the CPU. Previous work [19] has shown that GPUs can process range queries even faster.

Figure 2 illustrates the execution steps used in [19] to process a range query on the GPU. Initially, the compressed bit vectors are stored on the GPU. When the GPU receives a query, the required bit vectors are sent to the **Decompressor**. The decompressed columns are then sent to the **Query Engine** where the query is processed, and the result is sent to the CPU.

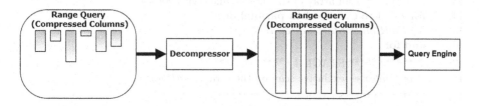

Fig. 2. Main components used to process WAH range queries on a GPU.

Using NVIDIA's compute unified device architecture (CUDA), Nelson et al. [19] presented three parallel reduction-based methods for the query engine: column-oriented access (COA), row-oriented access (ROA), and a hybrid approach. COA performs the reduction on columns, and ROA performs the reduction across single rows. In the hybrid approach, GPU threads are grouped into blocks, and thread blocks are tiled into grids to cover the query data. The blocks then perform a reduction on their data. This approach makes the most efficient use of the GPU memory system. Specifically, it utilizes both the coalesced memory accesses of COA and the use of shared memory for processing along rows of ROA; the hybrid was found to be the fastest method in their experimental study. For the remainder of the paper, we will only be considering the hybrid approach for the query engine though our improvements would benefit all approaches.

The work of this paper focuses on the decompressor component of the above approach. Algorithm 1 presents a procedure for the decompressor unit. It was designed by Andrezejewski and Wrembel [1] and modified in [19] to decompress multiple columns in parallel. The input to the algorithm is a compressed bit vector, *CompData*, the size of the compressed data, *CSize*, and the size of the decompressed data, *DSize*. The output is the corresponding decompressed bit vector, *DecompData*. The algorithm itself comprises five stages; the stages execute sequentially, but the work within stages is processed in parallel.

Stage 1 (lines 2–9) generates an array *DecompSizes* which has the same number of elements as *CompData*. At the end of **Stage 1**, each element in

---

Algorithm 1. Parallel decompression of compressed data

1: **procedure** DECOMP(*Compressed_BitVector CompData,CSize,DSize*)
2:        ************* STAGE 1 *************
3:        **for** $i \leftarrow 0$ to $CSize - 1$ **in parallel do**
4:            **if** $CompData(i)_{63} = 0b$ **then**
5:                $DecompSizes[i] \leftarrow 1$
6:            **else**
7:                $DecompSizes[i] \leftarrow$ the value of *len* encoded on bits $CompData(i)_{0 \rightarrow 61}$
8:            **end if**
9:        **end for**
10:       ************* STAGE 2 *************
11:       $StartingPoints \leftarrow$ exclusive scan on the array $DecompSizes$
12:       ************* STAGE 3 *************
13:       $EndPoints$ is an array of size $DSize$ filled with zeroes
14:       **for** $i \leftarrow 1$ to $CSize - 1$ **in parallel do**
15:           $EndPoints[StartingPoints[i] - 1] \leftarrow 1$
16:       **end for**
17:       ************* STAGE 4 *************
18:       $WordIndex \leftarrow$ exclusive scan on the array $EndPoints$
19:       ************* STAGE 5 *************
20:       **for** $i \leftarrow 0$ to $DSize - 1$ **in parallel do**
21:           $tempWord \leftarrow CompData[WordIndex[i]]$
22:           **if** $tempWord_{63} = 0b$ **then**
23:               $DecompData[i] \leftarrow tempWord$
24:           **else**
25:               **if** $tempWord_{62} = 0b$ **then**
26:                   $DecompData[i] \leftarrow 0_{64}$
27:               **else**
28:                   $DecompData[i] \leftarrow 0_1 + 1_{63}$
29:               **end if**
30:           **end if**
31:       **end for**
32:       **return** $DecompData$ ▷ contains a decompressed bit vector of $CompData$
33: **end procedure**

$DecompSizes$ will hold the number of words being represented by the atom with the same index in $CompData$. This is accomplished by creating a thread for each atom in $CompData$. If an atom is a literal, its thread assigns 1 to the appropriate index in $DecompSizes$ (line 5). If the atom is a fill, the thread assigns the number of words compressed by the atom (line 7).

Stage 2 (line 11) executes an exclusive scan (parallel element summations) on $DecompSizes$ storing the results in $StartingPoints$. $StartingPoints[i]$ contains the total number of decompressed words compressed into $CompData[0]$ to $CompData[i - 1]$, inclusive. $StartingPoints[i] * 63$ is the number of the bitmap row first represented in $CompData[i]$. Stage 3 (lines 13–16) creates an array of zeros, *EndPoints*. The length of *EndPoints* equals the number of words in the decompressed data. A 1 is assigned to *EndPoints* at the location of $StartingPoints[i] - 1$ for $i < |StartingPoints|$. In essence, each 1 in *EndPoints* represents where a heterogeneous chunk was found in the decompressed data by the WAH compression algorithm. Note that each element of *StartingPoints* can be processed in parallel.

Stage 4 (line 18) performs an exclusive scan over *EndPoints* storing the result in *WordIndex*. *WordIndex[i]* provides the index to the atom in *CompData* that contains the information for the *ith* decompressed word.

Stage 5 (lines 20–31) contains the final for-loop, which represents a parallel processing of every element of *WordIndex*. For each element in *WordIndex*, the associated atom is retrieved from *CompData*, and its type is checked. If *CompData[WordIndex[i]]* is a WAH literal atom (MSB is a zero), then it is placed directly into *DecompData[i]*. Otherwise, *CompData[WordIndex[i]]* must be a fill atom. If it is a fill of zeroes (second MSB is a zero), then 64 zeroes are assigned into *DecompData[i]*. If it is a fill of ones, a word consisting of 1 zero (to account for the flag bit) and 63 ones is assigned to *DecompData[i]*. The resulting *DecompData* is the fully decompressed bitmap.

4   Memory Use Strategies

We explored memory-focused strategies to accelerate GPU query processing: 1) data structure reuse, 2) metadata storage, and 3) employing a preallocated memory pool. Descriptions of each strategy are provided below.

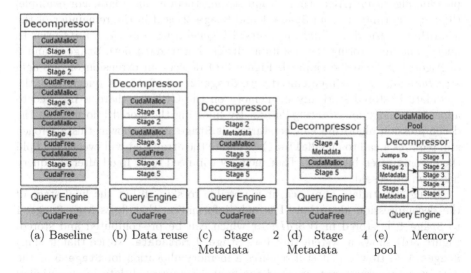

(a) Baseline   (b) Data reuse   (c) Stage 2 Metadata   (d) Stage 4 Metadata   (e) Memory pool

Fig. 3. Various implementations of a GPU-WAH system specialized for range queries.

Figure 3(a) depicts the steps required for our baseline implementation of Algorithm 1. As shown, this implementation requires five `cudaMalloc()` calls and four `cudaFree()` calls in the decompressor and an additional `cudaFree()` after the query engine has finished. Each `cudaMalloc()` is allocating an array needed in the following algorithmic stage. The CUDA library only supports synchronous memory allocation and deallocation. Synchronous memory operations combined with data dependencies in Algorithm 1 make memory operations a limiting factor for decompression.

Data structure reuse- We can reduce the number of CUDA memory calls by reusing data structures. The arrays created in **Stage** 1 and **Stage** 2 of Algorithm 1, *DecompSizes* and *StartingPoints*, are both the length of the compressed data. By performing an in-place exclusive scan on *DecompSizes*, meaning the results of the scan are saved back to *DecompSizes*, we no longer need to create *StartingPoints*. Similarly, we can perform an in-place scan on *EndPoints* in **Stage** 4. Moreover, we can reuse *EndPoints* for *DecompData*. After the data is read from *EndPoints* (line 21), the results of the writes in line 26 and line 28 can be written back to *EndPoints* without loss of data. Figure 3(b) shows the steps of implementation with data structure reuse. As shown, it only requires two calls to `cudaMalloc()`, one before **Stage** 1 and another before **Stage** 2. It requires a call to `cudaFree()` after **Stage** 3 is finished with *DecompSizes* and the final `cudaFree()` after the query engine has finished with the decompressed data saved in *EndPoints*. By careful reuse of data structures, we reduce the number of CUDA memory calls from 10 to 4.

Storing Metadata- Further memory management and even some processing stages can be skipped by pre-generating intermediate results of the decompression algorithm (Algorithm 1) and storing them as metadata. For example, the only information from **Stage** 1 and **Stage** 2 used in the remainder of the algorithm is stored in *StartingPoints*. By generating *StartingPoints* prior to query-time and storing the results as **Stage** 2 **metadata** both **Stage** 1 and 2 of Algorithm 1 can be skipped. Figure 3(c) depicts an extension of our *data structures reuse* system enhanced with **Stage** 2 **metadata**. At query-time, the metadata is stored statically in memory on the GPU, so there is no need to allocate memory for *StartingPoints*. By injecting the stored information, the decompression algorithm can be started at **Stage** 3. As shown, this approach still requires a call to `cudaMalloc()` to create the array that will eventually hold the decompressed bit vector. That memory will need to be freed after the query has been processed.

Using a metadata approach, it is possible to skip all but the final stage of the decompression algorithm. The only information that flows from **Stage** 4 to **Stage** 5 is stored in *WordIndex* which can be pre-computed and stored. Figure 3(d) shows a system that uses **Stage** 4 **metadata**. Notice that it skips **Stages** 1-4. However, it still requires a memory allocation for **Stage** 5 as the data structure reuse system saved the final decompressed data in the original *WordIndex* array. Now *WordIndex* is stored as metadata and overwriting it would slow the performance of subsequent queries as they would no longer have access to stored information. The `cudaMalloc()` call in Fig. 3(d) is allocating memory for a structure that will hold the fully decompressed data. This memory will need to be freed after the query is completed.

Any speedup realized by our metadata approaches is achieved at the cost of a larger memory footprint. To reduce the space requirements of our implementation, we explore the effects of using 32-bit and 64-bit integer types to store `Stage 2` and `Stage 4 metadata`. Our version of the decompression algorithm expected the WAH compression to be aligned with a 64-bit CPU system word size. However, `Stage 2 metadata` contains the total number of decompressed words compressed from $CompData[0]$ to $CompData[i-1]$, for some non-zero index i. The largest possible element is equal to the number of system words comprising the decompressed bit vector. Hence, for decompressed bitmaps containing less than $(2^{32}-1) \times 64$ rows `Stage 2 metadata` can be a 32-bit datatype. Essentially, this type-size reduction would make the `Stage 2 metadata` half the size of the compressed bitmap.

For each decompressed word w in a bit vector, `Stage 4 metadata` stores an index into $CompData$ where w is represented in compressed format. In essence, `Stage 4 metadata` maps decompressed words to their compressed representations. As long as the compressed bit vector does not contain more than $(2^{32}-1)$ atoms, a 32-bit data type can be used for `Stage 4 metadata`. This type reduction makes `Stage 4 metadata` half the size of the decompressed data. Note that storing `Stage 4 metadata` using 64-bit integer types would require the same memory footprint as the fully decompressed bitmap. In this case, it would be advantageous to store just the decompress bitmap and circumvent the entire decompression routine.

Memory Pool- A common approach to avoid the overhead of `cudaMalloc()` and `cudaFree()` is to create a preallocated static memory pool (e.g., [14,21,25]). We create a memory pool tailored to the bitmap that is stored on the GPU. A hashing function maps thread-ids to positions in preallocated arrays. The arrays are sized to accommodate a decompressed bit vector of the bitmap stored on the GPU. Threads lock their portion of the array during processing. The array is released back to the pool at the end of query processing. All available GPU memory that is not being used to store the bitmap and metadata is dedicated to the memory pool. This design will lead to a query failure if the memory requirements are too large. This limitation motivates future work that will explore methods for distributing massive indices across multiple GPUs.

Figure 3(e) shows the design of our fully enhanced GPU-WAH range query system. The use of a memory pool removes the need to invoke CUDA memory calls. As shown, the memory pool can be used in conjunction with both of our metadata strategies to circumvent stages of the decompression algorithm. It can also be used as a standalone solution.

5   Experiments

In this section, we describe the configuration of our testing environment and the process that was used to generate our results. All testing was executed on a machine running Ubuntu 16.04.5 LTS, equipped with dual 8-core Intel Xeon E5-2609 v4 CPUs (each at 1.70 GHz) and 322 GB of RAM. The CPU side of the system was written in C++ and compiled with GCC v5.4.0. The GPU components were developed using CUDA v9.0.176 and run on an NVIDIA GeForce GTX 1080 with 8 GB of memory.

We used the following data sets for evaluation. They are representative of the type of read-only applications (*e.g.*, scientific) that benefit from bitmap indexing.

- BPA – contains measurements reported from 20 synchrophasors (measures magnitude and phase of AC waveform) deployed by Bonneville Power Administration over the Pacific Northwest power grid [4]. Data from each synchrophasors was collected over approximately one month. The data arrived at a rate of 60 measurements per second and was discretized into 1367 bins. We use a 7, 273, 800 row subset of the measured data.
- linkage – contains anonymous records from the Epidemiological Cancer Registry regarding the German state of North Rhine-Westphalia [20]. The data set contains 5, 749, 132 rows and 12 attributes. The 12 attributes were discretized into 130 bins.
- kddcup – contains data obtained from the 1999 Knowledge Discovery and Data Mining competition. These data describe network flow traffic. The set contains 4, 898, 431 rows and 42 attributes [15]. Continuous attributes were discretized into 25 bins using Lloyd's Algorithm [16], resulting in 475 bins.
- Zipf – contains data generated using a Zipf distribution. This is the only synthetic data set on which we tested. A Zipf distribution represents a clustered approach to discretization, which can capture the skew of dense data in a bitmap. With the Zipf distribution generator, the probability of each bit being assigned to 1 is: $P(k, n, skew) = (1/k^{skew})/\sum_{i=1}^{n}(1/i^{skew})$ where n is the number of bins determined by cardinality, k is their rank (bin number: 1 to n), and the parameter $skew$ characterizes the exponential skew of the distribution. Increasing $skew$ increases the likelihood of assigning 1s to bins with lower rank (lower values of k) and decreases the likelihood of assigning 1s to bins with higher rank. We set $n = 10$ and $skew = 2$ for 10 attributes, which generated a data set containing 100 bins (*i.e.,* ten attributes discretized into ten bins each) and 32 million rows. This is the same synthetic data set used in Nelson et al. [19].

We tested multiple configurations of additional enhancement strategies for query execution. These configurations are comprised of three classes of options:

1. Data structure reuse
2. Metadata: None, 32-bit Stage 2, 64-bit Stage 2, 32-bit Stage 4, and fully decompressed columns.
3. Memory pool usage: used or unused.

We tested all valid combinations of these options on each of the four data sets. Due to the mutually exclusive nature of the metadata storage options, this results in 10 augmented configurations plus the baseline approach.

All tests used range queries of sizes 64 columns. To obtain representative execution times for each query configuration we repeated each test 6 times. The execution time of the first test is discarded to remove transient effects, and the arithmetic mean of the remaining 5 execution times is recorded. We used the average to calculate our performance comparison metric, $speedup = t_{base}/t$, where t_{base} is the execution time of the baseline for comparison and t is the execution time of the test of interest. The baseline we used for all speedup calculations was the implementation of the decompression algorithm from [19] (a slightly modified version of the algorithm presented in [1]).

6 Results

Here we present the results obtained from the experiments described in the previous section. We first discuss the impact of memory requirements. We then present results for data structure reuse, metadata, data type size, and memory pool strategies that were described in Sect. 4.

The performance provided by some of the techniques in this paper comes at the cost of additional memory costs, which are shown in Fig. 4. Relative to standard storage requirements, the storage requirements when using 32-bit **Stage 2** metadata, 64-bit **Stage 2 metadata**, and 32-bit **Stage 4 metadata**, are 1.5×,

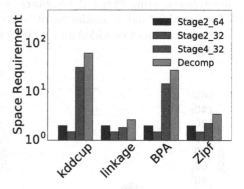

Fig. 4. Metadata memory space requirements relative to the baseline approach storing only compressed bitmaps. Note, the vertical axis is logarithmic.

2×, and an average of 12.8×, respectively.

The speedup provided by reuse of data structures to eliminate memory operations is shown in Fig. 5. Eliminating the overhead of many memory operations enhanced performance by a maximum of 8.53× and 5.43×, on average. The kddcup and BPA data sets exhibited greater speedup than the linkage and Zipf datasets due to the relative compressibility of these data sets and its effect on the decompression routine.

Performance enhancement provided by the use of a memory pool is shown in Fig. 6(a). This enhancement consistently provided an average of 24.4× speedup across all databases and a maximum speedup of 37.0×.

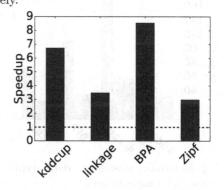

Fig. 5. Speedup provided by data structure reuse.

Incorporating metadata also provided consistent results as can been seen in Fig. 6(b). Using `Stage 2` metadata provided an average of 15.1× speedup. `Stage 4` metadata is more beneficial with an average of 20.5× speedup.

Varying data type size yielded negligible performance enhancement. When a memory pool was not used, as shown in Fig. 6(b), there was no observable performance difference between 32-bit and 64-bit data types. On average, their separation was less than 0.234× speedup. When a memory pool was used, as shown in Fig. 6(c), there was a performance boost when using 32-bit data types with an average improvement of 9.24× over 64-bit types.

Using a combination of metadata, data type size, and memory pool techniques produced the greatest performance benefit, as seen in Fig. 6(c). Across all databases, using `Stage 2 metadata` and a memory pool provided an average 37.7× speedup and a maximum of 58.8× speedup. Using `Stage 4 metadata` and a memory pool provided an average 113× speedup and a maximum of 166× speedup.

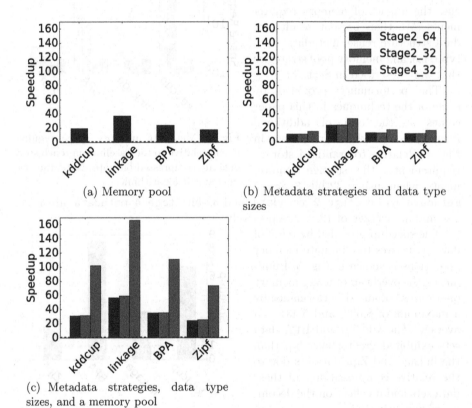

(a) Memory pool

(b) Metadata strategies and data type sizes

(c) Metadata strategies, data type sizes, and a memory pool

Fig. 6. Shown are performance results for a) memory pool usage, b) different metadata strategies and data type sizes, and c) different metadata strategies, data type sizes, and a memory pool. The dashed horizontal line indicates a speedup of 1. Figures b) and c) share a legend.

7   Discussion of Results

The performance provided by data reuse is dependant on the compressibility of the data set. Data sets with greater compressibility exhibit stronger performance relative to those with less compressibility. This is because data sets with less compressibility incur more global memory accesses on the GPU.

Storing the results of the first exclusive scan as 32-bit metadata instead of 64-bit not only saved storage space but also provided faster execution times (23.4% faster, on average). On NVIDIA GPUs, 32-bit integer operations are faster than 64-bit because the integer ALUs are natively 32-bits. 64-bit operations are performed using combinations of 32-bit operations.

When combining metadata and memory pool strategies, the attained speedup was greater than the sum of the speedup of each individual strategy. When only using metadata, the final stage can not begin until the necessary memory is allocated. When only using a memory pool, the final stage can not begin until the subsequent stage is completed. Combining the methods removes both bottlenecks and allows Stage 5 to execute almost immediately.

Although it has the highest storage cost, using fully decompressed columns as metadata reduces execution time because the decompression routine is completely avoided. Figure 7 shows the performance enhancement provided by using fully decompressed columns as "metadata". This option is only reasonable for small databases or GPUs with large storage space. This strategy provided a maximum of 699× speedup and an average of 411× speedup.

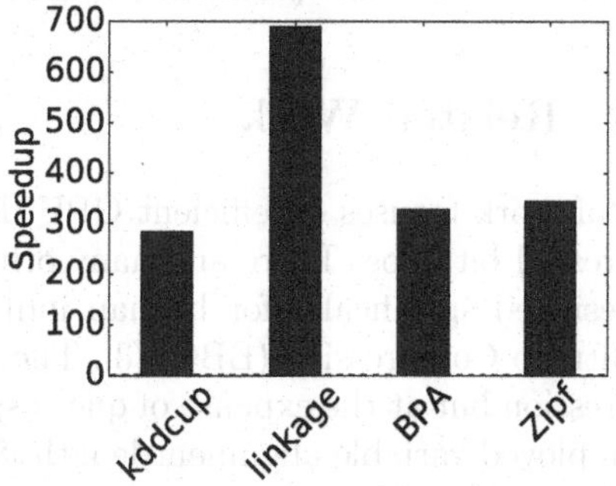

Fig. 7. Speedup provided by using decompressed bit vectors as "metadata".

Figure 8 shows execution profiles when using (a) data structure reuse, (b) 32-bit Stage 2 metadata without a memory pool, (c) 32-bit Stage 4 metadata without a memory pool, (d) only a memory pool, and 32-bit Stage 2 and 32-bit Stage 4 metadata with a memory pool in (e) and (f), respectively.

Data structure reuse (shown in Fig. 8(a)) eliminated three of five allocation/free pairs providing an average of 5.43× speedup. Profiles using Stage 2 and Stage 4 metadata are shown in Fig. 8(b) and (c), respectively. Both provide a noticeable reduction in execution time as each eliminate a memory allocation and free pair. The major cost of memory operations remains a dominant factor so the difference between Stage 2 and Stage 4 metadata use is limited.

The profile when using only a memory pool is shown in Fig. 8(d). The memory pool removes the overhead of memory operations providing a greater reduction in execution time than pure metadata strategies. Strategies combining a memory pool with Stage 2 or Stage 4 metadata are shown in Fig. 8(e) and (f), respectively. These combination strategies provide the benefits of both strategies: short-circuiting to a mid-point of the decompression routine and removing the overhead of GPU memory operations.

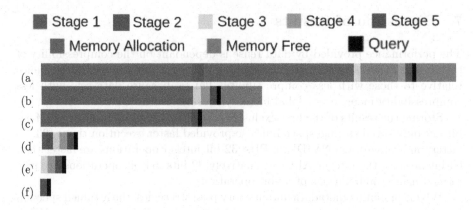

Fig. 8. A representative (albeit approximate) view of execution profiles of six identical query executions on the *linkage* database with varying enhancement strategy configurations. Query execution progresses from left to right. Longer bars correspond to longer execution times. While execution profiles for other databases exhibited slight variations, their interpretations remained consistent.

8   Related Work

Our work focuses on efficient GPU decompression and querying of WAH compressed bitmaps. There are many other hybrid run-length compression schemes designed specifically for bitmap indices. One of the earliest was Byte-aligned Bitmap Compression (BBC) [3]. The smaller alignment can achieve better compression but at the expense of query speed [23]. Other compression schemes have employed variable alignment length [8,13]. These approaches try to balance the trade-offs between compressing shorter runs and increasing query processing time. Others use word alignment but embed metadata in fill atoms that improve compression or query speed [7,10,12,22]. These techniques were developed for execution on the CPU, though they could be ported to the GPU by altering the decompressor component of the GPU system described above. We believe that many, if not all, of these compression techniques on the GPU would benefit from a variation of our metadata and memory pool enhancements.

To the best of our knowledge, we are first to use metadata to efficiently decompress WAH bitmap indices on the GPU. However, other work has explored the benefits of memory pools on GPU in a variety of applications. For example, Hou et al. [14] used a specialized memory pool to create kd-trees in the GPU. Their approach allowed them to process larger scenes on the GPU than previous work. Wang et al. [21] used a preallocated memory pool to reduce the overhead of large tensor allocations/deallocations. Their approach produced speedups of 1.12× to 1.77× over the use of `cudaMalloc()` and `cudaFree()`. The work of Simin et al. [25] is similar to our work in that they use a memory pool to increase the query processing of R-trees on GPU's. We were unable to find any work that used a GPU memory pool specifically designed for use with bitmap indices.

As mentioned above, our work extends the works of Andrzejewski and Wrembel [1,2] and Nelson et al. [19]. Andrzejewski and Wrembel introduced WAH and PLWAH [10] compression and decompression algorithms for GPUs as well as techniques to apply bitwise operations to pairs of bit vectors. Their decompression algorithm details a parallel approach for a decompressing a single WAH or PLWAH compressed bit vector. Nelson et al. modified Andrzejewski and Wrembel's decompression algorithm to apply it to multiple bit vectors in parallel. They then presented multiple algorithms for executing bitmap range queries on the GPU. Our experimental study used their most efficient range query implementation. As the work in this paper improves the efficiency of WAH bitmap decompression on the GPU, it represents a significant enhancement to approaches presented by Andrzejewski and Wrembel's and Nelson et al.

9 Conclusion and Future Works

In this paper, we present multiple techniques for accelerating WAH range queries on GPUs: data structure reuse, storing metadata, and incorporating a memory pool. These methods focus on reducing memory operations or removing repeated decompression work. These techniques take advantage of the static nature of bitmap indexing schemes and the inherent parallelism of range queries.

We conducted an empirical study comparing these acceleration strategies to a baseline GPU implementation. The results of our study showed that the data reuse, metadata, and memory pool strategies provided average speedups of $5.4\times$, $17.8\times$, and $24.4\times$, respectively. Combining these techniques provided an average of $75.4\times$ speedup. We also found that storing the entire bitmaps as accessible metadata on the GPU resulted in an average speedup of $411\times$ by eliminating the need for decompression altogether. This option is only feasible for configurations with small databases or GPUs with large storage space.

In future work, comparing energy consumption of the above approaches may prove interesting. We would also like to investigate executing WAH queries using multiple GPUs. Using multiple GPUs would provide additional parallelism and storage capabilities. Furthermore, since WAH compression is designed for CPU style processing, future studies could investigate new compression schemes that are potentially better fit for the GPU architecture.

References

1. Andrzejewski, W., Wrembel, R.: GPU-WAH: applying GPUs to compressing bitmap indexes with word aligned hybrid. In: Bringas, P.G., Hameurlain, A., Quirchmayr, G. (eds.) DEXA 2010. LNCS, vol. 6262, pp. 315–329. Springer, Heidelberg (2010). https://doi.org/10.1007/978-3-642-15251-1_26
2. Andrzejewski, W., Wrembel, R.: GPU-PLWAH: GPU-based implementation of the PLWAH algorithm for compressing bitmaps. Control Cybern. **40**, 627–650 (2011)
3. Antoshenkov, G.: Byte-aligned bitmap compression. In: Proceedings DCC 1995 Data Compression Conference, p. 476. IEEE (1995)

4. Bonneville power administration. http://www.bpa.gov
5. Bradlow, E., Gangwar, M., Kopalle, P., Voleti, S.: The role of big data and predictive analytics in retailing. J. Retail. **93**, 79–95 (2017)
6. Chambi, S., Lemire, D., Kaser, O., Godin, R.: Better bitmap performance with roaring bitmaps. Softw. Pract. Exp. **46**(5), 709–719 (2016)
7. Colantonio, A., Di Pietro, R.: CONCISE: compressed 'n' composable integer set. Inf. Process. Lett. **110**(16), 644–650 (2010)
8. Corrales, F., Chiu, D., Sawin, J.: Variable length compression for bitmap indices. In: Hameurlain, A., Liddle, S.W., Schewe, K.-D., Zhou, X. (eds.) DEXA 2011. LNCS, vol. 6861, pp. 381–395. Springer, Heidelberg (2011). https://doi.org/10.1007/978-3-642-23091-2_32
9. Davenport, T., Dyche, J.: Big data in big companies. Technical report, International Institute for Analytics (2013)
10. Deliège, F., Pedersen, T.B.: Position list word aligned hybrid: optimizing space and performance for compressed bitmaps. In: International Conference on Extending Database Technology, EDBT 2010, pp. 228–239 (2010)
11. Erevelles, S., Fukawa, N., Swaynea, L.: Big data consumer analytics and the transformation of marketing. J. Bus. Res. **69**, 897–904 (2016)
12. Fusco, F., Stoecklin, M.P., Vlachos, M.: NET-FLi: on-the-fly compression, archiving and indexing of streaming network traffic. VLDB **3**(2), 1382–1393 (2010)
13. Guzun, G., Canahuate, G., Chiu, D., Sawin, J.: A tunable compression framework for bitmap indices. In: 2014 IEEE 30th International Conference on Data Engineering, pp. 484–495. IEEE (2014)
14. Hou, Q., Sun, X., Zhou, K., Lauterbach, C., Manocha, D.: Memory-scalable GPU spatial hierarchy construction. IEEE Trans. Vis. Comput. Graph. **17**(4), 466–474 (2011)
15. Lichman, M.: UCI machine learning repository (2013). http://archive.ics.uci.edu/ml
16. Lloyd, S.: Least squares quantization in PCM. IEEE Trans. Inf. Theor. **28**(2), 129–137 (1982)
17. Marr, B.: Starbucks: using big data, analytics and artificial intelligence to boost performance. Forbes, May 2018. https://www.forbes.com/sites/bernardmarr/2018/05/28/starbucks-using-big-data-analytics-and-artificial-intelligence-to-boost-performance/#5784902e65cd
18. McAfee, A., Brynjolfsson, E.: Big data: the management revolution. Harvard Bus. Rev. **90**, 61–68 (2012)
19. Nelson, M., Sorenson, Z., Myre, J., Sawin, J., Chiu, D.: GPU acceleration of range queries over large data sets. In: Proceedings of the 6th IEEE/ACM International Conference on Big Data Computing, Application, and Technologies (BDCAT 2019), pp. 11–20 (2019)
20. Sariyar, M., Borg, A., Pommerening, K.: Controlling false match rates in record linkage using extreme value theory. J. Biomed. Inf. **44**(4), 648–654 (2011)
21. Wang, L., et al.: SuperNeurons: dynamic GPU memory management for training deep neural networks. In: Proceedings of the 23rd ACM SIGPLAN Symposium on Principles and Practice of Parallel Programming, pp. 41–53 (2018)
22. Wu, K., Otoo, E.J., Shoshani, A., Nordberg, H.: Notes on design and implementation of compressed bit vectors. Technical report, LBNL/PUB-3161, Lawrence Berkeley National Laboratory (2001)
23. Wu, K., Otoo, E.J., Shoshani, A.: Compressing bitmap indexes for faster search operations. In: Proceedings 14th International Conference on Scientific and Statistical Database Management, pp. 99–108. IEEE (2002)

24. Wu, K., Otoo, E.J., Shoshani, A.: Optimizing bitmap indices with efficient compression. ACM Trans. Database Syst. **31**(1), 1–38 (2006)
25. You, S., Zhang, J., Gruenwald, L.: Parallel spatial query processing on GPUs using r-trees. In: Proceedings of the 2nd ACM SIGSPATIAL International Workshop on Analytics for Big Geospatial Data, pp. 23–31 (2013)
26. Zaker, M., Phon-Amnuaisuk, S., Haw, S.C.: An adequate design for large data warehouse systems: bitmap index versus b-tree index. Int. J. Comput. Commun. **2**, 39–46 (2008)

An Effective and Efficient Re-ranking Framework for Social Image Search

Bo Lu[1(✉)], Ye Yuan[2], Yurong Cheng[2], Guoren Wang[2], and Xiaodong Duan[1]

[1] School of Computer Science and Engineering, Dalian Minzu University, Dalian, China
{lubo,Duanxd}@dlnu.edu.cn
[2] School of Computer Science and Technology, Beijing Institute of Technology, Beijing, China
yuanye@mail.neu.edu.cn, yrcheng@bit.edu.cn, wanggrbit@126.com

Abstract. With the rapidly increasing popularity of social media websites, large numbers of images with user-annotated tags are uploaded by web users. Developing automatic techniques to retrieval such massive social images attracts much attention of researchers. The method of social image search returns top-k images according to several keywords input by users. However, the returned results by existing methods are usually irrelevant or lack of diversity, which cannot satisfy user's veritable intention. In this paper, we propose an effective and efficient re-ranking framework for social image search, which can quickly and accurately return ranking results. We not only consider the consistency of visual content of images and semantic interpretations of tags, but also maximize the coverage of the user's query demand. Specifically, we first build a social relationship graph by exploring the heterogeneous attribute information of social networks. For a given query, to ensure the effectiveness, we execute an efficient keyword search algorithm over the social relationship graph, and obtain top-k relevant candidate results. Moreover, we propose a novel re-ranking optimization strategy to refine the candidate results. Meanwhile, we develop an index to accelerate the optimization process, which ensures the efficiency of our framework. Extensive experimental conducts on real-world datasets demonstrate the effectiveness and efficiency of proposed re-ranking framework.

1 Introduction

With the rapid development of the Web technology and the popularity of online social media sharing websites, such as *Flickr* and *Instagram*, large amounts of media data appear on the Internet at an alarming rate. The development of effective and efficient automatic media search approach has become an increasingly important requirement [1]. As an important branch of social media websites, the social image sharing websites allow users not only uploading different types of image data and sharing them with others, but also annotating their image

Fig. 1. Illustration of different social image search methods. For different user queries given in (a) and (b), (a) Example of tag-based social image search, (b) Example of relevance based ranking of social image search. Note that the red dotted line box indicates user's desired results, the blue dotted line box indicates actual returned results. (Color figure online)

content with free tags. For a given social image search task, intuitively, it generally enables users to formulate queries by keywords, and return beneficial search results in response to queries.

In the past decade, most research communities make dedicated effort to social image search and analysis on the following aspects:

(1) Tag-based social image search [2–6]. Since social image data are often associated with user contributed tags, these tags can be used to index the social image data to facilitate users' query. However, the accuracy and validity of existing tag-based social image search methods usually cannot meet the users' query intention. This is because the existing methods only consider whether the keywords are contained in the tags, but ignore whether the semantic of tags is the same with the keywords. For example, in Fig. 1(a), the desired search results of users are various types of dogs by given a query *"dog"*, but the returned results of search may include a number of distinct objects and scenes, such as *"hot dog"*, *"dog house and food"*, and even *"books about dog"*.

(2) Relevance-based social image ranking [7–14]. The relevance-based methods considered whether the semantic interpretation of tags can exactly express the visual content of images. However, they failed to consider the query semantic of a given set of keywords. When the users input several keywords, they usually would like to query such results expressed by the hold keyword sets, rather than obtain several results which presenting each keyword individually. For example in Fig. 1(b), a query consists of *"man, car, racing, champion"*. It is observed that the user actually wants to obtain the images of persons who won in a car racing competition. We present the illustration

in the red dotted line box of Fig. 1(b). Unfortunately, the returned results by the existing methods can only return images matching individual keywords, such as those in the blue dotted line box of Fig. 1(b).

Challenges. As mentioned in the above, there are some challenges in the task of social image search. (1) How to effectively return ranking results which is able to exactly match the semantic-level interpretation of images with users' query. (2) How to efficiently return results when performing queries and matches in massive social image data sets.

Our Approaches. To solve these problems, we propose an effective and efficient re-ranking framework for social image search, which not only considers the consistency of visual content of images and semantic relationships among tags, but also maximizes the coverage of the user's query demand. Our framework allows users to perform social image queries in a natural and simple way. The users only need to input a set of query keywords, the proposed framework can effectively return top-k image results that satisfy users' query intention. Meanwhile, there exist several techniques to ensure the efficiency. Specifically, our framework includes two phases.

Keyword Search over Social Relationship Graph. We first explore the heterogeneous attribute information from social image network to build a Social Relationship Graph, which is an undirected graph with weight of edge. The relationship of users in social network helps to improve the effectiveness of social image search. For a given user query, we execute an efficient keyword search algorithm over the social relationship graph. Answers to a user query are modeled as minimal trees covered by the keyword nodes in the social relationship graph. Each answer tree is assigned a relevance score as a candidate result and ranked in a decreasing order.

Re-ranking Optimization Strategy. In view of the inconsistency between the visual content of images and the semantic interpretations of tags, we present a re-ranking optimization strategy to refine the candidate results from keyword search. Firstly, the entire social image dataset is represented as Region Adjacency Graph (RAG) [15], and the images from candidate results are processed in the same way. Secondly, we use an automatic image region annotation method to annotate the RAGs of images from the candidate results. We hope to obtain the regional nodes, which are closely related to the semantic representation of the query keywords. We then obtain a set of subgraphs of RAGs corresponding to images from the candidate results, and we execute a subgraph matching algorithm over RAGs over the whole social image dataset. Particularly, we use closure tree to index the RAGs of social image dataset for accelerating the optimization process.

To summarize, our contributions as follows:

(1) We propose an effective and efficient re-ranking framework for social image search.
(2) We use keyword search method to solve the problem of matching the semantics of the query keywords and social images, and provide a candidate result set.

(3) We propose a novel re-ranking optimization strategy to refine the candidate results obtained from keyword search.

The remainder of this paper is organized as follows. We introduce the proposed keyword search method over social relationship graph in Sect. 2. In Sect. 3, we describe our re-ranking optimization strategy in detail. In Sect. 4, we report our experiment results. Finally, we make a conclusion in Sect. 5.

2 Keyword Search over Social Relationship Graph

In this section, we introduce the first step of our framework. Firstly, we review the process of building a social relationship graph. Then, we elaborate the query and answer model using an effective keyword search algorithm.

2.1 Overview of Building the Social Relationship Graph

The multimodal information available on social image network contains rich user-generated contents, including shared images, user-annotated tags, comments, etc. Besides these explicit information about images, the relationship among users can be leveraged to improve the effectiveness of social image search. For example, users in social image networks can create and join interested groups where users share images of interests and comments with each other. Thus, modeling the relationship among such user is beneficial to return more relevant and accurate image results in the task of social image search.

In order to build a social relationship graph, an important problem is to measure the social relationship among users by exploring heterogeneous social metadata. We first present the definition of social relationship graph as follows:

Definition 1 (*Social Relationship Graph*). *Given a collection of social users N, a social relationship graph G is defined as a three-tuple $G = (V, E, W)$, where V is a set of user nodes in N, $E \in V \times V$ is a set of edges between user nodes, W is a set of correlation weights between user nodes, which represents the social relationship amongst users.*

Specifically, a user node is defined as a five-tuple $n_i = (P, T, C, U, I)$, where P is a collection of images uploaded by user n_i, T is a set of tags associated with each image $p_i \in P$, C is a collection of comments of P, $U \in N$ is a collection of users who appear in the contact list of user n_i, I is a collection of interest groups joined by user n_i.

Here, we use a kernel-based learning rank framework [17] for inferring the social relationship of users. The method includes two learning stages. Firstly, multiple metadata is represented as proximity graphs by defining various kernel functions. Secondly, all proximity graphs are weighted by the kernel target alignment principle [18]. We further combine these multiple proximity graphs by learning the optimal combination.

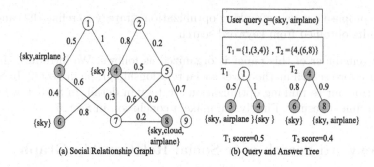

Fig. 2. Example of query and answer trees

2.2 Query and Answer Model

Given the social relationship graph G and a user query $q = (k_1, ..., k_m)$ which contains m keywords, we need to locate keyword nodes of matching search keywords. For each keyword k_i in the query, we find a set of nodes S_i that are relevant to k_i.

Answers to a user query are modeled as minimal trees, we name them as answer trees, which contain at least one node from each S_i. We formulate the definition as follows:

Definition 2 (Answer Tree). *Given a query $q = (k_1, ..., k_m)$ and a social relationship graph G, an answer tree of q is a pair $\{r, S_i\}$, where r is the root of the answer tree and S_i is a set of keyword nodes that contain query terms k_i. r and S_i satisfying: (1) each keyword node contains corresponding keyword; (2) there exists a shortest path of G from r to keyword nodes in S_i.*

Figure 2 illustrates an example of a query and its answer trees. Figure 2(a) shows a social relationship graph with weights. Simplicity, we indicate the tags of user nodes which contain the query keywords, such as (3, 4, 6, 8) respectively. Each weight represents the social relationship strength among users. Figure 2(b) shows answer trees corresponding to the given query $q = (sky, airplane)$.

Each answer tree is assigned a relevance score, which is presented in a decreasing order. We normalize the relevance score in the range $[0, 1]$. We obtain the normalized edge score $Escore(e)$ of an edge by dividing the edge weight by w_{min}, w_{min} is the minimum edge weight in the social relationship graph. The e of edge score is defined as $Esocre(e) = \log(1 + w(e)/w_{min})$. The overall relevance score of an answer tree is defined as $1/(1 + \sum_e Escore(e))$. As shown in Fig. 2 (a), the minimum edge weight w_{min} is 0.2. Thus, the relevance scores of the answer trees in Fig. 2(b) are 0.5 and 0.4 respectively.

2.3 Searching for the Optimal Answer Trees

In this section, we present an efficient keyword search algorithm which offers a heuristic solution for incrementally finding the optimal answer trees as candidate results.

Keyword Search Algorithm. The idea of proposing keyword search algorithm is to find a set of common nodes in the social relationship graph, from which there exists a shortest path to at least one node in each keyword node set. We define an answer tree, in which the common node is the root of the answer tree and the corresponding keyword nodes are the leaves of the answer tree.

Given a query $q = (k_1, ..., k_m)$, we find the set of keyword nodes S_i which are relevant to the keyword k_i. For all query keywords, we formulate it as $S = \bigcup_{i=1}^{m} S_i$. Then, the keyword search algorithm concurrently runs $|S|$ copies of Dijkstra's single source shortest path method by using iterators. All the Dijkstrs's single source shortest path methods traversal the social relationship graph G in a reverse direction. The nodes visited by each copy of Dijkstra's algorithm are called a cluster. The shortest path information is used to guide the iterators. Specifically, we use a cursor to traversal each keyword node and give the equidistance expansion order in each cluster, which guarantees an optimal keyword search. When an iterator for keyword node v visits a node u, it finds a shortest path from u to the keyword node v.

In order to find the root nodes, given a node v, we maintain a nodelist $v.L_i$ for each search keyword k_i, where $v.L_i \subset S_i$. For example, if an iterator started from a keyword node $u \in S_i$, visiting node v, we generate a cross product of node u with the rest of the nodelists $\{u \times \prod_{i \neq j} v.L_i\}$ and each cross product tuple corresponds to an answer tree rooted at node v. When all answer trees are generated, we insert node u into the list $v.L_i$.

Since the relevance of an answer tree is computed by using the edge weights, it causes that the answer trees may not be generated in decreasing order. Therefore, we maintain a small fixed-size heap for the generated answer trees in a heuristic way. The heap is ordered by the relevance of the answer trees. Particularly, if we want to add a new answer tree when the heap is filled, we output the answer tree with the highest relevance score and replace it in the heap. When all iterations finish, the preserved answer trees in the heap can be output as candidate results with a decreasing order.

3   Re-ranking Optimization Strategy

In this section, we introduce the second step of our framework. Recall that in the first step, we obtain a candidate set of results set of results using the method of keyword search over the social relationship graphs, which solve the problem of matching the semantics of queries and social image. However, some results in the candidate sets may only match the query keywords, but they do not satisfy the users' query semantics. Meanwhile, some images without tags matching the keywords that meet the users' query semantics may not be returned. Thus, we further refine the candidate results by an effective re-ranking optimization strategy for improving the performance of social image search.

3.1 Preprocessing

In order to obtain more accurate image results, we analyzed and preprocessed the visual content of social images collection (e.g., *pixels*, *color*, etc.). Note that the social images collection include all the images of user nodes of the social relationship graph. We divide these images into homogeneous color regions by using region segmentation algorithm called EDISON (Edge Detection and Image Segmentation System) [19]. The relationships among segmented regions can be represented as a Region Adjacency Graph (RAG) [20], which is defined as follows.

Definition 3 (*Region Adjacency Graph*). *Given an image $p_i \in P$, a region adjacency graph of p_i, is a four-tuple $G(p_i) = (V, E, \vartheta, \zeta)$, where V is a set of nodes for the segments regions of p_i; $E \in V \times V$ is a set of spatial edges between adjacent nodes of p_i;*
   ϑ is a set of functions generating node attributes;
   ζ is a set of functions generating spatial edge attributes.

A node ($v \in V$) corresponds to a region, and a spatial edge ($e \in E$) represents a spatial relationship between two adjacent nodes (regions). The node attributes are size (number of pixels), color and location (centroid) of corresponding region. The spatial edge attributes indicate relationships between two adjacent nodes such as spatial distance and orientation between centroids of two regions.

Particularly, we use an automatic image region annotation method [21,22] to annotate the RAGs of images from the candidate results. Because we want to retain the regional nodes that are closely related to the semantic representation of the query keywords. We then obtain a set of subgraph of RAGs corresponding to image from candidate results.

3.2 Region Adjacency Graph Closure

As mentioned above, after the preprocessing process, we need to execute the subgraph matching algorithm over RAGs of the whole social image dataset. Our purpose is to find more image results that can match the user's query semantics. Note that these image results may not be found by keyword search in the first step, due to the semantic ambiguity of tags. However, there is a situation cannot be ignored that the number of generated RAGs is huge, it will cause that the computational cost of subgraph matching algorithm is very expensive. Thus, we use the indexing mechanism for improving the efficiency of our framework. Specifically, we index all RAGs of the social image dataset by closure-tree, the closure-tree can capture the similar structural information of each RAG [16]. We present some fundamental definition as follows, and we first formally define the region adjacency graph closure. And then introduce the building process of closure-tree in Sect. 3.3.

We denote that a RAG is an undirected graph G by (V, E) as shown in Fig. 3, where V is a vertex set and E is an edge set. Vertices and edges have attributes denoted by $attr(v)$ and $attr(e)$. All RAGs corresponding to the images is a set of graph $D = (G_1, ..., G_n)$.

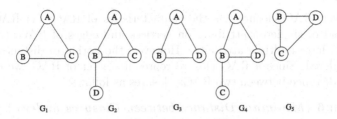

Fig. 3. Example of RAGs

Given two RAGs and a mapping ϕ, if we take an element-wise union of the two graphs, we can get a new graph, where the attribute of each vertex and each edge is a union of the constituent attribute value. The graph captures the structural information of each graph, and serves as a bounding container. This leads to the concept of region adjacency graph closure.

Definition 4 (*Vertex Closure and Edge Closure*). *The closure of a set of vertices (edges) is a generalized vertex (edge) whose attribute is the union of the attribute values of the vertices (edges).*

In particular, a vertex (or edge) closure may contain the special value ε corresponding to a dummy.

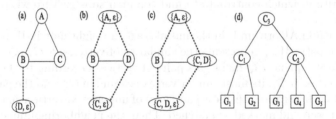

C1=RAG closure(G1, G2)   C2=RAG closure(G3, G4, G5)   C3=RAG closure(C1, C2)   A Closure-tree of RAG closure (C1, C2, C3)

Fig. 4. Example of RAG closure and closure-tree

Definition 5 (*RAG Closure under ϕ*). *The closure of two RAGs G_1 and G_2 under a mapping ϕ is a generalized graph (V, E) where V is the set of vertex closures of the corresponding vertices and E is the set of edge closures of the corresponding edges. This is denoted by $closure(G_1, G_2)$.*

The closure of two graphs depends on the underlying graph mapping. We usually leave this mapping implicit. In the case of multiple graphs, we can compute the closure incrementally. Figure 4(a), (b), and (c) show that the closures of the RAGs from Fig. 3. The dotted edges represent the closure of a dummy and a non-dummy edge.

Moreover, RAG closure has the characteristics of RAG: (1) RAG closure only instead of singleton attribute on vertices and edges of RAG, (2) a RAG closure can have multiple attributes. However, the notion of distance needs to be reconsidered. Since a RAG closure represents a set of RAG, we define the minimum distance between two RAGs closures as follows:

Definition 6 (*Minimum Distance between Closures under* ϕ). *The minimum distance between two RAGs closures G_1 and G_2 under a mapping ϕ is defined as follows:*

$$d_\phi(G_1, G_2) = \sum_{v \in V_1^*} d_{\min}(v, \phi(v)) + \sum_{e \in E_1^*} d_{\min}(e, \phi(e)) \tag{1}$$

where the $d_{\min}$ distances are obtained using the underlying distance measure for vertices and edges.

3.3 Building Process of Closure-Tree

In view of above mentioned, we need to find the graph mapping of all RAGs closure for generating the closure-tree. Figure 4(d) shows an example of a Closure-tree for the RAG closures C_1, C_2, and C_3.

The key issue of finding the graph mapping is to capture the common substructures among RAG closure. In this case, the calculation of finding graph mapping is very expensive. In our work, we present an efficient graph mapping algorithm, which contribute to find common substructure with the way of heuristics.

As shown in Algorithm 1, in the initial stage, a weight matrix W is computed where each entry $W_{u,v}$ represents the similarity of vertex $u \in G_1$ and vertex $v \in G_2$. A priority queue PQ maintains pairs of vertices according to their weights. For each vertex in G_1, its most similar vertex is found in G_2, and the pair is added to PQ. At each iteration, the best pair (u, v) of unmatched vertices in the priority queue is chosen and marked as matched. Then, the neighboring unmatched pairs of (u, v) are assigned higher weights, thus increasing their chance of being chosen. The iterations continue until all vertices in graph G_1 have been matched.

After the closure-tree is generated by finding the graph mapping, all RAGs in the leaf node of closure-tree corresponding to the entire social image dataset. The closure-tree captures the entire structure information of all images. The common ancestor of RAGs correspond to the closure which defined the similarity of RAGs (Definition 4 and 5). Let C be a common ancestor node of RAG G_1 and RAG G_2, if the distance (height) from common ancestor node C to its leaf nodes (RAG G_1 and RAG G_2) is smaller, then the RAGs have a larger similarity, vice versa.

The time complexity of the algorithm can be computed as follows. Let n be the number of vertices and d be the maximum degree of vertices. The initial computation of matrix W and insertions into the priority queue take $O(n^2)$ time, assuming uniform distance measure. In each iteration, the algorithm removes one pair from and inserts at most d^2 unmatched pairs into the priority queue. Totally, there are $O(n)$ iterations. Thus, the time complexity is $O(nd^2 \log n)$.

Algorithm 1: Graph Mapping Algorithm

---

1 Compute the initial similarity matrix W for G_1 and G_2;
2 **for** *each* $u \in G_1$ **do**
3 $\quad$ Find v_m such that $W_{u,v_m} = \max\{W_{u,v}|v \in G_2\}$
4 $\quad$ $PQ.Insert(W_{u,v_m}, <u, v_m>)$
5 $\quad$ $mate[u] := v_m$ /* best mate of u */
6 $\quad$ $wt[u] := W_{u,v_m}$ /* best weight of u */
7 $\quad$ **while** *PQ is not empty* **do**
8 $\quad\quad$ $<u, v> := PQ.dequeue()$
9 $\quad\quad$ **if** *u and v is matched* **then**
10 $\quad\quad\quad$ Find v_m such that $W_{u,v_m} = \max\{W_{u,v}|v \in G_2, v\,is\,unmatched\}$
11 $\quad\quad\quad$ $PQ.Insert(W_{u,v_m}, <u, v_m>), mate[u] := v_m, wt[u] := W_{u,v_m}$

12 $\quad$ Mark $<u, v>$ as matched, Let N_u, N_v be the neighbors of u, v
13 $\quad$ **for** *each* $u' \in N_u$, *u' is unmatched* **do**
14 $\quad\quad$ **for** *each* $v' \in N_v$, *v' is unmatched* **do**
15 $\quad\quad\quad$ Add weights to $W_{u',v'}$
16 $\quad\quad\quad$ **if** $W_{u',v'} \succ wt[u']$ **then**
17 $\quad\quad\quad\quad$ $mate[u'] := v', wt[u'] := W_{u',v'}$
18 $\quad\quad\quad\quad$ **if** $wt[u']$ *has changed* **then**
19 $\quad\quad\quad\quad\quad$ $PQ.Insert(wt[u'], <u', mate[u']>)$

20 return all matches

3.4 Re-ranking Optimization Strategy over Closure-Tree

When all RAGs closure are merged into the closure tree by finding the graph mapping, we execute the optimization strategy of refinement over closure-tree.

We describe in detail the entire process of optimization as follows.

Given a set of subgraph of RAGs corresponding to the candidate results, and generated closure-tree by all RAGs of entire social image dataset.

The steps of re-ranking optimization strategy over closure-tree include:

(1) We conduct a subgraph matching algorithm over closure-tree, the similarity between the subgraph and the RAGs in the leaf node of closure-tree can be calculated.

(2) To obtain more desired results, we consider three cases in the re-ranking optimization process, which are *True Positive, False Negative, False positive* respectively. Specifically, we set a threshold λ, which is a user-specific parameter, used to judge the similarity of the returned results, and decide whether the returned image results will be used as the final results. Note that the threshold λ can be seen as the height from a RAG G (leaf node) to closure of RAG G (common ancestor node).

As shown in Fig. 5, given Top-3 answer trees, for each answer tree, such as Top-1, we assume that G_1 and G_2 are the returned image results from the keyword nodes of Top-1 answer tree. When two RAGs G_1 and G_2 are located

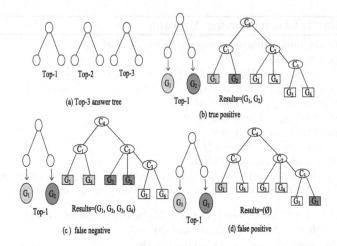

Fig. 5. Example of three cases. Given Top-3 answer trees, for each answer tree, such as Top-1, there are three cases corresponding to the Closure-tree. (Color figure online)

according to the leaf node of closure-tree calculated by subgraph matching algorithm, we mark them with the color of green and yellow respectively.

True Positive: For the Top-1 answer tree, if G_1 and G_2 belong to a common ancestor closure and satisfy $d(G_1, G_2) \leq \lambda$, then, the images corresponding to the G_1 and G_2 can be used as the final results which will be returned to the users. The true positive case can be observed in Fig. 5(b).

False Negative: For the Top-1 answer tree, if G_1 and G_2 do not belong to a common ancestor closure, but satisfy

$$\max\{d_{\min}(G_1, closure(G_1)), d_{\min}(G_2, closure(G_2))\} \leq \lambda \qquad (2)$$

then G_1, G_2, G_3 and G_4 can be returned as the final results to the user, as shown in Fig. 5(c).

In the false negative case, G_3 and G_4 are also returned to the users. Because G_1 and G_4 belong to a common ancestor closure C_1, which means the images corresponding to G_1 and G_4 have a high similarity. Likewise, G_2 and G_3 are able to be interpreted in the same way. Thus, if G_1 and G_2 are the final results, then the G_3 and G_4 are also the final results. This case demonstrate that the images corresponding to G_3 and G_4 are not be found by keyword search in the first stage, but they are indeed the results which should be returned to the users.

False Positive: For the Top-1 answer tree, if $d(G_1, G_2) \succ \lambda$, then we discard the G_1 and G_2, as shown in Fig. 5(d).

The false positive case shows that G_1 and G_2 are actually not satisfy the request of the user's query, even though they are used as the returned image results by keyword search. In this case, we remove the G_1 and G_2 from the top-k candidate results.

For the Top-k candidate results, we refine them by using the above mentioned method of re-ranking optimization to obtain desirable final image results.

4   Performance Evaluations

4.1   Experiments Setup

In this section, we introduce the setup of the experiments, including the data preparation, experimental environment.

We evaluate our approach on a set of real social images that are collected from Flickr *API*. Moreover, we collect heterogeneous data of multiple modalities, such as *tags, comments, friendships amongst users, interest group*, etc. These information is used to build the social relationship graph. All the associated metadata are downloaded in the XML format. We obtain a social image collection consisting of 3,375,000 images with 276,500 associated tags. In order to model the relationship of users, we obtain 5000 users and 78,590 contact list, which start from a random user as seed and expand the crawling according to its friend list in a breadth-first search manner.

4.2   Evaluation Criteria

- **Evaluation set for social image search.** We create a ground truth set as follows. We select 20 diverse tags which are most popular query tags by statistics of Flickr, as the query keywords to execute the keyword search over social relationship graph. The query keywords include *cat, sky, mountain, automobile, water, flower, bird, tree, sunset, sea, airplane, beach, boat, bridge, tiger, dog, lion, butterfly, horse, street* respectively.
- **Measurement.** We use Precision@k, Recall@k, Mean Average Precision (MAP) respectively as the performance evaluation metric of our framework. Specifically, for the proportion of relevant instances in the top k retrieved results, we report precision@k and Recall@k for each query, when $k = 10$, 20, 30, 50, 80, 100. Mean average precision is used to evaluate the overall performance.
- **Method Comparison.** We compare the following social image search methods:

(1) Relevance based Ranking social image search (RR): the method takes the relevance and diversity into account by exploring the content of images and their associated tags [23].

(2) Relevance based Ranking with semantic representation (RRS): the method use both visual information of images and semantic information of associated tags for social image search [24].

(3) Hypergraph learning based social image search (HLS): the method conduct joint Learning which investigate the bag of visual words, accomplish the relevance estimation with a hypergraph learning approach [25].

(4) Social image search based on the User-Image-Tag model (UIT): the method constructs the user-image-tag tripartite graphs to calculate the image similarity [26].

Experiment Results. We conduct experiment with 10 out of 20 query by compare five methods, which is RR, RRS, HLS, UIT and our method, respectively. As shown in

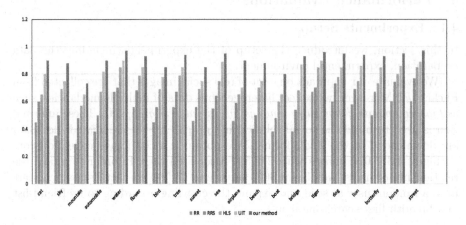

Fig. 6. The comparison of MAP measurement of different approaches for 20 query keywords.

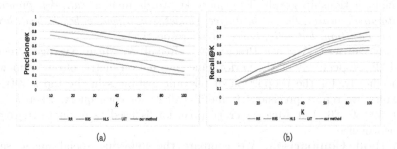

Fig. 7. The precision and recall of top-k results of our framework.

Figure 6 shows the comparison of MAP measurement of five approaches for 20 query keywords. We observe that our method outperforms the other methods by improving the performance 30.8% in terms of MAP.

We compared our method with others method with precision@k and recall@k. According to the experimental results, our method has the highest precision and recall compared to others in the top-k returned results. In Fig. 7(a), our method has a very smooth decline in precision. Particularly, the precision@100 still has 65% accuracy, it is much higher than others method. Meanwhile, the recall@100 in Fig. 7(b)is as high as 80%, which shows that our proposed re-rank optimization strategy can effectively return all possible results.

Query: cat, Top 10 ranked image

Query: mountain, Top 10 ranked image

Query: sky, Top 10 ranked image

Query: flower, Top 10 ranked image

Query: bird, Top 10 ranked image

Query: sunset, Top 10 ranked image

Query: tree, Top 10 ranked image

Query: sea, Top 10 ranked image

Query: water, Top 10 ranked image

Query: automobile, Top 10 ranked image

Fig. 8. The top 10 results for 10 out of 20 queries by using our framework.

In Fig. 8, we demonstrate the top-10 results for 10 out 20 queries by using our method, the image results are ranked in light of their relevance scores. If two images have the same score, they are ranked according to image ids in descending order.

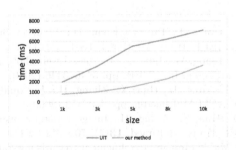

Fig. 9. Example of running time vs social image data size.

We analyzed the efficiency between our method and UIT method, because the size of the Flickr datasets used in UIT method is almost the same as ours. Figure 9 shows the running time versus social image dataset size for the our method and UIT method. From the results, we can observe that with the increase of image datasets size, the time cost for the two methods all increase. However, the running time of our method has not increased much by matching from 1000 to 5000 images in social image dataset, it proves that we build the closure tree can indeed accelerate the optimization process, which ensures the efficiency of our framework.

5 Conclusion

In this paper, we propose an effective and efficient re-ranking framework for social image search, which can quickly and accurately return ranking results. Specifically, we build a social relationship graph by exploring the heterogeneous attribute information of social networks. For a given query, to ensure the effectiveness, we execute an efficient keyword search algorithm over the social relationship graph, and obtain top-k relevant candidate results. Moreover, we propose a novel re-ranking optimization strategy to refine the candidate results. Meanwhile, we develop an index to accelerate the optimization process, which ensures the efficiency of our framework. Extensive experimental conducts on real-world datasets demonstrate the effectiveness and efficiency of proposed re-ranking framework. In the future work, we keep studying the problem of social image search in depth, we will use the technology of deep learning to further learning the visual characteristics and semantic interpretation of social images.

Acknowledgements. Bo Lu is supported by the NSFC (Grant No. 61602085), Ye Yuan is supported by the NSFC (Grant No. 61932004, N181605012), Yurong Cheng is supported by the NSFC (Grant No. 61902023, U1811262) and the China Postdoctoral Science General Program Foundation (No. 2018M631358).

References

1. Wu, Y., Cao, N., Gotz, D., Tan, Y., Keim, D.: A survey on visual analytics of social media data. IEEE Trans. Multimed. **18**(11), 2135–2148 (2016)
2. Chen, L., Xu, D., Tsang, W., Luo, D.: Tag-based web photo retrieval imapproved by batch mode re-tagging. In: CVPR, pp. 3440–3446 (2010)
3. Liu, D., Wang, M., Yang, L., Hua, X., Zhang, H.: Tag quality improvement for social images. In: ACM Multimedia, pp. 350–353 (2009)
4. Liu, D., Yan, D., Hua, S., Zhang, H.: Image retagging using collaborative tag propagation. IEEE Trans. Multimed. **13**(4), 702–712 (2011)
5. Zhu, G., Yan, S., Ma, Y.: Image tag refinement towards low-rank, content-tag prior and error sparsity. In: ACM Multimeida, pp. 461–470 (2010)
6. Yang, K., Hua, X., Wang, M., Zhang, H.: Tag tagging: towards more descriptive keywords of image content. IEEE Trans. Multimed. **13**(4), 662–673 (2011)
7. Gao, Y., Wang, M., Zha, Z., Shen, J., Li, X.: Visual-textual joint relevance learning for tag-based social image search. IEEE Trans. Image Process. **22**(1), 363–376 (2013)
8. Seah, B., Bhowmick, S., Sun, A.: PRISM: concept-preserving social image search results summarization. PVLDB **8**(12), 1868–1871 (2015)
9. Huang, F., Zhang, X., Li, Z., He, Y., Zhao, Z.: Learning social image embedding with deep multimodal attention networks. In: ACM Multimedia, pp. 460–468 (2017)

10. Dao, M., Minh, P., Kasem, A., Nazmudeen, M.: A context-aware late-fusion app-
    roach for disaster image retrieval from social media. In: ICMR, pp. 266–273 (2018)
11. Chen, Y., Tsai, Y., Li, C.: Query embedding learning for context-based social
    search. In: CIKM, pp. 2441–2444 (2018)
12. Wu, B., Jia, J., Yang, Y., Zhao, P., Tian, Q.: Inferring emotional tags from social
    images with user demographics. IEEE Trans. Multimed. **19**(7), 1670–1684 (2017)
13. Zhang, J., Yang, Y., Tian, Q., Liu, X.: Personalized social image recommendation
    method based on user-image-tag model. IEEE Trans. Multimed. **19**(11), 2439–2449
    (2017)
14. Lu, D., Liu, X., Qian, X.: Tag-based image search by social re-ranking. IEEE Trans.
    Multimed. **18**(8), 1628–1639 (2016)
15. Tremeau, A., Colantoni, P.: Regions adjacency graph applied to color image seg-
    mentation. IEEE Trans. Image Process **9**(4), 735–744 (2000)
16. He, H., Singh, A.: Closure-tree: an index structure for graph queries. In: ICDE,
    pp. 38–49 (2006)
17. Zhuang, F., Mei, T., Steven, C., Hua, S.: Modeling social strength in social media
    community via kernel-based learning. In: ACM Multimedia, pp. 113–122 (2011)
18. Cortes, C., Mohri, M., Rostamizadeh, A.: Two-stage learning kernel algorithms.
    In: ICML, pp. 239–246 (2010)
19. Comanicu, D., Meer, P.: Mean shift: a robust approach toward feature space anal-
    ysis. IEEE Trans. Pattern Anal. Mach. Intell. **24**(5), 603–619 (2002)
20. Lee, J., Hwang, S.: STRG-Index: spatio-temporal region graph indexing for large
    video databases. In: SIGMOD, pp: 718–729 (2005)
21. Yuan, J., Li, J., Zhang B.: Exploiting spatial context constraints for automatic
    image region annotation. In: ACM Multimedia, pp. 595–604 (2007)
22. Wang, Y., Yuan, Y., Ma, Y., Wang, G.: Time-dependent graphs: definitions, appli-
    cations, and algorithms. Data Sci. Eng. **4**(4), 352–366 (2019). https://doi.org/10.
    1007/s41019-019-00105-0
23. Wang, M., Yang, K., Hua, X., Zhang, H.: Towards relevant and diverse search of
    social images. IEEE Trans. Multimed. **12**(8), 829–842 (2010)
24. Yang, K., Wang, M., Hua, X.S., Zhang, H.J.: Tag-based social image search: toward
    relevant and diverse results. In: Hoi, S., Luo, J., Boll, S., Xu, D., Jin, R., King, I.
    (eds.) Social Media Modeling and Computing, pp. 25–45. Springer, London (2011).
    https://doi.org/10.1007/978-0-85729-436-4_2
25. Gao, Y., Wang, M., Luan, H., Shen, C.: Tag-based social image search with visual-
    text joint hypergraph learning. In: ACM Multimedia, pp. 1517–1520 (2017)
26. Zhang, J., Yang, Y., Tian, Q., Zhuo, L.: Personalized social image recommendation
    method based on user-image-tag model. IEEE Trans. Multimed. **19**(11), 2439–2449
    (2017)

HEGJoin: Heterogeneous CPU-GPU Epsilon Grids for Accelerated Distance Similarity Join

Benoit Gallet$^{(\boxtimes)}$ ⓘ and Michael Gowanlock ⓘ

School of Informatics, Computing, and Cyber Systems, Northern Arizona University,
Flagstaff, AZ 86011, USA
{benoit.gallet,michael.gowanlock}@nau.edu

Abstract. The distance similarity join operation joins two datasets (or tables), A and B, based on a search distance, ϵ, ($A \ltimes_\epsilon B$), and returns the pairs of points (p_a, p_b), where $p_a \in A$ and $p_b \in B$ such that the distance between p_a and $p_b \leq \epsilon$. In the case where $A = B$, then this operation is a similarity self-join (and therefore, $A \bowtie_\epsilon A$). In contrast to the majority of the literature that focuses on either the CPU or the GPU, we propose in this paper Heterogeneous CPU-GPU Epsilon Grids Join (HEGJOIN), an efficient algorithm to process a distance similarity join using both the CPU and the GPU. We leverage two state-of-the-art algorithms: LBJOIN for the GPU and SUPER-EGO for the CPU. We achieve good load balancing between architectures by assigning points with larger workloads to the GPU and those with lighter workloads to the CPU through the use of a shared work queue. We examine the performance of our heterogeneous algorithm against LBJOIN, as well as SUPER-EGO by comparing performance to the upper bound throughput. We observe that HEGJOIN consistently achieves close to this upper bound.

Keywords: Heterogeneous CPU-GPU computing · Range query · Similarity join · SUPER-EGO

1 Introduction

Distance similarity searches find objects within a search distance ϵ from a set of query points (or feature vectors). These searches are extensively used in database systems for fast query processing. Due to the high memory bandwidth and computational throughput, GPUs (Graphics Processing Units) have been used to improve database performance (e.g., similarity joins [8], high dimensional similarity searches [19], and indexing methods for range queries [2,16,17,21,24]). Despite the GPU's attractive performance characteristics, it has not been widely utilized to improve the throughput of modern database systems. Consequently, there has been limited research into concurrently exploiting the CPU and GPU to improve database query throughput. In this paper, we propose a hybrid CPU-GPU algorithm for computing distance similarity searches that combines two

ⓒ Springer Nature Switzerland AG 2020
Y. Nah et al. (Eds.): DASFAA 2020, LNCS 12114, pp. 372–388, 2020.
https://doi.org/10.1007/978-3-030-59419-0_23

highly efficient algorithms designed for each architecture. Note the use of the CUDA terminology throughout this paper.

In database systems, distance similarity searches are typically processed using a join operation. We focus on the distance similarity self-join, defined as performing a distance similarity search around each object in a table $(A \bowtie_\epsilon A)$. The method employed in this paper could be used to join two different tables using a semi-join operation $(A \bowtie_\epsilon B)$, where A is a set of query points, and B is a set of entries in the index. Throughout this paper, while we refer to the self-join, we do not explore the optimizations applicable only to the self-join, and so leave the possibility to semi-join instead.

The complexity of a brute force similarity search is $O(|D|^2)$, where D is the dataset/table. Thus, indexing methods have been used to prune the search to reduce its complexity. In this case, the *search-and-refine* strategy is used, where the *search* of an indexing structure generates a set of candidate points that are likely to be within ϵ of a query point, and the *refine* step reduces the candidate set to the final set of objects within ϵ of the query point using distance calculations.

The distance similarity search literature typically focuses on either low [7, 8, 11, 15] or high [19] dimensional searches. Algorithms designed for low dimensionality are typically not designed for high dimensionality (and vice versa). The predominant reason for this is due to the *curse of dimensionality* [5, 15], where index searches become more exhaustive and thus tend to degrade into a brute force search as the dimensionality increases. In this work, we focus on low-dimensional exact searches that do not dramatically suffer from the curse of dimensionality. Since the cost of the distance calculation used to refine the candidate set increases with dimensionality, low-dimensional searches are often memory-bound, as opposed to compute-bound in high-dimensions. The memory-bound nature of the algorithm in low-dimensionality creates several challenges that may hinder performance and limit the scalability of parallel approaches.

As discussed above, there is a lack of heterogeneous CPU-GPU support for database systems in the literature. This paper proposes an efficient distance similarity search algorithm that uses both the CPU and the GPU. There are two major CPU-GPU similarity search algorithm designs, described as follows:

- **Task parallelism:** Assign the CPU and GPU particular tasks to compute. For example, Kim and Nam [18] compute range queries using the CPU to search an R-TREE while the GPU refines the candidate set, while Shahvarani and Jacobsen [23] use the CPU and the GPU to search a B-TREE, and then use the CPU to refine the candidate set.

- **Data parallelism:** Split the work and perform both the search and refine steps on each architecture independently, using an algorithm suited to each architecture. To our knowledge, while no other works have used the data-parallel approach in a hybrid CPU-GPU distance similarity search algorithm, we found that this design has been used by Gowanlock [12] for k Nearest Neighbor (kNN) searches. In this paper, we focus on the *data-parallel* approach because it allows us to assign work to each architecture based on the workload of each query point. In modern database systems, this approach

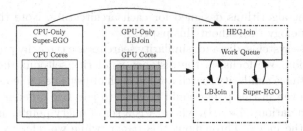

Fig. 1. Representation of how we combine SUPER-EGO and LBJOIN by using a single work queue to form HEGJOIN.

allows us to exploit all available computational resources in the system, which maximizes query throughput.

Our algorithm leverages two previously proposed independent works that were shown to be highly efficient: the GPU algorithm (LBJOIN) by Gallet and Gowanlock [11] and the CPU algorithm (SUPER-EGO) presented by Kalashnikov [15]. Figure 1 illustrates how these algorithms work together through the use of a single shared work queue. By combining these two algorithms, we achieve better performance on most experimental scenarios than CPU-only or GPU-only approaches. This paper makes the following contributions:

- We combine state-of-the-art algorithms for the CPU and GPU (Sect. 3).
- We propose an efficient double-ended work queue (deque) that assigns work on-demand to the CPU and GPU algorithms. This allows both architectures to be saturated with work while achieving low load imbalance (Sect. 4).
- By using the work queue, we split the work between the CPU and GPU by allowing the CPU (GPU) to compute the query points with the lowest (highest) workload. This exploits the GPU's high computational throughput.
- We optimize SUPER-EGO to further improve the performance of our hybrid algorithm. We denote this optimized version of SUPER-EGO as SEGO-NEW.
- We evaluate the performance using five real-world and ten exponentially distributed synthetic datasets. We achieve speedups up to 2.5× (11.3×) over the GPU-only (CPU-only) algorithms (Sect. 5).

The rest of the paper is organized as follows: we present background material in Sect. 2, and conclude the paper in Sect. 6.

2   Background

2.1   Problem Statement

Let D be a dataset in d dimensions. Each point in D is denoted as q_i, where $i = 1, ..., |D|$. We denote the jth coordinate of $q_i \in D$ as $q_i(j)$, where $j = 1, ..., d$. Thus, given a distance threshold ϵ, we define the distance similarity search of a query point q as finding all points in D that are within this distance ϵ to q. We also define a candidate point $c \in D$ as a point whose distance to q is evaluated. Similarly to related work, we use the Euclidean distance. Therefore, the similarity join finds pairs of points $(q \in D, c \in D)$, such that $dist(q,c) \le \epsilon$, where $dist(q,c) = \sqrt{\sum_{j=1}^{d}(q(j) - c(j))^2}$. All processing occurs in-memory. While we consider the case where the result set size may exceed the GPU's global memory capacity, we do not consider the case where the result set size may exceed the platform's main memory capacity.

2.2   Related Work

We present relevant works regarding the distance similarity join. Since the similarity join is frequently used as a building block within other algorithms, the literature regarding the optimization of the similarity join is extensive. However, the vast majority of existing literature aims at improving performance using either the CPU or the GPU, and rarely both. Hence, literature regarding heterogeneous CPU-GPU similarity join optimizations remains relatively scarce. The search-and-refine strategy (Sect. 1) largely relies on the use of data indexing methods that we describe as follows.

Indexes are used to prune searches. Given a query point q and a distance threshold ϵ, indexes find the candidate points that are likely to be within a distance ϵ of q. Also, the majority of the indexes are designed for a specific use, whether they are for low or high dimensional data, for the CPU, for the GPU or for both architectures. We identify different indexing methods, including those designed for the CPU [3,4,6,7,9,14,15,22], the GPU [2,13,17], or both architectures [12,18,23]. As our algorithm focuses on the low dimensionality distance similarity search, we focus on presenting indexing methods also designed for lower dimensions. Since indexes are an essential component of distance similarity searches, identifying the best index for each architecture is critical to achieve good performance, especially when using two different architectures. Furthermore, although our heterogeneous algorithm leverages two previously proposed works [11,15] that both use a grid indexing for the CPU and the GPU, we discuss in the following sections several other indexing methods based on trees.

CPU Indexing: In the literature, the majority of indexes designed for the CPU to index multi-dimensional data are based on trees, such as the kD-TREE [6], the QUAD TREE [10] or the R-TREE [14]. The B-TREE [3] is designed to index 1-dimensional data. All these indexes are designed for range queries and can be used for distance similarity searches. Grid indexes such as the Epsilon Grid Order (EGO) [7,15] have been designed for distance similarity joins. We discuss this EGO index that we leverage in Sect. 3.2.

GPU Indexing: Similarly to CPU indexes, index-trees have been specially optimized to be efficient on the GPU. The R-TREE has been optimized by Kim and Nam [17] and the B-TREE by Awad et al. [2], both designed for range queries. As an example of the optimizations they make to the R-TREE and the

B-TREE, both works mostly focus on removing recursive accesses inherent to tree traversals or on reducing threads divergence. We present the grid index proposed by Gowanlock and Karsin [13] designed for distance similarity joins and that we leverage in Sect. 3.1.

CPU-GPU Indexing: Kim and Nam [18] propose an R-TREE designed for range queries that uses task parallelism. The CPU searches the internal nodes of the tree, while the GPU refines the objects in the leaves. Gowanlock [12] instead elects to use two indexes for data parallelism to compute kNN searches: the CPU uses a kD-TREE while the GPU uses a grid, so both indexes are suited to their respective architecture.

3 Leveraged Work

In this section, we present the previously proposed works we leverage to design HEGJOIN. Therefore, we use LBJOIN [11] for the GPU and SUPER-EGO [15] for the CPU, two state-of-the-art algorithms on their respective platforms. The GPU[1] and CPU[2] algorithms are publicly available.

3.1 GPU Algorithm: LBJoin

The GPU component of HEGJOIN is based on the GPU kernel proposed by Gallet and Gowanlock [11]. This kernel also uses the grid index and the batching scheme by Gowanlock and Karsin [13]. This work is the best distance similarity join algorithm for low dimensions that uses the GPU (there are similar GPU algorithms but they are designed for range queries, see Sect. 2).

Grid Indexing: The grid index presented by Gowanlock and Karsin [13] allows the query points to only search for candidate points within its 3^d adjacent cells (and the query points' own cell), where d is the data dimensionality. This grid is stored in several arrays: *(i)* the first array represents only the non-empty cells to minimize memory usage, *(ii)* the second array stores the cells' linear id and a minimum and maximum indices of the points, *(iii)* the third array corresponds to the position of the points in the dataset and is pointed to by the second array. Furthermore, the threads within the same warp access neighboring cells in the same lock-step fashion, thus avoiding thread divergence. Also, note that we modify their work and now construct the index directly on the GPU, which is much faster than constructing it on the CPU as in the original work.

Batching Scheme: Computing the ϵ-neighborhood of many query points may yield a very large result set and exceed the GPU's global memory capacity. Therefore, in Gowanlock and Karsin [13], the total execution is split into multiple batches, such that the result set does not exceed global memory capacity.

---

[1] https://github.com/benoitgallet/self-join-hpbdc2019, last accessed: Feb. 27th, 2020.
[2] https://www.ics.uci.edu/~dvk/code/SuperEGO.html, last accessed: Feb. 27th, 2020.

The number of batches that are executed, n_b, are defined by an estimate of the total result set size, n_e, and a buffer of size n_s, which is stored on the GPU. The authors use a lightweight kernel to compute n_e, based on a sample of D. Thus, they compute $n_b = n_e/n_s$.[3] The buffer size, n_s, can be selected such that the GPU's global memory capacity is not exceeded. The number of queries, n_q^{GPU}, processed per batch (a fraction of $|D|$) are defined by the number of batches as follows: $n_q^{GPU} = |D|/n_b$. Hence, a smaller number of batches will yield a larger number of queries processed per batch.

The total result set is simply the union of the results from each batch. Let R denote the total result set, where $R = \bigcup_{l=1}^{n_b} r_l$, where r_l is the result set of a batch, and where $l = 1, 2, \ldots, n_b$.

The batches are executed in three CUDA *streams*, allowing the overlap of GPU computation and CPU-GPU communication, and other host-side tasks (e.g., memory copies into and out of buffers), which is beneficial for performance.

Sort by Workload and Work Queue: The sorting strategy proposed by Gallet and Gowanlock [11] sorts the query points by non-increasing workload. The workload of a query point is determined by the sum of candidate points in its own cell and its 3^d adjacent cells in the grid index. This results in a list of query points sorted from most to least workload, which is then used in the work queue to assign work to the GPU's threads. The consequence of sorting by workload and of using this work queue is that threads within the same warp will compute query points with a similar workload, thereby reducing intra-warp load imbalance. This reduction in load imbalance, compared to their GPU reference implementation [11], therefore reduces the overall number of periods where some threads of the warp are idle and some are computing. This yields an overall better response time than when not sorting by workload. This queue is stored on the GPU as an array, and a variable is used to indicate the head of the queue. In this paper, we store this queue on the CPU's main memory to be able to share the work between the CPU and the GPU components of HEGJOIN.

GPU Kernel: The GPU kernel [11] makes use of a grid index, the batching scheme, as well as the sorting by workload strategy and the work queue presented above. Moreover, we configure the kernel [11] to use a single GPU thread to process each query point ($|D|$ threads in total). Thus, each thread first retrieves a query point from the work queue using an atomic operation. Then, using the grid index, the threads search for their non-empty neighboring cells corresponding to their query point, and iterate over the found cells. Finally, for each point within these cells, the algorithm computes the distance to the query point and if this distance is $\leq \epsilon$, then the key/value pair made of the query point's id and the candidate point's id is added to the result buffer r of the batch.

---

[3] In this section, for clarity, and without the loss of generality, we describe the batching scheme assuming all values divide evenly.

3.2 CPU Algorithm: Super-EGO

Similarly to our GPU component, the CPU component of our heterogeneous algorithm is based on the efficient distance similarity join algorithm, SUPER-EGO, proposed by Kalashnikov [15]. We present its main features as follows.

Dimension Reordering: The principle of this technique is to first compute a histogram of the average distance between the points of the dataset and for each dimension. A dimension with a high average distance between the points means that points are more spread across the search space, and therefore fewer points will join. The goal is to quickly increase the cumulative distance between two points so it reaches ϵ with fewer distance calculations, allowing the algorithm to short-circuit the distance calculation and continue computing the next point.

EGO Sort: This sorting strategy sorts the points based on their coordinates in each dimension, divided by ϵ. This puts spatially close points close to each other in memory, and serves as an index to find candidate points when joining two sets of points. This sort was originally introduced by Böhm et al. [7].

Join Method: The SUPER-EGO algorithm joins two partitions of the datasets together, and recursively creates new partitions until it reaches a given size. Then, since the points are sorted based on their coordinates and the dimensions have been reordered, two partitions are compared only if their first point is within ϵ from each other. If they are not, then subsequent points will not join either and the join of the two partitions is aborted. The join is thus made directly on several partitions of the input datasets.

Parallel Algorithm: SUPER-EGO also adds parallelism to the original EGO algorithm, using PTHREADS and a producer-consumer scheme to balance the workload between threads. When a new partition is recursively created, if the size of the queue is less than the number of threads (i.e., some threads have no work), the newly created partitions are added to the work queue to be shared among the threads. This ensures that no threads are left without work to compute.

4 Heterogeneous CPU-GPU Algorithm: HEGJoin

We present the major components of our heterogeneous CPU-GPU algorithm, HEGJOIN, as well as improvements made to the leveraged work (Sect. 3).

4.1 Shared Work Queue

As mentioned in Sect. 3.1, we reuse the work queue stored on the GPU that was proposed by Gallet and Gowanlock [11], which efficiently balances the workload between GPU threads. However, to use the work queue for the CPU and the GPU components, we must relocate it to the host/CPU to use it with our CPU algorithm component. Because the GPU has a higher computational throughput than the CPU, we assign the query points with the most work to the GPU, and those with the least work to the CPU. Similarly to the shared work queue

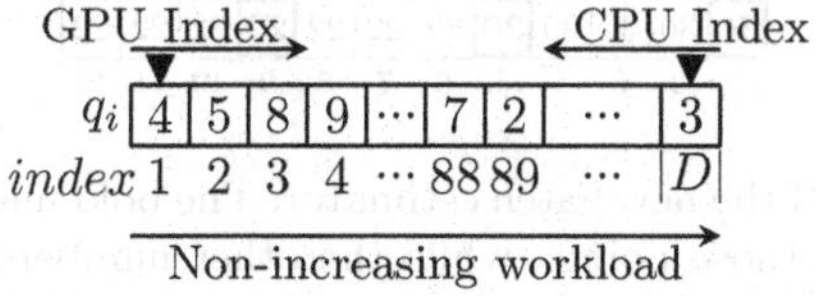

GPU Index→   ←CPU Index

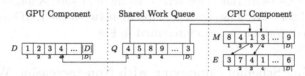

Non-increasing workload→

Fig. 2. Representation of our deque as an array. The numbers q_i are the query points id, the triangles are the starting position of each index, and the arrows above it indicate the indices progression in the deque.

GPU Component   Shared Work Queue   CPU Component

D | 1 | 2 | 3 | 4 | ... | $|D|$ | Q | 4 | 5 | 8 | 9 | ... | 3 | M | 8 | 4 | 1 | 3 | ... | 9 |

E | 3 | 7 | 4 | 1 | ... | 6 |

Fig. 3. Illustration of an input dataset D, the shared deque sorted by workload Q, the input dataset EGO-sorted E and the mapping M between Q and E. The numbers in D, Q, and E correspond to query point ids, while the numbers in M correspond to their position in E. The numbers below the arrays are the indices of the elements.

proposed by Gowanlock [12] for the CPU-GPU kNN algorithm, the query points need to be sorted based on their workload, as detailed in Sect. 3.1. However, while query points' workload in Gowanlock [12] is characterized by the number of points within each query point's cell, we define here the workload as the number of candidate points within all adjacent cells. Our sorting strategy is more representative of the workload than in Gowanlock [12], as it yields the exact number of candidates that must be filtered for each query point (Fig. 2).

Using this queue with the CPU and the GPU requires modifying the original work queue [11] to be a double ended-queue (deque), as well as defining a deque index for each architecture. Since the query points are sorted by workload, we set the GPU's deque index to the beginning of the deque (greatest workload), and to the end of the deque for the CPU's index (smallest workload). Therefore, the GPU's workload is configured to decrease while CPU's workload increases, as their respective index progresses in the deque. Also, note that while n_q for the CPU (n_q^{CPU}) is fixed, n_q for the GPU (n_q^{GPU}) varies based on the dataset characteristics and on ϵ (Sect. 3.1).

We assign query points from the deque to each architecture, described as follows: (a) We set the GPU's deque index to 1 and the CPU's deque index to $|D|$; (b) We create an empty batch if the GPU's index and the CPU's index are at the same position in the deque, and the program terminates; (c) To assign query points to the GPU, we create and assign a new batch to the GPU, and increase GPU's deque index; (d) To assign query points to the CPU, we create and assign a new batch to the CPU, and decrease CPU's deque index.

As described in Sect. 3, HEGJOIN uses two different sorts: sorting by workload (Sect. 3.1) and SUPER-EGO's EGO-sort (Sect. 3.2). However, as these two

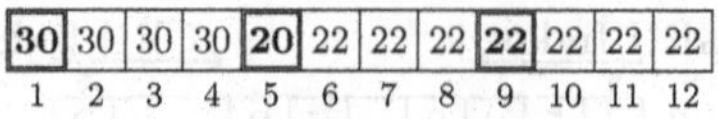

30	30	30	30	**20**	22	22	22	**22**	22	22	22

1 2 3 4 5 6 7 8 9 10 11 12

Fig. 4. Representation of the new batch estimator. The bold numbers are the estimated number of neighbors of those points, while the other numbers are inferred, based on the maximum result between the two closest estimated points shown in bold.

strategies sort following different criteria, it is not possible to first sort by workload then to EGO-sort (and vice-versa), as the first sort would be overwritten by the second sort. We thus create a mapping between the EGO-sorted dataset and our shared work queue, as represented in Fig. 3.

4.2 Batching Scheme: Complying with Non-Increasing Workload

A substantial issue arises when combining the batching scheme and the sorting by workload strategy (Sect. 3.1). As the batch estimator creates batches with a fixed number of query points, and because the query points are sorted by workload, the original batching scheme creates successive batches with a non-increasing workload. Hence, as the execution proceeds, the batches take less time to compute and the overhead of launching many kernels may become substantial, especially as the computation could have been executed with fewer batches.

We modify the batching scheme (Sect. 3.1) to accommodate the sorting by workload strategy, and that we represent in Fig. 4. While still estimating a fraction of the points, the rest of the points get a number of neighbors inferred from the maximum value of the two closest estimated points (to overestimate and avoid buffer overflow during computation). Adding the estimated and the inferred numbers of neighbors yields an estimated result set size n_e. We then create the batches so they have a consistent result set size r_l close to the buffer size n_s. As the number of estimated neighbors should decrease (as their workload decreases), the number of query points per batch increases. Furthermore, we set a minimum number of batches to $2 \times n^{streams}$, where $n^{streams} = 3$ is the number of CUDA streams used. Therefore, the GPU can only initially be assigned up to half of the queries in the work queue. This ensures that the GPU is not initially assigned too many queries, which would otherwise starve the CPU of work to compute.

4.3 GPU Component: HEGJoin-GPU

The GPU component of our heterogeneous algorithm, which we denote as HEGJOIN-GPU and that we can divide into two parts (the host and the kernel), remains mostly unchanged from the algorithm proposed by Gallet and Gowanlock [11] and presented in Sect. 3.

Regarding the host side of our GPU component, we modify how the kernels are instantiated to use the shared work queue presented in Sect. 4.1. Therefore, as the original algorithm was looping over all the batches (as given by the batch estimator, presented in Sect. 3), we loop while the shared deque returns a valid batch to execute (Sect. 4.1).

In the kernel, since the work queue has been moved to the CPU, a batch corresponds to a range of queries in the deque whose interval is determined when taking a new batch from the queue, and can be viewed as a "local queue" on the GPU. Therefore, the threads in the kernel update a counter local to the batch to determine which query point to compute, still following the non-increasing workload that yields a good load balancing between threads in the same warp.

4.4 CPU Component: HEGJoin-CPU

The CPU component of HEGJOIN is based on the SUPER-EGO algorithm proposed in [15] and presented in Sect. 3.2. We make several modifications to SUPER-EGO to incorporate the double ended queue we use, and we also optimize SUPER-EGO to improve its performance.

As described in Sect. 3.2, SUPER-EGO uses a queue and a producer-consumer system for multithreading. We remove this system and replace it with our shared double-ended queue. Because the threads are continuously taking work from the shared deque until it is empty, the producer-consumer originally used becomes unnecessary, as the deque signals SUPER-EGO when it is empty.

Table 1. Summary of the real-world datasets used for the experimental evaluation. $|D|$ denotes the number of points and d the dimensionality.

Dataset	$|D|$	d	Dataset	$|D|$	d	Dataset	$|D|$	d
$SW2DA$	1.86M	2	$SW2DB$	5.16M	2	$SDSS$	15.23M	2
$SW3DA$	1.86M	3	$SW3DB$	5.16M	3			

The original SUPER-EGO algorithm recursively creates sub-partitions of contiguous points on the input datasets until their size is suited for joining. As one of the partitions is now taken from our deque, which is sorted by workload, it no longer corresponds to a contiguous partition of the input dataset. Thus, we loop over the query points of the batch given by the deque to join it with the other points in the partition. This optimization requires the use of the mapping presented in Sect. 4.1 and illustrated in Fig. 3.

SUPER-EGO uses QSORT from the C standard library to EGO-sort, and we replace it by the more efficient and parallel BOOST::SORT::SAMPLE_SORT algorithm, a stable sort from the Boost C++ library. This allows SEGO-NEW to start its computation earlier than SUPER-EGO would, as it is faster than QSORT. We use as many threads to sort as we use to compute the join.

Finally, in contrast to the original SUPER-EGO algorithm, this new version of SUPER-EGO is now capable of using 64-bit floats instead of only 32-bit floats.

5   Experimental Evaluation

5.1   Datasets

We use real-world and exponentially distributed synthetic datasets (using $\lambda = 40$), spanning 2 to 8 dimensions. The real-world datasets we select are the *Space Weather* datasets (*SW-*) [20], composed of 1.86M or 5.16M points in two dimensions representing the latitude and longitude of objects, and adding the total number of electrons as the third dimension. We also use data from the *Sloan Digital Sky Survey* dataset (*SDSS*) [1], composed of a sample of 15.23M galaxies in 2 dimensions. A summary of the real-world datasets is given in Table 1.

Additionally, we use synthetic datasets made of 2M and 10M points spanning two to eight dimensions. These datasets are named using the dimensions and number of points: *Expo3D2M* is a 3-dimensional dataset of 2M points. We elect to use an exponential distribution as this distribution contains over-dense and under-dense regions, which are representative of the real-world datasets we select (Table 1). Finally, exponential distributions yield high load imbalance between the points, and should thus be more suited to outline a load imbalance between the processors, an important aspect of HEGJOIN, than uniform distributions.

5.2   Methodology

The platform we use to run our experiments is composed of 2 × Intel Xeon E5-2620v4 with 16 total cores, 128 GiB of RAM, equipped with an Nvidia Quadro GP100 with 16 GiB of global memory. The code executed by the CPU is written in C/C++, while the GPU code is written using CUDA. We use the GNU compiler and use the O3 optimization flag for all experiments.

We summarize the different implementations we test as follows. For clarity, we differentiate between similar algorithm components since they may use slightly different experimental configurations. For example, we make the distinction between the CPU component of HEGJOIN (HEGJOIN-CPU) and the original SUPER-EGO algorithm due to slight variations in their configurations.

LBJoin is the GPU algorithm proposed by Gallet and Gowanlock [11], uses 3 GPU streams (managed by 3 CPU threads), 256 threads per block, $n_s = 5 \times 10^7$ key/value pairs, 64-bit floats, and n_q^{GPU} is given by the batch estimator presented in Sect. 4.2. **Super-EGO** is the CPU algorithm developed by Kalashnikov [15], uses 16 CPU threads (on 16 physical cores) and 32-bit floats. **SEGO-New** is our optimized version of SUPER-EGO as presented in Sect. 4.4, using 16 CPU threads, 64-bit floats and the sorting by workload strategy. **HEGJoin-GPU** is the GPU component of HEGJOIN, using the same configuration as LBJOIN and our shared work queue (Sect. 4.1). **HEGJoin-CPU** is the CPU component of HEGJOIN, using the same configuration as SEGO-NEW and our shared work queue, with $n_q^{CPU} = 1,024$ (Sect. 4.1). Finally, **HEGJoin** is our heterogeneous algorithm using the shared work queue (Sect. 4.1) to combine HEGJOIN-CPU and HEGJOIN-GPU.

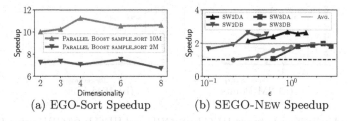

(a) EGO-Sort Speedup   (b) SEGO-New Speedup

Fig. 5. (a) Speedup to EGO-Sort our synthetic datasets using SAMPLE_SORT from the Boost library over QSORT from the C standard library. $S = 0$–9.39K and $S = 0$–1.99K on the 2M and 10M points datasets, respectively. (b) Speedup of SEGO-New over Super-EGO on the SW- real-world datasets.

LBJoin, Super-EGO, SEGO-New and HEGJoin are standalone, and thus compute all the work. HEGJoin-CPU and HEGJoin-GPU, as part of our HEGJoin algorithm, each compute a fraction of the work.

The response times presented are averaged over three trials, and include the end-to-end computation time, i.e., the time to construct the grid index on the GPU, sort by workload, reorder the dimensions and to EGO-sort, and the time to join. Note that some of these time components may overlap (e.g., EGO-sort and GPU computation may occur concurrently).

5.3 Selectivity

We report the selectivity (defined by Kalashnikov [15]) of our experiments as a function of ϵ. We define the selectivity $S = (|R| - |D|)/|D|$, where R is the result set and D is the input dataset. Across all experiments, selectivity ranges from 0 on the *Expo8D2M* and *Expo8D10M* datasets to 13, 207 on the *SW3DA* dataset. We include the selectivity when we compare the performance of all algorithms.

5.4 Results

Performance of SEGO-New: We evaluate the optimized version of Super-EGO and denoted as SEGO-New. The major optimizations include a different sorting algorithm, sorting by workload strategy and work queue (Sect. 4.4).

We evaluate the performance of EGO-sort using the parallel SAMPLE_SORT algorithm from the C++ Boost library over the QSORT algorithm from the C standard library. SAMPLE_SORT is used by SEGO-New (and thus by HEGJoin), while QSORT is used by Super-EGO. Figure 5(a) plots the speedup of SAMPLE_SORT over QSORT on our synthetic datasets. We observe an average speedup of 7.18× and 10.55× on the 2M and 10M points datasets, respectively. Note that we elect to use the SAMPLE_SORT as the EGO-sort needs to be stable.

Figure 5(b) plots the speedup of SEGO-New over Super-EGO on the *SW*-real-world datasets. SEGO-New achieves an average speedup of 1.97× over Super-EGO. While SEGO-New uses 64-bit floats, Super-EGO only uses 32-bit floats and is thus advantaged compared to SEGO-New. We explain this

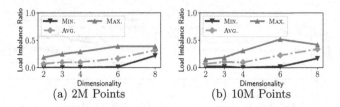

(a) 2M Points   (b) 10M Points

Fig. 6. Load imbalance between HEGJOIN-CPU and HEGJOIN-GPU using HEGJOIN across our synthetic datasets spanning $d = 2$–8. $S = 157$–$9.39K$ on the 2M points datasets (a), and $S = 167$ – $1.99K$ on the 10M points datasets (b).

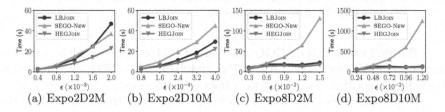

(a) Expo2D2M   (b) Expo2D10M   (c) Expo8D2M   (d) Expo8D10M

Fig. 7. Response time vs. ϵ of LBJOIN, SEGO-NEW and HEGJOIN on 2M and 10M point synthetic datasets in 2-D and 8-D. S is in the range (a) 397–9.39K, (b) 80–1.99K, (c) 0–157 and (d) 0–167.

overall speedup by using SAMPLE_SORT over QSORT, and the sorting by workload strategy with the work queue. Therefore, SEGO-NEW largely benefits from balancing the workload between its threads and from using the work queue.

Evaluating the Load Balancing of the Shared Work Queue: In this section, we evaluate the load balancing efficiency of our shared work queue, which can be characterized as the time difference between the CPU and the GPU ending their respective work. Indeed, a time difference close to 0 indicates that both CPU and GPU components of HEGJOIN ended their work at a similar time, and therefore that their workload was balanced.

Figure 6 plots the ratio of load imbalance as the time difference between the CPU (HEGJOIN-CPU) and the GPU (HEGJOIN-GPU) ending their respective work over the total response time of the application on our exponentially distributed synthetic datasets spanning 2 to 8 dimensions with (a) 2M and (b) 10M points. While we can observe a relatively high maximum load imbalance, these cases arise when the selectivity, and so the workload, are low. As the workload increases our deque becomes more efficient, and the load imbalance is reduced.

Performance of HEGJoin: We now compare the overall response time vs. ϵ of HEGJOIN, LBJOIN, and SEGO-NEW. Note that we decide to use SEGO-NEW instead of SUPER-EGO as it performs consistently better (Fig. 5(b)).

Figure 7 plots the response time vs. ϵ of HEGJOIN, LBJOIN and SEGO-NEW on (a) *Expo2D2M*, (b) *Expo2D10M*, (c) *Expo8D2M* and (d) *Expo8D10M*. We select these datasets as they constitute the minimum and maximum in terms

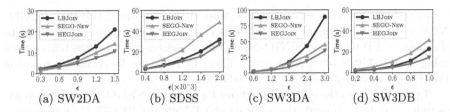

Fig. 8. Response time vs. ϵ of LBJOIN, SEGO-NEW and HEGJOIN on the (a) *SW2DA*, (b) *SDSS*, (c) *SW3DA* and (d) *SW3DB* real-world datasets. S is in the range (a) 295–5.82K, (b) 1–31, (c) 239–13.20K and (d) 33–2.13K.

Table 2. Query throughput (queries/s) of LBJOIN, SEGO-NEW, the upper bound of LBJOIN plus SEGO-NEW, HEGJOIN, and the performance ratio between HEGJOIN and the upper bound across several datasets.

Dataset	ϵ	S	LBJOIN	SEGO-NEW	Upper Bound	HEGJOIN	Perf. Ratio
Expo2D2M	2.0×10^{-3}	9,392	42,589	53,996	96,585	87,681	0.91
Expo4D2M	1.0×10^{-2}	9,262	14,745	13,557	28,302	26,382	0.93
Expo8D2M	1.5×10^{-2}	157	88,968	15,264	104,232	115,674	1.11
Expo2D10M	4.0×10^{-4}	1,985	340,252	221,288	561,540	451,875	0.80
Expo4D10M	4.0×10^{-3}	1,630	142,816	49,704	192,520	217,297	1.13
Expo8D10M	1.2×10^{-2}	167	77,507	8,055	85,562	90,654	1.06
SW2DA	1.5×10^{0}	5,818	88,749	130,942	219,691	176,574	0.80
SDSS	2.0×10^{-3}	31	485,508	314,771	798,834	567,086	0.71
SW3DA	3.0×10^{0}	13,207	20,930	41,143	62,073	53,093	0.86

of size and dimensionality among our synthetic datasets, and we observe similar results on the intermediate datasets of different dimensionality and size. Hence, on such datasets, while SEGO-NEW performs relatively well on the *Expo2D2M* dataset (Fig. 7(a)) and even better than LBJOIN when $1.6 < \epsilon$, HEGJOIN performs better than LBJOIN and SEGO-NEW in all subfigures (Fig. 7(a)–(d)). However, as dimensionality increases and as the performance of SEGO-NEW decreases in 8 dimensions (Fig. 7(c) and (d)), combining it with LBJOIN does not significantly improve performance. Thus, as SEGO-NEW (and therefore as HEGJOIN-CPU as well) do not scale well with dimensionality, the performance of HEGJOIN relies nearly exclusively on the performance of its GPU component HEGJOIN-GPU. HEGJOIN achieves a speedup of up to 2.1× over only using LBJOIN on the *Expo2D2M* dataset (Fig. 7(a)), and up to 11.3× over SEGO-NEW on the *Expo8D10M* dataset (Fig. 7(d)).

Figure 8 plots the response time vs. ϵ of LBJOIN, SEGO-NEW and HEGJOIN on our real-world datasets. Similarly to Fig. 7, as these datasets span 2 and 3 dimensions, the performance of SEGO-NEW is better than LBJOIN on *SW2DA* and *SW3DA* (Fig. 8(a) and (c), respectively). Thus, using HEGJOIN always improves performance over only using LBJOIN or SEGO-NEW. We achieve a speedup of up to 2.5× over using LBJOIN on the *SW3DA* dataset (Fig. 8(c)), and up to 2.4× over SEGO-NEW on the *SDSS* dataset (Fig. 8(b)).

From Figs. 7 and 8, we observe that using HEGJOIN over LBJOIN or SEGO-NEW is beneficial. Typically, in the worst case, HEGJOIN performs similarly to the best of LBJOIN and SEGO-NEW, and consistently performs better than using just LBJOIN or SEGO-NEW. Thus, there is no disadvantage to using HEGJOIN instead of LBJOIN or SEGO-NEW.

Table 2 presents the query throughput for LBJOIN, SEGO-NEW, HEGJOIN, as well as an upper bound (the addition of LBJOIN and SEGO-NEW respective throughput), and the ratio of the throughput HEGJOIN achieves compared to this perfect throughput. The query throughput corresponds to the size of the dataset divided by the response time of the algorithm, as shown in Figs. 7 and 8. We observe a high performance ratio, demonstrating that we almost reach a performance upper bound. Moreover, we also observe that on the *Expo4D10M* and the 8-D datasets, we achieve a ratio of more than 1. We explain this by the fact that LBJOIN's throughput includes query points with a very low workload, thus increasing its overall throughput compared to what HEGJOIN-GPU achieves. Similarly, SEGO-NEW's throughput includes query points with a very large workload, thus reducing its overall throughput compared to what HEGJOIN-CPU achieves. When combining the two algorithms, we have the GPU computing the query points with the largest workload and the CPU the points with the smallest workload. The respective throughput of each component should, therefore, be lower for the GPU and higher for the CPU, than their throughput when computing the entire dataset. Moreover, performance ratios lower than 1 indicate that there are several bottlenecks, including memory bandwidth limitations, with the peak bandwidth potentially reached when storing the results from the CPU and the GPU. We particularly observe this on low dimensionality and for low selectivity, as it yields less computation and a higher memory pressure than in higher dimensions or for higher selectivity (Figs. 7 and 8). We confirm this by examining the ratio of kernel execution time over the time to compute all batches, using only the GPU. Focusing on the datasets with the minimum and maximum performance ratio from Table 2, we find that *SDSS* has a kernel execution time ratio of 0.16, while *Expo4D10M* a kernel execution time ratio of 0.72. Hence, most of the *SDSS* execution time is spent on memory operations, while *Expo4D10M* execution time is mostly spent on computation. When executing HEGJOIN on *SDSS* (and other datasets with low ratios in Table 2), we observe that the use of the GPU hinders the CPU by using a significant fraction of the total available memory bandwidth.

6   Conclusion

The distance similarity join transitions from memory- to compute-bound as dimensionality increases. Therefore, the GPU's high computational throughput and memory bandwidth make the architecture effective at distance similarity searches. The algorithms used in HEGJOIN for the CPU and GPU have their own performance advantages: SEGO-NEW (LBJOIN) performs better on lower (higher) dimensions. By combining these algorithms, we exploit more computational resources, and each algorithm's inherent performance niches.

To enable these algorithms to efficiently compute the self-join, we use a double-ended queue that distributes and balances the work between the CPU and GPU. We find that HEGJOIN achieves respectable performance gains over the CPU- and GPU-only algorithms, as HEGJOIN typically achieves close to the upper bound query throughput. Thus, studying other hybrid CPU-GPU algorithms for advanced database systems is a compelling future research direction.

Acknowledgement. This material is based upon work supported by the National Science Foundation under Grant No. 1849559.

References

1. Alam, S., Albareti, F., Prieto, C., et al.: The eleventh and twelfth data releases of the sloan digital sky survey: final data from SDSS-III. Astrophys. J. Suppl. Ser. **219**, 12 (2015)
2. Awad, M.A., Ashkiani, S., Johnson, R., Farach-Colton, M., Owens, J.D.: Engineering a High-performance GPU B-Tree. In: Proceedings of the 24th Symposium on Principles and Practice of Parallel Programming, pp. 145–157 (2019)
3. Bayer, R., McCreight, E.M.: Organization and maintenance of large ordered indexes. Acta Informatica **1**(3), 173–189 (1972)
4. Bellman, R.: Adaptive Control Processes: A Guided Tour. Princeton University Press, Princeton (1961)
5. Bellman, R.: Adaptive Control Processes: A Guided Tour. Princeton University Press, Princeton (1961)
6. Bentley, J.L.: Multidimensional binary search trees used for associative searching. Commun. ACM **18**(9), 509–517 (1975)
7. Böhm, C., Braunmüller, B., Krebs, F., Kriegel, H.P.: Epsilon grid order: an algorithm for the similarity join on massive high-dimensional data. In: Proceedings of the ACM SIGMOD International Conference on Management of Data, pp. 379–388 (2001)
8. Böhm, C., Noll, R., Plant, C., Zherdin, A.: Index-supported similarity join on graphics processors, pp. 57–66 (2009)
9. Comer, D.: The ubiquitous B-tree. ACM Comput. Surv. **11**(2), 121–137 (1979)
10. Finkel, R.A., Bentley, J.L.: Quad trees: a data structure for retrieval on composite keys. Acta Informatica **4**(1), 1–9 (1974)
11. Gallet, B., Gowanlock, M.: Load imbalance mitigation optimizations for GPU-accelerated similarity joins. In: Proceedings of the 2019 IEEE International Parallel and Distributed Processing Symposium Workshops, pp. 396–405 (2019)
12. Gowanlock, M.: KNN-joins using a hybrid approach: exploiting CPU/GPU workload characteristics. In: Proceedings of the 12th Workshop on General Purpose Processing Using GPUs, pp. 33–42 (2019)
13. Gowanlock, M., Karsin, B.: Accelerating the similarity self-join using the GPU. J. Parallel Distrib. Comput. **133**, 107–123 (2019)
14. Guttman, A.: R-trees: a dynamic index structure for spatial searching. SIGMOD Rec. **14**(2), 47–57 (1984)
15. Kalashnikov, D.V.: Super-EGO: fast multi-dimensional similarity join. VLDB J. **22**(4), 561–585 (2013)

16. Kim, J., Jeong, W., Nam, B.: Exploiting massive parallelism for indexing multi-dimensional datasets on the GPU. IEEE Trans. Parallel Distrib. Syst. **26**(8), 2258–2271 (2015)
17. Kim, J., Kim, S.G., Nam, B.: parallel multi-dimensional range query processing with R-trees on GPU. J. Parallel Distribut. Comput. **73**(8), 1195–1207 (2013)
18. Kim, J., Nam, B.: Co-processing heterogeneous parallel index for multi-dimensional datasets. J. Parallel Distrib. Comput. **113**, 195–203 (2018)
19. Lieberman, M.D., Sankaranarayanan, J., Samet, H.: A fast similarity join algorithm using graphics processing units. In: 2008 IEEE 24th International Conference on Data Engineering, pp. 1111–1120 (2008)
20. MIT Haystack Observatory: Space Weather Datasets. ftp://gemini.haystack.mit.edu/pub/informatics/dbscandat.zip. Accessed 27 Feb 2020
21. Prasad, S.K., McDermott, M., He, X., Puri, S.: GPU-based parallel R-tree construction and querying. In: 2015 IEEE International Parallel and Distributed Processing Symposium Workshops, pp. 618–627 (2015)
22. Sellis, T., Roussopoulos, N., Faloutsos, C.: The R+-tree: a dynamic index for multi-dimensional objects. In: Proceedings of the 13th VLDB Conference, pp. 507–518 (1987)
23. Shahvarani, A., Jacobsen, H.A.: A Hybrid B+-tree as solution for in-memory indexing on CPU-GPU heterogeneous computing platforms. In: Proceedings of the International Conference on Management of Data, pp. 1523–1538 (2016)
24. Yan, Z., Lin, Y., Peng, L., Zhang, W.: Harmonia: a high throughput B+tree for GPUs. In: Proceedings of the 24th Symposium on Principles and Practice of Parallel Programming, pp. 133–144 (2019)

String Joins with Synonyms

Gwangho Song[1], Hongrae Lee[2], Kyuseok Shim[1(✉)], Yoonjae Park[1],
and Wooyeol Kim[1]

[1] Seoul National University, Seoul, South Korea
{ghsong,yjpark,wykim}@kdd.snu.ac.kr, kshim@snu.ac.kr
[2] Google, Mountain View, USA
mr.hongrae.lee@gmail.com

Abstract. String matching is a fundamental operation in many applications such as data integration, information retrieval and text mining. Since users express the same meaning in a variety of ways that are not textually similar, existing works have proposed variants of Jaccard similarity by using synonyms to consider semantics beyond textual similarities. However, they may produce a non-negligible number of false positives in some applications by employing set semantics and miss some true positives due to approximations. In this paper, we define new match relationships between a pair of strings under synonym rules and develop an efficient algorithm to verify the match relationships for a pair of strings. In addition, we propose two filtering methods to prune non-matching string pairs. We also develop join algorithms with synonyms based on the filtering methods and the match relationships. Experimental results with real-life datasets confirm the effectiveness of our proposed algorithms.

Keywords: String matching · String join with synonyms

1 Introduction

There is a wide range of applications where string matching is considered as a fundamental operation. Such applications include information retrieval [4], text mining, [9,12], data integration and deduplication [6,13,17,18,22]. Many studies for string matching rely on the string similarity measures such as edit-distance and Jaccard similarity (e.g., [5,10,11,14,15,21,23]). Since such measures focus on syntactic similarity without recognizing the semantics, they fail to match two strings that are textually very dissimilar but have the same meaning, or match two strings incorrectly that are textually very similar but have different semantics.

Several studies [2,15,20] use synonyms to consider semantics beyond textual similarities. In [2], semantic relationship between strings is modeled as transformation rules, and a variant of Jaccard similarity called *JaccT* is proposed. To compute *JaccT* between a pair of strings, it transforms both strings with all possible rules and then measures the maximum Jaccard similarity between

© Springer Nature Switzerland AG 2020
Y. Nah et al. (Eds.): DASFAA 2020, LNCS 12114, pp. 389–405, 2020.
https://doi.org/10.1007/978-3-030-59419-0_24

the transformed strings. Another variant called *Selective-Jaccard* proposed in [15] expands each string by appending applicable synonyms selectively to maximize Jaccard similarity between the expanded strings. As shown in [20], when a sequence of tokens has multiple synonyms, *JaccT* and *Selective-Jaccard* have low precision due to *topic drift*. For example, if a token CA has multiple synonyms Canada and California, the value of *JaccT* between Canada and California is 1, even though they have different semantics. Thus, the *pkduck* similarity was proposed in [20] and it uses the maximum Jaccard similarity between a transformed string of one string and the other string.

The proposed join algorithms with such similarity measures miss some true positives since they produce approximate results only due to NP-hardness of computing the similarities. Furthermore, they yield a non-negligible number of false positives by ignoring the order of tokens. Motivated by the observations, we propose new match relationships between strings with synonym rules. Two strings are considered to be equivalent if one of them is exactly transformed to the other by applying synonym rules. If we look at the general area of string matching, there are many application scenarios where the order of tokens is important and false positives from set semantics are not desirable. For example, William Gates is equivalent to Bill Gates, given that Bill has a synonym William. However, Warren William (who is an actor) and William Warren (who is a baseball player) are not matched. Note that the Jaccard-based measures in [2,15,20] match them incorrectly. Our goal is to provide an alternative angle on the problem of string matching with synonyms and develop the techniques that can consider the order of tokens, complementing prior work.

Our contributions are summarized as follows. We first define the novel match relationships for a pair of strings with synonyms by considering the order of tokens in a string and develop the *MATCH-DP* algorithm based on dynamic programming to determine the match relationships for a pair of strings. Based on the filter-verification framework [11], we next develop the novel filtering methods that prune a large number of non-matching string pairs by using the properties of the match relationships. We also propose several algorithms to efficiently perform the string join with synonyms by using the *MATCH-DP* algorithm as well as both filtering methods. Finally, we conduct a performance study with real-life datasets to show the effectiveness of our algorithms.

2   Related Work

The string similarity join problems have been extensively studied in [2,5,10, 11,14,15,20,21,23]. To exploit the semantic relationships such as synonyms and abbreviations, the set-based similarity measures *JaccT* [2], *Selective-Jaccard* [15] and *pkduck* [20] are proposed. Although computing *JaccT* is NP-hard, it takes polynomial time using maximum-matching if both left-hand and right-hand sides of every rule are single tokens [2]. Furthermore, since computing *Selective-Jaccard* and *pkduck* are also NP-hard, the polynomial time approximation algorithms are used to compute both measures in [15] and [20], respectively. *JaccT* and

Selective-Jaccard have low precision due to topic drift [20]. Thus, *pkduck* uses the maximum Jaccard similarity between a derived string of one string and the other string. Furthermore, in [20], the signature set of a string for the prefix filtering is calculated in polynomial time, while the signature set is computed in exponential time in [2] and [15].

Such set-based similarity measures may produce a non-negligible number of false positive results for some applications by ignoring the order of tokens. In contrast to them, our work cares about the order of tokens. Furthermore, our proposed join algorithms produce the exact join results in polynomial time, while the join algorithms with *Selective-Jaccard* and *pkduck* may miss matching pairs due to approximation.

Similar to our proposed algorithms in this paper, many works treat strings as sequences. While an algorithm for top-k similarity search with edit distance constraints is proposed in [9], the problem of similarity join using the edit distance is studied in [14]. Recently, many works employ sequence-to-sequence neural network models to learn semantic representations of word sequences [7,19].

String similarity join algorithms with a syntactic or a semantic similarity measure generally utilize the filter-verification framework [5,11,14,15,20,21,23]. Similarly, our proposed algorithms also use the filter-verification framework to efficiently process the string join operation.

3 Preliminaries

We first provide the definitions of strings and synonym rules. We next present the match relationships and the problem of the string join with the synonym rules.

3.1 Strings and Synonym Rules

A *string* s is a sequence of tokens $\langle s(1), \ldots, s(|s|) \rangle$ where $s(i)$ is the i-th token, and $|s|$ is the number of tokens in s. For a string s, $s[i,j]$ is the substring $\langle s(i), \ldots, s(j) \rangle$ with $1 \leq i \leq j \leq |s|$. When $i > j$, $s[i,j]$ is the empty string and we denote it by ε. For a pair of strings s and t, the concatenation of s and t is denoted by $s \oplus t$. In addition, if t is a suffix of s, the prefix $s[1, |s| - |t|]$ of s is represented by $s \ominus t$ (e.g., $\langle$San, Francisco, Airport$\rangle \ominus \langle$Francisco, Airport$\rangle = \langle$San$\rangle$).

We model synonyms as rules that define the transformation of a string into another string, as done in [2,20]. A *rule* $r = r.lhs \rightarrow r.rhs$ represents that a string $r.lhs$ in a string can be replaced with a string $r.rhs$ without changing its meaning. The strings $r.lhs$ and $r.rhs$ in the rule r are called the left-hand side (LHS) and right-hand side (RHS), respectively. We assume that there is always the *self-rule* $\langle a \rangle \rightarrow \langle a \rangle$ for every token a in a string. When $r.lhs$ is a substring of a string s, we say r is *applicable* to s. For instance, the rule $r_7 = \langle$NYU$\rangle \rightarrow \langle$New, York, Univ$\rangle$ is applicable to the string $s_1 = \langle$NYU, Library$\rangle$. Furthermore, we say that r is *applied* to s when $r.lhs$ in s is replaced by $r.rhs$.

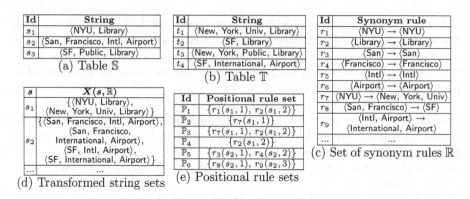

Id	String
s_1	⟨NYU, Library⟩
s_2	⟨San, Francisco, Intl, Airport⟩
s_3	⟨SF, Public, Library⟩

(a) Table $\mathbb{S}$

Id	String
t_1	⟨New, York, Univ, Library⟩
t_2	⟨SF, Library⟩
t_3	⟨New, York, Public, Library⟩
t_4	⟨SF, International, Airport⟩

(b) Table $\mathbb{T}$

Id	Synonym rule
r_1	⟨NYU⟩ → ⟨NYU⟩
r_2	⟨Library⟩ → ⟨Library⟩
r_3	⟨San⟩ → ⟨San⟩
r_4	⟨Francisco⟩ → ⟨Francisco⟩
r_5	⟨Intl⟩ → ⟨Intl⟩
r_6	⟨Airport⟩ → ⟨Airport⟩
r_7	⟨NYU⟩ → ⟨New, York, Univ⟩
r_8	⟨San, Francisco⟩ → ⟨SF⟩
r_9	⟨Intl, Airport⟩ → ⟨International, Airport⟩
...	...

s	$X(s, \mathbb{R})$
s_1	{⟨NYU, Library⟩, ⟨New, York, Univ, Library⟩}
s_2	{⟨San, Francisco, Intl, Airport⟩, ⟨San, Francisco, International, Airport⟩, ⟨SF, Intl, Airport⟩, ⟨SF, International, Airport⟩}
...	...

(d) Transformed string sets

Id	Positional rule set
$\mathbb{P}_1$	$\{r_1(s_1, 1), r_2(s_1, 2)\}$
$\mathbb{P}_2$	$\{r_7(s_1, 1)\}$
$\mathbb{P}_3$	$\{r_7(s_1, 1), r_2(s_1, 2)\}$
$\mathbb{P}_4$	$\{r_2(s_1, 2)\}$
$\mathbb{P}_5$	$\{r_3(s_2, 1), r_4(s_2, 2)\}$
$\mathbb{P}_6$	$\{r_8(s_2, 1), r_9(s_2, 3)\}$

(e) Positional rule sets

(c) Set of synonym rules $\mathbb{R}$

Fig. 1. Strings sets $\mathbb{S}$ and $\mathbb{T}$ with a rule set $\mathbb{R}$

If there are multiple occurrences of $r.lhs$ in s, the rule r can be applied to different places in s. To denote a specific application, a *positional rule* of s, denoted by $r(s, p)$, represents that the rule r is applicable at the position p of the string s. We refer to the integer range $[p, p + |r.lhs| - 1]$ as the *range of the positional rule* $r(s, p)$. For a rule set $\mathbb{R}$, $\mathbb{R}(s)$ is the set of every positional rule of s and $\mathbb{R}(s, i)$ is the set of every positional rule in $\mathbb{R}(s)$ whose LHS is a suffix of $s[1, i]$. The set $\mathbb{R}(s)$ can be computed in $\Theta(|s| + |\mathbb{R}(s)|)$ time by the AC algorithm [1]. We represent all synonym rules applicable to a string s by a positional rule set $\mathbb{R}(s)$.

Since applying the positional rules with any overlapping range results in an inconsistent interpretation, we only consider those without any overlapping range of them. A positional rule set $\mathbb{P} = \{r_1(s, p_1), \ldots, r_{|\mathbb{P}|}(s, \ p_{|\mathbb{P}|})\} \subseteq \mathbb{R}(s)$ is *valid*, if $p_i + |r_i.lhs| = p_{i+1}$ for all $1 \le i < |\mathbb{P}|$. Furthermore, the *range of* $\mathbb{P}$ is an integer range $[p_1, p_{|\mathbb{P}|} + |r_{|\mathbb{P}|}.lhs| - 1]$ and the RHS of $\mathbb{P} = \{r_1(s, p_1), \ldots, r_{|\mathbb{P}|}(s, p_{|\mathbb{P}|})\}$ is $r_1.rhs \oplus \ldots \oplus r_{|\mathbb{P}|}.rhs$. A valid positional rule set $\mathbb{P}$ of s is *fully valid*, if its range is $[1, |s|]$. For a string s, if we obtain a transformed string s' by applying a fully valid positional rule set $\mathbb{P}$ of s, we denote it by $s \xrightarrow{\mathbb{P}} s'$. We use $X(s, \mathbb{R})$ to represent the set of all transformed strings of s by using $\mathbb{R}$.

Example 1. For the string $s_2 = $ ⟨San, Francisco, Intl, Airport⟩ and the rule set $\mathbb{R}$ in Fig. 1, $\mathbb{R}(s_2) = \{r_3(s_2, 1), r_4(s_2, 2), r_5(s_2, 3), r_6(s_2, 4), r_8(s_2, 1), r_9(s_2, 3)\}$ and $\mathbb{R}(s_2, 2) = \{r_4(s_2, 2), r_8(s_2, 1)\}$. A positional rule set $\mathbb{P}_5 = \{r_3(s_2, 1), r_4(s_2, 2)\}$ of s_2 is valid, since $1 + |r_3.lhs| = 1 + 1 = 2$. The range of $\mathbb{P}_5$ is an integer range $[1, 2]$. Note that $\mathbb{P}_5$ is not fully valid. Furthermore, $RHS(\mathbb{P}_5) = r_3.rhs \oplus r_4.rhs = $ ⟨San, Francisco⟩. The positional rule set $\mathbb{P}_6$ is another valid positional rule set of s_2. Actually, $\mathbb{P}_6$ is fully valid since its range is $[1, 4]$. By applying $\mathbb{P}_6$ to s_2, s_2 is transformed into $s'_2 = $ ⟨SF, International, Airport⟩ (i.e., $s_2 \xrightarrow{\mathbb{P}_6} s'_2$). Figure 1(d) shows the set of all transformed strings of s_2.

3.2   Problem Definition

When matching two strings with synonyms, we can consider the synonyms in a single side only or both sides of the strings, depending on the application context. Thus, we first define the *smatch* and *bmatch* relationships with synonyms. To prevent topic drift, we adopt the idea of the bidirectional equivalence used in *pkduck* [20] rather than that in [15]. Using the match relationships, we next define the problem of the string join with synonyms.

Definition 1 *(The match relationships). Given a set of synonym rules* $\mathbb{R}$, *a string s is matched to a string t, written* $s \mathbin{\underline{\lessgtr}} t$, *if s can be transformed to t. Such a string pair is called a* smatch *pair. When a string s cannot be matched to a string t, we denote it by* $s \mathbin{\underline{\nlessgtr}} t$. *Note that* $s \mathbin{\underline{\lessgtr}} t$ *does not imply* $t \mathbin{\underline{\lessgtr}} s$. *We also refer to a pair of strings s and t as a* bmatch *pair, if* $s \mathbin{\underline{\lessgtr}} t$ *or* $t \mathbin{\underline{\lessgtr}} s$.

Definition 2 *(The string join with synonyms). Consider a pair of string sets* $\mathbb{S}$ *and* $\mathbb{T}$ *with a rule set* $\mathbb{R}$. *The problem is to find the set of all bmatch pairs in* $\mathbb{S} \times \mathbb{T}$. *We denote the set of all bmatch pairs in* $\mathbb{S} \times \mathbb{T}$ *by* $\mathbb{S} \bowtie_{\mathbb{R}} \mathbb{T}$.

Let $\mathbb{S} \ltimes_{\mathbb{R}} \mathbb{T}$ denote the set of all smatch pairs in $\mathbb{S} \times \mathbb{T}$. Since $\mathbb{S} \bowtie_{\mathbb{R}} \mathbb{T} = (\mathbb{S} \ltimes_{\mathbb{R}} \mathbb{T}) \cup (\mathbb{T} \ltimes_{\mathbb{R}} \mathbb{S})$ holds by the definition of the bmatch, if we have an algorithm to compute $\mathbb{S} \ltimes_{\mathbb{R}} \mathbb{T}$, we can calculate $\mathbb{S} \bowtie_{\mathbb{R}} \mathbb{T}$ by computing $(\mathbb{S} \ltimes_{\mathbb{R}} \mathbb{T}) \cup (\mathbb{T} \ltimes_{\mathbb{R}} \mathbb{S})$. Thus, we focus on the algorithms for calculating $\mathbb{S} \ltimes_{\mathbb{R}} \mathbb{T}$ in this paper.

4   String Matching with Synonyms

In this section, we develop the algorithms to determine the smatch relationship for two strings with a rule set $\mathbb{R}$. For a pair of strings s and t, if s is matched to t, there is a rule r transforming a suffix of s to a suffix of t such that $s \ominus r.lhs \mathbin{\underline{\lessgtr}} t \ominus r.rhs$. Consider a smatch pair of strings $s_2 = \langle$San, Francisco, Intl, Airport$\rangle$ and $t_4 = \langle$SF, International, Airport$\rangle$. The suffix $\langle$Intl, Airport$\rangle$ of s_2 can be transformed to the suffix $\langle$International, Airport$\rangle$ of t_4 by applying the rule $r_9 = \langle$Intl, Airport$\rangle \rightarrow \langle$International, Airport$\rangle$. In addition, the prefix $\langle$San, Francisco$\rangle$ of s_2 and the prefix $\langle$SF$\rangle$ of t_4 have the smatch relationship since $\langle$San, Francisco$\rangle$ is transformed to $\langle$SF$\rangle$ by applying $r_8 = \langle$San, Francisco$\rangle \rightarrow \langle$SF$\rangle$.

Based on the observation, we present the optimal substructure [8] of the smatch relationship. Let $m[i, j]$ be a boolean value which is true if $s[1, i] \mathbin{\underline{\lessgtr}} t[1, j]$, and false otherwise. We define $m[i, j]$ recursively as follows. If $i = j = 0$, since $s[1, 0] = t[1, 0] = \varepsilon$, $m[i, j]$ is true. If $(i = 0$ and $j > 0)$ or $(i > 0$ and $j = 0)$, $m[i, j]$ is false since one is an empty string and the other is not. Meanwhile, when $i > 0$ and $j > 0$, $m[i, j]$ is true if there exists a positional rule $r(s, p) \in \mathbb{R}(s, |s|)$ whose RHS is a suffix of $t[1, j]$ and $m[i - |r.lhs|, j - |r.rhs|]$ is true. Otherwise, $m[i, j]$ is false. Thus, the recursive equation becomes

$$m[i, j] = \begin{cases} \text{TRUE} & \text{if } i = 0, j = 0, \\ \text{FALSE} & \text{if } (i = 0 \text{ and } j > 0) \text{ or } (i > 0 \text{ and } j = 0), \\ \text{TRUE} & \text{if } i > 0, j > 0 \text{ and } \exists r(s, p) \in \mathbb{R}(s, i) \text{ s.t.} \\ & \quad \text{(i) } r.rhs \text{ is a suffix of } t[1, j] \text{ and} \\ & \quad \text{(ii) } m[i - |r.lhs|, j - |r.rhs|] \text{ is TRUE}, \\ \text{FALSE} & \text{otherwise.} \end{cases} \quad (1)$$

The *MATCH-DP* algorithm computes $m[i,j]$s by the recurrence equation (1) using dynamic programming, which takes $O(|t|^2 \cdot |\mathbb{R}(s)|)$ time to computing all $m[i,j]$s.

5 Filtering Methods

A common paradigm in the string joins is the filter-verification framework [11]. In the filtering step, it prunes a large number of non-smatch pairs and generates a set of candidate smatch pairs. In the verification step, expensive computation of the smatch relationships for the candidate string pairs is conducted to get the final smatch pairs. Based on the framework, we present two filtering methods.

5.1 The Length Filtering

For a string s and a rule set $\mathbb{R}$, the *maximum* and *minimum transform lengths* of s, denoted by $L_{\max}(s, \mathbb{R})$ and $L_{\min}(s, \mathbb{R})$, are the maximum and minimum lengths of the transformed strings from s, respectively. If $L_{\max}(s, \mathbb{R})$ (respectively, $L_{\min}(s, \mathbb{R})$) is smaller (respectively, larger) than the size of another string t, s cannot be matched to t. Thus, we develop a dynamic programming algorithm to efficiently compute them.

Assume that a rule r is used to transform a suffix of a substring $s[1,i]$ to get a transformed string with the maximum length. Then, we have $L_{\max}(s[1,i], \mathbb{R}) = L_{\max}(s[1, i-|r.lhs|], \mathbb{R}) + |r.rhs|$. Similarly, assume that a rule r' can transform a suffix of a substring $s[1,i]$ to get the minimum transform length. Then, we obtain $L_{\min}(s[1,i], \mathbb{R}) = L_{\min}(s[1, i-|r'.lhs|], \mathbb{R}) + |r'.rhs|$. Thus, the problem of finding $L_{\max}(s, \mathbb{R})$ and $L_{\min}(s, \mathbb{R})$ exhibits the optimal substructure [8]. Let $M[i]$ and $m[i]$ represent $L_{\max}(s[1,i], \mathbb{R})$ and $L_{\min}(s[1,i], \mathbb{R})$, respectively. Their optimal substructure properties give the following recurrences:

$$M[i] = \max_{r \in \mathbb{R}(s,i)} M[i - |r.lhs|] + |r.rhs|, m[i] = \min_{r \in \mathbb{R}(s,i)} m[i - |r.lhs|] + |r.rhs|.$$

Note that we can compute $L_{\max}(s, \mathbb{R})$ and $L_{\min}(s, \mathbb{R})$ in $O(|\mathbb{R}(s)|)$ time.

Length Filtering: To exploit string lengths to reduce the number of candidate pairs, we derive the following lemma to prune non-smatch pairs of strings.

Lemma 1. *For a pair of strings s and t with a rule set $\mathbb{R}$, if $L_{\max}(s, \mathbb{R}) < |t|$ or $|t| < L_{\min}(s, \mathbb{R})$, we have $s \npreceq t$ (i.e., the pair of s and t is a non-smatch pair).*

Proof. Assume that $s \preceq t$. Since $t \in X(s, \mathbb{R})$, we know $L_{\min}(s, \mathbb{R}) \le |t| \le L_{\max}(s, \mathbb{R})$. If $L_{\max}(s, \mathbb{R}) < |t|$, or $|t| < L_{\min}(s, \mathbb{R})$, we have $L_{\max}(s, \mathbb{R}) < |t| \le L_{\max}(s, \mathbb{R})$ or $L_{\min}(s, \mathbb{R}) \le |t| < L_{\min}(s, \mathbb{R})$ which contradict $s \preceq t$. Thus, we have $s \npreceq t$.

Given strings s and t, if $L_{\max}(s, \mathbb{R}) < |t|$ or $|t| < L_{\min}(s, \mathbb{R})$, we conclude $s \npreceq t$ without checking the smatch relationship.

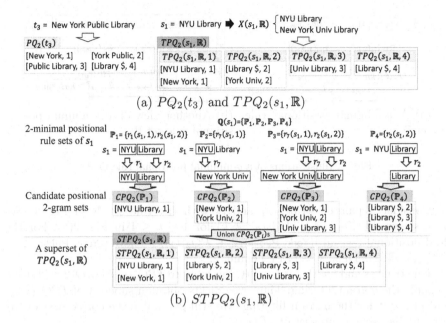

(a) $PQ_2(t_3)$ and $TPQ_2(s_1, \mathbb{R})$

(b) $STPQ_2(s_1, \mathbb{R})$

Fig. 2. An example of $PQ_2(t_3)$, $TPQ_2(s_1, \mathbb{R})$ and $STPQ_2(s_1, \mathbb{R})$

5.2   The q-gram Filtering

For a string s and a positive integer q, let $\tilde{s}$ be the padded string of s by adding $(q-1)$ special tokens, denoted by '$\$$', at the end of s. We call $g = \tilde{s}[k, k+q-1]$ the k-th q-gram of s with $1 \le k \le |s|$ and refer to the value k as its *position*. The k-th positional q-gram of the string s is also denoted by $[g, k]$. If a string s is matched to a string t with a rule set $\mathbb{R}$, every k-th positional q-gram of the string t should be the k-th positional q-gram of a transformed string of s.

Let us introduce the definitions to be used to illustrate the pruning by positional q-grams. $PQ_q(s)$ is the set of all positional q-grams of s, $TPQ_q(s, \mathbb{R})$ is the set of all positional q-grams appearing in a transformed string of s and $TPQ_q(s, \mathbb{R}, k)$ is the set of all k-th positional q-grams in $TPQ_q(s, \mathbb{R})$. As Figure 2(a) shows, the set of all positional 2-grams of the string t_3 (i.e. $PQ_2(t_3)$) is $\{[\langle\text{New, York}\rangle, 1], [\langle\text{York, Public}\rangle, 2], [\langle\text{Public, Library}\rangle, 3], [\langle\text{Library}, \$\rangle, 4]\}$. We also show $TPQ_2(s_1, \mathbb{R}, k)$ with $1 \le k \le 4$ and their union $TPQ_2(s_1, \mathbb{R})$.

Pruning by Positional q-grams: The following lemma allows us to prune non-smatch pairs by using the positional q-grams.

Lemma 2. *For a pair of strings s and t with a rule set $\mathbb{R}$, we have $s \npreceq t$, if there is a position k with $1 \le k \le |t|$ such that $[t[k, k+q-1], k] \notin TPQ_q(s, \mathbb{R}, k)$.*

Proof. We will prove the lemma by providing its contrapositive. If s is matched to t, the string t belongs to $X(s, \mathbb{R})$ and thus the k-th positional q-gram $[\tilde{t}[k, k+q-1], k]$ of t is contained in $TPQ_q(s, \mathbb{R}, k)$ for every $1 \le k \le |t|$.

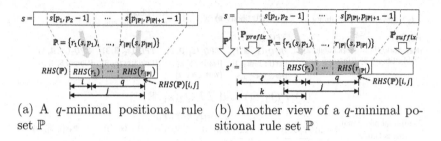

(a) A q-minimal positional rule set $\mathbb{P}$ (b) Another view of a q-minimal positional rule set $\mathbb{P}$

Fig. 3. Illustrations of a q-minimal positional ruleset $\mathbb{P}$

Consider a pair of strings s_1 and t_3 with the rule set $\mathbb{R}$ in Fig. 1. We have shown $PQ_2(t_3)$ and $TPQ_2(s_1, \mathbb{R}, k)$ for $1 \leq k \leq 4$ in Fig. 2(a). For the 2-nd positional 2-gram $[\langle \text{York, Public} \rangle, 2] \in PQ_2(t_3)$, we have $[\langle \text{York, Public} \rangle, 2] \notin TPQ_2(s_1, \mathbb{R}, 2)$. Thus, $s_1 \not\gtrsim t_3$ by Lemma 2.

For strings s and t with a rule set $\mathbb{R}$, we compute $TPQ_q(s, \mathbb{R})$ in $O(2^{|\mathbb{R}(s)|})$ time to apply the q-gram filtering. Thus, we quickly obtain a superset of $TPQ_q(s, \mathbb{R})$ and utilize it for the q-gram filtering. We still guarantee the correctness of the pruning by using a superset of $TPQ_q(s, \mathbb{R})$.

A q-minimal Positional Rule Set: To obtain a superset of $TPQ_q(s, \mathbb{R})$, we apply only the minimal-sized *valid* positional rule sets $\mathbb{P}_1, ..., \mathbb{P}_m$ of a string s such that every q-gram of all transformed strings of s appears in one of the $RHS(\mathbb{P}_i)$s. We then produce the positional q-grams by annotating the obtained q-grams with the possible positions.

For a string s with a rule set $\mathbb{R}$, let $\mathbb{P} = \{r_1(s, p_1), ..., r_{|\mathbb{P}|}(s, p_{|\mathbb{P}|})\}$ be a valid positional rule set of s. Figure 3(a) illustrates the relationship between $s[p_i, p_{i+1} - 1]$ and $RHS(r_i)$ for a string s, $\mathbb{P}$ and $RHS(\mathbb{P})$. Assume that the first and last tokens of $RHS(\mathbb{P})[i, j]$ appear in $r_1.rhs$ and $r_{|\mathbb{P}|}.rhs$, respectively. We represent the q-gram by a shaded rectangle in Fig. 3(a). Then, we call such a $\mathbb{P}$ a *q-minimal positional rule set* of s.

Definition 3. *For a string s with a rule set $\mathbb{R}$, consider a valid positional rule set $\mathbb{P} = \{r_1(s, p_1), ..., r_{|\mathbb{P}|}(s, p_{|\mathbb{P}|})\}$ of the string s such that $p_1 < p_2 < \cdots < p_{|\mathbb{P}|}$. Then, $\mathbb{P}$ is a q-minimal positional rule set if there is a q-gram $RHS(\mathbb{P})[i, j]$ such that $1 \leq i \leq |r_1.rhs|$ and $|RHS(\mathbb{P})| - |r_{|\mathbb{P}|}.rhs| < j = i + q - 1 \leq |r_1.rhs| + q - 1$.*

Example 2. Consider the string s_1 with its valid positional rule set $\mathbb{P}_3 = \{r_7(s_1, 1), r_2(s_1, 2)\}$ in Fig. 1. For the 2-gram $RHS(\mathbb{P}_3)[3, 4] = \langle \text{Univ, Library} \rangle$, the tokens Univ and Library appear in $r_7.rhs$ and $r_2.rhs$, respectively. Thus, $\mathbb{P}_3$ is a 2-minimal positional rule set of s_1 by Definition 3.

Generation of a Superset of $TPQ_q(s, \mathbb{R})$: For a string s and its transformed string s' obtained by applying a fully valid positional rule set $\mathbb{P}'$ to s, Fig. 3(b) shows the relationship between a positional q-gram $[g, k]$ in s' and a q-minimal positional rule set $\mathbb{P} \subseteq \mathbb{P}'$. Since the transform length ℓ of the substring $s[1, p_1 - 1]$

ranges from $L_{\min}(s[1, p_1 - 1], \mathbb{R})$ to $L_{\max}(s[1, p_1 - 1], \mathbb{R})$, $RHS(\mathbb{P})$ contains the positional q-gram $[g, k]$ if g is the i-th q-gram of $RHS(\mathbb{P})$ and $L_{\min}(s[1, p_1 - 1], \mathbb{R}) + i \leq k \leq L_{\max}(s[1, p_1 - 1], \mathbb{R}) + i$. Thus, we obtain the following lemma.

Lemma 3. *Consider a string s, a rule set $\mathbb{R}$ and a positional q-gram $[g, k]$ in a transformed string of s. Then, there always exists a q-minimal positional rule set $\mathbb{P} = \{r_1(s, p_1), \ldots, r_{|\mathbb{P}|}(s, p_{|\mathbb{P}|})\}$ with a position $1 \leq i \leq |r_1.rhs|$ satisfying that (1) $g = RHS(\mathbb{P})[i, i + q - 1]$ and (2) $L_{\min}(s[1, p_1 - 1], \mathbb{R}) + i \leq k \leq L_{\max}(s[1, p_1 - 1], \mathbb{R}) + i$.*

Due to the lack of space, we briefly sketch how to prove the lemma. For a positional q-gram in a transformed string of the string s, there is a fully valid positional rule set $\mathbb{P}'$ such that $s \xrightarrow{\mathbb{P}'} s'$. We can find a subset of $\mathbb{P}'$ satisfying both conditions in the lemma, which is indeed a q-minimal positional rule set of s.

Example 3. For a string $s_1 = \langle$NYU, Library$\rangle$ and a positional rule set $\mathbb{P}_4$ in Fig. 1, note that $RHS(\mathbb{P}_4)[1, 2]$ contains the 4-th 2-gram $\langle$Library, $\$\rangle$ of the transformed string s_1'. In addition, $L_{\min}(s_1[1, 1], \mathbb{R}) + 1 = 2 \leq 4$ and $4 \leq L_{\max}(s_1[1, 1], \mathbb{R}) + 1 = 4$. Thus, $\mathbb{P}_4$ is a 2-minimal positional rule set of s_1.

If we generate all possible positional q-grams in $RHS(\mathbb{P})$ satisfying both conditions in Lemma 3 for every possible q-minimal positional rule set $\mathbb{P}$ of a string s, we do not miss any positional q-gram in $TPQ(s, \mathbb{R})$. For a q-minimal positional rule set $\mathbb{P} = \{r_1(s, p_1), \ldots, r_{|\mathbb{P}|}(s, p_{|\mathbb{P}|})\}$ of s, let $CPQ_q(\mathbb{P})$ be the set of all positional q-grams $[RHS(\mathbb{P})[i, i + q - 1], k]$ with every pair of i and k such that $1 \leq i \leq |r_1.rhs|$ and $L_{\min}(s[1, p_1 - 1], \mathbb{R}) + i \leq k \leq L_{\max}(s[1, p_1 - 1], \mathbb{R}) + i$. Then, by Lemma 3, $STPQ_q(s, \mathbb{R}) = \bigcup_{\mathbb{P} \in \mathbb{Q}(s)} CPQ_q(\mathbb{P})$ is a superset of $TPQ_q(s, \mathbb{R})$, where $\mathbb{Q}(s)$ is the set of all q-minimal positional rule sets of the string s. For a string $s_1 = \langle$NYU, Library$\rangle$ with $\mathbb{Q}(s_1) = \{\mathbb{P}_1, \mathbb{P}_2, \mathbb{P}_3, \mathbb{P}_4\}$, we show the steps of computing $STPQ_2(s_1, \mathbb{R})$ in Fig. 2(b). We first compute $CPQ_2(\mathbb{P}_i)$ with $1 \leq i \leq 4$ for the string s_1 (e.g., $CPQ_2(\mathbb{P}_4) = \{[\langle$Library, $\$\rangle, 2], [\langle$Library, $\$\rangle, 3], [\langle$Library, $\$\rangle, 4]\}$). We next compute $STPQ_2(s_1, \mathbb{R})$ by taking the union of $CPQ_2(\mathbb{P}_j)$s.

To efficiently compute $STPQ_q(s, \mathbb{R})$, we define a *required q-minimal positional rule* set $\mathbb{P} = \{r_1(s, p_1), \ldots, r_{|\mathbb{P}|}(s, p_{|\mathbb{P}|})\}$ of s if there is no q-minimal positional rule set $\mathbb{P} \cup \{r_{|\mathbb{P}|+1}(s, p_{|\mathbb{P}|+1})\}$ with $p_{|\mathbb{P}|} < p_{|\mathbb{P}|+1}$. We generate all positional q-grams in $CPQ_q(\mathbb{P})$ not for every q-minimal positional rule set $\mathbb{P}$ but for every required q-minimal positional rule set $\mathbb{P}$ of the string s. We next present the *GenSTPQ* algorithm which enumerates every required q-minimal positional rule set $\mathbb{P}$ of the string s and outputs the union of the $CPQ_q(\mathbb{P})$s as $STPQ_q(s, \mathbb{R})$.

The *GenSTPQ* Algorithm: The pseudocode of the *GenSTPQ* algorithm is shown in Algorithm 1. At each position $1 \leq i \leq |s|$ of a string s, it calls *GenSTPQaux*($\mathbb{P}$) with $\mathbb{P} = \{r(s, p)\}$ for each positional rule $r(s, p) \in \mathbb{R}(s, i)$ (lines 2-4). The invocation of *GenSTPQaux*($\mathbb{P}$) produces $CPQ_q(\mathbb{P})$ for every required q-minimal positional rule set $\mathbb{P}$ such that $\mathbb{P}$ contains $r(s, p)$ and the range of $\mathbb{P}$ starts at p. During execution of *GenSTPQaux*($\mathbb{P}$), it adds all the positional q-grams in the $CPQ_q(\mathbb{P})$s to the set $STPQ$. After the for loop in *GenSTPQ* finishes, it returns $STPQ = STPQ_q(s, \mathbb{R})$.

Algorithm 1: GenSTPQ

Input: a string s, a set of rules $\mathbb{R}$
Output: $STPQ_q(s, \mathbb{R})$

1 $STPQ = \emptyset$
2 **for** $i = 1$ *to* $|s|$ **do**
3   **foreach** $r(s, p) \in \mathbb{R}(s, i)$ **do**
4          $GenSTPQaux(\{r(s, p)\})$
5 **return** $STPQ$

6 **Function** GenSTPQaux($\mathbb{P}$)
7      $[p_1, p_2] = range(\mathbb{P})$
8      $r_1(s, p_1) =$ positional rule with the smallest position in $\mathbb{P}$
9   **if** $p_2 \geq |s|$ *or* $|RHS(\mathbb{P})| - |r_1.rhs| \geq q - 1$ **then**
10         $STPQ = STPQ \cup CPQ_q(\mathbb{P})$
11   **else**
12         $\mathbb{R}' = \{r(s, p) \in \mathbb{R}(s) | p = p_2 + 1\}$
13   **foreach** $r(s, p_2 + 1) \in \mathbb{R}'$ **do**
14             $GenSTPQaux(\mathbb{P} \cup \{r(s, p_2 + 1)\})$

Upon calling $GenSTPQaux$ with $\mathbb{P}$ as input, we let $r_1(s, p_1)$ be the positional rule with the smallest position in $\mathbb{P}$ and $[p_1, p_2]$ be the range of $\mathbb{P}$ (lines 7-8). If $p_2 \geq |s|$ or $|RHS(\mathbb{P})| - |r_1.rhs| \geq q - 1$, since $\mathbb{P}$ is a required q-minimal positional rule set, we compute $CPQ_q(\mathbb{P})$ and add all positional q-grams in $CPQ_q(\mathbb{P})$ to $STPQ$ (lines 9-10). Otherwise, we recursively call $GenSTPQaux(\mathbb{P} \cup \{r'(s, p_2 + 1)\})$ for every positional rule $r'(s, p_2 + 1) \in \mathbb{R}(s)$ (lines 12-14).

6   Join Algorithms with Synonyms

For a pair of string sets $\mathbb{S}$ and $\mathbb{T}$, we present our proposed string join algorithms with synonyms based on the smatch relationship.

The *JOIN-NV* Algorithm: Given a pair of string sets $\mathbb{S}$ and $\mathbb{T}$, the *JOIN-NV* algorithm generates every transformed string s' of each string $s \in \mathbb{S}$ with respect to a given rule set $\mathbb{R}$, and outputs the pair (s, t) if s' is the same as a string $t \in \mathbb{T}$. To check efficiently whether a transformed string s' exists in $\mathbb{T}$ or not, it builds a hash table from all strings in $\mathbb{T}$ and utilizes the hash table to check whether a transformed string s' exists in $\mathbb{T}$ or not.

The *JOIN-BK* Algorithm: The *JOIN-BK* algorithm checks the smatch relationship from each string $s \in \mathbb{S}$ only for the remaining strings $t \in \mathbb{T}$ after applying both filtering methods in Sect. 5. For the pairs of strings $s \in \mathbb{S}$ and $t \in \mathbb{T}$ which are not pruned by the filtering methods, we check whether $s \gtrless t$.

To apply the q-gram filtering, it selects the best K positions $P(t) = \{p_1, \ldots, p_K\}$ for each string $t \in \mathbb{T}$ such that the estimated number of strings in $\mathbb{S}$ to be checked for the smatch relationship to t is minimized. Let $CSTPQ_q[g, k]$ be the number of strings $s \in \mathbb{S}$ such that $[g, k] \in STPQ_q(s, \mathbb{R}, k)$. *JOIN-BK* chooses the K positions with the smallest $CSTPQ_q[g,k]$s as $P(t)$, where g is the k-th q-gram of t. For a pair of strings $s \in \mathbb{S}$ and $t \in \mathbb{T}$, if there is a position $k \in P(t)$ such that $\tilde{t}[k, k+q-1] \notin STPQ_q(s, \mathbb{R}, k)$, JOIN-BK prunes the pair. Note that JOIN-BK has to compute $STPQ_q(s, \mathbb{R}, k)$s with $1 \leq k \leq L_{\mathsf{max}}(s, \mathbb{R})$ for every string $s \in \mathbb{S}$. However, since it is expensive to get $CSTPQ_q[g, k]$ by computing $STPQ_q(s, \mathbb{R}, k)$ for every string $s \in \mathbb{S}$, we compute $CSTPQ_q[g, k]$ from a sample $\mathbb{S}_\phi$ of $\mathbb{S}$ with a sample ratio of ϕ. Using samples to speed up may reduce the power of the positional q-gram filtering, but it does not affect the correctness of the join result.

The JOIN-FK Algorithm: While JOIN-BK finds the best K positions to apply the q-gram filtering, the JOIN-FK algorithm simply selects the first K positions of strings without additional computation. That is, $P(t) = \{1, ..., \min(|t|, K)\}$.

The JOIN-HB Algorithm: When strings in $\mathbb{S}$ have a small number of transformed strings, JOIN-NV is a clear winner due to its simplicity. However, if they have at least a fair number of transformed strings, either JOIN-FK or JOIN-BK becomes the best performer depending on the distribution of the applicable rules over the positions of strings in $\mathbb{S}$. To use such observations, we develop the hybrid algorithm JOIN-HB which utilizes the three join algorithms. It first splits $\mathbb{S}$ into two disjoint subsets $\mathbb{S}_1(\theta)$ and $\mathbb{S}_2(\theta) = \mathbb{S} - \mathbb{S}_1(\theta)$ such that the number of transformed strings of every string in $\mathbb{S}_1(\theta)$ is at most a given split threshold θ. After such partitioning, we apply JOIN-NV to join between $\mathbb{S}_1(\theta)$ and $\mathbb{T}$, and apply JOIN-FK or JOIN-BK based on the estimation of execution times to join between $\mathbb{S}_2(\theta)$ and $\mathbb{T}$.

To select the best threshold θ quickly, we use random samples $\mathbb{S}'$ and $\mathbb{T}'$ of $\mathbb{S}$ and $\mathbb{T}$, respectively. We first sort the strings $s \in \mathbb{S}'$ with non-decreasing order of $|X(s, \mathbb{R})|$. However, since it is very expensive to compute $|X(s, \mathbb{R})|$ for every string $s \in \mathbb{S}'$, we instead utilize an upper bound of $|X(s, \mathbb{R})|$ which can be quickly computed by using dynamic programming. We next consider the binary partition based on every split position in the sorted list. For each binary partition, we estimate the execution time of JOIN-HB by assuming that the left partition (i.e., $\mathbb{S}'_1(\theta)$) is processed by JOIN-NV and the right partition (i.e., $\mathbb{S}'_2(\theta)$) is processed by the one with the smaller execution time between JOIN-BK and JOIN-FK. After examining every split position, the position with the smallest estimated execution time of JOIN-HB is selected as the split threshold. Due to the lack of space, we omit the details of JOIN-HB.

7 Experiments

We empirically evaluated the performance of our algorithms using several real-life datasets. All experiments were performed on the machine with Intel(R) Core(TM) i3-3220 3.3 GHz CPU and 8 GB of main memory. All algorithms were implemented in Java 1.8. We ran all algorithms three times and reported the average execution times. We do not report the execution time of an algorithm if it did not finish in four hours or ran out of memory.

7.1 Tested Algorithms

We implemented the *JOIN-NV*, *JOIN-FK*, *JOIN-BK*, and *JOIN-HB* algorithms and compared them with the following state-of-the-art algorithms [14,15,20].

Pass-Join: This is the state-of-the-art algorithm for the string join with edit distance proposed in [14]. We downloaded the C++ source code available at https://github.com/TsinghuaDatabaseGroup/Similarity-Search-and-Join and re-implemented it in Java for a fair comparison. Since it does not utilize the synonyms, we modified the algorithm to enumerate all transformed strings of a string $s \in \mathbb{S}$ and output the smatch pairs (s,t) if there is a transformed string whose edit distance to a string $t \in \mathbb{T}$ is zero.

SI-Join: This is the join algorithm with *Selective-Jaccard* similarity in [15].

Pkduck-Join: It is the join algorithm with *pkduck* similarity in [20].

Since we consider exact match only, we set the minimum threshold to one for *SI-Join* and *Pkduck-Join*. We do not compare our algorithms with the neural network-based models [7,19] since they may fail to find all matching pairs.

7.2 Datasets

To compare the quality of join results by the string join algorithms, we use the following real-life datasets.

NAME: We use 578,979 names of people from Freebase [3].

PLAYER: It contains 30,168 names of football players in www.foot-balldb.com.

UNIV: It consists of 29,935 names of universities and colleges in Wikipedia. The redirection links are used to identify the same entity for the groundtruth and to get synonym rules for all pairs of names with the same redirection link.

CONF: This dataset contains 3,081 conference names with 792 synonym rules used in [15]. We also generated 96 synonym rules for the pairs of frequently used words and their abbreviations (e.g., db → database and database → db).

DRUG: It consists of 36,509 drug names available in http://polysearch.ca. We collected synonyms of the strings from http://www.druglib.com. In total, there are 1,544,568 synonyms rules.

AOL: There are 1M strings from AOL search queries with 215K synonym rules obtained from WordNet[16].

SPROT: It has protein and gene names with their synonyms gathered from Swiss-Prot dataset (http://www.uniprot.org).

USPS: This is a dataset of 1M addresses with 284 synonym rules used in [15].

Table 1. Statistics for the real-life datasets

Dataset		NAME	PLAYER	UNIV	CONF	DRUG	AOL	SPROT	USPS
# of records		578,979	30,168	29,935	3,081	36,509	1,000,000	466,158	1,000,000
# of distinct tokens		250,855	16,501	18,094	945	34,162	228,032	299,424	69,349
# of rules		284	284	39,798	888	1,544,568	215,672	353,577	284
Length of records	Avg	2.19	2.01	3.35	5.93	1.33	4.12	11.51	6.73
	Max	17	4	14	13	19	19	42	15
# of applicable rules per record	Avg	0.19	0.16	2.48	5.32	48.16	10.62	0.28	2.47
	Max	6	4	25	15	1,056	190	179	7

Table 2. Quality of join results

Dataset	JOIN-HB			Pkduck-Join			SI-Join		
	P	R	F1	P	R	F1	P	R	F1
$NAME \bowtie_R PLAYER$	1.0000	0.5866	0.7394	0.9881	0.5867	0.7363	0.9881	0.5867	0.7363
$UNIV \bowtie_R UNIV$	1.0000	1.0000	1.0000	0.9953	0.8792	0.9336	0.9915	0.8856	0.9356
$CONF \bowtie_R CONF$	1.0000	0.6299	0.7729	1.0000	0.6064	0.7550	1.0000	0.6289	0.7722
$DRUG \bowtie_R DRUG$	1.0000	0.3424	0.5101	1.0000	0.3061	0.4687	-	-	-

For each dataset, we generate all synonym rules obtained by the transitivity and use them together as its synonym rules for our experiments. Since *AOL*, *SPROT* and *USPS* datasets are large, we use them to study the scalability of our proposed algorithms. For quality comparison, we utilize the other smaller datasets with which we can obtain the groundtruth of matching pairs. To compute $NAME \bowtie_R PLAYER$, we use 284 synonym rules obtained from the United States Postal Service websites (www.usps.com). We also extracted the birth dates and occupations of people in both datasets to decide whether a pair of names in the join results indicates the same person. The statistics for all the datasets are provided in Table 1.

7.3 Quality of Join Results

We report the precision (P), recall (R) and F_1-score (F1) in Table 2. Since both *JOIN-HB* and *Pass-Join* produce the same join results, we omit those for *Pass-Join*. For all datasets, the F_1-scores of our *JOIN-HB* are always higher than all the other algorithms. Both *Pkduck-Join* and *SI-Join* show equal or lower precision than *JOIN-HB* since they ignore the order of tokens. Furthermore, except for $NAME \bowtie_R PLAYER$, the recall of *JOIN-HB* is higher than those of the others due to the use of approximate algorithms to compute the similarity values. Since there are people whose first name and last name are switched around in *PLAYER* dataset, *JOIN-HB* misses a few positive pairs in $NAME \bowtie_R PLAYER$. Since some strings in *DRUG* have a large number of applicable rules and *SI-Join* enumerates an exponential number of expansions for each string to build the SI-index for pruning, *SI-Join* did not finish within 4 hours. To give an idea of what false positive pairs are found by *Pkduck-Join* and *SI-Join*, we also provide some of them in Table 3.

Table 3. Examples of false positive pairs

Dataset	$NAME \bowtie_R PLAYER$	$UNIV \bowtie_R UNIV$
False positive pairs	James Daniel (1/17/1953, football coach)	Wa Ying College (Quarry Hill, HK)
	Dan James (8/10/1937, football player)	Ying Wa College (Cheung Sha Wan, HK)
	(with a rule daniel → dan)	
	Sam Michael (5/30/1989, actor)	Maris Stella College (Negombo, Sri Lanka)
	Michael Samson (2/17/1973, football player)	Stella Maris College (Tamil Nadu, India)
	(with a rule samson → sam)	
	Scott James (7/6/1994, snowboarder)	Lake Forest College (Illinois, US)
	James Scott (3/28/1952, football player)	Forest Lake College (Queensland, Australia)

In a nutshell, *Pkduck-Join* and *SI-Join* yield incorrect join results by employing set semantics and approximating the similarities. Thus, our algorithms can be used effectively for the applications that care about the order of tokens and prefer not to have false positives from set semantics.

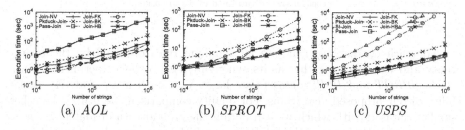

(a) *AOL*   (b) *SPROT*   (c) *USPS*

Fig. 4. Varying the number of strings $|\mathbb{S}|$ and $|\mathbb{T}|$

7.4   Efficiency of Join Algorithms

To study the performance of our proposed algorithms, we utilize *AOL*, *SPROT* and *USPS* datasets. Note that *USPS* dataset has a small number of rules, while the other datasets have a large number of rules. The default size of the datasets is 10^5. We set the parameters $K = 1$ and $q = 2$ as default values since those values of the parameters generally result in the best performance as we will show later. In addition, we set ϕ to 0.01 as the default value since the performances of *JOIN-BK* and *JOIN-HB* are generally the best and insensitive around the value.

Scalability Analysis: Figure 4 shows the execution times of the join algorithms. For *AOL* which has 10.62 applicable rules on the average, *JOIN-BK* shows the best performance, and *JOIN-FK* and *JOIN-HB* are in the second best group. The reason is that *JOIN-BK* chooses the best K positions for the strings $t \in \mathbb{T}$ to reduce a large number of invocations to *MATCH-DP*. Meanwhile, *JOIN-HB* is the best one and *JOIN-BK* is the second best one for *SPROT*. Even though there are a few strings with a large number of applicable rules, *JOIN-NV* performs worse due to such strings. However, for *USPS*, *JOIN-NV* shows

Table 4. Comparison of the filtering methods (*AOL*)

Filtering method	JOIN-FK		JOIN-BK	
	Time (sec)	V-ratio	Time (sec)	V-ratio
No filtering	6653.88	1.00000000	6600.18	1.00000000
LF	3113.64	0.50981044	3326.29	0.50981044
PQF$_{TPQ}$	335.86	0.00056044	458.23	0.00001496
PQF$_{STPQ}$	6.54	0.00056044	7.99	0.00001497
LF+PQF$_{TPQ}$	273.27	0.00029505	459.05	0.00001300
LF+PQF$_{STPQ}$	4.70	0.00029505	8.14	0.00001302

the best performance since the number of applicable rules to the strings $s \in \mathbb{S}$ is small. *JOIN-BK* and *JOIN-HB* perform similarly to *JOIN-NV* and belong to the second best group.

As a summary, *JOIN-BK* almost always outperforms *JOIN-FK* and the state-of-the-art algorithms. *JOIN-NV*, *Pass-Join* and *SI-Join* are inefficient since they have to enumerate all transformed strings of every string in $\mathbb{S}$. In addition, *JOIN-HB* is generally slightly slower than *JOIN-BK* due to the overhead of selecting the split threshold θ. However, it always belongs to the second best group for all the datasets since it properly selects *JOIN-NV* when it is faster than *JOIN-BK* or *JOIN-FK*. Thus, *JOIN-HB* is the best choice in practice.

Effectiveness of Filtering Methods: We compare the execution times and filtering powers of the *JOIN-FK* and *JOIN-BK* algorithms with various combinations of the proposed filtering methods on *AOL*. Note that we did not use sampling for *JOIN-BK* here to measure the filtering power exactly by removing the side effect of sampling. In Table 4, LF and PQF stand for the length filtering and the q-gram filtering, respectively. In addition, PQF$_{TPQ}$ and PQF$_{STPQ}$ represent the q-gram filtering by utilizing $TPQ_q(s, \mathbb{R})$ and $STPQ_q(s, \mathbb{R})$, respectively. To compare the filtering power, we measured the ratio of the remaining pairs after the filtering step (*V-ratio*). As Table 4 shows, our q-gram filtering method filters out a significant number of non-smatch pairs quickly. Furthermore, the q-gram filtering by using $STPQ_q(s, \mathbb{R})$s instead of $TPQ_q(s, \mathbb{R})$s speeds up the filtering step up to 58 times at the cost of a marginal reduction of the filtering power. We omit the results for *SPROT* and *USPS* datasets since the results are similar to the result of *AOL*.

8 Conclusion

We introduce the problem of the string join with synonyms to find string pairs with the same meaning. We next propose the efficient dynamic programming algorithm to compute the proposed match relationships for a pair of strings, and present the length filtering and q-gram filtering to prune non-matching string pairs. We also propose the join algorithms by applying the filtering methods. Experimental results confirm the effectiveness of our algorithms.

Acknowledgements. This research was supported by Next-Generation Information Computing Development Program through the National Research Foundation of Korea (NRF) funded by the Ministry of Science and ICT (NRF-2017M3C4A7063570).

References

1. Aho, A.V., Corasick, M.J.: Efficient string matching: an aid to bibliographic search. Commun. ACM **18**(6), 333–340 (1975)
2. Arasu, A., Chaudhuri, S., Kaushik, R.: Transformation-based framework for record matching. In: ICDE, pp. 40–49. IEEE, Cancun (2008)
3. Bollacker, K., Evans, C., Paritosh, P., Sturge, T., Taylor, J.: Freebase: a collaboratively created graph database for structuring human knowledge. In: Proceedings of the 2008 ACM SIGMOD International Conference on Management of Data, pp. 1247–1250. ACM (2008)
4. Carpineto, C., Romano, G.: A survey of automatic query expansion in information retrieval. ACM Comput. Surv. (CSUR) **44**(1), 1 (2012)
5. Chaudhuri, S., Ganti, V., Kaushik, R.: A primitive operator for similarity joins in data cleaning. In: ICDE, p. 5. IEEE, Atlanta (2006)
6. Chu, X., Ilyas, I.F., Koutris, P.: Distributed data deduplication. PVLDB **9**(11), 864–875 (2016)
7. Chung, J., Kastner, K., Dinh, L., Goel, K., Courville, A.C., Bengio, Y.: A recurrent latent variable model for sequential data. In: Advances in Neural Information Processing Systems, pp. 2980–2988 (2015)
8. Cormen, T.H., Leiserson, C.E., Rivest, R.L., Stein, C.: Introduction to Algorithms, vol. 6. MIT Press, Cambridge (2001)
9. Deng, D., Li, G., Feng, J., Li, W.S.: Top-k string similarity search with edit-distance constraints. In: ICDE, pp. 925–936. IEEE (2013)
10. Gravano, L., Ipeirotis, P.G., Jagadish, H.V., Koudas, N., Muthukrishnan, S., Srivastava, D.: Approximate string joins in a database (almost) for free. In: VLDB, vol. 1, pp. 491–500. VLDB, Rome (2001)
11. Jiang, Y., Li, G., Feng, J., Li, W.: String similarity joins: an experimental evaluation. PVLDB **7**(8), 625–636 (2014)
12. Kim, Y., Shim, K.: Efficient top-k algorithms for approximate substring matching. In: ACM SIGMOD, pp. 385–396 (2013)
13. Konda, P., et al.: Magellan: toward building entity matching management systems. Proc. VLDB Endowment **9**(12), 1197–1208 (2016)
14. Li, G., Deng, D., Wang, J., Feng, J.: PASS-JOIN: a partition-based method for similarity joins. PVLDB **5**(3), 253–264 (2011)
15. Lu, J., Lin, C., Wang, W., Li, C., Wang, H.: String similarity measures and joins with synonyms. In: ACM SIGMOD, New York, USA, pp. 373–384 (2013)
16. Miller, G.A.: Wordnet: a lexical database for English. Commun. ACM **38**(11), 39–41 (1995)
17. Mudgal, S., et al.: Deep learning for entity matching: a design space exploration. In: Proceedings of the 2018 International Conference on Management of Data, pp. 19–34. ACM (2018)

18. Naumann, F., Herschel, M.: An introduction to duplicate detection. Synth. Lect. Data Manag. **2**(1), 1–87 (2010)
19. Sutskever, I., Vinyals, O., Le, Q.V.: Sequence to sequence learning with neural networks. In: Advances in Neural Information Processing Systems, pp. 3104–3112 (2014)
20. Tao, W., Deng, D., Stonebraker, M.: Approximate string joins with abbreviations. PVLDB **11**(1), 53–65 (2017)
21. Wang, J., Li, G., Feng, J.: Can we beat the prefix filtering? An adaptive framework for similarity join and search. In: ACM SIGMOD, Scottsdale, Arizona, USA, pp. 85–96 (2012)
22. Whang, S.E., Menestrina, D., Koutrika, G., Theobald, M., Garcia-Molina, H.: Entity resolution with iterative blocking. In: ACM SIGMOD, pp. 219–232 (2009)
23. Xiao, C., Wang, W., Lin, X., Yu, J.X., Wang, G.: Efficient similarity joins for near-duplicate detection. ACM Trans. Database Syst. **36**(3), 15 (2011)

Efficient Query Reverse Engineering
Using Table Fragments

Meiying Li and Chee-Yong Chan$^{(\boxtimes)}$

National University of Singapore, Singapore, Singapore
{meiying,chancy}@comp.nus.edu.sg

Abstract. Given an output table T that is the result of some unknown query on a database D, Query Reverse Engineering (QRE) computes one or more target query Q such that the result of Q on D is T. A fundamental challenge in QRE is how to efficiently compute target queries given its large search space. In this paper, we focus on the QRE problem for PJ$^+$ queries, which is a more expressive class of queries than project-join queries by supporting antijoins as well as inner joins. To enhance efficiency, we propose a novel query-centric approach consisting of table partitioning, precomputation, and indexing techniques. Our experimental study demonstrates that our approach significantly outperforms the state-of-the-art solution by an average improvement factor of 120.

Keywords: Query Reverse Engineering · Query processing

1 Introduction

Query Reverse Engineering (QRE) is a useful query processing technique that has diverse applications including data analysis [4,18], query discovery [3,14,15, 19,20], query construction [9], and generating query explanations [5–7,11,17]. The fundamental QRE problem can be stated as follows: given an output table T that is the result of some unknown query on a database D, find a target query Q such that the result of Q on D (denoted by $Q(D)$) is equal T.

Research on the QRE problem can be categorized based on two orthognal dimensions. The first dimension relates to the class of queries being reverse-engineered, and the different query fragments that have been investigated for the QRE problem include select-project-join (SPJ) queries [9,17–19], top-k queries [11,13,14], aggregation queries [16,18], and queries with quantifiers [1]. The second dimension concerns the relationship between the target query result $Q(D)$ and the given output table T. One variant of the QRE problem [8,16,18,20] requires finding *exact target queries* such that $Q(D) = T$. Another variant of the QRE problem [9,14,15] relaxes the requirement to finding *approximate target queries* such that $Q(D) \supseteq T$. Solutions designed for the first problem variant could be adapted for the second easier problem variant.

In this paper, we focus on the exact target query variant of the QRE problem for the class of Project-Join (PJ) queries, which is a fundamental query fragment [3,14,15,20]. Specifically, we address two key challenges that arise from

© Springer Nature Switzerland AG 2020
Y. Nah et al. (Eds.): DASFAA 2020, LNCS 12114, pp. 406–422, 2020.
https://doi.org/10.1007/978-3-030-59419-0_25

the efficiency-expressiveness tradeoff for this QRE problem. The first challenge is the issue of search efficiency; i.e., how to efficiently find a target query from the large search space of candidate queries. Note that the QRE problem is NP-hard even for the simple class of PJ queries [2]. All the existing solutions for QRE are based on a *generate-and-test paradigm* that enumerates the space of candidate queries and validates each candidate query Q to check if the result of Q on D is T. All the existing approaches [8,20] generate a candidate query in two main steps. The first step enumerates the candidate projection tables for the query (i.e., the tables from which the output columns are projected), and the second step enumerates the candidate join paths to connect the projection tables. We refer to these approaches as *path-centric approaches*.

To address the efficiency of searching for candidate queries, two main strategies have been explored. The first strategy focuses on the design of heuristics to efficiently prune the search space [8,20]. The second strategy is to reduce the search space by considering only a restricted class of PJ queries; for example, the recent work on FastQRE considers only *Covering PJ* (CPJ) queries [8] which is a strict subset of PJ queries. However, even with the use of pruning heuristics, the performance of the state-of-the-art approach for PJ queries, STAR, could be slow for some queries; for example, STAR took 700s to reverse engineer the target query Q3 in [20][1].

The second challenge is the issue of supporting more expressive PJ queries. All existing approaches support PJ/CPJ queries without negation; i.e., the join operators in the PJ/CPJ queries are limited to inner joins and do not support antijoins. Thus, more expressive PJ queries such as "find all customers who have not placed any orders" can not be reversed engineered by existing approaches. Supporting antijoins in PJ queries is non-trivial as a straightforward extension of the generate-and-test paradigm to simply enumerate different combinations of join operators for each candidate PJ query considered could result in an exponential increase in the number of candidate queries. Indeed, the most recent QRE approach for PJ queries [8] actually sacrifices the support for expressive PJ queries to improve search efficiency over STAR [20].

Thus, the problem of whether it is possible to efficiently reverse engineer more expressive PJ queries remains an open question. In this paper, we address this problem by presenting a novel approach named QREF (for QRE using Fragments). Our approach's key idea is based on horizontally partitioning each database table into a disjoint set of fragments. using a set of PJ^+ queries. A PJ^+ *query* is a more general form of PJ queries that supports antijoins in addition to inner joins. By partitioning each table into fragments such that each fragment is associated with a PJ^+ query, QREF has the three advantages. First, in contrast to the existing path-centric approaches for generating candidate queries, QREF is a *query-centric approach* where given a set of candidate projection tables, it enumerates a set of candidate projection table queries (which are PJ^+ queries) for each projection table, and a candidate query is formed by merging a set of projection table queries. By judicious precomputation of the PJ^+ queries associated with subsets

---

[1] In contrast, our approach took 3s to reverse engineer this query (Sect. 6).

of table fragments, QREF is able to efficiently enumerate candidate projection table queries without the need to enumerate candidate join paths. Second, by indexing the PJ$^+$ queries associated with the table fragments, QREF is able to efficiently validate a candidate query Q by (1) accessing only relevant fragments instead of entire relations, and (2) eliminating some of the join computations in Q that are already precomputed by the table fragments. Third, QREF is able to reverse-engineer PJ$^+$ queries, which is a larger and more expressive class of queries than the PJ/CPJ queries supported by existing approaches [8, 20].

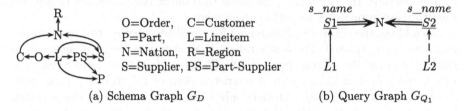

(a) Schema Graph G_D (b) Query Graph G_{Q_1}

Fig. 1. (a) Schema graph G_D for TPC-H database D with relation names abbreviated. (b) Query graph for query $Q_1 = \pi_{S1.s\_name, S2.s\_name}((S1 \ltimes L1) \bowtie N \bowtie (S2 \triangleright L2))$.

In summary, this paper proposes a novel query-centric approach to solve the QRE problem for the target class of PJ$^+$ queries based on table partitioning, precomputation, and indexing techniques. Although we have developed these techniques specifically for the class of PJ queries considered in this paper, our techniques could be adapted to optimize the QRE problem for other query fragments as well. To the best of our knowledge, our work is the first to reverse engineer queries with negation (specifically queries with antijoins).

2   Background

2.1   Query Classes

Given a database D consisting of a set of relations $\{R_1, \cdots, R_n\}$, we define its **schema graph** to be a directed graph $G_D = (V, E)$, where each node in $v_i \in V$ represents relation R_i, and each directed edge $(v_i, v_j) \in E$ represents a foreign-key relationship from R_i to R_j. In case if there are multiple foreign-key relationships from a relation to another relation, the directed edges would also be labeled with the names of the pair of foreign-key and primary-key attribute names. Figure 1a shows the schema graph for the TPC-H schema.

We now define all the fragments of PJ queries that have been studied in the past (i.e., PJ, PJ$^-$, CPJ) as well as the most general fragment that is the focus of this paper (i.e., PJ$^+$) These four fragments are related by the following strict containment relationship: $CPJ \subset PJ^- \subset PJ \subset PJ^+$.

A **Project-Join (PJ) query** is of the form $\pi_\ell(E)$, where E denote a join expression involving innerjoins ($\bowtie$), with all the join predicates based on foreign-key joins[2], and ℓ denote a list of attributes. A **Project-Join$^+$ (PJ$^+$) query** is a more general form of PJ queries that also supports left-antijoins ($\triangleright$) in addition to innerjoins (and left-semijoins ($\ltimes$)). Thus, the class of PJ$^+$ queries are strictly more expressive than PJ queries as every PJ query is also a PJ$^+$ query.

A PJ$^+$query Q can be represented by a **query graph** G_Q, where each relation in Q is represented by a node in G_Q, and each join between a pair of relations R_i and R_j in Q is represented by an edge between their corresponding nodes as follows: a innerjoin, left-semijoin, and left-antijoin, are denoted respectively, by a double-line, single-line, and dashed-line edges. The edges between nodes are directed following the edge directions in the schema graph.

To distinguish multiple instances of the same relation, we append a unique number to the relation instance names for convenience. If the projection list in Q includes some attribute from a relation R_i, then we refer to R_i (resp. the node representing R_i) as a **projection table** (resp. **projection node**). Each projection node in G_Q is underlined and annotated with a list of its projection attributes.

The non-projection nodes in a query graph are further classified into connection nodes and filtering nodes. Given a path of nodes $p = (v_1, \cdots, v_m)$ in G_Q, where the end nodes v_1 and v_m in p are the only two projection nodes in p, each $v_i, i \in (1, m)$ is a **connection node**. A node that is neither a projection nor a connection node is a **filtering node**.

A key feature of our approach is to horizontally partition each database table using a subclass of PJ$^+$ queries that we refer to as **filtering PJ$^+$ queries (FPJ$^+$)**. A FPJ$^+$ query is a PJ$^+$ query that has exactly one projection table and using only left-semijoins and/or left-antijoins. Thus, each FPJ$^+$ query consists of exactly one projection node and possibly some filtering nodes without any connection node.

A **Project-Join$^-$ (PJ$^-$) query** is a restricted form of PJ query that does not contain any filtering nodes. A **Covering Project Join (CPJ) query** is a restricted form of PJ$^-$ query with the constraint that for any two connection nodes v_i and v_j that correspond to two instances of the same table, v_i and v_j must be along the same path that connects a pair of projection nodes. Given a query Q, we define its size, denoted by $|Q|$, to be the number of nodes in its query graph.

Example 1. Figure 1b shows the query graph G_{Q_1} for the PJ$^+$ query Q_1 (on the TPC-H schema) that returns the names of all pairs of suppliers ($sname_1$, $sname_2$) who are located in the same country, where $sname_1$ has supplied some lineitem, and $sname_2$ has not supplied any lineitem. In G_{Q_1}, $L1$ and $L2$ are filtering nodes, N is a connection node, $S1$ and $S2$ are projection nodes, and

---

[2] Although our experiments focus on queries with foreign-key joins (similar to all competing approaches [8,20]), our approach can be easily extended to reverse engineer PJ queries with non-foreign key join predicates. The main extension is to explicitly annotate the database schema graph with additional join edges.

$|Q_1| = 5$. If N and its incident edges are removed from G_{Q_1}, then we end up with two FPJ$^+$ queries $\underline{S1} \ltimes L1$ and $\underline{S2} \rhd L2$. Note that Q_1 is not a PJ query due to the antijoin $\underline{S2} \rhd L2$; if the antijoin in Q_1 is replaced with a semijoin (i.e., $\underline{S2} \ltimes L2$), then the revised query Q_1' becomes a PJ query as the semijoins in Q_1' are effectively equivalent to innerjoins. Note that Q_1' is not a CPJ query due to the filtering nodes $L1$ and $L2$; if these filtering nodes are removed from Q_1', then the resultant query is a CPJ query.   □

3   Our Approach

In order to address the efficiency challenge of supporting PJ$^+$ queries, which is a even more expressive class of queries supported by STAR [20], we design a new approach, termed QREF, that applies table partitioning, precomputation and indexing techniques to improve the performance of the QRE problem. To the best of our knowledge, our approach is the first to utilize such techniques for the QRE problem and our work is also the first to support negation (in the form of antijoins) in target queries.

Query-Centric Approach. In contrast to existing path-centric QRE solutions, QREF is a *query-centric* approach that treats each PJ$^+$ query as a combination of FPJ$^+$ queries. Referring to the example target PJ$^+$ query Q_1 in Fig. 1b, Q_1 can be seen as a combination of two FPJ$^+$ queries, $Q_{S_1} = \pi_{s\_name}(\underline{S} \ltimes L \ltimes N)$ and $Q_{S_2} = \pi_{s\_name}(\underline{S} \rhd L \ltimes N)$. The target query Q_1 can be derived by merging these two FPJ$^+$ queries via the common table *Nation*.

We illustrate our approach using a running example of a simplified TPC-H database D consisting of the four tables shown in Figs. 2a–d; for convenience, we have indicated a *tid* column in each database table to show the tuple identifiers. In this example, our goal is to try to reverse engineer a PJ$^+$ query from the output table T shown in Fig. 2e.

Candidate Projection Tables. The first step in QREF, which is a necessary step in all existing approaches, is to enumerate possible *candidate projection tables* that could be joined to compute the output table T. Intuitively, for a set PT of table instances to be candidate projection tables for computing T, each of the columns in T must at least be contained in some table column in PT; otherwise, there will be some tuple in T that cannot be derived from PT. In our running example, since $\pi_{pkey,brand}(P) \supseteq \pi_{C_1,C_2}(T)$ and $\pi_{name}(S) \supseteq \pi_{C_3}(T)$, one possible set of candidate projection tables to compute T is $PT_1 = \{P(pkey,brand), S(name)\}$. We say that $P(pkey,brand)$ *covers* $T(C_1,C_2)$, and $P.pkey$ ($P.brand$, resp.) is a *covering attribute* of $T.C_1$ ($T.C_2$, resp.). Another set of candidate projection tables is $PT_2 = \{P(pkey), P(brand), S(name)\}$ which differs from PT_1 in that two instances of the table P are used to derive the output columns C_1 and C_2 separately. Figure 2k shows the covering attributes for each of the three columns in T; for example, there are two covering attributes (i.e., *P.pkey* and *PS.pkey*) for $T.C_1$. Figure 2l shows all the possible sets of candidate projection tables for T.

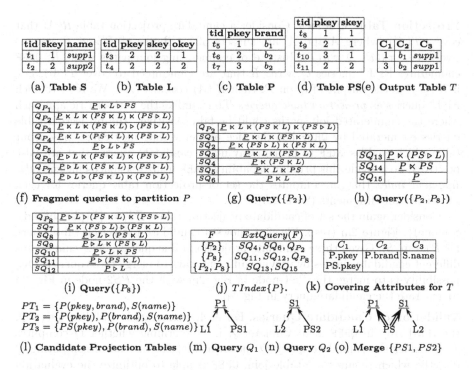

Fig. 2. Running example: simplified TPC-H database (Tables S, L, P & PS) with $\alpha = 2$

Candidate Queries. Given a set of candidate projection tables PT, the second step in QREF is to enumerate candidate queries that join the projection tables in PT. For example, given $PT_1 = \{P(pkey,brand), S(name)\}$, one possibility is to join the two projection tables via the intermediate table L to produce the candidate query Q_1: $\pi_{pkey,brand,name}(\underline{P} \bowtie_{pkey} L \bowtie_{skey} \underline{S})$. To avoid the performance issues with the path-centric approach, QREF uses a novel and more efficient *query-centric approach* to enumerate candidate queries that is based on a combination of three key techniques: table partitioning, precomputation, and indexing.

Table Fragments and Fragment Queries. QREF horizontally partitions each database table into a set of pairwise-disjoint fragments using a set of FPJ$^+$ queries. As an example, Fig. 2f shows a set of 8 FPJ$^+$ queries $\{Q_{P_1}, \cdots, Q_{P_8}\}$ to partition the *Part* table into 8 *table fragments*: $P = P_1 \cup \cdots \cup P_8$. Each table fragment P_i is the result of the FPJ$^+$ query Q_{P_i}, and we refer to each such query as a *fragment query*. In our running example, all the fragments of P are empty except for $P_2 = \{t_6\}$ and $P_8 = \{t_5, t_7\}$. To control the partitioning granularity, QREF uses a *query height parameter*, denoted by α, which determines the maximum height of each fragment query graph. Observe that the set of fragment queries in Fig. 2l are all the possible FPJ$^+$ queries w.r.t. P with a maximum height of 2.

Projection Table Queries. Consider a candidate projection table $R(A)$ that covers $T(A')$, where A and A' are subsets of attributes of R and T, respectively; i.e., $\pi_A(R) \supseteq \pi_{A'}(T)$. Instead of enumerating join paths w.r.t. R, QREF enumerates FPJ$^+$ queries Q w.r.t. R that are guaranteed to cover $T(A')$; i.e., if $R' \subseteq R$ is the output of Q on R, then $R'(A)$ covers $T(A')$. We refer to such FPJ$^+$ queries as *projection table queries*. Thus, unlike the path-centric approach where the enumerated join paths could be false positives, the projection table queries enumerated by QREF has the nice property that they cover the output table which enables QREF to prune away many candidate queries that are false positives. To provide this pruning capability, QREF utilizes a space-efficient *table fragment index* that precomputes the set of projection table queries for each subset of table fragments (Sect. 4.3).

Consider again the set of candidate projection tables $PT_1 = \{$P(pkey,brand), S(name)$\}$. Figure 2m (resp. Fig. 2n) shows a candidate projection table query for P (resp. S). Given these candidate projection table queries, QREF enumerates different candidate queries by varying different ways to merge these queries. One possibility is to merge the $PS1$ node in Fig. 2m with the $PS2$ node in Fig. 2n to generate the candidate query in Fig. 2o.

Validation of Candidate Queries. For each enumerated candidate query Q, the third step in QREF is to validate Q; i.e., determine whether Q computes T. Consider the candidate query Q shown in Fig. 2o. Instead of computing Q directly, which requires a 5-table join, QREF is able to optimize the evaluation of Q by making use of the table fragments. Specifically, Q is evaluated as $Q' = \underline{P'} \bowtie PS' \bowtie \underline{S'}$ where P' refers to the table fragment $\underline{P} \ltimes PS \rhd L$, PS' refers to the table fragment $\underline{PS} \ltimes (P \rhd L) \ltimes (S \rhd L)$, and S' refers to table fragment $\underline{S} \ltimes PS \rhd L$. To efficiently identify relevant table fragments for query evaluation, QREF builds an index on the fragment queries associated with the fragments of each table (Sect. 4.2).

Overall Approach. Our query-centric approach consists of an offline phase and an online phase. The offline phase partitions the database tables into fragments, precomputes candidate projection table queries from the table fragments, and build indexes on the table fragments and fragment queries. The online phase processes QRE requests at runtime, where each QRE request consists of an output table to be reverse-engineered.

4   Table Fragments and Indexes

In this section, we present QREF's offline phase which partitions each database table into fragments, precomputes candidate projection table queries, and builds indexes on the table fragments and query fragments to enable efficient query reverse engineering.

4.1   Partitioning Tables into Fragments

In this section, we explain how each database table is horizontally partitioned into fragments using a set of FPJ$^+$ queries. To control the granularity of the partitioning, QREF uses a **query height parameter**, denoted by α, so that the height of each fragment query graph is at most α. Thus, partitioning a table with a larger value of α will result in more table fragments.

We use $Frag(R) = \{R_1, \cdots, R_m\}$ to denote the set of m fragments created by the partitioning of a table R, and use Q_{R_i} to denote the fragment query that computes R_i from R. Each of the tables in the database is partitioned iteratively using FPJ^+ queries with increasing query height from 1 to α based on the join relationships among the tables. which are specified by the join edges in the database schema graph.

To make the discussion concrete, we consider the partitioning of three tables PS, P, and L from the schema graph G_D shown Fig. 1(a). Recall that we consider only foreign-key joins in G_D for simplicity, but the approach described can be generalized to other types of joins so long as they are captured in G_D. Each table R in G_D will be partitioned based on the set of foreign-key relationships to R. For the tables $\{PS, P, L\}$ being considered, the foreign-key join relationships are $\{PS \leftarrow L, P \leftarrow L, P \leftarrow PS\}$. In the first iteration, $PS = PS_1 \cup PS_2$ is partitioned into two fragments with the following FPJ^+ queries: $Q_{PS_1} = PS \ltimes L$ and $Q_{PS_2} = PS \rhd L$. Similarly, $P = P_1 \cup P_2 \cup P_3 \cup P_4$ is partitioned into four fragments with the following FPJ^+ queries: $Q_{P_1} = P \ltimes L \ltimes PS$, $Q_{P_2} = P \ltimes L \rhd PS$, $Q_{P_3} = P \rhd L \ltimes PS$, and $Q_{P_4} = P \rhd L \rhd PS$. Since there is no foreign-key relationship into L, L will not be partitioned at all. In the second iteration, the fragments PS_1 and PS_2 will not be further partitioned since L was not partitioned in the first iteration. On the other hand, since PS was partitioned into PS_1 and PS_2 in the first iteration, these will be used to further partition $P_1 = P_{1,1} \cup P_{1,2} \cup P_{1,3}$ into three fragments using the following FPJ^+ queries:

$$Q_{P_{1,1}} = P \ltimes L \ltimes PS_1 \ltimes PS_2 = P \ltimes L \ltimes (PS \ltimes L) \ltimes (PS \rhd L)$$
$$Q_{P_{1,2}} = P \ltimes L \ltimes PS_1 \rhd PS_2 = P \ltimes L \ltimes (PS \ltimes L) \rhd (PS \rhd L)$$
$$Q_{P_{1,3}} = P \ltimes L \rhd PS_1 \ltimes PS_2 = P \ltimes L \rhd (PS \ltimes L) \ltimes (PS \rhd L)$$

Similarly, fragment P_3 will be further partitioned into three fragments using PS_1 and PS_2. However, the fragments P_2 and P_4 will not be further partitioned because the tuples in them do not join with any tuple from PS. Thus, if $\alpha = 2$, the relation P will be partitioned by QREF into a total of 8 fragments as shown in Fig. 2f.

The partitioning of tables into fragments using FPJ^+ queries with a height of at most α can be performed efficiently by applying α-bisimulation techniques [12] on the database. Furthermore, the techniques from [12] can be applied to efficiently maintain the database fragments when the database is updated.

4.2 Fragment Query Index

For each table R, QREF builds a **fragment query index**, denoted by $QIndex(R)$, to index all the fragment queries used for partitioning R. These indexes are used for the efficient validation of candidate queries (to be explained in Sect. 5.3). Specifically, given an input FPJ^+ query Q w.r.t. table R, the fragment query index $QIndex(R)$ will return a list of the identifiers of all the fragments of R that match Q; i.e., the result of the query Q on the database is the union of all the fragments of R that are returned by the index search on $QIndex(R)$ w.r.t. Q.

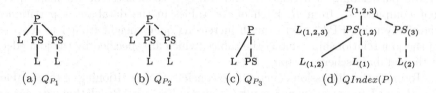

(a) Q_{P_1} (b) Q_{P_2} (c) Q_{P_3} (d) $QIndex(P)$

Fig. 3. Fragment queries & fragment query index $QIndex(P)$ for relation P

Each $QIndex(R)$ is a trie index that indexes all the root-to-leaf paths in the fragment query graphs of R. As an example, Fig. 3d shows $QIndex(P)$, the fragment query index for a relation P, which has been partitioned into a set of three non-empty fragments $\{P_1, P_2, P_3\}$ with their fragment queries shown in Figs. 3a, 3b, and 3c[3]. Each node in $QIndex(P)$ represents a relation, with P as the root node of $QIndex(P)$. Similiar to the edges in a query graph, each edge in $QIndex(P)$ represents either a semijoin (sold line) or an antijoin (dashed line). Each node N in $QIndex(P)$ stores a list of fragment identifiers (indicated by the node's subscript in Fig. 3d) identifying the subset of fragments of $Frag(P)$ that match the PJ^+ query represented by the single-path in $QIndex(P)$ that starts from the root node of $QIndex(P)$ to N. As an example, for the second child node PS of P in $QIndex(P)$, its list of fragment identifiers is $(1,2)$ indicates that only fragments P_1 and P_2 (but not P_3) match the PJ^+ query $\underline{P} \ltimes PS$.

4.3 Table Fragment Index

The second type of index built by QREF for each table R is the **table fragment index**, denoted by $TIndex(R)$. For each non-empty subset of fragments of R, $F \subseteq Frag(R)$, $TIndex(R)$ stores a set of FPJ^+ queries (with a maximum height of α) that computes F from R. We denote this set of queries by $Query(F)$. The table fragment index for R, $TIndex(R)$, is used to efficiently enumerate the set of candidate projection table queries for a projection table R (to be discussed in Sect. 5.2).

Given $F \subseteq Frag(R)$, we now explain how to derive $Query(F)$. The derivation algorithm uses the the following three rewriting rules (S1 to S3) and additional knowledge of whether any fragments in F are empty. In the following rules, each E_i denote some FPJ^+ query.

---

[3] Note that this example is not related to the example in Fig. 2.

S1. $(E_1 \bowtie E_2) \cup (E_1 \triangleright E_2) = E_1$

S2. $(E_2 \circ E_1) \cup (E_3 \circ E_1) = (E_2 \cup E_3) \circ E_1, \circ \in \{\bowtie, \triangleright\}$

S3. $(E_1 \bowtie E_2) \cup (E_1 \bowtie E_3) = E_1 \bowtie (E_2 \cup E_3)$

Example 2. Referring to our example database in Fig. 2, all the fragments of table P are empty except for $P_2 = \{t_6\}$ and $P_8 = \{t_5, t_7\}$; i.e., $P = P_2 \cup P_8$. Consider the set $F = \{P_7, P_8\}$. Observe that it is not possible to rewrite $Q_{P_7} \cup Q_{P_8}$ into a single FPJ$^+$ query using the above rules. However, since both fragments P_5 and P_6 are empty, we have $Q_{P_7} \cup Q_{P_8} = \bigcup_{i=5}^{8} Q_{P_i}$ which can now be rewritten into the FPJ$^+$ query $P \triangleright L$. Thus, $P \bowtie L \in Query(F)$. In addition, it is now possible to simplify Q_{P_8} in six ways. First, since $P_5 = \underline{P} \triangleright L \triangleright PS = \underline{P} \triangleright L \triangleright (PS \bowtie L) \triangleright (PS \triangleright L)$ is empty, $Q_{P_8} = Q_{P_8} \cup Q_{P_5}$ can be simplified using rule S1 to $SQ_8 = \underline{P} \triangleright L \triangleright (PS \bowtie L)$. Second, since P_4 is empty, $Q_{P_8} = Q_{P_8} \cup Q_{P_4}$ can be first rewritten using rule S2 to $(\underline{P} \bowtie L \cup \underline{P} \triangleright L) \triangleright (PS \bowtie L) \bowtie (PS \triangleright L)$ and further simplified using rule S1 to $SQ_7 = \underline{P} \triangleright (PS \bowtie L) \bowtie (PS \triangleright L)$.

Similarly, Q_{P_8} can be simplified to four other queries SQ_9, SQ_{11}, SQ_{10}, and SQ_{12}; the definitions of these queries are shown in Fig. 2i. The queries in $Query(\{P_2\})$ and $Query(\{P_2, P_8\})$ are shown in Figs. 2g and 2h, respectively. □

As the queries in $Query(F)$ are derived from the fragment queries, each query in $Query(F)$ also inherits the maximum query height of α from the fragment queries. Since there could be many queries in $Query(F)$, QREF uses a concise representation of $Query(F)$ to optimize both its storage as well as the efficiency of enumerating candidate projection table queries from $Query(F)$. Specifically, instead of storing $Query(F)$, QREF stores only the subset consisting the minimal and maximal queries (corresponding to the simplest and most complex queries), denoted by $ExtQuery(F)$ (for the extremal queries in $Query(F)$). The details of this optimization are given elsewhere [10] due to space constraint. Thus, the table fragment index for R_i, $TIndex(R_i)$, is a set of $(F, ExtQuery(F))$ entries where $F \subseteq Frag(R_i)$, $F \neq \emptyset$, and $ExtQuery(F) \neq \emptyset$. Figure 2j shows the table fragment index for relation P in the running example. Note that in the implementation of $TIndex(R_i)$, each F is represented by a set of fragment identifiers.

5   Reverse Engineering PJ$^+$ Queries

In this section, we present the online phase of QREF which given an output table T consisting of k columns $C_1, \cdots, C_k$ and a database D, generates a target PJ$^+$ query Q such that $Q(D) = T$. QREF optimizes this QRE problem by using the table fragments, fragment query index, and table fragment index computed in the offline phase in three main steps.

First, given D and T, QREF enumerates a set of candidate projection tables for computing T from D. Second, for each set of candidate projection tables $PT = \{P_1, \cdots, P_k\}$, $k \geq 1$, QREF enumerates a set of candidate projection queries $PQ = \{Q_1, \cdots, Q_k\}$, where each Q_i is a FPJ$^+$ query w.r.t. P_i. Third, given a set of candidate projection queries PQ, QREF enumerates candidate queries by merging the queries in PQ. For each enumerated candidate query Q, QREF performs a query validation to check whether Q could indeed compute T from D. The algorithm terminates when either QREF finds a target query to compute T or all the enumerated candidate queries fail the validation check.

5.1   Generating Candidate Projection Tables

The first step in QREF is to generate a set candidate projection tables PT that could compute the given output table $T(C_1, \cdots, C_k)$. This is a common task shared by all existing approaches that starts by identifying the set of covering attributes for each column C_i in the output table T. Based on the identified covering attributes for T, the next step is to enumerate all the sets of candidate projection tables. Since simpler target queries are preferred over more complex ones (in terms of query size), QREF enumerates the candidate projection table sets in non-descending order of the number of projection tables; for example, in Fig. 2l, the candidate set PT_1 will be considered before PT_2 and PT_3.

For each candidate projection table being enumerated; e.g.., $P(pkey, brand)$ to cover $T(C_1, C_2)$, it is necessary to validate that indeed $\pi_{pkey,brand}(P) \supseteq \pi_{C_1,C_2}(T)$. QREF optimizes this validation step by first performing a partial validation using a small random sample of tuples from T to check whether these tuples could be projected from the candidate projection table. If the partial validation fails, QREF can safely eliminate this candidate from further consideration.

5.2   Generating Candidate Queries

The second step in QREF is to generate candidate queries for each set $PT = \{R_1, \cdots, R_m\}$ of candidate projection tables enumerated by the first step. Recall that each R_i covers some subset of columns in T (i.e., $R_i(A_i)$ covers $T(A_i')$ for some attribute lists A_i and A_i') such that every column in T is covered.

Our query-centric approach performs this task using two main steps. First, QREF enumerates a set PQ_i of candidate projection table queries for each candidate projection table R_i in PT. Second, for each combination of candidate projection table queries $CPQ = \{Q_1, \cdots, Q_m\}$, where each $Q_i \in PQ_i$, QREF enumerates candidate queries that can be generated from CPQ using different ways to merge the queries in CPQ.

Generating Projection Table Queries. To generate candidate projection table queries for a projection table $R_i \in PT$, QREF uses the table fragment index $TIndex(R_i)$. Specifically, for each $(F, ExtQuery(F))$ entry in $TIndex(R_i)$, where $F \subseteq Frag(R_i)$, QREF determines whether $F(A_i)$ covers $T(A_i')$. If so, each query in $ExtQuery(F)$ is a candidate projection table query for R_i.

Example 3. Consider the set of candidate projection tables $PT_1 = \{P(pkey, brand), S(name)\}$, where $P(pkey, brand)$ covers $T(C_1, C_2)$ and $S(name)$ covers $T(C_3)$. Among the three index entries in $TIndex(P)$ (shown in Fig. 2j), $F(pkey, brand)$ covers $T(C_1, C_2)$ when $F = \{P_8\}$. Therefore, queries in $ExtQuery(\{P_8\})$ are candidate projection table queries for P.   □

Merging Projection Table Queries. Consider a set of candidate projection tables $PT = \{R_1, \cdots, R_m\}$ to compute T, where QREF has enumerated a projection table query Q_i for each R_i. We now outline how QREF enumerates different candidate queries from $\{Q_1, \cdots, Q_m\}$ by different ways to merge these queries. To merge the queries $\{Q_1, \cdots, Q_m\}$ into a candidate query Q, QREF first identifies a query node v_i in each query Q_i such that all the v_i's can all be merged into a single node v to connect all the Q_i's into a single query Q. We refer to each of the v_i's as a **mergeable node**, and to the combined query Q as a **merged query**. The details of the query merging processing are described elsewhere [10].

Example 4. Consider the merging of two projection table queries: $Q_1 = \underline{P1} \ltimes PS1 \rhd L1$ and $Q_2 = \underline{S1} \ltimes PS2 \rhd L2$, whose query graphs are shown in Fig. 2m and Fig. 2n. There is only one choice of mergeable nodes, i.e. $\{PS1, PS2\}$. The merged query generated by merging $\{PS1, PS2\}$ is shown in Fig. 2o, where the merged node is denoted by PS and all the semijoin edges along the path connecting the projection nodes $S1$ and $P1$ are converted to inner join edges. $\square$

Generating Candidate Queries. Each merged query Q generated by QREF is a candidate query, and Q itself could be used to derive possibly additional candidate queries by combining different subsets of mergeable nodes in Q. Thus, for a given set of candidate projection tables $PT = (R_1, \cdots, R_m)$ to compute T, QREF enumerates different candidate queries w.r.t. PT by applying different levels of enumeration. First, QREF enumerates different projection table queries for each R_i. Second, for each combination of projection table queries $PQ = \{Q_1, \cdots, Q_m\}$, QREF enumerates multiple merged queries from PQ via different choices of mergeable nodes. Third, for each merged query Q, QREF derives additional candidate queries from Q by enumerating different ways to further merge the query nodes in Q. As simpler target queries (with fewer tables) are preferred over more complex target queries, QREF uses various heuristics to control the enumeration order to try to generate simpler candidate queries earlier. For example, for enumerating combinations of projection table queries $PQ = \{Q_1, \cdots, Q_m\}$, QREF uses the sum of the query size (i.e., $\sum_{i=1}^{m} |Q_i|$) as a heuristic metric to consider those queries with smaller size earlier.

5.3 Validation of Candidate Queries

For each candidate query Q enumerated, QREF needs to validate whether Q could indeed compute T. In this section, we explain how QREF optimizes this validation process with the help of the table fragments and the fragment query indexes. The query validation process consists of two steps. First, QREF performs a less costly *partial validation* of Q by randomly choosing a small sample of tuples from the output table T and checking if these tuples are contained in Q's result. If the outcome is conclusive (i.e., one of the chosen tuples can not be produced by Q), then QREF concludes that Q is not a target query and the validation process terminates. On the hand, if the partial validation is not conclusive, QREF will resort to a more costly *full validation* to compute the result of Q on the database and check whether the result is equal to T.

QREF is able to optimize the evaluation of Q using table fragments for two reasons. First, some of joins with filter nodes in Q can be eliminated as they are already precomputed in the table fragments. As an example, consider the query Q_1 shown in Fig. 1b which involves a 5-table join: two projection nodes (S1 and S2), one connection node (N), and two filter nodes (L1 and L2). For this query, QREF is able to eliminate the join $\underline{S1} \ltimes L1$ by accessing the fragments of table S that satisfy FPJ$^+$ query expression $\underline{S} \ltimes L$ (instead of the entire table S). Similarly, QREF is able to eliminate the join $\underline{S2} \rhd L2$ by accessing the fragments of table S that satisfy the FPJ$^+$ query expression $\underline{S} \rhd L$.

Second, for joins that cannot be eliminated (i.e., joins with projection/connection nodes), QREF can optimize the evaluation by accessing only relevant table fragments instead of entire tables. Referring again to the query Q_1 in Fig. 1b, consider the table N. Instead of accessing the entire table N, QREF only need to access the fragments of table N that match the FPJ$^+$ query Q' $=\underline{N} \ltimes (S1 \ltimes L1) \ltimes (S2 \rhd L2)$ assuming $\alpha = 2$, and the matching fragments for Q' can be efficiently identified using the fragment index $QIndex(N)$. Specifically, QREF searches the trie index $QIndex(N)$ twice using the two root-to-leaf paths in Q' ($\underline{N} \ltimes (S \ltimes L1)$ and $\underline{N} \ltimes (S \rhd L2)$)). By intersecting the two lists of fragment identifiers associated with the paths' leaf nodes, QREF can efficiently identify the set of matching table fragments in N. Depending on the selectivity of Q', this alternative evaluation plan that retrieves only relevant table fragments could be more efficient than a full table scan.

6   Experiments

In this section, we present the results of a performance evaluation of our proposed approach, QREF. Our study compares QREF against our main competitor STAR [20] as STAR is designed for PJ queries, which is the closest query class to the PJ$^+$ queries supported by QREF[4]. Our results show that QREF generally outperforms STAR by an average improvement factor of 120 and up to a factor of 682.

We conducted four experiments using the same TPC-H benchmark database as [20], whose database size is 140 MB generated using Microsoft's skewed data generator for TPC-H. All the experiments were performed on an Intel Xeon E5-2620 v3 2.4 GHz server running Centos Linux 3.10.0 with 64 GB RAM, and the database was stored on a separate 1TB disk using PostgreSQL 10.5.

Experiments 1 and 2 are based on the same set of queries used by STAR [20], where Experiment 1 consists of 22 PJ queries (TQ1 to TQ22) modified from the 22 TPC-H benchmark queries, and Experiment 2 consists of 6 queries

---

[4] We did not compare against FastQRE [8] for two reasons. First, FastQRE supports only CPJ queries which are even more restrictive than PJ queries. Second, the code for FastQRE is not available, and its non-trivial implementation requires modification to a database system engine to utilize its query optimizer's cost model for ranking candidate queries.

(Q1 to Q6) created by [20]. Experiment 3 consists of 5 queries (QQ1 to QQ5) shown in Fig. 4a, which were created by us to showcase the strengths of QREF compared to STAR. Experiment 4 is designed to evaluate the effectiveness of QREF's fragment-based query validation; the results are given in [10] due to space constraint.

For each query Q, we first execute Q on the database to compute its result T and then measure the time taken to reverse engineer Q from T using each of the approaches. To be consistent with the experimental timings reported in [20], the execution timings for all approaches exclude the following components: preprocessing time and the validation of the last candidate query. Each of the timings (in seconds) reported in our experiments is an average of five measurements. If an approach did not complete after running for one hour, we terminate its execution and indicate its timing value by "E". For each of our experimental queries, both STAR and QREF are able to find the same target query if they both complete execution within an hour.

QID	Query	Projection
QQ1	$S \bowtie (N \triangleright C)$	S.name, N.name
QQ2	$C \ltimes (N \ltimes S) \triangleright O$	C.name
QQ3	$(L1 \bowtie S1) \bowtie N \bowtie (S2 \ltimes L2)$	S1.name, S2.name
QQ4	$(C \bowtie N) \bowtie (S \ltimes L)$	N.name, S.name
QQ5	$N \bowtie C \bowtie O \bowtie L$	C.name, O.orderdate, L.shipdate

(a) Expt. 3: Queries

QID	1	2	3	4	5	6	7	8	9	10	11
STAR	8.2	E	38.6	22.42	E	8.42	E	E	E	E	34.5
QREF	0.28	0.5	1.78	0.45	16.32	0.2	40.1	24.6	0.4	0.97	2.89
f	29.3	-	21.7	49.8	-	42.1	-	-	-	-	11.9

QID	12	13	14	15	16	17	18	19	20	21	22
STAR	27.27	2.2	16.6	82.21	80.09	48.4	14.0	E	E	E	4.87
QREF	0.04	2	0.38	0.23	0.5	0.41	2.11	0.57	0.8	294	0.13
f	681.8	1.1	43.7	357.4	160.2	118	6.6	-	-	-	37.5

(b) Expt. 1: Timing Results

QID	1	2	3	4	5	6	7	8	9	10	11
STAR	1	-	20	4	-	1	-	-	-	-	5
QREF	1	1	1	1	3	1	101	92	1	1	1

QID	12	13	14	15	16	17	18	19	20	21	22
STAR	6	1	11	225	11	8	9	-	-	-	6
QREF	1	1	1	1	1	1	1	1	1	263	1

(c) Expt. 1: # of Candidate Queries

QID	Q1	Q2	Q3	Q4	Q5	Q6
STAR.time	21.86	56.76	T	51.91	180.22	311.5
QREF.time	0.54	1.15	3.03	1.47	16.49	10.91
f	40.48	49.36	-	35.31	10.93	28.55
STAR.#cand	4	4	3	17	15	15
QREF.#cand	1	1	2	9	5	6

(d) Expt. 2: Timing Results

QID	QQ1	QQ2	QQ3	QQ4	QQ5
STAR	E	E	E	24.74	20.9
QREF	0.018	0.11	4.12	0.64	0.052
f	-	-	-	38.7	401.9

(e) Expt. 3: Timing Results

number	1	2	3	4
STAR.time	25.36	127.7	E	E
QREF.time	0.05	0.13	0.2	0.2
f	507.2	982.3	-	-
STAR.#cand	40	250	-	-
QREF.#cand	1	1	1	1

(f) Expt. 3: Effect of # Covering attributes

Fig. 4. Experimental results (timings in seconds)

Both QREF and STAR were implemented in C++[5]. We run approach STAR with its default settings and turn on all its optimization techniques. For QREF, we set $\alpha = 2$ and always use table fragments for query validation. QREF took

---

[5] The code for STAR was based on a version obtained from the authors of [20].

slightly under a minute to partition each database table into fragments, and build all the table fragment and fragment query indexes. The number of non-empty fragments created by QREF for tables Orders, Nation, Part, Region, Supplier, LineItem, PartSupp, and Customer are 2, 9, 3, 4, 3, 1, 5, and 6, respectively.

Experiment 1. This experiment compares the performance of QREF and STAR to reverse engineer the 22 TPC-H queries. The speedup factor f is defined to be the ratio of STAR's timing to QREF's timing. Figure 4b compares the execution time. STAR was unable to find queries TQ2, TQ5, TQ6, TQ7, TQ8, TQ9, TQ10, TQ19, TQ20, and TQ21 within one hour. The reason is that STAR is very memory intensive, and its executions ran out of memory for these queries. This behaviour of STAR was also observed by [8]. Specifically, TQ5, TQ7, TQ8, and TQ21 are large queries involving between 6 to 8 tables. STAR did not perform well for these queries as it spent a lot of time generating and validating complex candidate queries. In contrast, QREF could find all these complex queries even though it took longer compared to the timings for the other simpler queries.

For queries TQ2, TQ9, TQ10, TQ19, and TQ20, QREF outperformed STAR significantly for these queries. One common property among these queries is that they all have many output columns: the number of output columns for queries TQ2, TQ9, TQ10, TQ19 and TQ20 are 12, 7, 11, 8, and 7, respectively. For query TQ2, it even has 5 output columns that are projected from the same Supplier table. For the remaining queries where both STAR and QREF can find the target queries, QREF always outperformed STAR; the minimum, average, and maximum speedup factors of QREF over STAR for these queries are 1.12, 120, and 682.1, respectively. Figure 4c compares the number of candidate queries generated by STAR and QREF for the 22 queries in Experiment 1. The results clearly demonstrate the effectiveness of QREF's query-centric approach to enumerate candidate queries compared to STAR's path-centric approach.

Experiment 2. This experiment (shown in Fig. 4d) compares the performance using queries Q1-Q6 in [20]. Observe that QREF outperforms STAR by a minimum, average, and maximum speedup factor of 11, 33, and 49, respectively. Similar to Experiment 1, QREF also generated fewer candidate queries compared to STAR.

Experiment 3. This experiment (shown in Fig. 4e) compares the performance using the five queries in Fig. 4a. Both queries QQ1 and QQ2 consist of an antijoin, and STAR was unable to reverse engineer such queries as expected; in contrast, QREF was able to find these target queries very quickly. Queries QQ3 and QQ4 both consist of filtering nodes, and the results show that QREF was able to significantly outperform STAR for target queries containing filtering nodes. In particular, for query QQ3, STAR's path-centric approach actually requires generating join paths of up to length 3 for it to find the target query; however, this entails enumerating a large search space of join paths and the execution of STAR was terminated after running for one hour and generating 232 candidate queries without finding the target query.

Finally, Fig. 4f examines the effect of the number of covering attributes in a table using four variants of query QQ5. The first variant has only one column

C.name from the Customer table, and the additional variants are derived by progressively adding more projection columns (C.acctbal, C.address, and C.phone) from Customer table. Thus, the ith variant projects i columns from Customer table. Observe that as the number of covering attributes increases, both the execution time and the number of generated candidate queries increased significantly for STAR; indeed, when the number increases to more than 2, STAR had to enumerate hundreds of candidate queries and its executions did not terminate after running for more than an hour. In contrast, the performance of QREF is less negatively affected by the number of covering attributes.

7   Conclusions

In this paper, we focus on the QRE problem for PJ$^+$ queries, which is a more expressive class of queries than PJ queries by additionally considering antijoins as well as inner joins (and semijoins). To enhance efficiency, we propose a novel query-centric approach that is based on horizontally partitioning each database into fragments using FPJ$^+$ queries. By associating each table fragment with a FPJ$^+$ fragment query, our approach is amenable to using efficient precomputation and index techniques to both generate candidate projection table queries as well as validate candidate queries. Our experimental study demonstrates that our approach significantly outperforms the state-of-the-art solution (for PJ queries without antijoins) by an average improvement factor of 120.

Acknowledgements. We would like to thank Meihui Zhang for sharing the code of STAR. This research is supported in part by MOE Grant R-252-000-A53-114.

References

1. Abouzied, A., Angluin, D., Papadimitriou, C., Hellerstein, J.M., Silberschatz, A.: Learning and verifying quantified Boolean queries by example. In: PODS (2013)
2. Arenas, M., Diaz, G.I.: The exact complexity of the first-order logic definability problem. ACM TODS **41**(2), 13:1–13:14 (2016)
3. Bonifati, A., Ciucanu, R., Staworko, S.: Learning join queries from user examples. ACM TODS **40**, 1–38 (2016)
4. Das Sarma, A., Parameswaran, A., Garcia-Molina, H., Widom, J.: Synthesizing view definitions from data. In: ICDT (2010)
5. Gao, Y., Liu, Q., Chen, G., Zheng, B., Zhou, L.: Answering why-not questions on reverse top-k queries. PVLDB **8**, 738–749 (2015)
6. He, Z., Lo, E.: Answering why-not questions on top-k queries. In: ICDE (2012)
7. He, Z., Lo, E.: Answering why-not questions on top-k queries. TKDE **26**, 1300–1315 (2014)
8. Kalashnikov, D.V., Lakshmanan, L.V., Srivastava, D.: FastQRE: fast query reverse engineering. In: SIGMOD (2018)
9. Li, H., Chan, C.Y., Maier, D.: Query from examples: an iterative, data-driven approach to query construction. PVLDB **8**, 2158–2169 (2015)
10. Li, M., Chan, C.Y.: Efficient query reverse engineering using table fragments. Technical report (2019)

11. Liu, Q., Gao, Y., Chen, G., Zheng, B., Zhou, L.: Answering why-not and why questions on reverse top-k queries. VLDB J. **25**, 867–892 (2016)
12. Luo, Y., Fletcher, G.H.L., Hidders, J., Wu, Y., Bra, P.D.: External memory k-bisimulation reduction of big graphs. In: ACM CIKM, pp. 919–928 (2013)
13. Panev, K., Michel, S., Milchevski, E., Pal, K.: Exploring databases via reverse engineering ranking queries with paleo. PVLDB **13**, 1525–1528 (2016)
14. Psallidas, F., Ding, B., Chakrabarti, K., Chaudhuri, S.: S4: top-k spreadsheet-style search for query discovery. In: SIGMOD (2015)
15. Shen, Y., Chakrabarti, K., Chaudhuri, S., Ding, B., Novik, L.: Discovering queries based on example tuples. In: SIGMOD (2014)
16. Tan, W.C., Zhang, M., Elmeleegy, H., Srivastava, D.: Reverse engineering aggregation queries. PVLDB **10**, 1394–1405 (2017)
17. Tran, Q.T., Chan, C.Y.: How to conquer why-not questions. In: SIGMOD (2010)
18. Tran, Q.T., Chan, C.Y., Parthasarathy, S.: Query by output. In: SIGMOD (2009)
19. Weiss, Y.Y., Cohen, S.: Reverse engineering SPJ-queries from examples. In: PODS (2017)
20. Zhang, M., Elmeleegy, H., Procopiuc, C.M., Srivastava, D.: Reverse engineering complex join queries. In: SIGMOD (2013)

Embedding Analysis

Decentralized Embedding Framework for Large-Scale Networks

Mubashir Imran[1], Hongzhi Yin[1(✉)], Tong Chen[1], Yingxia Shao[2],
Xiangliang Zhang[3], and Xiaofang Zhou[1]

[1] School of Information Technology and Electrical Engineering,
The University of Queenslands, Brisbane, Australia
m.imran@uq.net.au, {h.yin1,tong.chen}@uq.edu.au, zxf@itee.uq.edu.au
[2] Beijing University of Posts and Telecommunications, Beijing, China
shaoyx@bupt.edu.cn
[3] King Abdullah University of Science and Technology, Thuwal, Saudi Arabia
xiangliang.zhang@kaust.edu.sa

Abstract. Network embedding aims to learn vector representations of vertices, that preserve both network structures and properties. However, most existing embedding methods fail to scale to large networks. A few frameworks have been proposed by extending existing methods to cope with network embedding on large-scale networks. These frameworks update the global parameters iteratively or compress the network while learning vector representation. Such network embedding schemes inevitably lead to a high cost of either high communication overhead or sub-optimal embedding quality. In this paper, we propose a novel *decentralized large-scale network embedding framework* called *DeLNE*. As the name suggests, DeLNE divides a network into smaller partitions and learn vector representation in a distributed fashion, avoiding any unnecessary communication overhead. Our proposed framework uses Variational Graph Convolution Auto-Encoders to embed the structure and properties of each sub-network. Secondly, we propose an embedding aggregation mechanism, that captures the global properties of each node. Thirdly, we propose an alignment function, that reconciles all sub-networks embedding into the same vector space. Due to the parallel nature of DeLNE, it scales well on large clustered environments. Through extensive experimentation on realistic datasets, we show that DeLNE produces high-quality embedding and outperforms existing large-scale network embeddings frameworks, in terms of both efficiency and effectiveness.

Keywords: Network embedding · Distributed system · Auto-encoder · Embedding alignment

---

H. Yin—Contributing equally with the first author.

© Springer Nature Switzerland AG 2020
Y. Nah et al. (Eds.): DASFAA 2020, LNCS 12114, pp. 425–441, 2020.
https://doi.org/10.1007/978-3-030-59419-0_26

1   Introduction

Learning continuous low-dimensional vector representations of nodes in a network has recently attracted substantial research interest. Data from networks such as E-commerce platforms, scholarly libraries, social media, medicine, service providers [2,5,24,25] etc. in its raw form is not directly applicable to emerging Machine Learning (ML) approaches, as they requires low-dimensional vector representations for computation. Formulating vector representations of these data networks, which can be utilized as an input to many ML systems, is identified as "Network Embedding" [1]. High-quality network embedding is imperative for accurately performing network inference tasks, e.g. link prediction, node classification, clustering, visualization and recommendation. State-of-the-art network embedding methods like node2vec, Line, DeepWalk and SDNE [3,17,20,22] effectively preserve the network structure properties. However, in the case of large-scale networks, especially the dynamically growing networks, being confronted with more sophisticated network structure, these state-of-the-art network embedding methods fall short on both computational efficiency and embedding quality.

To tackle the crucial task of efficiently producing network embedding on large networks, some large-scale network embedding frameworks have been proposed. These frameworks mostly perform the task by: 1) *coarsening* the network (i.e. dividing it into smaller chunks), 2) *sharing the embedding parameters* and 3) utilizing *matrix multiplication* to reduce data dimensions. Yet, existing network coarsening techniques can hardly guarantee the embedding quality, as the degree of coarsening increases [12], due to loss of information (i.e. local structure). Sharing of global parameters introduces immense communication overhead, while matrix multiplication-based approaches incur high memory cost. These frameworks are further described in Sect. 5.

With the increasing availability of large-scale distributed systems and cloud-based resources, another possible approach to address the large-scale network embedding problem is to carry out representation learning in a decentralized fashion. Decentralized scheme is beneficial for effective utilization of cloud-based resources. It is faster to generate node embedding due to parallel computation of gradients, while avoiding synchronization of global parameters and being easily scalable in the case growing networks. But before we can enjoy the benefits of decentralized scheme, we must solve the following three challenges imposed: 1) producing quality partitions, 2) preserving network properties in the absence of global parameters, and 3) aligning the distributively learned node representations into the same vector space.

Commonly, large networks can be expressed as a collection of smaller well-defined communities [13], interconnected through border nodes, known as *anchor nodes* [8]. For example, in an academic network, if a scholar collaborates frequently with researchers from two different research areas, this scholar can be regarded as the anchor node between two research communities. Anchor nodes act as bridges in order to diffuse information between communities [8].

Motivated by the fact that *large networks consist of numerous communities* that share a variety of *anchor nodes*, we present *Decentralized Large-scale Network Embedding Framework (DeLNE)*. DeLNE learns node representations from large-scale networks in a distributed fashion. Our proposed model aims to address the limitations of existing techniques by avoiding excessive communication overhead caused by parameter sharing. DeLNE identifies the optimal number of divisions for a given large-scale network, based on min-edge cut. It partitions the network into multiple well-defined communities with high neighborhood densities. Our framework then learns node embedding of each partition independently in a distributed environment. It then realigns the learned node representations into the same embedding space, by learning an alignment function. DeLNE stands out in its high scalability, even if the size of any partition grows or new sub-networks are introduced to the network. The main contributions of this paper are threefold:

- We advance the existing network embedding methods to cope with very large-scale networks in a parallel and distributed fashion. This allows for efficient computation on very large networks while preserving properties and structural information.
- In DeLNE, we present an aggregation function that captures the global properties of the sub-network, by actively fusing the information of neighbour nodes. We present an embedding alignment scheme that refines the distributed embedding onto the same embedding space.
- We conduct extensive experimentation on large-scale networks, to evaluate the effectiveness and efficiency of DeLNE. Experiments suggest that DeLNE outperforms the existing state-of-the-art and large-scale network embedding techniques.

2 Preliminaries

2.1 Notations and Definitions

In this section, we first explain the notations adopted throughout the paper (Table 1). For the ease of understanding, we also formally define the key technical terms in this paper as follows.

Definition 1 *(Super Network): a network $\mathcal{G}$ constructed by collapsing its nodes (i.e. $\{\forall v \in \mathcal{G}\}$) using Heavy Edge Matching [12]. A single node in $\mathcal{G}_S$ represents numerous nodes in the original $\mathcal{G}$.*

Definition 2 *(k^{th}-Order Walk): augmenting the neighbourhood S of a vertex $(v \in \mathcal{G})$, from $d = 1$ down to $d = k$. This augmentation is carried out in a breadth-first (BFS) search manner.*

Definition 3 *(Anchor Node): the border nodes (v_a) of a partitioned network. An anchor node v_a has an edge e_a (i.e. anchor link) to the adjacent sub-network, connecting one partition to another.*

Table 1. The notations adopted in the paper

Notation	Definition
$\mathcal{G}$	An undirected network
$\mathcal{V}$	A set of vertices belonging to network $\mathcal{G}$
$\mathcal{E}$	A set of edges belonging to network $\mathcal{G}$
$\mathcal{F}$	A set of features associated with $\mathcal{G}$
v_i	i^{th} vertex $\in \mathcal{V}$
e_i	i^{th} edge $\in \mathcal{E}$
d	Depth of network $\mathcal{G}$ from a vertex v
$\mathcal{G}_S$	**Super network:** Compressed representation of complete network
$\{\mathcal{G}_1, \mathcal{G}_2, ..\mathcal{G}_k\}$	A k way partitioned network
v_a	Anchor node (a node having an edge in adjacent partition)
$\mathbf{A}$	An adjacency matrix of the network $\mathcal{G}$
$\mathbf{Z}_i$	Embedding of sub-network i
$\mathbf{z}$	An embedded vector

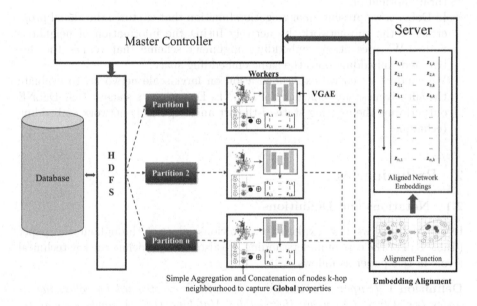

Fig. 1. The overview of DeLNE framework.

2.2 Problem Definition

Given a large-scale network $\mathcal{G} = (\mathcal{V}, \mathcal{E}, \mathcal{F})$ having dimensions $(\mathbf{d} \ll |\mathcal{V}|)$, we aim to learn node representations of a network $\mathcal{G}$ for its k partitions $\{\mathcal{G}_1, \mathcal{G}_2, ..\mathcal{G}_k\}$ in a distributed fashion. We also want to learn an alignment function $M : \mathcal{L}^{|\mathbf{z}_s - \mathbf{z}_t|}$, that aligns (partitioned network using anchor nodes $\mathbf{z}_s$ of source graph to $\mathbf{z}_t$ of target graph) and produces consistent node embeddings of each sub-network. It should also preserve both local and global network structures, as well as the node properties.

3   Methodology

To enable parallel computing for large-scale network embedding, DeLNE implements a master worker framework, consisting of four main phases: 1) *network partitioning*, 2) local property-preserving *node embedding*, 3) global property-preserving *embedding aggregation*, and 4) *embedding refinement*, i.e. mapping sub-networks into the same embedding space. Figure 1 presents an overview of the DeLNE architecture. Firstly, the master controller partitions the network using a non-overlapping network partitioning algorithm [9]. This divides the network into k high-quality partitions. These partitions take the shape of communities, based on their structural properties. Two neighborhood partitions are connected via anchor nodes v_a. A copy of anchor node and its first-order neighbourhood is kept at both sides of the connecting partitions. Master controller assigns each partition to a unique worker. Each worker employs a Variational Graph Auto-encoder (VGAE) to learn low-dimensional node representations dedicated to a sub-network and a property aggregation function to capture global perspective of the sub-network. Finally, we align the node embedding into the same space by learning an alignment function supervised by the observed anchor nodes.

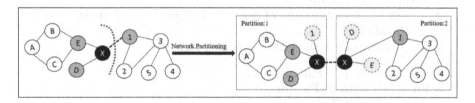

Fig. 2. Overlapping network partitioning by modifying multilevel graph partitioning algorithm. Node **X** is an anchor node connecting both partition-1 and partition-2. 1st order neighbour hood of **X** i.e. node **D** and **E** from partition-1 and node **1** from partition-2 are copied to the adjoining partitions.

3.1   Network Partitioning

An important property exhibited by very large networks is their community structures property. This translates the network into number of smaller modules called clusters [13]. Each large network consists of several smaller communities that share inherit characteristics or interests, e.g. people with interest in a common sport like football or basketball, followers of a particular political party or opinion etc. It is therefore critical to partition a network, such that nodes sharing similar characteristics are grouped together, against those nodes which exhibit distinct characteristics. In other words, we aim to increase intra-cluster connectivity and decrease the inter-cluster connectivity.

In order to partition the network, we propose a variant of multilevel graph partitioning algorithm (METIS) [6]. As shown in Algorithm 1, multilevel partitioning approach partitions a given graph in three phases, namely coarsening, partitioning and uncoarsening. During the coarsening phase, the algorithm iteratively applies heavy-edge matching (HEM) to compute the edge matching, i.e. finding a contraction between the vertex with the lowest degree and the vertex with the highest weight (v_1, v_2), such that weight between them is maximized. Since the size of the corset graph is small, multilevel graph partitioning algorithm efficiently apply graph growing heuristic to compute partitioning. Boundary Kernighan-Lin (BKLR) algorithm is used to iteratively refine the corset partitions. DeLNE does not involve sharing of any parameters or communication between the partitions during the learning phase. Rather, we learn the node embedding of each sub-network independently and then map them into the same space. To align the sub-networks and preserve better structural properties of the anchor nodes, we make use of the non-overlapping partitioning (see Fig. 2). Anchor nodes (e.g.. node **X** in Fig. 2) are copied to the linked partition (doted line in Fig. 2 represents such link), along with its 1st-order neighbourhood.

ALGORITHM 1: Multilevel Graph Partitioning

Input: $\mathcal{G} = (\mathcal{V}, \mathcal{E}), k$
Output: $[\mathcal{G}_1, \mathcal{G}_2, ...\mathcal{G}_k]$

1 **while** $|v \in \mathcal{G}| > k$ **do**
2   sort in increasing order of degree$(v \in \mathcal{G})$ **for** $\forall\ v \in \mathcal{G}$ **do in Parallel**
3   collapse v with neighbour having highest w

4 $[\mathcal{G}_{S1}, \mathcal{G}_{S2}, ...\mathcal{G}_{Sk}] \leftarrow$ Partition$(\mathcal{G}_S)$
5 **for** $\forall\ i \leftarrow k$ **do in Parallel**
6      $\mathcal{G}_i \leftarrow$ BKLR$(\mathcal{G}_{Si})$
7   **if** *Neighbour(v* $: (v \in \mathcal{G}_i) \notin \mathcal{G}_i)$ **then**
8         $\mathcal{G}_i \cup 1^{st}$order_neighbourhood$(v)$

9 **return** $[\mathcal{G}_1, \mathcal{G}_2, ...\mathcal{G}_k]$

3.2 Base Embedding

Our low-dimensional vector representations should effectively preserve both network structure (for supporting the network reconstruction tasks) as well as network properties (for network inference tasks). Commonly used network embedding models include linear models e.g. matrix factorization [21,23], skip-gram models [3,17,20] and non-linear Graph Convolution Networks (GCNs) [1,4,7] models. Traditional linear models can be classified as shallow models as they only captures first-order connections. Skip-gram models aim to learn embeddings from linear space, leading to limited expressiveness. Non-linear models e.g. GCNs address the previous problems, but heavily rely on high-quality labeled data and hardly adapt to unsupervised learning. While working with large-scale networks, due to their dynamics and node diversity, it is very difficult to associate high-quality and consistent labels with each individual node. To capture the non-linearity and preserve high-quality embeddings in the absence of labeled data, in DeLNE we employ Variational Graph Auto-Encoder (VGAE) [7], to support unsupervised learning task.

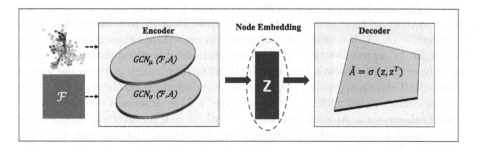

Fig. 3. VGAE, having two layer GCN as encoder and a inner product decoder.

Variational auto-encoder (VAE) modifies traditional auto-encoders by taking in a distribution $q\phi(\mathbf{z}|v)$ rather than data points. By doing so the model is able to adapt with and infer on unseen data. VAE encoder accepts data point v and generates the means and standard deviation of the Gaussian distribution. The lower-dimensional vector representation $\mathbf{z}$ is sampled using this distribution, $q\phi(\mathbf{z}|v)$. The decoder takes in the embedding $\mathbf{z}$ and produces output using variational approximation $p\theta(v|\mathbf{z})$.

To apply the idea of VAE on a network, we deploy a VGAE, that is able to generate as well as predict unseen links and the structure. VGAE consists of two parts: 1)An inference model (encoder) and 2) A generative model (decoder). The inference model takes an adjacency matrix $\mathbf{A}$ and feature matrix $\mathcal{F}$ as its input. It generates the mean and standard deviation through a function θ. As show in Fig. 3, this encoder is constructed using two layers of GCN, given as:

$$\mu = \text{GCN}_\mu(\mathbf{A}, \mathcal{F}) \tag{1}$$

$$\log \sigma = \text{GCN}_\sigma(\mathbf{A}, \mathcal{F}) \tag{2}$$

By combining the two layers together, we get:

$$\text{GCN}(\mathbf{A}, \mathcal{F}) = \tilde{\mathbf{A}} \; Relu(W_0 \mathcal{F} \tilde{\mathbf{A}}) W_1 \tag{3}$$

where, W_0 and W_1 are the weight matrices of each layer. The generative model consists of the inner product among the latent variables and is given as:

$$p(\mathbf{A}|\mathbf{Z}) = \Pi_{i=1}^n \Pi_{j=1}^n p(\mathbf{A}_{(i,j)}|\mathbf{z}_i, \mathbf{z}_j) \tag{4}$$

where $\mathbf{A}_{(i,j)} \in \mathbf{A}$ and $p(\mathbf{A}_{(i,j)} = 1|\mathbf{z}_i, \mathbf{z}_j) = \sigma(\mathbf{z}_i^T \mathbf{z}_j)$. Here $\sigma(\cdot)$ is a sigmoid function. The loss function for the optimization task, consists of two parts: 1) Reconstruction loss between the input and reconstructed adjacency and 2) Similarity loss, defined by KL-divergence.

$$L = -KL[q(Z|\mathcal{F}, \mathbf{A}||p(Z))] + E_{q(Z|\mathcal{F},\mathbf{A})}[\log(p(\mathbf{A}|\mathbf{Z}))] \tag{5}$$

3.3 Embedding Aggregation

Each sub-network exhibits a unique global prospective, depending on its structure. If two sub-networks have similar structures, they should also resemble each other in terms of their global characteristics [19]. In order to preserve this global perspective of a sub-network, we introduce an embedding aggregation function. This function aggregates the vector representations of vertices, that are k^{th} order neighbour of v in a breath first search (BFS) fashion, much in the same way as Weisfeiler-Lehman algorithm [19]. The resultant aggregation (i.e. $\hat{\mathbf{z}}_1 + \hat{\mathbf{z}}_2... + \hat{\mathbf{z}}_d$) is then concatenated with $\mathbf{z}$, as shown in Fig. 4. This process is repeated for the entire sub-network. At each level of the depth, a weight, inversely proportional to the depth is multiplied with aggregation. The aggregation and concatenation of each vertex is given as:

$$\hat{\mathbf{z}} = \mathbf{z} \oplus \sum_{l=1}^{d} w_l (\sum_{i=1}^{n} \mathbf{z}_{i,l}) \tag{6}$$

where, $\mathbf{z}$ is the base embedding of vertex v, d is the depth, n are the number of vertices at d and $w_l = \frac{1}{d}$.

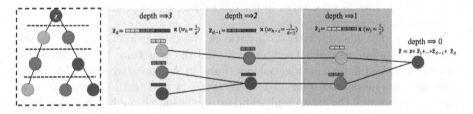

Fig. 4. Global aggregation augments walks from each node down to depth d, in a BFS fashion.

The benefit of this approach is that, for vertices having similar neighbourhood structures, their aggregated embeddings will be close to each other in the vector space. Correspondingly, if two sub-networks are alike in their structures, they will have highly similar node embeddings. In our experiments, we aggregated node embeddings down to $d = 4$.

3.4   Embedding Refinement

The key challenge faced while learning network representation in a distributed system is reconciling of sub-networks embeddings into the same embedding space. Directly projecting the embeddings within k partitions into a unified vector space is a straightforward approach. However, it is impractical when dealing with large-scale networks as the huge amount of parameters used for embeddings makes it highly demanding on computational resources. So, in what follows, we describe an innovative embedding alignment approach.

As discussed earlier, a large network consists of numerous smaller communities [13]. These communities are interconnected with each other through an edge e_B, connecting a common (anchor) node v_a [8]. For example, in a social network person X can be a member of both a hockey and a football fan club, making him the anchor node. In our framework we have partitioned the network in an overlapping fashion, such that we keep a copy of anchor node v_a as v_s, u_t (source and target) at both connected partitions of the network. For example, in Fig. 2 node $\mathbf{X}$ can be considered as v_s in partition-1 and v_t in partition-2. Embeddings of each sub-network are constructed independently and possess separate latent representation. Hence they require alignment. Anchor nodes (v_s, v_t, having presence in multiple sub-networks) act as a bridge to align one embedding space with the other. Therefore with the help of these anchor nodes we translate one embedding space into the other embedding space, such that the anchor nodes are aligned. As a result of this translation process, all nodes in both sub-networks are aligned with the alignment of the anchor nodes. It is important to note that first-order neighbourhood of an anchor node is copied to each partition (Fig. 2). This helps in preserving similar local properties of the same (anchor) nodes belonging to different partitions, during the network embedding phase.

Embedding Alignment: Assume we have a network partitioned into k sub-networks and learned node embeddings $\mathbf{Z}_1, \mathbf{Z}_2, ...\mathbf{Z}_k$ of the respective sub-networks. Without loss of generality, let us assume that we want to align embeddings $\mathbf{Z}_1, \mathbf{Z}_2, ...\mathbf{Z}_k$ into the same embedding space, say $\mathbf{Z}_0$. The purpose here is to learn the mapping function M, supervised by anchor nodes $v_0^s, v_k^t \in J$, having embeddings $\mathbf{z}_0^s, \mathbf{z}_k^t$. Here J is a collection of marked anchor links. Motivated by Matrix Translation and Supervised Embedding Space Matching [14,15] and given $\mathbf{z}_0^s \in \mathbf{Z}_0$, our mapping function projects $\mathbf{z}_0^s \in \mathbf{Z}^0$ and $\mathbf{z}_k^t \in \mathbf{Z}^t$ into same space. Alignment Loss (L_a) is given as:

$$L_a = \sum_{u_0^s, u_k^t \in J} \| M(\mathbf{z}_0^s; \mathbf{w}) - \mathbf{z}_k^t \|_F \qquad (7)$$

where $\mathbf{w}$ is the vector of weights, such that $\mathbf{z}_0^s \times w$ effectively approximates $\mathbf{z}_k^t$, while $\| \cdot \|_F$ represents Frobenius norm, that gives the distance between source embeddings and target embedding. The parameters of the mapping function M can be obtained by minimizing the loss function given in Eq. (7), using scholastic gradient descent (SGD). To capture non-linear relationship between v_0^s and v_k^t, we employ Multi-Layer Perceptron, having sigmoid σ activation at each layer.

On receiving the embeddings from the first two workers, whose sub-networks are connected, we align the node embeddings into the same vector space. As more workers report their learned embeddings they are aligned into the initial space in parallel. Even if we only focus on mapping the anchor nodes, the learned mapping function is able to map the whole embedding space. This comes from the fact that the embeddings of nodes in the same network are considered to be in the same space. As we align the anchor nodes, the mapping function is able to align the whole embedding space as well.

4   Experiments

In this section we report our experimental findings on several large-scale datasets to showcase the effectiveness of DeLNE. Particularly, we aim to answer the following research questions:

- How is the **embedding quality** of DeLNE compared with state-of-the-art embedding methods for medium-sized networks?
- When handling large-scale networks, is DeLNE able to effectively **preserve network properties** compared with other large-scale network embedding frameworks?
- How **efficient** is DeLNE compared with existing large-scale embedding frameworks?

4.1   Experimental Environment

We carry out our experiments on two Linux-based servers. The master controller server has 256-GB of RAM, 40 CPU cores and NVIDIA GPU (GeForce GTX 1080 Ti), is utilized as the master node. Worker nodes are deployed as separate virtual machines on another server with 1024 GB RAM and 80 CPU cores.

4.2   Dataset

We test and compare our framework on five real-world datasets from different domains, as presented in Table 2. Wiki [23] contains 2,405 documents, having 17,981 links and 19 classes. Github [18] is a social network consisting of github developers, where edges represents following relationship. BlogCatalog [26] is a directory of social blog that manages the bloggers and their blogs. The YouTube [16] dataset consists of friendship groups and consists of over 1.1 million users and 4.9 million links. Flickr [16] dataset consists of edges that are formed between images shared between friends, submitted to similar galleries or groups etc.

Table 2. Statistics of datasets used in our experiments.

Dataset	Wiki	Github	BlogCatalog	Youtube	Flickr
#Nodes	2,405	37,700	88,784	1,134,890	1,715,255
#Edges	17,98	289,003	4,186,390	4,945,382	22,613,981
#Class labels	19	2	39	47	20

4.3 Experimental Settings

To adjust the hyper-parameters, we adjust different settings of DeLNE and evaluate the performance on YouTube network. It is a large-scale network, having a large number of labeled nodes, making node classification task harder. For this reason YouTube network is widely adopted to test the performance by large-scale network embedding frameworks [11,12,27,29]. To identify the optimal number of anchor nodes, we conduct our experiments by altering the number of anchor nodes, kept at each partition. These anchor nodes are selected based on a decreasing order of their degrees. As indicated in Table 3, keeping 500 bridging nodes as anchor nodes outperforms other settings. We also run our experiments by altering the number of network partitioning (γ). For medium scale networks, we adjust $\gamma = 10,20$ and 30. For Large graphs, we set $\gamma = 100,150$ and 200 (Table 4).

Table 3. Effectiveness and efficiency of DeLNE for different setting of anchor nodes, on YouTube dataset.

Number of anchor nodes	0	500	1000	2000
Accuracy	0.27	**0.51**	0.47	0.42
Training time	2.06	**2.36**	2.85	3.55

Table 4. Running time for network embedding on YouTube dataset. For DeLNE running time refers to the sum of training time at each cluster, global properties aggregation time and Embedding Alignment Time.

Embedding framework	GPU	Clusters	Training time (min)
LINE	–	1	92.2
DeepWalk	–	1	107.8
SDNE	1	1	Memory error
COSINE (DeepWalk)	–	1	10.83
COSINE (LINE)	–	1	4.17
GraphVite	1	1	4.12
DeLNE-100	1	100	2.36
DeLNE-150	1	150	**2.30**
DeLNE-200	1	200	**2.30**

4.4 Baselines

Given below are the state-of-the-art network embedding methods, we employ as baseline to compare effectiveness and efficiency of DeLNE. Experiments are conducted using both computationally expensive embedding methods as well as large-scale network embedding frameworks.

- **DeepWalk** [17] constructs node sequences by employing random walks to learn embeddings.
- **LINE** [20] optimizes the objective function of edge reconstruction in-order to construct embeddings. parameters were set as;
- **SDNE** [22] preserves non-linear local and global properties using a semi-supervised deep model.
- **COSINE** [27] learns embeddings on top of existing state-of-the-art embedding methods, using parameter sharing on partitioned networks.
- **SepNE** [11] is a flexible, local and global properties preserving network embedding algorithm which independently learns representations for different subsets of nodes.
- **GraphVite** [29] is a CPU/GPU hybrid system, that augments random walks on CPU in a parallel fashion. On GPU's it trains node embedding simultaneously by deploying state-of-the-art embedding methods.

4.5 Effectiveness of DeLNE as Compared to Traditional Embedding Methods

To test the effectiveness of our proposed model we deploy multi-label classification task. The main idea is how well a given model can predict whether a node belongs to a certain community (label) or not. We test our model against state-of-the-art embedding models using medium sized datasets i.e. Wiki, Github and BlogCatalog. As for larger datasets, which are computationally expensive and slower to train on state-of-the-art models, we employ large-scale embedding frameworks.

Table 5. Results of Multi-label prediction on Wiki, Github and BlogCatalog (BC), as compared with state-of-the-art embedding methods (micro-averaged F1 scores).

Labeled nodes	Wiki		Github		BlogCatalog	
	10%	90%	10%	90%	10%	90%
DeepWalk	0.3265	0.4357	0.7417	0.7387	0.1578	0.2030
Line 1st order	0.3265	0.4357	0.7630	0.7689	0.1467	0.1989
Line 2nd order	0.3307	0.3307	0.7392	0.7469	0.1139	0.1533
SDNE	0.5044	0.6237	0.7766	0.7745	0.1734	0.1857
DeLNE-10	**0.6575**	**0.6596**	0.8397	0.8651	0.2151	0.2254
DeLNE-20	0.6486	0.6502	**0.8405**	**0.8711**	**0.2405**	0.2556
DeLNE-30	0.6331	0.6408	0.8331	0.8625	0.2403	**0.2728**

As Table 5 shows, our method out performs state-of-the-art network methods for all three datasets. This is because the structure of partitions plays an important role in identifying the node labels i.e, if the partitions form dense communities, better performance will be observed from our framework. For the smallest dataset, i.e. Wiki, DeLNE-10 outperforms all. DeLNE-20 works best for medium sized dataset.

4.6 Effectiveness of DeLNE as Compared to Large-Scale Embedding Frameworks

To compute the effectiveness of our model for large-scale datasets, we predict labels for networks consisting of millions of nodes and billion of edges. To compare our results with state-of-the-art methods we employ large-scale embedding frameworks, since these datasets are computationally expensive.

Both for Youtube and Flickr networks, as shown in Table 6 and Table 7, DeLNE outperforms SepNE, COSINE and GraphVite. For Flickr network, Deep-Walk and LINE settings of GraphVite slightly outperforms DeLNE. But as the number of labeled nodes are increased, DeLNE surpasses all the other frameworks.

4.7 Time Efficiency

Here we compare DeLNE with other large-scale network embedding methods regarding training efficiency. The time in Table 5, depicts training time plus aggregation time (maximum of all the workers) and network alignment time. It should be mentioned here that we do not compare time of SepNE, since it focuses on learning embeddings of certain components of a network and invites a sizeable computation overhead encase of embedding the entire network. By running efficiency tests on two large-scale datasets (YouTube and Flickr), we show our frame work outperforms all the base lines. We can also see that varying γ in $\{100, 150, 200\}$, for DeLNE, does not improve efficiency too much (Fig. 5).

Table 6. Results of Multi-label prediction on Youtube (micro-averaged F1 scores).

Labeled nodes	1%	3%	5%	10%
SepNE	0.2253	0.3361	0.3620	0.3882
COSINE (DeepWalk)	0.3650	0.4110	0.4230	0.4400
COSINE (LINE)	0.3631	0.4160	0.4271	0.4441
GraphVite (LINE)	0.3836	0.4217	0.4444	0.4625
GraphVite (DeepWalk)	0.3741	0.4212	0.4447	0.4639
DeLNE	**0.4132**	**0.4360**	**0.4892**	**0.5130**

Table 7. Results of Multi-label prediction on Flickr (micro-averaged F1 scores).

Labeled nodes	1%	3%	5%	10%
SepNE	0.4269	0.4468	0.4562	0.4623
COSINE (DeepWalk)	0.4040	0.4140	0.4190	0.4230
COSINE (LINE)	0.4080	0.4180	0.4240	0.429
GraphVite (LINE)	0.6103	**0.6201**	**0.6259**	0.6305
GraphVite (DeepWalk)	**0.6125**	0.6144	0.6216	0.6298
DeLNM-100	0.5933	0.6136	0.6201	**0.6312**

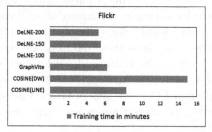

Fig. 5. A: Total Time in minutes taken to compute embeddings for Youtube dataset.
B: Total Time in minutes taken to compute embeddings for Flickr dataset.

5 Related Work

State-of-the-art embedding methods can be divided in to two categories, namely structure-preserving and property-preserving embedding methods. Structure preserving methods such as node2vec [3] and DeepWalk [17] samples k-order random walks on each node of a network. These walks are treated as sentences and Skip Gram is applied to obtain embedding. LINE [20] captures first and second-order proximity to learn embedding. SDNE [22], not only captures second-order proximity, but also non-linear structure using deep neutral networks. Property preserving embedding methods, such as MMDW[21] and TADW[23] deploy matrix factorization-based techniques to embed networks.

Large-scale network embedding frameworks, such as MILE [12], coarsens a very large network to perform network embedding and refine the embedding iteratively. This mechanism captures the global properties and transfers them down through refinement. A key limitation of this technique lies in it's ability to deal with the increasing network size. In this approach, the quality of embedding produced suffers, as the size of the network increases. In addition, refining embedding down from a super-node to the original nodes, is computationally expensive for large networks. Another approach to embed very large networks is parameter sharing. This technique is utilized by COSINE, GraphVite and Py-Torch-BigGraph (PGB) [10, 27, 29]. These frameworks adopt non-overlapping network partitioning methods to create smaller partitions with distinct vertices.

They update the global parameter, at each iteration of learning phases. A limitation to these approaches is there dependence on the bus-bandwidth. As the network size increases, the communication cost of updating the parameters over multiple iterations, aggravates.

Another variant network embedding method, Gaussian Random Projection (GRP), applied by RandNE [28] preserves high order structure using distributed multiplication, on the proximity matrix S. However, computing S on a single machine still remains expensive, for large networks. In parallel settings, this technique requires each computing machine to keep a complete copy of a network's adjacency matrix. This requirement is not viable when embedding a very large network.

In contrast, DeLNE utilizes parallelism and distributive computation to increase efficiency, for large-scale networks. It preservers non-linearity using VGAE while preserving global properties through an aggregation function. DeLNE also aligns the embedding into same vector space, to insure the consistency of sub-networks embedding.

6   Conclusion

In this paper, we presented a novel decentralized large-scale network embedding framework, called DeLNE, which divides a network into multiple dense partitions and performs node embedding in a parallel manner over distributed servers. Our proposed framework uses Variational Graph Convolution Auto-Encoders to embed structure, as well as local and global properties of each sub-network into the vector space. In order to construct consistent embeddings of the entire partitioned network, while avoiding parameter sharing overhead, we learn a network alignment function. The alignment function maps the node embeddings received from distributed servers onto the same embedding space. Through extensive experimentation on real world datasets, we show that DeLNE produce high quality embedding and outperforms state-of-the-art as well as large-scale network embedding frameworks in terms of efficiency and effectiveness.

Acknowledgement. This work is supported by Australian Research Council (Grant No. DP190101985, DP170103954) and National Natural Science Foundation of China (Grant No. U1936104 and 61702015).

References

1. Chen, H., Yin, H., Chen, T., Nguyen, Q.V.H., Peng, W.C., Li, X.: Exploiting centrality information with graph convolutions for network representation learning. In: 2019 IEEE 35th International Conference on Data Engineering (ICDE), pp. 590–601 (2019)
2. Chen, H., Yin, H., Wang, W., Wang, H., Nguyen, Q.V.H., Li, X.: PME: projected metric embedding on heterogeneous networks for link prediction. In: Proceedings of the 24th ACM SIGKDD International Conference on Knowledge Discovery & Data Mining, pp. 1177–1186 (2018)

3. Grover, A., Leskovec, J.: node2vec: scalable feature learning for networks. In: Proceedings of the 22nd ACM SIGKDD International Conference on Knowledge Discovery and Data Mining, pp. 855–864. ACM (2016)
4. Hamilton, W., Ying, Z., Leskovec, J.: Inductive representation learning on large graphs. In: Advances in Neural Information Processing Systems, pp. 1024–1034 (2017)
5. Imran, M., Akhtar, A., Said, A., Safder, I., Hassan, S.U., Aljohani, N.R.,: Exploiting social networks of Twitter in altmetrics big data. In: 23rd International Conference on Science and Technology Indicators (STI 2018) (2018)
6. Karypis, G., Kumar, V.: Metis-unstructured graph partitioning and sparse matrix ordering system, version 2.0 (1995)
7. Kipf, T.N., Welling, M.: Variational graph auto-encoders. In: NIPS Workshop on Bayesian Deep Learning (2016)
8. Kong, X., Zhang, J., Yu, P.S.: Inferring anchor links across multiple heterogeneous social networks. In: Proceedings of the 22nd ACM International Conference on Information & Knowledge Management, pp. 179–188. ACM (2013)
9. LaSalle, D., Patwary, M.M.A., Satish, N., Sundaram, N., Dubey, P., Karypis, G.: Improving graph partitioning for modern graphs and architectures. In: Proceedings of the 5th Workshop on Irregular Applications: Architectures and Algorithms, p. 14. ACM (2015)
10. Lerer, A., Wu, L., Shen, J., Lacroix, T., Wehrstedt, L., Bose, A., Peysakhovich, A.: PyTorch-BigGraph: a large-scale graph embedding system. In: Proceedings of The Conference on Systems and Machine Learning (2019)
11. Li, Z., Zhang, L., Song, G.: SepNE: bringing separability to network embedding. In: Proceedings of the AAAI Conference on Artificial Intelligence, vol. 33, pp. 4261–4268 (2019)
12. Liang, J., Gurukar, S., Parthasarathy, S.: MILE: a multi-level framework for scalable graph embedding. arXiv preprint arXiv:1802.09612 (2018)
13. Malliaros, F.D., Vazirgiannis, M.: Clustering and community detection in directed networks: a survey. Phys. Rep. **533**(4), 95–142 (2013)
14. Man, T., Shen, H., Huang, J., Cheng, X.: Context-adaptive matrix factorization for multi-context recommendation. In: Proceedings of the 24th ACM International on Conference on Information and Knowledge Management, pp. 901–910. ACM (2015)
15. Man, T., Shen, H., Liu, S., Jin, X., Cheng, X.: Predict anchor links across social networks via an embedding approach. IJCAI **16**, 1823–1829 (2016)
16. Mislove, A., Marcon, M., Gummadi, K.P., Druschel, P., Bhattacharjee, B.: Measurement and analysis of online social networks. In: Proceedings of the 7th ACM SIGCOMM Conference on Internet measurement, pp. 29–42. ACM (2007)
17. Perozzi, B., Al-Rfou, R., Skiena, S.: DeepWalk: online learning of social representations. In: Proceedings of the 20th ACM SIGKDD International Conference on Knowledge Discovery and Data Mining, pp. 701–710. ACM (2014)
18. Rozemberczki, B., Allen, C., Sarkar, R.: Multi-scale attributed node embedding (2019)
19. Shervashidze, N., Schweitzer, P., Leeuwen, E.J.v., Mehlhorn, K., Borgwardt, K.M.: Weisfeiler-Lehman graph kernels. J. Mach. Learn. Res. **12**, 2539–2561 (2011)
20. Tang, J., Qu, M., Wang, M., Zhang, M., Yan, J., Mei, Q.: LINE: Large-scale information network embedding. In: Proceedings of the 24th International Conference on World Wide Web, pp. 1067–1077. International World Wide Web Conferences Steering Committee (2015)

21. Tu, C., Zhang, W., Liu, Z., Sun, M., et al.: Max-margin DeepWalk: discriminative learning of network representation. IJCAI **2016**, 3889–3895 (2016)
22. Wang, D., Cui, P., Zhu, W.: Structural deep network embedding. In: Proceedings of the 22nd ACM SIGKDD International Conference on Knowledge Discovery and Data Mining, pp. 1225–1234. ACM (2016)
23. Yang, C., Liu, Z., Zhao, D., Sun, M., Chang, E.: Network representation learning with rich text information. In: Twenty-Fourth International Joint Conference on Artificial Intelligence (2015)
24. Yin, H., Zou, L., Nguyen, Q.V.H., Huang, Z., Zhou, X.: Joint event-partner recommendation in event-based social networks. In: 2018 IEEE 34th International Conference on Data Engineering (ICDE), pp. 929–940. IEEE (2018)
25. Yin, H., Wang, Q., Zheng, K., Li, Z., Yang, J., Zhou, X.: Social influence-based group representation learning for group recommendation. In: 2019 IEEE 35th International Conference on Data Engineering (ICDE), pp. 566–577. IEEE (2019)
26. Zafarani, R., Liu, H.: Social computing data repository at ASU (2009). http:// socialcomputing.asu.edu
27. Zhang, Z., Yang, C., Liu, Z., Sun, M., Fang, Z., Zhang, B., Lin, L.: COSINE: compressive network embedding on large-scale information networks. arXiv preprint arXiv:1812.08972 (2018)
28. Zhang, Z., Cui, P., Li, H., Wang, X., Zhu, W.: Billion-scale network embedding with iterative random projection. In: 2018 IEEE International Conference on Data Mining (ICDM), pp. 787–796. IEEE (2018)
29. Zhu, Z., Xu, S., Tang, J., Qu, M.: GraphVite: a high-performance CPU-GPU hybrid system for node embedding. In: The World Wide Web Conference, pp. 2494–2504. ACM (2019)

SOLAR: Fusing Node Embeddings and Attributes into an Arbitrary Space

Zheng Wang$^{(\boxtimes)}$, Jian Cui, Yingying Chen, and Changjun Hu

Department of Computer Science and Technology,
University of Science and Technology Beijing, Beijing, China
wangzheng@ustb.edu.cn

Abstract. Network embedding has attracted lots of attention in recent years. It learns low-dimensional representations for network nodes, which benefits many downstream tasks such as node classification and link prediction. However, most of the existing approaches are designed for a single network scenario. In the era of big data, the related information from different networks should be fused together to facilitate applications. In this paper, we study the problem of fusing the node embeddings and incomplete node attributes provided by different networks into an arbitrary space. Specifically, we first propose a simple but effective inductive method by learning the relationships among node embeddings and the given attributes. Then, we propose its transductive variant by jointly considering the node embeddings and incomplete attributes. Finally, we introduce its deep transductive variant based on deep AutoEncoder. Experimental results on four datasets demonstrate the superiority of our methods.

Keywords: Data fusion · Social network analysis · Data mining

1 Introduction

Nowadays, people are living in a connected world where information networks are ubiquitous [32]. Typical information networks are social networks, citation networks, e-commerce networks, and the World Wide Web. Recently, the task of network embedding, that consists of learning vector representations for nodes, has attracted increasing attention. Many advanced network embedding methods have emerged such as Deepwalk [21] and LINE [28], which have been proven to enhance numerous network analysis tasks such as node classification [21] and link prediction [13].

Motivated by the benefits, lots of studies [15,35] further consider more complex networks, like the networks with node attributes. However, despite the evident potential benefits, the knowledge of how to actually exploit the node embeddings that social network sites offer is still at its very preliminary stages.

Existing network embedding algorithms are mostly designed for a single network scenario, i.e., they only consider the information from the same network.

© Springer Nature Switzerland AG 2020
Y. Nah et al. (Eds.): DASFAA 2020, LNCS 12114, pp. 442–458, 2020.
https://doi.org/10.1007/978-3-030-59419-0_27

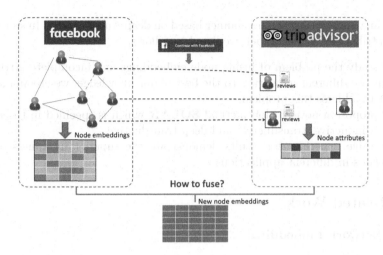

Fig. 1. Illustration of the studied problem, i.e., fusing the node embeddings and attributes provided by two different networks.

Actually, multiple network scenario also deserves lots of attention. As reported in the "Social Media Use in 2018" [27], the median American daily use at least three different social platforms. In the era of big data, it would be better to fuse the knowledge of different networks to facilitate applications. As illustrated in Fig. 1, we can login the review site TripAdvisor with our Facebook ID. As node embeddings not only contain substantial knowledge of users but also protect the privacy [25], Facebook could provide them to TripAdvisor to facilitate its applications, like the cold-start recommendation.

In this study, we focus on a "multiple network" scenario by studying the problem of fusing the node embeddings and attributes provided by two different networks. Notably, unlike the traditional fusion problem, the data in our problem is incomplete, i.e., only a few nodes have attributes. This has practical significance, since a small part of active users contribute the majority of social media content [6]. As illustrated in Fig. 1, on the one hand, TripAdvisor users could write reviews, inspiring us to consider data fusion to obtain the new node embeddings. On the other hand, we also notice that only a few TripAdvisor users write reviews.

In this paper, we propose **SOLAR** which can fu**S**e the inc**O**mp**L**ete network d**A**ta into an a**R**bitrary space. Specifically, we first introduce its inductive version which first learns a fusion model from those nodes with attributes and then utilizes the learned model to obtain the fusion results of nodes without attributes. After that, to obtain high-quality fusion results, we propose its shallow transductive version by jointly considering these two learning parts. Finally, motivated by the recent success of deep neural networks, we further extend the shallow transductive method to a deep one. In particular, we conduct the data fusion in a multi-task learning manner based on deep AutoEncoder. In summary, the contributions of this paper are listed as follows:

1. We study the problem of fusing node embeddings and incomplete attributes from two different networks. To the best of our knowledge, we are among the first to address this problem.
2. We propose a novel fusion method **SOLAR** which is specified in three versions: inductive, transductive and deep transductive.
3. Extensive experimental results demonstrate the superior usefulness of our methods in different applications.

2 Related Work

2.1 Network Embedding

There has been a lot of interest in learning useful node representations from networks automatically [8]. Some earlier works (like LLE [22] and IsoMAP [26]) first construct the affinity network based on the feature vectors (e.g., the K-nearest neighbor graph of data), then perform the generalized eigenvalue decomposition on the constructed network, and finally adopt the resulted eigenvectors as the network representations. Recently, to capture the semantic information of networks, DeepWalk [21] combines random walk and skip-gram to learn network representations. Another well-known work is LINE [28,31] which preserves both local and global network structure. More recently, a lot of studies consider the network embedding with side information. For example, by proving DeepWalk is equivalent to matrix factorization, [35] presents text-associated DeepWalk (TADW). CENE [29] jointly models structure and text information by regarding text content as a special kind of node. GCN [15] uses deep neural networks to utilize the labeled data in a network. RSDNE [34] and RECT [33] further consider the zero-shot label setting.

The most related to our method is the network embedding methods (like TADW and CENE) which utilize node attributes. However, all these methods are still designed for a single network scenario rather than a multiple network scenario. The proposed methods in this paper fill this gap.

2.2 Data Fusion

Data fusion techniques process data obtained from various sources, in order to improve the quality of the measured information for intelligent systems. Data fusion has proven useful in many disciplines. For instance, in bioinformatics, jointly analyzing multiple datasets describing different organisms improves the understanding of biological processes [2]. In information retrieval, fusing the retrieval results from multiple search engines would significantly improve the retrieval performance [14]. We refer to [4] and [7] for a comprehensive survey.

However, little work considers the fusion of incomplete data or network embedding data. This work is the first study of this problem.

3   Notations and Problem Statement

Throughout this paper, matrices are written as capital letters and vectors are denoted as lowercase letters. For an arbitrary matrix M, m_i denotes the i-th row of matrix M, M_{ij} denotes the (i,j)-th entry of M, $M_{i:j}$ denotes a block matrix containing rows i through j, and $M_{k,i:j}$ denotes the columns i through j in the k-th row of matrix M.

Network embedding aims to map the network data into a low dimensional latent space, where each node is represented as a low-dimensional vector. With the above notations and introduction, the studied problem in this paper is defined as follows:

Problem 1 (Fusion with node embeddings and incomplete attributes). Suppose we are given the network embedding data $U \in \mathbb{R}^{n \times d}$, where n is the network node number and the i-th row of U (denoted as u_i) is a d-dimensional embedding vector of node i. Additionally, another network further provides the attributes of the first l $(l < n)$ nodes: $\{a_1, ..., a_l\}$, where $a_i \in \mathbb{R}^m$ is the attribute vector and m is the attribute feature number. Our goal is to fuse the given node embeddings and known attributes into another h-dimensional space.

Note that: unlike existing network embedding methods, the original network structure is unknown in our problem.

4   Our Solution

In this section, we first briefly review the principle of the technique used in data fusion, and then present three data fusion methods for the studied problem.

4.1   Preliminaries

Data fusion can be typically formulated as jointly factorizing matrices [1]:

$$\min_{W,V,Z} \quad ||D^{(1)} - ZW||_2^2 + \alpha ||D^{(2)} - ZV||_2^2 \tag{1}$$

where $D^{(1)} \in \mathbb{R}^{n \times d}$ and $D^{(2)} \in \mathbb{R}^{n \times m}$ are two coupled data matrices obtained from different data sources; $W \in \mathbb{R}^{h \times d}$ and $V \in \mathbb{R}^{h \times m}$ are two functions; $Z \in \mathbb{R}^{n \times h}$ which is shared by both factorizations can be seen as the fusion results. Here, h indicates the fusion dimension, and α is a balancing parameter.

Intuitively, the first term and the second term in Eq. 1 ensures the fusion results Z should maintain the information coming from the first data source $D^{(1)}$ and the second data source $D^{(2)}$, respectively. This formulation can be easily extended to factorization of multiple matrices coupled in different modes, so as to handle the case of more sources.

4.2   The Proposed Inductive Method

As mentioned before, the primary challenge of the studied problem is the incompleteness of the node attributes. More specifically, only a small fraction of nodes have attributes in practical applications. We address this challenge in two steps. Similar to the method formulated in Eq. 1, we first train a matrix factorization model only on the complete data (i.e., the nodes have attributes). Supposing the attribute information matrix is $A \in \mathbb{R}^{n \times m}$ whose first l rows are the known node attributes $\{a_1, ..., a_l\}$. We can specify the factorization model as follows:

$$\min_{W,V,Z_{1:l}} \quad ||U_{1:l} - Z_{1:l}W||_2^2 + \alpha||A_{1:l} - Z_{1:l}V||_2^2 \tag{2}$$

where $U_{1:l}$ stands for the node embeddings of the first l nodes, $A_{1:l}$ stands for the node attributes of the first l nodes, $Z_{1:l}$ stands for the fusion results of the first l nodes, and α is a balancing parameter. Unlike the basic fusion model in Eq. 1, this formulation only learns the fusion model on the nodes with attributes. In other words, the data $U_{l+1:n}$ and $A_{l+1:n}$ are not used in this formulation.

By solving the problem in Eq. 2, we can obtain the fusion results (i.e., $Z_{1:l}$) of the nodes with attributes, and two functions $W \in \mathbb{R}^{h \times d}$ and $V \in \mathbb{R}^{h \times m}$. Intuitively, W can be seen as a mapping function that maps the fusion results to node embeddings. As the original network embedding data U is complete, we can reuse this map function to predict the fusion results of the nodes without attributes. Specifically, for those $\{l+1, ..., n\}$-th nodes with no attributes, we can generate their fusion results by solving the equation:

$$\min_{Z_{l+1:n}} \quad ||U_{1+1:n} - Z_{l+1:n}W||_2^2 \tag{3}$$

where $U_{1+1:n}$ stands for the node embeddings of the rest $n - l$ nodes, and $Z_{l+1:n}$ is the learned fusion results of the rest $n - l$ nodes. Intuitively, this model utilizes the pre-trained model (i.e., function W learned in Eq. 2) to infer the fusion results of the rest nodes.

Optimization. The problem in Eq. 3 has a closed-form solution: $Z_{l+1:n} = (U_{l+1:n}W')(WW')^{-1}$. As such, we finally obtain the fusion results $Z \in \mathbb{R}^{n \times h}$ for all nodes. Note, the matrix multiplication result WW' may be singular (or nearly singular), making it difficult to invert. Iterative solvers can be used to overcome this problem. In particular, supposing the loss of Eq. 3 is $\mathcal{J}_{ind}$, we iteratively update $Z_{l+1:n}$ as $Z_{l+1:n} = Z_{l+1:n} - \eta \frac{\partial \mathcal{J}_{ind}}{\partial Z_{l+1:n}}$, where $\frac{\partial \mathcal{J}_{ind}}{\partial Z_{l+1:n}} = 2(-U_{l+1:n}W' + Z_{l+1:n}WW')$ is the derivative function, and η is the learning rate.

4.3   The Proposed Transductive Method

As shown in Sect. 4.2, we can first learn a fusion model from those nodes with attributes and then utilize the learned model to obtain the fusion results of nodes without attributes. Based on the idea of transductive learning, we can exploit knowledge both from the nodes with or without attributes simultaneously for model learning. This can be formulated by the following optimization problem:

$$\min_{Z,Y,W,V} \mathcal{J}_{transd} = ||U - ZW||_2^2 + \alpha ||Y - ZV||_2^2$$

$$\text{s.t. } y_i = a_i, \ i = 1, ..., l \tag{4}$$

where $Y \in \mathbb{R}^{n \times m}$ denotes the predicted attributes, i.e., its i-th row (denoted as y_i) is the predicted attribute vector of node i; and α is a balancing parameter.

The first term in this method exploits the embedding information of all nodes. The second term utilizes the knowledge of the nodes with or without attributes, indicating that the fusion results should maintain the attribute information of all nodes. In addition, the imposed constraint in the second term guarantees the predicted attributes to be consistent with the known ones. Comparing to the inductive method (formulated in Eq. 2), this transductive one further utilizes the knowledge of those $\{l + 1, ..., n\}$-th nodes who have no attributes.

The above learning framework has one important characteristic which is also the key difference from existing data fusion methods. Specifically, the fused data is incompleted and can be learned in the fusion process. As such, our method can jointly exploit the knowledge of complete and incomplete fusion part, which is an ideal property for transductive learning.

Optimization. The objective function of SOLAR$_T$ (i.e., Eq. 4) is not a convex function, which might be difficult to solve by the conventional optimization tools. To address this, we adopt alternating optimization [3] which iteratively updates one variable with fixing the rest variables. In particular, we can update the variables in Eq. 4 iteratively, as follows.

Update rule of W. When the other variables are fixed, it is straight-forward to obtain the derivative of $\mathcal{J}_{transd}$ w.r.t. W:

$$\frac{\partial \mathcal{J}_{transd}}{\partial W} = 2(-Z'U + Z'ZW) \tag{5}$$

Therefore, we can update W as $W = W - \eta \frac{\partial \mathcal{J}_{transd}}{\partial U}$, where η is the learning rate.

Update rule of Y. When the other variables are fixed, we can obtain the partial derivative of $\mathcal{J}_{transd}$ w.r.t. Y as:

$$\frac{\partial \mathcal{J}_{transd}}{\partial Y} = 2\alpha(Y - ZV) \tag{6}$$

We can update Y as $Y = Y - \eta \frac{\partial \mathcal{J}_{transd}}{\partial Y}$. After that, to satisfy the constraint in Eq. 4, we reset:

$$y_i = a_i, \ i = 1, ..., l \tag{7}$$

Update rule of V and Z. When the other variables are fixed, it is straight-forward to obtain the derivative of $\mathcal{J}_{transd}$ w.r.t. V:

$$\frac{\partial \mathcal{J}_{transd}}{\partial V} = 2\alpha(-Z'Y + Z'ZV) \tag{8}$$

Similarly, when the other variables are fixed, we can obtain the derivative of $\mathcal{J}_{transd}$ w.r.t. Z:

$$\frac{\partial \mathcal{J}_{transd}}{\partial Z} = 2(-UW' + ZWW') + 2\alpha(-YV' + ZVV') \tag{9}$$

Therefore, we can update V and Z as $V = V - \eta\frac{\partial \mathcal{J}_{transd}}{\partial V}$ and $Z = Z - \eta\frac{\partial \mathcal{J}_{transd}}{\partial Z}$, respectively.

We can iterate the above four refinements until convergence or the maximum number of iterations is reached, so as to get the final fusion results Z.

4.4   The Proposed Deep Transductive Method

AutoEncoder (AE) [23] is one of the most popular deep models used in unsupervised feature learning. This kind of model uses artificial neural networks to learn representations by minimizing the reconstruction loss between the input and the reconstructed output. As illustrated in Fig. 2(a), given a set of n input samples $\{x_1, ..., x_n\}$, AE solves the following problem:

$$\min_\theta \sum_{i=1}^{n} ||x_i - f_{dec}(f_{enc}(x_i))||_2^2 \tag{10}$$

where $f_{enc}()$ is the encoder function which transforms the input data to latent representations, and $f_{dec}()$ is the decoder function which transforms the latent representations back to the initial input at the output layer; θ is the trainable parameter set of these two functions. Obviously, $f_{enc}(x_i)$ would return the learned representation of input x_i.

In our problem, optimizing Eq. 10 is intractable, as our fusion data is incomplete. To address this, we modify AE as follows. We first randomly assign attributes to the nodes without attributes (i.e., randomly initialize $A_{l+1:n}$). Then for each node i, we can obtain its complete feature vector as $\hat{x}_i = u_i \sqcup a_i$, where $\sqcup$ stands for the concatenation operation. After that, we minimize the reconstruction loss of the "real" input data (i.e., the given node embeddings and known attributes). This yields the following optimization problem:

$$\min_\theta \sum_{i=1}^{n} ||\psi_i \odot (\hat{x}_i - f_{dec}(f_{enc}(\hat{x}_i)))||_2^2 \tag{11}$$

where ψ_i is the i-th row of a binary matrix $\Psi \in \mathbb{R}^{n \times (d+m)}$, in which the matrix block $\Psi_{i,d+1:d+m} = 1$ if node i has attributes and 0 otherwise. Intuitively, the

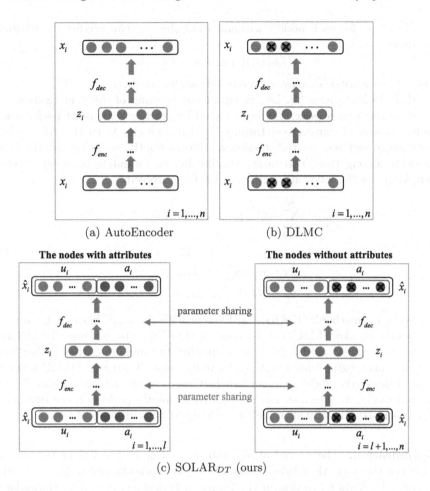

Fig. 2. Structures of (a) AutoEncoder, (b) DLMC, and (c) SOLAR$_{DT}$, where x_i (or $\hat{x}_i$) is the feature vector of an input sample, z_i is the learned latent representation, and a node with the symbol "×" is a missing entry. We utilize plate notation to indicate that there are many copies of the neural network (one for each sample).

Hadamard product with Ψ makes the missing node attributes have no contribution to the reconstruction loss. The network structure of our deep method SOLAR$_{DT}$ is shown in Fig. 2(c).

Let z_i denote the encoding vector of $\hat{x}_i$, i.e., $z_i = f_{enc}(\hat{x}_i)$. We can rewrite Eq. 11 as:

$$\min_{\theta} \sum_{i=1}^{n} ||\psi_i \odot (\hat{x}_i - f_{dec}(z_i))||_2^2 \tag{12}$$

$$\text{s.t. } z_i = f_{enc}(\hat{x}_i), \ i = 1, ..., n$$

Therefore, for each node i without attributes, we can predict its attribute vector as:

$$a_i = [f_{dec}(z_i)]_{d+1:d+m}, \ i = l+1, ..., n \tag{13}$$

where $[\cdot]_{j:k}$ denotes an array of vector values: $[e|e = [\cdot]_i, \ i = j, ..., k]$.

Multi-task learning [9], i.e., simultaneous learning of different tasks under some common constraint, has been tested in practice with good performance under various circumstances. Inspired by this, we can learn the AE model's parameters and recover the missing attributes jointly, so as to use the commonality among these two tasks. Specifically, we formulate our deep method SOLAR$_{DT}$ as the following multi-task learning problem:

$$\min_{\theta, A_{l+1:n}} \sum_{i=1}^{n} ||\psi_i \odot (\hat{x}_i - f_{dec}(z_i))||_2^2$$

$$\text{s.t.} \ z_i = f_{enc}(\hat{x}_i), \ i = 1, ..., n \tag{14}$$

$$\hat{x}_i = u_i \sqcup a_i, \ i = 1, ..., n$$

$$a_i = [f_{dec}(z_i)]_{d+1:d+m}, \ i = l+1, ..., n$$

Note, our method SOLAR$_{DT}$ is different with those AE based data completion methods, like DLMC [10]. As illustrated in Fig. 2(b), although DLMC also adopts a deep AE model, its aims to predict the missing entries rather than learning meaningful representations for data fusion. Therefore, DLMC needs to learn a deep AE model first, and then it predicts the missing entries by the learned model. In contrast, our method could jointly perform these two tasks, which is instructionally effective for the studied data fusion problem.

Optimization. The optimization problem in Eq. 14 consists of two sets of variables: the deep AE network parameters θ and node attributes $A_{l+1:n}$. Similar to the optimization solution in the proposed transductive method (formulated in Eq. 4 in Sect. 4.3), we also adopt an alternating optimization paradigm that iteratively updates one variable with the remaining variables fixed.

Learning θ. The AE network parameter set θ can be efficiently optimized via standard back-propagation (BP) algorithm. We utilize the automatic differentiation techniques in PyTorch.

Learning $A_{l+1:n}$. When the AE network has been trained, we can directly get the reconstructed attributes of all nodes. Therefore, for those nodes without attributes, we can update their attributes by Eq. 13.

5 Experiments

Datasets. To comprehensively evaluate the effectiveness of our methods, we use four widely used citation network datasets [18, 24]: Citeseer, Cora, Wiki, and Pubmed. In these networks, nodes are documents and edges are the citation rela-

Table 1. Datasets

Name	Citeseer	Cora	Wiki	Pubmed
#nodes	3,312	2,708	2,405	19,717
#edges	4,732	5,429	17,981	44,338
#classes	6	7	17	3
#attributes	3,703	1,433	4,973	500

tionships among them. Node attributes are bag-of-words vector text features. The detailed statistics of the datasets are summarized in Table 1.

Experimental Setting. As illustrated in Fig. 1, the node embeddings and attributes are provided by different networks. To align with this setting, we obtain these two kinds of data as follows. Firstly, we get the original node embeddings by the famous network embedding method LINE [28]. Then, we randomly select some nodes and provide them with attributes. After that, we employ different fusion methods to fuse these two kinds of data. Finally, we evaluate the fusion results on different tasks. Note, we also try other network embedding methods in Sect. 5.3.

Baseline Methods. Since this incomplete data fusion problem has not been previously studied, there is no natural baseline to compare with. To facilitate the comparison, we both set the attribute dimension $m = d$ and the final fusion embedding dimension $h = d$. This setting enables us to directly compare our methods with the original node embeddings and attributes. In particular, we compare the following three methods:

1. LINE. We adopt LINE (which preserves both first-order and second-order proximities) to obtain the original node embeddings.
2. Attributes. We use the zero-padded attributes, since only a few nodes have attributes.
3. NaiveAdd. We naively "add" the vectors of the given node embeddings and the zero-padded attributes. This method can be seen as a simple fusion strategy and will get d-dimensional fusion results.

In this comparison, we test three versions of our method: SOLAR$_I$ (the inductive version proposed in Sect. 4.2), SOLAR$_T$ (the transductive version proposed in Sect. 4.3), and SOLAR$_{DT}$ (the deep transductive version proposed in Sect. 4.4).

Parameters. Following LINE, we set the node embedding dimension to 128. Additionally, like [35], we reduce the dimension of attribute vectors to 128 by applying SVD decomposition [12] on the bag-of-words features.

We set the parameter $\alpha = 1$ in our SOLAR$_T$. For our deep method SOLAR$_{DT}$, we employ a two-layer AE model with 128 units in each hidden layer. In this method, we apply dropout ($p = 0.5$) only on the encoder output. Besides, we use Xavier initialization [11] for all variables and adopt the activation function RELU [20] on all layers except the last layer of the decoder.

Table 2. Node classification results (Micro-F1)

		LINE	Attributes	NaiveAdd	SOLAR$_I$	SOLAR$_T$	SOLAR$_{DT}$
Citeseer	10%	0.3928	0.2378	0.3968	0.4166	0.4213	**0.4256**
	30%	0.3928	0.3096	0.4149	0.4568	0.4592	**0.4789**
	50%	0.3928	0.4203	0.4565	0.4991	0.5025	**0.5409**
Cora	10%	0.6095	0.3022	0.5898	0.6029	0.6296	**0.6493**
	30%	0.6095	0.3297	0.5836	0.6107	0.6337	**0.6652**
	50%	0.6095	0.4114	0.5922	0.6300	0.6595	**0.6903**
Wiki	10%	0.5367	0.1838	0.5344	0.5427	0.5510	**0.5662**
	30%	0.5367	0.2752	0.5233	0.5584	0.5681	**0.5972**
	50%	0.5367	0.4073	0.5653	0.5889	0.5944	**0.6346**
Pubmed	10%	0.6467	0.4167	0.6378	0.6316	0.6628	**0.6731**
	30%	0.6467	0.4737	0.6348	0.6335	0.6738	**0.6879**
	50%	0.6467	0.5790	0.6511	0.6451	0.6844	**0.7349**

Table 3. Node classification results (Macro-F1)

		LINE	Attributes	NaiveAdd	SOLAR$_I$	SOLAR$_T$	SOLAR$_{DT}$
Citeseer	10%	0.3285	0.1145	0.3322	0.3482	0.3520	**0.3598**
	30%	0.3285	0.2228	0.3459	0.3864	0.3865	**0.4185**
	50%	0.3285	0.3586	0.3879	0.4269	0.4289	**0.4637**
Cora	10%	0.5441	0.0663	0.5187	0.5567	0.5952	**0.6128**
	30%	0.5441	0.1130	0.5072	0.5591	0.5913	**0.6255**
	50%	0.5441	0.2683	0.5366	0.5794	0.6216	**0.6544**
Wiki	10%	0.3467	0.0368	0.3324	0.3547	0.3529	**0.3727**
	30%	0.3467	0.1462	0.3382	0.3638	0.3756	**0.3967**
	50%	0.3467	0.2865	0.3633	0.3922	0.3847	**0.4273**
Pubmed	10%	0.5478	0.2393	0.5261	0.5343	0.5730	**0.6086**
	30%	0.5478	0.3787	0.5450	0.5610	0.6038	**0.6530**
	50%	0.5478	0.5452	0.6017	0.5839	0.6282	**0.7188**

5.1 Node Classification

Node classification is the most common task for evaluating the learned node representations [21, 28]. We treat the final fusion results as node representations, and use them to classify each node into a set of labels. The main experimental procedure is the same as [21]. Specifically, we first randomly sample a fraction of the labeled nodes as the training data and the rest as the test data. For Citeseer, Cora and Wiki, we fix the label rate in the classifiers to 10%. As Pubmed is a much larger dataset with fewer classes, we set the percentage of labeled data to 1%. Additionally, we vary the ratio of nodes with attributes among 10%,

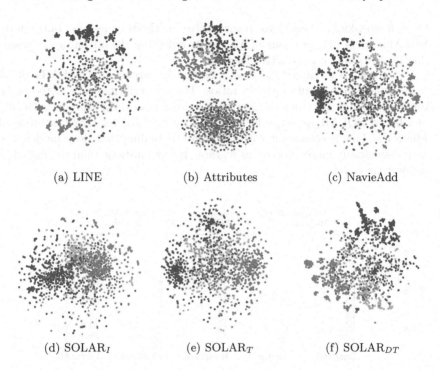

(a) LINE   (b) Attributes   (c) NavieAdd

(d) SOLAR$_I$ (e) SOLAR$_T$ (f) SOLAR$_{DT}$

Fig. 3. Visualization on Citeseer. Every point stands for one document. Color of a point indicates the document class. Red color is IR, Blue color is DB, and Green color is HCI. (Color figure online)

30%, and 50% on all datasets. Then we adopt one-vs-rest logistic regression classifiers to train the classifier. To be fair, we follow [28] to normalize all node representation vectors to unit length before evaluation. Finally, we report the average performance of 20 repeated runs.

Tables 2 and 3 present the classification results in terms of Micro-F1 and Macro-F1 [19], respectively. From these results, we have the following observations and analysis.

- Firstly, NaiveAdd, the simplest fusion strategy, outperforms both LINE and Attributes. This is in line with our expectation that data fusion always helps in achieving better performance compared to the non-fusion methods.
- Secondly, all our methods (including SOLAR$_I$, SOLAR$_T$ and SOLAR$_{DT}$) perform much better than baseline methods. For instance, on Citeseer with 50% attributes, SOLAR$_I$, which performs worst in our three methods, still outperforms LINE by 9.2%, Attributes by 7.8%, and NaiveAdd by 4.3% in terms of Micro-F1. Additionally, the improvement margins of our two transductive methods SOLAR$_T$ and SOLAR$_{DT}$ over these baselines are more obvious. These results clearly demonstrate the effectiveness of our fusion methods.
- Thirdly, the proposed two transductive methods (i.e., SOLAR$_T$ and SOLAR$_{DT}$) consistently outperform our inductive method SOLAR$_I$.

Overall speaking, these two transductive methods generally outperform SOLAR$_I$ by 5–12% in terms of Micro-F1, reflecting the advantage of transductive learning over inductive learning.

- Lastly, the deep version SOLAR$_{DT}$ consistently outperforms SOLAR$_T$ on all four datasets across all attribute ratios. The outperformance is noted to be especially significant when the attribute ratio increases. On average, with 50% attributes, SOLAR$_{DT}$ outperforms SOLAR$_T$ by 7.5% relatively in terms of Micro-F1. This is consistent with the previous finding that deep models can be exponentially more effective at learning representations than their shallow counterparts [16].

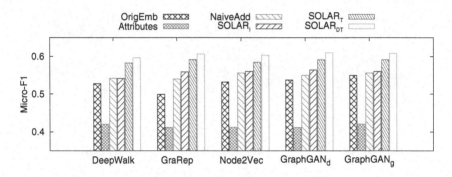

Fig. 4. Node classification results of more network embedding methods on Citeseer.

5.2   Visualization

In this part, we visualize the fusion results for qualitative comparison. Following [28], we use t-SNE [17] to visualize the fused node representations. For simplicity, we choose the first dataset Citeseer and test the case with 50% attributes. For a clear comparison, we only visualize the nodes from three research topics.

As shown in Fig. 3(a–b), the results of LINE (i.e., the original node embeddings) and original attributes are not satisfactory because their points belonging to different categories are heavily mixed together. For the simplest fusion method NaiveAdd, the result looks better. However, the boundaries of each group are not very clear. In contrast, as shown in Fig. 3(d–f), the results of our three methods form three main clusters, which are better than those of compared baselines. Especially, our deep transductive method SOLAR$_{DT}$ shows more meaningful layout. Specifically, each cluster is linearly separable with another cluster in the figure. Intuitively, this experiment demonstrates that our methods can obtain more meaningful fusion results.

5.3   More Network Embedding Baselines

In this part, we evaluate the performance of our methods based on more network embedding methods. In particular, we further test another four network embedding methods:

- DeepWalk [21] learns node embeddings by applying skip-gram model on the random walks of networks.
- GraRep [5] obtains node embeddings by computing high-order proximity and adopts SVD to reduce their dimensionality.
- Node2Vec [13] utilizes a biased random walk strategy based on DeepWalk to effectively explore neighborhood structure.
- GraphGAN [30] is a deep neural network based method that employs adversarial training to learn node embeddings. Its discriminator function (denoted as $GraphGAN_d$) and generator function (denoted as $GraphGAN_g$) could separately obtain node embeddings.

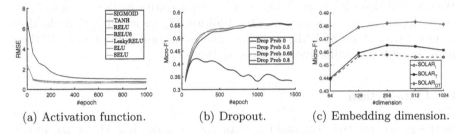

(a) Activation function.   (b) Dropout.   (c) Embedding dimension.

Fig. 5. Effects of various settings in $SOLAR_{DT}$ on Citeseer.

Without loss of generality, we choose 50% nodes with attributes and still fix the label rate to 10%. For convenience, we adopt "OrigEmb" to stand for the original node embeddings obtained by various network embedding methods. Figure 4 reports the performance on Citeseer. We can clearly find that all our three methods (including $SOLAR_I$, $SOLAR_T$ and $SOLAR_{DT}$) consistently outperform baselines significantly. Additionally, we also find that our transductive methods always perform better than the inductive one. Especially, our deep transductive method $SOLAR_{DT}$ always obtains the best performance.

5.4   Various Setting of $SOLAR_{DT}$

Effects of the Activation Types. To explore the effects of different activation functions [11], we test some popular choices in deep learning: SIGMOID, TANH, "rectified linear unit" (RELU), RELU6, LeakyRELU, "exponential linear units" (ELU) and "scaled exponential linear units" (SELU). We apply activation functions on a two-layer AutoEncoder with 128 units in each hidden layer. Except for the last layer of the decoder, the chosen activation function is applied in all layers. As shown in Fig. 5(a), except SIGMOID function, the RSME values of most activation functions show similar tends and results. In the experiments, we find that SIGMOID activation function performs badly in all evaluation tasks. In contrast, the other activation functions all achieve similar better performance.

Effects of Dropout. Dropout is usually applied to avoid the overfitting of deep models. In our method SOLAR$_{DT}$, we only add dropout on the encoder output. We also try to apply dropout on every layer, but that does not improve the performance. Figure 5(b) shows the experimental results. We can clearly find that the model overfits when the dropout rate is set to 0. On the contrary, the overfitting problem is largely prevented when the dropout is applied, i.e., with the dropout value 0.5, 0.65 and 0.85. These experimental results clearly demonstrate the advantages of dropout.

Effects of Embedding Dimension. Figure 5(c) shows the effects of the embedding dimension varied in {64, 128, 256, 512}, with 30% nodes having attributes. We can find that all our methods are not very sensitive to the dimension size, especially when this size exceeds 128. Additionally, we can find that the deep transductive method SOLAR$_{DT}$ always achieves the best performance. All these findings verifies the effectiveness of our methods.

6   Conclusion

In this paper, we study the problem of fusing the node embeddings and incompleted attributes provided by different networks into an arbitrary space. We develop two shallow methods: one is an inductive method and the other is a transductive method. Additionally, we also provide a deep transductive method. Extensive experiments conducted on four real-world datasets demonstrate the effectiveness of our methods. In the future, we will consider more types of related information from more different networks and resources.

Acknowledgment. This work is supported in part by National Natural Science Foundation of China (No. 61902020), Macao Youth Scholars Program (No. AM201912), China Postdoctoral Science Foundation Funded Project (No. 2018M640066), Fundamental Research Funds for the Central Universities (No. FRF-TP-18-016A1), and National Key R&D Program of China (No. 2019QY1402).

References

1. Acar, E., et al.: Structure-revealing data fusion. BMC Bioinform. **15**(1), 239 (2014)
2. Alter, O., Brown, P.O., Botstein, D.: Generalized singular value decomposition for comparative analysis of genome-scale expression data sets of two different organisms. PNAS **100**(6), 3351–3356 (2003)
3. Bezdek, J.C., Hathaway, R.J.: Some notes on alternating optimization. In: Pal, N.R., Sugeno, M. (eds.) AFSS 2002. LNCS (LNAI), vol. 2275, pp. 288–300. Springer, Heidelberg (2002). https://doi.org/10.1007/3-540-45631-7_39
4. Bleiholder, J., Naumann, F.: Data fusion. ACM Comput. Surv. **41**(1), 1–41 (2009)
5. Cao, S., Lu, W., Xu, Q.: GraRep: learning graph representations with global structural information. In: CIKM, pp. 891–900. ACM (2015)
6. Cheng, X., Li, H., Liu, J.: Video sharing propagation in social networks: measurement, modeling, and analysis. In: INFOCOM, pp. 45–49. IEEE (2013)

7. Clark, J.J., Yuille, A.L.: Data Fusion for Sensory Information Processing Systems, vol. 105. Springer, New York (2013)
8. Cui, P., Wang, X., Pei, J., Zhu, W.: A survey on network embedding. TKDE **31**, 833–852 (2018)
9. Evgeniou, T., Pontil, M.: Regularized multi-task learning. In: SIGKDD, pp. 109–117. ACM (2004)
10. Fan, J., Chow, T.: Deep learning based matrix completion. Neurocomputing **266**, 540–549 (2017)
11. Glorot, X., Bengio, Y.: Understanding the difficulty of training deep feedforward neural networks. In: AISTATS, pp. 249–256 (2010)
12. Golub, G.H., Reinsch, C.: Singular value decomposition and least squares solutions. In: Bauer, F.L., Householder, A.S., Olver, F.W.J., Rutishauser, H., Samelson, K., Stiefel, E. (eds.) Handbook for Automatic Computation. Die Grundlehren der mathematischen Wissenschaften, vol. 186, pp. 134–151. Springer, Heidelberg (1971). https://doi.org/10.1007/978-3-642-86940-2_10
13. Grover, A., Leskovec, J.: node2vec: scalable feature learning for networks. In: SIGKDD, pp. 855–864. ACM (2016)
14. Hsu, D.F., Taksa, I.: Comparing rank and score combination methods for data fusion in information retrieval. Inf. Retr. **8**(3), 449–480 (2005)
15. Kipf, T.N., Welling, M.: Semi-supervised classification with graph convolutional networks. In: ICLR (2017)
16. LeCun, Y., Bengio, Y., Hinton, G.: Deep learning. Nature **521**, 436–444 (2015)
17. Maaten, L., Hinton, G.: Visualizing data using t-SNE. JMLR **9**, 2579–2605 (2008)
18. McCallum, A.K., Nigam, K., Rennie, J., Seymore, K.: Automating the construction of internet portals with machine learning. Inf. Retr. **3**(2), 127–163 (2000)
19. Michalski, R.S., Carbonell, J.G., Mitchell, T.M.: Machine Learning: An Artificial Intelligence Approach. Springer, Berlin (2013)
20. Nair, V., Hinton, G.E.: Rectified linear units improve restricted Boltzmann machines. In: ICML, pp. 807–814 (2010)
21. Perozzi, B., Al-Rfou, R., Skiena, S.: DeepWalk: online learning of social representations. In: SIGKDD, pp. 701–710. ACM (2014)
22. Roweis, S.T., Saul, L.K.: Nonlinear dimensionality reduction by locally linear embedding. Science **290**(5500), 2323–2326 (2000)
23. Rumelhart, D.E., Hinton, G.E., Williams, R.J.: Learning internal representations by error propagation. Technical report, California Univ San Diego La Jolla Inst for Cognitive Science (1985)
24. Sen, P., Namata, G., Bilgic, M., Getoor, L., Galligher, B., Eliassi-Rad, T.: Collective classification in network data. AI Mag. **29**(3), 93 (2008)
25. Shakimov, A., et al.: Vis-a-Vis: privacy-preserving online social networking via virtual individual servers. In: COMSNETS, pp. 1–10. IEEE (2011)
26. Silva, V.D., Tenenbaum, J.B.: Global versus local methods in nonlinear dimensionality reduction. In: NIPS, pp. 721–728 (2003)
27. Smith, A., Anderson, M.: Social media use in 2018. Website (2018). https://www.pewresearch.org/internet/2018/03/01/social-media-use-in-2018/
28. Tang, J., Qu, M., Wang, M., Zhang, M., Yan, J., Mei, Q.: LINE: Large-scale information network embedding. In: WWW, pp. 1067–1077. ACM (2015)
29. Tu, C., Liu, H., Liu, Z., Sun, M.: CANE: context-aware network embedding for relation modeling. In: ACL, vol. 1, pp. 1722–1731 (2017)
30. Wang, H., et al.: GraphGAN: graph representation learning with generative adversarial nets. In: AAAI, pp. 2508–2515 (2018)

31. Wang, Q., Wang, Z., Ye, X.: Equivalence between line and matrix factorization. arXiv preprint arXiv:1707.05926 (2017)
32. Wang, Z., Wang, C., Pei, J., Ye, X., Philip, S.Y.: Causality based propagation history ranking in social networks. In: IJCAI, pp. 3917–3923 (2016)
33. Wang, Z., Ye, X., Wang, C., Cui, J., Yu, P.S.: Network embedding with completely-imbalanced labels. TKDE (2020)
34. Wang, Z., Ye, X., Wang, C., Wu, Y., Wang, C., Liang, K.: RSDNE: exploring relaxed similarity and dissimilarity from completely-imbalanced labels for network embedding. In: AAAI, pp. 475–482 (2018)
35. Yang, C., Liu, Z., Zhao, D., Sun, M., Chang, E.Y.: Network representation learning with rich text information. In: IJCAI, pp. 2111–2117 (2015)

Detection of Wrong Disease Information Using Knowledge-Based Embedding and Attention

Wei Ge[1], Wei Guo[1,2]($\boxtimes$), Lizhen Cui[1,2], Hui Li[1,2], and Lijin Liu[1]

[1] School of Software, Shandong University, Jinan, China
gw_ol@foxmail.com
[2] Joint SDU-NTU Centre for Artificial Intelligence Research (C-FAIR),
Shandong University, Jinan, China
{guowei,clz,lih}@sdu.edu.cn, liulijin@mail.sdu.edu.cn

Abstract. International Classification of Diseases (ICD) code has always been an important component in electronic health record (EHR). The coding errors in ICD have an extremely negative effect on the subsequent analysis using EHR. Due to some diseases been viewed as a stigma, doctors, despite having made the right diagnosis and prescribed the right drugs, would choose some diseases that symptom similarity instead of the real diseases to help patients, such as using febrile convulsions instead of epilepsy. In order to detect the wrong disease information in EHR, in this paper, we propose a method using the structured information of medications to correct the code assignments. This approach is novel and useful because patients' medications must be carefully prescribed without any bias. Specifically, we employ the Knowledge-based Embedding to help medications to get better representation and the Self-Attention Mechanism to capture the relations between medications in our proposed model. We conduct experiments on a real-world dataset, which comprises more than 300,000 medical records of over 40,000 patients. The experimental results achieve 0.972 in the AUC score, which outperforms the baseline methods and has good interpretability.

Keywords: Knowledge-based embedding · Self-attention · Wrong disease information detection

1 Introduction

In healthcare, electronic health records (EHRs) are widely accepted to store patient's medical information [9]. The information contained within these EHRs not only provides direct health information about patients but also is used to monitor hospital activities for medical billing [20]. Each EHR is typically accompanied by a set of metadata codes defined by the International Classification of Diseases (ICD), which presents a standardized way of indicating diagnosis and procedures performed by doctors. Many researchers have conducted an in-depth exploration of EHR data based on ICD coding, such as tracking patients with

© Springer Nature Switzerland AG 2020
Y. Nah et al. (Eds.): DASFAA 2020, LNCS 12114, pp. 459–473, 2020.
https://doi.org/10.1007/978-3-030-59419-0_28

specific diseases [11], clinical decision support [25] and modeling the patient state [6].

However, before using big data to analyze, data must be accurate. If not, inaccurate data may cause many errors [9]. Inconsistencies in medical characteristics and disease names are also an important factor in analysis errors, so data consistency is becoming a major social problem. We here refer to data consistency as the consistency of medical concepts, specifically between the name of the disease and the medical characteristics. Some data's medical characteristics differ from the record. Because certain diseases are normally linked to stigma, doctors tend to use other similar diseases instead to avoid such stigma feeling.

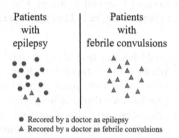

Fig. 1. Due to some social stigma, some doctors will help patients with epilepsy to conceal the truth of their illness and record it as febrile convulsions in the EHR.

A typical coding error caused by stigma is epilepsy cases in China. Epilepsy incidence is generally calculated in terms of the fraction of the number of newly discovered epilepsy patients in the population of 100,000 people per year. According to the World Health Organization [12], the incidence rate in high-income countries is 49, and in the low-income and middle-income countries, the number is stunned 139. In contrast, the number in China is 28.9. Based on the above data, we can see that the incidence of epilepsy in China is lower than the incidence of foreign countries. But the fact is not true. Epilepsy is one type of common central nervous system disorders. Despite advances in the control of epileptic seizures, individuals with epilepsy continue to experience stigma and discrimination [14,24]. International Epilepsy Bureau (IBE) survey shows that many parents or patients in China tend to conceal their illness [26]. So in EHR data, febrile convulsions or other related diseases are often used instead of epilepsy as the diagnostic results in China, as shown in Fig. 1. This will result in inconsistencies in medical characteristics and disease names, as well as the inconsistency of medical concepts. Many existing methods use clinical records as the only source of information for ICD coding. These methods normally use natural language processing to extract patients' symptoms, medication, disease description and other related information from clinical records. And thus the inconsistency of medical concepts is inevitable.

The patient's medications information is closely related to the patient's disease and personal condition, and each medication can be uniquely identified.

Therefore, medication information provides a promising way to auto-detect the coding errors embedded in EHR.

There are two main challenges in this approach. The first challenge is the imbalanced data distribution in EHR. The prescription rate of different medications is quite different. For two medications having the same efficacy, it may be difficult to obtain corresponding representation for the one with fewer occurrences. The other challenge is that the mixture of disease information in medication information. The relationship between medications and diseases is highly complex, and may often affect by certain factors of patients themselves (e.g., historical illness, age, gender, etc.). It is not trivial to capture such intertwined relationships.

To address these challenges, we propose a model called KEAM, which uses **K**nowledge-based **E**mbedding and self-**A**ttention **M**echanism to find the discrepancy between prescription and diagnosis, and correct coding errors. We use the knowledge guided embedding method to obtain precise and comprehensive representation, especially the ones with fewer prescriptions. In addition, we use the self-attention mechanism to obtain the representation of medication interrelationships for different patients.

We summarize our main contributions in this paper as follows:

- We propose KEAM, a simple yet robust model to effectively learn the representation matrix of medication combinations and be used in code assignment tasks, and the model has good interpretability for predicting results.
- KEAM incorporates medical knowledge into the embedding representation of medications and achieves effective embedding representation of medications. At the same time, KEAM uses an attention mechanism to get a 2-D matrix representation of medication combinations, with each row of the matrix attending on a different part of the medication.
- We experimented on a large-scale real-world EHR dataset comprising patient's medication information and diagnostic code. Specifically, we conducted epilepsy recognition experiments. The experimental results show that the predictive performance of KEAM is significantly better than that of baseline methods. Moreover, thanks to the incorporated self-attention mechanism, we are able to identify the specific medications that are deemed important in epilepsy diagnosis.

The rest of this paper is organized as follows. In the following section, we review the related work. We describe the details of our method in Sect. 3. Experiments and results analysis are given in Sect. 4. Then we conclude this paper with future work in Sect. 5.

2 Related Work

Code Assignment. In order to accomplish code assignments using computational methods, the majority of prior studies relied on the discharge summary records and normally employ machine learning methods. Perotte et al. [17] used an improved SVM to predict ICD code, wherein the hierarchical structure of ICD was taken into consideration.

Recently, mainstream research is employing deep learning approaches to perform the task of code assignment. Baumel et al. [4] classified ICD-9 diagnosis codes using an improved GRU, which employs two attention mechanisms in hierarchical sentence and word. Mullenbach et al. [15] made interpretable predictions using an improved CNN with an attention mechanism to improve both the accuracy and the interpretability of the predictions. However, in the discharge summary, if doctors help patients to cover up their epilepsy by providing similar diagnoses such as febrile convulsion, it is difficult to obtain the characteristics associated with predicting epilepsy in the discharge summary.

Scheurwegs et al. [19,20] studied the prediction method combining structured data and unstructured data. In their experiments, no significant difference was found between using medications (28.7%) and the discharge summary (30.2%) in code assignments. So we aim to build a model to use patients' medication information for code assignments, which can solve the errors in the discharge summary, for the patients' medications must be carefully prescribed without any bias.

Attention. Attention-based neural networks was firstly studied in Natural Language Processing (NLP), and in no time, attracted enormous interest in many other areas, such as multiple object recognition [2], neural machine translation [3], diagnosis prediction task [8,10].

RETAIN [8] used a reverse time attention mechanism to model event sequences, the attention mechanism makes RETAIN accurate and interpretable. Dipole [10] model used attention-based bidirectional RNN to make diagnosis. Attention mechanisms used in Dipole allows us to reasonably interpret the prediction results.

The traditional attention is designed to calculate relations between source and target. For capture relations of intra-sentence, a variation of attention mechanism, self-attention was firstly outlined in Vaswani et al. [23] There are two main schemes of self-attention schemes, i.e., dot-product attention and additive attention. The dot-product attention uses a parallel matrix to compute and thus has less time consumption. In order to obtain the representation of medications interrelationships of different options, we employ a self-attention mechanism.

Embedding. In recent years, medical concepts representation has witnessed great thrive. The earliest idea of word embeddings dated back to 1988, which proposed by Rumelhart et al. [18]

It is simple that represents words using one-hot embedding, but it exits some shortcomings, such as sparse vector and loses semantically. To address the issues, Mikolov et al. [13] proposes two models, CBOW and Skip-gram, which take context into consideration. The recent work of medical concepts representation mostly follows the same idea. Tran et al. [22] embedded medical concepts with a non-negative restricted Boltzmann machine in an EMR data set. Choi et al. [7]

proposed med2vec to learn representations of both features and visits in EMR data based on the Skip-gram model.

In order to utilize word sememes to capture exact meanings of a word, Zhiyuan Liu et al. [16] proposed a knowledge-based embedding, which use a CBOW model to learn the hierarchical data We adopt the similarity idea that uses a bi-lstm to learn the hierarchical data.

3 Method

This section introduces the detailed dynamics of our proposed model called KEAM, which uses Knowledge-based Embedding and self-attention Mechanism. Specifically, we employ the Knowledge-based Embedding to help medications to get better representation and the Self-Attention Mechanism to capture the relations between medications in our proposed model.

In the following sections, we first introduce the way we use Knowledge-based embedding to learn the hierarchical data. Then we discuss the novel model KEAM that we proposed to predict the code is an error or not. Finally, we choose some specific options to discuss the interpretability of KEAM.

3.1 Data Representation

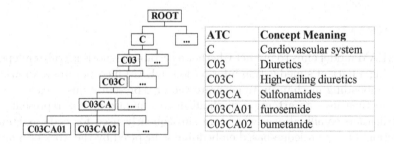

Fig. 2. Hierarchically structure of ATC

ATC	Concept Meaning
C	Cardiovascular system
C03	Diuretics
C03C	High-ceiling diuretics
C03CA	Sulfonamides
C03CA01	furosemide
C03CA02	bumetanide

Assuming that C_d denote the number of unique medications codes, according to one-hot encoding, each medication can be represented by a binary vector of C_d dimension. Therefore, a medication D_i can be expressed as a vector $x_i \in \{0,1\}^{|C_d|}$, and only the ith dimension is 1. If there are n kinds of medications prescribed by the doctor in one patient visit, the medication information of the visit can be expressed as a matrix of size $n \times C_d$.

Each medication code can be mapped to a node in Anatomical Therapeutic Chemical (ATC)[1], is hierarchically organized that each medication can be traced

---

[1] https://en.wikipedia.org/wiki/Anatomical_Therapeutic_Chemical_Classification_System.

with its parent categories. Figure 2 shows the structure of ATC. Leaf nodes represent specific medications, such as C03CA01, non-leaf nodes represent different subgroups, such as C03CA and C03C. If two medications belong to the same parent node, such as both C03CA01 and C03CA02 belong to C03CA, this indicates that they belong to the same chemical or pharmacological group.

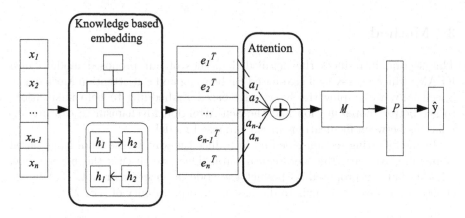

Fig. 3. The Proposed KEAM Model.

3.2 KEAM

Our KEAM model aims to predict whether a certain diagnosis is truly epilepsy by analyzing a patient's medication information. KEAM incorporates a knowledge-based embedding mechanism to generate vector representations of each medication and then uses an attention mechanism to generate matrix representations of medication combinations, which are subsequently used for code assignment prediction. Our knowledge-based embedding is inspired by the work of Shang et al. [21] and uses ATC-coded tree structure, which alleviates the issue of having imbalanced data distribution. The attention mechanism is to represent this medication information in matrix form, and different row represents different types of medication combinations information. Figure 3 shows the system architecture of our proposed KEAM. The dynamics of each building block are introduced as follows.

Embedding. We use the inner structure of ATC to help leaf nodes learn to embed by using the information of parent and sibling nodes. In our method, as shown in Fig. 4, only the information of leaf node and parent node is used as input of bi-directional GRU (Bi-GRU), which avoids the incorporation of too much information on other types of medications and ensures the specificity of medication expression. Bi-GRU consists of a forward GRU and a backward

GRU. They read the input sequence from forward and backward respectively. We get the medication representation vector by joining the hidden states of the two GRUs.

Each leaf node x_i has a basic representation r_i, which can be obtained by the following formula:

$$r_i = W_r x_i + b_i \qquad (1)$$

where the weight matrices $W_r \in \mathbb{R}^{m \times |C_d|}$ and bias term $b_r \in \mathbb{R}^m$ are the parameters to be learned. We use W_r and bias vector b_r to embed one-hot vector x_i. The representation of x_i's parent node p_i is calculated by x_i and its sibling nodes, as follows:

$$p_i = \frac{\sum_{x_k \in p_i} r_k}{count(x_k \in p_i)} \qquad (2)$$

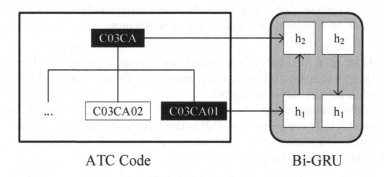

ATC Code   Bi-GRU

Fig. 4. Knowledge-based embedding

Thus, we can get the representation of x_i's parent node, and then use the x_i's basic representation r_i and the parent node representation p_i as the input of Bi-GRU to generate the final representation of medication x_i.

$$e_i = Bi\text{-}GRU[p_i, r_i], \qquad (3)$$

where e_i is a $2u$-dimensional representation vector, which is the final representation of medication x_i, u denotes the dimensions of hidden state for each GRU of Bi-GRU. There are a total of n kinds of medications in this visit, we note all the e_i as E, who have the size n-by-$2u$.

$$E = [e_1, e_2, \ldots, e_{n-1}, e_n] \qquad (4)$$

Attention. There may be many kinds of medications in one visit, they have different effects on different diseases. Therefore, we use the matrix E as the embedding representation of medications in a visit. Our goal is to encode variable-length medicines into fixed-size embedding. We use the attention mechanism to combine n medications.

The calculation process is as follows:

$$a_i = W_s e_i{}^T \tag{5}$$

where $W_s \in \mathbb{R}^{l \times 2u}$ is the parameter to be learned, l denotes the number of rows of the medication interaction representation matrix. For each medication x_i, we can get a weight vector a_i, which contains the weight information of the medication belonging to different types of medications.

$$A = softmax(a_1, a_2, \ldots, a_{n-1}, a_n) \tag{6}$$

The softmax() ensures all the computed weights sum up to 1. The expression matrix of medication combination can be obtained by multiplying A with embedding matrix E.

$$M = AE \tag{7}$$

M is the representation matrix of the medication combination for this visit. Each row of M represents different types of medicines.

Prediction. The final prediction module will output a probability by using the medication combination representation matrix M. We combine a fully connected layer with a sigmoid layer to predict.

The fully connected layer uses relu as the activation function, we use M as the input of a fully connected layer. P is the h-dimensional vector of the output from the fully connected layer. Then we put P into the final sigmoid layer to get a probability between 0 and 1.

$$\hat{y} = sigmoid(W_p^T P + b_p) \tag{8}$$

$W_p \in \mathbb{R}^h$, $b_p \in \mathbb{R}$ are the parameters to be learned in the sigmoid layer.

Loss Function. We use the cross entropy between the ground truth y and the predicted $\hat{y}$ to calculate the loss for all the visits as follows:

$$Loss = \sum_{i=1}^{N} -y_i log(\hat{y}_i) - (1 - y_i)log(1 - \hat{y}_i) \tag{9}$$

3.3 Interpretability in KEAM

Interpretability is much desirable in predictive models, especially in the medical field. By using the attention mechanisms, we can analyze the weight of each medication in the medication interaction representation. By summing all the attention vectors and normalizing the weight vectors with respect to 1, the embedding representation of medication combinations can be analyzed. Because the attention mechanism gives a general view of the main concerns of embedding. We can identify the medications that need to be considered and the medications that are skipped. By analyzing medications with high weights, we can explain the predicted results.

4 Experiment

In this section, we compare the performance of our proposed KEAM model against LR, DT, RF, Med2Vec [7] on a real-world dataset.

4.1 Experimental Setting

Data Description. This dataset was collected from a certain city in China, which includes both diagnoses and prescriptions. The dataset contains more than 300,000 admission records of 40,000 patients during the last five years. Figure 5 shows the age distribution of the dataset. More dataset information is reported in Table 1.

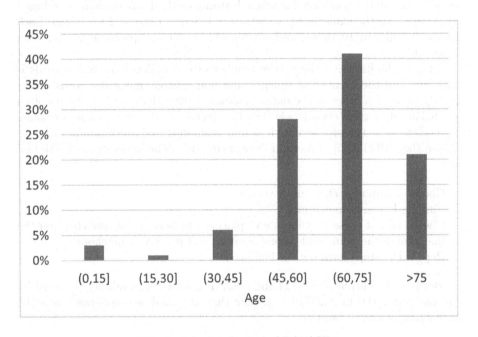

Fig. 5. Age distribution of DATASET

Baselines. We choose the following baseline models for performance comparison: including Logistic Regression (LR), Decision Tree (DT) and Random Forest (RF). These models have been widely used in classification problems, and known as being able to provide highly interpretable for prediction results. We also use a neural network model Med2Vec, which can efficiently learn medical code representations and improve the accuracy of prediction.

Table 1. Statistics of EHR dataset

# of patients	35308
# of visits	314933
Avg. # of visits per patient	8.919
# of unique diagnosis codes	1331
# of unique medication codes	1126
Avg. # of medication codes per visit	5.92

Implementation Details. All of the neural network models are implemented using tensorflow [1]. We use Adadelta [27] and mini-batch of 100 to train the models. And all the traditional machine learning methods are implemented based on sklearn[^2]. In the experiment we set, $m = 128, u = 128, r = 20, h = 256$. Each approach will run 10 times, and we report the average performance for each approach.

Despite the dataset is large, the number of real epilepsy medical records is only 504, and the number of epilepsy and non-epilepsy medical records differs greatly from each other in the dataset, almost 1:900. Medical datasets are often not balanced in their class labels [18]. But luckily in this area, there are many methods to solve the problem, such as undersampling and oversampling. And we choose the SMOTE [5], a method of oversampling. The main steps of SMOTE are:

- Choose a minority class input vector
- Find its k nearest neighbors
- Choose one of these neighbors and place a synthetic point anywhere on the line joining the point under consideration and its chosen neighbor
- Repeat the steps until data is balanced

Because the output of KEAM and KEAM- model is a possibility, we consider epilepsy when the probability is higher than 0.5, and non-epilepsy when the probability is lower than 0.5.

Evaluation Measures. In order to evaluate the predicted results of the model, we use accuracy to verify the predicted results of the model. To ensure the correct recognition of epilepsy, we compare the common indicators of recall, precision, and F1. Because the proportion of epilepsy patients and non-epilepsy patients in the experiment is quite different, in order to better show the recognition of epilepsy patients, we also calculated the AUC of each model.

---

[^2]: https://scikit-learn.org.

Table 2. The results of epilepsy recognition

	LR	DT	RF	Med2Vec	KEAM
Accuracy	0.9100	0.8833	0.9366	0.8966	**0.9543**
Recall	0.7580	0.7096	0.7338	0.7297	**0.8776**
Precision	0.7213	**0.9479**	0.7605	0.8307	0.8144
F1 score	0.7154	0.6029	0.8272	0.7769	**0.8448**
AUC	0.8538	0.8191	0.8616	0.9212	**0.9720**

4.2 Results and Discussion

Table 2 shows the results of epilepsy recognition experiments. The accuracy values of all methods are higher than 0.95, mainly because most of the dataset are non-epileptic diseases, and most of these test samples can be well distinguished.

DT model has a high precision, which is mainly due to the fact that the DT model assigns high weight to the antiepileptics, such as topiramate and phenobarbital. DT cannot capture the non-linear relationship and it is difficult to analyze the combination information of different medications. Therefore, the recall of the DT model is low. LR and RF consider medication combinations to some extent, so recall is higher than the DT model. However, due to the corresponding relationship between medications and diseases is complex. These two models fail to analyze this complex relationship very well, and most evaluation measures are lower than the model based on neural networks.

From the results, we can observe that almost all the evaluation measures of the model based on the neural networks are higher than those of the traditional machine learning methods, which indicates that the neural network model learns the information of different medication combinations spontaneously in the training process. AUC can indicate whether the model can recognize epileptic medication information. The higher the AUC, the higher the discrimination between epileptic medication use and non-epileptic medication use. Although traditional machine learning methods can find some important medications for identifying epilepsy, the combination of multiple medications can not be well analyzed, so the AUC value is low. Med2Vec and our model use a neural network to mine the corresponding relationship between the combination of information of different medications and diseases, which can better distinguish epileptic medications from non-epileptic medications. Our model learns medication combination information better through a knowledge-based embedding module and attention module helps to identify epilepsy diseases and achieves the best performance.

4.3 Interpretability

We use the analysis method mentioned in Sect. 3. C (Interpretability in KEAM) to analyze some medical records in the dataset. Three epilepsy cases and three non-epilepsy cases were randomly selected from the test set. Details are given in Table 3.

Table 3. Information of Analytical examples. Bold represents the top three medications with the highest weights

No	Medicine (ATC code)	Label
1	**glycerol(A06AX01)**, glucose(V06DC01), **thrombin(B02BD30)**, acetylcysteine(V03AB23), levetiracetam(N03AX14), oxcarbazepine(N03AF02), **carbamide(D02AE01)**	Epilepsy
2	**glycerol(A06AX01)**, glucose(V06DC01), **thrombin(B02BD30)**, acetylcysteine(V03AB23), valproic acid(N03AG01), carbamide(D02AE01), sodium chloride(B05CB01), **phenobarbital(N03AA02)**	Epilepsy
3	**glycerol(A06AX01)**, glucose(V06DC01), acetylcysteine(V03AB23), simvastatin(C10AA01), mannitol(B05BC01), felodipine(C08CA02), **carbamide(D02AE01)**, sodium chloride(B05CB01) **phenobarbital(N03AA02)**	Epilepsy
4	diazepam(N05BA01), **valproic acid(N03AG01)**, propranolol(C07AA05), **scopolamine(N05CM05)**, haloperidol(N05AD01), erythromycin(J01FA01), buspirone(N05BE01), citalopram(N06AB04), loratadine(R06AX13), clozapine(N05AH02), lithium(N05AN01), clonazepam(N03AE01), lorazepam(N05BA06), **aniracetam(N06BX11)**, quetiapine(N05AH04)	Non-epilepsy
5	**haloperidol(N05AD01)**, **amitriptyline(N06AA09)**, **scopolamine(N05CM05)**, perphenazine(N05AB03), clozapine(N05AH02)	Non-epilepsy
6	**cyclophosphamide(L01AA01)**, glucose(V06DC01), potassium chloride(B05XA01), glyceryl trinitrate(C01DA02), benzylpenicillin(J01CE01), **pyridoxine (vit B6)(A11HA02)**, ascorbic acid (vit C)(A11GA01), **lidocaine(C01BB01)**, isosorbide mononitrate(C01DA14), sodium chloride(B05CB01)	Non-epilepsy

By analyzing the attention vector of epileptic medications, we can find that the weight of epileptic medications such as phenobarbital (see Patients 2 and 3 in Table 3) is much higher than the average level of other medications. In addition to discovering that some medications are specifically prescribed for epilepsy, we identify some special medication combinations, such as glycerol and thrombin, which are also important to identify epilepsy.

The attention mechanism of KEAM takes into account all the medicines in the process of weight allocation, rather than single medicines. When the combination of medications does not conform to the pattern of epilepsy, although there are antiepileptic medications in the medications, it can still correctly identify the non-epilepsy patients. The accuracy of the model prediction results is ensured. Although Patients 4 and 5 were prescribed with antiepileptic medications, other psychoanalytic was also prescribed, which indicates that the patients actually suffer from some other psychiatric diseases rather than epileptic. Analysis of weights associated with medications of 3 non-epilepsy cases found that the weight of all medications is similar, no medication weight is significantly higher than that of other medications, this indicates that no pattern of epilepsy medication has been found in this 3 cases.

Compared with the baseline models such as logistic regression and decision tree, our method can assign a weight to each medication in the current medication mix through the overall situation of medication information, and make medication analysis more comprehensive and accurate. This can more effectively distinguish epilepsy from other similar psychiatric diseases and improve the accuracy of the model prediction.

5   Conclusion

We propose the KEAM model, which can be used to correct the codes disguised. This model can obtain the representation of medication combinations by using knowledge-based embedding and attention mechanisms. Knowledge-based embedding helps each medication get a representation, attention mechanism help generate medication combinations representation. Experiments on the real-world EHR dataset indicates the effectiveness of the KEAM. Moreover, attention is used to help identify important medications or medication combinations, which provides good explainability. We also find some future work directions, one is to extend the recognition of single disease to multi-disease recognition; the other is to add clinical records information, combining two parts of information to further improve the accuracy of the model in code assignment.

Acknowledgment. This work is partially supported by the National Natural Science Foundation of China under Grant (NSFC) No. 91846205, Major Science and Technology Innovation Projects of Shandong Province No. 2018YFJH0506, No. 2018CXGC0706.

References

1. Abadi, M., et al.: TensorFlow: large-scale machine learning on heterogeneous distributed systems. arXiv preprint arXiv:1603.04467 (2016)
2. Ba, J., Mnih, V., Kavukcuoglu, K.: Multiple object recognition with visual attention. arXiv preprint arXiv:1412.7755 (2014)
3. Bahdanau, D., Cho, K., Bengio, Y.: Neural machine translation by jointly learning to align and translate. arXiv preprint arXiv:1409.0473 (2014)
4. Baumel, T., Nassour-Kassis, J., Cohen, R., Elhadad, M., Elhadad, N.: Multi-label classification of patient notes: case study on icd code assignment. In: Workshops at the Thirty-Second AAAI Conference on Artificial Intelligence (2018)
5. Chawla, N.V., Bowyer, K.W., Hall, L.O., Kegelmeyer, W.P.: SMOTE: synthetic minority over-sampling technique. J. Artif. Intell. Res. **16**, 321–357 (2002)
6. Choi, E., Bahadori, M.T., Schuetz, A., Stewart, W.F., Sun, J.: Doctor AI: predicting clinical events via recurrent neural networks. In: Machine Learning for Healthcare Conference, pp. 301–318 (2016)

7. Choi, E., et al.: Multi-layer representation learning for medical concepts. In: Proceedings of the 22nd ACM SIGKDD International Conference on Knowledge Discovery and Data Mining, pp. 1495–1504 (2016)

8. Choi, E., Bahadori, M.T., Sun, J., Kulas, J., Schuetz, A., Stewart, W.: RETAIN: an interpretable predictive model for healthcare using reverse time attention mechanism. In: Advances in Neural Information Processing Systems, pp. 3504–3512 (2016)

9. Cimino, J.J.: Improving the electronic health record—are clinicians getting what they wished for? Jama **309**(10), 991–992 (2013)

10. Ma, F., Chitta, R., Zhou, J., You, Q., Sun, T., Gao, J.: Dipole: diagnosis prediction in healthcare via attention-based bidirectional recurrent neural networks. In: Proceedings of the 23rd ACM SIGKDD International Conference on Knowledge Discovery and Data Mining, pp. 1903–1911 (2017)

11. Mai, S.T., et al.: Evolutionary active constrained clustering for obstructive sleep apnea analysis. Data Sci. Eng. **3**(4), 359–378 (2018)

12. Megiddo, I., Colson, A., Chisholm, D., Dua, T., Nandi, A., Laxminarayan, R.: Health and economic benefits of public financing of epilepsy treatment in India: an agent-based simulation model. Epilepsia **57**(3), 464–474 (2016)

13. Mikolov, T., Chen, K., Corrado, G., Dean, J.: Efficient estimation of word representations in vector space. arXiv preprint arXiv:1301.3781 (2013)

14. Morrell, M.J.: Stigma and epilepsy. Epilepsy Behav. **3**(6), 21–25 (2002)

15. Mullenbach, J., Wiegreffe, S., Duke, J., Sun, J., Eisenstein, J.: Explainable prediction of medical codes from clinical text. arXiv preprint arXiv:1802.05695 (2018)

16. Niu, Y., Xie, R., Liu, Z., Sun, M.: Improved word representation learning with sememes. In: Proceedings of the 55th Annual Meeting of the Association for Computational Linguistics (Volume 1: Long Papers), pp. 2049–2058 (2017)

17. Perotte, A., Pivovarov, R., Natarajan, K., Weiskopf, N., Wood, F., Elhadad, N.: Diagnosis code assignment: models and evaluation metrics. J. Am. Med. Inform. Assoc. **21**(2), 231–237 (2014)

18. Rahman, M.M., Davis, D.: Addressing the class imbalance problem in medical datasets. Int. J. Mach. Learn. Comput. **3**(2), 224 (2013)

19. Scheurwegs, E., Cule, B., Luyckx, K., Luyten, L., Daelemans, W.: Selecting relevant features from the electronic health record for clinical code prediction. J. Biomed. Inform. **74**, 92–103 (2017)

20. Scheurwegs, E., Luyckx, K., Luyten, L., Daelemans, W., Van den Bulcke, T.: Data integration of structured and unstructured sources for assigning clinical codes to patient stays. J. Am. Med. Inform. Assoc. **23**(e1), e11–e19 (2016)

21. Shang, J., Hong, S., Zhou, Y., Wu, M., Li, H.: Knowledge guided multi-instance multi-label learning via neural networks in medicines prediction. In: Asian Conference on Machine Learning, pp. 831–846 (2018)

22. Tran, T., Nguyen, T.D., Phung, D., Venkatesh, S.: Learning vector representation of medical objects via EMR-driven nonnegative restricted Boltzmann machines (eNRBM). J. Biomed. Inform. **54**, 96–105 (2015)

23. Vaswani, A., et al.: Attention is all you need. In: Advances in Neural Information Processing Systems, pp. 5998–6008 (2017)

24. West, M.D., Dye, A.N., McMahon, B.T.: Epilepsy and workplace discrimination: population characteristics and trends. Epilepsy Behav. **9**(1), 101–105 (2006)

25. Yang, C., He, B., Li, C., Xu, J.: A feedback-based approach to utilizing embeddings for clinical decision support. Data Sci. Eng. **2**(4), 316–327 (2017)

26. Yu, P., Ding, D., Zhu, G., Hong, Z.: International bureau for epilepsy survey of children, teenagers, and young people with epilepsy: data in china. Epilepsy Behav. **16**(1), 99–104 (2009)
27. Zeiler, M.D.: ADADELTA: an adaptive learning rate method. arXiv preprint arXiv:1212.5701 (2012)

Tackling MeSH Indexing Dataset Shift with Time-Aware Concept Embedding Learning

Qiao Jin[1], Haoyang Ding[1], Linfeng Li[1,2], Haitao Huang[3], Lei Wang[1(✉)], and Jun Yan[1]

[1] Yidu Cloud Technology Inc., Beijing, China
`lei.wang01@yiducloud.cn`
[2] Institute of Information Science, Beijing Jiaotong University, Beijing, China
[3] The People's Hospital of Liaoning Province, Shenyang, Liaoning, China

Abstract. Medical Subject Headings (MeSH) is a controlled thesaurus developed by the National Library of Medicine (NLM). MeSH covers a wide variety of biomedical topics like diseases and drugs, which are used to classify PubMed articles. Human indexers at NLM have been annotating the PubMed articles with MeSH for decades, and have collected millions of MeSH-labeled articles. Recently, many deep learning algorithms have been developed to automatically annotate the MeSH terms, utilizing this large-scale MeSH indexing dataset. However, most of the models are trained on all articles non-discriminatively, ignoring the temporal structure of the dataset. In this paper, we uncover and thoroughly characterize the problem of MeSH indexing dataset shift (MeSHIFT), meaning that the data distribution changes with time. MeSHIFT includes the shift of input articles, output MeSH labels and annotation rules. We found that machine learning models suffer from performance loss for not tackling the problem of MeSHIFT. Towards this end, we present a novel method, time-aware concept embedding learning (TaCEL), as an attempt to solve it. TaCEL is a plug-in module which can be easily incorporated in other automatic MeSH indexing models. Results show that TaCEL improves current state-of-the-art models with only minimum additional costs. We hope this work can facilitate understanding of the MeSH indexing dataset, especially its temporal structure, and provide a solution that can be used to improve current models.

Keywords: Medical Subject Headings · Machine learning · Natural language processing · Dataset shift · Text classification

1 Introduction

Machine learning models can perform specific tasks on new input after being trained by past experiences. Most training algorithms estimate the test objectives from the training data, so it's presumed that the dataset for training and

Y. Nah et al. (Eds.): DASFAA 2020, LNCS 12114, pp. 474–488, 2020.
https://doi.org/10.1007/978-3-030-59419-0_29

the one for test should have similar data distribution. However, many biomedical machine learning datasets have been collected for decades, and the distribution of data changes along with time. Because algorithms are usually tested by the contemporary data, such shift of dataset can cause training-test discrepancy for machine learning.

MeSH indexing is a typical example: In the 1960s, the National Library of Medicine (NLM) started developing Medical Subject Headings (MeSH) thesaurus[1], a controlled vocabulary used to describe the topics of biomedical texts. PubMed articles are manually indexed with MeSH terms by well-trained annotators at NLM. Since manual annotation is expensive and time-consuming, NLM developed MetaMap [1], a rule-based software for automatically mapping freetexts to MeSH concepts, to facilitate the indexing process.

In the past several decades, more than 13M MeSH-labeled articles have been accumulated, which is an ideal training set for machine learning algorithms due to its large scale. To this end, BioASQ holds annual challenge on MeSH indexing [20], where the task is to predict the annotated MeSH terms given the abstract of articles. A number of machine learning based automatic MeSH indexing models have been proposed, such as DeepMeSH [15], AttentionMeSH [7] and MeSH-ProbeNet [24].

However, nearly all current automatic MeSH indexing models are trained on the historical data non-discriminatively, ignoring the temporal structure of the dataset. It undermines the full capacity of these algorithms, since we found that the data distribution is very time-dependent, which we denote as MeSH indexing dataset shift (MeSHIFT). Specifically, the data distribution changes with time in three different ways: (1) the distribution of articles change, because interests of researchers change; (2) the distribution of MeSH terms change, because the NLM keep updating the MeSH thesaurus each year; (3) the annotation rules change over time, which means the same article might be annotated with different MeSH terms in different years, due to the change of annotators and annotation rules.

The phenomenon is sometimes defined as dataset shift [17] or concept drift [21] when training-test discrepancy is caused by the fact that the training data is collected in a non-stationary stream. The problem has been discussed in the task of spam filtering [4], social network analysis [3], customer preferences [10] and financial data analysis [19]. Most current works tackle this problem by ensemble methods [11,23], like Streamed Ensemble Algorithm [18], Accuracy Weighted Ensemble [22] and Accuracy Updated Ensemble [2]. However, these methods generally require more computational resources and are hard to integrate to other models.

To solve the problem of MeSHIFT, we propose Time-aware Concept Embedding Learning (TaCEL), a novel method providing modules that can be used to improve current MeSH indexing models. TaCEL includes three sub-modules: (1) time-dependent projections to solve the problem of shifting input abstracts by conditioning the input representations on time; (2) a learnable year embedding matrix that further conditions the context representations on time; and (3) a

---

[1] https://www.nlm.nih.gov/mesh/.

dynamic masking mechanism explicitly designed to control which MeSH labels to learn at each year.

The overall architecture of this paper is shown in Fig. 1: we first describe the task of automatic MeSH indexing and the datasets we use in the paper in Sect. 2. Then we thoroughly characterize the problem of MeSHIFT, which includes the shift of input abstracts, MeSH labels and the annotation mapping in Sect. 3. We introduce the proposed TaCEL method to tackle MeSHIFT in Sect. 4. Finally, we discuss and conclude the main findings of this paper in Sect. 5 and 6.

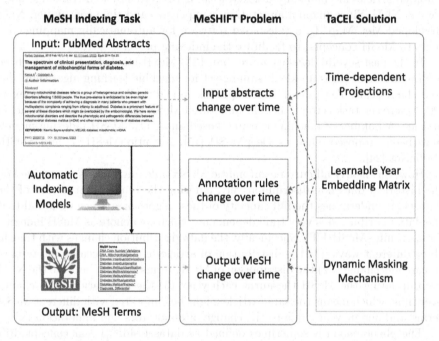

Fig. 1. This figure shows the overall architecture of the paper. Our contributions include (1) we introduce and characterize the problem of MeSHIFT, which has long been ignored and (2) we propose a novel method, TaCEL to tackle this problem. The example abstract is from [8].

2 Task and Data

In this section, we will first describe the MeSH indexing task. Then we will introduce two datasets we used in the paper: (1) the complete dataset which is used to evaluate the utility of TaCEL; and (2) the diagnostic dataset which is used to characterize the problem of MeSHIFT.

2.1 The MeSH Indexing Task

Formally, the task of automatic MeSH indexing is: given the title, abstract, published journal and year of an article, predict the MeSH terms that can be used to describe this article. The ground truth labels are derived from human annotations.

In general, most input information resides in the abstract[2], which we denote as **x** and

$$\mathbf{x} = \{w_1, w_2, w_3, ..., w_L\} \tag{1}$$

where w_i is the i-th token in the abstract and L is the length. We denote the MeSH labels as **y** and

$$\mathbf{y} \in \{0, 1\}^{|\mathcal{M}|} \tag{2}$$

where $\mathcal{M}$ is the set of all possible MeSH terms.

Following the standards in BioASQ [20], we evaluate the MeSH indexing models by Micro-F1 (MiF) value, which is a harmonic average between Micro-Precision (MiP) and Micro-Recall (MiR):

$$\mathrm{MiF} = \frac{2 \cdot \mathrm{MiP} \cdot \mathrm{MiR}}{\mathrm{MiP} + \mathrm{MiR}} \tag{3}$$

where

$$\mathrm{MiP} = \frac{\sum_{i=1}^{N_a} \sum_{j=1}^{|\mathcal{M}|} y_j^i \cdot \widehat{y}_j^i}{\sum_{i=1}^{N_a} \sum_{j=1}^{|\mathcal{M}|} \widehat{y}_j^i} \tag{4}$$

$$\mathrm{MiR} = \frac{\sum_{i=1}^{N_a} \sum_{j=1}^{|\mathcal{M}|} y_j^i \cdot \widehat{y}_j^i}{\sum_{i=1}^{N_a} \sum_{j=1}^{|\mathcal{M}|} y_j^i} \tag{5}$$

In these equations, i is indexed for articles and j is indexed for MeSH terms. N_a is the number of articles in the development or test set. y_j^i and $\widehat{y}_j^i$ are both binary variables to denote whether MeSH term j is annotated to article i in ground-truth and prediction, respectively.

To study the time-dependency structure of the MeSH indexing dataset, we binned the PubMed articles into years[3], and denote the publication year of an article as t. Conventionally, people use the most recent data for training. For example, if the most recent T years of data is used, the articles from year $t_c - T + 1$ to year t_c will be included in the training set. t_c is the current year where development and test data are sampled.

2.2 Complete Dataset

We use the 2018 version of PubMed dump from BioASQ[4], and included all articles from the 1979 to 2018 in the complete dataset. Figure 2 shows the distribution of training article numbers on each year, where there is an increasing trend as expected.

The complete dataset is used to test the utility of TaCEL method in real settings. For this, we incorporate the TaCEL modules in the state-of-the-art

---

[2] In this paper, we use "abstracts" and "articles" interchangeably.

[3] Since year is the minimum unit in the MeSH indexing dataset.

[4] http://participants-area.bioasq.org/general_information/Task7a/.

MeSH indexing models and train the hybrid model on the complete dataset. We randomly sample 10k and 50k articles published in year 2018 to build the development and test sets, respectively.

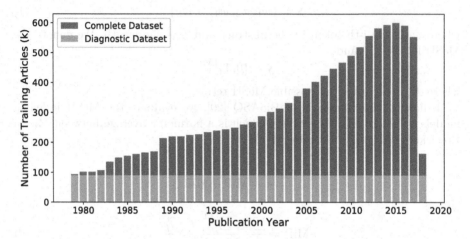

Fig. 2. This figure shows that the distribution of training article numbers.

2.3 Diagnostic Dataset

As is shown in Fig. 2, in the complete MeSH indexing dataset, there are different numbers of articles in different years, which might lead to different priors in downstream analyses. To mitigate the effects of uneven time distribution and better characterize the problem of MeSHIFT, we built a standard dataset that has the same number of articles published in each year.

For this, we randomly sampled 100k articles each year from 1979 to 2018. 90k of the abstracts consist of the training set and 10k consist of the development set for each year. In addition, we randomly sampled another 50k articles with MeSH labels from year 2018 to build the test set.

3 MeSH Indexing Dataset Shift

In this section, we first introduce the diagnostic model that is used to characterize MeSHIFT, and present our main findings. We then show that machine learning models suffer from not tackling the temporal structure of MeSH indexing dataset, which is the consequence of MeSHIFT.

3.1 Diagnostic Model

In order to conduct large-scale experiments, we use a lightweight diagnostic model for the analyses of MeSHIFT. In the diagnostic model, we use term frequency inverse document frequency (TF-IDF) statistics $\mathbf{a} \in \mathbb{R}^{|\mathcal{V}|}$ to represent the input abstract, where $\mathcal{V}$ is the vocabulary. We apply a feed-forward neural network layer to get the hidden states $\mathbf{h} \in \mathbb{R}^{D_h}$:

$$\mathbf{h} = \text{ReLU}(\mathbf{Wa} + \mathbf{b}) \tag{6}$$

where d_h is the hidden dimension. $\mathbf{W}$ and $\mathbf{b}$ are the weight matrix and bias, respectively. ReLU stands for Rectified Linear Unit [13], which serves as a non-linear activation function.

Then we apply an output layer:

$$\mathbf{y} = g(\mathbf{W}'\mathbf{h} + \mathbf{b}') \tag{7}$$

where $\mathbf{y}$ is the output. $\mathbf{W}'$ and $\mathbf{b}'$ are the weight matrix and bias. g depends on usage of the diagnostic model: (1) in the task of publication year prediction, we use a linear output layer where g is an identity function. (2) in the task of MeSH prediction, we use a sigmoid output layer where g is a sigmoid function.

3.2 Findings

We will formally define three kinds of data distribution shifts that compose of MeSHIFT, and provide thorough analyses to characterize each of them.

Definition 1. *The input abstracts change over time if $p(\mathbf{x}|t_a) \neq p(\mathbf{x}|t_b)$ when $t_a \neq t_b$.*

To reveal such shift, we train the diagnostic model to predict the publication year of the abstract, and compare the performance with random guesses. If $p(\mathbf{x}|t_a) = p(\mathbf{x}|t_b)$ for any given t_a and t_b (i.e.: there is no input abstract shift), the classifier won't outperform the random guess since there is not time-dependent structures to learn. We use mean square error (MSE) to train and evaluate the predictions, and show the results in Table 1.

Table 1. The publication year prediction diagnostic task.

Time Span	#Years	MSE (diagnostic model)	MSE (random)
1979–2018	40	**138.6**	266.5
1989–2018	30	**78.0**	149.8
1999–2018	20	**32.7**	66.5
2009–2018	10	**8.6**	16.5

Results in Table 1 show that the diagnostic model performs much better than random guess for predicting the publication year of articles. This indicates that the distribution of input abstracts changes over time, which might come from the fact that scientific research interests are dependent on time.

Definition 2. *The MeSH labels change over time if $p(\mathbf{y}|t_a) \neq p(\mathbf{y}|t_b)$ when $t_a \neq t_b$.*

The change of MeSH labels can be decomposed into: 1. Some MeSH terms are added and some are deleted at certain year; and 2. For existing MeSH terms, the proportions of articles annotated by them change each year. In Fig. 3, we visualize the frequency of several typical MeSH terms at different years. Clearly, there are at least 4 basic patterns: (i) some MeSH terms have a peak in frequency-time curve. For example, the MeSH term "Molecular Sequence Data" was peaked at 1995 and the usage keeps decreasing to 0 in the year 2018; (ii) some MeSH terms don't exist until its introduction and the usage keeps growing, like "Young Adult"; (iii) the frequency of some MeSH terms is stable, e.g.: "Adolescent"; and (iv) the frequency of some MeSH terms keeps increasing or decreasing over time. For example, the usage of "Retrospective Studies" keeps increasing and the usage of "Liver" keeps decreasing.

The shift of MeSH proportions might be related to the fact that annotation mapping also changes over time, which is discussed below.

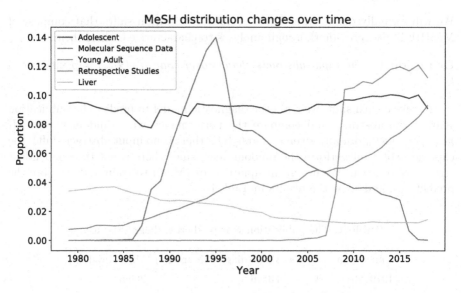

Fig. 3. This figure shows that the distribution of MeSH labels change over time.

Definition 3. *The annotation mapping changes over time if $p(\mathbf{y}|\mathbf{x}, t_a) \neq p(\mathbf{y}|\mathbf{x}, t_b)$ when $t_a \neq t_b$.*

This means even the same article can be annotated with different MeSH terms at different years. Obviously, addition and deletion of MeSH terms will cause the same article to be annotated with different labels in different years.

Besides, we also found that the annotation rules of MeSH terms change over time. For example, the MeSH term "Molecular Sequence Data" is now used to "index only for general articles on sequence data", but it was used to "indexed for articles which contained databank accession numbers for sequences deposited in a molecular sequence databank" from 1988 to 2016[5]. The annotation rule of "Molecular Sequence Data" changes in 2016, which partially explains its dramatic change of prevalence in Fig. 3.

In Fig. 4, we show the distribution of created year and last revision year of MeSH. From this figure, we can learn that: (1) There were major introductions of new MeSH terms in 1999 and around 2003; and (2) Most annotations rules of the MeSH terms have been changed after 2015, which means that labels of these MeSH terms before 2015 were annotated by different rules, and machine learning models might suffer from learning these instances which have noisy labels.

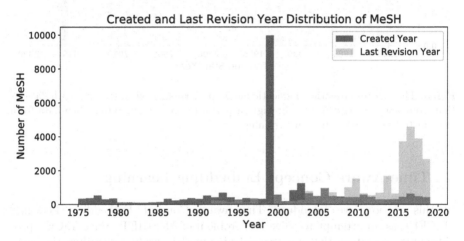

Fig. 4. This figure shows the created and last revision year distribution of MeSH terms.

3.3 Consequences of MeSHIFT

Ideally, the performance of a machine leaning model improves with more training data. However, it's not true in the MeSH indexing task, as a consequence of MeSHIFT.

In Fig. 5, we showed the test set performance of the diagnostic model when trained on data starting from different years. The diagnostic model doesn't perform well if it uses all data to train (i.e.: the model uses all data from 1979). When it uses less data which starts from more recent years, its performance increases instead of decreasing. The diagnostic model achieves the highest performance when only the data from 2016, 2017 and 2018 is used.

The peaked performance curve can be explained by the trade-off between the quantity and quality of training data: as more data is used, the distribution

---

[5] https://meshb.nlm.nih.gov/record/ui?ui=D008969.

of the dataset is more different from the development and test dataset. Most annotation rules have been revised after 2015 as shown in Fig. 4, which might be the reason why training by data before 2015 decreases the model performance.

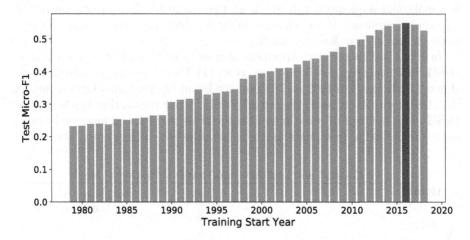

Fig. 5. This figure shows the diagnostic model performance when training with dataset that from year x to year 2018. The highest performance is highlighted when data from year 2016, 2017, 2018 is used for training.

4 Time-Aware Concept Embedding Learning

In this section, we introduce Time-aware Concept Embedding Learning (TaCEL), as an attempt to solve the problem of MeSHIFT. Since TaCEL provides plug-in modules that improve indexing models, we first introduce the backbone model and then propose the TaCEL modules. The model architecture is shown in Fig. 6. In the end, we show large-scale experiments with state-of-the-art MeSH indexing models and TaCEL modules in the complete dataset.

4.1 Backbone Model

We use a similar architecture to state-of-the-art MeSH indexing model, MeSH-ProbeNet [24] as the backbone model to test on the full dataset. For a given input abstract $\mathbf{x} = \{w_1, w_2, w_3, ..., w_L\}$, we first embed them to $\{\mathbf{v_1}, \mathbf{v_2}, \mathbf{v_3}, ..., \mathbf{v_L}\}$, where $\mathbf{v_i} \in \mathbb{R}^{D_e}$ is the pre-trained biomedical word embeddings [12] of the word w_i.

Then we use a bidirectional LSTM (BiLSTM) [6] to encode the words:

$$\tilde{\mathbf{x}} = \{\mathbf{h_1}, \mathbf{h_2}, \mathbf{h_3}, ..., \mathbf{h_L}\} = \text{BiLSTM}(\{\mathbf{v_1}, \mathbf{v_2}, \mathbf{v_3}, ..., \mathbf{v_L}\}) \tag{8}$$

where $\mathbf{h_i}$ is the hidden state of $\mathbf{v_i}$.

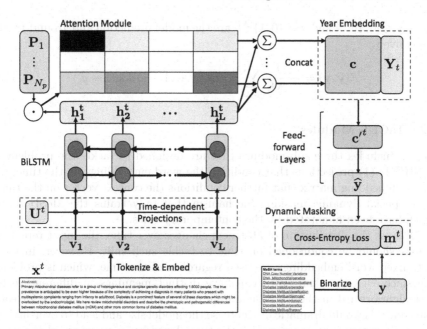

Fig. 6. Architecture of the Backbone Model and the TaCEL modules. TaCEL modules are highlighted by red dashed boxes. The example abstract and corresponding MeSH terms are from [8]. (Color figure online)

A probe embedding matrix $\mathbf{P} \in \mathbb{R}^{N_p \times D_p}$ is maintained where N_p is the number of probes and D_p is the dimension of probes. Each probe is designed to capture a specific aspect of the abstracts. The probes interact with the BiLSTM hidden states of the input by an attention mechanism:

$$\mathbf{A} = \text{Softmax}(\mathbf{P}\widetilde{\mathbf{x}}) \in \mathbb{R}^{N_p \times L} \tag{9}$$

where Softmax normalizes the sum of dot products between probe embeddings and hidden states along each input axis to one.

We multiply the attention weights with the hidden states and concatenate the results to get a flattened context vector:

$$\mathbf{c} = \text{Concat}(\mathbf{A}\widetilde{\mathbf{x}}^T) \in \mathbb{R}^{N_p D_p} \tag{10}$$

The context vector is then passed through a feed-forward layer:

$$\mathbf{c}' = \text{ReLU}(\mathbf{W}\mathbf{c} + \mathbf{b}) \tag{11}$$

where $\mathbf{W}$ and $\mathbf{b}$ are the weight matrix and bias, respectively $\mathbf{c}'$ is used to predict all the MeSH labels via sigmoid output units. We use Adam optimizer [9] to minimize the binary cross-entropy loss between the predicted MeSH probabilities and the ground truth labels:

$$\theta_{\text{backbone}} \leftarrow \text{argmin}_\theta \mathcal{L}(\widehat{\mathbf{y}}, \mathbf{y}) \tag{12}$$

where $\widehat{\mathbf{y}} \in [0,1]^{|\mathcal{M}|}$ and $\mathbf{y} \in \{0,1\}^{|\mathcal{M}|}$ are the model prediction and ground-truth, respectively.

$$\mathcal{L}(\widehat{\mathbf{y}}, \mathbf{y}) = -\frac{1}{|\mathcal{M}|} \sum_{i \in \mathcal{M}} y_i \log(\widehat{y}_i) + (1 - y_i)\log(1 - \widehat{y}_i) \qquad (13)$$

4.2 TaCEL Modules

TaCEL includes three sub-modules that are designed to tackle the problem of MeSHIFT: (1) projections that condition the word embeddings on the time; (2) a year embedding matrix that further conditions the context vector on the time; (3) a special dynamic masking mechanism that only trains the MeSH terms which have been annotated by the contemporary rules.

Because the distribution of abstracts changes over time, the input representations should be conditioned on time to tackle such shift. However, in most cases fixed word embeddings are used regardless of the time, which is problematic since even the same word can have distinct meanings in different years. A straightforward method is to maintain one word embedding matrix for each year, but this would dramatically increase the model size and lead to over-fitting. As an alternative, we parameterize the time-dependency structure of the word embeddings by linear projections:

$$\mathbf{E}^t = \mathbf{E}\mathbf{U}^t \in \mathbb{R}^{|\mathcal{V}| \times D_e} \qquad (14)$$

where $\mathbf{E}^t$ denotes the word embedding matrix of year t, $\mathbf{E} \in \mathbb{R}^{|\mathcal{V}| \times D_e}$ is the pre-trained biomedical word embedding matrix and $\mathbf{U}^t$ is the learnable projection matrix for year t. During training, the $\mathbf{U}^t$ is intialized with an identity matrix. We aim the time-dependent projection module at solving the shift of input abstracts.

In the backbone model, the context vector $\mathbf{c}$ is calculated by an attention-based contextual module. TaCEL further conditions $\mathbf{c}$ on time by concatenating it with a learnable year embedding, getting $\mathbf{c}^t$:

$$\mathbf{c}^t = \text{Concat}(\mathbf{c}, \mathbf{Y}_t) \in \mathbb{R}^{N_p D_p + D_h} \qquad (15)$$

where $\mathbf{Y} \in \mathbb{R}^{N_y \times D_h}$ is the learnable year embedding matrix, and N_y is the number of different publication years in the training set. $\mathbf{c}^t$ is then passed through a feed-forward layer, resulting in $\mathbf{c}'^t$. $\mathbf{c}'^t$ is used to classify the MeSH labels in year t. The year embedding module has the potentials to solve all three MeSHIFT problems.

Unlike other tasks with concept drift problems, we have access to the MeSH indexing logs which kept track of all annotation rule changes. Therefore, we can explicitly know whether a specific MeSH term was annotated by the same rule as it is in the current year. Based on such information, we add a dynamic mask $\mathbf{m}$ in the loss function:

$$\mathcal{L}^t(\widehat{\mathbf{y}}, \mathbf{y}) = -\frac{1}{|\mathcal{M}|} \sum_{i \in \mathcal{M}} m_i^t(y_i \log(\widehat{y}_i) + (1 - y_i)\log(1 - \widehat{y}_i)) \qquad (16)$$

where the dynamic mask $\mathbf{m}^t \in \{0,1\}^{|\mathcal{M}|}$ is a binary vector. $m_i^t = 1$ denotes that MeSH term i was annotated in year t by the same rule as it is in current year, and $m_i^t = 0$ denotes that MeSH term i underwent a change of annotation rule between year t and current year. The dynamic masking mechanism is designed to solve the shift of MeSH labels and annotation rules.

4.3 Experiments

The models are implemented in PyTorch [14]. We use the 2018 version of MeSH thesaurus where $|\mathcal{M}|$ is 28863. The input abstract will be cut to 384 tokens if it's longer than 384. The vocabulary size $|\mathcal{V}|$ is 131072, and we use the word embedding size D_e of 200. We use $N_p = 24$ probes which has D_p of 256, and the hidden size D_h is also 256. One Tesla P40 GPU is used for training and inference. We set the batch size of 512 and the Adam learning rate is 0.0005. Early stopping training strategy is used to prevent over-fitting [25].

For the thresholds used to convert probabilities to predictions, we searched for an optimal global threshold for all MeSH terms on the development set, and use it during inference on the test set. The threshold ranges from 0.25 to 0.35 in most experiments.

We conducted experiments of the backbone model, the backbone model with full TaCEL and different ablations on the complete dataset. The results[6] are shown in Table 2. Clearly, TaCEL improves the backbone model by a large margin (from 63.18 to 66.05 in MiF). Each of the TaCEL ablation also surpasses the backbone model as expected. In general, the more TaCEL modules added to the backbone model, the better the performance.

The performance loss from the ablation of each TaCEL module is slightly different: Dynamic masking might be the most important module because ablation of dynamic masking is the most destroying setting with two modules, and only incorporating the dynamic masking increases the performance the most (from 63.18 to 65.00 in MiF). Similarly, we can infer that the projection module is more important than the year embedding module. In conclusion, while all TaCEL modules are useful, ablation studies show that dynamic masking module > projection module > year embedding module in terms of importance.

---

[6] We didn't compare with the BioASQ challenge results for several reasons: (1) labels of the challenge test sets are not publicly available; (2) submitted results are generated by model ensembles. In the experiments, we use the challenge winner system, MeSHProbeNet [24], as a strong baseline (i.e. the Backbone Model).

Table 2. Performance of different models on the test set, where Micro-F is the main metric. B.M.: Backbone Model; p: projection module; y: year embedding module; m: dynamic masking module.

Model	Micro-P (%)	Micro-R (%)	Micro-F (%)
Backbone Model [24]	66.79	59.93	63.18
B.M. w/ TaCEL (p)	65.95	62.26	64.05
B.M. w/ TaCEL (y)	69.28	58.75	63.58
B.M. w/ TaCEL (m)	66.68	62.97	64.77
B.M. w/ TaCEL (p, y)	66.66	63.41	65.00
B.M. w/ TaCEL (p, m)	69.72	62.23	65.76
B.M. w/ TaCEL (y, m)	66.74	64.01	65.35
B.M. w/ TaCEL (p, y, m)	**69.88**	**62.61**	**66.05**

5   Discussion

Models and algorithms have been the focus of machine learning research, but tasks and data are equally important, especially for applications in biomedical domain. In this work, we showed that after thoroughly studying and analysing the structure of the dataset, we can achieve significant performance improvements by introducing modules that tackle commonly ignored aspects of the dataset. Specifically, we found that the MeSH indexing dataset has a dataset shift problem, which we called MeSHIFT. In addition, we designed a novel TaCEL method to tackle MeSHIFT, which improves the current state-of-the-art model by a large margin.

There are several interesting future directions to explore based on this work: 1. In this work, we only used simple linear projections or concatenations to model the time dependency as a demonstration. Using more complicated models to parametrize TaCEL has the potentials to further improve the model performance; 2. Currently, we cannot disentangle the effects of different MeSHIFT aspects on the results, so we are unable to quantitatively evaluate how well different TaCEL modules tackle different aspects of MeSHIFT. It would be interesting to analyze the decoupled effects; 3. Motivated by recent successes of large-scale pre-training models in natural language processing like ELMo [16] and BERT [5], incorporating the time-dependent structure of the dataset into a pre-training task is also a promising direction to explore and 4. The proposed methods are only tested on a biomedical domain specific task, and their utility in more general scenarios remains to be studied.

6   Conclusion

In conclusion, we revealed and thoroughly characterized the problem of MeSHIFT in the MeSH indexing dataset. Using the diagnostic dataset and model, we formally defined and proved that MeSHIFT includes the shifts of: (a) input abstracts due to the shift of scientific research interests; (b) MeSH labels due to the addition and deletion of MeSH terms each year; and (c) annotation mapping due to the changes of annotation rules.

We also proposed a novel method, TaCEL, as an attempt to solve the problem of MeSHIFT. TaCEL has three modules: 1. the time-dependent projections which condition the input representations on time; 2. the year embedding module which further conditions the context representations on time; and 3. the dynamic masking mechanism which dynamically determining which MeSH labels to learn at each year. We aim the module 1. of TaCEL at solving the aspect (a), the module 2. at solving all aspects and module 3. at solving aspects (b) and (c) of MeSHIFT. Our experiments show that TaCEL improves the current state-of-the-art model by a large margin and each of its module proves to be useful.

We hope this work can facilitate understanding of the temporal structures in the MeSH indexing dataset, and encourage other models to improve by tackling the dataset shift problem in similar scenarios.

References

1. Aronson, A.R.: Effective mapping of biomedical text to the UMLs Metathesaurus: the MetaMap program. In: Proceedings of the AMIA Symposium, p. 17. American Medical Informatics Association (2001)
2. Brzeziński, D., Stefanowski, J.: Accuracy updated ensemble for data streams with concept drift. In: Corchado, E., Kurzyński, M., Woźniak, M. (eds.) HAIS 2011. LNCS (LNAI), vol. 6679, pp. 155–163. Springer, Heidelberg (2011). https://doi.org/10.1007/978-3-642-21222-2_19
3. Costa, J., Silva, C., Antunes, M., Ribeiro, B.: Concept drift awareness in twitter streams. In: 2014 13th International Conference on Machine Learning and Applications, pp. 294–299. IEEE (2014)
4. Delany, S.J., Cunningham, P., Tsymbal, A., Coyle, L.: A case-based technique for tracking concept drift in spam filtering. In: Macintosh, A., Ellis, R., Allen, T. (eds.) SGAI 2004, pp. 3–16. Springer, London (2004). https://doi.org/10.1007/1-84628-103-2_1
5. Devlin, J., Chang, M.W., Lee, K., Toutanova, K.: BERT: pre-training of deep bidirectional transformers for language understanding. arXiv preprint arXiv:1810.04805 (2018)
6. Hochreiter, S., Schmidhuber, J.: Long short-term memory. Neural Comput. 9(8), 1735–1780 (1997)
7. Jin, Q., Dhingra, B., Cohen, W., Lu, X.: AttentionMeSH: simple, effective and interpretable automatic mesh indexer. In: Proceedings of the 6th BioASQ Workshop A Challenge on Large-Scale Biomedical Semantic Indexing and Question Answering, pp. 47–56 (2018)
8. Karaa, A., Goldstein, A.: The spectrum of clinical presentation, diagnosis, and management of mitochondrial forms of diabetes. Pediatr. Diab. 16(1), 1–9 (2015)
9. Kingma, D.P., Ba, J.: Adam: a method for stochastic optimization. arXiv preprint arXiv:1412.6980 (2014)
10. Koren, Y.: Collaborative filtering with temporal dynamics. In: Proceedings of the 15th ACM SIGKDD International Conference on Knowledge Discovery and Data Mining, pp. 447–456. ACM (2009)

11. Krawczyk, B., Minku, L.L., Gama, J., Stefanowski, J., Woźniak, M.: Ensemble learning for data stream analysis: a survey. Inf. Fusion **37**, 132–156 (2017)
12. Moen, S., Ananiadou, T.S.S.: Distributional semantics resources for biomedical text processing. In: Proceedings of LBM (2013)
13. Nair, V., Hinton, G.E.: Rectified linear units improve restricted boltzmann machines. In: Proceedings of the 27th International Conference on Machine Learning (ICML 2010), pp. 807–814 (2010)
14. Paszke, A., et al.: Pytorch: an imperative style, high-performance deep learning library. In: Advances in Neural Information Processing Systems, pp. 8024–8035 (2019)
15. Peng, S., You, R., Wang, H., Zhai, C., Mamitsuka, H., Zhu, S.: DeepMeSH: deep semantic representation for improving large-scale mesh indexing. Bioinformatics **32**(12), i70–i79 (2016)
16. Peters, M.E., et al.: Deep contextualized word representations. arXiv preprint arXiv:1802.05365 (2018)
17. Quionero-Candela, J., Sugiyama, M., Schwaighofer, A., Lawrence, N.D.: Dataset Shift in Machine Learning. The MIT Press, Cambridge (2009)
18. Street, W.N., Kim, Y.: A streaming ensemble algorithm (sea) for large-scale classification. In: Proceedings of the Seventh ACM SIGKDD International Conference on Knowledge Discovery and Data Mining, pp. 377–382. ACM (2001)
19. Sun, J., Li, H.: Dynamic financial distress prediction using instance selection for the disposal of concept drift. Expert Syst. Appl. **38**(3), 2566–2576 (2011)
20. Tsatsaronis, G., et al.: An overview of the BIOASQ large-scale biomedical semantic indexing and question answering competition. BMC Bioinform. **16**(1), 138 (2015)
21. Tsymbal, A.: The problem of concept drift: definitions and related work. Comput. Sci. Dep. Trinity College Dublin **106**(2), 58 (2004)
22. Wang, H., Fan, W., Yu, P.S., Han, J.: Mining concept-drifting data streams using ensemble classifiers. In: Proceedings of the Ninth ACM SIGKDD International Conference on Knowledge Discovery and Data Mining, pp. 226–235. ACM (2003)
23. Woźniak, M., Graña, M., Corchado, E.: A survey of multiple classifier systems as hybrid systems. Inf. Fusion **16**, 3–17 (2014)
24. Xun, G., Jha, K., Yuan, Y., Wang, Y., Zhang, A.: MeSHProbeNet: a self-attentive probe net for mesh indexing. Bioinformatics **35**, 3794–3802 (2019)
25. Yao, Y., Rosasco, L., Caponnetto, A.: On early stopping in gradient descent learning. Constr. Approximation **26**(2), 289–315 (2007)

Semantic Disambiguation of Embedded Drug-Disease Associations Using Semantically Enriched Deep-Learning Approaches

Janus Wawrzinek$^{(\boxtimes)}$ ⓘ, José María González Pinto ⓘ, Oliver Wiehr ⓘ,
and Wolf-Tilo Balke ⓘ

IFIS TU-Braunschweig, Mühlenpfordstrasse 23, 38106 Brunswick, Germany
{wawrzinek,pinto,wiehr,balke}@ifis.cs.tu-bs.de, o.wiehr@tu-bs.de

Abstract. State-of-the-art approaches in the field of neural-embedding models (NEMs) enable progress in the automatic extraction and prediction of semantic relations between important entities like active substances, diseases, and genes. In particular, the *prediction property* is making them valuable for important research-related tasks such as hypothesis generation and drug-repositioning. A core challenge in the biomedical domain is to have *interpretable* semantics from NEMs that can distinguish, for instance, between the following two situations: a) drug x *induces* disease y and b) drug x *treats* disease y. However, NEMs alone cannot distinguish between associations such as treats or induces. Is it possible to develop a model to learn a *latent representation* from the NEMs capable of such disambiguation? To what extent do we need domain knowledge to succeed in the task? In this paper, we answer both questions and show that our proposed approach not only succeeds in the *disambiguation* task but also advances current growing research efforts to find real predictions using a sophisticated retrospective analysis.

Keywords: Digital libraries · Association mining · Information extraction · Neural embeddings · Semantic enrichment

1 Introduction

Today's digital libraries have to manage the exponential growth of scientific publications [5], which results in faster-growing data holdings. To illustrate the effects of this growth, consider as an example Sara, a young scientist from the pharmaceutical field who wants to find drugs related to "Diabetes" to design a new hypothesis that might link an existing drug with "Diabetes" that has not yet been discovered (not published in a paper). Indeed, this is a complex information need, and in this context, a term-based search in the digital library PubMed leads to ~39,000 hits for the year 2019 alone. Due to these data amounts, Sara will have to dedicate considerable time to analyse each paper and take some other steps to satisfy her information need. Given this complicated situation, we believe that this problem makes innovative access paths beyond term-based searches necessary.

One of the most effective ways to help users like Sara is based on the automatic extraction of entity relations that are embedded in literature, i.e. such as those that exist

© Springer Nature Switzerland AG 2020
Y. Nah et al. (Eds.): DASFAA 2020, LNCS 12114, pp. 489–504, 2020.
https://doi.org/10.1007/978-3-030-59419-0_30

between drugs and diseases [10, 12]. Considering pharmaceutical research drug-disease associations (DDAs) play a crucial role because they are considered candidates for drug-repurposing [3]. The central idea behind drug-repurposing is to use an already known and well-studied drug for the treatment of another disease. In addition, drug-repurposing generally leads to lower risk in terms of adverse side effects [3]. However, it is not only the therapeutic application that is of interest, but also whether a drug induces a disease and may therefore be life-threatening for patients [8, 15].

From this motivation, numerous computer-based approaches have been developed in recent years which attempt not only to extract but also to predict DDAs. Here, for the majority of the methods, entity-similarities form the basis [7, 8, 15]. For example, one of the most important assumptions is that drugs with a similar chemical (sub) structure also have similar (therapeutic) properties [14]. Moreover, while structural similarity is extremely useful for screening, it does not capture other important semantic features.

Scientific literature is one of the primary sources in the investigation of new drugs [10], which is why newer approaches use Neural Embedding Models (NEMs) to calculate linguistic or lexical similarities between entities in order to deduce their properties, semantics, and relationships [1, 9]. The use of NEMs in this area is based on a (context) hypothesis [1], where words which share numerous similar surrounding word-contexts are spatially positioned closer to each other in a high dimensional space. This property leads to the fact, that with increasing similarity (i.e. cosine-similarity) also a possible semantic relationship between two entities can be deduced.

This *contextualization* property can be used to predict complex chemical properties decades in advance [9, 16] or novel therapeutic drug-properties [24]. However, using NEMs to disambiguate the type (i.e. "treats", "induces") of a drug-disease association is a challenging task; after all, what *semantic* is behind a cosine similarity between such entity-pairs? All we know is that they appear in similar contexts, but we do not know how to *interpret* it. Despite NEMs being an excellent foundation for several tasks, little is known about how to find –if it exists- a feature space within the NEMs that allow us to disambiguate associations between drug-disease pairs such as treats or induces. In this paper, we hypothesize that such disambiguation is possible. In particular, we propose to apply deep learning to find the *latent feature space* from the NEMs and, thus, disambiguate associations between pharmaceutical entities.

In summary, the questions that guided our research were the following:

- *RQ1*: Do word embeddings contain the information of the *type* of an embedded drug-disease association, and if yes, to which extent?
- *RQ2*: Distance in the embedding space has an effect on semantics [16]. In this context we want to answer what is the impact of the distance between entities in the embedding space in the disambiguation task?
- *RQ3*: Does domain-specific knowledge have an impact to uncover the embedding space needed to disambiguate predicted drug-disease associations?

To answer the first two questions, we investigate different Deep-Learning models and compare the results with a baseline-approach. We show that distance has indeed an effect on accuracy-quality, but not all models are affected by it in the same strength. Hereafter, we propose different semantic enriched Deep-Leaning models and using a retrospective analysis we can show, that our semantic enriched models lead to improved results in disambiguating the DDA-Type ("treats", "induces") for *real* DDA predictions.

2 Related Work

Entities like active substances, diseases, genes, and their interrelationships are of central interest for bio-medical digital libraries [4]. In this context, manual curation is a key-point that guarantees a high quality in today's digital libraries. Arguably, one of the best bio-medical databases for curated associations is the Comparative Toxicogenomics Database (CTD). In the case of DDAs, information about the specific type of a DDA association is curated in the CTD, i.e., either a drug is used to treat a disease, or it induces a disease. Due to the high quality, we use the manually curated drug-disease associations from the CTD as ground truth in our work.

On the one side, manual curation leads to the highest quality, but on the other side it is also time-consuming and often tends to be incomplete [18]. Considering these problems, numerous *entity-centric* methods have been developed for the automatic extraction and prediction of DDAs [3, 8, 15]. Entity specific similarities form the basis for these approaches, e.g., in the case of active substances similar therapeutic properties can be inferred [8] by calculating a molecular/chemical similarity between different drugs. In addition to these *entity-centric* approaches, which mostly require information from specialized databases [15], also literature-based methods exist, e.g., like the co-occurrence approach [19] that can be used for detection and prediction of DDAs. Usually, with this approach, entities in the documents are first recognized using Named Entity Recognition. Afterwards, in a next step, a co-occurrence of different entities in documents is counted. The hypothesis is that if two entities co-occur in documents, then a relationship between them can be assumed. Besides, it can be assumed that with an increasing number of co-occurrences, the probability of an association also increases [19]. In our work, we use the co-occurrence approach in combination with a retrospective analysis to determine DDA predictions. Newer literature-based techniques in the field of neural embedding models use state-of-the-art approaches like Word2Vec [6, 11] to efficiently learn and predict semantic associations based on word contexts [20]. In comparison to a co-occurrence approach, entities do not necessarily have to occur in the same document but rather can have similar word contexts. Therefore, we use the Word2Vec (Skip-gram) implementation from the open source Deep-Learning-for-Java[1] library in our investigation. Recent work in the field of NLP shows that the use of word embeddings in various NLP classification tasks outperforms previous (classical) approaches [13]. That finding led to an increasing interest using word embeddings also in the bio-medical field [2]. For example, in the work of Patrick et al. [24], the authors are using word embeddings as features to predict a therapeutic effect of drugs on a specific group of diseases. In our case, we not only try to predict a therapeutic association but also whether a drug can cause a disease in the sense of a side effect. Furthermore, we do not limit the body of documents or our evaluation to certain pharmaceutical entities, but consider all DDAs curated in the CTD.

---

[1] https://deeplearning4j.org/.

3 Problem Formulation and Methodology

In this section, we define the problem and provide definitions to accomplish our goal: a deep-learning based approach to disambiguate semantic relations between entities in the biomedical field. Hereafter, we present our proposed method.

3.1 Semantic Classification of Drug-Disease Relationships

Problem Definition: Given a neural embedding model M over a suitable collection of pharmaceutical documents, two sets of n-dimensional entity embeddings $Drugs_M$ and $Diseases_M$ can be learned by embedding techniques (e.g., Word2Vec), such that:

- $Drugs_M$ collects all entity representations where the respective entities correspond to drugs identified by some controlled vocabulary (e.g., MeSH identifier)
- $Diseases_M$ collects all entity representations corresponding to diseases again identified by some controlled vocabulary.

Then, given a set *Rel* of labels $r_1,...,r_k$ for possible and clearly disambiguated semantic relationships between drugs and diseases (e.g., *treats, induces*, etc.), the *semantic classification problem of drug-disease relationships* means to learn a classifier $f : \mathbb{R}^n \times \mathbb{R}^n \to Rel$ with $(e_i, e_j) \to r_m$, where $e_i \in Drugs_M$, $e_j \in Diseases_M$, and $r_m \in Rel$.

Since there is a variety of alternatives to learn such a classification function from data in a supervised fashion, and we will explore some in this work. In the following, we present our approach as well as briefly describe the used deep-learning models and provide the details of their implementation in our experiments.

3.2 Methodology

Our method consists of the following six steps (Fig. 1):

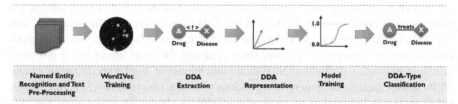

| Named Entity Recognition and Text Pre-Processing | Word2Vec Training | DDA Extraction | DDA Representation | Model Training | DDA-Type Classification |

Fig. 1. Method overview: We start with NER and text pre-processing, followed by Word2Vec training. Afterward, we extract Drug-Disease Associations from the embedding space. As next, we generate different DDA representations and afterward we use them for training a deep-learning model in order to disambiguate a DDA-type.

Steps 1–2. Text Pre-processing and NEM Learning. Our document corpus consists of scientific publications from the medical field. First, we remove stop-words and apply Named Entity Recognition to identify drug- as well as disease entities in the documents.

After this initial pre-processing we train a Neural Embedding Model (i.e., Word2Vec) on this corpus resulting in a matrix of entity embeddings.

Step 3. Drug-Disease Association Extraction. Using a k-Nearest-Neighbor approach we extract for a given drug-entity the k-Nearest-Disease-Neighbors. For example, for $k = 10$ we will extract ten DDAs, where the association-type i.e. "treats" or "induces" is unknown.

Step 4. DDA-Representations. In our investigations we create different DDA representations for model training. For example, given a pair of entity-vectors $<e_i, e_j>$ we create a (DDA) representation by i.e. concatenating or averaging the two vectors. Using taxonomic information from pharmaceutical classifications systems, we also create a *semantic* representation of a DDA.

Steps 5–6. Model Training and DDA-Type Classification. Next, we use the different representations to train a deep-learning model to classify a DDA-type. In this context, we investigate the following two deep-learning models:

- *Multilayer Perceptron (MLP):* The multilayer perceptron represents one of the most straightforward architectures available. Fixed length input vectors are handed over from layer to layer sequentially, and no recursion is used.
- *Convolutional Neural Networks (CNN):* Convolutional neural networks (CNN) are known best for their use on image data, such as object detection, handwritten character classification, or even face recognition. Recently, they have also shown to achieve state of the art results on natural language tasks. Goodfellow, I. et al. [21] have emphasized that three essential ideas motivate the use of CNNs in different machine learning tasks, including our disambiguation task: sparse interactions, parameter sharing, and equivariant representations. For our task, sparse interactions allow for learning automatically -without manual feature engineering- patterns from d-dimensional spaces; parameter sharing influences computation storage requirements; equivariant representation allows for robustness in the patterns learned.

4 Experimental Investigation

In this section, we will first describe our pharmaceutical text corpus and the necessary experimental set-up decisions. Afterward, we define for each research-question the quality criteria that our proposed models should fulfill, followed by our evaluation.

Experimental Setup

Evaluation Corpus. In the biomedical field PubMed[2] is one of the most comprehensive digital libraries. For the most publications a full-text access is not available and therefore we collected only abstracts for our experiments. Furthermore, all collected abstracts were published between 1900-01-01 and 2019-06-01. Word embedding algorithms usually train on single words, resulting in one vector per word and not per entity. This is a problem, because disease and drug names often consist of several words (e.g., ovarian cancer). Therefore, we first use PubTator[3] to identify the entities in documents. Afterward, we place a unique identifier at the entity's position in the text.

---

[2] https://www.ncbi.nlm.nih.gov/pubmed/.

[3] https://www.ncbi.nlm.nih.gov/CBBresearch/Lu/Demo/PubTator/.

Retrospective Analysis Evaluation Corpora. In order to detect predicted DDAs using a retrospective analysis, we divide our evaluation corpus into two corpora: 1900.01.01–1988.31.12 (1989 corpus) and 1900.01.01–2019.06.01 (2019 corpus). Each corpus contains only the documents for the respective time period.

Query Entities. As query entities for the evaluation, we selected all Drugs from the *DrugBank*[4] collection, which can be also be found (using a MeSH-Id) in CTD Database as well as in the pre-processed documents. Therefore, our final document set for evaluation contains ~29 million abstracts for ~1700 drugs. As ground truth, we selected for each drug all manually curated drug-disease associations from CTD, resulting in a data set of 33541 inducing and 18664 therapeutic drug-disease associations.

Experiment Implementation and Parameter Settings

Text Pre-processing. In an initial document pre-processing step, we removed stop-words and performed stemming using Lucene's[5] *Porter Stemmer* implementation. Here we made sure that the drug and disease identifiers were not affected.

Word Embeddings. After document pre-processing, word embeddings were created with DeepLearning4J's *Word2Vec*[6] implementation. A larger window-size can lead to improved results in learning (pharmaceutical) associations [2, 17]. Therefore, to train the neural embedding model, we set the word window size in our investigations to 50. Further, we set the layer size to 200 features per word, and we used a minimum word frequency of 5 occurrences.

Similarity-Measure. To measure a similarity between drug and disease embeddings we choose cosine similarity in all experiments. A value of 1 between two vectors means a perfect similarity, and the value 0 means a maximum dissimilarity.

Model Training and Evaluation Settings: For all experiments, cross-validation is applied by creating ten identical models, where each of them is trained on a randomly selected and balanced data set, containing 50% inducing associations and 50% therapeutic associations. The selected data set is then randomly split into 90% training data, and 10% test data in a stratified way, meaning both training and test set will also consist of 50% inducing associations and 50% therapeutic associations. As a measure for the performance of the model, the average test accuracy of the ten models on each test data set is measured. For the neural networks, the average of the maximum accuracy overall epochs of all models will serve as the measure for comparison.

4.1 Quality Criteria

First, we investigate whether, and to what extent, word-embeddings are suitable for learning the DDA type (i.e. "treats", "induces"). In addition, we investigate if latent

---

[4] https://www.drugbank.ca/.

[5] https://lucene.apache.org/.

[6] https://deeplearning4j.org/word2vec.

features can be learned and therefore dimensions where semantics of an association type are probably expressed. In this context, the following quality criterion should be fulfilled to answer *RQ1*:

1. *Disambiguation Suitability:* If latent features exist in an entity-vector representation, that indicate the type of a DDA, then Machine-Learning approaches that are able to weight certain features higher should lead to better results (i.e. increased classification accuracy) compared to methods that asses all features equally. In this context, a sufficiently good quantitative result (i.e. high accuracy) should be achieved.

Distance between two embedded entities can affect the quality of a DDA predictions [16]. Therefore, we investigate whether distance can also affect the quality of a DDA-type prediction. In this context, the following criterion should be fulfilled to answer *RQ2*:

2. *Disambiguation Stability*: As the distance between a drug and a disease increases, the accuracy disambiguating a DDA should not decrease *substantially*. This would further indicate that a certain latent subspace has been learned by the Deep-Learning approaches. Therefore, with increasing distance between a drug and a disease vector-representation, always a sufficiently good quantitative result (i.e. high classification accuracy) should be achieved.

In our last investigation, we propose semantically enhanced deep learning models that can determine the type of (real) DDA predictions. In this context, the following criterion should be fulfilled:

3. *Prediction Accuracy with Semantic Enhancement:* So far, we have only evaluated the classification of existing DDAs on current datasets, i.e. DDAs and their association type are available in the CTD and in addition the DDAs can be found in publications. However, our primary interest is to classify the type of (real) DDA predictions. For this task, our proposed semantic-enhanced deep learning models should lead to improved results compared to modes trained without semantic information.

4.2   General Suitability of Word Embeddings to Disambiguate Drug-Disease Associations

In our first experiment, we investigate RQ1 and the hypothesis that latent information about the type of a DDA is encoded in certain areas of the vectors, hence certain dimensions, and can be learned using deep learning approaches. In this context, we will first describe the used dataset followed by our Baseline description. Afterward, we describe the used Deep-Learning models as well as their implementation details followed by our experiments.

Experimental Data-Set.   In this experiment, we use all drug-disease vectors that can also be found as a curated drug-disease association in CTD. Thus, our data set contains 33541 inducing associations as well as 18664 therapeutic associations. Since the classes are not strongly skewed, representative results can be expected when training on a balanced data set. Therefore, our data-set contains 18644 therapeutic as well as the same number of induce associations.

Baseline Construction. The work of Lev et al. [13] demonstrates that vector pooling techniques, applied to Word2Vec vectors, can outperform various literature-based algorithms in different NLP tasks and therefore can be seen as a method to construct strong baselines. As an example, a common pooling technique is the calculation of a mean vector $\vec{v}$ for N different vectors with:

$$\vec{v} = \frac{1}{N} \sum_{i=1}^{n} x_i$$

In our case, we use a drug vector and a disease vector for the pooling approach. However, since this '*mean*'-pooling approach, resulting in 200-dimensional vectors, blurs part of the information contained in the entity vectors, two more approaches will be tested. For the '*concat*'-pooling, both vectors are concatenated, resulting in 400-dimensional association vectors. Finally, for the 'stack'-polling, the drug and disease vectors are stacked, resulting in association matrices of the shape 200×2. Afterward, using our new vector representations, we train a scikit-learn's Support Vector Classifier (SVC) on our data-set to learn the drug-disease association types "treats" and "induces". As for the SVC-parameters, a degree of 3 has proven to be the best choice in our experiments. In addition, the kernel is a radial basis function and for all other parameters we used the default values.

Next, we describe the investigated Deep-Learning models and their implementation details.

Multilayer Perceptron (MLP). Using Keras' Sequential class, this model consists of three densely connected layers of decreasing size. The first two layers use a reactive linear unit as the activation function. The final layer uses a sigmoid function to produce the binary classification output and the loss function used is a binary cross-entropy loss function. The optimizer is Adam [22], with a learning rate of $1e-4$ and the batch size is set to 8.

Convolutional Neural Network (CNN). This model is built on Keras' Sequential class as well and also uses a single-neuron sigmoid layer as the final layer. The hidden layers describe an underlying CNN architecture. A dropout [23] layer with a rate of 0.1 is applied to combat overfitting. The kernel size is set to value of 3 for the 'mean' and 'concat' vectors and 3×2 for the 'stack' vectors. Similar to the MLP, the loss function is a binary cross-entropy loss function, and the optimizer is Adam [22] with a learning rate of $1e-4$. The batch size is eight, as well.

For a comparison, the two deep-learning models were trained with the different pooling vector-representations, where the "stack" vectors were exclusively used for training the CNN. The average Accuracies in a 10-fold cross validation are presented in Table 1:

Table 1. Accuracies achieved with different models and pooling-approaches. Best values in bold.

	Mean	Concat	Stack
SVC	71.84	**73.84**	–
MLP	80.75	**81.64**	–
CNN	80.40	**81.61**	81.54

Results and Results Interpretation. We can observe in Table 2 that both deep learning approaches learned a feature space capable of disambiguating between "treats" and "induces". Moreover, the concatenation of the vectors delivered the best results overall, achieving more than 81% of accuracy. Compared to the baseline, we can recognize an increase in accuracy of up to ~8%. In summary, we obtained empirical evidence that answers our first research question: deep learning models can find a latent space to disambiguate associations between pharmaceutical entities.

However, further investigation is needed to assess their performance. In particular, given the findings of [16] regarding the impact of the distance in the embedding space between entities to find associations, we would like to test the *stability* of our proposed methods in the following section.

4.3 Distance Relationship and Learning

In our next experiment, we investigate the fact that with increasing entity distance, the accuracy quality using k-NN approaches can decrease [16]. Therefore, we assume that with increasing distance also the semantic disambiguation accuracy (SDA) might show the same characteristic, and thus we lose the information for classifying a DDA Type.

Evaluation Dataset. To test this assumption, we select for each drug the k-nearest disease neighbors (k-NDNs), where $k = 10, 20, 50$. With increasing k also the distance between two entity-vectors will increase. In addition, we test only with DDAs that are curated in the CTD i.e. the association type in known. This selection results in data sets of the following sizes:

Table 2. Evaluation datasets for distance related evaluation.

k-NDNs	10	20	50
Inducing	688	1172	2264
Therapeutic	1939	3044	4812

Since the amount of training data will probably influence a model's accuracy, each model is trained and tested on data sets of equal size to achieve comparable results. Each subset of each data set will, therefore, contain exact 688 inducing as well as

688 therapeutic associations. We use the same models as in the previous experiment in combination with a concatenation approach. The results achieved with a 10-fold cross validation are presented in Table 3:

Table 3. Investigation of the influence of distance for DDA-type disambiguation. Accuracies achieved for different k-Nearest-Disease-Neighbors sets.

k-NDN	10	20	50
SVC	77.32	72.97	73.26
MLP	82.61	80.58	77.17
CNN	81.38	79.93	79.93

Results and Result Interpretations. We can observe in Table 3 that the SVC accuracy has a declining tendency with an increasing number of NDNs. The effect from 10 to 50 NDNs reaches up to ~4% for the SVC. The MLP model shows similar results with a decreasing rate in accuracy of ~5%. Finally, with a change of ~2% the CNN model shows a rather stable performance. This is a somewhat surprising result, given that the dataset used is small, which tends to lead to more volatile results in the model's accuracy. A possible explanation of the stability of the CNN model could be found in the rationale behind using CNNs in machine learning tasks. In Goodfellow, I. et al. [36], in particular, two properties from CNNs: sparse interactions and equivariant representation. For our task, the results confirm that sparse interactions allow for learning automatically - without manual feature engineering- patterns from d-dimensional spaces and equivariant representation allows for robustness in the patterns learned.

Given our experimental findings, we can claim that the performance of the CNN model is not affected by the distance in the original embedding space. Thus, our model has learned a robust latent representation that succeeds to disambiguate associations between entities. In the next section, we build on our findings to assess the impact of incorporating domain knowledge by introducing a hybrid neural network architecture and performing a sophisticated retrospective analysis to assess model performance when facing real DDA predictions.

4.4 Impact of Domain Knowledge to Disambiguate Predicted Drug-Disease Associations

In our previous sections, we have empirically proven that DDAs extracted from the year 2019 can indeed be classified with higher accuracy. In this section, we will verify that this is not only true for already discovered/existing DDAs, but also for real predictions using retrospective analysis. Moreover, we answer our third research question in the difficult task of real predictions by introducing domain knowledge into the models. In this context, we first describe how we identify real DDA predictions. Hereafter, we present an approach to create a semantic representation of a DDA using medical classification systems. Afterward, we present our proposed semantically enriched Deep-Learning models for DDA type disambiguation and compare the results with all previously tested models. With this retrospective approach, we want to simulate today's situation, where we have on the one hand possible DDA predictions and in addition we have access to rich taxonomic data that can be used as a source for semantic information.

Evaluation Dataset for Retrospective Analysis. To detect real predictions, first we train our Word2Vec model with the historical corpus (Publication date < 1989). Next, we extract DDAs from the resulting embedded space using a k-Nearest-Neighbor approach. Afterward, using a co-occurrence approach [19], we first check if a DDA does not exists, i.e., does not appear in at least three publications in our historical corpus. Then we check if a non-existing DDAs will appear in the actual corpus 2019 and/or can also be found in CTD. With this approach, we identify DDA predictions within the k-NDNs sets of each drug (where k = 10, 20, 50). In addition, we identify and train our proposed models on existing DDAs (DDA appears in documents where publication date < 1989) and test these models with the predicted DDAs sets. This yields to the data sets shown in Table 4.

Table 4. Number of real predicted as well as existing DDAs extracted using a k-Nearest-Disease-Neighbors (k-NDN) approach.

	k-NDNs/DDA-type	10	20	50
Test (predicted)	Inducing	88	168	402
	Therapeutic	91	195	474
Train (existing)	Inducing	687	1161	2311
	Therapeutic	1408	2182	3535

Entity Specific Semantic Information. In order to semantically enrich drug as well as disease entities, we use (medical) classification systems as a source. Considering pharmaceutical entities, there are a couple of popular classification systems such as the Medical Subject Headings (MeSH) Trees[7] or the Anatomical Therapeutic Chemical (ATC) Classification System[8]. The ATC subdivides drugs (hierarchically) according to their anatomical, therapeutic/pharmacologic, and chemical features. For example, the cancer related drug *'Cisplatin'* has the class label *'L01XA01'*. In this context the first letter indicates the anatomical main group, where *'L'* stands for 'Antineoplastic and immunomodulating agents'. The next level consists of two digits *'01'* expressing the therapeutic subgroup 'Antineoplastic agents'. Each further level classifies the object even more precisely, until the finest level usually uniquely identifies a drug. We use the ATC-Classification system exclusively for drug-entities. To collect semantic information also

---

[7] https://www.nlm.nih.gov/mesh/intro_trees.html.
[8] https://www.whocc.no/atc_ddd_index/.

for diseases as well as for drugs we use the *Medical Subject Headings (MeSH)*. MeSH is a controlled vocabulary with a hierarchical structure and serves as general classification system for biomedical entities (Table 5).

Table 5. Example: classes in different classification systems for the drug 'Cisplatin'.

Classification system	Assigned classes
ATC	L01XA01
MeSH trees	D01.210.375, D01.625.125, D01.710.100

Semantically Enriched DDA Representations. In our previous experiments, we have generated a DDA vector representation by concatenation of a drug vector with a disease vector. Afterwards, we trained different approaches with these DDA representations. In order to additionally use semantic information from classification systems, we also need a DDA vector representation using the MeSH trees and the ATC classes. For this task, we proceed as follows: MeSH trees consist of a letter in the beginning, followed by groups of digits from 0 to 9, separated by dots (Table 5). The leading letter signifies one of 16 categories, e.g., C for diseases and D for drugs and chemicals. Since the categories will not be mixed in any data sets, this letter can be omitted in our approach. Furthermore, the dots separating the levels provide no additional information, since each level is of the same length. After removing the dots and leading letters, a string of digits remains. To use this string as meaningful input to the neural networks, we need two more steps. First, the strings need to be of equal length. We achieved it through the padding of all strings, which are not of the maximum occurring length with zeros. Finally, since we will have a sparse representation, we use a Keras embedding layer to build a latent (dense) representation for each MeSH Tree. The structure of the ATC class labels consists of a string, containing both letters and digits. To prepare the class-strings for processing in a neural network, we mapped each character to a unique number. The resulting vector of numbers is then padded with zeros in the same way as the PubMed Vectors to achieve equal lengths. The resulting vectors can then be processed through an embedding layer to build a latent representation as well.

Semantically Enriched Deep-Learning Models. At this point we have two semantically different DDA representations. To achieve meaningful processing of these two fundamentally different information sources, we introduce a new type of neural network architecture. We will refer to it hereafter as a "hybrid network". Herein, a hybrid network is a neural network that is split at the input level and the first hidden layer and then merged into a single network through a merging layer, which is followed by further hidden layers and, finally, the output layer. By using the previously investigated MLP and CNN networks we propose and evaluate the following three hybrid architectures:

Hybrid MLP (Fig. 2a) - This architecture consists of two symmetrical densely connected networks that merge into one densely connected network through a concatenation layer.

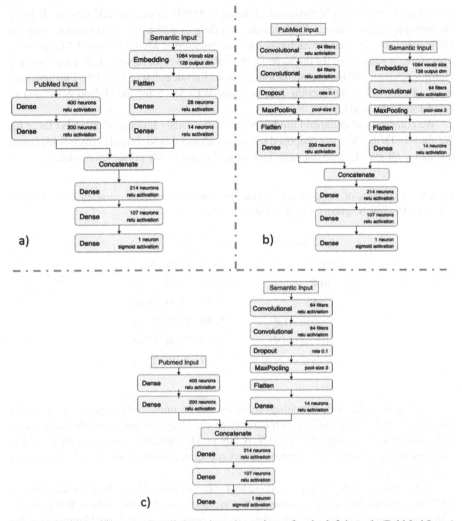

Fig. 2. Hybrid Architectures Detail Overview. As an input for the left branch (PubMed Input) Word2Vec DDA representations (Concatenated vectors) were used and for the right branch (Semantic Input) we used the semantic DDA representations.

Hybrid CNN (Fig. 2b) - The lower part of this network is identical to the Hybrid MLP. The two upper branches, however, have been replaced by CNNs.

Hybrid CNN & MLP (Fig. 2c) - This architecture is a mix of the previously introduced hybrid networks. The semantic input is pre-processed through a CNN, while PubMed vectors are pre-processed through an MLP.

Model Evaluation and Result Interpretation. For the one branch of a hybrid network, we will use the NEM (PubMed) vector representation of a DDA as input and for the other branch, we will use the semantic DDA representation as input. In this context, drugs, as well as diseases, can be assigned to multiple MeSH trees and ATC classes. If there are multiple MeSH trees or ATC classes for a drug or a disease, an additional semantic representation is created for each possible combination of MeSH trees and ATC classes. Therefore, in such cases for the same DDA (PubMed) vector representation we can have multiple semantic DDA representations as input for our enhanced models. Furthermore, to avoid overfitting the learning rate is adjusted to 5e−5 for the Hybrid MLP and to 2e−5 for the Hybrid CNN.

We compare our enhanced models with all previously introduced models (Table 6). We show in Table 6 that our proposed semantic enhanced models lead to overall improved results for all investigated datasets, including the previously presented models.

Table 6. Accuracies of the different approaches achieved for DDAs extracted from different k-Nearest-Disease-Neighbors sets. Best values in bold.

k-NDNs	10	20	50
SVC	66.93	70.54	71.53
MLP	74.83	76.25	77.05
CNN	75.74	74.79	76.44
Hybrid MLP	**76.80**	79.14	77.44
Hybrid CNN	75.99	**80.60**	**77.65**
Hybrid CNN & MLP	75.34	79.46	77.21

Compared to the SVC baseline, Accuracy can be improved by up to 10% (Hybrid CNN, k = 20). Moreover, in comparison to the deep-learning models we achieve improvements of Accuracy by up to 6% (CNN vs. Hybrid CNN). We conclude from this experiment that semantic information allows substantial improvements in the prediction of the DDA type.

Furthermore, CNNs are less sensitive to distances (Table 3) and, besides, the Hybrid-CNN leads to the best average values. We conclude from this observation, that (hybrid) CNNs are generally better suited for the classification of the DDA type. Compared to the previous experiments (Table 3, results for SVC, MLP and CNN) the accuracy-results are generally lower although the number of training samples is comparable (see Table 2 and 4). We therefore connect this effect mainly to the different sizes of training corpora that were used for the Word2Vec training. For example, the historical corpus of 1989 is many times smaller than the corpus of 2019, which can lead to averagely worse contextualized entities after Word2Vec training. We conclude from this, that if the contextualization quality decreases, this also hurts classification accuracy.

5   Conclusion and Future Work

In this paper, we addressed the central question of finding interpretable semantics from Neural Embedding Models (NEMs) to disambiguate associations such as "treats" and "induces" between pharmaceutical entities. To do so, we explored the use of deep learning models to learn a latent feature space from the NEMs and performed an in-depth analysis of the performance of the models.

We found that the deep learning models are stable in the sense that they learned a latent feature space that successfully delivers an accuracy of up to 80%. Moreover, we proposed Hybrid Deep-Learning models that incorporate domain knowledge. With our retrospective analysis, we showed that these models are robust and lead to improved performance for real DDA predictions.

References

1. Gefen, D., et al.: Identifying patterns in medical records through latent semantic analysis. Commun. ACM **61**(6), 72–77 (2018)
2. Chiu, B., Crichton, G., Korhonen, A., Pyysalo, S.: How to train good word embeddings for biomedical NLP. In: Proceedings of the 15th Workshop on Biomedical Natural Language Processing, pp. 166–174, 2016 August
3. Chiang, A.P., Butte, A.J.: Systematic evaluation of drug–disease relationships to identify leads for novel drug uses. Clin. Pharmacol. Ther. **86**(5), 507–510 (2009)
4. Herskovic, J.R., Tanaka, L.Y., Hersh, W., Bernstam, E.V.: A day in the life of PubMed: analysis of a typical day's query log. J. Am. Med. Inform. Assoc. **14**(2), 212–220 (2007)
5. Larsen, P.O., Von Ins, M.: The rate of growth in scientific publication and the decline in coverage provided by Science Citation Index. Scientometrics **84**(3), 575–603 (2010)
6. Baroni, M., Dinu, G., Kruszewski, G.: Don't count, predict! a systematic comparison of context-counting vs. context-predicting semantic vectors. In: Proceedings of the 52nd Annual Meeting of the Association for Computational Linguistics (Volume 1: Long Papers), vol. 1, pp. 238–247 (2014)
7. Gottlieb, A., Stein, G.Y., Ruppin, E., Sharan, R.: PREDICT: a method for inferring novel drug indications with application to personalized medicine. Mol. Syst. Biol. **7**(1), 496 (2011)
8. Zhang, W., et al.: Predicting drug-disease associations based on the known association bipartite network. In: 2017 IEEE International Conference on Bioinformatics and Biomedicine (BIBM), pp. 503–509. IEEE, 2017 November
9. Tshitoyan, V., et al.: Unsupervised word embeddings capture latent knowledge from materials science literature. Nature **571**(7763), 95 (2019)
10. Agarwal, P., Searls, D.B.: Can literature analysis identify innovation drivers in drug discovery? Nat. Rev. Drug Disc. **8**(11), 865 (2009)
11. Mikolov, T., Chen, K., Corrado, G., Dean, J.: Efficient estimation of word representations in vector space. arXiv preprint arXiv:1301.3781 (2013)
12. Dudley, J.T., Deshpande, T., Butte, A.J.: Exploiting drug–disease relationships for computational drug repositioning. Brief. Bioinf. **12**(4), 303–311 (2011)
13. Lev, G., Klein, B., Wolf, L.: In defense of word embedding for generic text representation. In: Biemann, C., Handschuh, S., Freitas, A., Meziane, F., Métais, E. (eds.) NLDB 2015. LNCS, vol. 9103, pp. 35–50. Springer, Cham (2015). https://doi.org/10.1007/978-3-319-19581-0_3
14. Keiser, M.J., et al.: Predicting new molecular targets for known drugs. Nature **462**(7270), 175 (2009)
15. Lotfi Shahreza, M., Ghadiri, N., Mousavi, S.R., Varshosaz, J., Green, J.R.: A review of network-based approaches to drug repositioning. Brief. Bioinform. **19**, 878–892 (2017)
16. Wawrzinek, J., Balke, W.-T.: Measuring the semantic world – how to map meaning to high-dimensional entity clusters in PubMed? In: Dobreva, M., Hinze, A., Žumer, M. (eds.) ICADL 2018. LNCS, vol. 11279, pp. 15–27. Springer, Cham (2018). https://doi.org/10.1007/978-3-030-04257-8_2

17. Hill, F., Reichart, R., Korhonen, A.: Simlex-999: evaluating semantic models with (genuine) similarity estimation. Comput. Linguist. **41**(4), 665–695 (2015)
18. Rinaldi, F., Clematide, S., Hafner, S.: Ranking of CTD articles and interactions using the OntoGene pipeline. In: Proceedings of the 2012 BioCreative Workshop, April 2012
19. Jensen, L.J., Saric, J., Bork, P.: Literature mining for the biologist: from information retrieval to biological discovery. Nat. Rev. Genet. **7**(2), 119 (2006)
20. Mikolov, T., Yih, W.T., Zweig, G.: Linguistic regularities in continuous space word representations. In: Proceedings of the 2013 Conference of the North American Chapter of the Association for Computational Linguistics: Human Language Technologies, pp. 746–751 (2013)
21. Wick, Christoph: Deep Learning. Nature **521**(7553), 436–444 (2016). MIT Press, 800
22. Kingma, D.P., Ba, J.: Adam: a method for stochastic optimization. CoRR. abs/1412.6980 (2014)
23. Hinton, G.E., et al.: Improving neural networks by preventing co-adaptation of feature detectors (2012)
24. Patrick, M.T., et al.: Drug repurposing prediction for immune-mediated cutaneous diseases using a word-embedding–based machine learning approach. J. Invest. Dermatol. **139**(3), 683–691 (2019)

Recommendation

Heterogeneous Graph Embedding for Cross-Domain Recommendation Through Adversarial Learning

Jin Li[1], Zhaohui Peng[1(✉)], Senzhang Wang[2], Xiaokang Xu[1], Philip S. Yu[3], and Zhenyun Hao[1]

[1] School of Computer Science and Technology, Shandong University, Qingdao, China
{ljin,xuxiaokang,haozhyun}@mail.sdu.edu.cn, pzh@sdu.edu.cn
[2] School of Computer Science and Technology,
Nanjing University of Aeronautics and Astronautics, Nanjing, China
szwang@nuaa.edu.cn
[3] Department of Computer Science, University of Illinois at Chicago, Chicago, USA
psyu@uic.edu

Abstract. Cross-domain recommendation is critically important to construct a practical recommender system. The challenges of building a cross-domain recommender system lie in both the data sparsity issue and lacking of sufficient semantic information. Traditional approaches focus on using the user-item rating matrix or other feedback information, but the contents associated with the objects like reviews and the relationships among the objects are largely ignored. Although some works merge the content information and the user-item rating network structure, they only focus on using the attributes of the items but ignore user generated contents such as reviews. In this paper, we propose a novel cross-domain recommender framework called ECHCDR (Embedding content and heterogeneous network for cross-domain recommendation), which contains two major steps of content embedding and heterogeneous network embedding. By considering the contents of objects and their relationships, ECHCDR can effectively alleviate the data sparsity issue. To enrich the semantic information, we construct a weighted heterogeneous network whose nodes are users and items of different domains. The weight of link is defined by an adjacency matrix and represents the similarity between users, books and movies. We also propose to use adversarial training method to learn the embeddings of users and cross-domain items in the constructed heterogeneous graph. Experimental results on two real-world datasets collected from Amazon show the effectiveness of our approach compared with state-of-art recommender algorithms.

Keywords: Recommender system · Heterogeneous network · Deep learning

© Springer Nature Switzerland AG 2020
Y. Nah et al. (Eds.): DASFAA 2020, LNCS 12114, pp. 507–522, 2020.
https://doi.org/10.1007/978-3-030-59419-0_31

1   Introduction

Cross-domain recommendation has attracted considerable research attention in both academic and industry communities. In real application scenarios of a recommender system, the users' ratings on products of different domains are usually unevenly distributed. The ratings in one domain (e.g. book) can be scarce, while the ratings in another domain (e.g. movie) can be rich. Cross-domain recommendation aims to leverage all available data of multiple domains to generate better recommendations on target domain.

Currently, cross-domain recommendation methods can be categorized into collaborative filtering (CF) based methods [1], content-based methods [2] and graph-based methods [3,4]. The CF-based methods have been widely studied due to their impressive performance in single-domain recommender system [5]. To address the data sparsity issue, some works improve CF-based methods by employing side information, such as the properties of users/items, the behavior history of users and the relationship between users. Ren *et al.* [6] incorporated the similarity vector of users into user-item rating matrix. Mirbakhsh *et al.* [7] mixed user-item rating matrix with cluster algorithm and Xin *et al.* [2] proposed a cross-domain collaborative filtering with review text. Wang *et al.* [8] proved the influence of neighborhood.

Content-based methods combining with deep learning techniques have gained significant performance improvement [9,10]. Various deep learning methods are used for the task of content-based recommendation [11,12]. These methods aim to learn a mapping function across domains to address the data sparsity issue. Relations among objects, which includes user-user, user-item and item-item relationship, can find the potential interest of users and improve the recommendation performance. Although the above mentioned methods can address the data sparsity issue to some extent, the relationships among the objects cannot be effectively leveraged.

Graph-based methods focus on learning the user and item embedding for recommendation from the user-item network structure and semantics. DeepWalk [13] learned feature vectors for vertices from a corpus of random walks. Besides, some methods try to merge content information and network structure, such as HIN2vec [14] and knowledge graph. Although these methods merge content information and network structure, they only focus on items' properties but ignore users' reviews. Reviews can reflect users' concerns, which influence recommender effectiveness.

A major limitation of the above methods is that they do not fully consider content information and network structure from the perspective of user centered. In effect, synthesizing each kind of information will benefit to the effectiveness of cross-domain recommendation. However, it is challenging to make better recommendations fusing various kinds of contents and structures. First, it is challenging to merge content information, such as users' reviews and items' properties, and network structure from different objects and relationships. Second, it is also challenging to efficiently train the model to enhance semantics and minimize computational complexity.

In this paper, we construct a weighted heterogenous graph to more comprehensively fuse both the content information and network structure information. An adversarial learning algorithm is also proposed to learn user and item embeddings in the graph for cross-domain recommendation. More specifically, we extract content information from users' reviews and items' titles and construct an adjacency matrix according to rating matrix to get the adjacency representation of them. Then we construct a heterogeneous network whose nodes include users and cross-domain items nodes to solve above problems. The contents and adjacency representation are concatenated as the initial vector for them and GANs is used to train the proposed framework. For the generator, we construct a tree by BFS (Breadth-First-Search) algorithm to decrease the cost of generator computation and capture relationships among users, books and movies. We design experiment to make top-n recommendation for both cross-domain and single-domain users. The results demonstrate that our approach outperforms state-of-art baselines. The contributions of this work are summarized as follows.

- We present a hybrid model that takes into account content information including both reviews and titles, adjacency relationship and network structure, which is suitable for cross-domain recommendation;
- We propose generative adversarial nets method to perform heterogeneous network embedding and apply it to cross-domain recommender system;
- We evaluate our model over real datasets to show the effectiveness of the proposed model.

2   Related Work

Our work is closely related to content based recommender systems, and heterogeneous graph embedding. We review related works from the two aspects.

2.1   Content Based Recommender Systems

Content based recommender systems share in common a means for describing the items that may be recommended, a means for creating a profile of the user that describes the types of items the user likes, and a means of comparing items to the user profile to determine what to recommend [2]. Based on that data, we then use them to make suggestions to the user. TF/IDF are used in content based recommender systems to determine the correlation of a document/item/movie etc. Besides, many shallow methods can serve for content based recommender systems, such as K-means, decision tree and so on.

Along with the expansion of the data scale, shallow methods do not work well when a large volume of content information is available. Researchers adopt deep learning model in content based recommender system. Wang *et al.* [15] proposed a recommender named CAMO, which employed a multi-layer content encoder for simultaneously capturing the semantic information of multitopic and word order. Multi-view deep learning approach [16] combined different domains into a single model to improve the recommendation quality across all the domains, as well as having a more compact and a semantically richer user latent feature vector. Tag2word [17] focused on the content based tag recommendation due to its wider applicability. Although these methods fully tap the content potential, the relationships between users and items are ignored. Besides, reviews play a critical role in content information of users that Fu *et al.* [18] and Song *et al.* [19] utilized them to make recommendation.

2.2 Heterogeneous Graph Embedding

Graph embedding is an effective and efficient way to solve the graph analytics problem, which converts the graph data into a comprehensible view. Cai *et al.* [20] systematically categorized the graph embedding techniques and applications based on problem setting and summarize the challenges.

Compared with homogeneous graphs, heterogeneous graphs are more complex as they have multiple types of nodes and links. The research focuses on exploring global consistency among different types of objects and dealing with the imbalance of them. Searching suitable meta-path or meta-graph [3,4] is effective to capture the complex relationships in a heterogeneous graph. Node2vec [21] preserved network neighborhoods to learn feature vectors. Metapath2vec and Metapath2vec++ used different treatment for different semantic types. Meta-graph developed Metapath2vec and incorporated more complex semantics relationships into the heterogeneous graph and different embedding vectors are learned through different meta graphs. Kim *et al.* [22]. proposed a generative adversarial network model to discover relationships between different domains. Traditional recommender systems try to make use of the rich semantics provided by the heterogeneous information and treat them disparately.

The recent research of HIN has some novel methods, Hu et al. [23] used adversarial learning method to generate fake samples. Zhang *et al.* [11] jointly considered heterogeneous structural information as well as contents information for node embedding effectively. Su *et al.* [24] investigated the contagion adoption behavior under a set of interactions among users and contagions, which are learned as latent representations.

3   The ECHCDR Model

The framework of the proposed model ECHCDR is shown in Fig. 1. Here we use the two domains (movies and books) as an example to illustrate our model. As shown on the left of Fig. 1, we first extract text features from high-score reviews and high-score titles to represent user, movie and book respectively. We also construct an adjacency matrix according to rating matrix to get adjacency representation of them. As shown on the top-right of Fig. 1, we construct an heterogeneous network whose nodes are users, books, and movies, and concentrate content and adjacency representation as their initial representation vectors. To make the network more densely connected, we evolve it according to adjacency matrix. As the discriminator of GANs needs to distinguish that relationship is user-item, user-user, or item-item, we set weight in relationship from the adjacency matrix. As shown on the lower-right of Fig. 1, we use GANs to train the proposed framework. Besides, we construct a tree by Breadth First Search (BFS) algorithm to decrease the cost of generator computation and capture relationships among users, books and movies. Next, we will introduce the framework in detail.

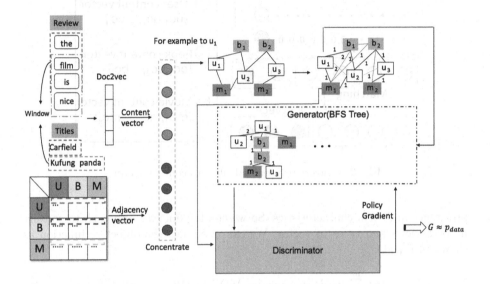

Fig. 1. Architecture of ECHCDR framework.

3.1 Text Feature Learning for Users and Items

In this section, we introduce the method of text feature learning from reviews of users U and titles of books B and movies M. $U = \{u_1, u_2, \ldots, u_i\}$, $B = \{b_1, b_2, \ldots, b_j\}$ and $M = \{m_1, m_2, \ldots, m_k\}$ represent the set of users, books and movies respectively. We select the reviews given by the users who give 4 or 5 scores to the item as the content representation for books and movies. Similarly, we choose the titles of books and movies that users give 4 or 5 scores as users' content representation because these items can better represent user interest. Let $R = \{r_1, r_2, \ldots, r_m\}$ be all the reviews and $T = \{t_1, t_2, \ldots, t_n\}$ be all the titles, where m, and n represent the total number of reviews and titles, respectively. We define $uc_i = \{t_{i1}, t_{i4}, t_{i5}, \ldots\}$ as the titles of u_i's favorite items and $bc_j = \{r_{j2}, r_{j6}, r_{j7}, \ldots\}$, $mc_k = \{r_{k3}, r_{k8}, r_{k9}, \ldots\}$ as the received reviews of books and movies respectively. The different subscripts means each user has different numbers of titles and each item has different numbers of reviews.

Doc2vec [25] represents each paragraph by a dense vector which is trained to predict words in the paragraph. Besides, the order of each word decides the user

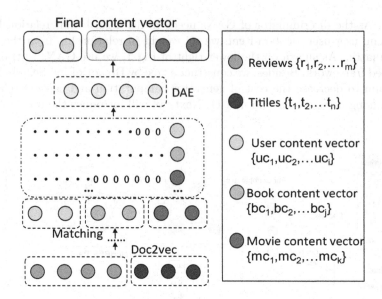

Fig. 2. Convert reviews and titles to content vector.

preference and item characteristics. So we employ doc2vec approach to convert users' reviews and titles to vector. We can maximize the objective function as shown in Eq. (1):

$$O = \sum_{\nu=1}^{m} \log P\big(w_{-l} : w_l \mid r_\nu\big) \| \sum_{u=1}^{n} \log P\big(w_{-l} : w_l \mid t_u\big) \qquad (1)$$

where w is a word in the text information of r_ν, t_u ($r_\nu, t_u \in R, T$), and l is the window size of the word sequence. After optimizing this objective, we can get the mapped vector for every review and title. Then we integrate these vectors by matching corresponding objects and get the content vectors with different dimensions for users, books and movies. To represent the features of all the items and users in a unified way, we adopt filling-zero method to unify it to the highest dimension. After that, we use Denoising auto-encoder (DAE) to extract content vector. DAE first encodes the input (preserves the information about the input), and then undoes the effect of a corruption process stochastically applied to the input of the auto-encoder, which has better results compared with traditional AutoEncoder. After a feature extraction process with DAE, we get the final content vector $\{uc_i, \ldots\}, \{bc_j, \ldots\}, \{mc_k, \ldots\}$ for each user, book and movie as shown in Fig. 2.

3.2 Adjacency Matrix Construction with Rating Matrix

As shown in the far left of Fig. 3, we stack users, books and movies to build an adjacency matrix among them based on the cross-domain symmetric rating matrix and all the weights are initialized to zeros. Then if one user gives high rating score (4 or 5) to the item, we set the corresponding user-item weight in the adjacency matrix as the rating score. Similarly, the other weights' initialization are as following:

- *Item-Item Weight:* If one book or movie is given high score by the same user, the weight between them will be marked 1. The weight between a book and a movie will be marked as n if there are n common users giving high scores on both items.
- *User-User Weight:* If two users give high score for the same item, the weight between them will be marked 1. More items that are rated with high scores by the two users will lead to larger weight between the two users.

Following the rules mentioned above, we obtain the adjacency matrix. The new matrix reflects the nearest neighbor and relevance degree of each object. Through extracting one row or column of the cross-domain matrix, we can get the adjacency vector for each user (U), book (B) and movie (M). However, the adjacency vectors are extremely sparse and most elements are zero. To solve this problem, we employ DAE to learn a low dimensional representation vector from the initial adjacency vector and get final adjacency vector $\{ua_i, \ldots\}, \{ba_j, \ldots\}, \{ma_k, \ldots\}$. The complete framework is shown in Fig. 3.

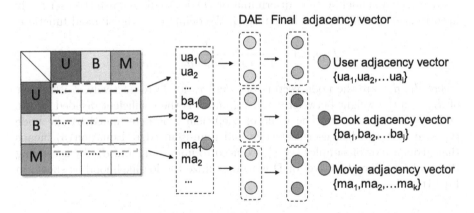

Fig. 3. From rating matrix to adjacency matrix.

3.3   Training HIN with Generative Adversarial Networks

After that, we build a heterogeneous dense graph according to the adjacency matrix that we add the item-item and user-user links. The matrix value is corresponding to the link weight in the heterogeneous graph. We denote the graph as $\mathcal{G} = (\mathcal{V}, \mathcal{E})$, where $\mathcal{V} = \{u_1, \ldots, u_i, b_1, \ldots, b_j, m_1, \ldots, m_k\}$ represents the nodes and $\mathcal{E} = \{e_{pq}\}_{p,q \in \mathcal{V}}$ represents the set of edges. For simplicity, we let v represent the $\{u_1, \ldots, u_i, b_1, \ldots, b_j, m_1, \ldots, m_k\}$. Inspired by the idea of GANs, we train two models simultaneously and regard weight as the parameter for discriminative model. When the GANs model converges, we learn the suitable representations for users, books and movies.

- Generator $G(v|v_c; \theta)$: It tries to fit the true connectivity distribution $p_{data}(v|v_c)$ that the user really likes and obtain the most similar vertices that connected with v_c (users vertices) from books and movies to deceive the discriminator.
- Discriminator $D(v, v_c; \phi)$: It tries to distinguish the item that user really likes or the generator selects.

Formally, G and D are playing two-player *minimax* game with value function $J(G, D)$ as following:

$$\min_{\theta} \max_{\phi} = \sum_{c=1}^{v} \left(\mathbb{E}_{v \sim p_{data}}(.|v_c) \left[\log D(v, v_c; \phi) \right] + \mathbb{E}_{v \sim G(.|v_c; \theta)} \left[\log \left(1 - D(v, v_c; \phi) \right) \right] \right) \tag{2}$$

In each iteration, discriminator D is trained with positive samples that it has real edge with v_c and negative samples from generator G. Generator is updated by policy gradient using a reward function that can be dynamically updated by discriminator. Next, we introduce them in detail.

In a recommender system, discriminator D tries to distinguish the user really liked items from the items selected by G. We define D as the sigmoid function:

$$D(v, v_c) = \frac{1}{1 + exp(-\boldsymbol{d}_v {}^\top \boldsymbol{d}_{v_c} + \lambda w)} \tag{3}$$

where $\boldsymbol{d}_v$, $\boldsymbol{d}_{v_c}$ are the node vectors of v and v_c for D respectively, ϕ is the set of $\boldsymbol{d}_v$, w is the weight between v and v_c, and λ is the coefficient decided by the node types of v and v_c. The value of coefficient λ is determined by the node types: user-item edge, user-user edge and item-item edge. Equation (3) means that given a pair of samples (v, v_c), we need to update $\boldsymbol{d}_v$ and $\boldsymbol{d}_{vc}$ by ascending the gradient. The objective for D is to maximize the log-likelihood as shown in Eq. (4).

$$\nabla_{\phi} J(G, D) = arg \max_{\phi} \sum_{c=1}^{v} \left(\underbrace{\mathbb{E}_{v \sim p_{data}}(.|v_c) \left[\log D(v, v_c; \phi) \right]}_{v \sim p_{data}} + \underbrace{\mathbb{E}_{v \sim G(.|v_c; \theta)} \left[\log \left(1 - D(v, v_c; \phi) \right) \right]}_{v \sim G} \right) \tag{4}$$

By contrary, the generator G selects samples to deceive D and intends to minimize the objective $\mathbb{E}_{v \sim G(.|v_c; \theta)} \left[\log \left(1 - D(v, v_c; \phi) \right) \right]$. Inspired by the method of Wang et al. [26], we also perform Breadth First Search (BFS) on the heterogeneous graph starting from each vertex v_c, which will get a BFS-tree T_c rooted at v_c as shown in Fig. 1 for u_1. The relevance probability of a given v_c and a random vertex v can be calculated by Eq. (5). Conditional probability $p_{data}(v|v_c)$ means

Algorithm 1. Full Path from v_c to v

Require: BFS-tree T_c
Ensure:
  selected destination samples v_{des}
  $v_{cur} \Rightarrow v_c, v_{-1} \Rightarrow v_c$;
  while true **do**
    Randomly select v and compute $p(v|v_{cur})$ in Eq. (5);
    if $v = v_{-1}$ **then**
      $v_{des} = v_{cur}$;
      return v_{des} ;
    else
      $v_{-1} \Rightarrow v_{cur}, v_{cur} \Rightarrow v$;
    end if
  end while

---

the underlying true connectivity distribution for vertex v_c, which reflects v_c's connectivity preference distribution over all other direct neighbor vertices in $\mathcal{V}$.

$$p(v|v_c) = \frac{exp(\boldsymbol{g_v}^\top \boldsymbol{g_{v_c}})}{\sum\limits_{v_n \in \mathcal{N}(v_c)} exp(\boldsymbol{g_{v_n}}^\top \boldsymbol{g_{v_c}})} \tag{5}$$

where $\boldsymbol{g_v}$, $\boldsymbol{g_{v_c}}$ are the node vectors of vertices v and v_c for G respectively, and $\mathcal{N}(v_c)$ represents the direct neighbor vertices of v_c. So the generator $G(v|v_c; \theta)$ can be defined as follows:

$$G(v|v_c) = \big(\prod_{c=1}^{v} p(v|v_{-1}) p(v_{c+1}|v_c) \big) \tag{6}$$

We use BFS to build these trees, so the root v_c to the destination v's path is unique. To get the destination v, we describe the process in Algorithm 1. v_{c+1} is the second node, and v_{-1} is the prior node of v. This algorithm shows that if we visit one vertex twice, we optimize the vector of v and v_c by GANs. The optimization method uses policy gradient derived in Eq. (7), which is suitable for discrete data [27].

$$\begin{aligned}
&\nabla_\theta J(G, D) \\
&= \nabla_\theta \sum_{c=1}^{V} \mathbb{E}_{v \sim G(.|v_c)}[\log(1 - D(v, v_c))] \\
&= \sum_{c=1}^{V} \sum_{t=1}^{N} \nabla_\theta G(v_t|v_c)[\log(1 - D(v_t, v_c))] \\
&= \sum_{c=1}^{V} \sum_{t=1}^{N} G(v_t|v_c) \nabla_\theta G(v_t|v_c)[\log(1 - D(v_t, v_c))] \\
&= \sum_{c=1}^{V} \mathbb{E}_{v \sim G(.|v_c)}[\nabla_\theta \log G(v|v_c) \log(1 - D(v, v_c))]
\end{aligned} \tag{7}$$

With reinforcement learning terminology, the term $\log(1 + exp(\boldsymbol{g_v}^\top \boldsymbol{g_{v_c}}))$ acts as the reward for the policy gradient. The overall training process is summarized in Algorithm 2. The inputs of generator and discriminator are sampled in graph, and they play a *minimax* game to train node vectors. When the algorithm converges after several rounds of iterations, we can work out the score between any two nodes, which is computed by the score function f: $f = g_v^\top g_{v_c}$. Because we recommend books or movies for user, we only calculate the user-book score and user-movie score. Then we can make top-n recommendation according to the scores between items and the user. The higher score means, the higher rank of the recommendation for the user.

Algorithm 2. Minimax Game for Graph

Require: Initialize generator $G(v|v_c; \theta)$ and discriminator $D(v, v_c; \phi)$ by obtained node vectors
    Construct T_c for all $v_c, c \in \{1, 2, \ldots, i + j + k\}$
Ensure:
  repeat
    for g-steps **do**
      $G(v|v_c; \theta)$ recommends n vertices for each vertex v_c by Algorithm 1;
      Update generator parameters via policy gradient according to Eq. (5), (6) and (7);
    end for
    for d-steps **do**
      Use negative vertices from $G(v|v_c; \theta)$ and combine with given positive samples from training data for each vertex v_c;
      Training discriminator ϕ by Eq. (3) and (4);
    end for
  until converge

---

4   Experiments

4.1   Experiment Setup

We use the Amazon dataset[1] to evaluate the performance of our model. We consider a high score (4 or 5) means the user is interested in the item. We extract reviews text from active users. Active users refer to the users that have reviewed more than 20 items and the comment contains more than 10 words on average. Similarly, we select the active items that have more than 50 high-score reviews. Finally we obtain the dataset consisting of about 0.25 million reviews from 2792 users, 60730 books and 82844 movies. We divide the set of testing users into two categories: cross-domain users and single-domain users. Moreover, the cross-domain users and single-domain users have the same number of users in each domain:

---

[1] http://jmcauley.ucsd.edu/data/amazon/.

- Cross-domain users have reviews and rating scores in both domains. We remove the information in one domain as testing set and preserve complete information in the other domain as training set.
- Single-domain users have reviews and rating scores only in one domain. We randomly remove 20% reviews and rating scores of users as testing set and the remaining data is used as training set.

We conduct five-fold cross-validation in the experiment for all methods and report the average results. To further verify the performance of the model, we recommend books for cross-domain users and single-domain users respectively and then recommend movies for them.

4.2   Performance Evaluation

We mainly compare our approach with the following methods including CF-based methods, content-based methods and graph embedding methods.

- **PMF:** Probabilistic Matrix Factorization (PMF) [28] modeled latent factors of users and items by Gaussian distributions.
- **Cross-CBMF:** Cross-CBMF [7] extended matrix factorization to cross domain and utilized unobserved ratings.
- **Multi-view Deep Learning:** Multi-View deep learning approach [16] is a content based recommender model that maps users and items to a latent space where the similarity between users and items is maximized.
- **Metapath2vec:** Metapath2vec [3] is a graph embedding method and preserved both the structure and semantics for a given heterogeneous network.
- **EMCDR** [9]: It adopted matrix factorization to learn latent factors first and then utilized a multi-layer perceptron network to map the user latent factors from one domain to the other domain.

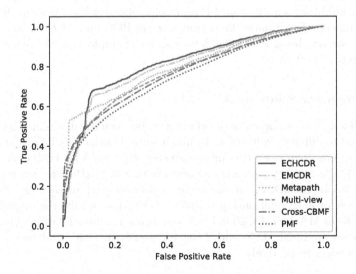

Fig. 4. ROC curves of various models on Amazon dataset

Table 1. AUC of different methods

Method	AUC (n = 5)	AUC (n = 10)	AUC (n = 20)
PMF	0.7544	0.7615	0.7695
Cross-CBMF	0.7512	0.7633	0.7698
Multi-view	0.7632	0.7726	0.7810
Metapath2vec	0.7823	0.7959	0.8011
EMCDR	0.7943	0.8026	0.8133
ECHCDR	**0.8026**	**0.8139**	**0.8165**

As our goal is to recommend items that users may like, we use two evaluation metrics as follows. **Normalized discounted cumulative gain (NDCG@n)**, which is a valid measure of ranking quality and **Mean Average Precision (MAP@n)**, which is a precision metric and defined as:

$$MAP = \frac{\sum_{u=1}^{i} AveP(u)}{U}$$

where U is the set of testing users and $AveP$ is the average precision. Here we set $n = \{5, 10, 20\}$ to evaluate the proposed approach. We also use more intuitive and comprehensive methods, receiver operating characteristic (ROC) curve and the area under ROC (AUC) as our evaluation metrics. The true positive rate (TPR) and false positive rate (FPR) used for generating ROC curves are defined as:

$$TPR = \frac{TP}{TP+FN} \quad FRR = \frac{FP}{FP+TN}$$

where TP represents we recommend an item that user likes, FN represents we do not recommend an item that user does not like, TN represents we recommend an item that user does not like, and FP represents one user likes an item that we do not recommend. To evaluate the performance of ROC and AUC among different methods, we use book recommendation as an example and the recommended book number is set to 10.

4.3   Parameter Study on λ

λ can influence the accuracy of discriminator, because the different relationships correspond to different value of λ. To find a suitable value setting of the parameter λ for discriminator, we conduct parameter study on λ for both cross-domain user and single domain-user, and the performance of MAP@n is shown in Fig. 5. Different λ can remarkably influence recommender performance and the influence degree for each relationship is different. Finally we take the average of the three peaks in MAP@5, 10, 20 in book and movie recommendation. The parameter λ is set to 10.32, 6.75, and 20.96 for user-item edge, user-user edge and item-item edge, respectively.

4.4   Results and Analysis

Experimental results of different methods are presented in Table 2 for cross-domain and single-domain users. One can see that our model achieves the best performance. PMF as a trivial method achieves the worst performance. Moreover, the results demonstrate that our model brings significant improvement compared to Metapath2vec that only considers graph network structure and Multi-view Deep Learning that only considers content information. When $n = 10$ for cross-domain users, the average NDCG of our model is 0.2948 in book domain, indicating that it outperforms the PMF model by 6.35%, the Cross-CBMF model

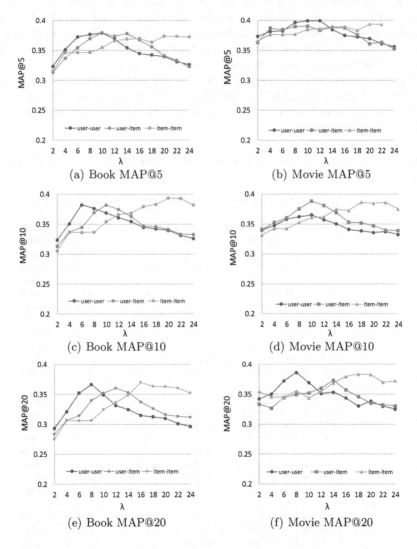

(a) Book MAP@5

(b) Movie MAP@5

(c) Book MAP@10

(d) Movie MAP@10

(e) Book MAP@20

(f) Movie MAP@20

Fig. 5. Effect of parameter λ on book and movie recommendation

by 5.44%, the Multi-view model by 3.0%, the Metapath2vec model by 1.97% and EMCDR model by 1.84%. The average MAP achieved by our model is 35.29%, which outperforms the PMF, Cross-CBMF, Multi-view, Metapath2vec and EMCDR models by 30.91%, 31.10%, 31.93%, 32.39% and 34.24%, respectively.

From the experiments, we can draw the following conclusions. First, performance on movie recommendation is better than that on the book recommendation in both cross-domain users and single-domain users. Second, compared with cross-domain users, recommendation for single-domain users usually achieves higher NDCG and MAP. This is mainly because the data quantity in the movie domain is better than the book domain. In addition, cross-domain users preference in one domain cannot be fully reflected and captured by other domains' feedback. Instead, the single-domain users regard the other domains as side information and retain the complete information in source domain, which enhances the recommender performance in source domain.

Figure 4 shows the ROC curves of PMF, Cross-CBMF, Multi-view deep learning, Metapath2vec, EMCDR, and ECHCDR. Table 1 shows the corresponding AUC of all compared methods. We can observe that our model obtains higher AUC than other methods.

Table 2. Overall performance of different methods for cross-domain/single-domain users

			PMF	Cross-CBMF	Multi-view	Metapath2vec	EMCDR	ECHCDR
Cross-domain users								
$n = 5$	Book	NDCG	0.2495	0.2593	0.2704	0.2824	0.2891	**0.3047**
		MAP	0.3153	0.3198	0.3278	0.3318	0.3556	**0.3674**
	Movie	NDCG	0.2613	0.2666	0.2816	0.2967	0.3054	**0.3175**
		MAP	0.3309	0.3389	0.3403	0.3437	0.3604	**0.3751**
$n = 10$	Book	NDCG	0.2313	0.2404	0.2648	0.2751	0.2764	**0.2948**
		MAP	0.3091	0.3110	0.3193	0.3239	0.3424	**0.3529**
	Movie	NDCG	0.2536	0.2581	0.2762	0.2855	0.2997	**0.3121**
		MAP	0.3216	0.3264	0.3358	0.3385	0.3532	**0.3641**
$n = 20$	Book	NDCG	0.2261	0.2243	0.2567	0.2602	0.2651	**0.2756**
		MAP	0.2976	0.3026	0.3087	0.3191	0.3351	**0.3483**
	Movie	NDCG	0.2473	0.2489	0.2665	0.2759	0.2816	**0.2965**
		MAP	0.3087	0.3135	0.3206	0.3217	0.3413	**0.3556**
Single-domain users								
$n = 5$	Book	NDCG	0.2625	0.2703	0.2844	0.2914	0.3081	**0.3217**
		MAP	0.3229	0.3308	0.3418	0.3507	0.3609	**0.3789**
	Movie	NDCG	0.2734	0.2876	0.2956	0.3087	0.3204	**0.3372**
		MAP	0.3439	0.3499	0.3543	0.3627	0.3754	**0.3975**
$n = 10$	Book	NDCG	0.2453	0.2550	0.2688	0.2841	0.2914	**0.3108**
		MAP	0.3171	0.3220	0.3333	0.3429	0.3494	**0.3689**
	Movie	NDCG	0.2566	0.2691	0.2752	0.2845	0.3033	**0.3211**
		MAP	0.3336	0.3374	0.3497	0.3555	0.3682	**0.3851**
$n = 20$	Book	NDCG	0.2391	0.2483	0.2537	0.2692	0.2800	**0.2976**
		MAP	0.3016	0.3136	0.3227	0.3381	0.3412	**0.3547**
	Movie	NDCG	0.2403	0.2598	0.2615	0.2747	0.2965	**0.3125**
		MAP	0.3207	0.3245	0.3326	0.3415	0.3563	**0.3716**

5   Conclusions

In this paper, we propose a novel approach for cross-domain recommendation. First, we integrate reviews text, item title and adjacency relationship and then extract features as their representation including content-vector and adjacency vector. Second, we build a weighted dense heterogeneous graph by the constructed adjacency matrix to alleviate the data sparsity and enrich semantics. Finally, we train it by GANs to generate the recommendation items. We evaluate our proposed model on the Amazon-Book and Amazon-Movie datasets. The experiments conducted on two types of users (cross-domain and single-domain users) demonstrate the superior performance of the model. In the future, we will extend our model to more domains, such as book, movie and music, which can further improve recommender effectiveness.

Acknowledgements. This work is supported by Industrial Internet Innovation and Development Project in 2019 of China, Shandong Provincial Key Research and Development Program (Major Scientific and Technological Innovation Project) (NO. 2019JZZY010105), NSF of Shandong, China (No. ZR2017MF065), NSF of Jiangsu, China (No. BK20171420). This work is also supported in part by US NSF under Grants III-1526499, III-1763325, III-1909323, and CNS-1930941.

References

1. Hernando, A., Bobadilla, J., Ortega, F.: A non negative matrix factorization for collaborative filtering recommender systems based on a bayesian probabilistic model. Knowl.-Based Syst. **97**, 188–202 (2016)
2. Xin, X., Liu, Z., Lin, C.-Y., Huang, H., Wei, X., Guo, P.: Cross-domain collaborative filtering with review text. In: IJCAI (2015)
3. Dong, Y., Chawla, N.V., Swami, A.: metapath2vec: scalable representation learning for heterogeneous networks. In: KDD (2017)
4. Zhao, H., Yao, Q., Li, J., Song, Y., Lee, D.L.: Meta-graph based recommendation fusion over heterogeneous information networks. In: KDD (2017)
5. Han, X., Shi, C., Wang, S., Yu, P.S., Song, L.: Aspect-level deep collaborative filtering via heterogeneous information networks. In: IJCAI (2018)
6. Ren, S., Gao, S., Liao, J., Guo, J.: Improving cross-domain recommendation through probabilistic cluster-level latent factor model. In: AAAI (2015)
7. Mirbakhsh, N., Ling, C.X.: Improving top-n recommendation for cold-start users via cross-domain information. In: TKDD (2015)
8. Wang, X., Peng, Z., Wang, S., Yu, P.S., Fu, W., Hong, X.: Cross-domain recommendation for cold-start users via neighborhood based feature mapping. In: Pei, J., Manolopoulos, Y., Sadiq, S., Li, J. (eds.) DASFAA 2018. LNCS, vol. 10827, pp. 158–165. Springer, Cham (2018). https://doi.org/10.1007/978-3-319-91452-7_11
9. Man, T., Shen, H., Jin, X., Cheng, X.: Cross-domain recommendation: an embedding and mapping approach. In: IJCAI (2017)
10. Chao-Yuan, W., Ahmed, A., Beutel, A., Smola, A.J., Jing, H.: Recurrent recommender networks. In: WSDM (2017)
11. Zhang, S., Yao, L., Sun, A., Tay, Y.: Deep learning based recommender system: a survey and new perspectives. ACM Comput. Surv. (CSUR) **52**, 1–38 (2019)

12. Li, S., Kawale, J., Fu, Y.: Deep collaborative filtering via marginalized denoising auto-encoder. In: CIKM (2015)
13. Perozzi, B., Al-Rfou, R., Skiena, S.: Deepwalk: online learning of social representations. In: KDD (2014)
14. Shi, C., Hu, B., Zhao, W.X., Yu, P.S.: Heterogeneous information network embedding for recommendation. In: TKDE (2018)
15. Wang, C., Zhou, T., Chen, C., Tianlei, H., Chen, G.: Camo: a collaborative ranking method for content based recommendation. In: AAAI (2019)
16. Elkahky, A.M., Song, Y., He, X.: A multi-view deep learning approach for cross domain user modeling in recommendation systems. In: WWW (2015)
17. Wu, Y., Yao, Y., Xu, F., Tong, H., Lu, J.: Tag2word: using tags to generate words for content based tag recommendation. In: CIKM (2016)
18. Fu, W., Peng, Z., Wang, S., Xu, Y., Li, J.: Deeply fusing reviews and contents for cold start users in cross-domain recommendation systems. In: AAAI (2019)
19. Song, T., Peng, Z., Wang, S., Fu, W., Hong, X., Yu, P.S.: Review-based cross-domain recommendation through joint tensor factorization. In: Candan, S., Chen, L., Pedersen, T.B., Chang, L., Hua, W. (eds.) DASFAA 2017. LNCS, vol. 10177, pp. 525–540. Springer, Cham (2017). https://doi.org/10.1007/978-3-319-55753-3_33
20. Cai, H., Zheng, V.W., Chang, K.C.-C.: A comprehensive survey of graph embedding: problems, techniques, and applications. In: TKDE (2018)
21. Grover, A., Leskovec, J.: node2vec: scalable feature learning for networks. In: KDD (2016)
22. Kim, T., Cha, M., Kim, H., Lee, J.K., Kim, J.: Learning to discover cross-domain relationships with generative adversarial networks. In: ICML (2017)
23. Hu, B., Fang, Y., Shi, C.: Adversarial learning on heterogeneous information networks. In: SIGKDD (2019)
24. Su, Y., Zhang, X., Wang, S., Fang, B., Zhang, T., Yu, P.S.: Understanding information diffusion via heterogeneous information network embeddings. In: Li, G., Yang, J., Gama, J., Natwichai, J., Tong, Y. (eds.) DASFAA 2019. LNCS, vol. 11446, pp. 501–516. Springer, Cham (2019). https://doi.org/10.1007/978-3-030-18576-3_30
25. Le, Q., Mikolov, T.: Distributed representations of sentences and documents. In: ICML (2014)
26. Wang, H., et al.: GraphGAN: graph representation learning with generative adversarial nets. In: AAAI (2018)
27. Wang, J., et al.: IRGAN: a minimax game for unifying generative and discriminative information retrieval models. In: SIGIR (2017)
28. Mnih, A., Salakhutdinov, R.R.: Probabilistic matrix factorization. In: NIPS (2008)

Hierarchical Variational Attention for Sequential Recommendation

Jing Zhao[1], Pengpeng Zhao[1(✉)], Yanchi Liu[2], Victor S. Sheng[3], Zhixu Li[1],
and Lei Zhao[1]

[1] Institute of Artificial Intelligence, School of Computer Science and Technology,
Soochow University, Suzhou, China
`jzhao1@stu.suda.edu.cn`, {`ppzhao,zhixuli,zhaol`}`@suda.edu.cn`
[2] NEC Labs America, Princeton, USA
`yanchi@nec-labs.com`
[3] Department of Computer Science, Texas Tech University, Lubbock, USA
`victor.sheng@ttu.edu`

Abstract. Attention mechanisms have been successfully applied in many fields, including sequential recommendation. Existing recommendation methods often use the deterministic attention network to consider latent user preferences as fixed points in low-dimensional spaces. However, the fixed-point representation is not sufficient to characterize the uncertainty of user preferences that prevails in recommender systems. In this paper, we propose a new **H**ierarchical **V**ariational **A**ttention **M**odel (HVAM), which employs variational inference to model the uncertainty in sequential recommendation. Specifically, the attention vector is represented as density by imposing a Gaussian distribution rather than a fixed point in the latent feature space. The variance of the attention vector measures the uncertainty associated with the user's preference representation. Furthermore, the user's long-term and short-term preferences are captured through a hierarchical variational attention network. Finally, we evaluate the proposed model HVAM using two public real-world datasets. The experimental results demonstrate the superior performance of our model comparing to the state-of-the-art methods for sequential recommendation.

Keywords: Recommender systems · Variational attention · User preferences

1 Introduction

Recommender systems have played an increasingly significant role in our daily life, especially in social media sites and e-commerce. Due to the inherent dynamics and uncertainty of user preferences and tastes, sequential recommendation has become an attractive topic in recommender systems. And a lot of efforts have been made to recommend the next item that the user might like according to his/her past interaction sequences.

© Springer Nature Switzerland AG 2020
Y. Nah et al. (Eds.): DASFAA 2020, LNCS 12114, pp. 523–539, 2020.
https://doi.org/10.1007/978-3-030-59419-0_32

Early approaches often use separate models to capture user's long-term and short-term preferences respectively and finally integrate them [5,6,13]. For example, Rendel et al. [13] proposed a method which contains both the Markov chain and the matrix factorization model, and then linearly combined them for the next basket recommendation. However, it is not enough to learn a static vector for each user to capture his/her long-term preferences. Furthermore, since fixing the weights of the different components linearly, these approaches have limited abilities to capture high-level user-item interactions.

Recently, attention mechanisms have been used to obtain user preferences and item features in recommender systems [3,10,21]. For example, Chen et al. [3] used the attention network to incorporate its components (frames or regions) to obtain representations of multimedia objects (video, text, or image), while similar attention mechanisms are employed to merge the interacted items to catch a user representation for recommendations. Ying et al. [21] proposed a two-layer hierarchical attention network to model the user's long-term and short-term preferences.

Despite their successes, the above models employ a deterministic attention network, which lacks the ability to model the uncertainty of user preferences. In recommender systems, user preferences may produce significant uncertainty for the following reasons: (1) Users may have rich and varied interests by nature. (2) Users may be affected by their surrounding environment. For example, user u dislikes science fiction films, but her boyfriend influences her, and then she slowly falls in love with science fiction films. (3) The sparseness of user data makes the depiction of user preferences full of uncertainty. Ying et al. [21] modeled the attention vector as a point in the low-dimensional feature space to represent the user preferences. We argue that the attention vector represented by this approach is not sufficient to express the uncertainty of user preferences. Because the attention vector (i.e., the user representation) is confined to a certain point in the low-dimensional space, there is no constraint of the error term, which may result in inaccurate recommendation results.

To address these problems, we put forward a novel **H**ierarchical **V**ariational **A**ttention **M**odel (HVAM) for sequential recommendation, which employs variational inference to model the attention vector as a random variable by imposing a probability distribution. Compared with the conventional deterministic counterpart, the stochastic units incorporated by HVAM allow multi-modal attention distributions. Compared with the point vectors, it is promising to serve the attention vector (i.e., user representation) with Gaussian distributions (i.e., mean and variance) to represent the uncertainty. By applying a Gaussian distribution to denote the attention vector, the mean and the variance need to retain different attributes to make such representations of great value. Specifically, the mean vector should reflect the position of the attention vector in the low-dimensional feature space, and the variance term should contain its uncertainty [23]. Moreover, we model the user's long- and short-term preferences through a nonlinear hierarchical variational attention network to obtain a mixed user representation. In this way, our model not only captures the user's long- and short-term preferences simultaneously but also has the sufficient ability to model the uncertainty of user preferences.

To summarize, our contributions are as follows:

- The proposed HVAM introduces a variational attention network, which has an excellent ability to capture the uncertainty of user preferences.
- To obtain high-level mixed user representations, we integrate the user's long- and short-term preferences by utilizing a hierarchical structure.
- We evaluate our HVAM model on two public real-world datasets. Our experimental results reveal that HVAM exceeds various baselines by a significant margin.

2   Related Work

In the context of collaborative filtering, researchers have made great efforts to capture temporal dynamics in past user behaviors. For example, Rendel et al. [13] proposed a model, which mixes matrix factorization and Markov chains to capture the user general interests and sequential patterns. The model takes into account personalized first-order transition probabilities between items, which are modeled to decompose the underlying tensor by user embeddings and item embeddings. Following this direction, researchers adopted different methods to extract user general interests (long-term preferences) and sequential patterns (short-term preferences) [6,17]. Wang et al. [6] integrated the similarity of the factored items with a Markov chain to capture the user's long- and short-term preferences. Feng et al. [5] and Chen et al. [4] exploited metric embedding to embed items into fixed points in the low-dimensional space for next new POI recommendation and playlist prediction. However, the above methods fix the weight of different components, so that it is difficult to model high-order relationships.

Recently, deep learning has attracted considerable attention and achieved significant progress. Sedhain et al. [15] exploited autoencoder to encode user preferences, which is also a common choice in the recommender systems. In addition, Recurrent Neural Networks (RNNs) have been widely applied to process sequential data [7,19,22]. Hidasi et al. [7] proposed a recurrent neural network model based on Gated Recurrent Units (GRUs) to predict the next item in a user session by utilizing the user's interaction histories. Moreover, hybrid approaches integrating deep learning and latent variable modeling have also attracted attention. For example, Sachdeva et al. [14] combined recurrent neural networks and variational autoencoders to model the temporal dynamics of user preferences upon different perspectives. Chatzis et al. [2] proposed a Bayesian version of GRU, which treated hidden units as random latent variables with some prior distribution, and inferred the corresponding posteriors for the recommendation. They all have demonstrated the potential for learning valid non-linear representations of user-item interactions. However, the items in a session do not strictly conform to the time order under various real-world circumstances, e.g., transactions in online shopping. RNN assumes that the change of temporal-dependent is monotonic, which means the current item or hidden state plays an increasingly significant role compared with the previous one. This assumption weakens the generating of the short-term interests of users.

To address the above problems, attention mechanisms have recently been studied thanks to the capability of capturing user preferences. Li et al. [9] proposed a hybrid model, which uses an attention mechanism to highlight the user's primary purpose in the current session. Xiao et al. [18] utilized attentional pooling networks to obtain high-order feature interaction automatically. Ying et al. [21] developed a two-layer attention network to get hybrid user representation. However, they all utilize the deterministic neural networks and hold attention vectors as point vectors in a latent low-dimensional continuous space. Our model addresses this problem by modeling the attention vectors as random variables through a Gaussian distribution. Therefore, the attention vector is represented by a density instead of a fixed point in the latent feature space. The variance of the Gaussian distribution of attention vector measures the uncertainty associated with the users' preference representation. Moreover, our model can well capture dynamic user general interests and sequential behaviors by the hierarchical variational attention architecture.

3   Hierarchical Variational Attention Model (HVAM)

In this section, we first define our next item recommendation problem and then review the variational inference (VI). Next, we display the overall layered architecture of HVAM and show how each layer works. Finally, we will introduce the optimization objective of HVAM.

3.1   Problem Definition

We denote a set of users as $\mathcal{U} = \{u_1, u_2, \ldots, u_M\}$ and an item set as $\mathcal{X} = \{x_1, x_2, \ldots, x_N\}$, where M and N represent the amount of users and items, respectively. We model implicit and sequential user-item feedback data. For each user $u \in \mathcal{U}$, the user u's sequential interactions (or transactions) are denoted as $S^u = \{S_1^u, S_2^u, \ldots, S_T^u\}$, where T represents the total time steps. $S_t^u \in \mathcal{X}$ is the t-th item set following the timestamps. For a fixed time t, the item set $S_{short}^u = S_t^u$ can express user u's short-term preferences, while the set of items before the timestamp t are denoted as $S_{long}^u = S_1^u \cup S_2^u \cup \ldots \cup S_{t-1}^u$, which can reflect user u's long-term preferences. Formally, given a user and his/her sequential interactions S^u, our main purpose is to recommend the next item the user might like through mining S^u.

3.2   Variational Inference

In this subsection, we briefly introduce variational inference (VI), an approach of approximating the probability density from machine learning. Many issues have applied VI (e.g., large-scale document analysis, computational neuroscience, and computer vision). Variational inference is an alternative method for Markov

Chain Monte Carlo (MCMC) sampling, which is commonly applied to approximate the posterior density in Bayesian models. Fortunately, the variational inference is often easier and faster to extend to large datasets than MCMC.

Given the observed variable, approximating the conditional density of latent variables is the primary purpose of variational inference. The main idea behind variational inference is to use optimization. Assume that a set of data points $Y = \{y_1, y_2, \ldots, y_d\}$, $P(Y)$ represents the true distribution of the data points Y. We can obtain Eq. 1 from Bayesian properties, where Z is the latent variable.

$$P(Y) = \frac{P(Y, Z)}{P(Z|Y)}. \tag{1}$$

However, the posterior distribution $P(Z|Y)$ is usually complicated and challenging to solve, so we consider a relatively simple distribution $Q(Z|Y)$ to approximate the posterior distribution. Take the logarithm of both sides of the Eq. 1 and introduce $Q(Z|Y)$ to the right of the equation,

$$\log P(Y) = \log P(Y, Z) - \log P(Z|Y). \tag{2}$$

$$\log P(Y) = \log \frac{P(Y, Z)}{Q(Z|Y)} - \log \frac{P(Z|Y)}{Q(Z|Y)}. \tag{3}$$

Then, solve the expected value of Eq. 3 under $Q(Z|Y)$,

$$\begin{aligned}
\int Q(Z|Y) \cdot \log P(Y) \mathrm{d}Z &= \int Q(Z|Y) \cdot \log \frac{P(Y, Z)}{Q(Z|Y)} \mathrm{d}Z \\
&- \int Q(Z|Y) \cdot \log \frac{P(Z|Y)}{Q(Z|Y)} \mathrm{d}Z.
\end{aligned} \tag{4}$$

The $P(Y)$ at the left side of the Eq. 4 is independent of Z, and the integral of $Q(Z|Y)$ for Z is 1; the extension of the right end of the equation yields:

$$\begin{aligned}
\log P(Y) &= \int Q(Z|Y) \cdot \log \frac{P(Y|Z) \cdot P(Z)}{Q(Z|Y)} \mathrm{d}Z + \int Q(Z|Y) \cdot \log \frac{Q(Z|Y)}{P(Z|Y)} \mathrm{d}Z \\
&= \int Q(Z|Y) \cdot \log P(Y|Z) \mathrm{d}Z - \int Q(Z|Y) \cdot \log \frac{Q(Z|Y)}{P(Z)} \mathrm{d}Z \\
&+ KL[Q(Z|Y)||P(Z|Y)] \\
&= E_{Z \sim Q}[\log P(Y|Z)] - KL[Q(Z|Y)||P(Z)] + KL[Q(Z|Y)||P(Z|Y)]
\end{aligned} \tag{5}$$

The Eq. 5 is very important for variational inference. We can observe that the $\log P(Y)$ is converted to the sum of ELBO (Evidence Lower Bound Objective) and KL divergence (Kullback-Leibler Divergence). We don't know the true distribution of the sample Y, but the objective truth does not change. In other words, $P(Y)$ and $\log P(Y)$ are unknown constants. The first two terms on the right side of the equation are called ELBO, which is a functional (i.e., a function on $Q(Z|Y)$). The last KL divergence on the right-hand side of the equation is non-negative, so the upper bound of ELBO is $\log P(Y)$. Minimizing the KL

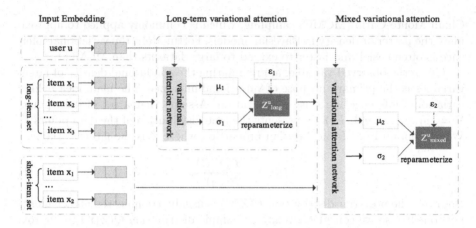

Fig. 1. The architecture of the hierarchical variational attention model.

divergence is our primary purpose, but $\log P(Z|Y)$ is challenging to know. Fortunately, the KL divergence and ELBO in the equation are in a reciprocal relationship. And then, minimizing the KL divergence is equivalent to maximizing ELBO. Thus, the inference problem is turned into an optimization problem by variational inference.

3.3 The Overall Architecture

Figure 1 illustrates the overall architecture of the HVAM. The HVAM model is based on a hierarchical variational attention network. Moreover, we model the attention vectors as random variables by imposing a probabilistic distribution to take into account the uncertainty of user preferences.

The core idea of HVAM is to obtain a valid mixed user expression so that when the user interacts with distinct candidate items, the user's interests can be more precisely characterized. The model HVAM is made up of three parts, input embedding, long-term variational attention, and mixed variational attention. In more detail, the bottom input embedding layer converts user sparse vectors and item sparse vectors from $\mathcal{U}$ and $\mathcal{X}$ respectively into low-dimensional dense vectors. Then, user embedding and item embedding from S_{long}^u are fed into the first variational attention network to acquire the mean and the variance of the attention vector. We use reparameterization techniques to get the user's long-term preference representation before time step t. To further integrate short-term preferences, the final mixed user representation combines long-term user representation with item embedding from the short-term item set S_{short}^u, where the mean and the variance are learned through another variational attention network. Next, we will present each component of the model HVAM.

3.4 Input Embedding

As shown in Fig. 1, the bottom input includes both user sparse vectors and item sparse vectors from the long-term item set S_{long}^u and the short-term item set S_{short}^u respectively. However, similar to discrete word symbols in natural language processing, one-hot encoded vectors (i.e., sparse vectors) are always high-dimensional and sparse. Simultaneously, they have minimal representation capability. This approach is not computationally efficient in a big dataset. Thence, we first construct two consecutive low-dimensional embeddings of users and items using two separate fully connected layers of weight matrices $U \in \mathbb{R}^{M \times K}$ and $X \in \mathbb{R}^{N \times K}$, respectively, where M and N represent the number of users and items, and K is the embedded dimension. Here, we only provide a one-hot representation of the user (or item) to the fully connected layer network. Then the network outputs corresponding embedding for the user i (or item j), represented by the $u_i \in \mathbb{R}^K$, i.e., the i-th row of U (or $x_j \in \mathbb{R}^K$, the j-th row of X).

3.5 Long-Term Variational Attention

Ideally, we believe a good recommender should have the ability to understand users' long-term preferences (e.g., what items users usually like), as well as account for exploring the short-term preferences (e.g., buying one item leads to wanting another) for recommendation [17]. Our model can well capture both users' general tastes and sequential behaviors. For each user, we argue that learning a static expression does not well reflect the evolution of long-term user preferences. Moreover, we hold the opinion that distinct items have distinct effects upon the identical user, and the identical item may produce distinct influences on the distinct users. The first half of the sentence is very well understood. The latter half of the sentence can be interpreted as user x buys an iphone11 for herself due to her taste, while user y buys iphone11 as a gift to her boyfriend or parents.

Attention mechanisms have been well employed in numerous fields, such as machine translation [11], document classification [20], etc. They have the intuition that people only focus on the essential parts of the target. In order to satisfy the above discussion, we will introduce the attention mechanism as well. We first need to get a deterministic attention vector, and then get the variational attention vector based on it. Specifically, the attention mechanism computes a probabilistic distribution by

$$v_{1i} = \delta(W_1 x_i + b_1), \tag{6}$$

$$\alpha_i = \frac{exp(u^T v_{1i})}{\sum_{m \in S_{long}^u} exp(u^T v_{1m})}, \tag{7}$$

where x_i represents the dense embedding vector of item $i \in S_{long}^u$, $W_1 \in \mathbb{R}^{K \times K}$ and $b_1 \in \mathbb{R}^{K \times 1}$ represent the weight and the bias of the model, respectively. $\delta(.)$ is the sigmoid function, and we utilize it to enhance nonlinear capability. We first feed each item x_i's ($i \in S_{long}^u$) dense low-dimensional embedding through a multi-layer perceptron (MLP) to acquire a pre-normalized score v_{1i}. Then, we use the user embedding u as the context vector to achieve the goal of personality (i.e., assigning different weights of same items to different users) and apply the softmax function to calculate the normalized similarity score α_i between u and v_{1i}. At last, the item embedding from S_{long}^u is summed by the attention scores to obtain the deterministic attention vector:

$$Z_{det1} = \sum_{i \in S_{long}^u} \alpha_i x_i. \tag{8}$$

As we discussed above, we argue that deterministic attention vectors are not sufficient to characterize the uncertainty of user preferences. Therefore, we introduce the variational attention vector Z_{long}^u. We propose two plausible prior distributions for Z_{long}^u. First, the simplest prior perhaps is the standard normal:

$$P_1(Z_{long}^u) = \mathcal{N}(0; I). \tag{9}$$

The central intuition is that standard normal distribution variables can generate even complex dependencies. Second, we observe that the attention vector has to be in the convex hull of the latent representation of the original input [1]. We impose a prior whose mean is the average of the item embeddings in the user's long-term item set, i.e., $\overline{X}$, making the prior non-informative,

$$P_2(Z_{long}^u) = \mathcal{N}(\overline{X}; I). \tag{10}$$

Inspired by variational inference, we model the posterior of $Q_1(Z_{long}^u | u, S_{long}^u)$ as a normal distribution $\mathcal{N}(\mu_1; \sigma_1^2)$, where the parameters μ_1 and σ_1^2 are obtained by a neural network, u represents a given user's embedding vector, and S_{long}^u represents the user's long-term item sets. For the mean μ_1, we apply an identity conversion, i.e., $\mu_1 = Z_{det1}$. The identity conversion is significant because it retains the spirit of "attention". As for the σ_1^2, we first convert Z_{det1} by a neural layer with tanh activation. Then, the resulting vector is linearly transformed, and executing the exp activation function is to ensure that the value is a positive number.

We can obtain the unbiased estimation of ELBO by sampling $Z_{long}^u \sim Q_1(Z_{long}^u | u, S_{long}^u)$ and optimize it by stochastic gradient descent. However, depending on the parameters μ_1 and σ_1^2, the sampling is an indeterminate function and is not differentiable. The solution called the "reparameterization trick" in [8] is to move the sampling to an input layer. In other words, instead of using the standard sampling method, we reparameterize Z_{long}^u to a function of μ_1 and σ_1 as follows:

$$Z_{long}^u = \mu_1 + \sigma_1 \cdot \varepsilon_1, \tag{11}$$

where ε_1 is a standard Gaussian variable that plays the role of introducing noises. In summary, we are able to smoothly acquire and optimize Z_{long}^u, which represents the user's long-term preferences.

3.6   Mixed Variational Attention

In the previous subsection, we have discussed the user's long-term preferences. However, the user's sequential behaviors, i.e., the short-term preferences, cannot be ignored. Short-term preferences are essential for predicting the next item. Nonetheless, many previous methods did not respond well to the impact of item properties on the next item prediction, which limits the performance of the model. Similar to modeling user long-term preferences, we also seek attention mechanism to obtain a specific attention vector and then get a corresponding variational attention vector. It first computes the importance of each item in the short-term item set S_{short}^u of a given user, and then aggregates the embedding of these items and the long-term preference representation Z_{long}^u to form the high-level user preference representation. Specifically,

$$v_{2j} = \delta(W_2 x_j + b_2),   \tag{12}$$

$$\beta_j = \frac{exp(u^T v_{2j})}{\sum_{n \in S_{long}^u \cup \{0\}} exp(u^T v_{2n})},   \tag{13}$$

where x_j represents the dense embedding vector of item $j \in S_{short}^u$ when $j > 0$ and $x_j = Z_{long}^u$ when $j = 0$, $W_2 \in \mathbb{R}^{K \times K}$ and $b_2 \in \mathbb{R}^{K \times 1}$ represent the weight and biase of the model, respectively. Similarly, after obtaining attention scores β_j, the item embedding from S_{short}^u and the long-term user representation Z_{long}^u is summed by the attention scores to obtain the mixed deterministic attention vector:

$$Z_{det2} = \beta_0 Z_{long}^u + \sum_{j \in S_{short}^u} \beta_j x_j,   \tag{14}$$

where β_0 is the weight of long-term user representation Z_{long}^u. In the same measure, we propose two plausible prior distributions for Z_{mixed}^u, i.e., $P_1(Z_{mixed}^u) = \mathcal{N}(0; I)$ and $P_2(Z_{mixed}^u) = \mathcal{N}(\overline{X}_2; I)$, where $\overline{X}_2$ represents the average of the item embeddings in the user's short-term item set and the long-term user representation Z_{long}^u. We model the posterior of $Q_2(Z_{mixed}^u | u, S_{short}^u, Z_{long}^u))$ as a normal distribution $\mathcal{N}(\mu_2; \sigma_2^2)$, where u represents a given user's embedding vector, and S_{short}^u represents the user's short-term item sets. Then, we use the method similar to the above Sect. 3.5 to get the parameter μ_2 and σ_2^2. Finally, we sample Gaussian noise $\varepsilon_2 \sim \mathcal{N}(0; I)$, and reparameterize Z_{mixed}^u (i.e., user mixed preference representation) to a function of μ_2 and σ_2 as follows:

$$Z_{mixed}^u = \mu_2 + \sigma_2 \cdot \varepsilon_2.   \tag{15}$$

To summarize, Z_{mixed}^u not only considers the dynamic characteristics in long- and short-term preferences but also distinguishes the items' contributions used to predict the next item. Moreover, modeling the attention vector as a random variable using a variational attention network, the attention vector can be represented by a density instead of a fixed point in latent feature space. Therefore, the variance of the Gaussian distribution of attention vector measures uncertainty associated to the users' preference representations.

3.7   Training Objective

Given the user's mixed preference representation Z^u_{mixed}, we use the traditional latent factor model to obtain his/her preferences for the candidate item set $\mathcal{X}$.

$$Y_u = Z^u_{mixed} \cdot X, \qquad (16)$$

where X represents the dense embedding of candidate item sets, $Y_u = Y_u(u, S^u_{long}, S^u_{short})$, Y_u is a function about u, S^u_{long} and S^u_{short}. Following Eq. 5, the overall training objective of our model HVAM with both the long-term variational attention vector Z^u_{long} and the mixed variational attention vector Z^u_{mixed} is to minimize:

$$\begin{aligned}
\mathcal{L}(\Theta) = &-E_{Z^u_{long}\sim Q_1, Z^u_{mixed}\sim Q_2}[\log P(Y_u|Z^u_{long}, Z^u_{mixed})] \\
&+ \lambda_{KL}(KL[Q_1(Z^u_{long}|u, S^u_{long})||P(Z^u_{long})] \\
&+ KL[Q_2(Z^u_{mixed}|u, S^u_{short}, Z^u_{long})||P(Z^u_{mixed})])
\end{aligned} \qquad (17)$$

where Θ denotes the set of model parameters, and λ_{KL} represents the annealing factor for regularising KL divergence. Specifically, the calculation of KL divergence is as follows:

$$KL[Q_1(Z^u_{long}|u, S^u_{long})||P(Z^u_{long})] = \frac{1}{2}\sum_k(-\log \sigma^2_{1k} + \mu^2_{1k} + \sigma^2_{1k} - 1), \qquad (18)$$

$$\begin{aligned}
KL[Q_2(Z^u_{mixed}|u, &S^u_{short}, Z^u_{long})||P(Z^u_{mixed})] \\
&= \frac{1}{2}\sum_k(-\log \sigma^2_{2k} + \mu^2_{2k} + \sigma^2_{2k} - 1),
\end{aligned} \qquad (19)$$

where k denotes the dimension of mean and variance. The primary purpose of our model HVAM is to obtain a user's preference representation and provide a sorted list of candidate items for the user instead of refactoring the input. Therefore, when optimizing the first term of ELBO, we utilize a pairwise ranking loss objective function according to the BPR optimization criterion [12].

4   Experiments

In this section, to evaluate our model HVAM, we will conduct experiments on two public real-world datasets. Firstly, we briefly describe our experimental settings, including datasets, evaluation metrics, and baselines. Next, we will evaluate the effectiveness of our model compared to baselines. Finally, we separately assess the impact of model components and hyper-parameters for model performance.

4.1   Experimental Settings

Datasets. We conduct our experiments on the Foursquare[1] dataset and the Gowalla[2] dataset. Foursquare and Gowalla are two widely used in location-based social networking sites (LBSNs), which contain many user-POI check-ins information. For the Foursquare dataset, we select check-ins records of users whose home is at California, collected from January 2010 to February 2011. For the Gowalla dataset, we focus on the data generated in the last seven months. Similar to [21], we generate a transaction by splitting user actions in one day on two datasets. Then, we first discard items that have been interacted by less than twenty users. Next, we filter out all transactions with fewer than three items and users with less than three transactions. After processing, Table 1 summarizes the statistics of the datasets.

Table 1. Datasets statistics.

Dataset	Foursquare	Gowalla
#user	12,565	5,623
#item	26,159	11,364
avg.transaction length	3.77	4.63
sparsity	99.85%	99.71%

Baselines. We compare our model HVAM with following seven state-of-the-art baseline models.

- **POP:** This method is simple where users are recommended the most popular (the largest number of interactions in the training set) items.
- **BPR** [12]: It is a traditional latent factor CF (collaborative filtering) model. To model the relative preference of users, BPR designs a pair-wise optimization method.
- **FPMC** [13]: A hybrid model for next-basket recommendation, which combines Markov chain and matrix factorization to capture users' general preferences and sequential behaviors.
- **GRU4REC** [7]: An RNN-based model for session-based recommendation, which contains GRUs and utilizes session-parallel minibatches as well as a pair-wise loss function for training.
- **HRM** [17]: This method generates a hierarchical user representation to mine sequential behaviors and general tastes. We employ the max pooling as the aggregation operation, because this works best.
- **SHAN** [21]: This method employs two layered attention networks to model user's general interests and sequential behaviors, respectively.

---

[1] https://sites.google.com/site/yangdingqi.
[2] http://snap.stanford.edu/data/loc-gowalla.html.

- **SVAE** [14]: A model for sequential recommendation, which combines recurrent neural networks and variational autoencoders to capture the temporal dynamics of user preferences.

Table 2. The overall performances of all models on Foursquare and Gowalla datasets.

Datasets	Models	MAP	Rec@10	Rec@20	NDCG@10	NDCG@20
Foursquare	POP	0.0150	0.0304	0.0504	0.0152	0.0203
	BPR	0.0259	0.0564	0.0996	0.0268	0.0376
	FPMC	0.0636	0.1272	0.1841	0.0710	0.0853
	GRU4REC	0.0558	0.1518	<u>0.2213</u>	0.0691	0.0869
	SVAE	0.0499	0.1000	0.1500	0.0563	0.0692
	HRM	0.0674	0.137	0.2151	0.0745	0.0942
	SHAN	<u>0.0741</u>	<u>0.1572</u>	0.2209	<u>0.0854</u>	<u>0.1025</u>
	HVAM-P_1	0.0957	0.1914	0.2642	0.1091	0.1273
	HVAM-P_2	**0.0959**	**0.1956**	**0.2690**	**0.1107**	**0.1276**
	Improv.	29.4%	24.43%	21.55%	29.63%	24.49%
Gowalla	POP	0.0487	0.0541	0.0806	0.027	0.0338
	BPR	0.0383	0.1012	0.1592	0.051	0.0655
	FPMC	0.0852	0.1599	0.2200	0.0956	0.1107
	GRU4REC	0.0872	0.1778	0.2438	0.0878	0.1044
	SVAE	<u>0.1105</u>	<u>0.2100</u>	<u>0.2650</u>	<u>0.1287</u>	<u>0.1425</u>
	HRM	0.0652	0.1414	0.2006	0.0763	0.0913
	SHAN	0.0752	0.1466	0.2065	0.0847	0.0989
	HVAM-P_1	0.1082	0.2268	0.3047	0.1271	0.1467
	HVAM-P_2	**0.1110**	**0.2351**	**0.3120**	**0.1311**	**0.1505**
	Improv.	0.45%	11.95%	17.74%	1.86%	5.61%

Metrics. To compare the performance of each model, we adopt three metrics widely used in previous works, including Mean Average Precision (MAP), Recall@N, and Normalized Discounted Cumulative Gain (NDCG@N). Average precision (AP) is called order sensitive recall. MAP is the average of the AP of all users. The second metric Recall evaluates the fraction of ground truth items that have been retrieved over the total amount of ground truth items. The NDCG is an evaluation index to measure the quality of ranking. The larger the values of three metrics, the better the performance. In our experiments, we set $N = \{10, 20\}$.

4.2   Comparison of Performance

We compare the performance of our model HVAM with seven state-of-the-art baselines regarding MAP, Recall, and NDCG on Foursquare and Gowalla datasets. Table 2 demonstrates the overall performance of all models, where bold fonts represent the best results among all methods and underlined fonts mean the best results among baselines. HVAM-P_1 means that the prior P_1 mentioned in Sect. 3 is used in both layers of variational attention networks, while HVAM-P_2 uses the prior P_2. From this table, we can get the following interesting observations.

(1) POP has the most unfavorable performance under all cases among all baselines. The reason is easy to understand. That is, POP always recommends the most popular items to users, which does not meet the needs of users well. Although the results of BPR are better than POP, it still performs poorly. All other methods outperform BPR on both datasets since BPR is a non-sequential recommendation method. These experimental results demonstrate that sequential information is helpful to enhance the real-world recommendation performance, where BPR ignores it. Moreover, our model HVAM achieves the best results in these baselines under all cases.

(2) SHAN is always better than HRM in all measurements on both datasets. This demonstrates that SHAN can generate high-level and complex nonlinear interactions for long- and short-term representations by the attention network, while HRM may lose much information through hierarchical maximum pooling operations. Moreover, our model gets better performance than SHAN. Besides, interestingly, we find that the results of SVAE are extremely unstable on two datasets. This may be because the Gowalla dataset is denser than the Foursquare dataset. Then, SVAE can tackle the uncertainty of user preferences better from the user's vast amount of interactive information through the variational autoencoder, to get better recommendation results. Furthermore, our model achieves better results than SVAE regardless of the dataset is sparse or dense. This demonstrates that the hierarchical variational attention network can not only better characterize the uncertainty of user preferences but also model the user's general interests and sequential patterns more accurately.

(3) Finally, it is clear to see that our model HVAM consistently outperforms all baselines, whether it uses prior P_1 or prior P_2 on all datasets with a large margin. The baselines include non-sequential methods (i.e., POP and BPR), traditional MC-based next item recommendation methods (i.e., FPMC), RNN-based method (i.e., GRU4REC), the state-of-the-art hierarchical representation methods (i.e., HRM and SHAN), and a recurrent version of the VAE (i.e., SVAE). This observation empirically verifies the superiority of our proposed HVAM. It can not only capture the uncertainty of user preferences through a hierarchical variational attention network but also better model users' hybrid representations. Moreover, we can observe that HVAM-P_2 usually performs better than HVAM-P_1. This may be because that a simplistic

prior like the standard Gaussian distribution will limit the ability of latent factors to a certain extent.

4.3 Influence of Components

To further explore the impact of the user's long- and short-term preferences, we propose two new simplified versions of the model, namely L-HVAM and S-HVAM. L-HVAM indicates that only the first layer of the hierarchical attention network employs variational attention. As a comparison, S-HVAM uses variational attention only when considering short-term user preferences. Owing to the space constraints, We only demonstrate the experimental results of the MAP, Rec@20, and NDCG@20.

Table 3. Influence of long- and short-term preferences.

Foursquare	HVAM-P_1			HVAM-P_2		
Metrics	MAP	Rec@20	NDCG@20	MAP	Rec@20	NDCG@20
L-HVAM	0.0911	0.2517	0.1205	0.0952	0.2603	0.1237
S-HVAM	0.0948	0.2556	0.1235	**0.0966**	0.2648	**0.1283**
HVAM	**0.0957**	**0.2642**	**0.1273**	0.0959	**0.2690**	0.1276
Gowalla	HVAM-P_1			HVAM-P_2		
Metrics	MAP	Rec@20	NDCG@20	MAP	Rec@20	NDCG@20
L-HVAM	0.1008	0.2866	0.1360	0.1009	0.2880	0.1353
S-HVAM	0.1052	0.2905	0.1413	0.1081	0.2969	0.1431
HVAM	**0.1082**	**0.3047**	**0.1467**	**0.1110**	**0.3120**	**0.1505**

Table 3 demonstrates the experimental results on the Foursquare dataset and the Gowalla dataset, where bold fonts indicate the best results. We can observe that although L-HVAM achieves relatively worse results, it still performs better than the baseline methods on both datasets under most cases. This indicates that modeling long-term preferences through variational attention is effective. Variational attention can well capture the evolution of users' general tastes and model uncertainty in user preferences. The result of S-HVAM is better than L-HVAM based on two prior assumptions, which reveals that the users' sequential behaviors are more significant on the next item recommendation task. Moreover, HVAM performs better than two single conversational models under most cases on the two datasets, whether it uses prior P_1 or prior P_2. It demonstrates that adding users' long-term preferences modeled by variational attention to S-HVAM is useful to make the next item recommendation because S-HVAM just combines the user's static general tastes and sequential information. Finally, we notice that using the prior P_2 is better than using the prior P_1 on all three methods (i.e., L-HVAM, S-HVAM or HVAM), which is consistent with our above experimental results. It verified our conjecture that the simplicity of a prior would limit the ability of latent factors to express to some extent once again.

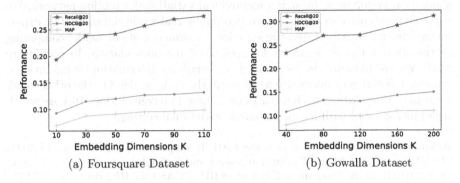

(a) Foursquare Dataset   (b) Gowalla Dataset

Fig. 2. The impact of embedding dimensions.

4.4   Influence of Hyper-parameter

HVAM has one crucial hyper-parameter, i.e., the embedding dimension K. It is not only related to the embedding dimension of users and items but also related to the MLP parameters in the variational attention network. For simplicity, the embedding dimension of the users and the items are the same. We will explore the influence of K on Foursquare and Gowalla datasets, respectively. We only show the experimental results of the MAP, Rec@20, and NDCG@20 due to the space constraints.

Figure 2 demonstrates the specific results on the Foursquare and the Gowalla datasets. We can observe that a more significant value of K can improve performance. This can be understood intuitively. That is, higher dimensions can embed better for users and items, and it will be more helpful to establish high-level factor interaction and capture the uncertainty of user preferences through the variational attention network. This situation is similar to the traditional latent factor model. In our experiments, we set the embedding dimension of the Foursquare dataset to 100 and the size of the Gowalla dataset to 200 to make a balance between the cost of calculation and the capacity of recommendations.

5   Conclusion

In this paper, we proposed a new model named Hierarchical Variational Attention Model (HVAM), which employs variational inference to model the attention vector as a random variable by imposing a Gaussian distribution. Explicitly, the attention vector is represented by a density rather than a fixed point in the latent feature space. The variance of the Gaussian distribution of the attention vector measures the uncertainty associated with the user's preference representation. On the other hand, the model captures the user's long-term and short-term preferences through a hierarchical variational attention network. We conducted extensive experiments on two public real-world datasets. The experimental results demonstrated the superior performance of our model comparing to the state-of-the-art methods for sequential recommendation. In our future work, we are planning to focus on using auxiliary information (e.g., reviews, images, and category information) to strengthen our model. On the other hand, we believe that distinct architectures (e.g., based on convolution [16]) are worth being investigated within a probabilistic variational setting.

Acknowledgments. This research was partially supported by NSFC (No. 61876117, 61876217, 61872258, 61728205), Suzhou Science and Technology Development Program (SYG201803), Open Program of Key Lab of IIP of CAS (No. IIP2019-1) and PAPD.

References

1. Bahuleyan, H., Mou, L., Vechtomova, O., Poupart, P.: Variational attention for sequence-to-sequence models. arXiv preprint arXiv:1712.08207 (2017)
2. Chatzis, S.P., Christodoulou, P., Andreou, A.S.: Recurrent latent variable networks for session-based recommendation. In: DLRS@RecSys, pp. 38–45. ACM (2017)
3. Chen, J., Zhang, H., He, X., Nie, L., Liu, W., Chua, T.S.: Attentive collaborative filtering: Multimedia recommendation with item-and component-level attention. In: SIGIR, pp. 335–344 (2017)
4. Chen, S., Moore, J.L., Turnbull, D., Joachims, T.: Playlist prediction via metric embedding. In: KDD, pp. 714–722 (2012)
5. Feng, S., Li, X., Zeng, Y., Cong, G., Chee, Y.M., Yuan, Q.: Personalized ranking metric embedding for next new POI recommendation. In: IJCAI (2015)
6. He, R., McAuley, J.: Fusing similarity models with Markov chains for sparse sequential recommendation. In: ICDM, pp. 191–200. IEEE (2016)
7. Hidasi, B., Karatzoglou, A., Baltrunas, L., Tikk, D.: Session-based recommendations with recurrent neural networks. ICLR (2016)
8. Kingma, D.P., Welling, M.: Auto-encoding variational Bayes. arXiv preprint arXiv:1312.6114 (2013)
9. Li, J., Ren, P., Chen, Z., Ren, Z., Lian, T., Ma, J.: Neural attentive session-based recommendation. In: CIKM, pp. 1419–1428 (2017)
10. Liu, J., et al.: Attention and convolution enhanced memory network for sequential recommendation. In: Li, G., Yang, J., Gama, J., Natwichai, J., Tong, Y. (eds.) DASFAA 2019. LNCS, vol. 11447, pp. 333–349. Springer, Cham (2019). https://doi.org/10.1007/978-3-030-18579-4_20
11. Luong, M.T., Pham, H., Manning, C.D.: Effective approaches to attention-based neural machine translation. arXiv preprint arXiv:1508.04025 (2015)
12. Rendle, S., Freudenthaler, C., Gantner, Z., Schmidt-Thieme, L.: BPR: Bayesian personalized ranking from implicit feedback. In: UAI, pp. 452–461. AUAI Press (2009)
13. Rendle, S., Freudenthaler, C., Schmidt-Thieme, L.: Factorizing personalized Markov chains for next-basket recommendation. In: WWW, pp. 811–820. ACM (2010)

14. Sachdeva, N., Manco, G., Ritacco, E., Pudi, V.: Sequential variational autoencoders for collaborative filtering. In: WSDM, pp. 600–608. ACM (2019)
15. Sedhain, S., Menon, A.K., Sanner, S., Xie, L.: Autorec: autoencoders meet collaborative filtering. In: WWW, pp. 111–112 (2015)
16. Tang, J., Wang, K.: Personalized top-n sequential recommendation via convolutional sequence embedding. In: WSDM, pp. 565–573. ACM (2018)
17. Wang, P., Guo, J., Lan, Y., Xu, J., Wan, S., Cheng, X.: Learning hierarchical representation model for next basket recommendation. In: SIGIR, pp. 403–412 (2015)
18. Xiao, J., Ye, H., He, X., Zhang, H., Wu, F., Chua, T.S.: Attentional factorization machines: Learning the weight of feature interactions via attention networks. arXiv preprint arXiv:1708.04617 (2017)
19. Xu, C., et al.: Recurrent convolutional neural network for sequential recommendation. In: WWW, pp. 3398–3404 (2019)
20. Yang, Z., Yang, D., Dyer, C., He, X., Smola, A., Hovy, E.: Hierarchical attention networks for document classification. In: HLT-NAACL, pp. 1480–1489 (2016)
21. Ying, H., et al.: Sequential recommender system based on hierarchical attention network. In: IJCAI (2018)
22. Zhao, P., et al.: Where to go next: a spatio-temporal gated network for next poi recommendation. In: AAAI, vol. 33, pp. 5877–5884 (2019)
23. Zhu, D., Cui, P., Wang, D., Zhu, W.: Deep variational network embedding in wasserstein space. In: KDD, pp. 2827–2836. ACM (2018)

Mutual Self Attention Recommendation with Gated Fusion Between Ratings and Reviews

Qiyao Peng[1], Hongtao Liu[1], Yang Yu[1], Hongyan Xu[1], Weidi Dai[1(✉)], and Pengfei Jiao[1,2]

[1] College of Intelligence and Computing, Tianjin University, Tianjin, China
{qypeng,htliu,yuyangyy,hongyanxu,davidy}@tju.edu.cn
[2] Center for Biosafety Research and Strategy, Tianjin University, Tianjin, China
pjiao@tju.edu.cn

Abstract. Product ratings and reviews can provide rich useful information of users and items, and are widely used in recommender systems. However, it is nontrivial to infer user preference according to the history behaviors since users always have different and complicated interests for different target items. Hence, precisely modelling the context interaction of users and target items is important in learning their representations. In this paper, we propose a unified recommendation method with mutual self attention mechanism to capture the inherent interactions between users and items from reviews and ratings. In our method, we design a review encoder based on mutual self attention to extract the semantic features of users and items from their reviews. The mutual self attention could capture the interaction information of different words between the user review and item review automatically. Another rating-based encoder is utilized to learn representations of users and items from the rating patterns. Besides, we propose a neural gated network to effectively fuse the review- and rating-based features as the final comprehensive representations of users and items for rating prediction. Extensive experiments are conducted on real world recommendation datasets to evaluate our method. The experiments show that our method can achieve better recommendation performance compared with many competitive baselines.

Keywords: Recommender system · Mutual self attention · Rating prediction

1 Introduction

Recently recommender systems have been more and more important in many e-commerce platforms such as Amazon [27] in the age of information explosion. The key of recommender systems is to learn precise representations for users and items according to their historical behaviors (e.g., user-generated ratings and reviews). There are many recommendation methods proposed based on user

© Springer Nature Switzerland AG 2020
Y. Nah et al. (Eds.): DASFAA 2020, LNCS 12114, pp. 540–556, 2020.
https://doi.org/10.1007/978-3-030-59419-0_33

ratings, and Collaborative Filtering (CF) [21] is one of the most popular technologies [15]. CF methods can project user preference and item feature into a low-dimensional space for predicting rating score. For example, PMF [15] learns the latent features of users and items based on the rating matrix data via probabilistic matrix factorization. However these rating-based methods may suffer from the sparsity problem, which would limit ability of these methods on modelling user and item [2].

To alleviate the problem, many works start to utilize the reviews to enhance the representations of users and items, since the reviews accompanying with the ratings contain rich semantic information and can reveal user preference and item features fully such as [2,4,10,13,24,27]. For example, DeepCoNN [27] and NARRE [2] utilize Convolutional Neural Network (CNN) to extract features of reviews as the representations of users and items for rating prediction, nevertheless without considering the interactions between users and items. To capture the interaction information (i.e., context-awareness), MPCN [24] adopts pointer networks to learn user and target item features from words and reviews. DAML [10] utilizes Euclidean distance to learn the interactions between users and target items from reviews and ratings.

Though these methods have gained superior performance in recommendation, it is still difficult to understand the complex rating behaviors fully due to users usually have complicated interests for different target items, and there are some issues not thoroughly studied. First, the interactions between user reviews and item reviews are not exploited fully. For example, suppose that one of User A's reviews is "this phone is of high price but with good quality.", and another review on Item B is "the mobile phone is durable and has a long battery life." The word "quality" has strong relatedness with "durable". Besides, a word may interact with multiple words, i.e., "quality" also has semantic interactions with "good", "long", "life". Existing methods are usually incapable of involving this scenario effectively, for example DAML [10] with a Euclidean distance approach is limited by the linear interaction issue. Second, in addition to the review-based features above, the features derived from the rating matrix are also important for users and items, and they are complementary with each other. However, there is no effective method that unifies the features from both ratings and reviews. Most existing methods simply conduct concatenation or add operations over them [2,10,13], which is insufficient to integrate the two features fully.

To this end, in this paper we propose a Mutual Self Attention Recommendation model (MSAR) with a gated fusion layer between ratings and reviews, which aims to learn comprehensive representations of users and items in recommendation. There are three main modules in our method, i.e., a review encoder to learn representations of users and items from reviews, a rating-based encoder to learn representations from ratings, and a gated fusion layer to combine the two kinds of representations together. In our review encoder, for a given user-item pair we first use Convolutional Neural Networks to extract the semantic features of words in user/item review document respectively, and then propose a mutual self attention layer [25] to capture the underlying interaction information automatically between words in user and item review documents. At last, we can obtain the contextual review-based representations of the user and the target item via aggregating the interaction features of words of reviews. In our rating-based encoder, we derive the rating features from the user and item IDs, which can reveal the identity characteristic and rating pattern in rating matrix of users and items. In addition, we propose to apply a neural gated network to effectively fuse the review-based and rating-based features to generate the final comprehensive representations for the user and the target item. Afterwards, our model feeds the representations of user and item into the Latent Factor Model (LFM) [8] to predict the rating that the user would score towards the item.

The main contributions of this paper can be summarized as follows:

(1) We propose the mutual self-attention mechanism to learn the contextual representations for user and item by capturing the words interactions between user and item reviews.
(2) The neural gated network is applied to fuse the rating-based feature and the review-based feature, which can capture the complementary information of the two kinds features.
(3) Extensive experiments are conducted on four real-world recommendation datasets to evaluate the performance of our method, the results show the effectiveness of our model MSAR.

2 Related Works

Traditional collaborative filtering methods (CF) [16,20,21] are popular rating-based technologies, which are based on Matrix Factorization (MF) to learn latent features of users and items from the rating matrix [9,15,19]. For example, Mnih et al. [15] extended MF to Probabilistic Matrix Factorization (PMF), and Koren et al. [8] introduced bias information of users and items into MF. Though collaborative filtering recommender systems have achieved favourable performance, nevertheless suffering from data sparsity problem.

With the increase of interaction between users and system platforms, some auxiliary information related to users and items have been utilized for relieving the above problem, especially, reviews have been introduced to improve the performance of recommender systems. Early review-based works adopted topic modelling technology [17] incorporating with reviews to generate the latent factors for users and items. For example, Mcauley et al. [14] used LDA model to extract the semantic topics of reviews. Yang et al. [1] proposed TopicMF, which utilized MF and topic model respectively for learning latent features of users and items from ratings and reviews.

Recently, deep neural networks which have a strong ability to representation learning have yielded an immense success in review-based recommender systems. More and more works apply neural network to exploit review information for improving the recommender systems performance [2,4,5,10–13,22,24,26]. Deep-CoNN [27] combined all the reviews of users or items and then utilized Convolutional Neural Network (CNN) to learn the representations from reviews.

D-Attn [22] utilized local attention to provide user preferences or item properties and global attention for helping CNNs focus on the semantic meaning of
the whole review. ANR [4] proposed an aspect-based co-attention network to
extract different aspects information for estimating aspect-level user and item
importance from the review documents. However, DeepCoNN, D-Attn and ANR
ignore the rating-based features which can reflect rating patterns of users and
items. NARRE [2] considered the helpfulness of reviews and introduced review-
level attention to estimate the importance of different reviews for a user or an
item. Lu et al. [13] employed an attention-based GRU method that extracts
semantic features of users and items from ratings and reviews by optimizing
matrix factorization. Though have achieved improvement on recommender system, these method still learn the latent factors for users and items from user
reviews and item reviews respectively, which fail to capture the words interactions between user and item reviews.

A very recent method, DAML [10] exploited an attention layer with the
Euclidean distance to study the correlation between user and item reviews, which
achieved state-of-the-art performance in review-based recommendations. However, DAML only takes word interaction between user and item review documents into account via a linear method, which is not capable of fully capturing
effective interaction information. Besides, most existing methods fuse the review-
based feature and rating-based feature via simple add/concatenation operation
which would not learn the deep interaction features between them. In this paper,
our proposed method MSAR aims to learn comprehensive representations of
users and items from reviews and ratings in recommendation.

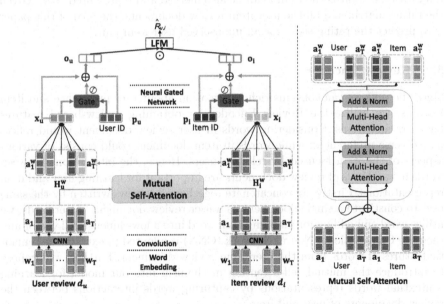

Fig. 1. Overview of our method MSAR, which learns the representation of user u or
item i from reviews and ratings.

3   Proposed Method

In this section, we will present our proposed method Mutual Self Attention Recommendation with Gated Fusion between Ratings and Reviews (MSAR) in detail. The overview of our approach is shown in Fig. 1. There are three main modules in our method, i.e., a review encoder with convolutional neural network and mutual self-attention layer to learn representations of words for modelling user and item by capturing interactions between words, a rating-based encoder to extract rating features from user-item rating matrix, and a gated fusion layer to integrate the rating-based and review-based features into a neural gated network, which can capture the nonlinear interactions between two kinds of latent features. Then a prediction layer is devised to estimate the final rating score (i.e., $\hat{R}_{u,i}$) of the model. In the following sections, we will cover the problem definition in this paper firstly, and then describe our method in detail.

3.1   Problem Definition

Rating prediction is a fundamental problem in recommendation. Suppose that there are a User set U, Item set I, and N is the number of users, and M is the number of items. We assume that the rating matrix $\mathbf{R} \in \mathcal{R}^{N \times M}$, where the entry R_{ui} indicates the rating score of user $u \in U$ towards item $i \in I$ if the user u has given a rating to the item i. For a user u, all reviews written by u can be merged and denoted by one word sequence $d_u = [w_1, w_2, \cdots, w_T]$ (i.e., user review document) where T is the number of words in d_u. Similarly, for an item i, all reviews towards the item i can be also merged and represented as d_i. Given the rating matrix and all the user item review documents, the goal of this paper is to predict the rating score for an unobserved user-item pair.

3.2   Review Encoder

Since the reviews can contains rich information about user preference and item features, we utilize the Review Encoder for obtaining review-based features for users and items. Considering words in user review document would reflect user diverse preferences, while words in item document could consist of various aspects of the item focused by relevant users. Hence, the interactions between words in two kinds of review documents are important for learning user and item representations. Firstly, we concatenate together all reviews written by the same user to constitute a single document as a user review document. Afterwards, we utilize word embedding for mapping each word into a low-dimensional vector and then the Convolutional Neural Network (CNN) is adopted to extract the semantic features of words from the user/item review document. Further, we propose to introduce the mutual self-attention mechanism into our model for learning contextual words representations by capturing words interactions between the review documents of user and item.

Word Embedding. As shown in Fig. 1, given the review document of user u, i.e., $d_u = [w_1, w_2, \cdots, w_T]$, which is composed of T words. The word embedding layer is utilized for projecting each word w_m in d_u to a vector $\mathbf{w_m} \in \mathcal{R}^{d_w}$ with d_w as dimension. Then the review document embedding matrix $\mathbf{D} \in \mathcal{R}^{T \times d_w}$ is formed by concatenating these words in order to preserve their appearances in the document and the order of words. In the same method, we can obtain the embedding matrix of item i review document d_i.

Convolution Operation. Afterwards, we utilize Convolutional Neural Networks (CNN) to extract the semantic feature of each word from the review document embedding matrix. Suppose there are K different convolution filters, denoted as $\mathbf{F} = \{\mathbf{f_1}, \mathbf{f_2}, \cdots, \mathbf{f_K}\}$, each filter is a parameter matrix $\mathbf{f_j} \in \mathcal{R}^{s \times d_w}$, where s is the filter window size. The convolution result of x-th filter over y-th window on the whole review document embedding matrix $\mathbf{D}$ is computed as:

$$c_{xy} = \text{ReLU}(\mathbf{D_{y:y+s-1}} * \mathbf{f_x} + b_x), \tag{1}$$

where $*$ is the convolution operating, b_x is the bias and ReLU is a nonlinear activation function. Then we can obtain the features produced by the K filters, stacked into a matrix $\mathbf{H} = [\mathbf{c_1}, \cdots, \mathbf{c_K}] \in \mathcal{R}^{T \times K}$. Thus the m-th row of $\mathbf{H}$ is the semantic feature of the m-th word in the review document, denoted as $\mathbf{a_m} \in \mathcal{R}^K$. In this work, we apply two different CNNs to process the user u review document and item i review document, and the outputs of two CNN are denoted as $\mathbf{H_u}$ and $\mathbf{H_i}$ respectively.

Mutual Self-attention Layer. As stated before, for a given user-item pair, the interaction information between words in user or item review document could be important for predicting rating score. Since the informativeness of a same word in user review should be dependent on different target items and vice versa, hence it is important to model this contextual interaction between users and items. To effectively capture word interaction information of a user-item pair, we propose to adopt a mutual self-attention layer to jointly process the review documents of the user-item pair. First of all, we combine the semantic features of user and item via concatenation, (i.e., $\mathbf{A} = [\mathbf{H_u}, \mathbf{H_i}]$, $\mathbf{A} \in \mathcal{R}^{2T \times K}$). Thereafter, inspired the superior performance of self attention in capturing the interactions between words in sentences [25], we utilize multi-head self attention mechanism to learn word importance by capturing their interactions between two review documents. The representation of the i_{th} word learned by the k_{th} attention head is computed as:

$$\alpha_{i,j}^k = \frac{\exp(\mathbf{a}_i^{\mathbf{T}} \mathbf{Q}_k^w \mathbf{a}_j)}{\sum_{n=1}^{2T} \exp(\mathbf{a}_i^{\mathbf{T}} \mathbf{Q}_k^w \mathbf{a}_n)}, \tag{2}$$

$$\mathbf{h}_{i,k}^w = \mathbf{V}_k^w (\sum_{j=1}^{2T} \alpha_{i,j}^k \mathbf{a}_j), \tag{3}$$

where $\mathbf{a}_i^{\mathbf{T}}$ and $\mathbf{a}_j$ are the semantic features of i_{th} and j_{th} words, $\mathbf{Q}_k^w$ and $\mathbf{V}_k^w$ are the parameters in the k_{th} self-attention head, and $\alpha_{i,j}^k$ indicates the interaction information between the i_{th} and j_{th} words. The multi-head representation $\mathbf{h}_i^w$ of the i_{th} word is the concatenation of the representations produced by h different self-attention heads, i.e., $\mathbf{h}_i^w = [\mathbf{h}_{i,1}^w; \mathbf{h}_{i,2}^w; ...; \mathbf{h}_{i,h}^w]$. We can obtain the weighted matrix consisting of word interaction information from $\mathbf{A}$ via mutual self-attention operation. Afterwards, we derive the user and item weighted word features via splitting the whole matrix, denoted as $\mathbf{H}_u^w$ and item review weighted matrix $\mathbf{H}_i^w$ as shown in Fig. 1.

Since different words have obtained different contextual attention weights in $\mathbf{H}_u^w$ and $\mathbf{H}_i^w$, we utilize mean-pooling operation to derived the final review document features, denoted as:

$$\mathbf{x_u} = \text{Mean}(\mathbf{H}_u^w), \mathbf{x_i} = \text{Mean}(\mathbf{H}_i^w). \tag{4}$$

$\mathbf{x_u}$ and $\mathbf{x_i}$ are the final review-based representations of the user u and the item i respectively.

3.3   Rating-Based Encoder

Though the mutual learning of different review documents can provide rich word interaction information to represent user preference and item characteristics, there are some users or items with very few or no reviews. Therefore, we design a rating-based encoder to extract the user and item latent feature from the rating patterns following previous works [2,10].

Specifically, we define an ID embedding for each user and item respectively, which can describe the intrinsic characteristics of users and items. We introduce a user ID embedding matrix $\mathbf{P_u} \in \mathcal{R}^{N \times d}$ and an item ID embedding matrix $\mathbf{P_i} \in \mathcal{R}^{M \times d}$ to indicate the rating-based features, where d is the dimension of the ID embedding. Then for a user u and a given item i, a embedding look-up layer is followed:

$$\mathbf{p_u} = \mathbf{P_u}(u), \mathbf{p_i} = \mathbf{P_i}(i), \tag{5}$$

where $\mathbf{p_u}$ is the rating-based feature (i.e., ID embedding) of the user u, and $\mathbf{p_i}$ is the rating-based feature of the item i.

3.4   Gated Fusion Layer

For the user u, $\mathbf{x_u}$ and $\mathbf{p_u}$ are from two heterogeneous information which would be complementary with each other. It is necessary to integrate them together for yielding better comprehensive representations. However, most existing methods adopt simple concatenation or add operations to fuse them, which may not effectively unify the rating-based features and review-based features.

To distill more valuable information precisely from the different features, as shown in Fig. 1, we design a gated fusion layer to integrate $\mathbf{x_u}$ and $\mathbf{p_u}$. Specifically, the gate layer is computed as:

$$\mathbf{g_u} = 1 - \sigma((\mathbf{W_g}\mathbf{x_u} + \mathbf{b_g}) \odot \mathbf{p_u}), \tag{6}$$

where $\sigma(\cdot)$ is the sigmoid function, $\odot$ denotes the element-wise multiplication, $\mathbf{W_g}$ and $\mathbf{b_g}$ are parameter matrix and bias respectively.

Considering that review-based feature contains rich semantic information and rating-based feature includes user inherent characteristics, we concatenate the interaction information $\mathbf{g_u}$, the review-based information $\mathbf{x_u}$ and the rating-based information $\mathbf{p_u}$ as the final representation $\mathbf{o_u}$ for user u, which is computed as:

$$\mathbf{o_u} = \mathbf{g_u} \oplus \mathbf{x_u} \oplus \mathbf{p_u}, \tag{7}$$

where $\oplus$ is the concatenation operator, we can also obtain latent factor $\mathbf{o_i}$ for item i in a similar way.

3.5   Rating Prediction Framework

In this section, we will report the final rating prediction process. Here, we combine the user representation $\mathbf{o_u}$ and the given item representation $\mathbf{o_i}$ as follows:

$$\mathbf{b_{u,i}} = (\mathbf{o_u} \odot \mathbf{o_i}), \tag{8}$$

where $\odot$ is the element-wise product. We apply the Latent Factor Model (LFM) [8] to predict the rating that the user u would score the item i. The rating $\hat{R}_{u,i}$ is computed by:

$$\hat{R}_{u,i} = \mathbf{W_r^T}(\mathbf{b_{u,i}}) + \mathbf{b_u} + \mathbf{b_i} + \mu, \tag{9}$$

where $\mathbf{W_r^T}$ denotes the parameter matrix of the LFM model, μ denotes the global bias, $\mathbf{b_u}$ and $\mathbf{b_i}$ are the corresponding bias of user u and item i individually.

To train the model parameters of our method MSAR, we need to specify a target function to optimize. Since the task in this work is rating prediction which is a regression problem, we take the square loss function as the objective function for parameter optimization:

$$L_{sqr} = \sum_{u,i \in \Omega} (\hat{R}_{u,i} - R_{u,i})^2, \tag{10}$$

where Ω denotes the set of instances in training datasets, and $R_{u,i}$ is the ground truth rating assigned by the user u to the item i.

4   Experiments

In this section, we conduct extensive experiments over four recommendation datasets to assess the performance of our proposed approach. Firstly, the datasets and experimental settings are introduced, and then experimental results and analysis are presented.

4.1   Datasets and Experimental Settings

Datasets. We utilize four real-world used recommendation datasets, **Office Products**, **Tools Improvement**, **Digital Music** and **Video Games**, which are selected from Amazon Review[1]. All datasets consist of user-item pairs with 5-score ratings (from 1–5) and reviews. Following the preprocessing approach [6], we delete stop words that appear in the review document with a frequency higher than 0.5 and corpus-specific stop words. User and Item review documents are generated by amputating long review documents or padding short review documents, which have 300 words. The detailed characteristics of the datasets are presented in Table 1 after preprocessing.

Table 1. Statistical details of the datasets.

Datasets	# users	# items	# ratings	density (%)
Digital music	5,540	3,568	64,666	0.327
Video games	24,303	10,672	231,577	0.089
Office products	4,905	2,420	53,228	0.448
Tools improvement	16,638	10,217	134,345	0.079

We randomly split each dataset into training set (90%) and testing set (10%), and we take (10%) from training set as validation set randomly for tuning our hyperparameters in our model. In addition, at least one interaction per user/item is included in the training set, and the target reviews are excluded in validation and testing sets from training set since they are not available in real-world recommendation.

Experimental Setting. We report our experimental configure in this section, which are tuned with the validation set. The word embedding matrix can be initialized via using word vectors that have been pre-trained on large corpora, such as Glove [18], and the embedding dimension is set to 300. The dimension of user or item rating-based feature is set to 32, the number of convolution filters (i.e., K) is 100 and the window size of CNN (i.e., l) is 3. The dropout ratio is 0.5 for alleviating the overfitting problem, the learning rate of Adam [7] optimizer is set to 0.002 for optimizing our model and the weight decay is 0.0001.

Evaluation Metric. The Mean Square Error (MSE) is adopted in our model as the evaluation metric following most review-based methods [2,27], which is denoted as:

$$MSE = \frac{\sum_{u,i \in \Omega}(\hat{R}_{u,i} - R_{u,i})^2}{|\Omega|} , \qquad (11)$$

---
[1] http://jmcauley.ucsd.edu/data/amazon/.

where $R_{u,i}$ is the ground truth rating, $\hat{R}_{u,i}$ is the rating score predicted by our model and Ω indicates the set of user-item pairs in testing sets. The lower MSE indicates the better performance of the models. We report the average MSE value of 5 times experiment results in our comparison.

Baselines. To evaluate the performance of our method MSAR, we compare MSAR with the several recent competitive baseline methods including:

- **RBLT** [23]: This method proposes a rating-boosted approach which utilizes rating-boosted reviews and rating scores for rating prediction and integrates MF and topic model.
- **DeepCoNN** [27]: The model trains convolutional representations for user and item via two parallel CNNs from reviews respectively and passed the concatenated embedding into a FM model for predicting the rating score.
- **D-Attn** [22]: This model is characterized by its usage of two forms of attentions (local and global) for obtaining an interpretable embedding for user or item and capturing the precise semantic features.
- **TARMF** [13]: A novel recommendation model utilizes attention-based recurrent neural networks to extract topical information from user and item review documents.
- **NARRE** [2]: The newly proposed method that introduces neural attention mechanism to select highly-useful reviews simultaneously and completes the rating prediction task.
- **MPCN** [24]: A neural model exploits a novel pointer-based learning scheme and enables not only noise-free but also deep word-level interaction between user and item.
- **ANR** [4]: A novel aspect-based neural recommendation method proposes to utilize a co-attention mechanism in the architecture to estimate aspect-level ratings in an end-to-end fashion.
- **DAML** [10]: DAML adopts local attention layer and mutual attention layer to select informative words and relevant semantic information between user and item review text for realizing the dynamic interaction of users and items.

Besides, we compare the rating-based method **PMF** [15], which learns user and item latent features only from rating matrix without review information via probabilistic matrix factorization with Gaussian distribution. Note that there are many more other recommendation models, such as HFT [14], JMARS [5], ConvMF [6], ALFM [3]. Since these works have been exceeded by the above baselines, we did not compare our method with the these approaches.

4.2   Performance Evaluation

Table 2 reports the results of all methods over the four datasets. We compare our method MSAR with the above baselines in terms of MSE and have the following observations. First of all, it is reasonable that the review-based approaches consistently outperform than the rating-based method (i.e., PMF) over all four

Table 2. MSE comparisons of our model **MSAR** (in bold) and baseline methods. The best performance of the baselines are underlined.

Method	Digital music	Office products	Tools improvement	Video games
PMF	1.206	1.092	1.566	1.672
RBLT	0.870	0.759	0.983	1.143
DeepCoNN	1.056	0.860	1.061	1.238
D-Attn	0.911	0.825	1.043	1.145
TARMF	0.853	0.789	1.169	1.195
NARRE	0.812	0.732	0.957	1.112
MPCN	0.903	0.769	1.017	1.201
ANR	0.867	0.742	0.975	1.182
DAML	0.813	0.705	0.945	1.165
MSAR	**0.802**	**0.689**	**0.928**	**1.106**

datasets. This is because rating matrix data is sparse and reviews contain rich information of user preference and product features, which would be more useful for understanding user rating behaviors than using a single rating score. This observation is consistent with many review-based works [2]. Second, the neural network methods (i.e., DeepCoNN, D-Attn, TARMF, NARRE, MPCN, ANR, DAML, MSAR) usually perform better than topic-based model (i.e., RBLT). The reason is that neural networks have powerful representation capabilities, which allow the neural network models to learn more semantic information from reviews in the non-linear way, the topic-based modeling approach may not fully capture the deep characteristics of reviews. Furthermore, we find out that the attention-based mechanisms (e.g., D-Attn, TARMF, NARRE, MPCN, ANR) generally perform better than those without attention (e.g., DeepCoNN, RBLT). We believe that this phenomenon is due to the presences of noise words and sentences in reviews, and the attention mechanism can help models focus on the more useful reviews and words, which will help model to learn the representations of users and items more accurately.

Third, as shown in Table 2, our method MSAR achieves the best MSE score over the four datasets and outperforms than the very recent method DAML. The reason may be that DAML adopts linear Euclidean distance approach, which would be incapable of fully capturing the complicated interaction information between user reviews and item reviews. And our model utilizes mutual self-attention mechanism to learn the review-based features consisting of interaction information for user and item by capturing words interactions between user and item review documents, which could yield a better understanding of user rating behaviors. Further more, the gated fusion layer is applied in our model for integrating the rating-based feature and review-based feature, which could extract non-linear interaction information for generating more knowledge about user preference and item features.

Overall the experimental results meet our motivation in Sect. 1, and demonstrate the effectiveness of our method in learning more precise and comprehensive representations for users and items in recommendation.

4.3   Analysis of Our Method

In this section, we conduct ablation experiments to evaluate the effect of different components in our model. We mainly study effect of mutual self-attention layer, and effect of our feature fusion module.

Effect of Mutual Self-Attention Layer. In our method, the mutual self-attention layer is utilized for capturing the words interactions between user and item review documents. In this section, we conduct ablation experiments for studying the impact of mutual self-attention layer over all datasets. We design one variant, which removes the mutual self-attention layer:

- MSAR-MS: removing the mutual self-attention layer from our model, i.e., utilizing mean-pooling operation on H_u and H_i to obtain user feature x_u and item feature x_i directly.

Table 3. Comparisons between MSAR and baseline Variant.

Variant	Digital music	Office products	Tools improvement	Video games
MSAR-MS	0.818	0.701	0.938	1.112
MSAR	**0.802**	**0.689**	**0.928**	**1.106**

The comparison results are presented in Table 3. We can find that removing mutual self-attention layer (i.e., MSAR-MS) would degrade the performance of the model over all datasets. This indicates that user preference and item characteristics with interaction information generated by mutual self-attention layer are favorable for rating prediction.

Effect of Gated Fusion Layer. The gated fusion layer is an important building block for MSAR since it captures some non-linear interaction information between review-based feature and rating-based feature. In this section, we conduct ablation and replacement experiments for studying the impact of gated fusion layer over four datasets. We design two variants, which removes the gated fusion layer and uses dot product operation to replace that:

- MSAR-GF: removing the gated fusion layer from our model, and the interaction information between rating- and review-based features is excluded from the representations of user and item, i.e., $o_u = x_u \oplus p_u$, where $\oplus$ is the concatenation operation.

- MSAR-dot: applying dot product operation to structure the mutual information between review-based feature and rating-based feature instead of the gated fusion layer, i.e., $\mathbf{o_u} = (\mathbf{x_u} \odot \mathbf{p_u}) \oplus \mathbf{x_u} \oplus \mathbf{p_u}$, where $\odot$ is the dot product operation.

Table 4. Comparisons between MSAR and baseline Variants.

Variant	Digital music	Office products	Tools improvement	Video games
MSAR-GF	0.815	0.707	0.936	1.141
MSAR-dot	0.813	0.695	0.932	1.137
MSAR	**0.802**	**0.689**	**0.928**	**1.106**

The comparison results are presented in Table 4. We can find that removing gated fusion layer (i.e., MSAR-GF) and replacing that with dot product (i.e., MSAR-dot) would degrade the performance of the model over all datasets. It is worth noting that MSAR-GF performs better on all datasets than MSAR-dot. This demonstrates that our proposed gated fusion layer is superior to dot product operation. The two variant results indicate that the gated fusion layer can effectively fuse review-based feature and rating-based feature as the final representations, which are favorable for rating prediction.

4.4 HyperParameter Analysis

In this section, we conduct hyper-parameters sensitivity analysis experiments to explore the effect of hyper-parameters on the performance of the model in this paper. We mainly analyze two important hyper-parameters over four datasets: the number of convolution filters in CNN, and the dimension of rating-based feature.

Dimension of Rating-Based Feature. Since the rating-based feature is an explicit feedback of user behavior, the dimension of that is important for representing user and item. We explore the effect of the dimension of Rating-Based feature with other parameters fixed. It is noticed that the lower MSE score expresses the better model performance. As shown in the left of Fig. 2, we can find that as the dimension of rating-based feature increases, the MSE first decreases, then reaches the lowest (i.e., the bset performance) when the dimension of rating-based feature is 32, and increases afterwards. We argue that when the dimension of rating-based feature is too small, the rating-based feature may not be able to capture the inherent preference for user and potential attributes of item. When the dimension is too large, the model may suffer from overfitting problem and the computational complexity would increase. Hence we set up the optimal dimension for rating based feature is 32.

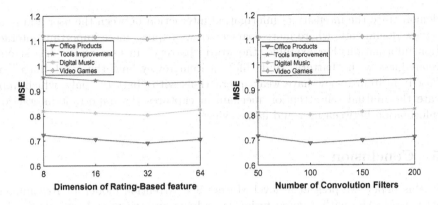

Fig. 2. The impact of hyper-parameters.

Number of Convolution Filters. Considering that the number of convolution filters might affect the ability of the convolutional neural network to extract semantic features from review documents. We analyze the influence of the Number of Convolution Filters on the model performance. From the right of Fig. 2, with the increase of the Convolution Filters Number, the MSE declines at first, and reaches the minimum value (i.e., the best performance) when the convolution filters number is 100, and then increases. When convolution filter number is too smaller, the convolution operator may not extract rich information of users and items from reviews. However, the model may suffer from overfitting problem if the number of convolution filter is too larger. Therefore we set the optimal Number of Convolution Filters to 100 regardless of different datasets.

4.5   Case Study

To indicate the effect of mutual self-attention more intuitively and explainable, we visualize the word interaction weight between user and item review for qualitative analysis.

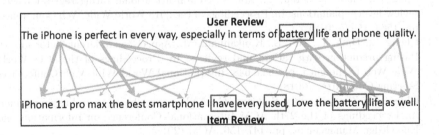

Fig. 3. Visualization of different word interaction weight.

As shown in Fig. 3, the upper review is belong to user review document and the bottom review is part of item review document. We utilize arrows to demonstrate the intensity of information interaction between the user and item review documents, where more coarser arrows express more important interaction information. For example, the word "battery" in user review has strong relatedness with "battery" and "life" in item review and the word "have" or "used" in item review have weak interactions with "battery" only, which indicate the mutual self-attention mechanism captures the important interaction information between user and item reviews.

5   Conclusion

In this paper, we present a novel Mutual Self Attention model for Recommendation (MSAR) with a gated fusion layer between ratings and reviews. In this model, we propose to utilize a mutual self attention layer to learn representations of users and items from reviews, which can automatically capture the underlying interaction information between user and item review documents. The rating-based features of users and items are derived from their ID embeddings, which can depict the rating characteristics of users and items. Besides, a fusion layer with neural gated network is introduced to merge rating- and review-based features effectively, instead of combining them simply. We conduct extensive experiments on four real-world recommendation datasets to evaluate our model, and the results demonstrate that our method can consistently outperform the existing state-of-the-art methods.

Acknowledgments. This work is surpported by the National Social Science Fund of China (15BTQ056), and the National Key R&D Program of China (2018YFC0832101).

References

1. Bao, Y., Fang, H., Zhang, J.: TopicMF: simultaneously exploiting ratings and reviews for recommendation. In: Twenty-Eighth AAAI Conference on Artificial Intelligence (2014)
2. Chen, C., Zhang, M., Liu, Y., Ma, S.: Neural attentional rating regression with review-level explanations. In: Proceedings of the 2018 World Wide Web Conference, pp. 1583–1592 (2018)
3. Cheng, Z., Ding, Y., Zhu, L., Kankanhalli, M.: Aspect-aware latent factor model: Rating prediction with ratings and reviews. In: Proceedings of the 2018 World Wide Web Conference, pp. 639–648. International World Wide Web Conferences Steering Committee (2018)
4. Chin, J.Y., Zhao, K., Joty, S., Cong, G.: ANR: aspect-based neural recommender. In: Proceedings of the 27th ACM International Conference on Information and Knowledge Management, pp. 147–156. ACM (2018)
5. Diao, Q., Qiu, M., Wu, C.Y., Smola, A.J., Jiang, J., Wang, C.: Jointly modeling aspects, ratings and sentiments for movie recommendation (JMARS). In: KDD, pp. 193–202 (2014)

6. Kim, D., Park, C., Oh, J., Lee, S., Yu, H.: Convolutional matrix factorization for document context-aware recommendation. In: RecSys, pp. 233–240. ACM (2016)
7. Kingma, D.P., Ba, J.: Adam: a method for stochastic optimization. In: International Conference on Learning Representations (2015)
8. Koren, Y., Bell, R., Volinsky, C.: Matrix factorization techniques for recommender systems. Computer **8**, 30–37 (2009)
9. Lee, D.D., Seung, H.S.: Algorithms for non-negative matrix factorization. In: Advances in Neural Information Processing Systems, pp. 556–562 (2001)
10. Liu, D., Li, J., Du, B., Chang, J., Gao, R.: DAML: dual attention mutual learning between ratings and reviews for item recommendation. In: Proceedings of the 25th ACM SIGKDD International Conference on Knowledge Discovery & Data Mining, pp. 344–352. ACM (2019)
11. Liu, H., Wang, Y., Peng, Q., Wu, F., Gan, L., Pan, L., Jiao, P.: Hybrid neural recommendation with joint deep representation learning of ratings and reviews. Neurocomputing **374**, 77–85 (2020)
12. Liu, H., et al.: NRPA: neural recommendation with personalized attention. In: Proceedings of the 42nd International ACM SIGIR Conference on Research and Development in Information Retrieval. ACM (2019)
13. Lu, Y., Dong, R., Smyth, B.: Coevolutionary recommendation model: mutual learning between ratings and reviews. In: Proceedings of the 2018 World Wide Web Conference, pp. 773–782. International World Wide Web Conferences Steering Committee (2018)
14. McAuley, J., Leskovec, J.: Hidden factors and hidden topics: understanding rating dimensions with review text. In: RecSys, pp. 165–172. ACM (2013)
15. Mnih, A., Salakhutdinov, R.R.: Probabilistic matrix factorization. In: NIPS, pp. 1257–1264 (2008)
16. Pan, R., et al.: One-class collaborative filtering. In: 2008 Eighth IEEE International Conference on Data Mining, pp. 502–511. IEEE (2008)
17. Papadimitriou, C.H., Raghavan, P., Tamaki, H., Vempala, S.: Latent semantic indexing: a probabilistic analysis. J. Comput. Syst. Sci. **61**(2), 217–235 (2000)
18. Pennington, J., Socher, R., Manning, C.: Glove: global vectors for word representation. In: Proceedings of the 2014 Conference on Empirical Methods in Natural Language Processing (EMNLP), pp. 1532–1543 (2014)
19. Rendle, S.: Factorization machines. In: ICDM, pp. 995–1000 (2010)
20. Salakhutdinov, R., Mnih, A., Hinton, G.: Restricted Boltzmann machines for collaborative filtering. In: Proceedings of the 24th International Conference on Machine Learning, pp. 791–798. ACM (2007)
21. Sarwar, B.M., Karypis, G., Konstan, J.A., Riedl, J., et al.: Item-based collaborative filtering recommendation algorithms. WWW **1**, 285–295 (2001)
22. Seo, S., Huang, J., Yang, H., Liu, Y.: Interpretable convolutional neural networks with dual local and global attention for review rating prediction. In: Proceedings of the Eleventh ACM Conference on Recommender Systems, pp. 297–305. ACM (2017)
23. Tan, Y., Zhang, M., Liu, Y., Ma, S.: Rating-boosted latent topics: understanding users and items with ratings and reviews. IJCAI **16**, 2640–2646 (2016)
24. Tay, Y., Luu, A.T., Hui, S.C.: Multi-pointer co-attention networks for recommendation. In: Proceedings of the 24th ACM SIGKDD International Conference on Knowledge Discovery & Data Mining, pp. 2309–2318. ACM (2018)

25. Vaswani, A., et al.: Attention is all you need. In: Advances in Neural Information Processing Systems, pp. 5998–6008 (2017)
26. Wang, X., et al.: Neural review rating prediction with hierarchical attentions and latent factors. In: Li, G., Yang, J., Gama, J., Natwichai, J., Tong, Y. (eds.) DAS-FAA 2019. LNCS, vol. 11448, pp. 363–367. Springer, Cham (2019). https://doi.org/10.1007/978-3-030-18590-9_46
27. Zheng, L., Noroozi, V., Yu, P.S.: Joint deep modeling of users and items using reviews for recommendation. In: WSDM, pp. 425–434 (2017)

Modeling Periodic Pattern with Self-Attention Network for Sequential Recommendation

Jun Ma[1], Pengpeng Zhao[1](✉), Yanchi Liu[2], Victor S. Sheng[3], Jiajie Xu[1], and Lei Zhao[1]

[1] Institute of Artificial Intelligence, School of Computer Science and Technology, Soochow University, Suzhou, China
jma0@stu.suda.edu.cn, {ppzhao,xujj,zhaol}@suda.edu.cn
[2] NEC Labs America, Princeton, USA
yanchi@nec-labs.com
[3] Department of Computer Science, Texas Tech University, Lubbock, USA
victor.sheng@ttu.edu

Abstract. Repeat consumption is a common phenomenon in sequential recommendation tasks, where a user revisits or repurchases items that (s)he has interacted before. Previous researches have paid attention to repeat recommendation and made great achievements in this field. However, existing studies rarely considered the phenomenon that the consumers tend to show different behavior periodicities on different items, which is important for recommendation performance. In this paper, we propose a holistic model, which integrates **G**raph Convolutional Network with **P**eriodic-**A**ttenuated **S**elf-**A**ttention **N**etwork (GPASAN) to model user's different behavior patterns for a better recommendation. Specifically, we first process all the users' action sequences to construct a graph structure, which captures the complex item connection and obtains item representations. Then, we employ a periodic channel and an attenuated channel that incorporate temporal information into the self-attention mechanism to model the user's periodic and novel behaviors, respectively. Extensive experiments conducted on three public datasets show that our proposed model outperforms the state-of-the-art methods consistently.

Keywords: Sequential recommendation · Self-attention network · Periodic pattern

1 Introduction

Recommender systems play an essential role in people's daily life. In many real-world applications, recommender systems help users alleviate the problem of information overload and recommend products that users are likely interested in. At the same time, repeat recommendation is a significant component of recommender systems as the repetitive behaviors frequently exhibit in our daily

© Springer Nature Switzerland AG 2020
Y. Nah et al. (Eds.): DASFAA 2020, LNCS 12114, pp. 557–572, 2020.
https://doi.org/10.1007/978-3-030-59419-0_34

life. For example, we may visit the same breakfast shop every day and buy a toothbrush every three months. It is important for recommender systems to make repeat recommendations at a particular time, for both business benefits and customer satisfaction.

Most sequential recommendation methods mainly concentrate on predicting the unknown/novel items to target users but ignore the importance of consumers' repeat behaviors. Only a few researchers paid attention to repeat recommendation and introduced mathematical probability models to predict users' repeat behaviors. The Negative Binomial Distribution (NBD) model [5] was used to model customers' repeat behaviors for consumer brands. Then, Anderson et al. [1] proposed the importance of *recency*, which indicates that people have a tendency to repurchase the items recently consumed. Afterward, Benson et al. [2] integrated time, distance, and quality factors into repeat recommendation and made a significant achievement. However, the existing studies ignore the importance of the users' periodic behaviors for repeat recommendation. More specifically, most of the users' repeat behaviors are not random. A user tends to show different behavior periodicities for different items, depending on the items' life cycles and user preferences, such as weekly milk purchase, monthly toothpaste replacement, and a continuous loop of music. Therefore, the repeat recommendation is supposed to be provided with a periodic frequency. In consideration of the temporal accuracy of repeat recommendation, we further refine the concept of repeat as periodic.

In order to recommend proper items at the right time, we propose a holistic model, **G**raph Convolutional Network with **P**eriodic-**A**ttenuated **S**elf-**A**ttention Network (GPASAN), to predict both novel and consumed items simultaneously. In GPASAN, we first encode all users' action sequences as a large *item-item* graph. And we employ a multi-layer Graph Convolutional Network (GCN) to capture the complex item transitions and obtain accurate item representations. Then, we design a Periodic-Attenuated Self-Attention Network (PASAN) to model the user's periodic and novel behavior patterns with a periodic channel and an attenuated channel, respectively. To be specific, the periodic channel is used to model the user's periodic behavior pattern. We apply a periodic activation function to process the time difference between items in the same sequence, by which the time intervals are translated into cyclical influence that changes dynamically with time. We combine it with the self-attention process so as to transmit the impact information of different items for better attention weights computing. In the attenuated channel, we assume that the user has a tendency to interact with the novel items next time. Thus, the influence of previous items is decaying with time. We employ an attenuated function to process the time intervals between sequential items and integrate them into the self-attention mechanism. It's worth noting that all parameter settings are item-specific, as the lengths of life cycle and attenuation rates of influence vary among items. Finally, we adaptively balance the preference orientation of users' periodic and novel behaviors in different scenarios for an accurate recommendation. In summary, the main contributions of our work are listed as follows:

- To the best of our knowledge, it is the first effort that emphasizes the importance of user's periodic pattern on different items for repeat recommendation.
- We propose a holistic model called GPASAN, which applies a multi-layer graph convolutional network to model the complex transactions between items in a sequence and integrates with the periodic-attenuated self-attention mechanism to model the user's periodic and novel behavior patterns.
- We conduct extensive experiments on three real-world datasets, which demonstrate that our model outperforms state-of-the-art methods significantly.

2   Related Work

2.1   Sequential Recommendation

Sequential recommendation is one of the typical applications of recommender systems that predict the next item with users' previous implicit feedback [23]. Conventional recommendation methods include Matrix Factorization (MF) [11] and Markov Chains (MC) [17]. MF factorizes a user-item rating matrix into low-rank matrices. This method captures the user's general interests based on the user's whole interaction history but ignores the information of the whole interactive sequence. MC based methods predictes a user's next behavior based on the previous one. Yap et al. [24] introduced a competence score measure in personalized sequential pattern mining for next-item recommendation. In this way, MC is able to model the sequence information, but the limitation is that it is difficult to capture users' long-term preferences. Later, deep-learning-based methods achieved great success in modeling sequential data. Hidasi et al. [7] applied a gated recurrent unit (GRU) as a special variant of RNN to model session representation and made great improvements over traditional methods. Then, Tan et al. [18] further developed an RNN model by using a data augmentation technique to improve the model performance. In recent years, Graph Convolutional Network (GCN) has been proposed to learn the representation of graph-structured data. GC-MC [3] employs GCN to model the complex connection between users and items and obtained accurate representations. PinSage, an industrial solution, applies multiple graph convolution layers on the *item-item* graph for image recommendation and achieves a great success. On the other hand, attention mechanisms attract considerable attention recently, because of its surprising effect in various domains [13,22]. Li et al. [12] proposed a model called NARM, which employs the attention mechanism on RNN to capture the user's main purpose. Then, Vaswani et al. [19] proposed a sequence-to-sequence attention-based method called Transformer, which achieves the state-of-the-art performance and efficiency on machine translation tasks. Soon after, Kang et al. [9] introduced the self-attention mechanism in sequential recommendation and made a great success. However, all of the methods above don't consider users' periodic behaviors, which is essential to analyse users' consumption habits.

2.2   Repeat Recommendation

Repeat recommendation is an important component of sequential recommendation. In many life scenarios, people often interact with the items, which have been visited before, e.g., a user may listen to the same music several times on loop. As a consequence, a number of studies have taken efforts to analyze users' repeat behaviors. The early explorations of repeat recommendation applies a marginal utility [21] in economics to evaluate the possibility that the user will purchase a specific item. However, it didn't apply to most scenarios. Later, Anderson et al. [1] proposed the importance of recency, which indicates that users tend to interact with items recently visited. Furthermore, Benson et al. [2] and Cai et al. [4] introduced the item factor into repeat recommendation and obtained a good performance. Then, Ren et al. [14] divided users' behaviors in repeat and exploration modes with a repeat-explore mechanism to predict different items for session-based recommendation. More recently, Wang et al. [20] emphasized the importance of the item-specific effect of previous consumed items and introduced Hawkes Process into Collaborative Filtering (CF) for a better recommendation.

Nonetheless, existing studies ignore the periodic consumption of users, which means that the consumer may interact with the same item multiple times at similar intervals. Furthermore, the user tends to show different behavior cycle lengths on different items. With that in mind, we design a periodic-attenuated self-attention network, which considers users' periodic and novel behaviors simultaneously and models the item-specific influential change. We also employ a multi-layer graph convolutional network to capture the complex transitions of graph items and introduce a time difference matrix to overcome the lack of time perception in self-attention mechanism. To the best of our knowledge, it is the first attempt to incorporate users' periodic and novel behaviors into the self-attention network for sequential recommendation.

3   Proposed Method

3.1   Problem Definition

The task of sequential recommendation is predicting the next item, which the user will interact with when given his/her history action sequence. Here, we give a formulation of the problem.

Let $U = \{u_1, u_2, ..., u_{|U|}\}$ denote a set of all users and $I = \{i_1, i_2, ..., i_{|I|}\}$ denote a set of all unique items, where $|U|$ and $|I|$ are the number of users and items, respectively. For each user's interaction history, we denote an action sequence as $S^u = \{i_1^u, i_2^u, ..., i_t^u, ...i_{|S^u|}^u\}$, where i_t^u denotes the item that the user interacts at time t $(1 \leq t \leq |S^u|)$. The goal of our task is to predict the next item i_{t+1}^u when given a user's previous action sequence $S_t^u = \{i_1^u, i_2^u, ..., i_t^u\}$.

3.2   The Architecture of GPASAN

The workflow of GPASAN is illustrated in Fig. 1, which consists of three layers. In the Graph Convolutional Network (GCN) Embedding Layer, the historical

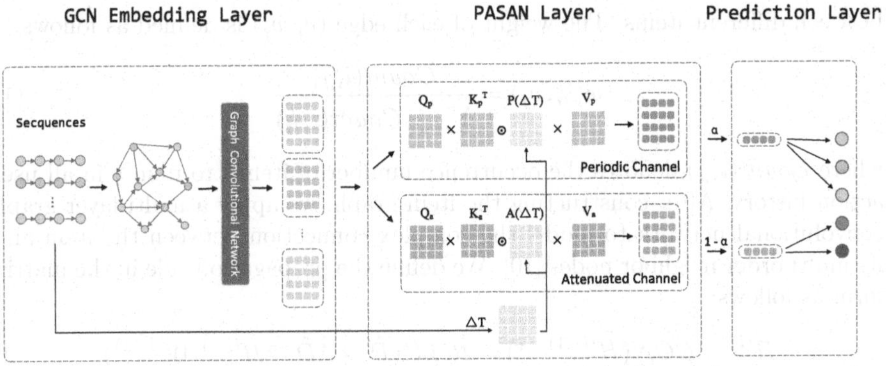

GCN Embedding Layer   PASAN Layer   Prediction Layer

Fig. 1. The architecture of the GPASAN model, which integrates graph convolutional network with periodic-attenuated self-attention network.

interactions of all users are aggregated together to construct a directed item graph. Then, for more accurate item representations, a graph convolution network is introduced to capture the complex transitions of graph items. In the Periodic-Attenuated Self-Attention Network (PASAN) Layer, we design a two-channel self-attention network to model the different behavior patterns of users. To overcome the drawback of the self-attention mechanism, which ignores the time interval information between items, we construct a ΔT matrix to record the time intervals between items for each action sequence and encode it with different activation functions for each channel, respectively. In the periodic channel, we consider the user has a tendency of periodic behavior next time, so we design a periodic activation function $P(\cdot)$ for ΔT to model the periodic fluctuation of visited item influence. We multiply $P(\Delta T)$ matrix by matrices Q_p, K_p^T and V_p to update the hidden vectors of sequence items. Then, we extract the vector of the last item as the periodic user representation for prediction. In attenuated channel, we consider the possibility of the user's novel behavior next time, so we design a decreasing activation function $A(\cdot)$ for ΔT. The influence of visited items is attenuated by time. After that, we multiply $A(\Delta T)$ matrix by matrices Q_a, K_a^T and V_a to get the attenuated user representation for prediction. As a user has a clear consumption orientation at a specific time, we adaptively assign different weights to the two channels so as to model the user's behavior disposition in different scenarios. In the Prediction Layer, we multiply periodic and attenuated user vectors by visited and unvisited item representations in the candidate set, respectively, to get the ranking scores.

GCN Embedding Layer. Given the item sequences of all users, we first transform the sequence information into a large directed graph $G(V, E)$, where the directed edges (E) in the graph indicate the interactive order of item nodes (V). Then, we assign weights for the edges to distinguish the relational degree between different items. The weight of each edge (v_i, v_j) is defined as follows:

$$w(v_i, v_j) = \frac{Count(i, j)}{\sum_{k=1}^{|I|} Count(i, k)} \tag{1}$$

where $Count(i, j)$ denotes the occurrence number of item i to item j in all user action history. After constructing the item graph, we apply a multi-layer graph convolutional network to model the complex connection between the item and its multi-order neighbor nodes [10]. We define the propagation rule in the matrix form as follows:

$$H^{(l)} = GCN(H^{(l-1)}, \bar{A}) = ReLU(\hat{D}^{-\frac{1}{2}} \hat{A} \hat{D}^{-\frac{1}{2}} H^{(l-1)} W^{(l-1)}) \tag{2}$$

where $H^{(l)} \in R^{|I| \times d}$ are the representations of items after l steps of embedding propagation. $\bar{A}$ is the adjacency matrix of $G(V, E)$, and d is the size of hidden units. H^0 is set as the initial embedding of all items. $W^{(l-1)} \in R^{d \times d}$ is a weight matrix and $ReLU(\cdot)$ is ReLU activation function. $\hat{D}^{-\frac{1}{2}} \hat{A} \hat{D}^{-\frac{1}{2}}$ represents the Laplacian matrix for the item graph. We formulate $\hat{A}$ and $\hat{D}$ as follows:

$$\hat{A} = \bar{A} + C \tag{3}$$

$$\hat{D}_{ii} = \sum_j \hat{A}_{ij} \tag{4}$$

where C denotes an identity matrix. If the graph node doesn't have a self-joining edge, the value of the adjacency matrix $\bar{A}$ at the diagonal position is 0. So, we add an identity matrix C to introduce the feature of node self-joining, which is important in feature extraction. $\hat{D}$ is a diagonal degree matrix, where the i-th diagonal element is the degree of node i, and j denotes the item nodes connected with i. By implementing multi-layer matrix-form propagation (considering multi-order neighborhoods), we get the new item representation matrix $H \in R^{|I| \times d}$.

PASAN Layer. As the traditional self-attention mechanism ignores the importance of time intervals between items, and single-channel self-attention can not model multiple behavior patterns of users, we propose an improved method called Periodic-Attenuated Self-Attention Network (PASAN). First, we transform the item sequence $S^u = \{i_1^u, i_2^u, ..., i_t^u, ...i_{|S^u|}^u\}$ into a fixed-length sequence, which considers the most recent n actions. If the sequence length is less than n, we add "padding" items to the front. For position embedding, we inject a learnable position embedding $P \in R^{|I| \times d}$. Finally, we get a sequence embedding $\hat{S}$ as follows:

$$\hat{S} = \begin{bmatrix} h_1 + p_1 \\ h_2 + p_2 \\ ... \\ h_n + p_n \end{bmatrix} \tag{5}$$

where $\hat{S} \in R^{n \times d}$, $h_n \in H$, and $p_n \in P$. Considering the influence of time interval, we construct a ΔT matrix for each action sequence, which records the time difference between an item and the items interacted after it. We ignore the item intervals from the back item to the front, so the ΔT is an upper triangular matrix. We define the $\Delta T \in R^{n \times n}$ of an n-length sequence as follows:

$$
\Delta T = \begin{bmatrix} \Delta t_{11} & \Delta t_{12} & \cdots & \Delta t_{1n} \\ & \Delta t_{22} & \cdots & \Delta t_{2n} \\ & & \cdots & \cdots \\ & & & \Delta t_{nn} \end{bmatrix} \tag{6}
$$

Different from the conventional self-attention network, we design a periodic channel and an attenuated channel to model the user's different behavior patterns (i.e., periodic behaviors and novel behaviors). We will elaborate on them in the next two sections.

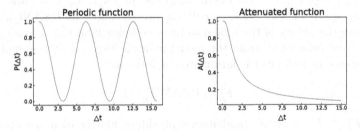

Fig. 2. An illustration of $P(\Delta t)$ and $A(\Delta t)$.

Periodic Channel. In this channel, we consider that a user has a tendency to interact with the items, which (s)he visited before. So, the influence of the visited items is periodic variation, rather than decaying with time. In order to model the lifetime effect of visited items, we design a periodic activation function, which is formulated as follows:

$$
P(\Delta t) = \frac{1}{2}cos(m_p \times \Delta t) + \frac{1}{2} \tag{7}
$$

where $cos(\cdot)$ is the cosine function, and $m_p \in R$ is a learnable variable, which is designed to learn the effect cycle of the item. Specially, we assign a different m_p for each item, as the life cycle of influence varies with each other (e.g., the lifetime of a box of milk may be a day, and the lifetime of a toothbrush is about three months). Furthermore, we set the coefficients and bias terms to be $\frac{1}{2}$ to distribute the value of $P(\Delta t)$ between 0 and 1. The illustration of $P(\cdot)$ is shown in Fig. 2. After obtaining the $P(\Delta T)$ matrix, we combine it with the self-attention mechanism to update the item vectors.

$$
E_p = softmax\left(\frac{Q_p K_p^T}{\sqrt{d}} \odot P(\Delta T)\right) V_p \tag{8}
$$

$$Q_p = \hat{S}W_p^Q, \quad K_p = \hat{S}W_p^K, \quad V_p = \hat{S}W_p^V \tag{9}$$

where $\hat{S} \in R^{n \times d}$ is the sequence item embedding matrix, and $W_p^Q, W_p^K, W_p^V \in R^{d \times d}$ are the projection matrices, which are used to make the model more flexible. The scale factor $\sqrt{d}$ is to avoid overly large values. The $\odot$ symbol denotes Hadamard product (element-wise product). Intuitively, $P(\Delta t)$ can be viewed as a filter gate, which is used to filter the important influencing factors in the weight matrix of each sequence. If the value of $P(\Delta t)$ is 1, we keep the whole information of the hidden unit. If the value is 0, we will discard all. To overcome the linear limitation of the self-attention mechanism, we apply a point-wise feed-forward network with a *ReLU* activation function to endow the model, which is defined as follows:

$$F_p = ReLU(E_p W_{p1} + b_{p1})W_{p2} + b_{p2} \tag{10}$$

where $W_{p1}, W_{p2} \in R^{d \times d}$ are weight matrices, and $b_{p1}, b_{p2} \in R^d$ are bias. For simplicity, we define the whole calculation process above as $PSAN(\cdot)$. In order to improve the ability of the model to learn complex item transitions, we apply a multi-layer network to train the model multiple times. The item embeddings in a sequence at *b-th* layer is formulated as follows:

$$F_p^b = PSAN(F_p^{b-1}) \tag{11}$$

where $F_p^b \in R^{n \times d}$ is the final item embedding matrix of a sequence. After multi-layer self-attention learning, F_p^b can fully capture the long-term interdependent between the items of a sequence. Finally, following [9], we define the final sequence representation of the periodic channel as the last clicked-item embedding of F_p^b, which can be noted as $F_p^b[n]$.

Attenuated Channel. In this channel, we think about the behavior pattern that a user is inclined to interact with a novel item next time, which (s)he has not visited before. Therefore, the influence of previous items will be decaying with time. In order to model the attenuated lifetime of different items, we design an attenuated activation function as follows:

$$A(\Delta t) = tanh\left(\frac{1}{m_a \times (\Delta t + 0.0001)}\right) \tag{12}$$

where $tanh(\cdot)$ is the tanh function, and $m_a \in R$ is a learnable variable, which is applied to control the decaying rate of item influence. Specially, we assign a specialized m_a for each item, as the various decaying rate of different item impact. In addition, we add 0.0001 for Δt to prevent the case that the denominator is 0. Finally, we apply a tanh function to distribute the value of $A(\Delta t)$ between 0 and 1, as an oversize value will destroy the balance of the two channels. The illustration of $A(\cdot)$ is shown in Fig. 2. After obtaining the $A(\Delta T)$ matrix, we combine it with the self-attention mechanism to update the item vectors as follows:

$$E_a = softmax\left(\frac{Q_a K_a^T}{\sqrt{d}} \odot A(\Delta T)\right) V_a \tag{13}$$

$$Q_a = \hat{S}W_a^Q, \quad K_a = \hat{S}W_a^K, \quad V_a = \hat{S}W_a^V \tag{14}$$

where $\hat{S} \in R^{n \times d}$ is the sequence item embedding matrix. In order to make the model learn the different representations of the same sequence in two different channels, we employ W_a^Q, W_a^K, $W_a^V \in R^{d \times d}$, which differ from that in the periodic channel, to project the item embedding matrix into different hidden space. In addition, similar to $P(\Delta t)$, the $A(\Delta t)$ can be considered as a filter gate, which decides the proportion of weight information that can be passed to V_a. To enhance the expressive ability of the model, we use a nonlinear activation function $ReLU$ to encode E_a as follows:

$$F_a = ReLU(E_a W_{a1} + b_{a1})W_{a2} + b_{a2} \tag{15}$$

where W_{a1}, $W_{a2} \in R^{d \times d}$, and b_{a1}, $b_{a2} \in R^d$. For simplicity, we define the whole calculation process above as $ASAN(\cdot)$. For the multi-layer training in the attenuated channel, we formulate it as follows:

$$F_a^b = ASAN(F_a^{b-1}) \tag{16}$$

where $F_a^b \in R^{n \times d}$ is the final item embedding matrix. Finally, we define the final sequence representation of the attenuated channel as the last clicked-item embedding of F_a^b, which can be noted as $F_a^b[n]$.

Prediction Layer. Different from traditional prediction methods, which calculate the candidate item scores with one user representation, we employ $F_p^b[n]$ and $F_a^b[n]$ as user vectors of different behavior patterns, which refer to periodic behaviors and novel behaviors. Considering that a user has only one behavior tendency at time t, we adaptively assign different weight coefficients to the two channels respectively as follows:

$$f_p = \alpha F_p^b[n], \quad f_a = (1 - \alpha)F_a^b[n] \tag{17}$$

where $\alpha \in R^{1 \times d}$ is a learnable weight coefficient, which is designed to control the diverse importance of two channels in different scenarios. f_p and f_a are the user representations of the periodic channel and the attenuated channel after assigning weights. Finally, we define the score calculation of the candidate item as follows:

$$\begin{cases} \hat{y}_p = I(i_j \in S^u)f_p^T h_j \\ \\ \hat{y}_a = I(i_j \notin S^u)f_a^T h_j \end{cases} \tag{18}$$

$$\hat{y} = \hat{y}_p + \hat{y}_a \tag{19}$$

where $\hat{y}_p$ denotes the visited item scores and $\hat{y}_a$ denotes the unvisited item scores in candidate set. $I(\cdot)$ is the indicator function, S^u denotes the set of items visited by the user. f_p^T and f_a^T are the transpose vectors of f_p and f_a, and h_j refers to the item vector of candidate item i_j. In general, we calculate the candidate item scores that the user has visited with f_p and the unvisited item score in $\hat{y}_p$ will be 0. Also, we only compute the unvisited item scores with f_a. Finally, we add them together and get the final scores of all candidate items for ranking.

For user modeling, following [6,8,9], we consider the user's previous interacted items, and induce implicit user vectors from visited item embeddings f_p and f_a, as we find that an explicit user embedding doesn't improve the model performance. That is probably because after multi-layer GCN and PASAN training, the item embeddings have adequately captured the preferences information of the users.

3.3   Network Learning

For model optimization, as we employ two implicit user vectors to calculate different item scores, we design a specialized cross entropy loss function to optimize the user representations of two channels respectively:

$$\begin{cases} L_p = -\sum \left(\log(\sigma(\hat{y}_p^+)) + \log(1 - \sigma(\hat{y}_p^-))\right) \\ L_a = -\sum \left(\log(\sigma(\hat{y}_a^+)) + \log(1 - \sigma(\hat{y}_a^-))\right) \end{cases} \tag{20}$$

$$L = L_p + L_a \tag{21}$$

where $\hat{y}_p^+$ and $\hat{y}_a^+$ denote the prediction scores of positive items. The difference is that $\hat{y}_p^+$ only contains scores of visited items and $\hat{y}_a^+$ contains unvisited item scores. $\hat{y}_p^-$ and $\hat{y}_a^-$ are negative sample scores. L is the loss sum of two channels. To summarize, we define the loss function of the two channels separately, and add them together for optimization. In this way, both user representations of two channels are more adequately optimized compared with sharing the same loss function.

4   Experiment

4.1   Experimental Setup

Datasets. We conduct the experiments on different datasets (i.e., TaoBao[1] and BrightKite[2]). Specifically, the TaoBao dataset is released by IJCAI SocInf'16 Contest-Brick-and-Mortar Store Recommendation[3], which includes the users' browse and purchase data accumulated on Tmall.com, Taobao.com and the app Alipay between July 1, 2015 and November 30, 2015. In our experiment, we take the 1/8 proportion of all purchase and click records as the TaoBao Buy and TaoBao Click datasets, respectively, considering the vast quantity of the whole dataset. In addition, BrightKite is a widely used LSBN dataset, which contains massive implicit feedbacks from the users' POI check-ins. To remove rare cases, we filter out the user sequences whose length is less than 5 and items

---

[1] https://tianchi.aliyun.com/dataset/dataDetail?dataId=53.

[2] https://snap.stanford.edu/data/loc-Brightkite.html.

[3] https://tianchi.aliyun.com/competition/entrance/231532/information.

Table 1. Statistics of experiment datasets.

Dataset	TaoBao Buy	TaoBao Click	BrightKite
# of users	75,282	60,762	35,099
# of items	189,671	634,061	85,308
# of interactions	1,168,594	4,397,421	3,604,962
repeat ratio	9.17%	13.77%	90.04%
sparsity	99.99%	99.99%	99.88%
avg.len	15.52	72.37	102.71

that appear less than 5 times for three datasets. In our experiments, we take the last interacted item of each user for testing, the penultimate one for validating, and the rest for training. The statistics of the filtered datasets are shown in Table 1.

Baselines. We select the following commonly used state-of-the-art baselines to compare with our proposed model GPASAN.

- **POP:** This method always recommends the items that have the most frequent occurrence.
- **BPR-MF**[15]**:** This model optimizes a pairwise ranking objective function and combines with matrix factorization.
- **FPMC**[16]**:** This method combines matrix factorization and markov chains to capture users' long-term preference and item-to-item transitions.
- **GRU4REC**[7]**:** This model employs GRU for session-based recommendation, which utilizes a session-parallel mini-batch training process and a ranking-based loss function.
- **SASRec**[9]**:** This is a state-of-the-art method that utilizes self-attention network for sequential recommendation.
- **SLRC**[20]**:** This method utilizes collaborative filter (CF) and Hawkes Process to model the user's preference, which considers the two temporal characteristics of items' lifetime. $SLRC_{BPRMF}$, $SLRC_{Tensor}$ and $SLRC_{NCF}$ are three variants of SLRC with different CF methods.
- **GPASAN-G:** This is a variant of GPASAN, which removes the multi-layer graph convolutional network and initializes all item embeddings with a random uniform initializer.
- **GPASAN-P:** This is a variant of GPASAN, which removes the periodic channel and considers all user behaviors in the attenuated channel.
- **GPASAN-A:** This is a variant of GPASAN, which removes the attenuated channel and calculates the candidate item scores with the user representation from the periodic channel.

Evaluation Metrics. We use NDCG and Recall to evaluate all compared methods, which are introduced as follows.

Table 2. The performance of GPASAN and baselines on three datasets.

Method	TaoBao Buy				TaoBao Click				BrightKite			
	@5		@10		@5		@10		@5		@10	
	NDCG	Recall	NDCG	Recall	NDCG	Recall	NDCG	Recall	NDCG	Recall	NDCG	Recall
POP	0.2094	0.2506	0.2300	0.2925	0.3205	0.3851	0.3642	0.4374	0.5630	0.5942	0.5860	0.6317
BPRMF	0.4182	0.4802	0.4411	0.5501	0.5269	0.5979	0.5406	0.6550	0.7143	0.7609	0.7229	0.7872
FPMC	0.4555	0.5526	0.4863	0.6275	0.5459	0.6256	0.5612	0.6817	0.6927	0.7369	0.7018	0.7651
GRE4REC	0.4721	0.5673	0.5064	0.6254	0.5570	0.6426	0.5729	0.7010	0.7366	0.7889	0.7458	0.8065
SASRec	0.5291	0.6053	0.5545	0.6739	0.5947	0.6889	0.6177	0.7490	0.7639	0.8005	0.7739	0.8210
$SLRC_{BPRMF}$	0.4922	0.5509	0.5275	0.6252	0.5637	0.6433	0.5817	0.6999	0.7778	0.8083	0.7844	0.8286
$SLRC_{Tensor}$	0.5156	0.5710	0.5409	0.6493	0.5830	0.6634	0.6095	0.7253	0.7658	0.7923	0.7722	0.8119
$SLRC_{NCF}$	0.4678	0.5241	0.4921	0.5997	0.5301	0.6151	0.5542	0.6799	0.7768	0.8044	0.7824	0.8219
GPASAN-G	0.5398	0.6134	0.5683	0.6849	0.6115	0.6975	0.6344	0.7510	0.7712	0.8057	0.7800	0.8296
GPASAN-P	0.5810	0.6610	0.6090	0.7346	0.6511	0.7374	0.6752	0.7917	0.7386	0.7893	0.7422	0.8013
GPASAN-A	0.5697	0.6331	0.5899	0.7065	0.6345	0.7043	0.6518	0.7682	0.7699	0.8087	0.7706	0.8214
GPASAN	**0.6076**	**0.6818**	**0.6304**	**0.7523**	**0.6774**	**0.7539**	**0.6973**	**0.8154**	**0.7910**	**0.8241**	**0.8002**	**0.8498**
Improvement	14.84%	12.64%	13.69%	11.63%	13.91%	9.44%	12.89%	8.87%	1.70%	1.95%	2.01%	2.56%

NDCG@K is the normalized discounted cumulative gain. The higher items are ranked in the recommendation list top K, the higher NDCG scores will be, and it is a normalized score of range [0,1].

Recall@K is a widely used metric in the recommendation algorithm, which represents the proportion of correctly recommended items in a top K list. The higher scores indicate better performance.

Parameter Settings. In our experiment, we set the dimension of item vectors to be 128 and the dropout rate to be 0.1. The default number of PASAN layers is 2, and default GCN layers number is 5. Besides, we initialize weight matrices with a random uniform initializer and set the maximal sequence length as 20, 80, and 120 for TaoBao Buy, Taobao Click, and BrightKite. When constructing the candidate set for testing, we take 1 positive case and 99 negative cases on TaoBao Buy and Taobao Click. Following [20], we take 1 positive case and 4999 negative cases on BrightKite for a fair comparison. All of the experiments are conducted on an Nvidia Titan V GPU and 256G memory capacity.

4.2   Comparison of Performance

To demonstrate the performance of GPASAN, we compare it with other state-of-the-art methods and the experimental results are illustrated in Table 2. Moreover, We calculate the improvement ratio relative to the best baseline on three datasets without considering three variants of GPASAN. From our experimental results, we have the following observations.

First, the most traditional and straightforward algorithm POP performs poorly, as it just considers the popularity of items without the user's history and personalized customization. The performance of BPRMF varies on different datasets, and the lower sparsity ratio is conducive to model capacity. Besides, FPMC and GRU4REC, which employ Markov chain and GRU respectively to

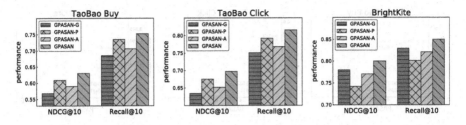

Fig. 3. Performance comparisons between GPASAN and its variant models.

capture the users' sequence information, perform better than BPRMF under most scenarios. It indicates that the sequential feature is an important factor for the next-item recommendation.

Second, the state-of-the-art methods SASRec and SLRC achieve a great performance on three datasets. SASRec utilizes the self-attention mechanism to capture the global and local connection with high parallelism effectively. However, it is lack of time perception and has no single consideration for repeat behaviors. That is why SLRC performs better on BrightKite, which has a high repeat ratio. But on sparse and low repeat ratio dataset TaoBao, SLRC performs poorer. That is because the Hawkes process of SLRC contributes less with a low repeat ratio, and the model performance mainly relies on the CF-based methods.

Finally, our proposed model GPASAN outperforms all methods mentioned above on three datasets, which indicates the importance of considering the user's behavior periodicities on different items. As GPASAN adopts an adaptive way to distinguish the importance of two behavior patterns respectively in different scenarios, our model gets the best performance on both low repeat and high repeat datasets. In addition, we retain the high parallelism of the self-attention mechanism and overcome the lack of time awareness. As a result, GPASAN achieves an excellent performance while maintaining high efficiency.

4.3 Influence of Components

To verify the effect of the different parts in GPASAN, we compare GPASAN with three variants GPASAN-G, GPASAN-P and GPASAN-A, which remove GCN layer, periodic channel or attenuated channel, respectively. We conduct experiments on three datasets and show their results in Fig. 3. It is obvious that the loss of any component will degrade the performance of the model. Specifically, the graph convolutional network plays an important role on sparse datasets (i.e., TaoBao Buy and TaoBao Click). On e-commerce dataset, it is objectively impossible for each customer to interact with everything, the customers' purchase/click records must only be a small fraction of a large number of items. Therefore, it is difficult to model users' accurate preference facing numerous unknown items. With the help of GCN, which constructs a large item graph with all users' action sequences, GPASAN can integrate information from the users with similar consumer preferences and learn more accurate item representations. In addition, as the low repeat ratio on TaoBao Buy and TaoBao Click, the absence of the

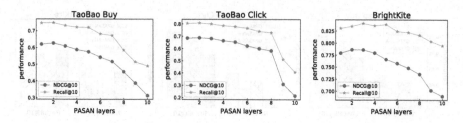

Fig. 4. Effect of the number of periodic-attenuated self-attention network layers.

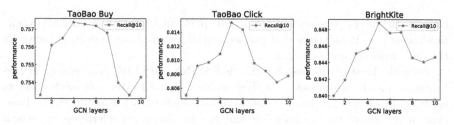

Fig. 5. Effect of the number of graph convolutional network layers.

attenuated channel causes a higher performance loss for the model compared with the lack of the periodic channel. Conversely, on BrightKite, which has a higher repeat ratio and density, the periodic channel plays greater roles than the attenuated channel and the GCN layer. That is because, on check-in datasets, people tend to visit the same place with the same frequency and periodicity. The lack of periodic channel will steer the model in the wrong direction. Also, the GCN layer contributes less as the model can capture the user preferences by dense user interaction independently. To the end, the complete model GPASAN achieves the best performance on three datasets.

4.4 Influence of Hyper-parameters

Impact of PASAN Layers. To enhance the capabilities of learning the complex relationships between different items, we employ the multi-layer periodic-attenuated self-attention network. In order to find out the appropriate number of layers for PASAN, we set the GCN layers to be 5 and vary the PASAN layers from 1 to 10. As shown in Fig. 4, with the increment of network layers several times, GPASAN achieves a better performance. However, when the network goes deeper, the performance of the model begins to degrade on three datasets. This is because the increment of the model capacity will lead to overfitting, and the training process becomes unstable. In addition, more layers lead to more time for training. Considering the balance of effectiveness and efficiency, we set the default PASAN layer number to be 2 for experiments.

Impact of GCN Layers. In order to consider the multi-order neighborhood information propagation, we utilize a multi-layer graph convolutional network. We set the PASAN layer to be 2 and vary the GCN layers from 1 to 10 so as to investigate a proper GCN layer number. As the performance difference with diverse GCN layers is not as large as PASAN layers, we only show the performance of Recall@10 on three datasets for more accurate observation. As illustrated in Fig. 5, the model achieves a peak performance, when we consider 4 to 6-order neighborhood of a node (4 to 6 GCN layers). Then, the model performance decreases quickly when the network goes deeper. It is because that the overfitting problem is serious with many layers, and containing the neighbor information from 4 to 6-order is enough to capture the complex relationship between items.

5 Conclusion

In this paper, we emphasized the importance of modeling user's periodic behavior pattern for repeat recommendation and proposed a novel model called GPASAN. In our model, we first employed a graph convolutional network to capture the complex transitive connection between items and obtain accurate item representations. Then, we developed a periodic-attenuated self-attention network to model user's periodic and novel behavior patterns, respectively. In order to obtain the time-varying influence of previous interacted items in different patterns, we constructed a periodic function and an attenuated function to process the time intervals between items and integrated the items' influence information into a multi-layer self-attention network to form accurate user representations for a better recommendation. Extensive experiments conducted on the three benchmark datasets demonstrate the superiority of GPASAN. As the future work, we will take more item features into consideration, such as images and item description text, which can describe the user's preferences more accurately, to improve the performance of GPASAN.

Acknowledgments. This research was partially supported by NSFC (No. 61876117, 61876217, 61872258, 61728205), Suzhou Science and Technology Development Program (SYG201803), Open Program of Key Lab of IIP of CAS (No. IIP2019-1) and A Project Funded by the Priority Academic Program Development of Jiangsu Higher Education Institutions(PAPD).

References

1. Anderson, A., Kumar, R., Tomkins, A., Vassilvitskii, S.: The dynamics of repeat consumption. In: WWW, pp. 419–430. ACM (2014)
2. Benson, A.R., Kumar, R., Tomkins, A.: Modeling user consumption sequences. In: Proceedings of the 25th International Conference on World Wide Web, pp. 519–529. International World Wide Web Conferences Steering Committee (2016)
3. van den Berg, R., Kipf, T.N., Welling, M.: Graph convolutional matrix completion. arXiv preprint arXiv:1706.02263 (2017)
4. Cai, R., Bai, X., Wang, Z., Shi, Y., Sondhi, P., Wang, H.: Modeling sequential online interactive behaviors with temporal point process. In: Proceedings of the 27th ACM CIKM, pp. 873–882. ACM (2018)

5. Ehrenberg, A.S.: The pattern of consumer purchases. J. Roy. Stat. Soc.: Ser. C (Appl. Stat.) **8**(1), 26–41 (1959)
6. He, R., McAuley, J.: Fusing similarity models with Markov chains for sparse sequential recommendation. In: 2016 IEEE 16th ICDM, pp. 191–200. IEEE (2016)
7. Hidasi, B., Karatzoglou, A., Baltrunas, L., Tikk, D.: Session-based recommendations with recurrent neural networks. arXiv preprint arXiv:1511.06939 (2015)
8. Kabbur, S., Ning, X., Karypis, G.: FISM: factored item similarity models for top-n recommender systems. In: Proceedings of the 19th ACM SIGKDD International Conference on Knowledge Discovery and Data Mining, pp. 659–667. ACM (2013)
9. Kang, W.C., McAuley, J.: Self-attentive sequential recommendation. In: 2018 IEEE ICDM, pp. 197–206. IEEE (2018)
10. Kipf, T.N., Welling, M.: Semi-supervised classification with graph convolutional networks. arXiv preprint arXiv:1609.02907 (2016)
11. Koren, Y., Bell, R., Volinsky, C.: Matrix factorization techniques for recommender systems. Computer **8**, 30–37 (2009)
12. Li, J., Ren, P., Chen, Z., Ren, Z., Lian, T., Ma, J.: Neural attentive session-based recommendation. In: Proceedings of the 2017 ACM on Conference on Information and Knowledge Management, pp. 1419–1428. ACM (2017)
13. Luo, A., et al.: Adaptive attention-aware gated recurrent unit for sequential recommendation. In: Li, G., Yang, J., Gama, J., Natwichai, J., Tong, Y. (eds.) DASFAA 2019. LNCS, vol. 11447, pp. 317–332. Springer, Cham (2019). https://doi.org/10.1007/978-3-030-18579-4_19
14. Ren, P., Chen, Z., Li, J., Ren, Z., Ma, J., de Rijke, M.: Repeatnet: A repeat aware neural recommendation machine for session-based recommendation. arXiv preprint arXiv:1812.02646 (2018)
15. Rendle, S., Freudenthaler, C., Gantner, Z., Schmidt-Thieme, L.: BPR: Bayesian personalized ranking from implicit feedback. In: Proceedings of the Twenty-Fifth UAI, pp. 452–461. AUAI Press (2009)
16. Rendle, S., Freudenthaler, C., Schmidt-Thieme, L.: Factorizing personalized Markov chains for next-basket recommendation. In: WWW, pp. 811–820. ACM (2010)
17. Shani, G., Heckerman, D., Brafman, R.I.: An MDP-based recommender system. J. Mach. Learn. Res. **6**(Sep), 1265–1295 (2005)
18. Tan, Y.K., Xu, X., Liu, Y.: Improved recurrent neural networks for session-based recommendations. In: DLRS, pp. 17–22. ACM (2016)
19. Vaswani, A., et al.: Attention is all you need. In: Advances in Neural Information Processing Systems, pp. 5998–6008 (2017)
20. Wang, C., Zhang, M., Ma, W., Liu, Y., Ma, S.: Modeling item-specific temporal dynamics of repeat consumption for recommender systems. In: WWW, pp. 1977–1987. ACM (2019)
21. Wang, J., Zhang, Y.: Utilizing marginal net utility for recommendation in e-commerce. In: Proceedings of the 34th International ACM SIGIR Conference on Research and Development in Information Retrieval, pp. 1003–1012. ACM (2011)
22. Xu, C., et al.: Graph contextualized self-attention network for session-based recommendation. In: Proceedings of 28th International Joint Conference Artificial Intelligence (IJCAI), pp. 3940–3946 (2019)
23. Xu, C., et al.: Recurrent convolutional neural network for sequential recommendation. In: WWW, pp. 3398–3404 (2019)
24. Yap, G.-E., Li, X.-L., Yu, P.S.: Effective next-items recommendation via personalized sequential pattern mining. In: Lee, S., Peng, Z., Zhou, X., Moon, Y.-S., Unland, R., Yoo, J. (eds.) DASFAA 2012. LNCS, vol. 7239, pp. 48–64. Springer, Heidelberg (2012). https://doi.org/10.1007/978-3-642-29035-0_4

Cross-Domain Recommendation with Adversarial Examples

Haoran Yan[1], Pengpeng Zhao[1(✉)], Fuzhen Zhuang[2,6], Deqing Wang[3],
Yanchi Liu[4], and Victor S. Sheng[5]

[1] Institute of Artificial Intelligence, Soochow University, Suzhou, China
hryan@stu.suda.edu.cn, ppzhao@suda.edu.cn
[2] Key Lab of IIP of CAS, Institute of Computing Technology, Beijing, China
zhuangfuzhen@ict.ac.cn
[3] School of Computer Science and Engineering, Beihang University, Beijing, China
dqwang@buaa.edu.cn
[4] NEC Labs America, Princeton, USA
yanchi@nec-labs.com
[5] Department of Computer Science, Texas Tech University, Lubbock, USA
victor.sheng@ttu.edu
[6] University of Chinese Academy of Sciences, Beijing 100049, China

Abstract. Cross-domain recommendation leverages the knowledge from relevant domains to alleviate the data sparsity issue. However, we find that the state-of-the-art cross-domain models are vulnerable to adversarial examples, leading to possibly large errors in generalization. That's because most methods rarely consider the robustness of the proposed models. In this paper, we propose a new Adversarial Cross-Domain Network (ACDN), in which adversarial examples are dynamically generated to train the cross-domain recommendation model. Specifically, we first combine two multilayer perceptrons by sharing the user embedding matrix as our base model. Then, we add small but intentionally worst-case perturbations on the model embedding representations to construct adversarial examples, which can result in the model outputting an incorrect answer with a high confidence. By training with these aggressive examples, we are able to obtain a robust cross-domain model. Finally, we evaluate the proposed model on two large real-world datasets. Our experimental results show that our model significantly outperforms the state-of-the-art methods on cross-domain recommendation.

Keywords: Cross-domain recommendation · Adversarial examples · Sparse data

1 Introduction

Due to the information explosion, recommendation systems play a vital role in today's world. Among various recommendation techniques, collaborative filtering (CF), which aims at providing a personalized list of items to a user by

© Springer Nature Switzerland AG 2020
Y. Nah et al. (Eds.): DASFAA 2020, LNCS 12114, pp. 573–589, 2020.
https://doi.org/10.1007/978-3-030-59419-0_35

learning her prior preferences, is the most popular and widely used method with many variants, like matrix factorization (MF) and neural collaborative filtering [1]. However, existing collaborative filtering approaches still suffer from the data sparsity issue. In this paper, we focus on the cross-domain recommendation, which transfers knowledge from relevant domains to alleviate the data sparsity issue in recommendation.

During the development of cross-domain recommendation, many cross-domain recommendation approaches have appeared. The collective matrix factorization (CMF) [7] is a typical approach, which jointly factorizes user-item interaction matrices in two domains by sharing the user latent factors. Cross-Domain CF with factorization machines (FM) [15] extends FM [16] by applying factorization on the merged domains. The collaborative cross networks (Conet) [8] introduces cross connections between two neural networks to transfer knowledge across domains. With various state-of-the-art approaches proposed, cross-domain recommendation has achieved great success in real applications.

However, the existing cross-domain recommendation methods rarely consider the robustness of a cross-domain model, which is both sufficient and necessary for generalizability [5]. Although transferring knowledge from auxiliary domains can improve the model performance, we find the robustness of cross-domain model is still not satisfying compared with single-domain model. For example, we add small but intentional perturbations on the parameters of a cross-domain model and find that the model's performance drops a lot. The poor robustness implies possibly large errors in generalization [12]. There are several reasons for this. First, cross-domain recommendation technique is often used to alleviate the data sparsity issue, so the dataset used for cross-domain recommendation is much sparser than the general recommendation task. It is more difficult to learn robust parameter representations from sparser data. Another reason for the problem is the complexity of the cross-domain model. Compared with the single-domain model, the cross-domain model is more complicated. Complex model is more vulnerable to noise [9]. In addition, when we transfer knowledge from the source domain to the target domain, we may transfer useless noise instead of useful knowledge [2]. It is a challenge to transfer knowledge due to the noise from auxiliary domain [8], which may destroy the robustness of the model.

In this paper, we develop a new Adversarial Cross-Domain Network (ACDN), where adversarial examples are dynamically generated to train a deep cross-domain recommendation model. To benefit from deep learning and keep our discussion uncluttered, we combine two multilayer perceptrons by sharing the user embedding matrix as our base model. Different from adversarial examples in the computer vision domain, it is meaningless to apply noise to discrete features in the recommendation domain, which may change their semantics. So we add intentional perturbations which are essentially a kind of gradient noise on the embedding representations to generate adversarial examples. And then we train our model with these adversarial examples. They help the model to transfer more robust representations instead of useless noise. In this way, our model alleviates the data sparsity issue and learns robust parameters. The contributions of this work can be summarized as follows:

- We point out that the cross-domain recommendation model suffers from the poor robustness issue and we are the first to address this issue.
- We propose an Adversarial Cross-Domain Network to introduce adversarial examples to enhance the robustness of the cross-domain recommendation and thus improve its generalization performance.
- We compare our model with the state-of-the-art methods and verify the superiority of our model through the quantitative analysis on two large real-world datasets.

2   Preliminaries

2.1   Notation

Given a source domain $\mathcal{S}$ and a target domain $\mathcal{T}$, in which users $\mathcal{U}$ (size $m = |\mathcal{U}|$) are shared, let I_S (size $n_S = |I_S|$) and I_T (size $n_T = |I_T|$) denote the item sets of $\mathcal{S}$ and $\mathcal{T}$, respectively. We define user-item interactions matrix $\boldsymbol{R}_T \in \mathbb{R}^{m \times n_T}$ from user's implicit feedback in $\mathcal{T}$, where the element r_{ui} is 1 if user u and item i have an interaction and 0 otherwise. Similarly, we define user-item interactions matrix $\boldsymbol{R}_S \in \mathbb{R}^{m \times n_S}$ from user's implicit feedback in $\mathcal{S}$, where the element r_{uj} is 1 if user u and item j have an interaction and 0 otherwise. We rank the items through the predicted scores:

$$\hat{r}_{ui} = f(u, i|\Theta), \tag{1}$$

where Θ are parameters of the model, and f is the user-item interaction function. The function is dot production for MF techniques:

$$\hat{r}_{ui} = \boldsymbol{P}_u^T \boldsymbol{Q}_i, \tag{2}$$

and the latent vectors of users and items are $\Theta = \{\boldsymbol{P}, \boldsymbol{Q}\}$, in which $\boldsymbol{P} \in \mathbb{R}^{m \times d}$, $\boldsymbol{Q} \in \mathbb{R}^{n \times d}$ and d is the size of dimension. Neural collaborative filtering methods utilize neural networks to parameterize interaction function f:

$$\hat{r}_{ui} = f(\boldsymbol{v}_{ui}|\boldsymbol{P}, \boldsymbol{Q}, \theta_f) = \phi_o(\phi_L(...(\phi_1(\boldsymbol{v}_{ui})))), \tag{3}$$

where the projections of user and item are merged as the input $\boldsymbol{v}_{ui} = [\boldsymbol{v}_u, \boldsymbol{v}_i]$. $\boldsymbol{v}_u = \boldsymbol{P}^T \boldsymbol{x}_u$ and $\boldsymbol{v}_i = \boldsymbol{Q}^T \boldsymbol{x}_i$ are the projections of user u and item i, which are consisted of the embedding matrices $\boldsymbol{P} \in \mathbb{R}^{m \times d}$, $\boldsymbol{Q} \in \mathbb{R}^{n \times d}$ and one-hot encodings $\boldsymbol{x}_u \in \{0,1\}^m, \boldsymbol{x}_i \in \{0,1\}^n$. In a multilayer feedforward network, ϕ_o and ϕ_l ($l \in [1, L]$) can compute the output and hidden layers, and θ_f is the weight matrices and biases.

2.2   Base Model

MLP++ [8] is adopted as our base model, which combines two MLPs through sharing the user embedding matrix. The source and target domains are modelled by two neural networks, and the performance can be improved through transferring knowledge between two domains. The MLP++ model is shortly recapitulated as follows.

$$\Theta_t = \{P, Q_t, \theta_{f_t}\}, \Theta_s = \{P, Q_s, \theta_{f_s}\}, \qquad (4)$$

where the item embedding matrices are denoted by Q and the corresponding domain is specified with the subscript, and P denotes the user embedding matrix. θ_f are the weights and biases in the multilayer feedforward neural network.

$$L_0 = - \sum_{(u,i) \in \mathbf{R}^+ \cup \mathbf{R}^-} r_{ui} log \hat{r}_{ui} + (1 - r_{ui}) log(1 - \hat{r}_{ui}), \qquad (5)$$

where $\mathbf{R}^+$ is the positive examples and $\mathbf{R}^-$ is the randomly sampled negative examples. The base model's objective function is the joint losses of the source domain (L_s) and target domain (L_t), which can be instantiated by Eq. (5).

$$L_{MLP++}(\Theta) = L_t(\Theta_t) + L_s(\Theta_s), \qquad (6)$$

where the model parameters $\Theta = \Theta_t \cup \Theta_s$ and the user embeddings are shared by Θ_t and Θ_s.

3   Proposed Method

3.1   Adversarial Examples

Adversarial examples, which are constructed by adding small but intentional perturbations to the input examples, are originally proposed in the image classification [9,10]. Different from some random perturbations (i.e., Gaussian noise), the adversarial perturbations applied in the adversarial examples are intentionally designed, which can cause the model to give a wrong output with high confidence. In the image domain, adding small noise on an image will not change its visual content. However, in the recommendation domain, discrete ID features are often as the input of the model. If we add perturbations to the discrete ID features, the semantics of the input may change. For example, a user u reads a book i, and if we add noise on the user ID u, we may do not know who reads the book i. Similarly, if the book ID i is changed, the user u does not know which book she has read. Since it is irrational to add noise in the input layer, we instead apply the adversarial perturbations to the parameters of the underlying recommendation model [12]. We define the adversarial perturbations as the model parameter perturbations maximizing the overall loss function:

$$n_{adv} = \underset{\|n\| \le \epsilon}{\arg\max} \, L_{MLP++}(\theta_{emb} + n, \theta_f), \qquad (7)$$

where n is the perturbations on the embedding parameters $\theta_{emb} = \{\boldsymbol{P}, \boldsymbol{Q}_t, \boldsymbol{Q}_s\}$, ϵ controls the perturbations level, $\|.\|$ denotes the L_2 norm, and $\theta_f = \{\theta_{f_t}, \theta_{f_s}\}$ is the parameters in output and hidden layers. Essentially, the adversarial perturbations n_{adv} is a kind of gradient noise. In most neural network models, the gradient descent algorithm is used to minimize the loss function for training. The random perturbations are almost orthogonal to the gradient of the loss function in high dimensional space, which has little effect on the loss function. If input examples move along the direction of gradient ascent, the model's prediction may change a lot. Inspired by the fast gradient method in [10], we approximate the worst-case perturbations n_{adv} by linearizing $L_{MLP++}(\Theta)$ around θ_{emb}. With this approximate and the max-norm constraint, the solution of Eq. (7) is given by:

$$n_{adv} = \epsilon \frac{\nabla_{\theta_{emb}} L_{MLP++}(\Theta)}{\|\nabla_{\theta_{emb}} L_{MLP++}(\Theta)\|}, \tag{8}$$

where $\Theta = \{\theta_{emb}, \theta_f\}$ is the model parameters. Then we can get the representations of adversarial examples $\theta_{emb} + n_{adv}$, which can be used to train our model.

3.2 Adversarial Cross-Domain Network

Here, we first introduce the objection function of our model, and then the architecture of proposed model is described.

Because of the poor robustness in the cross-domain recommendation model, we train the cross-domain model with adversarial examples, which can enhance the model's robustness. Different form the image domain, it is irrational to add perturbations in the input layer in the recommendation domain. Instead, we apply the adversarial perturbations to the parameters of underlying recommendation model. To benefit from deep learning and keep our discussion uncluttered, we use a deep cross-domain model (MLP++) as our base model. The model is enforced to perform well even when attacked by adversarial examples for model's robustness. Therefore, adversarial examples are added to the MLP++ objective function to be minimized. The objective function of ACDN is defined as follows:

$$L_{ACDN}(\Theta) = L_{MLP++}(\theta_{emb}, \theta_f) + \lambda L_{MLP++}(\theta_{emb} + n_{adv}, \theta_f),$$
$$where \; n_{adv} = \underset{\|n\| \le \epsilon}{\arg\max} \, L_{MLP++}(\theta_{emb} + n, \theta_f), \tag{9}$$

where $\theta_{emb} + n_{adv}$ are the representations of adversarial examples, n_{adv} denotes the adversarial perturbations on the embedding parameters $\theta_{emb} = \{\boldsymbol{P}, \boldsymbol{Q}_t, \boldsymbol{Q}_s\}$, $\epsilon \ge 0$ controls the magnitude of the perturbations, $\theta_f = \{\theta_{f_t}, \theta_{f_s}\}$ is the output and hidden layers parameters and $\Theta = \{\theta_{emb}, \theta_f\}$ is the model parameters. In the formulation, $L_{MLP++}(\theta_{emb} + n_{adv}, \theta_f)$ can also be seen as a regularization term, and λ controls its strength.

The architecture of the proposed model is shown in Fig. 1, which consists of four modules: Input Layer, Embedding Layer, Hidden Layers and Output Layer. First, the Input Layer adopts the one-hot encoding to encode user-item

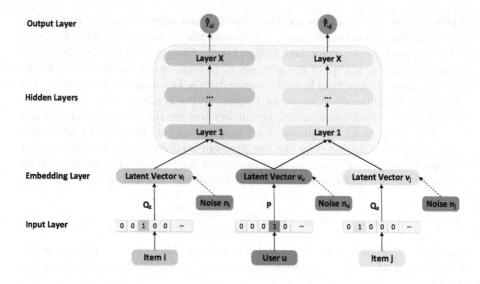

Fig. 1. The architecture of ACDN.

interaction indices. Second, the one-hot encodings are embedded into continuous representations and add adversarial perturbations on them to construct adversarial examples in the Embedding Layer. Third, we transform the representations to the final ones in the Hidden Layers. Finally, the Output Layer predicts the score with the final representations.

Input Layer: This module utilizes one-hot encoding to encode user-item interaction indices. It maps user u and item i, j into one-hot encodings $\boldsymbol{x}_u$, $\boldsymbol{x}_i$ and $\boldsymbol{x}_j$ respectively, which are as the input of the Embedding Layer.

Embedding Layer: It takes the one-hot encodings, and embeds them into the continuous representations $\boldsymbol{v}_u = \boldsymbol{P}^T \boldsymbol{x}_u$, $\boldsymbol{v}_i = \boldsymbol{Q}_t^T \boldsymbol{x}_i$ and $\boldsymbol{v}_j = \boldsymbol{Q}_s^T \boldsymbol{x}_j$. The adversarial perturbations are added on them to get the representations of the adversarial examples $\boldsymbol{v}'_u = \boldsymbol{v}_u + n_u$, $\boldsymbol{v}'_i = \boldsymbol{v}_i + n_i$ and $\boldsymbol{v}'_j = \boldsymbol{v}_j + n_j$, where n_u, n_i and n_j are the adversarial perturbations of $\boldsymbol{v}_u$, $\boldsymbol{v}_i$ and $\boldsymbol{v}_j$ in n_{adv}. We merge them as $\boldsymbol{v}_{ui} = [\boldsymbol{v}_u, \boldsymbol{v}_i]$, $\boldsymbol{v}_{uj} = [\boldsymbol{v}_u, \boldsymbol{v}_j]$, $\boldsymbol{v}'_{ui} = [\boldsymbol{v}'_u, \boldsymbol{v}'_i]$ and $\boldsymbol{v}'_{uj} = [\boldsymbol{v}'_u, \boldsymbol{v}'_j]$, and input them into the Hidden Layers.

Hidden Layers: It transforms the representations from the Embedding Layer to the final representations. To learn abstractive features of the input, the size of the higher layer is halved, whose configuration is $[64 \rightarrow 32 \rightarrow 16 \rightarrow 8]$.

Output Layer: It predicts the score, based on the final representations from the Hidden Layers. We use a softmax layer to achieve the output, which is the probability of a positive interaction in the input pair.

3.3   Adversarial Training

Here we introduce the training procedure of our model. First, we optimize MLP++ to initialize our model parameters Θ. Then the procedure of model training can be seen as minimizing the worst-case error when the embedding parameters are perturbed by the perturbations. That can be interpreted as learning to play an adversarial game:

$$\theta_{emb}, n = \arg\min_{\theta_{emb}} \max_{\|n\| \leq \epsilon} L_{MLP++}(\theta_{emb}, \theta_f) + \lambda L_{MLP++}(\theta_{emb} + n, \theta_f), \quad (10)$$

where the maximizing player is the process of achieving perturbations n against the model, and the minimizing player is the procedure for optimizing embedding parameters θ_{emb}. Specifically, the adversarial training procedure includes two steps. First step, the adversarial examples are dynamically generated during the training process, and we use Eq. (8) to construct the adversarial perturbations and apply them to the embedding parameters get adversarial examples. Second step, ACDN can be trained with the objection function Eq. (9). Stochastic gradient descent (SGD) and adaptive moment method (Adam) [13] can be used to optimize the model. The update equation is:

$$\Theta^{new} \leftarrow \Theta^{old} - \eta \nabla_\Theta L_{ACDN}(\Theta), \quad (11)$$

where η is the learning rate. The two steps are iterated until the model converges.

4   Experiment

The experimental settings are introduced first, and then the following research questions are answered via experiments.

RQ1: How is the robustness of ACDN compared with the base model MLP++?

RQ2: How does ACDN perform compared with the state-of-the-art recommendation methods?

RQ3: How is the ability of ACDN to reduce labelled data compared with the advanced model?

RQ4: How do the hyper-parameters affect the model performance, such as noise level ϵ and adversarial regularization strength λ?

Table 1. Datasets and Statistics

Dataset	#User	Target Domain			Source Domain		
		#Items	#Interactions	Density	#Items	#Interactions	Density
Mobile	23,111	14,348	1,164,394	0.351%	29,921	617,146	0.089%
Amazon	80,763	93,799	1,323,101	0.017%	35,896	963,373	0.033%

4.1   Experimental Settings

Dataset. Table 1 summarizes the statistics of the two real-world cross-domain datasets, which are collected by [8]. The first dataset Mobile contains data of user reading news and app installation. The app installation and news reading are the target and source domain, respectively. The second dataset Amazon contains the two largest categories, Books and Movies, where Books is the target domain and Movies is the source domain. On the two sparse datasets, we hope to transfer knowledge form the source domain to improve the performance of the target domain recommendation.

Baselines. To evaluate the performance of ACDN, we compare it with following methods:

- **BPRMF** Bayesian Personalized Ranking [14], which extends Matrix Factorization by optimizing the pairwise rank loss, is a popular collaborative filtering method.
- **MLP** Multi-Layer Perceptron [6], which uses neural networks to learn user-item nonlinear interaction function, is a popular neural collaborative filtering method.
- **CMF** Collective Matrix Factorization [7] jointly factorizes two domains' matrices and transfers knowledge across two domains via the shared user factors.
- **CDCF** Cross-Domain Collaborative Filtering [16] extends factorization machines [15] by applying factorization on the domains which are merged with the shared users.
- **MLP++** It combines two MLPs by sharing user embedding matrix [8], which is our base model described in Sect. 2.2.
- **CSN** Cross-Stitch Network [17] introduces a linear combination of the activation maps to connect two neural networks and transfers knowledge between two domains via the shared coefficient.
- **CoNet** Collaborative cross networks [8], which is an improved MLP++, introduces cross connections to connect two neural networks to enable dual knowledge transfer between two domains.
- **SCoNet** Sparse collaborative cross networks [8] enforces a sparse prior on the structure of the cross connections to enable each network to learn the intrinsic representations.

Implementations. We implement our model with tensorflow. Adam [13] optimizer is used to learn model parameters, in which batch size and learning rate η are fixed to 128 and 0.001, respectively. The embedding size is 32 and the configuration of hidden layers is $[64 \rightarrow 32 \rightarrow 16 \rightarrow 8]$. We search the noise level ϵ in $\{0.01, 0.05, 0.1, 0.5, 1, 1.5, 2\}$, and the adversarial regularization strength λ in $\{0.001, 0.01, 0.1, 10, 100\}$.

Evaluation Protocol. We use the leave-one-out method [6,18,19] to evaluate the item recommendation model. We randomly sample one interaction for each user as the validation item to determine the model hyper-parameters. For each user, one interaction is reserved as the test item and 99 items which are not interacted are randomly sampled as negative examples. We evaluate the model by ranking the 100 items. We adopt the Hit Ratio (HR), the Normalized Discounted Cumulative Gain (NDCG) and the Mean Reciprocal Rank (MRR) as the evaluation metrics and cut the ranked list at K = 10. HR, which measures whether the Top-K list has the test item, is defined as:

$$HR = \frac{1}{|\mathcal{U}|} \sum_{u \in \mathcal{U}} \delta(p_u \leq K), \tag{12}$$

where $\delta(\cdot)$ is indicator function, and p_u is hit position of test item for user u. NDCG and MRR are defined respectively as:

$$NDCG = \frac{1}{|\mathcal{U}|} \sum_{u \in \mathcal{U}} \frac{\log 2}{\log(p_u + 1)}, \tag{13}$$

$$MRR = \frac{1}{|\mathcal{U}|} \sum_{u \in \mathcal{U}} \frac{1}{p_u}. \tag{14}$$

Note that a higher value is better.

4.2 Robustness Comparison (RQ1)

We compare the robustness of our model ACDN and the base model MLP++ by applying adversarial perturbations at different levels to MLP++ and ACDN. Table 2 shows the robustness of MLP++ and ACDN against adversarial perturbations.

Table 2. The impact of adding adversarial perturbations to the base model MLP++ and our model ACDN. Note that the number (%) is the relative decrease of performance.

Dataset	Metric	$\epsilon = 1$		$\epsilon = 5$		$\epsilon = 10$	
		MLP++	ACDN	MLP++	ACDN	MLP++	ACDN
Mobile	HR	−1.18	-	−28.82	−1.32	−56.42	−6.52
	NDCG	−2.09	-	−38.20	−2.12	−70.77	−8.75
	MRR	−2.27	-	−37.63	−2.13	−68.11	−8.67
Amazon	HR	−11.10	-	−67.89	−7.34	−79.06	−27.41
	NDCG	−13.90	-	−76.70	−10.93	−85.80	−36.77
	MRR	−11.46	-	−59.11	−10.19	−64.18	−32.46

From Table 2, we can see that MLP++ is very sensitive to adversarial perturbations. For example, on Mobile, adding adversarial perturbations at the noise

level of 5 to MLP++ can decrease HR by 28.82%, which means that the model
has poor robustness. And the performance on Amazon decreases more than on
Mobile at the same noise level. For example, we add the same noise level of 5
to MLP++ trained on Mobile and Amazon, respectively. NDCG on Amazon
decreases 76.70% which is about twice as much as it on Mobile. We think that it
is caused by the sparser data in Amazon, both in the source and target domains.
When the target domain is sparse, it is difficult to train a robust model. And
because of the sparse source domain, we may transfer more useless noise instead
of useful knowledge, which destroys the model robustness. However, by training MLP++ with adversarial examples, our model ACDN is less sensitive to
adversarial perturbations compared to MLP++ both on Mobile and Amazon.
At a noise level of 1, our model's performance is almost unchanged where the
numbers are too small and we do not show them. At a noise level of 5, MLP++
decreases HR by 28.82%, while the number is only 1.32% for ACDN on Mobile.
On Amazon, ACDN decreases MRR by 10.19%, while the number is 59.1% for
MLP++ at a noise level of 5. The results show that our model ACDN is robust to
adversarial perturbations, which indicates that the generalization performance
of our model is good.

4.3   Performance Comparison (RQ2)

In this section, we compare our model with baselines and discuss the findings.
Table 3 shows the recommendation performance of different methods on the two

Table 3. Experimental results of different methods on two real-world datasets. Note
that the best results are boldfaced, the second best are marked with * and the results
of the base model are underlined.

Dataset	Mobile			Amazon		
Metric	HR	NDCG	MRR	HR	NDCG	MRR
BPRMF	0.6175	0.4891	0.4489	0.4723	0.3016	0.2971
CMF	0.7879	0.5740	0.5067	0.3712	0.2378	0.1966
CDCF	0.7812	0.5875	0.5265	0.3685	0.2307	0.1884
MLP	0.8405	0.6615	0.6210	0.5014	0.3143	0.3113
MLP++	<u>0.8445</u>	<u>0.6683</u>	<u>0.6268</u>	<u>0.5050</u>	<u>0.3175</u>	<u>0.3053</u>
CSN	0.8458	0.6733	0.6366	0.4962	0.3068	0.2964
Conet	0.8480	0.6754	0.6373	0.5167	0.3261	0.3163
SConet	0.8583*	0.6887*	0.6475*	0.5338*	0.3424*	0.3351*
ACDN	**0.8692**	**0.7026**	**0.6647**	**0.5760**	**0.3818**	**0.3602**
Improve-best	1.27%	2.20%	2.66%	7.91%	11.51%	7.49%
Improve-base	2.92%	5.13%	6.05%	14.06%	20.25%	17.98%

datasets under three ranking metrics. The improve-best and improve-base are the relative improvements compared with the best baseline SConet and the base model MLP++, respectively. We can see that ACDN is better than all of the baselines on two datasets.

On Mobile, ACDN improves more than 25% in terms of MRR compared with CMF and CDCF, which shows the effectiveness of neural network methods. And compared with MLP, ACDN achieves 7.04% improvements in terms of MRR, showing the benefits of knowledge transfer. Our model betters the base model MLP++ by 6.05% in terms of MRR, which shows that by training with adversarial examples, we can improve performance of the cross-domain recommendation model. As compared to Conet, an improved MLP++, which has cross connection units, ACDN achieves 4.30% improvements in terms of MRR. Although our base model MLP++ is a simple and shallow knowledge transfer approach, we can still improve it better than Conet through our method. Our model improves 2.66% in terms of MRR compared with the best baseline SConet which enforces a sparse prior on the cross connections structure to select the useful representations and ignore the noisy ones. It means that our model, trained with adversarial examples which are constructed by adding adversarial perturbations on the embedding representations, has a better ability to get robust representations instead of useless noise.

On Amazon, our model achieves 11.51% and 20.25% improvements in terms of NDCG compared with the best baseline SConet and the base model MLP++, respectively. However, CMF and CDCF show worse performance than single-domain model BPRMF. And a similar phenomenon occurs between CSN and its single-domain model MLP, where CSN decreases 4.79% in terms of MRR compared with MLP. These models are all cross-domain models transferring knowledge on Amazon. We find that ACDN achieves greater improvements on Amazon, but some cross-domain models perform worse than single-domain models on Amazon. We think that it is caused by data sparsity of Amazon.

The Amazon dataset is much larger but sparser than the Mobile dataset, which brings us opportunities as well as challenges. On the one hand, a sparser dataset means that we are more likely to enjoy the benefits of knowledge transfer. For example, the HR of MLP is 84.45% on the Mobile dataset. The best cross-domain baseline SConet can only improve 1.47% compared with MLP. Mobile's target domain is not as sparse as Amazon's, so we have obtained good performance without knowledge transfer. In other words, the sparser the dataset, the more we can benefit from knowledge transfer. On the other hand, transferring knowledge on sparser datasets is more difficult. CMF, CDCF and CSN perform worse than single-domain models on Amazon. Because we may transfer useless noise instead of useful representations from the source domain. Our model trained with adversarial examples can learn robust parameter representations which can alleviate noise issues on the sparse data and achieves good performance.

4.4   Reducing Labelled Data Comparison (RQ3)

Transfer learning can reduce the labelled data which is hard to collect. It is an important ability to reduce the labelled data for cross-domain recommendation model. In this section, we compare our model ACDN with the state-of-the-art cross-domain model SConet to see their ability to reduce labelled data.

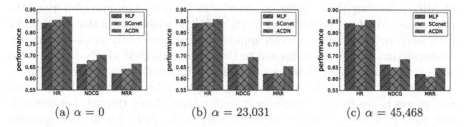

(a) $\alpha = 0$ \qquad (b) $\alpha = 23{,}031$ \qquad (c) $\alpha = 45{,}468$

Fig. 2. The performance of reducing α training examples on Mobile where α denotes the amount of reduction. Note that the results of MLP are fixed at $\alpha = 0$.

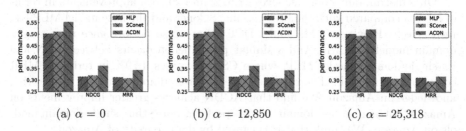

(a) $\alpha = 0$ \qquad (b) $\alpha = 12{,}850$ \qquad (c) $\alpha = 25{,}318$

Fig. 3. The performance of reducing α training examples on Amazon where α denotes the amount of reduction. Note that the results of MLP are fixed at $\alpha = 0$.

Figure 2 and Fig. 3 show the results of reducing training examples on Mobile and Amazon, respectively. We use MLP as the single-domain baseline and fix its results after it has been trained on the complete training examples. Then, we gradually reduce the number of training examples in the target domain for ACDN and SConet. When the cross-domain model uses reduced training examples and is still better than single-domain model MLP which uses the complete training examples, it means that transfer knowledge from source domain can reduce the labelled data. And the more training examples we can reduce, the stronger the ability of the cross-domain model to transfer knowledge from source domain.

On the Mobile dataset, each user has about 50 training examples on average. Figure 2(a) shows the performance when the models are trained with complete training examples. We first reduce one training example per user who has more than one training example to train SConet and ACDN. From Fig. 2(b), we can see that SConet achieves similar performance to MLP despite a reduction of 23,031 training examples and our model ACDN is better than both of them. Then, we reduce two training examples per user who has more than two training examples to train SConet and ACDN. Figure 2(c) shows that SConet is worse than MLP when 45,468 training examples are reduced. However, ACDN is still better than MLP. The results show that we can avoid collecting more than about 45,000 labelled data by using our model but SConet can only reduce about 25,000 labelled data. On the Amazon dataset, there are about 16 examples per user. Amazon is a much sparser dataset. We use the same setting as on the Mobile dataset. In Fig. 3, we find that SConet can reduce about 13,000 labelled data. However, our model ACDN can avoid collecting more than about 25,000 labelled data. The above experimental results show that ACDN is an effective way to alleviate the data sparsity issue and save the collecting data cost.

4.5 Hyper-parameter Influence (RQ4)

In this section, we explore the influences of noise level ϵ and strength of adversarial regularizer λ in our model ACDN. Due to the space limitation, we just show the results on Amazon dataset, and the results on the Mobile dataset are similar.

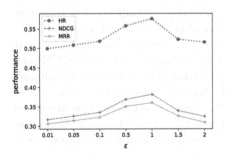

Fig. 4. Influence of noise level

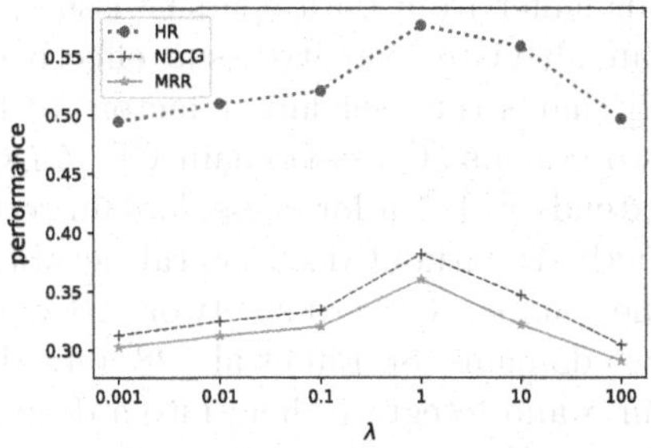

Fig. 5. Influence of adversarial regularization strength

Influence of Noise Level ϵ. We first fix λ to 1 and vary ϵ to explore the influence of noise level. We choose ϵ in $\{0.01, 0.05, 0.1, 0.5, 1, 1.5, 2\}$ to find the noise level which makes the performance of ACDN better. As is shown in Fig. 4, when ϵ is less than 0.1, ACDN behaves similarly to MLP++ and has minor improvements. Because the adversarial examples are almost equivalent to the original examples when the perturbations added to the parameters are too small. When ϵ is more than 1, the large ϵ doesn't work either. Too large perturbations may destroy the learning process of the model parameters. The optimal value is around 1, which is similar to ϵ on the Mobile dataset.

Influence of Adversarial Regularization Strength λ. Here we fix ϵ to 1 and vary λ to explore the influence of adversarial regularization strength. We search the λ from {0.001, 0.01, 0.1, 1, 10, 100} to optimize performance of our model. From Fig. 5, we can see that the optimal λ is 1. Whether λ is too large or too small, it will impair the performance of our model. It is worth mentioning that the Fig. 4 and Fig. 5 have similar changes. One possible explanation is that ϵ can also control the impact of the adversarial regularizer by control the noise level. Therefore, we can simply set λ to 1 to reduce the number of hyper-parameters.

5   Related Work

5.1   Collaborative Filtering

The collaborative filtering (CF) collects a user's prior preference history to provide her a personalized list of items [1]. Among the various CF methods, matrix factorization (MF) [3,4], which can learn from the linear interaction between latent factors of users and items, has been used in many recommendation methods [14,20–22]. BPRMF [14] is a prominent example, which optimizes a model through a user's relative preference over pairs of items. Due to the development of deep learning, there are many efforts for developing non-linear neural network models [6,8,23–27]. For example, neural collaborative filtering uses neural networks to learn the non-linear user-item interaction [6]. However, these CF methods suffer from data sparsity issues. Cross-domain recommendation technique can alleviate sparsity issue effectively. Collective matrix factorization(CMF) [7] shares the user latent factors to factorize user-item interaction matrices in two domain. Cross-Domain CF (CDCF) with factorization machines(FM) [15] extends FM [16] for cross-domain recommendation. These methods are shallow methods without deep neural networks. To transfer knowledge at a deep level, the collaborative cross networks (Conet) introduces cross connections to connect two domains [8]. Liu et al. [28] introduce aesthetic network to get aesthetic features and integrate them into a deep cross-domain network for item recommendation. However, these state-of-the-art recommendation methods rarely consider the robustness of the cross-domain recommendation model.

5.2   Adversarial Learning

The application of adversarial learning can be roughly divided into two classes: Generative Adversarial Networks(GAN) [30] and Adversarial Examples [9,10]. IRGAN [29] is the first to utilize GAN to unify the generative and discriminative information retrieval model. To improve pairwise ranking in recommendation, GAN is often used to generate difficult negative samples [33–35]. Wang et al. [33] adopt GAN to generate negative samples to optimize the neural memory networks for streaming recommendation. Geo-ALM [34] fuses geographic features and GAN used to sample negative examples for POI recommendation. Fan et al. [35] utilize GAN to generate negative items for item recommendation in item domain and use the bidirectional mapping approach to transfer the information of users across social domain. Different from GAN, adversarial examples are used to enhance the robustness of the model [9–12, 31, 32]. In [9, 10], it was found that a classier under normal supervised training process is vulnerable to adversarial examples. To solve the problem, adversarial examples are constructed to train the model. Miyato et al. [11] add adversarial perturbations to the word embeddings in the text domain. Wu et al. [32] extract relations in the multi-instance learning framework by applying adversarial training. Adversarial Personalized Ranking [12] minimizes the BPR objective function by utilizing adversarial examples. However, adversarial examples have not been used in collaborative cross-domain recommendation. To the best of our knowledge, we are the first to introduce adversarial examples to enhance the robustness of a cross-domain recommendation model and thus improve the generalization of the model.

6   Conclusion

In this paper, we find that the cross-domain recommendation model suffers from the poor robustness issue. To enhance the robustness of the cross-domain recommendation model, we propose a new Adversarial Cross-Domain Network (ACDN). We combine two multilayer perceptrons by sharing the user embedding matrix as our base model and construct adversarial examples by adding adversarial perturbations on the embedding representations to train it. We compare the robustness of our model and the base model at different noise levels and find that the model's robustness is improved a lot by training with adversarial examples. Experimental results demonstrate that ACDN achieves a good performance in terms of HR, NDCG and MRR, comparing with several competitive methods on two large real-world datasets. What's more, our model can reduce a lot of labelled data comparing with the state-of-the-art cross-domain recommendation model. In the future, we plan to combine adversarial examples and content information to further improve the performance of cross-domain recommendation.

Acknowledgments. This research was partially supported by NSFC (No. 61876117, 61876217, 61872258, 61728205), Suzhou Science and Technology Development Program (SYG201803), Open Program of Key Lab of IIP of CAS (No. IIP2019-1) and PAPD.

References

1. Lee, J., Lee, D., Lee, Y.C., et al.: Improving the accuracy of top-N recommendation using a preference model. Inf. Sci. **348**, 290–304 (2016)
2. Cantador, I, Cremonesi, P.: Tutorial on cross-domain recommender systems. In: RecSys, pp. 401–402. ACM (2014)
3. Koren, Y., Bell, R., Volinsky, C.: Matrix factorization techniques for recommender systems. Computer **42**(8), 30–37 (2009)

4. Mnih, A, Salakhutdinov, R.R.: Probabilistic matrix factorization. In: NIPS, pp. 1257–1264 (2017)
5. Xu, H., Mannor, S.: Robustness and generalization. Mach. Learn. **86**(3), 391–423 (2012)
6. He, X., Liao, L., Zhang, H., Nie, L., Hu, X., Chua, T.S.: Neural collaborative filtering. In: WWW, pp. 173–182. IW3C2 (2017)
7. Singh, A., Gordon, G.: Relational learning via collective matrix factorization. In: SIGKDD, pp. 650–658. ACM (2008)
8. Hu, G., Zhang, Y., Yang, Q.: CoNet: collaborative cross networks for cross-domain recommendation. In: CIKM, pp. 667–676. ACM (2018)
9. Szegedy, C, Zaremba, W, Sutskever, I, et al.: Intriguing properties of neural networks. http://arxiv.org/abs/1312.6199 (2013)
10. Goodfellow I J, Shlens J, Szegedy C.: Explaining and harnessing adversarial examples. http://arxiv.org/abs/1412.6572 (2014)
11. Miyato, T., Dai, A.M., Goodfellow, I.: Adversarial training methods for semi-supervised text classification. http://arxiv.org/abs/1605.07725 (2016)
12. He, X., He, Z., Du, X., et al.: Adversarial personalized ranking for recommendation. In: SIGIR, pp. 355–364. ACM (2018)
13. Kingma, D.P., Ba, J.: Adam: a method for stochastic optimization. arXiv preprint arXiv:1412.6980 (2014)
14. Rendle S, Freudenthaler C, Gantner Z, et al.: BPR: Bayesian personalized ranking from implicit feedback. In: UAI, pp. 452–461. AUAI Press (2009)
15. Loni, B., Shi, Y., Larson, M., Hanjalic, A.: Cross-domain collaborative filtering with factorization machines. In: de Rijke, M., et al. (eds.) ECIR 2014. LNCS, vol. 8416, pp. 656–661. Springer, Cham (2014). https://doi.org/10.1007/978-3-319-06028-6_72
16. Rendle, S.: Factorization machines with libFM. ACM Trans. Intell. Syst. Technol. **3**(3), 57 (2012)
17. Misra, I., Shrivastava, A., Gupta, A., et al.: Cross-stitch networks for multi-task learning. In: Proceedings of the IEEE Conference on Computer Vision and Pattern Recognition, pp. 3994–4003 (2016)
18. Bayer, I., He, X., Kanagal, B., et al.: A generic coordinate descent framework for learning from implicit feedback. In: WWW, pp. 1341–1350. IW3C2 (2017)
19. He, X., Zhang, H., Kan, M.Y., et al.: Fast matrix factorization for online recommendation with implicit feedback. In: SIGIR, pp. 549–558. ACM (2016)
20. Ning, X., Karypis, G.: SLIM: sparse linear methods for top-n recommender systems. In: IEEE 11th International Conference on Data Mining, pp. 497–506. IEEE (2011)
21. Shi, Y., Karatzoglou, A., Baltrunas, L., et al.: CLiMF: learning to maximize reciprocal rank with collaborative less-is-more filtering. In: RecSys, pp. 139–146. ACM (2012)
22. Kabbur, S., Ning, X., Karypis, G.: FISM: factored item similarity models for top-N recommender systems. In: SIGKDD, pp. 659–667. ACM (2013)
23. Dziugaite, G.K., Roy, D.M.: Neural network matrix factorization. http://arxiv.org/abs/1511.06443 (2015)
24. Chen, J., Zhang, H., He, X., et al.: Attentive collaborative filtering: multimedia recommendation with item-and component-level attention. In: Proceedings of the 40th International ACM SIGIR Conference on Research and Development in Information Retrieval, pp. 335–344. ACM (2017)
25. Zhao, P., Zhu, H., Liu, Y., et al.: Where to go next: a Spatio-temporal gated network for next POI recommendation. In: AAAI, pp. 5877–5884 (2019)

26. Xu, C., Zhao, P., Liu, Y., et al.: Recurrent convolutional neural network for sequential recommendation. In: WWW, pp. 3398–3404. IW3C2 (2019)
27. Zhang, T., Zhao, P., Liu, Y., et al.: Feature-level deeper self-attention network for sequential recommendation. In: IJCAI, pp. 4320–4326. AAAI Press (2019)
28. Liu, J., Zhao, P., Liu, Y., et al.: Deep cross networks with aesthetic preference for cross-domain recommendation. http://arxiv.org/abs/1905.13030 (2019)
29. Wang, J., Yu, L., Zhang, W., et al.: IRGAN: a minimax game for unifying generative and discriminative information retrieval models. In: SIGIR, pp. 515–524. ACM (2017)
30. Goodfellow, I., Pouget-Abadie, J., Mirza, M., et al.: Generative adversarial nets. In: NIPS, pp. 2672–2680 (2014)
31. Moosavi-Dezfooli, S.M., Fawzi, A., Fawzi, O., et al.: Universal adversarial perturbations. In: CVPR, pp. 1765–1773. IEEE (2017)
32. Wu, Y., Bamman, D., Russell, S.: Adversarial training for relation extraction. In: EMNLP, pp. 1778–1783. ACL (2017)
33. Wang, Q., Yin, H., Hu, Z., et al.: Neural memory streaming recommender networks with adversarial training. In: SIGKDD, pp. 2467–2475. ACM (2018)
34. Liu, W., Wang, Z.J., Yao, B., et al.: Geo-ALM: POI recommendation by fusing geographical information and adversarial learning mechanism. In: IJCAI, pp. 1807–1813. AAAI Press (2019)
35. Fan, W., Derr, T., Ma, Y., et al.: Deep adversarial social recommendation. http://arxiv.org/abs/1905.13160 (2019)

DDFL: A Deep Dual Function Learning-Based Model for Recommender Systems

Syed Tauhid Ullah Shah[1], Jianjun Li[1(✉)], Zhiqiang Guo[1], Guohui Li[2], and Quan Zhou[1]

[1] School of Computer Science and Technology,
Huazhong University of Science and Technology, Wuhan, China
{tauhidshah,jianjunli,georgeguo,quanzhou}@hust.edu.cn
[2] School of Software Engineering, Huazhong University of Science and Technology,
Wuhan, China
guohuili@hust.edu.cn

Abstract. Over the last two decades, latent-based collaborative filtering (CF) has been extensively studied in recommender systems to match users with appropriate items. In general, CF can be categorized into two types: matching function learning-based CF and representation learning-based CF. Matching function-based CF uses a multi-layer perceptron to learn the complex matching function that maps user-item pairs to matching scores, while representation learning-based CF maps users and items into a common latent space and adopts dot product to learn their relationship. However, the dot product is prone to overfitting and does not satisfy the triangular inequality. Different from latent based CF, metric learning represents user and item into a low dimensional space, measures their explicit closeness by Euclidean distance and satisfies the triangular inequality. In this paper, inspired by the success of metric learning, we supercharge metric learning with non-linearities and propose a Metric Function Learning (MeFL) model to learn the function that maps user-item pairs to predictive scores in the metric space. Moreover, to learn the mapping more comprehensively, we further combine MeFL with a matching function learning model into a unified framework and name this new model Deep Dual Function Learning (DDFL). Extensive empirical results on four benchmark datasets are conducted and the results verify the effectiveness of MeFL and DDFL over state-of-the-art models for implicit feedback prediction.

Keywords: Recommender systems · Collaborative filtering · Matching function learning · Metric function learning

---

Supported by the National Natural Science Foundation of China under Grant No. 61672252, and the Fundamental Research Funds for the Central Universities under Grant No. 2019kfyXKJC021.

Y. Nah et al. (Eds.): DASFAA 2020, LNCS 12114, pp. 590–606, 2020.
https://doi.org/10.1007/978-3-030-59419-0_36

1 Introduction

In recent years, with the rapid development of Internet technology, e-commerce, social media, and entertainment platforms, recommender systems (RSs) have been widely investigated. The job of a recommender system is to model user's preference on items based upon their past interactions, known as collaborative filtering (CF). To enrich RSs with such kind of capabilities, collaborative filtering based algorithms, specifically matrix factorization, have been used quite extensively in the literature [8,13,28]. Matrix factorization (MF) is the bedrock and most popular technique for CF based recommender systems. MF maps users and items into a common low-dimensional latent space and model the similarity between them by using the dot product or cosine similarity.

In the past few years, deep learning has achieved great success in computer vision, speech recognition and natural language processing [6,24]. Due to its excellent performance at representation learning, deep learning models have also been widely used in recommender systems. For example, He et al. [4] proposed a neural collaborative filtering (NCF) method that models the interaction between user and item latent factors with a multi-layer perceptron neural network for implicit feedback. In [29], by utilizing both explicit ratings and implicit feedback, a deep matrix factorization (DMF) model was proposed, which uses two distinct multi-layer perceptron (MLP) networks to learn user and item latent vectors and then models the user-item interaction by cosine similarity of the latent vectors. Recently, Deng et al. [3] categorized CF models into two types, i.e., representation learning-based CF and matching function learning-based CF, and proposed a Deep Collaborative Filtering (DeepCF) framework, which combines the strengths of these two types of CF models to achieve better performance.

Though deep latent-based methods have substantially improved the recommendation performance, they still need to utilize the dot product or cosine similarity to predict the unknown ratings or interactions. But as pointed out in a recent work [7], the inner product or cosine similarity does not satisfy the triangle inequality, which may limit the expressiveness of matrix factorization and result in sub-optimal solutions. In other words, though matrix factorization can capture user's general interests, it may fail to capture the fine-grained preference information when the triangle inequality is violated. To address this problem, collaborative metric learning (CML) [7] was proposed to encode user-item relationships and user-user/item-item similarity in a joint metric space, which results in superior accuracy over most existing CF methods. In [30], Zhang et al. proposed a new technique called factorized metric learning (FML) for both rating prediction and rating ranking. Instead of latent-vectors, they used low-dimensional distance vectors for both users and items and predicted the rating based on their explicit closeness.

In this paper, inspired by the success of metric learning on addressing the limitation of matrix factorization, as well as the superiority of deep metric learning on learning new metric space [14], we propose a metric function learning (MeFL) model. With MeFL, we first transform user-item interactions to user-item distances to get user's and item's dense distance representation, then utilize squared

Euclidean distance to derive a user-item distance vector, and finally stack multiple neural layers on the top of the distance vector to better learn the mapping between the user-item distance representation and the predictive score. Furthermore, to learn the mapping between user-item representation and predictive score more comprehensively, we combine MeFL with a matching function learning (MaFL) model, and term the joint model as Deep Dual Function Learning (DDFL). The proposed DDFL model can learn the mapping between user-item distance representation and predictive score, as well as the complex matching between user-item latent representation and the matching score jointly. Our main contributions can be summarized as follows.

- We propose a novel Metric Function Learning (MeFL) model that can effectively learn the complex mapping between user-item distance representation and predictive score, which satisfies the triangular inequality and yield to better recommendation performance.
- We propose a Deep Dual Function Learning (DDFL) model, which fuses MeFL and MaFL to learn both user and item distance vectors and latent features jointly for collaborative filtering.
- Extensive experiments are conducted on four publicly available datasets and the results demonstrate the effectiveness of our models over several state-of-the-art schemes.

2   Related Work

2.1   Collaborative Filtering-Based Recommendation

Modern recommender systems are mainly based on collaborative filtering (CF), where the idea is to utilize user-item past behavior to model user's preferences on items [22]. Among different CF models, latent factor-based CF models [1] are extensively adopted. Matrix Factorization (MF) [18] adopts latent vectors to represent users and items in a common low-dimensional space and use inner product to compute the preference score. Early literature on recommender systems mainly focused on explicit feedback, where users express his preference for items in the form of rating [12,20]. However, it is difficult to collect explicit ratings because most users do not incline to rate items. As a result, implicit feedback, e.g., mouse click, view, or purchase, outweighs the quantity of explicit feedback. The implicit collaborative filtering is formulated as an item-based recommendation task, which aims to recommend a list of items to users [2,5]. Implicit feedback based recommendation is a challenging task [2], where the key issue is to model the missing data [11]. Some early studies [9,19] adopt two distinct uniform weighting strategies, i.e., either treat all missing data as negative instances or sample negative instances from missing data.

Recently, due to the powerful learning abilities, deep learning attracted a vast interest in recommender systems. Restricted Boltzmann Machines (RBMS) was the first model attempting to use a two-layer network to model user's explicit feedback on items [21]. AutoRec [23] uses auto-encoder to determine the hidden structure for rating predictions. Collaborative Denoising Auto-encoders (CDAE) [26] further improves the performance of auto-encoders by utilizing both ratings and IDs. NueMF [4] replaces the dot product with a neural network to learn the matching function. It unifies generalized matrix factorization (GMF) and a multi-layer perceptron (MLP) under the NCF model. Deep matrix factorization (DMF) [29] follows a two pathway structure to map users and items into a low-dimensional space and utilize the inner product to compute recommendations. DeepCF [3] unifies the strengths of both matching function learning and representation learning under the CFNet framework. However, the representation part still resorts to the inner product, which is expensive in terms of computational time and does not satisfy the triangular inequality.

2.2 Metric Learning-Based Recommendation

Recently, there is also some work on employing metric learning for the recommendation task. Collaborative metric learning (CML) [7] replaces the widely adopted dot product with the Euclidean distance-based approach and satisfies the triangle inequality, which is essential to model user's preference for items. The user-item embedding is learned by maximizing the distance among users and their disliked items and also by minimizing the distance among the users and their desired items. Latent relational metric learning (LRML) [27] proposed a memory-based attention model for metric learning. LRML uses a centralized memory model to show the latent relationship between user and item. It parameterizes the global memories by a memory matrix, which is shared by user-item interaction data. Factorized Metric Learning (FML) [30] represents users and items as points in a low-dimensional space. FML first converts the similarity matrix to the distance matrix and then measure the explicit closeness between users and items by the squared-Euclidean distance.

Different from previous work, in this paper, we use a multi-layer perceptron to enrich metric learning with the ability of deep learning. Furthermore, we combine the strength of both matching function learning and our proposed metric function learning to bestow the joint model with great flexibility and faster computational speed while solving the triangular inequality and maintaining the ability to learn the low-rank relationships for implicit feedback.

3 Preliminaries

We consider a recommender system with M users $U = \{u_1, u_2, \ldots, u_M\}$ and N items $V = \{i_1, i_2, \ldots, i_N\}$. Compared with explicit feedback, such as ratings and reviews, implicit feedback can be tracked automatically and is much easier to obtain. Hence, in this paper, we focus on learning user-item interactions from implicit feedback. Following [4, 30], the user-item interaction matrix $\mathbf{Y} \in \mathbb{R}^{M \times N}$ is constructed as follows,

$$y_{ui} = \begin{cases} 1, & \text{if interaction } (u, i) \text{ is observed;} \\ 0, & \text{otherwise.} \end{cases} \tag{1}$$

Note here the observed user-item interaction $y_{ui} = 1$ only reflects user u's indirect preference on item i, it cannot tell how much u likes i. Likewise, $y_{ui} = 0$ also does not indicate that u does not like i, it may be the case that u have never seen i before. Due to the binary nature of implicit feedback, it only provides noisy signals about user's preference, which poses great challenge of lacking real negative instances for training. To address this problem, there are generally two ways, one is to treat all the unobserved interactions as weak negative instances [9,17], while the other one is sampling some negative instance from the unobserved interactions [4,17]. In this paper, following [4,17], we opt for the second method, i.e., randomly sample some unobserved interactions as negative instances.

Recommendation with implicit feedback can be formulated as a missing value prediction problem, which predicts the missing entries in the user-item interaction matrix $\mathbf{Y}$. The estimated missing values can then be utilized for item ranking. Model-based recommendation methods usually consider that data can be produced by an underlying model, which can be abstracted as learning,

$$\hat{y}_{ui} = f(u, i | \Theta) \tag{2}$$

where Θ, f and $\hat{y}_{ui}$ denote the model parameters, the interaction function (maps the model parameters to the prediction score) and the predicted score of the interaction between u and i, respectively. Different from explicit feedback, implicit feedback is binary, which makes predicting y_{ui} to be a binary classification problem. It is clear that solving such a problem cannot help rank and select the top ranking items for recommendation. To address this problem, the same as [3], we employ a probabilistic treatment for interaction matrix $\mathbf{Y}$ by assuming that y_{ui} obeys Bernoulli distribution,

$$P(y_{ui} = k | P_{ui}) = \begin{cases} p_{ui}, & k = 1; \\ 1 - p_{ui}, & k = 0 \end{cases} \tag{3}$$
$$= p_{ui}^k (1 - p_{ui})^{1-k},$$

where p_{ui} is the probability of y_{ui} being equal to 1, which can also be explained as the probability that user u is matched with item i. In other words, instead of modeling y_{ui} that is binary and discrete, we model p_{ui} (which is continuous) in our method. In this way, the binary classification problem is transformed to a matching score prediction problem, and the predicted score can be used for ranking items and finally making recommendations.

4   DDFL: Deep Dual Function Learning

4.1   Metric Function Learning (MeFL)

Regardless of the success of matrix factorization, the widely adopted dot product in MF-based recommendation models does not satisfy the triangular inequality, which is essential to model the fine-grained user-item similarity [7]. This may limit the expressiveness of the model and result in sub-optimal solutions. Different from matrix factorization, metric factorization [30] considers users and items

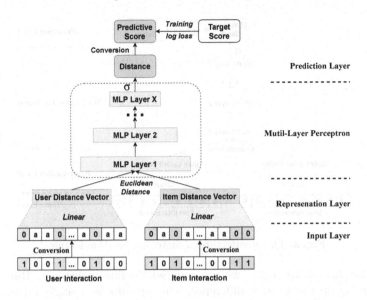

Fig. 1. The architecture of metric function learning.

as points in a low-dimensional coordinate system and models their relationship by using Euclidean distances. Specifically, it converts the implicit feedback to distance by using the following transformation,

$$d_{ui} = \alpha(1 - y_{ui}) \tag{4}$$

Since y_{ui} is equal to either 0 or 1, the distance $d_{ui} = 0$ if $y_{ui} = 1$, and $d_{ui} = \alpha$ if $y_{ui} = 0$, where $\alpha > 0$ is a user-defined distance factor.

Since we focus on implicit feedback in this work, we use the interaction matrix $\mathbf{Y}$ as input. Specifically, we use $\mathbf{v}_u^U = \mathbf{Y}_{u,*}$ to represent user u's interactions across all items, and use $\mathbf{v}_i^I = \mathbf{Y}_{*,i}$ to represent item i's interactions across all users. We first adopt the same operation as in Eq. (4) to transform the user and item interactions to user and item distances, denoted by $\mathbf{d}_u^U$ and $\mathbf{d}_i^I$, respectively. Considering that $\mathbf{d}_u^U$ and $\mathbf{d}_i^I$ are high dimensional and very sparse, we use a linear embedding layer to transform the sparse input into dense representation. Let $\mathbf{P} \in \mathbb{R}^{N \times K}$ and $\mathbf{Q} \in \mathbb{R}^{M \times K}$ denote the linear embedding matrices of users and items respectively, we have,

$$\begin{aligned} \mathbf{p}_u^d &= \mathbf{P}^T \mathbf{d}_u^U \\ \mathbf{q}_i^d &= \mathbf{Q}^T \mathbf{d}_i^I \end{aligned} \tag{5}$$

where $\mathbf{p}_u^d = (d_{u1}, d_{u2}, \ldots, d_{uK})$ and $\mathbf{q}_i^d = (d_{1i}, d_{2i}, \ldots, d_{Ki})$ denote the user and item distance vectors, respectively. From metric learning perspective, we calculate the distance vector by $\mathbf{p}_u^d$ and $\mathbf{q}_i^d$ as follows,

$$\mathbf{d}_0 = ((d_{u1} - d_{1i})^2, (d_{u2} - d_{2i})^2, \ldots, (d_{uK} - d_{Ki})^2) \tag{6}$$

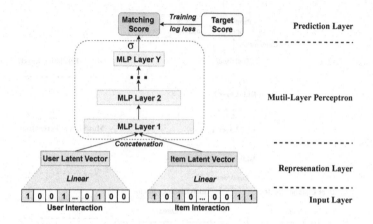

Fig. 2. The architecture of matching function learning.

Note that similar to [7,30], we use squared Euclidean distance. But, instead of summing up the squared difference, which results in a single value, we only calculate the squared difference to get the distance vector $\mathbf{d}_0$. Then, we can learn a better metric function non-linearly by adopting a deep neural network [14]. As shown in Fig. 1, we stack deep hidden layers on the top of $\mathbf{d}_0$ to better learn the metric function, endow non-linearity and further improve the flexibility of our model. By using an MLP, the distance between u and i can be modeled as,

$$\mathbf{d}_1 = a(\mathbf{W}_1^T \mathbf{d}_0 + \mathbf{b}_1),$$

$$......$$

$$\mathbf{d}_X = a(\mathbf{W}_X^T \mathbf{d}_{X-1} + \mathbf{b}_X),$$

$$\hat{d}_{ui} = \sigma(\mathbf{W}_{out}^T \mathbf{d}_X) \tag{7}$$

where $\mathbf{W}_x$, $\mathbf{b}_x$, $\sigma()$ and $\mathbf{d}_x$ denote the weight matrix, bias vector, sigmoid function and output of the x-the layer, respectively. $a()$ is the activation function and if not otherwise specified, we use ReLU, which is faster, plausible and less likely to overfitting. After obtaining $\hat{d}_{ui}$, the user-item distance score, we convert it to predictive score by,

$$\hat{y}_{ui} = 1 - \frac{\hat{d}_{ui}}{\alpha} \tag{8}$$

In summary, the MeFL model is implemented by Eq. (4)–(8).

4.2   Matching Function Learning (MaFL)

The matching function learning-based CF method learns the relationship between user-item representation and matching score by stacked neural layers. As deep neural networks have been proved to be capable of approximating any continuous function, such kind of method can better capture complex relationships. In fact, MaFL has been widely used in the literature, such as the MLP part in NeuMF [4], and the CFNet-ml component in DeepCF [3].

To make the paper self-contained, we briefly introduce MaFL. Figure 2 depicts the architecture of MaFL, where the input is the same as that in MeFL, i.e., $\mathbf{v}_u^U$ and $\mathbf{v}_i^I$. Let $\widetilde{\mathbf{P}} \in \mathbb{R}^{N \times K}$ and $\widetilde{\mathbf{Q}} \in \mathbb{R}^{M \times K}$ denote the linear embedding matrices of users and items, respectively. We first transform the sparse input into the dense user and item latent vectors as follows,

$$
\begin{aligned}
\mathbf{p}_u &= \widetilde{\mathbf{P}}^T \mathbf{v}_u^U \\
\mathbf{q}_i &= \widetilde{\mathbf{Q}}^T \mathbf{v}_i^I
\end{aligned}
\tag{9}
$$

Then, the MaFL model is defined as,

$$
\begin{aligned}
\mathbf{h}_0 &= \begin{bmatrix} \mathbf{p}_u \\ \mathbf{q}_i \end{bmatrix}, \\
\mathbf{h}_1 &= a(\mathbf{W}_1^T \mathbf{h}_0 + \mathbf{b}_0), \\
&\quad \ldots \ldots \\
\mathbf{h}_Y &= a(\mathbf{W}_Y^T \mathbf{h}_{Y-1} + \mathbf{b}_Y), \\
\hat{y}_{ui} &= \sigma(\mathbf{W}_{out}^T \mathbf{h}_Y)
\end{aligned}
\tag{10}
$$

where $\mathbf{W}_y$, $\mathbf{b}_y$ and $\mathbf{h}_y$ denote the weight matrix, bias vector and output of the y-th layer, respectively. $a()$ is the activation function and $\sigma()$ is the sigmoid function. Despite the simplicity of the concatenation operation, it allows us to maintain the information passed by previous layers and make full use of the non-linearity and high flexibility of the MLP structure.

4.3 Fusion and Learning

Fusion. We now introduce how to fuse MeFL and MaFL so that they can benefit each other. For MeFL, the model first calculates the distance score by learning the distance vector $\mathbf{d}_X$ and then converts the distance score to the predictive score by Eq. (9). Here, by Eq. (8), we can convert $\mathbf{d}_X$ to a predictive vector. For MaFL, the last layer of MLP is called the predictive vector, as well. In both cases, the predictive vectors can be viewed as the representation for the corresponding user-item pair. Since these two types of methods learn the predictive vectors from different perspectives and they have different advantages, the concatenation of the two predictive vectors will result in a more robust joint representation for the user-item pair. Suppose the predictive vector of MeFL and MaFL are $\mathbf{d}_X^{\mathrm{Me}}$ and $\mathbf{h}_Y^{\mathrm{Ma}}$, respectively, a widely used fusing strategy is to concatenate the learned representation and then feed it into a fully connected layer, then the output of the fusion model can be defined as,

$$
\hat{y}_{ui} = \sigma\left(\mathbf{W}_{comb}^T \begin{bmatrix} \mathbf{d}_X^{\mathrm{Me}} \\ \mathbf{h}_Y^{\mathrm{Ma}} \end{bmatrix}\right)
\tag{11}
$$

We call such a model DualFL, short for dual function learning.

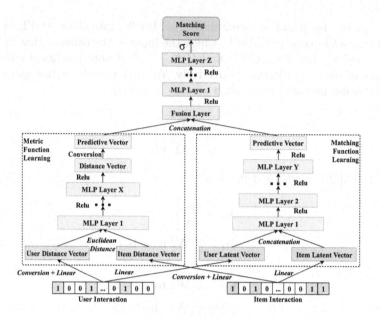

Fig. 3. The architecture of DDFL

Moreover, inspired by the non-linear learning characteristics of deep neural networks, we have an conjecture that it is possible to learn a better model for user-item interaction by stacking multiple layers on the top of the fusion. To verify our assumption, we first concatenate $\mathbf{d}_X^{\text{Me}}$ and $\mathbf{h}_Y^{\text{Ma}}$ to obtain,

$$\mathbf{f}_0 = (\begin{bmatrix} \mathbf{d}_X^{\text{Me}} \\ \mathbf{h}_Y^{\text{Ma}} \end{bmatrix}) \tag{12}$$

Then, we stack multiple deep neural layers on the top of $\mathbf{f}_0$ by,

$$\mathbf{f}_1 = a(\mathbf{W}_1^T \mathbf{f}_0 + \mathbf{b}_0),$$
$$\cdots \cdots$$
$$\mathbf{f}_Z = a(\mathbf{W}_Z^T \mathbf{f}_{Z-1} + \mathbf{b}_Z), \tag{13}$$
$$\hat{y}_{ui} = \sigma(\mathbf{W}_{out}^T \mathbf{f}_Z)$$

where $\mathbf{f}_z$ denotes the output of the z-th layer, and the meanings of other notions are the same as that in MaFL. We use the tower-like structure for the overall design of our network, where the bottom-most is the broadest and successive layers have a smaller amount of neurons. We call such a fusion model deep dual function learning (DDFL). Figure 3 illustrates the architecture of DDFL.

Learning. For parameters Θ, existing schemes estimate it by optimizing an objective function. Two types of objective functions, i.e., point-wise loss [5,9] and pair-wise loss [15,19,25], are commonly used in the literature. In this paper,

Table 1. Statistics of the four datasets

Dataset	#User	#item	#Interaction	Sparsity
ml-1m	6,040	3,706	1,000,209	0.9553
lastfm	1,741	2,665	69,149	0.9851
AMusic	1,776	12,929	46,087	0.9980
AKindle	7,795	30,171	372,867	0.9984

we use point-wise loss and leave the exploration of the pair-wise loss as future work. To learn model parameters, the squared loss is widely used in existing models [13,16,30] as a point-wise loss. For implicit CF, the squared loss is not a suitable choice, because it is derived by assuming the error between the given rating r_{ui} and the predicated rating $\hat{r}_{ui}$ obeys a normal distribution. This does not hold for an implicit feedback scenario since y_{ui} is discrete and binary. Hence, following [3], we use the binary cross-entropy function as the loss function,

$$L = -\sum_{(u,i) \in Y^+ \cup Y^-} y_{ui} \log \hat{y}_{ui} + (1 - y_{ui}) \log(1 - \hat{y}_{ui}) \tag{14}$$

where Y^+ and Y^- denote set of observed interactions and sampled negative instances, respectively.

To optimize the model, we use mini-batch Adam [10], which dynamically adjusts the learning rate and leads to very fast convergence. We uniformly sample the negative instances Y^- from the observed interactions in each iteration.

Pre-training. The pre-trained models can be used to initialize the ensemble framework, which can significantly change the performance and lead to faster convergence. Since DDFL is composed of MeFL and MaFL, we pre-train both models with random initialization until convergence and use their model parameters to initialize DDFL. As mentioned earlier, we train MeFL and MaFL with Adam. But instead of Adam, we optimize DDFL with vanilla SGD combined with the pre-trained parameters. This is due to the momentum-based nature of Adam, which makes it unsuitable for DDFL with pre-trained parameters.

5   Experiments

We conduct experiments to answer the following research questions **(RQs)**.

- **RQ1:** Do the proposed MeFL and DDFL models outperform the state-of-the-art implicit collaborative filtering methods?
- **RQ2:** How does pre-training impact the performance of DDFL?
- **RQ3:** Are deeper hidden layers helpful for the performance of DDFL?
- **RQ4:** How does DDFL perform with different values of hyper-parameters?

Table 2. Comparison results of different methods in terms of NDCG@10 and HR@10.

Datasets	Measures	Existing methods					Ours			Improv. of DDFL
		NeuMF	DMF	FML	MLP	DeepCF	MeFL	DualFL	DDFL	over DeepCF
ml-1m	HR	0.7205	0.6572	0.6935	0.7057	**0.7257**	0.7131	0.7289	**0.7383**	1.74%
	NDCG	0.4376	0.3764	0.4168	0.4271	**0.4432**	0.4327	0.4487	**0.4560**	2.89%
lastfm	HR	0.8892	0.8882	0.8626	0.8794	**0.8950**	0.8894	0.9018	**0.9156**	2.30%
	NDCG	0.6010	0.5820	0.5847	0.5852	**0.6190**	0.6012	0.6214	**0.6424**	3.78%
AMusic	HR	0.3885	0.3751	0.3830	0.4060	**0.4165**	0.3987	0.4257	**0.4412**	5.93%
	NDCG	0.2387	0.2022	0.2453	0.2323	**0.2526**	0.2521	0.2677	**0.2713**	7.40%
AKindle	HR	0.7543	0.7219	0.7368	0.7497	**0.7690**	0.7564	0.7784	**0.8018**	4.91%
	NDCG	0.4801	0.4232	0.4347	0.4810	**0.4928**	0.4828	0.5048	**0.5170**	5.11%

5.1   Datasets

We conduct experiments on four widely used datasets in recommender systems: MovieLens 1M (ml-1m)[1], Lastfm (lastfm)[2], Amazon music (AMusic) and Amazon Kindle (AKindle)[3]. We do not process the ml-1m dataset, as the provider has already preprocessed it. For the remaining datasets, we filter them to ensure that only users with at least 20 interactions and items with 5 interactions are retained. The statistics of the four datasets are reported in Table 1.

5.2   Compared Methods

As our model aims to learn the relationship between users and items, we mainly compare it with user-item methods, including,

- **NueMF** [4] fuses generalized matrix factorization with and a multi-layer perceptron. It treats the recommendation problem with implicit feedback as a binary classification problem and utilizes log loss as the loss function.
- **DMF** [29] is a neural network architecture, which maps users and items into a common latent low dimensional space. It also designs a new normalized cross-entropy loss that makes use of both explicit ratings and implicit feedback. We only utilize implicit feedback as input in our experiments.
- **FML** [30] represents user and item in a low-dimensional coordinate system with squared loss and uses the Euclidean distance to measure the similarity between user and item.
- **MLP** [3,4] is a method that adopts a concatenation operation with stacked neural layers to learn the user-item nonlinear relationship with log loss as the loss function.
- **DeepCF** [3] combines the strengths of matching function learning based CF and representation learning-based CF. The same as NeuMF, it uses implicit feedback as input and log loss as the loss function.

---

[1] https://grouplens.org/datasets/movielens/.
[2] http://trust.mindswap.org/FilmTrust/.
[3] http://jmcauley.ucsd.edu/data/amazon/.

Table 3. Performance of DDFL with/without pre-training.

Factors	ml-1m				Lastfm			
	Without pre-training		With pre-training		Without pre-training		With pre-training	
	HR	NDCG	HR	NDCG	HR	NDCG	HR	NDCG
8	0.7005	0.4275	0.7255	0.4432	0.8439	0.5475	0.8996	0.5985
16	0.7058	0.4308	0.7342	0.4498	0.8612	0.5667	0.9078	0.6199
32	0.7046	0.4327	0.7334	0.4536	0.8751	0.5744	0.9073	0.6324
64	0.7091	0.4366	**0.7383**	**0.4560**	0.8807	0.5995	**0.9156**	**0.6424**
Factors	Amusic				Akindle			
	Without pre-training		With pre-training		Without pre-training		With pre-training	
	HR	NDCG	HR	NDCG	HR	NDCG	HR	NDCG
8	0.3494	0.2306	0.4323	0.2538	0.7410	0.4501	0.7861	0.4989
16	0.3679	0.2204	0.4296	0.2647	0.7475	0.4584	0.7912	0.5048
32	0.3752	0.2272	0.4247	0.2602	0.7538	0.4659	0.7965	0.5104
64	0.3956	0.2306	**0.4422**	**0.2713**	0.7602	0.4728	**0.8018**	**0.5170**

5.3 Evaluation Protocols

To evaluate the performance of item recommendation, we adopt the widely used *leave-one-out* evaluation [2, 4]. We keep the latest interaction of every user as a test item and manipulate the remaining for training. As it takes excessive time to rank all items for every user during the evaluation, following [3, 4], we randomly sample 100 items that are not yet interacted by the user. For prediction, we further rank the test item. We employ two widely used metrics, Hit Ratio (HR) and Normalized Discounted Cumulative Gain (NDCG), to evaluate the ranking performance. For both measures, we truncated the ranked list at 10 for both evaluation metrics. By definition, the HR measures whether the test item exists in the top 10 list, and the NDCG measures the ranking, which gives higher scores to hits at top position ranks.

5.4 Implementation Details

We implemented DDFL with *Keras*[4] and *Tensorflow*[5]. For hyper-parameters, we report the model which performs the best on the test set. The models are trained until convergence with a maximum of 20 epochs for every dataset. We set the batch size to 256 and tune the learning rate for all models amongst {0.001, 0.005, 0.0001, 0.0005}. To train DDFL from scratch, we randomly initialized the model parameters with the Gaussian distribution (with a mean of 0 and standard deviation of 0.01). For FML, we use the released source code[6] and change the loss function to binary cross-entropy. We tune the number of dimensions amongst

---

[4] https://github.com/keras-team/keras.
[5] https://github.com/tensorflow/tensorflow.
[6] https://github.com/cheungdaven/metricfactorization.

Table 4. Performance of DDFL with different number of layers.

Dataset	Measure	1-Layer	2-Layer	3-Layer	4-Layer	5-Layer
ml-1m	HR	0.7114	0.7173	0.7229	0.7284	**0.7383**
	NDCG	0.4291	0.4368	0.4421	0.4474	**0.4560**
lastfm	HR	0.8894	0.8939	0.9007	0.8084	**0.9156**
	NDCG	0.6176	0.6229	0.6298	0.6334	**0.6426**
AMusic	HR	0.4016	0.4085	0.4168	0.4277	**0.4422**
	NDCG	0.2701	0.2654	0.2591	0.2671	**0.2713**
AKindle	HR	0.7728	0.7798	0.7856	0.7914	**0.8018**
	NDCG	0.4952	0.4934	0.4987	0.5059	**0.5170**

{32, 64, 128, 256} and report the best results. For NueMF, we follow the same tower-like architecture proposed in [4] with 3-fully connected layers for MLP. For DeepCF, we follow the same architecture[7] with the best results. For our model, the distance factor α is set to 1.0.

5.5 Experimental Results

Performance Comparison (RQ1). The comparison results of our models MeFL, DualFL and DDFL versus other baselines are summarized in Table 2, where the best results in both categories (DeepCF and DDFL) are in boldface. The results demonstrate the consistent performance improvements of our models over state-of-the-art methods on all four datasets in terms of HR and NDCG. As can be observed, the most significant performance gap between DDFL and DeepCF can reach up to 7.40% on the Amusic dataset. From Table 2, we can see that DDFL achieves the best results among all the evaluated methods, which demonstrates that combining MeFL and MaFL jointly indeed can lead to better performance. Moreover, it can be observed that MeFL always performs better than FML, illustrating the effectiveness of applying deep neural networks on metric learning. Finally, the consistent performance improvement of DDFL over DualFL also verifies our conjecture that stacking multiple layers on the top of the fusion results can lead to a better model for learning the mapping relationship.

Impact of Pre-training (RQ2). To show the impact of pre-training on DDFL, we conduct experiments to compare the performance of DDFL with and without pre-training. Specifically, we conduct the experiments with a 5 layer-model and the predictive factors are tuned amongst 8, 16, 32, 64. In the case of without pre-training, we use mini-batch-Adam with random initialization to evaluate DDFL. As shown in Table 3, DDFL with pre-training always achieves better performance than DDFL without pre-training on every dataset. Moreover, DDFL generates the best result with 64 predictive factors for every dataset. The results

---

[7] https://github.com/familyld/DeepCF.

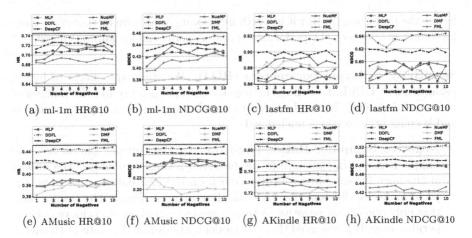

(a) ml-1m HR@10   (b) ml-1m NDCG@10   (c) lastfm HR@10   (d) lastfm NDCG@10

(e) AMusic HR@10   (f) AMusic NDCG@10   (g) AKindle HR@10   (h) AKindle NDCG@10

Fig. 4. Performance w.r.t. the number of negative samples per positive instance

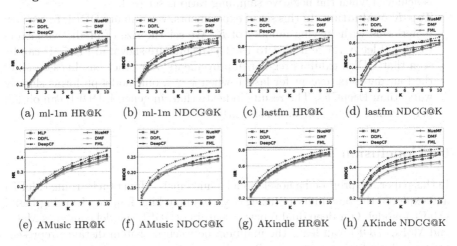

(a) ml-1m HR@K   (b) ml-1m NDCG@K   (c) lastfm HR@K   (d) lastfm NDCG@K

(e) AMusic HR@K   (f) AMusic NDCG@K   (g) AKindle HR@K   (h) AKinde NDCG@K

Fig. 5. Evaluation of Top-K item recommendation where K ranges from 1 to 10

demonstrate the usefulness and effectiveness of DDFL with pre-training and also suggest that the model can produce better results with more predictive factors, since it leads to the superior capability of representation.

Number of Layers in the Networks (RQ3). DDFL contains several deep hidden layers. Hence, it is essential to see whether using more layers is beneficial for it. To this end, we examine DDFL's performance with a various number of hidden layers. The results for both HR and NDCG with a different number of layers are shown in Table 4. As can be observed, stacking more layers on the top is indeed beneficial for the recommendation accuracy. The more layers, the better the performance. But it is worth mentioning that if we further stack more layers (after 5 layers), the performance of our model cannot improve much.

Sensitivity to Hyper-parameters (RQ4). We study the effects of hyper-parameters of DDFL, including the negative sampling ratio and the number of recommended items. We conduct experiments by varying these hyper-parameters and keeping other settings unchanged.

- **Negative Sampling Ratio.** When training the network parameters, we sample negative instances form the observed ratings. To analyze the impact of the negative sampling ratio for DDFL on the four datasets, we show the performance of different negative sampling ratios in Fig. 4. The results show that only one or two negative instances are not significant to achieve optimal performance. It is observed that our model outperforms the state-of-the-art in all the cases. The best performance on ml-1m for both HR@10 and NDCG@10 is achieved when the negative sampling ratio is set to 4.
- **Top-K evaluation.** In this set of experiments, we evaluate DDFL to see how it performs with varying number of K. We set the size of predictive factors to 64 and let K range from 1 to 10. The results are shown in Fig. 5. It can be seen that DDFL exhibits regular improvements over other baselines on all of the four datasets (except for lastfm when $k < 4$, our model performs slightly worse than DeepCF). The results indicate that, in general, DDFL can obtain performance improvement consistently.

6   Conclusion

In this paper, we proposed a novel model named metric function learning (MeFL) and explored the possibility of fusing it with a matching function learning-based (MaFL) model. Our deep dual function learning (DDFL) model combines MeFL and MaFL. DDFL can learn the mapping between user-item distance representation and predictive score, as well as the complex matching between user-item latent representation and the matching score jointly. DDFL not only exploits the benefits of metric learning but also takes the powerful abilities of deep learning into consideration. Moreover, our model points out the importance of fusing these two models, allowing the model to efficiently learn the low-rank relationship between users and items while satisfying the triangular inequality. Extensive experiments on four real-world datasets demonstrate the superior performance of our model over several state-of-the-art models. For future work, we plan to extend DDFL to model auxiliary information to improve the initial representation of both users and items for further performance improvement.

References

1. Aggarwal, C.C., Parthasarathy, S.: Mining massively incomplete data sets by conceptual reconstruction. In: Proceedings of ACM SIGKDD, pp. 227–232 (2001)

2. Bayer, I., He, X., Kanagal, B., Rendle, S.: A generic coordinate descent framework for learning from implicit feedback. In: Proceedings of WWW, pp. 1341–1350 (2017). https://doi.org/10.1145/3038912.3052694

3. Deng, Z., Huang, L., Wang, C., Lai, J., Yu, P.S.: DeepCF: a unified framework of representation learning and matching function learning in recommender system. In: Proceedings of AAAI, pp. 61–68 (2019). https://doi.org/10.1609/aaai.v33i01.330161

4. He, X., Liao, L., Zhang, H., Nie, L., Hu, X., Chua, T.: Neural collaborative filtering. In: Proceedings of WWW, pp. 173–182 (2017)

5. He, X., Zhang, H., Kan, M., Chua, T.: Fast matrix factorization for online recommendation with implicit feedback. In: Proceedings of ACM SIGIR, pp. 549–558 (2016). https://doi.org/10.1145/2911451.2911489

6. Hong, R., Hu, Z., Liu, L., Wang, M., Yan, S., Tian, Q.: Understanding blooming human groups in social networks. IEEE Trans. Multimedia 17(11), 1980–1988 (2015)

7. Hsieh, C., Yang, L., Cui, Y., Lin, T., Belongie, S.J., Estrin, D.: Collaborative metric learning. In: Proceedings of WWW, pp. 193–201 (2017)

8. Hu, Q., Zhao, Z., Wang, C., Lai, J.: An item orientated recommendation algorithm from the multi-view perspective. Neurocomputing 269, 261–272 (2017)

9. Hu, Y., Koren, Y., Volinsky, C.: Collaborative filtering for implicit feedback datasets. In: Proceedings of ICDM, pp. 263–272 (2008)

10. Kingma, D.P., Ba, J.: Adam: a method for stochastic optimization. In: Proceedings of ICLR (Poster) (2015). http://arxiv.org/abs/1412.6980

11. Koren, Y.: Factorization meets the neighborhood: a multifaceted collaborative filtering model. In: Proceedings of ACM SIGKDD, pp. 426–434 (2008)

12. Koren, Y.: Collaborative filtering with temporal dynamics. In: Proceedings of ACM SIGKDD, pp. 447–456 (2009)

13. Koren, Y., Bell, R.M., Volinsky, C.: Matrix factorization techniques for recommender systems. IEEE Comput. 42(8), 30–37 (2009)

14. Li, W., Huo, J., Shi, Y., Gao, Y., Wang, L., Luo, J.: Online deep metric learning. CoRR abs/1805.05510 (2018)

15. Mnih, A., Teh, Y.W.: Learning label trees for probabilistic modelling of implicit feedback. In: Proceedings of NIPS, pp. 2825–2833 (2012)

16. Ning, X., Karypis, G.: SLIM: sparse linear methods for top-N recommender systems. In: Proceedings of ICDM, pp. 497–506 (2011)

17. Pan, R., et al.: One-class collaborative filtering. In: Proceedings of ICDM, pp. 502–511 (2008)

18. Paterek, A.: Improving regularized singular value decomposition for collaborative filtering. In: Proceedings of KDD Cup and Workshop 2007, pp. 5–8 (2007)

19. Rendle, S., Freudenthaler, C., Gantner, Z., Schmidt-Thieme, L.: BPR: Bayesian personalized ranking from implicit feedback. In: Proceedings of UAI, pp. 452–461 (2009)

20. Rendle, S., Schmidt-Thieme, L.: Online-updating regularized kernel matrix factorization models for large-scale recommender systems. In: Proceedings of ACM RecSys, pp. 251–258 (2008)

21. Salakhutdinov, R., Mnih, A., Hinton, G.E.: Restricted Boltzmann machines for collaborative filtering. In: Proceedings of ICML, pp. 791–798 (2007)

22. Sarwar, B.M., Karypis, G., Konstan, J.A., Riedl, J.: Item-based collaborative filtering recommendation algorithms. In: Proceedings of WWW, pp. 285–295 (2001)

23. Sedhain, S., Menon, A.K., Sanner, S., Xie, L.: AutoRec: autoencoders meet collaborative filtering. In: Proceedings of WWW, pp. 111–112 (2015). https://doi.org/10.1145/2740908.2742726

24. Serban, I.V., Sordoni, A., Bengio, Y., Courville, A.C., Pineau, J.: Building end-to-end dialogue systems using generative hierarchical neural network models. In: Proceedings of AAAI, pp. 3776–3784 (2016)

25. Socher, R., Chen, D., Manning, C.D., Ng, A.Y.: Reasoning with neural tensor networks for knowledge base completion. In: Proceedings of NIPS, pp. 926–934 (2013)

26. Strub, F., Mary, J.: Collaborative filtering with stacked denoising autoencoders and sparse inputs. In: NIPS Workshop on Machine Learning for eCommerce, Montreal, Canada (2015)

27. Tay, Y., Tuan, L.A., Hui, S.C.: Latent relational metric learning via memory-based attention for collaborative ranking. In: Proceedings of WWW, pp. 729–739 (2018)

28. Wang, C., Deng, Z., Lai, J., Yu, P.S.: Serendipitous recommendation in E-commerce using innovator-based collaborative filtering. IEEE Trans. Cybern. **49**(7), 2678–2692 (2019)

29. Xue, H., Dai, X., Zhang, J., Huang, S., Chen, J.: Deep matrix factorization models for recommender systems. In: Proceedings of IJCAI, pp. 3203–3209 (2017). https://doi.org/10.24963/ijcai.2017/447

30. Zhang, S., Yao, L., Huang, C., Xu, X., Zhu, L.: Position and distance: recommendation beyond matrix factorization. CoRR abs/1802.04606 (2018). http://arxiv.org/abs/1802.04606

Zero-Injection Meets Deep Learning: Boosting the Accuracy of Collaborative Filtering in Top-N Recommendation

Dong-Kyu Chae, Jin-Soo Kang, and Sang-Wook Kim$^{(\boxtimes)}$

Hanyang University, Seoul, South Korea
{dongkyu,jensoo7023,wook}@hanyang.ac.kr

Abstract. *Zero-Injection* has been known to be very effective in alleviating the *data sparsity* problem in *collaborative filtering* (CF), owing to its idea of finding and exploiting *uninteresting items* as users' negative preferences. However, this idea has been only applied to the linear CF models such as SVD and SVD++, where the linear interactions among users and items may have a limitation in fully exploiting the additional negative preferences from uninteresting items. To overcome this limitation, we explore CF based on *deep learning* models which are highly flexible and thus expected to fully enjoy the benefits from *uninteresting items*. Empirically, our proposed models equipped with Zero-Injection achieve great improvements of recommendation accuracy under various situations such as basic top-N recommendation, long-tail item recommendation, and recommendation to cold-start users.

Keywords: Recommender systems · Collaborative filtering · Data sparsity · Zero-injection

1 Introduction

The *collaborative filtering* (CF) is one of the most widely used methods in top-N recommendation tasks [3,18]. CF relies on users' prior *preference* history, usually represented by a *rating matrix* where each cell indicates a *rating score* given by a *user* to an *item*. However, most users tend to rate only a small proportion of available items, which results in the so-called *data sparsity* problem [10]. Indeed, this problem has been a universal obstacle to most CF methods, making them face difficulty in learning preferences of users on items.

We note, in particular, that the recently proposed *Zero-Injection* [10,14,16] is arguably one of the most successful methods to address the sparsity problem in the context of pure CFs (i.e., recommendation tasks given only with the user-item interaction data). It carefully finds the *uninteresting items* (U-items in short, hereafter) that each user has not rated yet but is unlikely to prefer even if recommended to her. Then, it injects "0" rating to the identified U-items in the original rating matrix so as to exploit the U-items for each user additionally

© Springer Nature Switzerland AG 2020
Y. Nah et al. (Eds.): DASFAA 2020, LNCS 12114, pp. 607–620, 2020.
https://doi.org/10.1007/978-3-030-59419-0_37

in CF as her *negative* preferences. Empirically, several CF methods employed in [10,14] improved their accuracy significantly when applied to this enriched matrix.

Despite its promising idea of enriching the rating matrix by carefully chosen U-items, it has been applied only to the linear *matrix factorization* (MF) models such as SVD [12] and SVD++ [11], and heuristic user-based or item-based CFs [19]; applying Zero-Injection to recently proposed CF models based on *deep neural networks* (DNN) has not been explored yet. Even though SVD equipped with Zero-Injection has reported the state-of-the-art accuracy in [10] and [14], we believe that there are still more rooms to achieve further improvement by using DNN-based models with higher complexity beyond the linear models.

Recently, DNN has been very successful in various areas, such as computer vision [13], natural language processing [1], and speech recognition [7], and applying DNN to recommendation systems has also gained significant attention. Indeed, DNN's capability of discovering *non-linear* interactions of users' and items' latent factors has meaningfully improved the performance of recommendation in the literature, where notable work includes AutoRec [20], CDAE [24], and NCF [6]. Furthermore, DNN's high model capacity is generally known to allow for better accuracy when there exist more training data. We thus expect that DNN and Zero-Injection bring together a great synergy effect in CF tasks, thereby providing even more accurate results.

Along this line, this paper proposes three DNN-based CF models: they are on the basis of *Autoencoder*, one of the most popular DNNs in the field of CF. One of our models uses the basic form of Autoencoder, and the other two incorporate the *denoising* scheme and the *adversarial training* technique, respectively. Each model successfully embeds the idea of Zero-Injection in its own way. To validate the effectiveness of our proposed models, we conduct extensive experiments by using four real-world datasets and four evaluation metrics. The results demonstrate that our DNN-based CF models equipped with Zero-Injection are much more accurate than the previous linear models as well as several state-of-the-arts in this field. Notably, these results are consistently shown not only under basic top-N settings, but also under more-difficult settings: *long-tail* item recommendations and recommendations to *cold-start users*.

The rest of the paper is organized as follows. Section 2 explains how the U-items are chosen. Section 3 introduces our DNN-based CF models exploiting the chosen U-tems. Section 4 elaborates our experimental settings. Section 5 reports the results of our extensive experiments. Section 6 summarizes and concludes the paper.

2   Finding U-Items

Our proposed models exploit the notion of *uninteresting items* (U-items, in short), which is the core part of Zero-Injection [10]. Hence, we here elaborate what the U-items are and how we find them among the vast amount of unrated items.

U-items of a user are defined as the items with her low *pre-use preference*. Here, a user's pre-use preference on an item is her judgment on the item *before* using it. For example, a user's pre-use preferences for items that have already been rated by her are set to *one* (i.e., the highest score), since she must have bought the items that she might be interested in (i.e., having a very high pre-use preference). Then, the rest (unknown) pre-use preferences of hers on all unrated items should be inferred based on the observed pre-use preferences on rated items. This inference is naturally formulated as *one-class collaborative filtering* (OCCF) [9,24] by using the pre-use preference matrix $\mathbb{P} = (p_{ui})_{m \times n}$, where each entry p_{ui} is 1 if it is observed and empty otherwise, as training data. Among the inferred pre-use preferences, where each $\hat{p}_{ui}$ implies the degree of a user u's interestingness on an item i, the bottom θ percent of $\hat{p}_{ui}$ whose values are the *lowest* are selected; then, each corresponding i is regarded as u's U-item.

There have been numerous methods for OCCF; examples include WRMF [9], BPR [18], and CDAE [24]. Among them, we employ CDAE (*Collaborative Denoising Autoencoder*), which is a DNN-based latent factor model and has shown a state-of-the-art accuracy under OCCF setting. We first prepare $\mathbb{P}$ by converting all the ratings in $\mathbb{R}$ into 1 in it and leaving the rest unknown. Then, we train CDAE by using $\mathbb{P}$ as training data[1], and finally obtain the dense matrix $\hat{\mathbb{P}}$ that contains predictions of every $\hat{p}_{ui}$ value. Note that we have the opposite goal with CDAE: while its original goal is to find the most interesting items, our goal for using CDAE here is to find the items that each user is highly unlikely to be interested in. For each user u, we define N_u as a set of her U-items determined by CDAE.

3 Proposed Models

We now present our DNN-based CF models. Our models are based on Autoencoder [20,21,24], which is one of the most successful DNNs and has also been actively adopted in CF, recently. In Sect. 4.1, we show our Autoencoder-based model which exploits the U-items identified by CDAE as stated in the previous section. Then, in Sect. 4.2 and 4.3, we present two variations, where each incorporates the *denoising* scheme and the *adversarial training* technique, respectively. For simplicity, we denote the three proposed models as M1, M2, and M3, respectively.

3.1 Autoencoder-Based Model (M1)

Autoencoder is a feed-forward neural network trying to reconstruct it's the initial input data through learning hidden structures. For simplicity, suppose we have Autoencoder having one hidden layer with K hidden nodes. Let I_u and $\mathbf{r_u}$ as a

---

[1] Here, we omit the detailed descriptions for CDAE (e.g., its objective function and learning procedure) for simplicity; they can be found in [24].

set of items rated by a user u and her rating vector, respectively. Fed with an input $\bar{\mathbf{r}}_\mathbf{u} \in \mathbb{R}^N$, Autoencoder first maps it to a hidden layer by:

$$\mathbf{h} = \sigma(\mathbf{W}_1^\mathbf{T}\bar{\mathbf{r}}_\mathbf{u} + \mathbf{b}_1), \tag{1}$$

where $\bar{\mathbf{r}}_\mathbf{u}$ is user u's *modified* rating vector $\mathbf{r}_\mathbf{u}$ in such a way that $r_{uj} \leftarrow \delta$, $\forall j \in N_u$. Here, δ is a low rating value to be added for U-items, typically chosen among $\{0, 1, 2\}$, due to the belief that U-items would not satisfy u even if recommended to her. By doing so, $\bar{\mathbf{r}}_\mathbf{u}$ contains observed ratings as well as negative ratings on the identified U-items. Then, Autoencoder tries to reconstruct the initial input by:

$$\hat{\bar{\mathbf{r}}}_\mathbf{u} = \sigma(\mathbf{W}_2\mathbf{h} + \mathbf{b}_2) \tag{2}$$

The weight matrices ($\mathbf{W}_1$ and $\mathbf{W}_2$) and the bias vectors ($\mathbf{b}_1$ and $\mathbf{b}_2$) are trained by optimizing the loss function below:

$$\min_{\Theta} \Sigma_u \| (\bar{\mathbf{r}}_\mathbf{u} - \hat{\bar{\mathbf{r}}}_\mathbf{u}) \cdot \mathbf{e}_\mathbf{u} \|^2 + \frac{\alpha}{2} \| \Theta \|^2, \tag{3}$$

where the first term is the reconstruction loss and the second one is the so-called L2 regularization term that helps to avoid overfitting, controlled by α. In order to address the sparse nature of CF setting, $\mathbf{e}_\mathbf{u}$ is used as u's indicator function such that $e_{ui} = 1$ if $i \in I_u$ or $i \in N_u$, and 0 otherwise. Multiplying $\mathbf{e}_\mathbf{u}$ makes the network disregard the loss from missing entries in $\bar{\mathbf{r}}_\mathbf{u}$ (i.e., neither u's rated items nor uninteresting items) by zeroing out back-propagated values [21]. Θ denotes a set of model parameters, which can be learn by the stochastic gradient descent [8].

After the weights and biases are learned, Autoencoder is fed with $\bar{\mathbf{r}}_\mathbf{u}$ and reconstruct a dense vector containing rating predictions on all items in $I \setminus I_u$ for u. Finally the top-N items with the highest predicted ratings are recommended to u.

3.2 Variation Using Denoising Scheme (M2)

Our second model is based on another type of Autoencoder: *denoising Autoencoder*, which is one of the most popular variants of the Autoencoder. As its name implies, the denoising Autoencoder aims to force the hidden layer to discover latent factors more robust to noise. Towards this goal, it first *corrupts* the original input vectors in each training iteration, and then put them to Autoencoder. We note, in calculating the reconstruction loss, it compares the output values with the original input, rather than the corrupted input.

In order to make the (partially) corrupted version of rating vectors, we first define $\psi(0 < \psi < 100)$ as a tunable parameter indicating the percentage of input nodes to be corrupted. For corruption, we employ the widely-used *masking noise*, which randomly overwrites a fraction of observed values in the input vector with zero. For example, if there are 100 ratings in $\bar{\mathbf{r}}_\mathbf{u}$, including u's observed ratings as well as injected low ratings for her, and ψ is set to 20, then u's 20 ratings are randomly chosen and converted to zero. We denote each user's corrupted rating vector as $\bar{\mathbf{r}}_{\mathbf{u}}^{\mathbf{c}}$. Then, our model maps $\bar{\mathbf{r}}_{\mathbf{u}}^{\mathbf{c}}$ to the hidden layer, and maps the resulting vector to the output layer, as follows:

$$\hat{\bar{\mathbf{r}}}_{\mathbf{u}}^{\mathbf{c}} = \sigma(\mathbf{W}_2\sigma(\mathbf{W}_1^T\bar{\mathbf{r}}_{\mathbf{u}}^{\mathbf{c}} + \mathbf{b}_1) + \mathbf{b}_2) \tag{4}$$

Finally, when computing the reconstruction loss, we compare the output vector $\hat{\bar{\mathbf{r}}}_{\mathbf{u}}^{\mathbf{c}}$ with its original one, i.e., $\bar{\mathbf{r}}_{\mathbf{u}}$, rather than the corrupted input $\bar{\mathbf{r}}_{\mathbf{u}}^{\mathbf{c}}$. This denoising scheme helps our model to learn *more-robust* latent factors since it forces our model to make predictions as accurately as possible even though some part of input ratings are missing [24].

3.3 Variation Using Adversarial Training (M3)

Since its birth, *Generative Adversarial Network* (GAN) [4] has gained popularity due to its novelty in training DNNs and its great achievement in generating synthetic images. And very recently, applying the GAN framework to recommendation systems has also gained significant attention (e.g., IRGAN [23], CFGAN [2], and APR [5]). Instead of using the conventional reconstruction or cross-entropy loss, GAN trains DNN (shortly, $\mathcal{G}$) by employing a discriminative model (shortly, $\mathcal{D}$) additionally, and forces DNN to produce output data that could "fool" $\mathcal{D}$ that tries to classify whether the given data is real (i.e., from training data) or fake (i.e., from $\mathcal{G}$). Eventually, the completely trained $\mathcal{G}$ could produce *realistic data* which are almost indistinguishable from the original data.

In line with this trend, our last model incorporates adversarial training, with the expectation that the adversarial training makes our model to capture each user's *true preference distribution* over items, thus improving the recommendation accuracy. In the GAN framework, we have Autoencoder take a role of $\mathcal{G}$ and build $\mathcal{D}$ to train $\mathcal{G}$, so that Autoencoder is not trained by minimizing the reconstruction loss, but trained by maximizing the likelihood of deceiving $\mathcal{D}$. Formally, the objective function can be written as:

$$\min_{\Theta} \Sigma_u\left(ln(1 - \mathcal{D}(\hat{\mathbf{r}}_{\mathbf{u}} \odot \mathbf{e}_{\mathbf{u}}|\mathbf{r}_{\mathbf{u}})) + \frac{\beta}{2}\Sigma_{j \in N_u}(\hat{r}_{uj} - \delta)^2\right) + \frac{\alpha}{2}\| \Theta \|^2 \tag{5}$$

where $\mathbf{e}_{\mathbf{u}}$ is u's indicator function and $\mathcal{D}(.)$ indicates the estimated probability of given data coming from the ground truth. $\hat{\mathbf{r}}_{\mathbf{u}}$ is the output of Autoencoder, computed as:

$$\hat{\mathbf{r}}_{\mathbf{u}} = \sigma(\mathbf{W}_2\sigma(\mathbf{W}_1^T\mathbf{r}_{\mathbf{u}} + \mathbf{b}_1) + \mathbf{b}_2) \tag{6}$$

The above formulation can be regarded as *conditional GAN* [17], where each user's rating vector $\mathbf{r}_{\mathbf{u}}$ corresponds to a user-specific condition; Autoencoder generates output conditioned by $\mathbf{r}_{\mathbf{u}}$. Similarly, our $\mathcal{D}$ is trained by:

$$\max_{\Phi} \Sigma_u ln\mathcal{D}(\mathbf{r}_{\mathbf{u}}|\mathbf{r}_{\mathbf{u}}) + \Sigma_u ln(1 - \mathcal{D}(\hat{\mathbf{r}}_{\mathbf{u}} \odot \mathbf{e}_{\mathbf{u}}|\mathbf{r}_{\mathbf{u}})) - \frac{\alpha}{2}\| \Phi \|^2 \tag{7}$$

where Φ denotes a set of model parameters of $\mathcal{D}$.

Table 1. Statistics of datasets

Datasets	# users	# items	# ratings	Sparsity
Ciao	996	1,927	18,648	98.72%
Watcha	1,391	1,927	101,073	96.98%
ML100K	943	1,682	100,000	93.69%
ML1M	6,039	3,883	1,000,209	95.72%

In order to take the notion of U-items into account while training, the term $\frac{\beta}{2}\Sigma_{j \in N_u}(\hat{r}_{uj} - \delta)^2$ is used in Eq. (5). By minimizing this term as well, Autoencoder would focus on generating rating scores close to the low rating value δ chosen on the entries corresponding to U-items. The importance of this term is controlled by β.

4   Experimental Environment

We used four real-world datasets: Watcha[2], Ciao [22], Movielens 100K, and Movielens 1M[3] (ML100K and ML1M in short, hereafter). The detailed statistics of each dataset are shown in Table 1. Each data was randomly divided into 80% for training and the rest 20% for testing, where the ratings of 5 were regarded as our ground truth [10,14].

4.1   Evaluation Metrics

We evaluate the accuracy of top-N recommendation, where $N \in \{5, 20\}$. We employed four widely-used ranking metrics: *Precision, Recall, Normalized Discounted Cumulative Gain (nDCG)*, and *Mean Reciprocal Rank (MRR)* [10,14]. For simplicity, we denote each metric as *P@N*, *R@N*, *G@N*, and *M@N*, respectively.

4.2   Implementation Details

We performed a grid search to tune the hyper-parameters. Specifically, we varied the number of hidden layers of our Autoencoders with $\{1, 2, 3, 4, 5\}$, the number of hidden nodes per hidden layer with $\{50, 100, 150, 200, 300\}$, a learning rate with $\{0.001, 0.0005, 0.0001\}$, a percentage of U-items (θ) with $\{10, 30, 50, 70, 90\}$, and a low rating value for U-items (δ) with $\{0, 1, 2\}$. We also varied α and β with $\{0.01, 0.001, 0.0001\}$ and $\{1, 0.5, 0.1, 0.05\}$, respectively. For the corruption ratio ψ in M2, we used $\{20, 40, 60, 80\}$.

---

[2] http://watcha.net.
[3] https://grouplens.org/datasets/movielens/.

Table 2. Basic top-N recommendation on Ciao

Metrics	P@5	R@5	G@5	M@5	P@20	R@20	G@20	M@20
ItemKNN	.0067	.0145	.0101	.0185	.0059	.0441	.0219	.0252
SVD	.0187	.0392	.0341	.0506	.0112	.0857	.0485	.0571
PureSVD	.0278	.0626	.0492	.0662	.0159	.1278	.0678	.0694
Zero-Injection+SVD	.0347	.0689	.0571	.0795	.0210	.1772	.0903	.0927
M1\U-items (AutoRec)	.0219	.0492	.0363	.0495	.0130	.1139	.0592	.0643
M2\U-items (CDAE)	.0191	.0375	.0288	.0434	.0088	.0649	.0323	.0350
M3\U-items	.0212	.0471	.0341	.0459	.0124	.1047	.0522	.0508
M1	.0403	.0817	.0691	.0938	**.0221**	.1856	**.1009**	**.1096**
M2	**.0409**	**.0889**	**.0712**	.0949	.0219	.1862	.0972	.1015
M3	.0375	.0855	.0702	**.0962**	.0218	**.1874**	.0957	.0994

Table 3. Basic top-N recommendation on Watcha

Metrics	P@5	R@5	G@5	M@5	P@20	R@20	G@20	M@20
ItemKNN	.0087	.0096	.0131	.0213	.0085	.0373	.0210	.0298
SVD	.0344	.0449	.0497	.0887	.0226	.1167	.0705	.0985
PureSVD	.0573	.0844	.0826	.1274	.0389	.2259	.1303	.1509
Zero-Injection+SVD	.0672	.1048	.1002	.1551	.0425	.2434	.1439	.1721
M1\U-items (AutoRec)	.0346	.0564	.0502	.0795	.0248	.1501	.0809	.0968
M2\U-items (CDAE)	.0318	.0455	.0467	.0787	.0195	.1094	.0606	.0764
M3\U-items	.0331	.0519	.0462	.0734	.0203	.1264	.0700	.0867
M1	.0681	.1091	.1044	.1600	.0444	.2522	.1480	.1801
M2	**.0739**	**.1126**	**.1087**	**.1709**	**.0445**	**.2600**	**.1565**	**.1896**
M3	.0680	.1091	.1044	.1600	.0425	.2538	.1482	.1735

4.3 Situations

We evaluated our models not only under basic setting of top-N recommendation, but also under the following two more-challenging situations: *long-tail* items recommendation and recommendation to *cold-start users* [15]. In the first situation, we defined *top-head* popular items that received 30% of all ratings and the rest *long-tail* items, and then made each CF model recommend only the long-tail items to users. In the second situation, we randomly selected 200 users as cold-start users. Then, we also randomly selected five ratings for each cold-start user and discarded the rest of her ratings in training. We finally recommended items only to these cold-start users [14].

5 Results and Analyses

The results on our comparative experiments are reported in Tables 2, 3, 4, 5, 6, 7, 8, 9, 10, 11, 12 and 13. The key observations from the results are summarized as follows.

Table 4. Basic top-N recommendation on ML100K

Metrics	P@5	R@5	G@5	M@5	P@20	R@20	G@20	M@20
ItemKNN	.0372	.0332	.0450	.0613	.0337	.1174	.0686	.0830
SVD	.0931	.0853	.1159	.1951	.0524	.1852	.1386	.2119
PureSVD	.1466	.1717	.2019	.3104	.0835	.3704	.2573	.3330
Zero-Injection+SVD	.1429	.1621	.1927	.2914	.0837	.3390	.2379	.3077
M1\U-items (AutoRec)	.0946	.0937	.1253	.2135	.0552	.2265	.1522	.2184
M2\U-items (CDAE)	.0887	.0992	.1210	.2085	.0554	.2226	.1528	.2228
M3\U-items	.0806	.0901	.1117	.1892	.0451	.1748	.1304	.2072
M1	.1726	.2003	.2342	.3423	.0958	.3944	.2782	.3488
M2	**.1815**	**.2144**	**.2490**	**.3636**	**.0977**	.4140	**.3007**	**.3926**
M3	.1688	.2011	.2314	.3371	.0970	**.4205**	.2868	.3528

Table 5. Basic top-N recommendation on ML1M

Metrics	P@5	R@5	G@5	M@5	P@20	R@20	G@20	M@20
ItemKNN	.0361	.0229	.0272	.0668	.0329	.0795	.0541	.0870
SVD	.0418	.0281	.0370	.0866	.0321	.0836	.0616	.1068
PureSVD	.1485	.1200	.1558	.3114	.0940	.2797	.2166	.3400
Zero-Injection+SVD	.1856	.1486	.2265	.3613	.1117	.3202	.2567	.3821
M1\U-items (AutoRec)	.0785	.0572	.0772	.1220	.0688	.1195	.1024	.1561
M2\U-items (CDAE)	.1058	.0766	.1288	.2297	.0684	.1894	.1354	.2122
M3\U-items	.1035	.0750	.1222	.2149	.0645	.1822	.1395	.2321
M1	.2039	.1587	.2478	.3909	.1230	.3573	.2820	.4085
M2	**.2213**	**.1858**	**.2747**	**.4245**	**.1292**	**.3817**	**.3078**	**.4395**
M3	.1860	.1563	.2257	.3557	.1144	.3496	.2669	.3821

5.1 Impact of U-Items on DNN-based CF Models

In the tables, we examine the accuracy of our models trained *with* using U-items, compared with those trained *without* using U-items, denoted as M1\U-items, M2\U-items, and M3\U-items, respectively. For example, M3\U-items indicates the model that does not use the second term in Eq. (5); M1\U-items and M2\U-items indicate the models that use the original rating vector r_u as their inputs, rather than $\bar{r}_u$. We observe that our three models considering U-items provide significantly higher accuracy: improving 53.1% on average in terms of *Precision*@5 and 116.4% in terms of *nDCG*@20, respectively, compared to those only using $\mathbb{R}$.

5.2 Comparisons with State-of-the-arts

We also compare our models with state-of-the-art CF models in terms of accuracy. Zero-Injection+SVD is the most similar to ours: it trains SVD with exploiting U-items as extra information, as in our models. PureSVD considers *all*

Table 6. Recommendation to cold-start users on Ciao

Metrics	P@5	R@5	G@5	M@5	P@20	R@20	G@20	M@20
ItemKNN	.0064	.0133	.0099	.0124	.0058	.0418	.0210	.0246
SVD	.0164	.0332	.0246	.0298	.0097	.0785	.0379	.0386
PureSVD	.0185	.0390	.0296	.0401	.0115	.1019	.0493	.0488
Zero-Injection+SVD	.0236	.0499	.0454	.0679	.0138	.1118	.0625	.0689
M1\U-items (AutoRec)	.0174	.0338	.0248	.0337	.0100	.0815	.0372	.0398
M2\U-items (CDAE)	.0185	.0361	.0281	.0412	.0085	.0620	.0319	.0338
M3\U-items	.0205	.0418	.0312	.0445	.0090	.0776	.0430	.0507
M1	.0287	**.0620**	.0455	.0609	.0156	**.1425**	.0696	.0754
M2	**.0308**	.0613	.0474	.0638	**.0167**	.1360	**.0703**	**.0779**
M3	.0284	.0584	**.0503**	**.0726**	**.0167**	.1412	.0696	.0736

Table 7. Recommendation to cold-start users on Watcha

Metrics	P@5	R@5	G@5	M@5	P@20	R@20	G@20	M@20
ItemKNN	.0077	.0087	.0122	.0201	.0059	.0357	.0201	.0260
SVD	.0220	.0324	.0367	.0670	.0157	.0824	.0540	.0808
PureSVD	.0290	.0515	.0393	.0561	.0202	.1161	.0644	.0719
Zero-Injection+SVD	.0250	.0448	.0367	.0627	.0215	.1342	.0669	.0736
M1\U-items (AutoRec)	.0240	.0357	.0390	.0699	.0160	.0834	.0548	.0819
M2\U-items (CDAE)	.0312	.0445	.0450	.0759	.0190	.1084	.0583	.0670
M3\U-items	.0310	.0492	.0414	.0695	.0197	.1135	.0597	.0673
M1	.0350	.0586	.0437	.0647	.0237	.1471	.0711	.0754
M2	**.0410**	**.0653**	**.0556**	**.0864**	**.0270**	**.1588**	**.0748**	.0752
M3	.0335	.0528	.0493	.0791	.0233	.1359	.0747	**.0951**

unrated items as negative preferences and assigns "0" scores to them before training. Autorec [20] and CDAE [24] are also widely-used baselines; they are identical to M1\U-items and M1\U-items, respectively.

We observe that our models consistently and universally outperform those state-of-the-art CF models. In particular, our models beat Zero-Injection+SVD by a wide margin: the average improvements of accuracy with our M1, M2, and M3 over Zero-Injection+SVD are 12.8%, 20.6%, and 6.1%, respectively. These results verify that DNN provides a more powerful synergy effect with Zero-Injection+SVD than the conventional MF. Finally, the p-values of the paired sample t-test between our models and the best competitor demonstrate statistical significance of the improvements, where p-values are 0.0011, 0.0002, 0.0004, and 0.0003 in Ciao, Watcha, ML100K, and ML1M, respectively.

Next, we conducted another set of experiments to examine the accuracy changes of our models according to different key hyper-parameters: the percentage of U-items (θ) and the value of low ratings (δ). Figure 1 shows the results w.r.t. $MRR@5$ on Watcha; the other results, which showed similar trend, are

Table 8. Recommendation to cold-start users on ML100K

Metrics	P@5	R@5	G@5	M@5	P@20	R@20	G@20	M@20
ItemKNN	.0159	.0110	.0187	.0400	.0149	.0554	.0317	.0539
SVD	.0430	.0514	.0529	.0899	.0280	.1167	.0747	.1099
PureSVD	.0570	.0873	.0834	.1264	.0380	.1992	.1219	.1486
Zero-Injection+SVD	.0600	.0913	.0904	.1444	.0335	.1651	.1098	.1565
M1\U-items (AutoRec)	.0440	.0530	.0571	.1023	.0295	.1298	.0795	.1117
M2\U-items (CDAE)	.0610	.0861	.0889	.1473	.0367	.1640	.1064	.1500
M3\U-items	.0590	.0741	.0784	.1192	.0352	.1650	.1024	.1361
M1	.0720	.0951	.0965	.1473	.0430	.2131	.1318	.1660
M2	.0790	**.1094**	**.1074**	**.1647**	.0437	**.2194**	**.1376**	.1800
M3	**.0812**	.0964	.1048	.1645	**.0475**	.2027	.1354	**.1975**

Table 9. Recommendation to cold-start users on ML1M

Metrics	P@5	R@5	G@5	M@5	P@20	R@20	G@20	M@20
ItemKNN	.0284	.0213	.0242	.0525	.0240	.0741	.0515	.0744
SVD	.0390	.0264	.0333	.0857	.0237	.0630	.0577	.0902
PureSVD	.0620	.0645	.0849	.1487	.0360	.1295	.0923	.1610
Zero-Injection+SVD	.0600	.0708	.0820	.1232	.0455	.1545	.0972	.1503
M1\U-items (AutoRec)	.0620	.0540	.0687	.1213	.0395	.1190	.0883	.1488
M2\U-items (CDAE)	.0620	.0514	.0673	.1252	.0453	.1532	.0932	.1431
M3\U-items	.0640	.0600	.0721	.1242	.0422	.1389	.0932	.1549
M1	.0720	.0757	.0890	.1531	.0505	.1728	.1181	.1784
M2	**.0870**	**.0844**	**.1072**	**.1917**	**.0580**	**.2167**	**.1345**	**.1928**
M3	.0780	.0647	.0929	.1646	.0560	.1714	.1195	.1840

Table 10. Long-tail item recommendation on Ciao

Metrics	P@5	R@5	G@5	M@5	P@20	R@20	G@20	M@20
ItemKNN	.0065	.0142	.0096	.0180	.0058	.0432	.0215	.0243
SVD	.0072	.0115	.0113	.0186	.0048	.0353	.0176	.0194
PureSVD	.0131	.0213	.0192	.0315	.0088	.0547	.0308	.0392
Zero-Injection+SVD	.0219	.0400	.0337	.0494	.0110	.0851	.0478	.0529
M1\U-items (AutoRec)	.0078	.0130	.0116	.0188	.0048	.0323	.0181	.0231
M2\U-items (CDAE)	.0094	.0177	.0161	.0232	.0060	.0474	.0256	.0293
M3\U-items	.0050	.0098	.0083	.0120	.0034	.0218	.0097	.0105
M1	.0222	.0407	.0358	.0500	**.0125**	.0902	.0486	.0569
M2	.0228	.0414	.0375	.0563	.0119	**.0945**	.0500	.0594
M3	**.0234**	**.0457**	**.0405**	**.0598**	.0116	.0933	**.0535**	**.0631**

Table 11. Long-tail item recommendation on Watcha

Metrics	P@5	R@5	G@5	M@5	P@20	R@20	G@20	M@20
ItemKNN	.0071	.0078	.0070	.0106	.0083	.0370	.0181	.0195
SVD	.0204	.0249	.0265	.0461	.0134	.0666	.0393	.0554
PureSVD	.0377	.0487	.0504	.0844	.0246	.1257	.0745	.1011
Zero-Injection+SVD	.0439	.0564	.0594	.0990	.0272	.1348	.0860	.1168
M1\U-items (AutoRec)	.0212	.0262	.0266	.0446	.0135	.0689	.0389	.0506
M2\U-items (CDAE)	.0186	.0262	.0256	.0437	.0122	.0577	.0320	.0406
M3\U-items	.0187	.0221	.0236	.0439	.0113	.0587	.0326	.0454
M1	.0467	.0652	.0648	.1037	.0273	.1428	.0880	.1174
M2	**.0471**	.0647	.0639	.1054	.0272	.1425	.0884	.1208
M3	.0469	**.0666**	**.0656**	**.1086**	**.0275**	**.1443**	**.0897**	**.1214**

Table 12. Long-tail item recommendation on ML100K

Metrics	P@5	R@5	G@5	M@5	P@20	R@20	G@20	M@20
ItemKNN	.0261	.0197	.0269	.0515	.0196	.0910	.0539	.0723
SVD	.0312	.0251	.0353	.0642	.0228	.0760	.0487	.0742
PureSVD	.0753	.0733	.0977	.0647	.0449	.1653	.1208	.0839
Zero-Injection+SVD	.0751	.0697	.0948	.1595	.0448	.1614	.1148	.1762
M1\U-items (AutoRec)	.0324	.0256	.0428	.0643	.0235	.0804	.0509	.0737
M2\U-items (CDAE)	.0433	.0369	.0515	.0969	.0304	.1104	.0714	.1113
M3\U-items	.0271	.0261	.0324	.0570	.0187	.0721	.0440	.0627
M1	**.0906**	.0900	.1164	.1927	.0493	.1826	.1341	.2039
M2	.0850	.0853	.1102	.1790	.0492	.1821	.1313	.1931
M3	.0880	**.0944**	**.1198**	**.2035**	**.0504**	**.1881**	**.1422**	**.2221**

Table 13. Long-tail item recommendation on ML1M

Metrics	P@5	R@5	G@5	M@5	P@20	R@20	G@20	M@20
ItemKNN	.0329	.0211	.0271	.0584	.0182	.0427	.0492	.0805
SVD	.0255	.0137	.0282	.0574	.0159	.0371	.0282	.0641
PureSVD	.0923	.0596	.1083	.1992	.0469	.1156	.1154	.1999
Zero-Injection+SVD	.0926	.0643	.1123	.1402	.0522	.1334	.1112	.2059
M1\U-items (AutoRec)	.0231	.0136	.0258	.0522	.0211	.0366	.0329	.0706
M2\U-items (CDAE)	.0238	.0139	.0259	.0521	.0178	.0398	.0311	.0633
M3\U-items	.0246	.0138	.0277	.0574	.0167	.0375	.0303	.0625
M1	.1021	.0728	.1217	.2170	.0608	.1566	.1308	.2361
M2	**.1043**	**.0728**	**.1242**	**.2223**	**.0613**	**.1600**	**.1337**	**.2417**
M3	.0997	.0721	.1194	.2164	.0598	.1586	.1306	.2345

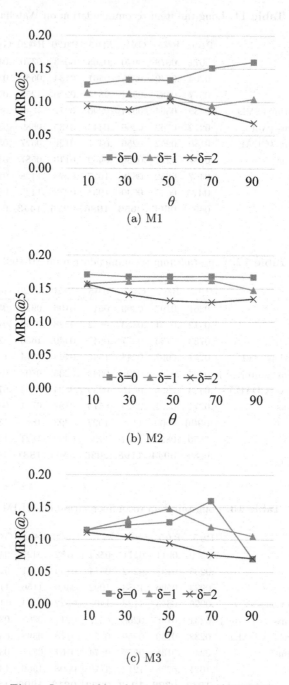

Fig. 1. Impact of key hyper-parameters ($MRR@5$).

omitted for simplicity. We observe that setting δ with lower values provides consistently higher accuracy: we thus confirm that considering U-items as severe negative preferences helps our models to accurately capture users' *relative preferences* on items. M1 and M3 exhibit accuracy quite sensitive to θ: they achieve their best accuracy when $\theta = 90$ and $\theta = 70$, respectively. In contrast, M2 shows the highest accuracy as well as the highest insensitivity against varying θ.

6   Conclusions and Future Work

This paper proposes three DNN-based CF models, each of which successfully exploits the idea of Zero-Injection in its own way. Through our extensive experiments, all of our models outperform SVD equipped with Zero-Injection, which demonstrates impressive synergy effect through combination of DNN and Zero-Injection. Among our proposed models, M2 showed the best accuracy as well as insensitivity against the ratio of U-items.

Acknowledgements. This research was supported by (1) Next-Generation Information Computing Development Program through the National Research Foundation of Korea (NRF) funded by the Ministry of Science, ICT (NRF-2017M3C4A7069440), (2) Basic Science Research Program through the National Research Foundation of Korea (NRF) funded by the Ministry of Education (No. 2019R1I1A1A01061588), and (3) Next-Generation Information Computing Development Program through the National Research Foundation of Korea (NRF) funded by the Ministry of Science and ICT (No. NRF-2017M3C4A7083678).

References

1. Bahdanau, D., Cho, K., Bengio, Y.: Neural machine translation by jointly learning to align and translate. arXiv preprint arXiv:1409.0473 (2014)
2. Chae, D.K., Kang, J.S., Kim, S.W., Lee, J.T.: CFGAN: a generic collaborative filtering framework based on generative adversarial networks. In: Proceedings of the 27th ACM International Conference on Information and Knowledge Management, pp. 137–146 (2018)
3. Chae, D.K., Lee, S.C., Lee, S.Y., Kim, S.W.: On identifying k-nearest neighbors in neighborhood models for efficient and effective collaborative filtering. Neurocomputing **278**, 134–143 (2018)
4. Goodfellow, I., et al.: Generative adversarial nets. In: Advances in Neural Information Processing Systems, pp. 2672–2680 (2014)
5. He, X., He, Z., Du, X., Chua, T.S.: Adversarial personalized ranking for recommendation. In: The 41st International ACM SIGIR Conference on Research and Development in Information Retrieval, pp. 355–364 (2018)
6. He, X., Liao, L., Zhang, H., Nie, L., Hu, X., Chua, T.S.: Neural collaborative filtering. In: Proceedings of the 26th International Conference on World Wide Web, pp. 173–182 (2017)
7. Hinton, G., et al.: Deep neural networks for acoustic modeling in speech recognition. IEEE Signal Process. Mag. **29**(6), 82–97 (2012)

8. Hinton, G.E., Salakhutdinov, R.R.: Reducing the dimensionality of data with neural networks. Science **313**(5786), 504–507 (2006)
9. Hu, Y., Koren, Y., Volinsky, C.: Collaborative filtering for implicit feedback datasets. In: Proceedings of the 2008 IEEE 8th International Conference on Data Mining, pp. 263–272 (2008)
10. Hwang, W.S., Parc, J., Kim, S.W., Lee, J., Lee, D.: Told you i didn't like it: exploiting uninteresting items for effective collaborative filtering. In: Proceedings of the IEEE 32nd International Conference on Data Engineering, pp. 349–360 (2016)
11. Koren, Y.: Factorization meets the neighborhood: a multifaceted collaborative filtering model. In: Proceedings of the 14th ACM SIGKDD International Conference on Knowledge Discovery and data Mining, pp. 426–434 (2008)
12. Koren, Y., Bell, R., Volinsky, C.: Matrix factorization techniques for recommender systems. Computer **42**(8), 30–37 (2009)
13. Krizhevsky, A., Sutskever, I., Hinton, G.E.: ImageNet classification with deep convolutional neural networks. In: Advances in Neural Information Processing Systems, pp. 1097–1105 (2012)
14. Lee, J., Hwang, W.S., Parc, J., Lee, Y., Kim, S.W., Lee, D.: l-injection: toward effective collaborative filtering using uninteresting items. IEEE Trans. Knowl. Data Eng. **31**(1), 3–16 (2019)
15. Lee, J., Lee, D., Lee, Y.C., Hwang, W.S., Kim, S.W.: Improving the accuracy of top-N recommendation using a preference model. Inf. Sci. **348**, 290–304 (2016)
16. Lee, Y.C., Kim, S.W., Lee, D.: gOCCF: graph-theoretic one-class collaborative filtering based on uninteresting items. In: Proceedings of the 2018 AAAI International Conference on Artificial Intelligence (2018)
17. Mirza, M., Osindero, S.: Conditional generative adversarial nets. arXiv preprint arXiv:1411.1784 (2014)
18. Rendle, S., Freudenthaler, C., Gantner, Z., Schmidt-Thieme, L.: BPR: Bayesian personalized ranking from implicit feedback. In: Proceedings of the 25th International Conference on Uncertainty in Artificial Intelligence, pp. 452–461 (2009)
19. Sarwar, B., Karypis, G., Konstan, J., Riedl, J.: Item-based collaborative filtering recommendation algorithms. In: Proceedings of the 10th International Conference on World Wide Web, pp. 285–295 (2001)
20. Sedhain, S., Menon, A.K., Sanner, S., Xie, L.: AutoRec: autoencoders meet collaborative filtering. In: Proceedings of the 24th International Conference on World Wide Web, pp. 111–112 (2015)
21. Strub, F., Mary, J.: Collaborative filtering with stacked denoising autoencoders and sparse inputs. In: NIPS Workshop on Machine Learning for eCommerce (2015)
22. Tang, J., Gao, H., Liu, H.: mTrust: discerning multi-faceted trust in a connected world. In: Proceedings of the 5th ACM International Conference on Web Search and Data Mining, pp. 93–102 (2012)
23. Wang, J., et al.: IRGAN: a minimax game for unifying generative and discriminative information retrieval models. In: Proceedings of the 40th International ACM SIGIR Conference on Research and Development in Information Retrieval, pp. 515–524 (2017)
24. Wu, Y., DuBois, C., Zheng, A.X., Ester, M.: Collaborative denoising auto-encoders for top-N recommender systems. In: Proceedings of the 9th ACM International Conference on Web Search and Data Mining, pp. 153–162 (2016)

DEAMER: A Deep Exposure-Aware Multimodal Content-Based Recommendation System

Yunsen Hong, Hui Li, Xiaoli Wang, and Chen Lin$^{(\boxtimes)}$ ⓘD

School of Informatics, Xiamen University, Xiamen, China
yshong@stu.xmu.edu.cn, {hui,xlwang,chenlin}@xmu.edu.cn

Abstract. Modern content-based recommendation systems have greatly benefited from deep neural networks, which can effectively learn feature representations from item descriptions and user profiles. However, the supervision signals to guide the representation learning are generally incomplete (i.e., the majority of ratings are missing) and/or implicit (i.e., only historical interactions showing implicit preferences are available). The learned representations will be biased in this case; and consequently, the recommendations are over-specified. To alleviate this problem, we present a Deep Exposure-Aware Multimodal contEnt-based Recommender (i.e., DEAMER) in this paper. DEAMER can jointly exploit rating and interaction signals via multi-task learning. DEAMER mimics the expose-evaluate process in recommender systems where an item is evaluated only if it is exposed to the user. DEAMER generates the exposure status by matching multi-modal user and item content features. Then the rating value is predicted based on the exposure status. To verify the effectiveness of DEAMER, we conduct comprehensive experiments on a variety of e-commerce data sets. We show that DEAMER outperforms state-of-the-art shallow and deep recommendation models on recommendation tasks such as rating prediction and top-k recommendation. Furthermore, DEAMER can be adapted to extract insightful patterns of both users and items.

Keywords: Neural recommender systems · Content-based recommendation · Deep generative model · Multi-task learning · Multi-modal learning

1 Introduction

Recommender systems have been widely used in e-commerce platforms such as Amazon and eBay. Recommender systems endow e-commerce platforms with the capability to deliver personalized filtering of innumerable consumption choices. Tackling the over-choice problem in online shopping brings both better user experience and higher enterprise revenue. Therefore, during the past decades,

ⓒ Springer Nature Switzerland AG 2020
Y. Nah et al. (Eds.): DASFAA 2020, LNCS 12114, pp. 621–637, 2020.
https://doi.org/10.1007/978-3-030-59419-0_38

recommender systems have become an indispensable part in e-commerce platforms and enormously impacted other domains, such as social networks and media websites [1].

Abundant content information is available on e-commerce platforms, including item meta-data, user profiles, user generated reviews, and other auxiliary data sources such as videos and images. To exploit content information, content-based methods, which is a main paradigm of recommender systems, have been extensively studied in the past. Content-based methods offer recommendations based on comparisons across contents of users and items. Comparisons can be done by either finding exact keywords or computing relevance scores on hand-crafted and/or learned content representations [29]. Recent breakthroughs of deep learning boost the performance of content-based methods, as deep neural networks are superior at deriving underlying representations without hand-crafted features [35]. Compared to the other paradigm of recommender systems, i.e., collaborative filtering methods which solely learn user preferences from user-item historical interactions, content-based methods are able to alleviate the cold-start problem effectively [1]. When a new user has zero or only a few ratings, content-based approaches can still generate reasonable recommendations by performing content-based comparisons. However, it is reported that content-based methods lack some diversity and serendipity, which leads to inferior performance to collaborative filtering methods when sufficient ratings are provided [29].

One possible reason for the inferior performance of content-based approaches is that the supervision signals which are used to guide the representation learning of content information are incomplete or implicit in general. Traditional content-based methods require users to label item documents by assigning a relevance score [1,29]. Such approaches are labor-consuming and thus not applicable. Other content-based methods, including shallow [18] and deep models [21] use ratings to measure the content relevance. However, rating signals are generally missing for a large portion of user-item pairs in practice. Moreover, content-based methods will learn over-specialized representations on items that users already like, as well as non-distinguishable representations on items that users have never rated. To cope with aforementioned issues, some contemporary content-based approaches consider binary interactions (i.e., implicit feedback), e.g., whether a user has rated, clicked, tagged or consumed an item. In particular, a positive interaction usually indicates positive preference, while a zero interaction does not indicate negative preference. Nevertheless, content-based methods using implicit feedback are still unable to learn accurate representations for user-item pairs with zero interactions.

Let us recall the expose-evaluate process which is typically adopted by users when using a recommender system. The process is shown in Fig. 1. A user will rate an item, only if the item is exposed to him/her. To further specify the process, we make the following assumptions: (1) The decision of exposure can be roughly interpreted as a quick content matching. It does not matter whether the occurrence of exposure is due to the pro-activeness of users or the intelligent decision from recommender systems. Exposure indicates preferences.

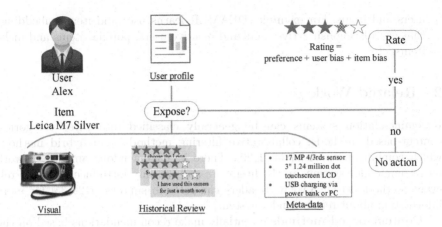

Fig. 1. Expose-evaluate process where an item is rated after it is exposed to a user

The user will interact (e.g., rate, consume or click) with an exposed item. (2) Once exposed, the item will be further evaluated (in many scenarios, it means the item will be consumed) and rated. A well-known and successful assumption is that the value of rating is an aggregation of user preference, user bias and item bias [13].

The expose-evaluate process inspires us to generate rating values based on interactions in learning content representations and improve the quality of content-based recommendation. We present a Deep Exposure-Aware Multimodal contEnt-based Recommender (i.e., DEAMER). DEAMER can jointly exploit rating and interaction signals via multi-task learning and it consists of two modules. The exposure generation module predicts the binary exposure based on representations extracted from multi-modal contents including item meta-data, item images, and historical reviews grouped by users and items. The predicted probability of an interaction, which is treated as the user preference over an item, is transmitted to the rating generation module which estimates the user-item relevance (i.e., rating) based on the user-item preference, user-specific and item-specific embeddings.

In summary, the contributions of this paper is three-fold:

- **A novel deep generative model.** Unlike previous methods which model ratings and interactions in parallel [22], DEAMER generates ratings from exposures. The content representations are first learned in the coarse-grained bottom module and then finer tuned in the top module.
- **A multi-modal multi-task framework.** Multi-modal content information (i.e., visuals and texts) and heterogeneous supervision signals (i.e., ratings and interactions) are utilized to improve recommendations in DEAMER.
- **Superior recommendation performance and a behavior analysis.** We experimentally verify, on a variety of e-commerce data sets, that DEAMER outperforms state-of-the-art shallow and deep recommendation models. Furthermore, we show that DEAMER is able to capture interesting patterns of

users and items. For example, DEAMER learns user and item embeddings that can distinguish active users and inactive users, popular items and niche items, etc.

2 Related Work

Recommendation systems can be generally classified into three categories: content-based methods, collaborative filtering methods, and hybrid methods which combine both of them [1,29]. There is a tremendous amount of work on recommender systems in the literature and we only introduce the most relevant methods to DEAMER. Readers can refer to surveys [1,29,35] for more information about recommender systems.

Content-based methods essentially make recommendations based on the degrees of resemblance between item content and user profiles. Existing works have investigated the mining of various content features [1,29] including social network [19], user grouping data [2,18], relationships in a graph [27], time-series information [17], locations [24], review text [5], just to name a few. The rapid development of deep neural network techniques have fostered modern content-based recommenders. Many deep neural networks have been leveraged to analyze different data modalities. For example, CNNs are often used for image feature extraction, CNNs and RNNs are usually necessary to deal with texts, and attention mechanism is commonly adopted for textual and tag information [35]. To learn the corresponding user and item representations over heterogeneous information, researchers recently turn to multi-modal learning [33,36] in order to incorporate multiple types of information sources (e.g., review text, product image and numerical rating) into one content-based recommendation model.

Collaborative filtering methods extract users with similar tastes as the target user from interactions/rating data only. In the fruitful literature of collaborative filtering methods, matrix factorization [13,16] and factorization machines [28] have exhibited superior performances in rating prediction task. However, it is problematic to only consider ratings, since the underlying rating matrix is *sparse*. When ratings are *Missing Not At Random* (MNAR) [26], the performance of collaborative filtering methods will further degrade. Therefore, many MNAR-targeted models have been proposed [23,26]. Recent advances of deep neural networks have also benefited collaborative filtering methods. For instance, NeuCF [9] generalizes matrix factorization with multi-layer perceptron. DeepFM [7] and xDeepFM [22] extend factorization machines by learning high-order feature interactions via a compressed interaction network. AutoRec [31] introduced autoencoder into collaborative filtering.

In many cases where a wider variety of data sources is available, one has the flexibility of using both content-based and collaborative filtering recommenders for the task of recommendation. Using **hybridization of different types of recommenders**, the various aspects from different types of recommender systems are combined to achieve the best performance. There is a surge of works on hybrid recommenders. HybA [6] is a hybrid recommender system for automatic playlist continuation which combines Latent Dirichlet Allocation and case-based reasoning. MFC [20] is a hybrid matrix factorization model which uses both ratings and the social structure to offer recommendations. Deep neural network has also been introduced into hybrid recommender systems. For instance, CFN [32] utilizes autoencoder and incorporates side-information to construct a hybrid recommender. aSDAE [3] is a hybrid model which combines additional stacked denoising autoencoder and matrix factorization together.

To the best of our knowledge, there exists methods modeling ratings and explicit interactions in parallel [22], but the dependency between ratings values and explicit user-item interactions has not been explored in deep neural network architecture for the recommendation task. In addition to generalizing and extending the shallow MNAR and exposure-aware collaborative filtering models, DEAMER learns the ratings through multi-modal content information. Thus the contents are assembled to provide more accurate recommendations than pure collaborative filtering or content-based methods.

3   DEAMER

In this section, we will elaborate on the proposed Deep Exposure-Aware Multimodal Content-based Recommender System (i.e., DEAMER).

Suppose that we have a set of users $\mathcal{U}$ and a set of items $\mathcal{V}$. For each user $u \in \mathcal{U}$ and each item $v \in \mathcal{V}$, the inputs $\mathbf{X}_u$ and $\mathbf{X}_v$ consists of user and item contents, respectively (they will be described below); the output contain an observed interaction $y_{u,v}$ and an observed rating $r_{u,v}$. The interaction is binary (i.e., $y_{u,v} \in \{0,1\}$, where 1 indicates that the user has rated/clicked/viewed the item, and 0 otherwise). The rating is numerical and normalized (i.e., $r_{u,v} \in [0,1]$). We aim to estimate proper parameters which can generate observations $y_{u,v}$, $r_{u,v}$ given $\mathbf{X}_u$ and $\mathbf{X}_v$ for $\forall u \in \mathcal{U}$ and $\forall v \in \mathcal{V}$.

The user content $\mathbf{X}_u$ is constructed by the user's historic reviews on different items. That is, we aggregate u's reviews and make $\mathbf{X}_u = (x_{u,1}, \cdots, x_{u,T_{(x,u)}})$, where $x_{u,i}$ is the i-th word and $T_{(x,u)}$ is the number of words in u's content. We gather item content $\mathbf{X}_v$ in three different modalities: meta-data, textual reviews and images (i.e., $\mathbf{X}_v = \{\mathbf{w}_v, \mathbf{m}_v, \mathbf{G}_v\}$). For each item v, we aggregate its reviews from different users as its textual reviews. Item reviews for item v is a sequence of words $\mathbf{w}_v = (w_{w,1}, \cdots, w_{w,T_{(w,v)}})$ where $w_{v,i}$ is the i-th word and $T_{(w,v)}$ is the number of words in v's reviews. We extract the name and textual description of item as its meta-data. The meta-data for item v is a sequence of words $\mathbf{m}_v = (m_{v,1}, \cdots, m_{v,T_{(m,v)}})$ where $m_{v,i}$ is the i-th word and $T_{(m,v)}$ is the number of words in v's meta-data. We use images which are associated with an item as its visual input. The image is represented as $\mathbf{G}_v \in \mathcal{R}^{N(G) \times N(G)}$ where $N(G)$ is the dimensionality of the image.

Figure 2 depicts the overall architecture of DEAMER. The inputs $\mathbf{X}_u$ and $\mathbf{X}_v = \{\mathbf{w}_v, \mathbf{m}_v, \mathbf{G}_v\}$ are first passed to the exposure generation module at the bottom which leverages real binary interaction to guide the exposure generation. Then, rating generation module on the top uses real rating value to generate

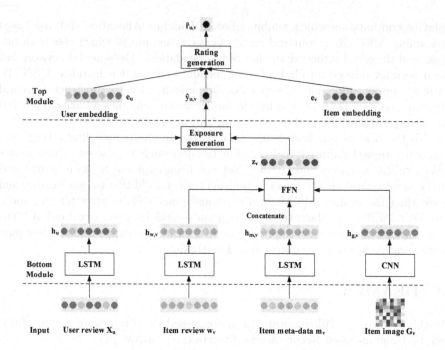

Fig. 2. Overall architecture of DEAMER

ratings. In the following, we will illustrate the exposure generation module and rating generation module of DEAMER in detail.

3.1 Exposure Generation Module

We extract feature representations from different modalities separately. For each pair of user u and item v, we use LSTM [10] to extract the features from $\mathbf{X}_u$, $\mathbf{w}_v$ and $\mathbf{m}_v$.

LSTM reads a sequence of tokens and generates hidden states one by one. For the t-th word n_t in each word sequence $(n_1, \dots, n_T)$ ($\{n_t, T\}$ can be one of $\{x_{u,t}, T_{(x,u)}\}$, $\{w_{v,t}, T_{(w,v)}\}$ and $\{m_{v,t}, T_{(m,v)}\}$), LSTM learns a hidden state $\mathbf{h}_t$. To be specific, LSTM operates on the hidden state $\mathbf{h}_{t-1}$ of the previous token n_{t-1} and the input vector $\mathbf{n}_t$ of current token n_t to get an input gate $\mathbf{i}_t$, a forget gate $\mathbf{f}_t$, an output gate $\mathbf{o}_t$ and the cell $\mathbf{g}_t$ for the t-th input:

$$\mathbf{i}_t = \sigma(\mathbf{W}_i[\mathbf{n}_t, \mathbf{h}_{t-1}] + \mathbf{b}_i) \tag{1}$$
$$\mathbf{f}_t = \sigma(\mathbf{W}_f[\mathbf{n}_t, \mathbf{h}_{t-1}] + \mathbf{b}_f)$$
$$\mathbf{o}_t = \sigma(\mathbf{W}_o[\mathbf{n}_t, \mathbf{h}_{t-1}] + \mathbf{b}_o)$$
$$\mathbf{g}_t = \phi(\mathbf{W}_g[\mathbf{n}_t, \mathbf{h}_{t-1}] + \mathbf{b}_g)$$

where $\mathbf{W}_i$, $\mathbf{W}_f$, $\mathbf{W}_o$ and $\mathbf{W}_g$ are the learnable weight matrices, and $\mathbf{b}_i$, $\mathbf{b}_f$, $\mathbf{b}_o$ and $\mathbf{b}_g$ are bias vectors. $\sigma(x) = \frac{1}{1+e^{-x}}$ is the sigmoid activation function, and $\phi(x) = \frac{x}{1+|x|}$ is the softsign activation function.

Then, LSTM computes the cell state and the hidden state for the t-th input:

$$\begin{aligned} \mathbf{c}_t &= \mathbf{f}_t \odot \mathbf{c}_{t-1} + \mathbf{i}_t \odot \mathbf{g}_t \\ \mathbf{h}_t &= \mathbf{o}_t \odot \phi(\mathbf{c}_t) \end{aligned} \qquad (2)$$

where $\odot$ indicates the Hadamard product.

We use the final hidden state $\mathbf{h}_T$ as the representation for user reviews, item reviews and item meta-data:

$$\mathbf{h}_u = \mathbf{h}_{T(x,u)}, \ \mathbf{h}_{w,v} = \mathbf{h}_{T(w,v)}, \ \mathbf{h}_{m,v} = \mathbf{h}_{T(m,v)} \qquad (3)$$

where $\mathbf{h}_u$, $\mathbf{h}_{w,v}$ and $\mathbf{h}_{m,v}$ are the representations learned from user u's reviews, item v's reviews and item v's meta-data, respectively.

We utilize CNN [15] to extract features from item visual content $\mathbf{G}_v$. In CNN, each neuron j in the convolutional layer uses a filter $K_j \in \mathcal{R}^{S \times S}$ on a slide window of size $S \times S$. We obtain a feature map f_j for each neuron j:

$$b_j = ReLU(\mathbf{G}_{1:N(G),1:N(G)} \star K_j), \qquad (4)$$

where $\star$ is the convolutional operator, and $ReLU(x) = \max\{0, x\}$ indicates the Rectified Linear Units (ReLU) activation function. Then we apply a max pooling operation of a slide window size $P \times P$ over the new feature map b_j, which is calculated by:

$$q_j = \max \begin{pmatrix} b_j & \cdots & b_{j+P-1} \\ \vdots & \ddots & \vdots \\ b_{j+(P-1)\cdot L} & \cdots & b_{j+(P-1)\cdot L+P-1} \end{pmatrix}, \qquad (5)$$

where $L = N(G) - S + 1$ represents the length of side of the new feature map after the convolutional operation.

In this framework, multiple filters are used to gain informative feature maps, resulting $\mathbf{Q} = flat\{q_1, \ldots, q_I\}$, where I denotes the number of kernels in the convolutional layer and $flat$ is the flatten operation. Then the image representation $\mathbf{h}_{g,v}$ can be obtained by:

$$\mathbf{h}_{g,v} = ReLU(\mathbf{W}_G \mathbf{Q}), \qquad (6)$$

where $\mathbf{W}_G$ is the learnable weight matrix.

Given the representations learned separately from heterogeneous content sources of item v, we first concatenate all the representations of item v:

$$\mathbf{h}_v^c = concat(\mathbf{h}_{w,v}, \mathbf{h}_{m,v}, \mathbf{h}_{g,v}), \qquad (7)$$

where $concat(\cdot)$ is the concatenation operation. The concatenated representation is then passed through a fully connected layer to further reduce its dimensionality and generate the representation $\mathbf{z}_v$ of item v:

$$\mathbf{z}_v = ReLU(\mathbf{W}_z \mathbf{h}_v^c), \qquad (8)$$

where $\mathbf{W}_z$ denotes the learnable weight matrix.

Finally, we use $\mathbf{h}_u$ and $\mathbf{z}_v$ to predict the probability of generating a positive interaction $\hat{y}_{u,v}$ between user u and item v:

$$\hat{y}_{u,v} = \sigma\big(\mathbf{W}_L(\mathbf{h}_u \odot \mathbf{z}_v)\big), \tag{9}$$

where $\mathbf{W}_L$ is the weight matrix in this layer.

3.2 Rating Generation Module

The rating generation module shown in the top of Fig. 2 takes the output $\hat{y}_{u,v}$ from exposure generation module, the user embedding vector $\mathbf{e}_u \in \mathcal{R}^{D_e}$ for user u and the item embedding vector $\mathbf{e}_v \in \mathcal{R}^{D_e}$ for item v as inputs where D_e is the dimensionality of embeddings.

DEAMER first concatenates the three inputs and then transform the concatenation to generate the rating. The procedure can be formally defined as:

$$\hat{r}_{u,v} = \sigma\Big(\mathbf{W}_R \cdot concate(\mathbf{e}_u, \hat{y}_{u,v}, \mathbf{e}_v)\Big), \tag{10}$$

where $\mathbf{W}_R$ denotes the weight matrix and σ is the sigmoid activation function.

3.3 Training

DEAMER optimizes its parameters by using multi-task learning and minimizing the following loss function:

$$\mathcal{L} = \sum_u \sum_v \{[y_{u,v} \log \hat{y}_{u,v} + (1 - y_{u,v}) \log(1 - \hat{y}_{u,v})] + \lambda[r_{u,v} - \hat{r}_{u,v}]^2\}, \tag{11}$$

where the first term is the interaction prediction loss, the second term is the rating prediction loss, and the loss coefficient λ is used to balance the two prediction tasks.

In the implementation, we pretrain the word embeddings of user reviews, item metadata and item reviews using Doc2vec [14] and set the pretrained embedding size for each word as 500. The pretrained item image embeddings is available from the data sets we used which will be illustrated in Sect. 4.1.

3.4 Discussion

The probabilistic matrix factorization (PMF) [30] assumes a rating is sampled from a Gaussian distribution:

$$p(r_{u,v}) = \mathcal{N}(r_{u,v}; \frac{1}{1 + \exp[-(\mathbf{uv} + b_u + b_v)]}, \beta^{-1}), \tag{12}$$

where $\mathbf{u}, \mathbf{v}$ can be considered as user and item embeddings, β^{-1} is the precision of the Gaussian distribution. Under this assumption, the likelihood $p(r_{u,v})$ is equivalent to the square loss function.

Table 1. Statistics of data sets

Data sets	# Users	# Items	# Ratings	RR (%)	MM (%)	MI (%)
Musica Instruments (MI)	1,429	900	10,261	0.798	0.22	1.00
Office Products (OP)	4,905	2,420	53,258	0.449	0.04	0.58
Digital Music (DM)	5,541	3,568	64,706	0.327	11.66	0.28
Sports and Outdoors (SO)	35,598	18,357	296,337	0.045	0.22	0.98
Health and Personal care (HP)	38,609	18,534	346,355	0.048	0.19	0.92

Given observations $r_{u,v}$ and $y_{u,v}$, the likelihood $p(r_{u,v}, y_{u,v})$ can be decomposed as:

$$p(r_{u,v}, y_{u,v}) = p(y_{u,v})p(r_{u,v}|y_{u,v}). \tag{13}$$

Therefore, we can see that DEAMER is essentially a generative model for two tasks:

1. The exposure generation module at the bottom of Fig. 2 estimates $p(y_{u,v})$ based on user and item contents. The first term of Eq. 11 is the negative likelihood of $p(y_{u,v})$ where $p(y_{u,v} = 1) = \sigma\big(\mathbf{W}_L(\mathbf{h}_u \odot \mathbf{z}_v)\big)$.
2. The rating generation module in the top of Fig. 2, which assesses $p(r_{u,v}|y_{u,v})$, is based on the assumption of PMF. But it is more expressive than standard PMF. Equation 10 generalizes Eq. 12 when $\mathbf{W}_R, \mathbf{W}_L$ are both unit matrices, $\mathbf{h}_u = \mathbf{u}, \mathbf{z}_v = \mathbf{v}, \mathbf{e}_u \in \mathcal{R}^{D_e}, \mathbf{e}_v \in \mathcal{R}^{D_e}$ and σ is the sigmoid function. The second term of Eq. 11 is the negative likelihood of PMF.
3. We can derive that the loss coefficient $\lambda = 2/\beta$ from Eqs. 11, 12 and 13.

4   Experiment

In this section, we conduct experiments to answer the following research questions:

1. Does DEAMER perform well on recommendation tasks, including rating prediction and top-k recommendation?
2. Do multi-modal learning and multi-task learning contribute to the recommendation of DEAMER?
3. How do the hyper-parameters affect the performance?
4. Can we discover interesting patterns by analyzing the user and item embeddings?

4.1   Experimental Setup

Data Sets. We adopt several public E-commerce data sets,[1] which are standard benchmarks in the recommendation community in our experiments. These data

---

[1] http://jmcauley.ucsd.edu/data/amazon/.

sets contain item descriptions, user reviews and ratings on the Amazon web store. We conduct experiments on five representative categories with different sizes and densities, as well as missing rate of meta-data and image. For each item, if there does not exist meta-data or an image, we regard it as missing. The statistics of the adopted data sets are listed in Table 1. We also provide the density–ratio of observed ratings (i.e., RR), the percentage of missing item meta-data (i.e., MM) and the percentage of missing item image (i.e., MI) of each data set.

Experimental Protocol. We use leave-one-out in both rating prediction and top-k recommendation tasks. In rating prediction, for each user in the data set, we holdout one rating randomly as test instance. In top-k recommendation, for each user, we use the holdout sample as the positive test instance and randomly sample 99 items that the user did not interact before as the negative test instances. The reported results are averaged over five runs.

Hyper-parameter Settings. Unless stated otherwise, we use the optimal hyper-parameters turned in the smallest MI data set for DEAMER. We leverage grid search, where the dimensionality of hidden states (i.e., $\mathbf{h}_u$, $\mathbf{h}_{w,v}$ and $\mathbf{h}_{m,v}$) D_h as well as dimensionality of embeddings D_e are searched in the range $\{32, 64, 128, 256\}$. The regularization weight and the dropout rate are tuned in the ranges of $\{0.1, 0.01, 0.001\}$ and $\{0.1, 0.25, 0.5\}$, respectively. The best performing D_h, D_e, regularization weight, and dropout rate are 256, 64, 0.01 and 0.25, respectively. In our training step, we set $\lambda = 1.2, 10, 15, 30, 40$ for data sets MI, OP, DM, SO and HP, respectively. We will report the impacts of different λ in Sect. 4.4. During training, for each user-item interaction training pair, we sample 5 items that this user has not interacted before. For CNN, we set the convolutional slide window size as 3×3 (i.e., $S = 3$) with slide step size 1×1, the max pooling size as 2×2 (i.e., $P = 2$) with step size 2×2, and the number of filters as 8. The Adam optimizer is employed with the initial learning rate, the training batch size and the max training epoch being 0.001, 256 and 80, respectively.

4.2   Analysis of Recommendation Performance

One advantage of DEAMER is that it can simultaneously predict rating values and make top-k recommendation. Thus, we conduct comparative studies on its regression performance (i.e., how close the predicted ratings are to the true ratings) and ranking performance (i.e., how close is the output recommendation list to the ranking list based on the true user feedback) in the following.

Rating Prediction. We first compare DEAMER with other state-of-the-art rating prediction models on the rating prediction task.

Table 2. Results of rating prediction with best performance in bold

Data set	UKNN	IKNN	NMF	logit-vd	AutoRec	DeepCoNN	DEAMER
MI	1.0348	1.0073	0.9558	1.0578	0.9178	0.9244	**0.9017**
OP	0.9646	0.9914	0.9308	0.9904	0.9269	0.8783	**0.8696**
DM	1.0683	1.0829	0.9973	1.0714	1.0010	1.0319	**0.9533**
SO	1.0683	1.0908	1.0087	1.8648	0.9981	0.9870	**0.9668**
HP	1.1929	1.1990	1.1208	1.8786	1.1275	1.0876	**1.0839**

Baselines. We compare with the conventional collaborative filtering models, probabilistic MNAR models, and deep neural network models:

1. UKNN [1]: the user-based collaborative filtering with Pearson correlation and $k = 50$.
2. IKNN [1]: the item-based collaborative filtering with cosine similarity and $k = 50$.
3. NMF [13]: nonnegative matrix factorization.
4. logit-vd [26]: a collaborative filtering method that assumes responses are *Missing Not At Random*.
5. AutoRec [31]: a deep model using autoencoder.
6. DeepCoNN [37]: a deep model that contains two parallel CNNs to extract latent factors from both user and item reviews.

Evaluation Metrics. We use Root Mean Square Error (RMSE) for evaluation.

Observations. We report the results of rating prediction in Table 2. We can find that DEAMER produces the best results on different data sets. DEAMER outperforms the probabilistic MNAR generative model logit-vd by more than 10% on all data sets. This is unsurprising, since DEAMER's deep architecture allows it to express the dependency structure between interactions and ratings. In addition, DEAMER achieves significantly lower RMSE than deep neural network recommendation models AutoRec and DeepCoNN . This shows the benefits of multi-modal learning and deep generative model over normal deep models.

Top-k Recommendation. We then assess the performance of DEAMER on the ranking task.

Baselines. We compare DEAMER with a wide range of state-of-the-art baselines including both shallow and deep models:

1. MostPopular: a method which outputs the most popular items as the recommendation list.

Table 3. Results of top-10 recommendation with best performance in bold

Methods		MostPopular	SVD++	FISM	NeuCF	DeepMF	ConvMF	CMN	DEAMER
MI	HR	0.3394	0.3471	0.3499	0.3478	0.3464	0.3674	0.3590	**0.4402**
	NDCG	0.1999	0.2081	0.2025	0.2058	0.2017	0.1718	0.2183	**0.2498**
	MRR	0.1571	0.1656	0.1575	0.1624	0.1573	0.1107	0.1749	**0.1924**
OP	HR	0.3321	0.4385	0.3382	0.4353	0.4112	0.4630	0.3823	**0.5739**
	NDCG	0.1725	0.2434	0.1741	0.2398	0.2280	0.2145	0.2082	**0.3350**
	MRR	0.1246	0.1842	0.1248	0.1805	0.1724	0.1374	0.1556	**0.2618**
DM	HR	0.3795	0.6979	0.3857	0.6694	0.6972	**0.7760**	0.7426	0.7553
	NDCG	0.2089	0.4525	0.2119	0.4044	0.4409	0.3956	**0.5279**	0.4813
	MRR	0.1569	0.3760	0.1591	0.3227	0.3615	0.2729	**0.4603**	0.3961
SP	HR	0.3911	0.5338	0.3957	0.4612	0.4344	0.5903	0.4722	**0.6668**
	NDCG	0.2262	0.3412	0.2284	0.2749	0.2448	0.2923	0.3057	**0.4222**
	MRR	0.1759	0.2816	0.1773	0.2176	0.1869	0.1974	0.2543	**0.3466**
HP	HR	0.3714	0.4992	0.3765	0.4427	0.4015	0.5121	0.4105	**0.6198**
	NDCG	0.2129	0.3172	0.2162	0.2675	0.2350	0.2557	0.2763	**0.4043**
	MRR	0.1647	0.2610	0.1674	0.2137	0.1841	0.1738	0.2349	**0.3376**

2. SVD++ [12]: a matrix factorization model that combines latent factor model and neighborhood model.
3. FISM [11]: an factorization matrix method that learns the item-item similarity of low dimensional latent factors.
4. NeuCF [9]: a deep collaborative filtering model that generalizes matrix factorization with multi-layer perceptron.
5. DeepMF [34]: a deep matrix factorization model that learns latent features of users and items using multi-layer perceptron.
6. ConvMF [8]: a deep model that uses an outer product to reconstruct the pairwise item correlations in the embedding space.
7. CMN [4]: a deep model that combines the global user and item embeddings and the local neighborhood-based structure with neural attention mechanism.

Evaluation Metrics. We adopt Hit Ratio (HR), Normalized Discounted Cumulative Gain (NDCG) and Mean Reciprocal Rank (MRR) as the metrics for the top-10 recommendation.

Observations. Table 3 illustrates the results of top-10 recommendation. From the results, we can conclude that DEAMER performs consistently well in terms of HR, NDCG and MRR, on different data sets. To be specific, it produces the best performances on four out of five data sets. On DM data set, although DEAMER does not produce the best results, it produces satisfying results (i.e., second best results) in terms of all evaluation metrics, while the best models ConvMF and CMN on DM data set do not perform best on other four data sets.

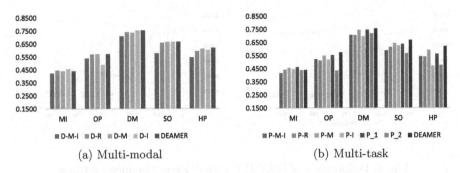

(a) Multi-modal   (b) Multi-task

Fig. 3. Performance of DEAMER with different multi-modal and multi-task settings

4.3 Effectiveness of Multi-modal Learning and Multi-task Learning

We further conduct experiments to show the effectiveness of multi-modal learning and multi-task learning used in DEAMER.

Figure 3(a) shows the impact of different modalities on the HR performance of DEAMER. User-item reviews are the major information source for content based recommendation and we denote the use of it as D. There are three additional modalities, namely item meta-data (M), item images (I) and user-item ratings (R). From Fig. 3(a), we can conclude that all the modalities contribute to the recommendation of DEAMER, as DEAMER using all modalities consistently performs better on all the data sets than the cases when some modalities are not used.

Furthermore, we compare DEAMER to an alternative multi-task learning model. DEAMER is a generative model which operates in a cascade manner. It is common to design a *parallel* multi-task learning model that uses the same representations to predict rating values and binary interactions. The parallel model implements an interaction layer and a rating prediction layer on a shared representation learning component. The representation component can be built on multi-modal data sources. Since the loss function of parallel model also consists of two parts, one is the square loss for rating values, the other is the cross-entropy loss for interactions. Similar to Eq. 11, the two parts are summated with λ. For a fair comparison, we use the same network structures. In Fig. 3(b), we compare the performance of the parallel alternative and DEAMER. The notations are similar as in Fig. 3(a). Additionally, P indicates the parallel architecture is used. For example, "P-I-M" uses LSTMs on user and item reviews, and the representations flow to a rating prediction layer and an interaction prediction layer parallely. We report the parallel multi-task model on all modalities in P_1 with $\lambda = 1.0$ for all data sets; and in P_2 with λ follows DEAMER's setting. From Fig. 3(b), we can observe that DEAMER outperform its parallel alternative which illustrates the power of its architecture.

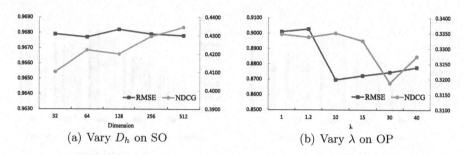

(a) Vary D_h on SO

(b) Vary λ on OP

Fig. 4. Performance of DEAMER with different parameter settings

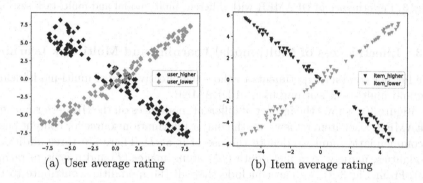

(a) User average rating

(b) Item average rating

Fig. 5. Distinguishing user and item embeddings for users and items with different average ratings

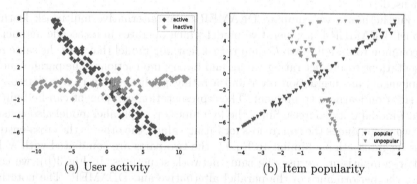

(a) User activity

(b) Item popularity

Fig. 6. Distinguishing user and item embeddings for different levels of user activity and item popularity

4.4 Impact of Hyper-parameters

In the following, we investigate how the hyper-parameters affect the performance of DEAMER. We report the change of RMSE and NDCG on the data set SO by varying $D_h \in \{32, 64, 128, 256, 512\}$ in Fig. 4(a) and on the data set OP by varying $\lambda \in \{1, 1.2, 10, 15, 30, 40\}$ in Fig. 4(b).

From Fig. 4(a), we can see that the dimensionality D_h of hidden states does not have a large impact on rating prediction. However, when D_h increases, the performance of top-k recommendation will be improved correspondingly. As the loss coefficient λ changes which is shown in Fig. 4(b), both RMSE and NDCG will be affected. The best overall performance of DEAMER can be achieved using $\lambda \in \{10, 15\}$ to balance rating prediction and top-k recommendation.

4.5 Visualization

After investigating the recommendation results of DEAMER, we discover that DEAMER is able to provide some insights on useful patterns of users and items.

We first compute the average rating for each user and item on data set MI and select 8% of the users/items with the highest average rating and 8% of the users/items with the lowest average rating. We project the embeddings of users and items into a two-dimensional surface with t-SNE [25] and plot the projection in Fig. 5. DEAMER clearly reveals user bias, i.e., a user tends to rate higher or lower in his/her preference, and item bias, i.e., items that are likely to receive higher or lower ratings.

We also plot the projected embeddings for 8% of the users/items with the most ratings and 8% of the users/items with the fewest ratings. Again, as depicted in Fig. 6, DEAMER can distinguish active users and inactive users, popular items and niche items.

5 Conclusion

In this work, we present DEAMER, a novel deep generative model that not only leans representations to generate the exposure, but also predict ratings simultaneously with the help of exposure. By using multi-modal learning and multi-task learning, DEAMER shows that it is possible to fully use heterogeneous information sources and achieve a better performance compared to the state-of-the-art approaches. In the future, we plan to explore the possibility of expanding the user modality used in DEAMER and making it more multivariate like the item modalities used in DEAMER. We also plan to replace the blocks for representation learning in DEAMER with other neural architectures to further improve its recommendation results.

Acknowledgement. This work was supported by the National Natural Science Foundation of China (no. 61702432, 61772209, 61972328), the Fundamental Research Funds for Central Universities of China (20720180070) and the international Cooperation Projects of Fujian in China (201810016).

References

1. Aggarwal, C.C.: Recommender Systems - The Textbook. Springer (2016)
2. Ding, D., Li, H., Huang, Z., Mamoulis, N.: Efficient fault-tolerant group recommendation using alpha-beta-core. In: CIKM. pp. 2047–2050 (2017)
3. Dong, X., Yu, L., Wu, Z., Sun, Y., Yuan, L., Zhang, F.: A hybrid collaborative filtering model with deep structure for recommender systems. In: AAAI. pp. 1309–1315 (2017)
4. Ebesu, T., Shen, B., Fang, Y.: Collaborative memory network for recommendation systems. In: SIGIR. pp. 515–524 (2018)
5. García-Durán, A., Gonzalez, R., Oñoro-Rubio, D., Niepert, M., Li, H.: Transrev: Modeling reviews as translations from users to items. arXiv Preprint (2018), https://arxiv.org/abs/1801.10095
6. Gatzioura, A., Vinagre, J., Jorge, A.M., Sànchez-Marrè, M.: A hybrid recommender system for improving automatic playlist continuation. In: IEEE Transactions on Knowledge and Data Engineering (2019). https://doi.org/10.1109/TKDE.2019.2952099
7. Guo, H., Tang, R., Ye, Y., Li, Z., He, X.: Deepfm: A factorization-machine based neural network for CTR prediction. In: IJCAI. pp. 1725–1731 (2017)
8. He, X., Du, X., Wang, X., Tian, F., Tang, J., Chua, T.: Outer product-based neural collaborative filtering. In: IJCAI. pp. 2227–2233 (2018)
9. He, X., Liao, L., Zhang, H., Nie, L., Hu, X., Chua, T.: Neural collaborative filtering. In: WWW. pp. 173–182 (2017)
10. Hochreiter, S., Schmidhuber, J.: Long short-term memory. Neural Computation 9(8), 1735–1780 (1997)
11. Kabbur, S., Ning, X., Karypis, G.: FISM: factored item similarity models for top-n recommender systems. In: KDD. pp. 659–667 (2013)
12. Koren, Y.: Factorization meets the neighborhood: a multifaceted collaborative filtering model. In: KDD. pp. 426–434 (2008)
13. Koren, Y., Bell, R.M., Volinsky, C.: Matrix factorization techniques for recommender systems. IEEE Computer 42(8), 30–37 (2009)
14. Lau, J.H., Baldwin, T.: An empirical evaluation of doc2vec with practical insights into document embedding generation. In: Rep4NLP@ACL. pp. 78–86 (2016)
15. LeCun, Y., Haffner, P., Bottou, L., Bengio, Y.: Object recognition with gradient-based learning. In: Shape, Contour and Grouping in Computer Vision. vol. 1681, p. 319 (1999)
16. Li, H., Chan, T.N., Yiu, M.L., Mamoulis, N.: FEXIPRO: fast and exact inner product retrieval in recommender systems. In: SIGMOD Conference. pp. 835–850 (2017)
17. Li, H., Liu, Y., Mamoulis, N., Rosenblum, D.S.: Translation-based sequential recommendation for complex users on sparse data. IEEE Trans. Knowl, Data Eng (2019)
18. Li, H., Liu, Y., Qian, Y., Mamoulis, N., Tu, W., Cheung, D.W.: HHMF: hidden hierarchical matrix factorization for recommender systems. Data Min. Knowl. Discov. 33(6), 1548–1582 (2019)
19. Li, H., Wu, D., Mamoulis, N.: A revisit to social network-based recommender systems. In: SIGIR. pp. 1239–1242 (2014)
20. Li, H., Wu, D., Tang, W., Mamoulis, N.: Overlapping community regularization for rating prediction in social recommender systems. In: RecSys. pp. 27–34 (2015)

21. Lian, J., Zhang, F., Xie, X., Sun, G.: Cccfnet: A content-boosted collaborative filtering neural network for cross domain recommender systems. In: WWW. pp. 817–818 (2017)
22. Lian, J., Zhou, X., Zhang, F., Chen, Z., Xie, X., Sun, G.: xdeepfm: Combining explicit and implicit feature interactions for recommender systems. In: KDD. pp. 1754–1763 (2018)
23. Liang, D., Charlin, L., McInerney, J., Blei, D.M.: Modeling user exposure in recommendation. In: WWW. pp. 951–961 (2016)
24. Lu, Z., Li, H., Mamoulis, N., Cheung, D.W.: HBGG: a hierarchical bayesian geographical model for group recommendation. In: SDM. pp. 372–380 (2017)
25. van der Maaten, L., Hinton, G.: Visualizing data using t-sne. JMLR **9**, 2579–2625 (2008)
26. Marlin, B.M., Zemel, R.S.: Collaborative prediction and ranking with non-random missing data. In: RecSys. pp. 5–12 (2009)
27. Qian, Y., Li, H., Mamoulis, N., Liu, Y., Cheung, D.W.: Reverse k-ranks queries on large graphs. In: EDBT. pp. 37–48 (2017)
28. Rendle, S.: Factorization machines with libfm. ACM TIST 3(3), 57:1–57:22 (2012)
29. Ricci, F., Rokach, L., Shapira, B. (eds.): Recommender Systems Handbook. Springer (2015)
30. Salakhutdinov, R., Mnih, A.: Probabilistic matrix factorization. In: NIPS. pp. 1257–1264 (2007)
31. Sedhain, S., Menon, A.K., Sanner, S., Xie, L.: Autorec: Autoencoders meet collaborative filtering. In: WWW. pp. 111–112 (2015)
32. Strub, F., Gaudel, R., Mary, J.: Hybrid recommender system based on autoencoders. In: DLRS@RecSys. pp. 11–16 (2016)
33. Wang, C., Niepert, M., Li, H.: LRMM: learning to recommend with missing modalities. In: EMNLP. pp. 3360–3370 (2018)
34. Xue, H., Dai, X., Zhang, J., Huang, S., Chen, J.: Deep matrix factorization models for recommender systems. In: IJCAI. pp. 3203–3209 (2017)
35. Zhang, S., Yao, L., Sun, A., Tay, Y.: Deep learning based recommender system: A survey and new perspectives. ACM Comput. Surv. 52(1), 5:1–5:38 (2019)
36. Zhang, Y., Ai, Q., Chen, X., Croft, W.B.: Joint representation learning for top-n recommendation with heterogeneous information sources. In: CIKM. pp. 1449–1458 (2017)
37. Zheng, L., Noroozi, V., Yu, P.S.: Joint deep modeling of users and items using reviews for recommendation. In: WSDM. pp. 425–434 (2017)

Recurrent Convolution Basket Map for Diversity Next-Basket Recommendation

Youfang Leng[1], Li Yu[1(✉)], Jie Xiong[1], and Guanyu Xu[2]

[1] School of Information, Renmin University of China, Beijing, China
{lengyoufang,buaayuli,xiongjiezk}@ruc.edu.cn
[2] Xuteli School, Beijing Institute of Technology, Beijing, China
xuguanyu25@gmail.com

Abstract. Next-basket recommendation plays an important role in both online and offline market. Existing methods often suffer from three challenges: information loss in basket encoding, sequential pattern mining of the shopping history, and the diversity of recommendations. In this paper, we contribute a novel solution called *Rec-BMap* ("**R**ecurrent **C**onvolution **B**asket **Map**"), to address those three challenges. Specifically, we first propose basket map, which encodes not only the items in a basket without losing information, but also static and dynamic properties of the items in the basket. A convolutional neural network followed by the basket map is used to generate basket embedding. Then, a Time-LSTM with time-gate is proposed to learn the sequence pattern from consumer's historical transactions with different time intervals. Finally, a deconvolutional neural network is employed to generate diverse next-basket recommendation. Experiments on two real-world datasets demonstrate that the proposed model outperforms existing baselines.

Keywords: Next-basket recommendation · Sequential recommendation · Basket map · Time-LSTM · Recurrent Neural Network

1 Introduction

In next basket recommendation task, given a sequence of shopping transactions for each user, where a transaction is a basket of items purchased at a point of time, the goal is to recommend a combination of items that user most wants to purchase next. It plays an important role in improving the user experience and increasing the shopping conversion rate, which has received extensive attention in both online and offline shopping [1,18,19,24–26].

Next-basket recommendation is fundamentally different from next-item recommendation, which usually treats a consumer's historical purchase behavior as a sequence and recommends the *next item* to him/her, often referred to as session-based recommendation [4,11,12,14,23]. In contrast, next-basket recommendation treats a consumer's *shopping baskets* as a sequence and recommends the *next set of items* for the consumer. The items in the set are

© Springer Nature Switzerland AG 2020
Y. Nah et al. (Eds.): DASFAA 2020, LNCS 12114, pp. 638–653, 2020.
https://doi.org/10.1007/978-3-030-59419-0_39

Fig. 1. The difference between next-item and next-basket recommendation.

usually associated and complementary to each other. For example, as shown in Fig. 1, when a consumer who is a photography enthusiast is searching for a photo printer, a next-item recommender system usually recommends a number of different printers for him/her, while a next-basket recommender system will recommend the most appropriate combination of items, such as printers, ink cartridges, printing paper, album, and even scissors.

The deep learning based next-item recommendation models, such as GRU4Rec [4], NARM [11] and STAMP [14], basically follows a general encoder-decoder framework with three steps: 1) **an encoder** converts an input sequence into a set of high-dimensional hidden representations, 2) which are then fed to **a sequential pattern learning module** to build the representation of the current state, 3) which will be transformed by **a decoder** component to produce the next item. Existing methods for next-basket recommendation usually apply next-item models directly with slight changes in those three steps [1, 24–26]. In this paper, we argue that it is unreasonable to directly apply next-item recommendation to next-basket recommendation due to the following three challenges faced by next-basket recommendation task:

Challenge 1: Information Loss in Basket Encoding. Existing methods typically use pooling to encode a basket. They use the average or maximum value of the item embeddings to represent a basket, which leads to information loss. For example, HRM [25] forms the basket representation by aggregating item embeddings from the last transaction. NN-Rec [24] uses the mean of all item embeddings included in a basket to represent it. In addition, a shopping basket usually contains rich attribute information, such as items, price, purchase amounts, which cannot be encoded by the pooling method.

Challenge 2: Long Sequence Dependencies of Different Time Intervals. The consumer's shopping history is usually a long sequence of transactions in which his/her shopping pattern is embedded. There can be different time intervals between any two consecutive baskets. These micro time interval information can well characterize a consumer's shopping habits. A good recommendation system should capture and take into consideration these long-term dependencies and time intervals simultaneously. Typically, there are two modeling paradigms for this problem. One is MC (Markov Chain) based methods, the other is RNN (Recurrent Neural Network) based methods. The former, such as FPMC [19] and HRM [25] only capture local sequential features between every two adjacent baskets. The latter, such as DREAM [26] use a standard RNN to capture global sequential features to make the best of all relations among sequential baskets. Nevertheless, all those methods have the same deficiency by ignoring the time interval between consecutive baskets.

Challenge 3: Diversity of Recommendation. Next-basket recommendation requires recommending a set of products that are complementary. Existing systems encode a user's shopping history into a latent vector and use the vector as the encoding of the next item, then generate the prediction of next basket by two ways [1,11,14,26]. One is to generate the probability of item occurrence through a softmax layer. Another is to calculate the similarity between the latent vector and all item vectors, then sort the above probabilities or similarities and take the top K as a basket. However, the items recommended by this top K method are often very similar to each other, even including the same products with different brands or types. This obviously does not meet the goal of next-basket recommendation, in which consumers prefer to recommend a printer and its accessories instead of recommending five different brands of printers in a basket. Therefore, generating a diverse recommendation for consumers is the third challenge that next-basket recommendation faces.

To address the above challenges, we propose a novel model, namely Recurrent Convolutional Basket **MAP** (*Rec-BMAP* for short), in this study. Specifically, Rec-BMap consists of three parts. The first part is Basket Map Learning, in which a multi-channel basket map is construct including item, price, purchase amount, then a Convolutional Neural Network (CNN) scans the multi-channel basket map to obtain the latent vector of the basket map, which is to address the challenge #1. The second part is Basket Sequence Learning, which is responsible for learning the long term dependencies of the baskets with different time interval by our proposed Time-LSTM to address the challenge #2. For the recommendation diversity of challenge #3, we replaced the method of softmax with a deconvolution method to directly restore the basket to ensure the diversity of recommendations. We have evaluated the proposed approach with two real world datasets, with existing methods MC and RNN based methods as baselines. The results showed that our method surpasses many traditional and deep learning based methods and achieves state-of-the-art.

Overall, the main contributions of this paper can be summarized as follows:

- We proposed **Basket Map** to encode a basket, as well as multiple channels to represent the properties of each item in the basket. The deconvolution is designed to decode the basket to ensure the diversity of the recommended basket. To the best of our knowledge, this is the first time that a basket has been encoded and decoded using this approach.

- We improved the LSTM unit by incorporating a time gate to capture the time interval between the baskets. The time gate can fine tune the input gate and forget gate at the same time.
- We proposed a Recurrent Convolution Basket Map model by integrating the encode/decode operation of the basket map and time-lstm layer.

In the rest of the paper, we will first introduce some related work in Sect. 2, then present the concept of basket map and the proposed Rec-BMAP model in Sect. 3. The experiments and results will be presented in Sect. 4. Finally, we will conclude this paper and discuss the future work in Sect. 5.

2   Related Work

This section will review the related work on next-basket recommendation. We categorize existing methods into three groups, including traditional methods, MLP based methods, and RNN based methods.

2.1   Traditional Methods

Typically, traditional recommendation methods can be divided into two types: general methods and serial methods.

General methods such as collaborative filtering (CF) is based on a user-item rating matrix extracted from the interaction history between users and items. For example, the Matrix Factorization (MF) approach [7] factorized a user-item matrix to estimate the user latent vector. Another approach is the neighborhood methods [13,20] which try to make recommendations based on item similarities calculated from the co-occurrences of items in baskets. Though these methods have proven to be effective and are widely employed, they break down the basic transaction unit (e.g., a basket) into multiple records (e.g., user-item interaction pairs) and miss the sequential features which contain the user's preference shift over time.

Sequential Methods. To address the above issues, sequential methods based on Markov chains are proposed which utilize sequential data to predict users' next action [19,21,28]. Zimdars et al. [28] proposed a sequential recommender algorithm based on Markov chains and investigated on how to extract sequential patterns to learn the next state using probabilistic decision-tree models. Shani et al. [21] presented a MDP (Markov Decision Processes) aiming to provide recommendations in a session-based manner and the simplest MDP boiled down to the first-order Markov chains where the next recommendation could be simply generated through the transition probabilities between items. Factorizing Personalized Markov Chains (FPMC) [19] was a hybrid model that combined the power of MF and MC to model both general interests and sequential behavior between every two adjacent baskets for next basket recommendation. Nevertheless, all the MC-based methods share the same deficiency in that those recommenders only modeled local sequential behaviors between every two adjacent actions and some of which may even be irrelevant.

2.2   MLP-Based Methods

Recently, some researchers apply the deep learning method to the next-basket recommendation research field, such as HRM [25] and NN-Rec [24]. HRM [25] employed a two-layer structure to construct a hybrid representation of users and items from last transaction: The first layer formed the transaction representation by aggregating item vectors from last transaction, while the second layer built the hybrid representation by aggregating the user vector and the transaction representation. Since the HRM model used information about the last basket only, Wang et al. (2015) further proposed the NN-Rec [24] model to improve the HRM model by concatenating the last K baskets. Those methods used a Multi-Layer Perception (MLP) approach that did not model all the user's shopping history and therefore cannot mine the global sequential behaviors.

2.3   RNN-Based Methods

Recurrent Neural Networks have proven to be very effective in modeling global sequence recently [9]. Inspired by recent advances in natural language processing techniques [22], some RNN based methods have been developed for next-basket recommendation [1,3,18,26]. DREAM [26] was the first model adopting RNN that not only learned user's dynamic interests over time, but also captured the global sequential pattern among baskets. However, existing RNN models ignores important item attributes such as product categories, prices, etc. [1]. Therefore, Bai et al. (2018) proposed an Attribute-aware Neural Attentive Model that used both an RNN to capture user's dynamic interests and a hierarchical structure to incorporate attribute information if items to improve the performance of next basket recommendation.

3   Proposed Model

In this section, we formulate the task of next-basket recommendation and then introduce the proposed **Rec-BMap** model in detail.

3.1   Problem Formulation

Given a set of users U and items I, let $u \in U$ denote a user and $i \in I$ denote an item. The numbers of users and items are denoted as $|U|$ and $|I|$, respectively. User u's purchase history sorted by time is a sequence of baskets $B^u = \{B_1^u, B_2^u, \ldots, B_t^u\}$, where $B_t^u \subseteq I$ is the basket that user u purchased at time t. As mentioned earlier, our model not only encodes the items in the basket, but also its attributes such as product category, price, and so on. Therefore, a basket can be extended to a set of tuples denoted as $B_t^u = \{< i_1, c_1, p_1, m_1 >, \ldots, < i_n, c_n, p_n, m_n >\}$, in which the four elements represent item, product category, price and purchase amount, respectively, means that user u purchased m_n units of items i_n in the category c_n at selling price of p_n. Based on the above annotations, our goal is to predict a set of items that u most likely to buy in the next basket B_{t+1}^u.

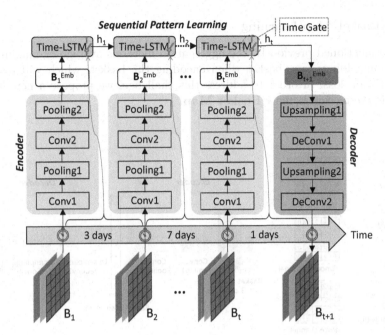

Fig. 2. An overview of the Recurrent Convolution Basket Map model

3.2   An Overview of REC-BMap Model

In this study, in order to solve the three challenges faced by the next-basket recommendation task, we proposed a **R**ecurrent **C**onvolution **B**asket *Map* model (**Rec-BMap**) for next-basket recommendation. Rec-BMap model is an encoder-decoder framework. Its basic idea is to build a latent representation of the history of one's previous baskets, on which next-basket recommendation will be generated. As shown in Fig. 2, the encoder converts an input basket sequence $B^u = \{B_1, B_2, \ldots, B_t\}$ into a set of high-dimensional latent representations $\hat{B}^u = \{B_1^{Emb}, B_2^{Emb}, \ldots, B_t^{Emb}\}$, which are fed into the sequential pattern learning component to build the representation of the current state h_t at time t. Finally, h_t is transformed by a decoder component to produce the recommendation of next basket B_{t+1}.

Based on the above basic idea, Rec-BMap consists of three main components, as show in Fig. 2: (1) A convolutional basket map learning encoder, which is designed to encode well-constructed basket maps (Sect. 3.3). (2) A sequential learning component, which equips LSTM with a time-gate is used to capture the global sequential pattern of one's all historical baskets (Sect. 3.4). (3) A deconvolution component, which commensurates with the basket encoder to decode and generate the diversity of next basket (Sect. 3.5). In the remainder of this chapter, we will first introduce these three components, then describe the design of the loss function and the algorithm for jointly training the Rec-BMap model (Sect. 3.6).

3.3   Basket Map Learning

As we mentioned previously, the aggregation methods average or maximize all item embeddings in a basket when encoding the basket, which will result in information loss. In order to address this challenge, we propose a new way to encode the basket, namely **Basket Map**.

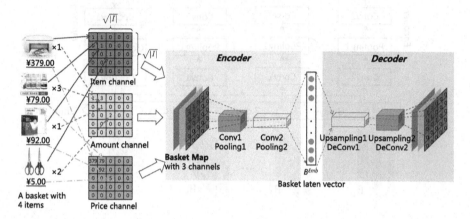

Fig. 3. Basket map construction process, and its convolutional encoder and deconvolutional decoder diagram.

Basket Map. As shown in the left part of Fig. 3, for the item set I with a total number of $|I|$, we first constructed a two-dimensional matrix of shape $\lceil\sqrt{|I|}\rceil \times \lceil\sqrt{|I|}\rceil$, with each element in the matrix representing an item. In order to better learn the category information better, we put the items of the same category together as much as possible. Then, for a user's basket B, the matrix element corresponding to an item contained in the basket B will be filled with 1, and the rest will be filled with 0. Just like an image has three channels (i.e., RGB) of the color space, we used two additional channels to indicate the purchase amount and the price of the corresponding item. So for multiple baskets in a user's shopping history, our task was converted to learn a sequence of basket map with different time intervals and then predicted the next one.

Convolution Basket Map. The Basket map organizes the items in the basket and their attribute information in a spatial structure. For example, a basket containing many baby products may activate an area in the basket map, while digital products will activate another area in the basket map. Convolutional neural network (CNN) has shown its powerful ability to hierarchically capture the spatial structural information [8]. Therefore, we use CNN to extract the rich regional features contained in the basket map. As shown in the encoder part of

Fig. 2 and Fig. 3, it contains two convolutional layers, each with a kernel size of 3×3 and filter sizes of 16 and 32 respectively. Each convolution layer is followed by an activation function ReLU [16] and a max pooling layer with a kernel size of 2×2. With such a convolution structure, we project each basket map into a latent space B^{Emb}.

3.4 Basket Sequence Learning

After constructing a basket into a spatial basket map, and using CNN to scan the basket map and embed it into a latent space, in order to model the basket map sequence, we resort to the Recurrent Neural Network (RNN), which has been proved effective when modeling sequential data, such as speech and language [9]. In traditional tasks such as language modeling, RNN usually only considers the sequential order of objects but not interval length. However, in a recommender system, time intervals between consumers' consecutive actions are of significant importance to capturing the relations of consumers' actions. The classical RNN architectures such as LSTM [5] are not good at modeling them [2,17,27]. So we propose a time-LSTM unit, which adds a time gate to LSTM and fine-tunes the input gate and forget gate through the time gate simultaneously.

Assuming consumer u'sbasket embedding sequence $S = \{B_1^{Emb}, \ldots, B_t^{Emb}\}$, the hidden state h_t of Time-LSTM as:

$$
\begin{aligned}
x_t &= B_t^{Emb} \\
h_t &= Time\text{-}LSTM(x_t, d_t, h_{t-1})
\end{aligned}
\tag{1}
$$

where the input x_t of Time-LSTM is basket embedding at current step. A time-gate is added in Time-LSTM unit to deal with time interval. So at time t, the hidden state h_t of Time-LSTM is determined by x_t, the time interval d_t and the previous hidden state h_{t-1} simultaneously. The internal structure of the Time-LSTM will be covered in the following subsection.

Time-LSTM Unit. Figure 4 presents the proposed Time-LSTM unit, which updates the equations of the standard LSTM [5] unit as follows:

$$
i_t = \sigma(W_i x_t + U_i h_{t-1}) \tag{2}
$$

$$
f_t = \sigma(W_f x_t + U_f h_{t-1}) \tag{3}
$$

$$
o_t = \sigma(W_o x_t + U_o h_{t-1}) \tag{4}
$$

$$
\hat{c}_t = tanh(W_c x_t + U_c h_{t-1}) \tag{5}
$$

$$
c_t = f_t \odot c_{t-1} + i_t \odot \hat{c}_t \tag{6}
$$

$$
h_t = o_t \odot tanh(c_t) \tag{7}
$$

where r_t and z_t are the reset gate and update gate respectively, and the final hidden state is h_t. Based on the updated Eqs. (2) to (7), we add one time gate $\mathbf{T_t}$ as:

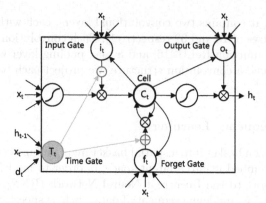

Fig. 4. The proposed Time-LSTM by equipping LSTM with a time-gate which can fine-tunes the input gate and forget gate simultaneously.

$$\mathbf{T_t} = \sigma(W_t x_t + U_t h_{t-1} + \mathbf{V_t d_t}) \tag{8}$$

where d_t is the dwell time and V_t is its weight matrix. Then the Eq. (6) can be rewritten as:

$$c_t = (f_t + \mathbf{T_t}) \odot c_{t-1} + (i_t - \mathbf{T_t}) \odot \hat{c}_t \tag{9}$$

The changes in the formula are highlighted in bold. The time gate T_m is helpful in two ways. First, as shown in Eq. (8), the time gate is affected not only by the input data, but also by the time interval. Second, as shown in Eq. (9), the time gate acts as a fine-tuning gate between the input gate and the forget gate. In LSTM [5], the input gate controls how much of the current input can enter the hidden state, while the forget gate controls how much of the previous information can enter the hidden state [9]. So when the time-gate has a positive effect on the input gate, it will adversely affect the forget gate accordingly. An intuitive understanding is that after a user has purchased a basket for a long time, the time-gate will reduce its impact on the next basket and increase the impact of other baskets.

3.5   Diversity Next-Basket Generation

The hidden state of each time step is derived through LSTM with the time gate. The last hidden state is used to generate the next basket prediction. There are two traditional ways to decode this hidden state. The first one, such as HRM [18], DREAM [26], ANAM [1], outputs the probability of the next item through a fully connected layer with the softmax activation function. The second one, such as NARM [11] and STAMP [14], uses a similarity layer to calculate the similarities between the hidden state and each candidate item. Finally, the probabilities or similarities are ranked and the top K items are selected to form the next basket recommendation for the consumer.

Regardless of probability or similarity, this hidden state is usually treated as the representation of the *next item* rather than a basket. Top K items obtained from the ranking list are usually similar, which leads to a lack of diversity in the recommended basket. To address this flaw, we propose an end-to-end method to generate the next basket directly. The unsupervised stacked convolutional auto-encoder (SCAE) [15] model uses stacked convolutional layers to compress an image, then uses the symmetric deconvolution layers to restore it. Inspired by SCAE, we designed supervised stacked deconvolutional layers to restore the basket map from the last hidden state of Time-LSTM directly, as shown in the decoder part of Fig. 2 and Fig. 3. The deconvolution decoder can learn parameters from the existing basket map during the training phase, and use the learned parameters to generate the next basket in the prediction phase. This approach ensures the diversity of the generated basket.

3.6 Objective Function and Jointly Learning

Given a consumer's shopping history, our model recurrently convolutes a basket map sequence by an end-to-end fashion to directly generate the next basket.

In order to optimize the three components (i.e., basket map learning, sequential pattern learning, and diversity next basket recommendation) at the same time, we designed a multi-objective loss function to jointly learn the weight parameters.

$$
\begin{aligned}
L = & \frac{1}{V} \sum_{m=1}^{V} (B_{t,m}^{Emb} - B_{t,m}^{Emb'})^2 \\
& - \frac{1}{|I|} \sum_{i=1}^{|I|} y_i log(\hat{y}_i) + (1 - y_i)log(1 - \hat{y}_i)
\end{aligned}
\tag{10}
$$

The first term of the loss function L represents the mean square error between latent vector of the basket output from the last step of Time-LSTM, in which V is the dimension of basket embedding. The second term is the binary cross-entropy error between the prediction basket map $\hat{y}$ and the ground truth y, in which $|I|$ is the total number of item set. An iterative stochastic gradient descent (SGD) optimizer is then performed to optimize the entire loss L.

4   Experiments

We conducted empirical experiments to assess the effectiveness of REC-BMap on next basket prediction based on the basket map. Specifically, we first prepared the datasets. Then we compared the proposed model with some state-of-the-art next-basket recommendation methods, including traditional methods and RNN-based models, as baselines.

4.1   Datasets

We conducted experiments on two real-world datasets, i.e., Ta-Feng and BeiRen. **Ta-feng**[1] dataset contained 4 months (November 2000 to February 2001) of shopping transactions from the Ta-feng grocery store. There were 8,266,741 transactions of 23,266 consumers that involved 23,812 products. **BeiRen** dataset was a public online e-commerce dataset released by BeiGuoRenBai[2], which contained 123,754 transactions of 34221 consumers that involved on 17,920 products.

Table 1. Statistics of datasets.

Statistics	Ta-Feng	BeiRen
Users	9,238	9,321
Items	7,982	5,845
Transactions	67,964	91,294
Category	1,074	1,205
Avg.transaction size	7.4	9.7
Avg.transaction per user	5.9	5.8

For both datasets, we removed the items that were purchased less than 10 times and the users who that has bought in total less than 10 items. The statistics of the two datasets after pre-processing is shown in Table 1. We took the last basket of each user as the testing data, the penultimate ones as the validation set to optimize parameters, and the remaining baskets as the training data. Same as [23], we used the data augmentation techniques that for an input baskets $B = [B_1, B_2, \ldots, B_n]$, we generate the sequences and corresponding labels $([B_1], B_2)$, $([B_1, B_2], B_3)$, $\ldots$, $([B_1, B_2, \ldots, B_{n-1}], B_n)$ for training and testing on both datasets.

4.2   Parameter Setup

We implemented a prototype of Rec-BMap on Pytorch[3]. The hyper parameters wre optimized via an extensive grid search on the datasets, and the best models were selected by early stopping based on the Recall@20 on the validation set. According to the averaged performance, the optimal parameters were empirically set as follows. The item embedding dimension were set to 100 and the Time-LSTM cell size is set to 200. For training, optimization was done using mini-batch gradient descent with the learning rate set to 0.001. The mini-batch size is fixed at 128.

---

[1] https://www.kaggle.com/chiranjivdas09/ta-feng-grocery-dataset.

[2] http://www.brjt.cn/.

[3] http://www.pytorch.org.

4.3 Baseline Methods

- **Top** is a naive model that always recommends the most popular items in the training set.
- **MC:** A Markov chain model that predicts the next purchase based on the last transaction of a user.
- **NMF** [10] is a collaborative filtering method, which applies non-negative matrix factorization over the user-item matrix.
- **FPMC** [19]: short for Factorizing personalized markov chains for next-basket recommendation. It is a hybrid model combining Markov chains and matrix factorizing, which can capture both sequential purchase behavior and general interests of users.
- **HRM** [25] is the first model to use neural network for next-basket recommendation. It is a two-level hierarchical representation model that captures a user's sequence behavior and general taste.
- **NNRec** [24] is an improvement of the HRM model that concatenates the last K baskets.
- **DREAM** [26] incorporates both customers' general preferences and sequential information by RNN.
- **ANAM** [1] uses RNN to capture the user's dynamic interests and a hierarchical structure for incorporating the attributes information of items to improve the performance of the next basket recommendation.

4.4 Evaluation Metrics

We adopted the widely used evaluation metrics for assessing the performance of next basket recommendation, i.e., F1-Score and Normalized Discounted Cumulative Gain (NDCG). A list of K items ($K = 5$), denoted by R^u, was generated for each user u, where R_i^u stands for the i_{th} item in the list.

- F1-Score: F1-score is the harmonic mean of precision and recall, which is a widely used metric for assessment of next basket recommendation [19].

$$Precision(B_{t_u}^u, R^u) = \frac{|B_{t_u}^u \bigcap R^u|}{|R^u|}$$

$$Recall(B_{t_u}^u, R^u) = \frac{|B_{t_u}^u \bigcap R^u|}{|B_{t_u}^u|} \tag{11}$$

$$F1\text{-}score = \frac{2 \times Precision \times Recall}{Precision + Recall}$$

- NDCG@K: Normalized Discounted Cumulative Gain (NDCG) is a ranking based measure that takes into account the order of recommendation items in the list [6]. It is formally defined as follows:

$$NDCG@k = \frac{1}{N_k} \sum_{j=1}^{k} \frac{2^{I^{(R_j^u \in B_{t_u}^u)}} - 1}{log_2(j+1)} \tag{12}$$

where $I(\cdot)$ is an indicator function and N_k is a constant which denotes the maximum give of $NDCG@k$ given $R(u)$.

Table 2. Performance comparisons of Rec-BMap vs. baseline methods over two benchmark datasets.

Methods	Ta-Feng		BeiRen	
	F1-score@5	NDCG@5	F1-score@5	NDCG@5
TOP	0.051	0.084	0.078	0.101
NMF	0.052	0.072	0.103	0.125
MC	0.053	0.074	0.091	0.109
FPMC	0.059	0.087	0.107	0.141
HRM	0.062	0.089	0.118	0.147
NNRec	0.063	0.105	0.105	0.148
DREAM	0.065	0.084	0.116	0.152
ANAM	0.146	0.190	0.132	0.164
Rec-BMap	**0.153**	**0.192**	**0.142**	**0.171**

4.5 Comparison Against Baselines

The performance results of Rec-BMap and baseline methods are shown in Table 2. We have the following observations from the results:

(1) Among all the traditional methods, TOP performed the worst because it always recommended the most popular item without using a user's personal purchase history information. NMF model breaks user's sequential transactions into a user-item matrix, and can learn the user's personalized preferences, so it performs better than Top. Contrast to the NMF which didn't consider the sequence information, MC and FPMC methods based on Markov Chain can capture local sequential pattern between adjacent baskets, outperforms slightly NMF. (2) The MLP-based approach uses the maximum or average aggregation method to encode the basket and uses a hierarchical multi-layer perceptron structure to obtain nonlinear interactions between adjacent baskets, so its performance is slightly better than to traditional linear methods. (3) While NN-Rec and HRM methods use only the local sequence information of the last 1 or K baskets, DREAM model which capture global sequential pattern by Recurrent Neural Network outperforms MLP methods. ANAM model not only acquires the global sequence pattern through LSTM, but also incorporates the category information of the item, which outperforms the existing deep learning-based approach. (4) The proposed Rec-BMap model replace the existing maximum or average aggregation method by a convolution-based basket map to encode and decode the basket, and uses the TIME-LSTM with time gate to obtain global sequence information. Its performance is better than all comparison methods.

Next, we went further by evaluating the proposed convolution basket map and TIME-LSTM separately, aiming to gain additional insights on the effectiveness of each.

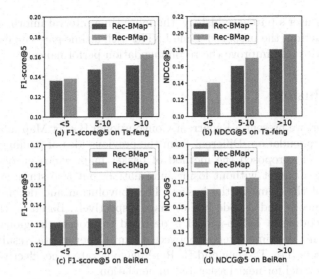

Fig. 5. The effect of time-gate under different basket lengths.

4.6 The Impact of Time-Gate

In order to test the validity of our proposed Time-LSTM, we changed the structure of the Rec-BMap model by removing the Time-gate in the LSTM. In order to better distinguish the two model structures, we denote the model with time-gate removed as Rec-BMap$^-$, and the complete model as Rec-BMap. Since we posit that the length of a consumer's purchase history has an impact on the recommender system performances and time-LSTM has the capacity of capturing such subtle information, we used both datasets and partitioned consumers into three groups: occasional consumers who had fewer than 5 purchase transactions in the data collection period, regular consumers who had between 5 and 10 purchase transactions, and frequent consumers who had more than 10 transactions. For measuring the impact of the complex long-term preference dynamics used in Rec-BMap$^-$ and Rec-BMap, we assessed their performances for these three consumer groups separately.

Figure 5 shows the performances with both datasets. Firstly, we pay attention to improvements over the length of user behaviors grows with the Ta-feng dataset. Rec-BMap achieved a significant improvement of F1-score@5 as the historical transaction lengths grow. For frequent consumers, Rec-BMap performs the best, with at least a 17.39% improvement of F1-score@5 than occasional consumers. Then, we turn to the comparison between Rec-BMap and Rec-BMap$^-$, taking NDCG@5 on Beiren as an example. It can be seen that Rec-BMap with a time gate outperformed than Rec-BMap$^-$ without a time gate. Interestingly, this difference increases as the length of the interaction increases. In summary, this comparative assessment provides two findings. First, Rec-BMap can learn the global sequential pattern from the a user's long-term interaction. Second, the longer a user's purchase history, the more time interval information hidden in the interaction, the proposed Time-LSTM with a time-gate can capture these micro-behavior and improve the recommendation performance.

5 Conclusion

In this paper, we present a Recurrent Convolutional Basket Map model for next basket recommendation. Different from the previous idea of pooling the basket, we innovatively proposed the method of basket map, which encodes not only the items in a basket without losing information, but also static and dynamic properties of the items in the basket. The convolution and deconvolution are then used encode and decode basket map, respectively. Based on the resulting basket embedding, a TIME-LSTM is designed to process the global sequences of baskets and the time interval between them. Experimental results with two public datasets (i.e., Ta-Feng and BeiRen) demonstrated the effectiveness of our Rec-BMap model for next basket recommendation.

References

1. Bai, T., Nie, J.Y., Zhao, W.X., Zhu, Y., Du, P., Wen, J.R.: An attribute-aware neural attentive model for next basket recommendation. In: The 41st International ACM SIGIR Conference on Research and Development in Information Retrieval, pp. 1201–1204 (2018)
2. Bogina, V., Kuflik, T.: Incorporating dwell time in session-based recommendations with recurrent neural networks. In: RecTemp@RecSys, pp. 57–59 (2017)
3. He, R., Kang, W.C., McAuley, J.: Translation-based recommendation. In: Proceedings of the 11th ACM Conference on Recommender Systems, pp. 161–169. ACM (2017)
4. Hidasi, B., Karatzoglou, A., Baltrunas, L., Tikk, D.: Session-based recommendations with recurrent neural networks. In: International Conference on Learning Representations (2016)
5. Hochreiter, S., Schmidhuber, J.: Long short-term memory. Neural Comput. 9(8), 1735–1780 (1997)
6. Järvelin, K., Kekäläinen, J.: IR evaluation methods for retrieving highly relevant documents. In: Proceedings of the 23rd Annual International ACM SIGIR Conference on Research and Development in Information Retrieval, pp. 41–48. ACM (2000)
7. Koren, Y., Bell, R., Volinsky, C.: Matrix factorization techniques for recommender systems. Computer 8, 30–37 (2009)
8. Krizhevsky, A., Sutskever, I., Hinton, G.E.: Imagenet classification with deep convolutional neural networks. Commun. ACM 60(6), 84–90 (2017)
9. LeCun, Y., Bengio, Y., Hinton, G.: Deep learning. Nature 521(7553), 436 (2015)
10. Lee, D.D.G., Seung, H.S.: Algorithms for non-negative matrix factorization. Adv. Neural Inf. Process. Syst. 13, 556–562 (2000)
11. Li, J., Ren, P., Chen, Z., Ren, Z., Lian, T., Ma, J.: Neural attentive session-based recommendation. In: Proceedings of the 2017 ACM on Conference on Information and Knowledge Management, pp. 1419–1428. ACM (2017)

12. Li, Z., Zhao, H., Liu, Q., Huang, Z., Mei, T., Chen, E.: Learning from history and present: next-item recommendation via discriminatively exploiting user behaviors. In: Proceedings of the 24th ACM SIGKDD International Conference on Knowledge Discovery & Data Mining, pp. 1734–1743. ACM (2018)
13. Linden, G., Smith, B., York, J.: Amazon.com recommendations: item-to-item collaborative filtering. IEEE Internet Comput. **7**(1), 76–80 (2003)
14. Liu, Q., Zeng, Y., Mokhosi, R., Zhang, H.: STAMP: short-term attention/memory priority model for session-based recommendation. In: Proceedings of the 24th ACM SIGKDD International Conference on Knowledge Discovery & Data Mining, pp. 1831–1839. ACM (2018)
15. Masci, J., Meier, U., Cireşan, D., Schmidhuber, J.: Stacked convolutional auto-encoders for hierarchical feature extraction. In: Honkela, T., Duch, W., Girolami, M., Kaski, S. (eds.) ICANN 2011. LNCS, vol. 6791, pp. 52–59. Springer, Heidelberg (2011). https://doi.org/10.1007/978-3-642-21735-7_7
16. Nair, V., Hinton, G.E.: Rectified linear units improve restricted Boltzmann machines. In: Proceedings of the 27th International Conference on Machine Learning, pp. 807–814 (2010)
17. Neil, D., Pfeiffer, M., Liu, S.C.: Phased LSTM: accelerating recurrent network training for long or event-based sequences. In: Advances in Neural Information Processing Systems, pp. 3882–3890 (2016)
18. Quadrana, M., Karatzoglou, A., Hidasi, B., Cremonesi, P.: Personalizing session-based recommendations with hierarchical recurrent neural networks. In: Proceedings of the 11th ACM Conference on Recommender Systems, pp. 130–137 (2017)
19. Rendle, S., Freudenthaler, C., Schmidt-Thieme, L.: Factorizing personalized Markov chains for next-basket recommendation. In: Proceedings of the 19th International Conference on World Wide Web, pp. 811–820. ACM (2010)
20. Sarwar, B.M., Karypis, G., Konstan, J.A., Riedl, J., et al.: Item-based collaborative filtering recommendation algorithms. WWW **1**, 285–295 (2001)
21. Shani, G., Heckerman, D., Brafman, R.I.: An MDP-based recommender system. J. Mach. Learn. Res. **6**, 1265–1295 (2005)
22. Sutskever, I., Vinyals, O., Le, Q.V.: Sequence to sequence learning with neural networks. In: Advances in Neural Information Processing Systems, pp. 3104–3112 (2014)
23. Tan, Y.K., Xu, X., Liu, Y.: Improved recurrent neural networks for session-based recommendations. In: Proceedings of the 1st Workshop on Deep Learning for Recommender Systems, pp. 17–22. ACM (2016)
24. Wan, S., Lan, Y., Wang, P., Guo, J., Xu, J., Cheng, X.: Next basket recommendation with neural networks. In: RecSys Posters (2015)
25. Wang, P., Guo, J., Lan, Y., Xu, J., Wan, S., Cheng, X.: Learning hierarchical representation model for nextbasket recommendation. In: Proceedings of the 38th International ACM SIGIR Conference on Research and Development in Information Retrieval, pp. 403–412 (2015)
26. Yu, F., Liu, Q., Wu, S., Wang, L., Tan, T.: A dynamic recurrent model for next basket recommendation. In: Proceedings of the 39th International ACM SIGIR Conference on Research and Development in Information Retrieval, pp. 729–732 (2016)
27. Zhu, Y., et al.: What to do next: modeling user behaviors by time-LSTM. In: IJCAI, pp. 3602–3608 (2017)
28. Zimdars, A., Chickering, D.M., Meek, C.: Using temporal data for making recommendations. In: Proceedings of the 17th Conference on Uncertainty in Artificial Intelligence, pp. 580–588. Morgan Kaufmann Publishers Inc. (2001)

Modeling Long-Term and Short-Term Interests with Parallel Attentions for Session-Based Recommendation

Jing Zhu, Yanan Xu, and Yanmin Zhu$^{(\boxtimes)}$

Department of Computer Science and Engineering, Shanghai Jiao Tong University,
Shanghai, China
{sjtu_zhujing,xuyanan2015,yzhu}@sjtu.edu.cn

Abstract. The aim of session-based recommendation is to predict the users' next clicked item, which is a challenging task due to the inherent uncertainty in user behaviors and anonymous implicit feedback information. A powerful session-based recommender can typically explore the users' evolving interests (i.e., a combination of his/her long-term and short-term interests). Recent advances in attention mechanisms have led to state-of-the-art methods for solving this task. However, there are two main drawbacks. First, most of the attention-based methods only simply utilize the last clicked item to represent the user's short-term interest ignoring the temporal information and behavior context, which may fail to capture the recent preference of users comprehensively. Second, current studies typically think long-term and short-term interests as equally important, but the importance of them should be user-specific. Therefore, we propose a novel Parallel Attention Network model (PAN) for Session-based Recommendation. Specifically, we propose a novel time-aware attention mechanism to learn user's short-term interest by taking into account the contextual information and temporal signals simultaneously. Besides,we introduce a gated fusion method that adaptively integrates the user's long-term and short-term preferences to generate the hybrid interest representation. Experiments on the three real-world datasets show that PAN achieves obvious improvements than the state-of-the-art methods.

Keywords: Attention mechanism · Behavior modeling · Session-based recommendation

1 Introduction

Recommender Systems play a significant role to provide personalized recommendations for different users in many application domains. Classical recommender systems typically utilize user historical interactions. In other words, user identity must be visible in each interaction record. However, in many application scenarios, the identities of users are unknown and the recommender system can

© Springer Nature Switzerland AG 2020
Y. Nah et al. (Eds.): DASFAA 2020, LNCS 12114, pp. 654–669, 2020.
https://doi.org/10.1007/978-3-030-59419-0_40

employ only the user behavior history during ongoing session. To solve this problem, session-based recommendation [1,14,23] is proposed to predict which item will be clicked by the user based on the sequence of the user's previous clicked items in the current session.

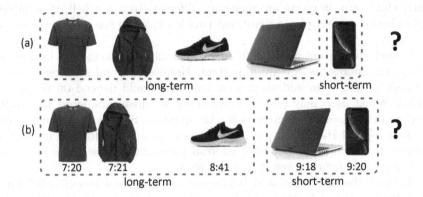

Fig. 1. The main difference between the existing attention-based methods and our method. (a): The existing attention-based methods distinguish long-term interest and short-term interest, but they neglect temporal signals and behavior context when identifying short-term interest. (b): Our method not only distinguishes long-term interest and short-term user interest but also considers temporal signals and contextual information to identify short-term interest.

Recently, the attention-based [1,8,16] methods have been largely developed in session-based recommendation. Compared with previous methods, attention-based models exhibit better capabilities to extract the user's interest (i.e., long-term interest or short-term interest) on user behavior data. Since the long-term interest and short-term interest of users are both significant for recommendation is confirmed by Jannach et al. [3]. Current attention-based methods [8,16,21] mainly model the user's long-term interest and short-term interests simultaneously. Although these studies have achieved state-of-the-art performance, there are also two main limitations not thoroughly studies.

First, most existing attention-based methods [8,16,21] focus more on modeling long-term interest, simply regard the embedding vector of last clicked item as short-term interest. However, they neglect two important information: the time interval between actions, and the user's behavior context. On the one hand, user's short-term interest is changing over time, because past interest might disappear and new interest will come out. Intuitively, a user tends to have similar interests within a short time gap, while a large time gap may cause user's interests to drift. The time interval between actions is an important signal to indicate the change of user's interest. On the other hand, user's behavior sequence is much more complex. Only last clicked item cannot represent the short-term interest efficiently. For example, The last two clicked items of user A and user B are

[*Apple Watch*, *MaBook Pro* 13] and [*Surface Pro* 7, *MacBook Pro* 13], respectively. Existing attention-based methods will model their short-term interest into the same, because they both clicked *MacBook Pro*13 finally. Nevertheless, according to the former actions, we can infer that user A may be interested in Apple series products, the current interest of user B should be laptops. Their short-term interests are relatively different. In a nutshell, time interval and behavior context play a significant part in exploring the short-term interest of a user.

Second, the current methods [8, 16, 23] typically treat long-term interest and short-term interest as equally important. However, in real scenarios, the significance of long-term and short-term interests should depend on the specific user. Different users may have various options for them. Therefore, how to fuse short-term and long-term interests is a key question. Simply merge, such as concatenate [6] and Hadamard product [8], does not account for any interactions between long-term and short-term interests latent features, which is insufficient for learning the hybrid interest representation.

To solve the aforementioned limitations, we propose a model called Parallel Attention Network (PAN) for Session-based Recommendation. The model adopts two parallel attention layers: a short-term attention layer and a long-term attention layer, which are used to model user's short-term and long-term interests, respectively. Since the user's short-term interest is the current interest that changes over time, a novel time-aware attention mechanism is proposed to learn user's short-term interest, which explicitly considers contextual information and temporal signals simultaneously to adaptively weighted aggregate most recent clicked items. Conversely, the user's long-term interest in the session has a rare probability to change over time [16], long-term attention layer extracts the user's long-term interest based on the whole sequential behavior characteristics. Then, we employ a gated fusion method to adaptively combine the long-term and short-term interests by taking into account the user's behavior characteristics. Eventually, the PAN makes recommendations based on the hybrid interest representation (i.e., the combination of long-term and short-term interests). Figure 1 concludes the main difference between the existing methods and our method. The main contributions of the proposed model are as follows:

- We propose a novel model PAN to enhance user's interest representation for session-based recommendation.
- We propose a novel time-aware attention mechanism, in which the attention weights are calculated by combining the embedding of last clicked item with the temporal signal (time interval).
- We introduce a gated fusion method to adaptively incorporate the long-term and short-term preferences according to the specific user.
- The proposed model is evaluated on three benchmark datasets, Experimental results show that PAN achieves better performance than the existing state-of-the-art methods, Further studies demonstrate the proposed time-aware attention and gated fusion method play an important role.

The remainder of this paper is organized as follows: we first discuss the related work in the session-based recommender systems in Sect. 2, and formulate the problem in Sect. 3. Then we illustrate our proposed model in Sect. 4. The experimental settings and results are presented in Sect. 5. Finally, we conclude this paper in Sect. 6.

2   Related Work

In this section, we briefly review the related work on session-based recommendation from the following three aspects.

2.1   Conventional Methods

Matrix factorization (MF) [4,5,13] is a general approach to recommender systems. The idea of MF is to simultaneously map user and item into a continuous and low-dimensional space. It is difficult to apply in session-based recommendation due to the lack of user's information. A natural solution is item-based neighborhood methods [15], in which item similarities are pre-computed by the co-occurrence in the same session. Because the essence of session-based recommendation is a problem of sequence modeling, these methods are not suitable for the sequential problems. Then, a series of methods that are based on Markov chains (MC) is proposed. MC predicts the next clicked item based on the previous clicked item by using sequence data. Shani et al. [14] utilize Markov decision processes (MDPs) to solve the problem. FPMC [14] combines the power of MF and MC for next-item recommendation. However, the major limitation of Markov-chain-based models is that they cannot consider the whole sequence information because the state space will become enormous when taking into account all possible clicks.

2.2   Recurrent-Network-based Methods

Recurrent neural network (RNN) has been applied successfully in natural language processing area [10,11], therefore, a variety of RNN-based methods have emerged for session-based recommendation. Hidasi et al. [1] propose GRU4REC, which is a first work that applies RNN into session-based recommendation and achieves dramatical improvements than conventional models. They employ gated recurrent unit as its core and utilize a novel pair-wise ranking loss function to optimize the model. Tan et al. [18] propose two techniques to enhance the performance of RNN-based methods, namely, data augmentation and a method to take shifts in the input data distribution into account. Although RNN-based methods have achieved significant improvements for session-based recommendation, these methods suffer two limitations due to the RNN's drawbacks. First, both learning and inference processes are time-consuming due to its inherently sequential nature precluding parallelization [19]. Second, RNN-based methods can model only single-way transitions between consecutive items and ignore the complex transitions among the context [21].

2.3 Attention-Mechanism-Based Methods

Attention mechanism has been shown to be effective in various tasks such as image caption [7, 22] and machine translation [2, 9]. For session-based recommendation, Li et al. [6] propose a neural attentive recommendation machine (NARM) to solve the problem. They utilize a hybrid encoder with an attention mechanism, which can capture the user's whole behavior characteristic and the main purpose. Then to improve the session representation by considering the user's interests drift, Liu et al. [8] propose STAMP to simultaneously model user's long-term interest and short-term interest. They employ the attention mechanism to capture the user's long-term interest and utilizes the embedding of last clicked item to represent user's short-term interests. A recent approach, ISLF [16] aims to capture the user's interest shift and latent factors simultaneously by an improved variational autoencoder and attention mechanism. However, these state-of-the-art attention-based methods ignore the importance of contextual information and temporal signals when learning short-term interest. In this study, we propose a time-aware attention mechanism that can take advantage of these two types of information.

3 Problem Statement

Session-based recommendation aims to predict which item the user would like to click next, only based upon his/her current sequential transaction data. Here we give a formulation of the session-based recommendation problem as below.

In session-based recommendation, let $\mathbf{V} = \{v_1, v_2, \ldots, v_{|V|}\}$ represents the set of all unique items that emerged in all sessions, called item dictionary. For each anonymous session, a click action sequence by unknown user can be denoted as a list $\mathbf{S} = [s_1, s_2, \ldots, s_N]$ ordered by timestamps, where s_i is a clicked item of the anonymous user at i-th timestamp within the session. $\mathbf{S}_n = [s_1, s_2, \ldots, s_n], 1 \leq n \leq N$ denotes a prefix of the action sequence truncated at n-th timestamp. Given session prefix $\mathbf{S}_n$, the goal of the session-based recommendation is to predict the next click (i.e., s_{n+1}). To be exact, the session-based recommendation task can be considered to learn a ranking model that generates a ranking list of all the possible items. Let $\hat{\mathbf{y}} = [\hat{y}_1, \hat{y}_2, \ldots, \hat{y}_{|V|}]$ denotes the output probability vector, where $\hat{y}_i$ corresponds to the probability of item v_i being clicked at next timestamp. The items with top-K values in $\hat{\mathbf{y}}$ will be the candidate items for recommendation.

4 The Proposed Model

In this section, we will introduce parallel attention network (PAN) model in detail. As illustrated in Fig. 2, there are two parts in our model: one is interest learning module; the other is interest fusion module.

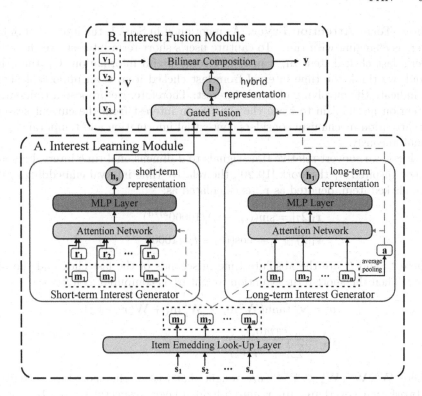

Fig. 2. The architecture of PAN model.

1. **Interest Learning Module.** The interest learning module consists of two components: short-term interest generator and long-term interest generator. The short-term interest generator utilizes a time-aware attention mechanism to learn short-term interest. The long-term interest generator extracts the long-term purpose within the session based on the behavior sequence representation.

2. **Interest Fusion Module.** We integrate the short-term and long-term interests by a gated fusion method in interest fusion module. Then a bi-linear similarity function is utilized to compute recommendation score for each candidate item.

4.1 Interest Learning Module

Embedding Layer. Given a session prefix $\mathbf{S}_n = [s_1, s_2, \ldots, s_n]$, which is composed of item IDs and ordered by time. In embedding layer, The model utilizes an item embedding matrix $\mathbf{I} \in \mathbb{R}^{|V| * d}$ to map each item s_i into the item vector $\mathbf{m}_i$, where d is the dimension of each item embedding. Then we stack these item embedding vectors together to obtain a set of high-dimensional latent representations $\mathbf{M} = [\mathbf{m}_1, \mathbf{m}_2, \ldots, \mathbf{m}_n]$.

Short-Term Attention Layer. As mentioned in Sect. 1, the interest of online users is changing with time. To capture user's short-term interest, we think the user's last clicked item $\mathbf{m}_n$ can reflect some useful information. On the other hand, we think the time interval from last clicked item is an important signal to indicate the envolving of user's interest. Therefore, we propose a time-aware attention mechanism to learn the short-term interest from the current session, the attention mechanism considers contextual information and temporal signals simultaneously.

First, we convert relative time signals to d-dimensional time interval embeddings. Inspired by the work [19,20], the relative time interval embedding of the i-th clicked item, denoted as $\mathbf{r}_i$, is calculated as:

$$
\begin{aligned}
\mathbf{r_i}[2j] &= \sin((t_n - t_i)/10000^{2j/d}) \\
\mathbf{r_i}[2j+1] &= \cos((t_n - t_i)/10000^{(2j+1)/d})
\end{aligned}
\tag{1}
$$

where t_n and t_i are the absolute time in seconds of the last click and the i-th click. Then the short-term attention weight α_s^i is computed as follows:

$$
\begin{aligned}
o_s^i &= \mathbf{v}_s^\top \tanh(\mathbf{W}_s^q \mathbf{m}_n + \mathbf{W}_s^k \mathbf{m}_i + \mathbf{W}_s^r \mathbf{r}_i + \mathbf{b}_s^c) \\
\alpha_s^i &= \frac{exp(o_s^i)}{\sum_{i=1}^k exp(o_s^i)}
\end{aligned}
\tag{2}
$$

where $\mathbf{W}_s^q, \mathbf{W}_s^k, \mathbf{W}_s^r \in \mathbb{R}^{d \times d}$ are weight matrices of the short-term attention network that covert $\mathbf{m}_n, \mathbf{m}_i, \mathbf{r}_i$ into a hidden layer, respectively; and $\mathbf{b}_s^c \in \mathbb{R}^d$ is the bias vector of the hidden layer; $\mathbf{v}_s \in \mathbb{R}^d$ is a weight vector to project the hidden layer output to a score o_s^i. Unlike a standard attention mechanism that is unaware of the time interval between the target item and last-click item, we inject a time interval embedding vector $\mathbf{r}_i$ to model the effect of different time intervals. We use tanh as the activation function of the hidden layer to enhance nonlinear capability. Then, we normalize the scores with a softmax function. As a result, we can compute the user's current interest as a sum of the item embeddings weighted by the attention scores as follows:

$$
\mathbf{c}_s = \sum_{i=1}^n \alpha_s^i \mathbf{m}_i
\tag{3}
$$

where $\mathbf{c}_s$ denotes user's short-term interest representation.

Long-Term Attention Layer. As mentioned in Sect. 1, the user's long-term interest within the current session is relatively difficult to change over time, so it is necessary to compute the user's attention on each clicked item. Previous studies [6,8] indicate that the user's overall behavior representation can provide useful information for capturing the user's long-term interest. Therefore, the long-term attention mechanism mainly considers overall behavior characteristics. Similar to STAMP [8], We first utilize the average of all the item vectors within the session to summarize the whole session behaviors.

$$\mathbf{a} = \frac{1}{n} \sum_{i=1}^{n} \mathbf{m}_i \tag{4}$$

where $\mathbf{a}$ denotes the whole sequential behavior representation. Then the user's long-term interest representation is computed as follows:

$$o_l^i = \mathbf{v}_l^\top \tanh(\mathbf{W}_l^q \mathbf{a} + \mathbf{W}_l^k \mathbf{m}_i + \mathbf{b}_l^c)$$
$$\alpha_l^i = \frac{exp(o_l^i)}{\sum_{i=1}^{k} exp(o_l^i)} \tag{5}$$
$$\mathbf{c}_l = \sum_{i=1}^{n} \alpha_l^i \mathbf{m}_i$$

where $\mathbf{c}_l$ denotes the user's long-term interest representation; α_l^i denotes the long-term attention score to the i-th click item in the session; $\mathbf{W}_l^q, \mathbf{W}_l^k \in \mathbb{R}^{d \times d}$ are weight matrices, $\mathbf{b}_l^c$ is a bias vector, and $\mathbf{v}_l$ is a weight vector.

Multilayer Perceptron (MLP). Similar to STAMP [8], we respectively abstract the long-term interest and short-term interest by two MLP networks, which both have one hidden layer. The network structure of MLP shown in Fig. 2 are identical to each other, but they have different learned parameters. The operation on the short-term interest $\mathbf{c}_s$ is defined as:

$$\mathbf{h}_s = f(\mathbf{W}_1^s \mathbf{c}_l + \mathbf{b}_1^s) \mathbf{W}_2^s + \mathbf{b}_2^s \tag{6}$$

where $\mathbf{W}_1^s, \mathbf{W}_2^s \in \mathbb{R}^{d \times d}$ are weight matrices; $f(\cdot)$ is a non-linear activation function (we found tanh has a better performance); and $\mathbf{h}_s$ denotes final short-term interest representation. The state vector $\mathbf{h}_l$ with regard to $\mathbf{c}_l$ can be computed similar to $\mathbf{h}_s$.

4.2   Interest Fusion Module

The *interest fusion module* adaptively combines the information from the short-term generator and long-term generator for recommendation. As mentioned in Sect. 1, The significance of long-term and short-term interests is uncertain, depending on the specific user. Since user identification is unknown in session-based recommendation, we conjecture that the whole behavior representation $\mathbf{a}$ within the session may provide some user information, Inspired by the repeat-Net [12], we employ a gated fusion method, which can balance the significance of user's short-term interest and long-term interest to adaptively construct the hybrid interest representation:

$$\mathbf{h} = \beta \mathbf{h}_s + (1 - \beta) \mathbf{h}_l \tag{7}$$

where the gate β is given by

$$\beta = \sigma(\mathbf{W}_s \mathbf{h}_s + \mathbf{W}_l \mathbf{h}_l + \mathbf{W}_a \mathbf{a} + \mathbf{b}) \tag{8}$$

where $\mathbf{h}$ is the hybrid representation vector of user's interest, and σ denotes sigmoid function. Similar to previous studies, then we exploit a bi-linear method to compute the recommendation score for each candidate item v_i:

$$\hat{z}_i = emb_i^\top \mathbf{B} \, \mathbf{h} \tag{9}$$

where emb_i^T is the corresponding embedding vector to v_i, $\mathbf{B} \in \mathbb{R}^{d*d}$ is a parameter matrix. The output $\hat{\mathbf{y}}$ is obtained by applying a softmax function to $\hat{\mathbf{z}} = [\hat{z}_1, \hat{z}_2, \ldots, \hat{z}_{|V|}]$.

$$\hat{\mathbf{y}} = \mathrm{softmax}(\hat{\mathbf{z}}) \tag{10}$$

where $\hat{\mathbf{y}} \in \mathbb{R}^{|V|}$ denotes the probability vector which is the output of PAN, and the utility of softmax function is to convert the recommendation score to a probabilistic distribution.

4.3 Objective Function

For any given session prefix $\mathbf{S}_n$, the goal of our model is to maximize the prediction probability of the actual next clicked item within the current session. Therefore, the loss function is defined as the cross-entropy of the prediction result $\hat{y}$:

$$\mathcal{L}(\hat{\mathbf{y}}) = -\sum_{i=1}^{|V|} \mathbf{y}_i \log(\hat{\mathbf{y}}_i) + (1 - \mathbf{y}_i) \log(1 - \hat{\mathbf{y}}_i) \tag{11}$$

where $\mathbf{y}$ denotes the one-hot encoding vector of the ground truth item. At last, we utilize the Back-Propagation algorithm to train the proposed PAN model.

5 Experiments

In this section, we conduct experiments with the aim of answering the following research questions:

RQ1 Can our proposed approach perform better than other competitive methods?

RQ2 Are the key components in PAN (i.e., time-aware attention layer, gated fusion) useful for improving recommendation results?

RQ3 How does session length affect the recommendation performance of our approach?

5.1 Experimental Settings

Datasets. We evaluate the proposed model and other models on two benchmark datasets, the first one is YOOCHOSE from RecSys'15 Challenge[1], which are composed of user clicks gathered from an e-commerce web site. Another is

---

[1] http://2015.recsyschallege.com/challege.html.

DIGENTICA from the CIKM Cup 2016[2], where we only use the transaction data.

For a fair comparison, following [1,6], we filter out all sessions with length 1, items with less than 5 occurrences in the datasets, and items in the test set which do not appear in the training set. Similar to [18], we use a sequence splitting preprocess. To be specific, for the input Session $\mathbf{S} = [s_1, s_2, \ldots, s_n]$, we generate a series of sequences and corresponding labels $([s_1], s_2)$, $([s_1, s_2], s_3)$, $\ldots$, $([s_1, s_2, \ldots, s_{n-1}], s_n)$ for both datasets, which proves to be effective. Because the training set of YOOCHOSE is quite large and training on the recent fractions yields better results than training on the entire fractions as per the experiments of [18], we use the recent fractions 1/64 and 1/4 of training sequences. The statistics of datasets are summarized in Table 1.

Table 1. Statistics of datasets

Dataset	YOOCHOSE 1/64	YOOCHOSE 1/4	DIGINETICA
# train	369,859	5,917,746	719,470
# test	55,898	55,898	60,858
# clicks	557,248	8,326,407	982,961
# items	17,745	29,618	43,097
avg.length	6.16	5.71	5.12

Baselines. For the purpose of evaluating the performance of PAN, we compare it with the following baselines:

- **Item-KNN** [15]: An item-to-item model, which recommends items that are similar to the previous clicked item in terms of cosine similarity.
- **BPR-MF** [13]: A commonly used matrix factorization method, which optimizes a pairwise ranking objective function via stochastic gradient descent.
- **FPMC** [14]: A classic hybrid model combining Markov chain and matrix factorization. In order to make it work on session-based recommendation, we treat each basket as a session.
- **GRU4REC** [1]: An RNN-based deep learning model for session-based recommendation, which employs a session-parallel mini-batch training process and a pair-wise loss function for training the model.
- **Time-LSTM** [23]: An improved RNN model, which proposes time gates to model the time intervals between consecutive items to capture user's long-term and short-term interests.
- **NARM** [6]: An RNN-based model, which utilizes a hybrid encoder with an attention mechanism to model the sequential behavior and the user's main purpose, then combines them to generate final representation.

---

[2] http://cikm2016.cs.iupui.edu/cikm-cup.

- **STAMP** [8]: A novel memory model, which captures the user's long-term preference from previous clicks and the current interest of last click in the session.
- **ISLF** [16]: A state-of-the-art model for session-based recommendation, which employs an improved variational autoencoder with an attention mechanism to capture the user's interest shift (i.e., long-term and short-term interest) and latent factors simultaneously.

Evaluation Metrics. To evaluate the recommendation performance of all models, we adopt the following metrics, which are widely used in other related works.

Recall@20: Recall is a widely used evaluation metric in session-based recommendation, which cannot take the rank of items into account. Recall@K indicates the proportion of test cases, which have the actual next clicked items among the top-K ranking list.

$$\text{Recall@}K = \frac{n_{hit}}{N} \tag{12}$$

where N is the number of the whole test samples, n_{hit} is the number of test samples which have the desired items in recommendation lists.

MRR@20: MRR is the average of reciprocal ranks of the actual next clicked item. MRR@20 will be 0 if the rank is above 20.

$$\text{MRR@}K = \frac{1}{N} \sum_{t \in G} \frac{1}{Rank(t)} \tag{13}$$

where G is the ground-truth set of the test cases. The MRR considers the order of the recommendation list, if the actual next clicked item is at the top of the recommendation list, the MRR value will be larger.

Parameter Settings. We select hyper-parameters on the validation set which is a random 10% subset of the training set. The dimensionality of latent vectors is searched in $\{25, 50, 100, 200\}$ and 100 is optimal. We choose Adam as the model learning algorithm to optimize these parameters, where the mini-batch size is searched in $\{64, 128, 256, 512\}$ and sets 128 finally. We set 0.001 as the initial value of the learning rate, and it will decay by 0.1 after every 10 epochs. Following previous method [8], All weighting matrices are initialized using a Gaussian distribution $N(0, 0.05^2)$. All the items embeddings are initialized randomly with a Gaussian distribution $N(0, 0.002^2)$. We use dropout [17] with drop ration $\rho = 0.5$. The model is written in Tensorflow and trained on an NVIDIA TITAN Xp GPU.

5.2   Performance Comparison (RQ1)

The overall performances of all contrast methods based on three datasets are shown in Table 2, with the best results highlighted in boldface. Please note that,

Table 2. The overall performance over three datasets

Method	YOOCHOSE 1/64		YOOCHOSE 1/4		DIGINETICA	
	Recall@20	MRR@20	Recall@20	MRR@20	Recall@20	MRR@20
Item-KNN	51.60	21.82	52.31	21.70	35.75	11.57
BPR-MF	31.31	12.08	3.40	1.57	5.24	1.98
FPMC	45.62	15.01	-	-	26.53	6.95
GRU4REC	60.64	22.89	59.53	22.60	29.45	8.33
Time-LSTM	67.78	28.12	69.14	28.96	47.13	15.22
NARM	68.32	28.63	69.73	29.23	49.70	16.17
STAMP	68.74	29.67	70.44	30.00	45.64	14.32
ISLF	69.32	**33.58**	71.02	32.98	49.35	16.41
PAN	**70.36**	31.97	**71.58**	**33.04**	**50.69**	**16.53**

as in [6], due to insufficient memory to initialize FPMC, the performance on Yoochoose 1/4 is not reported. From the table, we make the following observations from the results.

The performance of traditional methods such as Item-KNN, BPR-MF, and FPMC are not competitive, as they only outperform the naive POP model. These methods help justify the importance of taking the user's behavior context into consideration, as the results demonstrate that making recommendations only based on co-occurrence popularity, or employing a simple first-order transition probability matrix may not be sufficient.

Deep-learning-based methods, such as GRU4REC and NARM, outperform all conventional methods. This demonstrates the effectiveness of deep learning technology in this domain. Time-LSTM utilizes time interval to capture user's short-term interest and improves the performances of GRU4REC, which indicates that temporal signals are helpful to capture user's shifty interest. PAN significantly outperforms all baseline methods. Generally, PAN obtains improvements over the best baseline ISLF of 1.50%, 0.79%, and 2.72% in Recall@20 on the three datasets, respectively. Although both Time-LSTM and PAN taking temporal signals into account. We notice that PAN achieves consistent improvements over Time-LSTM. The reason is that RNN always models oneway transitions between consecutive items, neglecting the relations between other items in the session, however, attention mechanism can capture complex transitions within the entire session sequence. As to other baseline methods that also consider user's long- and short-term interests simultaneous (i.e., STAMP and ISLF), we find that PAN outperforms STAMP and ISLF on all datasets. This is because PAN learns user's current interest by taking into account the contextual information and temporal signals simultaneously. Conversely, STAMP and ISLF simply utilize the last clicked item, which may not be sufficient.

5.3 Effects of Key Components (RQ2)

Impact of Time-Aware Attention. In order to prove the effectiveness of the proposed time-aware attention mechanism, we introduce the following variant methods of PAN. (1) PAN-l excludes the short-term interest. (2) PAN-v utilizes a vanilla attention mechanism that excludes the time interval information to learn short-term interest.

Table 3. Performance comparison of PAN with different short-term representation

Method	YOOCHOSE 1/64		YOOCHOSE 1/4		DIGINETICA	
	Recall@20	MRR@20	Recall@20	MRR@20	Recall@20	MRR@20
PAN-l	67.21	28.03	68.88	31.13	46.97	13.06
PAN-v	69.72	30.72	71.33	32.99	49.02	15.20
PAN	**70.36**	**31.97**	**71.58**	**33.04**	**50.69**	**16.53**

Table 3 shows the experimental results of the Recall@20 and MRR@20 metrics on all datasets. We can see that PAN-v outperform PAN-l for all datasets. For example, for YOOCHOSE 1/64, the PAN-v improves Recall@20 and MRR@20 by 3.96% and 2.31%, which indicates that the effectiveness of considering the short-term interest representation when capturing the user's overall representation. We can also see that PAN outperforms PAN-v for all datasets, such as PAN improves Recall@20 and MRR@20 by 0.57% and 3.90% on YOOCHOSE 1/64, which proves that the temporal signals are conducive to capturing the user's short-term interest.

Impact of Fusion Operations. Next, we compare PAN with different fusion operations, i.e., concatenation, average pooling, and Hadamard product. As illustrated in Table 4, the introduced gated fusion method achieves better results than the other three fusion operations on two datasets. This indicates that the gated fusion method is more effective in modeling interactions between user's long-term and short-term interests. In most cases, PAN with Hadamard product and PAN with concatenation operation achieve similar performance and

Table 4. Performance comparison of PAN with different fusion operations

Method	YOOCHOSE 1/64		DIGINETICA	
	Recall@20	MRR@20	Recall@20	MRR@20
Average pooling	69.77	31.58	49.99	16.02
Hadamard product	70.18	31.79	50.13	16.37
Concatenation	70.11	31.82	50.35	16.37
Gated fusion	**70.36**	**31.97**	**50.69**	**16.53**

both outperform the average pooling on three datasets. This demonstrates that Hadamard product and concatenation operation have advantages over average pooling in modeling interactions between long-term and short-term interests latent features.

5.4 Influence of Different Session Lengths (RQ3)

We further analyze the impact of different session lengths on different methods. For a comparison, following [8], we partition sessions of Yoochoose 1/64 and Diginetica into two groups. Since the average length of sessions is almost 5, sessions with length greater than 5 are seen as long session, the remainder is called short sessions. The percentages of sessions belong to short sessions and long sessions are 70.1% and 29.9% on the YOOCHOSE 1/64, AND 76.4% and 23.6% on the Diginetica. We use Recall@20 and MRR@20 to compare the performance. The results of the methods are given in Fig. 3. We can see that all methods obtain worse performance in long group than to short group, which reveals the difficulty of handling long sequences in session-based recommendation. The performance of NARM changes greatly, The reasons for this phenomenon may be that NARM neglects the user's short-term interest. Similar to NARM, STAMP achieves better performance in the short sessions than long sessions. STAMP explains this difference based on repetitive click actions. It employs the attention mechanism, so duplicate items may be ignored when obtaining user's general interest. PAN is relative stable than NARM and STAMP.

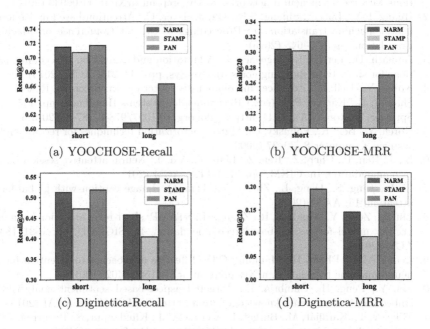

(a) YOOCHOSE-Recall   (b) YOOCHOSE-MRR

(c) Diginetica-Recall   (d) Diginetica-MRR

Fig. 3. Recall@20 and MRR@20 on different session length

6 Conclusion

In this paper, we propose a novel parallel attention network (PAN) for session-based recommendation to modeling user's short-term and long-term interests. We observe that user's short-term interest has great relations to temporal signal and behavior context, thus we propose a novel time-aware attention mechanism to learning the short-term interest. We further introduce a gated fusion method to adaptively incorporate the long-term and short-term interests according to the specific behavior context. Extensive experimental analysis on three real-world datasets shows that our proposed model PAN outperforms the state-of-the-art methods.

Acknowledgment. This research is supported in part by the 2030 National Key AI Program of China 2018AAA0100503 (2018AAA0100500), National Science Foundation of China (No. 61772341, No. 61472254), Shanghai Municipal Science and Technology Commission (No. 18511103002, No. 19510760500, and No. 19511101500), the Program for Changjiang Young Scholars in University of China, the Program for China Top Young Talents, the Program for Shanghai Top Young Talents, Shanghai Engineering Research Center of Digital Education Equipment, and SJTU Global Strategic Partnership Fund (2019 SJTU-HKUST).

References

1. Hidasi, B., Karatzoglou, A., Baltrunas, L., Tikk, D.: Session-based recommendations with recurrent neural networks. arXiv preprint arXiv:1511.06939 (2015)
2. Huang, P.Y., Liu, F., Shiang, S.R., Oh, J., Dyer, C.: Attention-based multimodal neural machine translation. In: Proceedings of the 1st Conference on Machine Translation, pp. 639–645 (2016)
3. Jannach, D., Lerche, L., Jugovac, M.: Adaptation and evaluation of recommendations for short-term shopping goals. In: RecSys, pp. 211–218. ACM (2015)
4. Koren, Y., Bell, R.: Advances in collaborative filtering. In: Ricci, F., Rokach, L., Shapira, B., Kantor, P.B. (eds.) Recommender Systems Handbook, pp. 145–186. Springer, Boston, MA (2011). https://doi.org/10.1007/978-0-387-85820-3_5
5. Koren, Y., Bell, R., Volinsky, C.: Matrix factorization techniques for recommender systems. Computer **8**, 30–37 (2009)
6. Li, J., Ren, P., Chen, Z., Ren, Z., Lian, T., Ma, J.: Neural attentive session-based recommendation. In: CIKM, pp. 1419–1428. ACM (2017)
7. Li, L., Tang, S., Deng, L., Zhang, Y., Tian, Q.: Image caption with global-local attention. In: AAAI (2017)
8. Liu, Q., Zeng, Y., Mokhosi, R., Zhang, H.: STAMP: short-term attention/memory priority model for session-based recommendation. In: SIGKDD, pp. 1831–1839. ACM (2018)
9. Luong, M.T., Pham, H., Manning, C.D.: Effective approaches to attention-based neural machine translation. arXiv preprint arXiv:1508.04025 (2015)
10. Ma, Y., Peng, H., Cambria, E.: Targeted aspect-based sentiment analysis via embedding commonsense knowledge into an attentive LSTM. In: AAAI (2018)
11. Mikolov, T., Karafiát, M., Burget, L., Černocký, J., Khudanpur, S.: Recurrent neural network based language model. In: 11th Annual Conference of the International Speech Communication Association (2010)

12. Ren, P., Chen, Z., Li, J., Ren, Z., Ma, J., de Rijke, M.: RepeatNet: a repeat aware neural recommendation machine for session-based recommendation. In: AAAI (2019)
13. Rendle, S., Freudenthaler, C., Gantner, Z., Schmidt-Thieme, L.: BPR: Bayesian personalized ranking from implicit feedback. In: Proceedings of the 25th Conference on Uncertainty in Artificial Intelligence, pp. 452–461. AUAI Press (2009)
14. Rendle, S., Freudenthaler, C., Schmidt-Thieme, L.: Factorizing personalized Markov chains for next-basket recommendation. In: WWW, pp. 811–820. ACM (2010)
15. Sarwar, B.M., Karypis, G., Konstan, J.A., Riedl, J., et al.: Item-based collaborative filtering recommendation algorithms. WWW **1**, 285–295 (2001)
16. Song, J., Shen, H., Ou, Z., Zhang, J., Xiao, T., Liang, S.: ISLF: interest shift and latent factors combination model for session-based recommendation. In: IJCAI, pp. 5765–5771 (2019)
17. Srivastava, N., Hinton, G., Krizhevsky, A., Sutskever, I., Salakhutdinov, R.: Dropout: a simple way to prevent neural networks from overfitting. J. Mach. Learn. Res. **15**(1), 1929–1958 (2014)
18. Tan, Y.K., Xu, X., Liu, Y.: Improved recurrent neural networks for session-based recommendations. In: Proceedings of the 1st Workshop on Deep Learning for Recommender Systems, pp. 17–22. ACM (2016)
19. Vaswani, A., et al.: Attention is all you need. In: NeurIPS, pp. 5998–6008 (2017)
20. Wang, Y., et al.: Regularized adversarial sampling and deep time-aware attention for click-through rate prediction. In: CIKM, pp. 349–358. ACM (2019)
21. Wu, S., Tang, Y., Zhu, Y., Wang, L., Xie, X., Tan, T.: Session-based recommendation with graph neural networks. In: AAAI (2019)
22. Xu, K., et al.: Show, attend and tell: neural image caption generation with visual attention. In: ICML, pp. 2048–2057 (2015)
23. Zhu, Y., et al.: What to do next: modeling user behaviors by time-LSTM. In: IJCAI, pp. 3602–3608 (2017)

Industrial Papers

Recommendation on Heterogeneous Information Network with Type-Sensitive Sampling

Jinze Bai[1], Jialin Wang[1], Zhao Li[2(✉)], Donghui Ding[2], Jiaming Huang[2], Pengrui Hui[2], Jun Gao[1(✉)], Ji Zhang[3], and Zujie Ren[3]

[1] The Key Laboratory of High Confidence Software Technologies, Department of Computer Science, Peking University, Beijing, China
gaojun@pku.edu.cn
[2] Alibaba Group, Hangzhou, China
lizhao.lz@alibaba-inc.com
[3] Zhejiang Lab, Hangzhou, China

Abstract. Most entities and relations for recommendation tasks in the real world are of multiple types, large-scale, and power-law. The heterogeneous information network (HIN) based approaches are widely used in recommendations to model the heterogeneous data. However, most HIN based approaches learn the latent representation of entities through meta-path, which is predefined by prior knowledge and thus limits the combinatorial generalization of HIN. Graph neural networks (GNNs) collect and generalize the information of nodes on the receptive field, but most works focus on homogeneous graphs and fail to scale up with regard to power-law graphs. In this paper, we propose a HIN based framework for recommendation tasks, where we utilize GNNs with a type-sensitive sampling to handle the heterogeneous and power-law graphs. For each layer, we adopt schema-based attention to output the distribution of sampling over types, and then we use the importance sampling inside each type to output the sampled neighbors. We conduct extensive experiments on four public datasets and one private dataset, and all datasets are selected carefully for covering the different scales of the graph. In particular, on the largest heterogeneous graph with 0.4 billion edges, we improve the square error by 2.5% while yielding a 26% improvement of convergence time during training, which verifies the effectiveness and scalability of our method regarding the industrial recommendation tasks.

Keywords: Heterogeneous information network · Graph neural networks · Recommendation

1 Introduction

Recommender systems, aiming to recommend products or service to users, play a significant role in various domains, such as e-commerce platforms and social networks [1,8,20]. Nowadays, various kinds of auxiliary data are incorporated

© Springer Nature Switzerland AG 2020
Y. Nah et al. (Eds.): DASFAA 2020, LNCS 12114, pp. 673–684, 2020.
https://doi.org/10.1007/978-3-030-59419-0_41

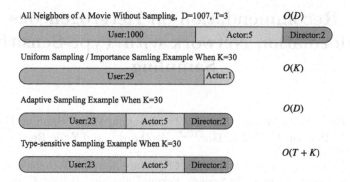

Fig. 1. The sampling varieties of GNNs. D denotes the degree, T denotes the number of types, K denotes the sample number.

into sizeable commercial recommender systems [15]. It is challenging to leverage such context information to improve recommendation performance effectively.

Heterogeneous information network (HIN) [13] contains multiple types of entities connected by different types of relations. It has been widely adopted in recommendation because of its flexibility in modeling various kinds of data. Most HIN based recommendation methods extract meta-paths or meta-graphs based features to improve the characterization of user-item interaction [21]. The meta-paths used in these approaches are predefined by prior knowledge in advance, which might limit combinatorial generalization of HIN. Besides, the meta-paths need to be made up manually with the growth of network schema. Hence, we leverage graph neural networks (GNNs) [2], which aggregate adjacent nodes and apply the neural message passing [4], to generate the high-level representation of nodes and edges in HIN.

However, GNNs suffer from the power-law issue. The high degree nodes tend to be involved in the receptive field and lead to uncontrollable neighborhood expansion across layers [7,9]. Even if we distribute the computation tasks among distinct nodes, the complexity of any flexible aggregator is proportional to the maximal degree of active nodes. That is, the power-law distribution of graphs limits the scalability of GNNs.

One common solution is sampling, which has been proven to be very effective on homogeneous graphs, but most works are hard to deply in the setting of heterogeneous and power-law graphs. GraphSAGE [6] and FastGCN [3] improve the aggregation by uniform sampling and fixed importance sampling respectively, which reduce the complexity of aggregation from $O(D)$ to $O(K)$ for a node or level, where D is the degree of nodes and K is the sample number. However, we also observe that the frequencies of edges across types follow a power law distribution, at least on our datasets. For example, as shown in Fig. 1, a movie may be watched by 1000 users and only has few neighbors about actors and directors. Hence, the uniform sampling and the fixed importance sampling are hard to capture the information about the minority or types (actors and directors), and this lead to slow convergence and inadequate performance improvement on heterogeneous graphs.

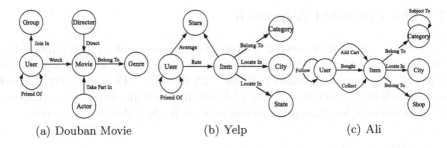

Fig. 2. Network schemas of heterogeneous information networks for datasets.

The adaptive sampling [7] and the stacked attention [16] handles the heterogeneous setting by computing the attention scores between the target node and all the neighbors, which gives trainable multinomial distribution over all neighbors. However, the complexity of full adaptive sampling is still $O(D)$ and suffers from the power-law issue owing to the cost of computing attention scores.

Heterogeneous graphs raise a dilemma between the power law of types and the power law of degrees. On the one hand, if we sample uniformly or in a way of importance sampling, some neighbors with a minority of types almost have few chances to be sampled owing to the power-law distribution of types. On the other hand, if we compute all unnormalized score by adaptive sampling or attention, the complexity limits the scalability owing to the power-law distribution of degrees.

To overcome the above limitations, we propose a type-sensitive sampling to balance between two aspects of power law. A schema-based attention mechanism produces the probability over types so that GNNs keep the complete combinatorial generalization on types. Moreover, the importance sampling inside types with the alias method reduces the complexity of aggregation of high degree nodes. Empirical studies on real-world datasets demonstrate the effectiveness of the proposed algorithm while benefiting from faster convergence time in the setting of heterogeneous and power-law graphs. We summarize the contributions of this paper as follows:

- We propose a HIN based model for recommendation tasks without manually predefined meta-paths, where we utilize GNNs with a proposed type-sensitive sampling to handle the heterogeneous and power-law graphs.
- The type-sensitive sampling focuses on the dilemma between two aspects of power law on heterogeneous graphs. It makes sampling adaptive over types by the attention mechanism and reduces the complexity of high degree nodes by importance sampling and the alias method.
- We conduct extensive experiments on four real-world public datasets and one industrial private dataset, whose results illustrate the fastest convergence time while outperforming the state-of-the-art methods in the setting of heterogeneous and power-law graphs. Our experiments verify the scalability and effectiveness regarding the industrial recommendation tasks on heterogeneous graphs with billion edges.

2   The Proposed Approach

Here we present a HIN based recommendation framework, which handles the heterogeneous and power-law graphs by graph neural networks with a proposed type-sensitive sampling, namely *TS-Rec*. TS-Rec is a HIN-based graph neural network whose layers consist of schema-based attention and importance sampling by alias method. We present these components in the following subsections.

2.1   HIN-Based Recommendation

The goal of HIN based recommendation is to predict the rating score $r_{u,i}$ given a user u and a item i. We define it as a matrix factorization form, namely,

$$r_{u,i} = \mathbf{h}^L(u)^T \mathbf{h}^L(i) + b_u + b_i \tag{1}$$

where $\mathbf{h}^l$ denotes the l-layer representation of nodes, b_u and b_i denote the trainable bias term. Then, the objective function is given by:

$$\mathcal{L} = \frac{1}{|\mathcal{D}_{train}|} \sum_{u,i \in \mathcal{D}_{train}} (r_{u,i} - \hat{r}_{u,i})^2 + \lambda \Theta(W) \tag{2}$$

where $\hat{r}_{u,i}$ is the ground truth that user u assigns to item i, $\mathcal{D}_{train}$ denotes the train set of rating records, $\Theta(W)$ denotes the regularization term, and λ is the regularization parameter.

For fusing the information in HIN, we adopt graph neural networks to produce the high-level representation of u and i with the fusion of heterogeneous information, denoted as $\mathbf{h}^L(u)$ and $\mathbf{h}^L(i)$ where L is the number of layers in graph neural networks.

2.2   Heterogeneous Graph Neural Networks

We adopt graph neural networks (GNNs) [2] to fuse the heterogeneous information, which aggregates adjacent nodes iteratively and applies the neural message passing [4], a generalized aggregate rule, to output the high-level representation of nodes and edges. For HIN, the feed-forward propagation for a single layer of GNNs can be defined as[1]

$$\mathbf{m}^l(v_i) = \frac{1}{D_{v_i}} \sum_{u_j \in nei[v_i]} M_\theta(\mathbf{h}^l(v_i), \mathbf{h}^l(u_j)))$$

$$\mathbf{h}^{l+1}(v_i) = U_\theta(\mathbf{h}^l(v_i), \mathbf{m}^l(v_i)) \tag{3}$$

where $\mathbf{h}^l(v_i)$ denotes the hidden representation of node v_i on layer l, $nei[v_i]$ denotes all neighbors and D_{v_i} denotes the degree of node v_i. A message function

---

[1] We omit the edge aggregation for convenience, while the conclusions in this paper can be generalized to the standard GNNs.

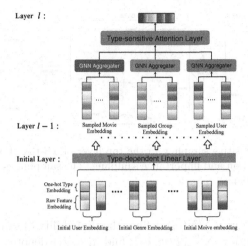

Fig. 3. The aggregation of high-level representation of nodes in TS-Rec.

$M(\cdot,\cdot)$ and an updating function $U(\cdot,\cdot)$ produce the upper hidden representation parameterized in terms of a generic θ. Note that, the mean aggregator of message function in Eq. (3) can be replaced by a sum aggregator. For recommendation tasks, we observe that the mean aggregator performs better than the sum aggregator.

The initial layer of GNNs, parametrized by a type-dependent weight matrix $\mathbf{W}_t^0 \in \mathbf{R}^{H \times H_t}$, is a linear transformation for mapping the feature of different types to a uniform space. And we convert the type of each node to a one-hot feature vector which is concatenated with the original feature:

$$\mathbf{h}^0(v_i) = \mathbf{W}_t^0[\mathrm{onehot}(t_{v_i})||\mathrm{feature}(v_i)] \tag{4}$$

where t_{v_i} denotes the type of node v_i, and $||$ denotes the concatenation operation. We illustrate the network in Fig. 3.

Graph neural networks bring a powerful capacity of combinatorial generalization for HIN. However, the power law of degree leads to expensive cost and uncontrollable neighborhood expansion. Meanwhile, the power law of types raises the issue that both uniform and importance sampling are hard to fuse the neighbors with a minority of type, which leads to slow convergence and inadequate performance. Hence, we introduce a *type-sensitive sampling* mechanism, which makes the sampling process efficient and sensitive to types.

2.3   Type-Sensitive Sampling

The type-sensitive sampling consists of a *schema-based attention* and an *importance sampling* with the alias method. In Eq. (3), the mean or sum of message passing function allows for the use of Monte Carlo approaches to estimate the result in a way of integral consistently, so we rewrite the mean aggregator to the expectation form, namely,

$$\mathbf{m}^l(v_i) = \mathbb{E}_{p(u_j|v_i)}[M(\mathbf{h}^l(v_i), \mathbf{h}^l(u_j))]) \tag{5}$$

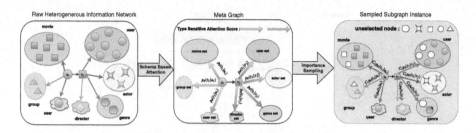

Fig. 4. The type-sensitive sampling in the top layer of TS-Rec. Each layer consists of schema-based attention and importance sampling. The sampling is similar with regard to all types in schema.

where $p(u_j|v_i) = a(v_i, u_j)/N(v_i)$ defines the probability of sampling the node u_j given v_i with $N(v_i) = \sum_{j=1}^{n} a(v_i, u_j)$, and $a(v_i, u_j)$ denotes the unnormalized score of sampling. Note that, if we adopt the sum aggregator instead of mean, the expectation should be multiplied by the $N(v_i)$ term in Eq. (5).

Then, we introduce the random variable for type and apply the Bayes formula:

$$\mathbf{m}^l(v_i) = \int \int p(u_j|t_k, v_i) M(\mathbf{h}^l(v_i), \mathbf{h}^l(u_j)) \, dt_k \, du_j$$

$$= \mathbb{E}_{p(t_k|v_i)} \left[\mathbb{E}_{p(u_j|t_k, v_i)} [M(\mathbf{h}^l(v_i), \mathbf{h}^l(u_j))] \right] \qquad (6)$$

where the conditional probability $p(t_k|v_i)$ defines the distribution over types given the node v_i. The number of types is much less than the number of nodes, so we calculate the outer expectation directly in Eq. (6), which is given by:

$$\mathbf{m}^l(v_i) = \sum_{t_k \in nei[v_i]} p(t_k|v_i) \mathbb{E}_{p(u_j|t_k, v_i)} [M(\mathbf{h}^l(v_i), \mathbf{h}^l(u_j))] \qquad (7)$$

The calculation of outer expectation in Eq. 7 is the *schema-based attention*, which could be defined as a bilinear attention function, namely,

$$p_\theta^l(t_k|v_i) = \frac{a_\theta^l(t_k, v_i)}{\sum_{t_k \in nei[v_i]} a_\theta^l(t_k, v_i)} \qquad \text{where } a_\theta^l(t_k, v_i) = \exp(\mathbf{h}(t_k)^T \mathbf{W}^l \mathbf{h}^0(v_i))$$

$$(8)$$

where $\mathbf{h}(t_k)$ denotes the layer-independent embedding for types, and $\mathbf{W}^l$ denotes the weight matrix on layer l. Note that, the nodes of lower layer are not available before sampling, so we can only use the original feature $\mathbf{h}^0(v_i)$ instead of $\mathbf{h}^l(v_i)$ in Eq. 8.

The schema-based attention calculates the parameterized distribution over types for each node and layer. The complexity is only $O(T)$ for each node by regarding the hidden size constant, and the implementation can be parallelized by masking efficiently. The attention is type-sensitive and graph neural networks could generalize the combination over the types' and nodes' representation.

For estimating the inner expectation in Eq. (6), we use the importance sampling to reduce the variance. Indeed, a good estimator should reduce the variance caused by the sampling process, since high variance usually impedes efficient training. Let $\mu(v_i, t_k) = \mathbb{E}_{p(u_j|t_k, v_i)}[M(\mathbf{h}^l(v_i), \mathbf{h}^l(u_j))]$, and we have

$$\mu(v_i, t_k) = \mathbb{E}_{q(u_j|t_k, v_i)}[\frac{p(u_j|t_k, v_i)}{q(u_j|t_k, v_i)}\mathbf{h}^l(u_j)] \qquad (9)$$

where $q(u_j|t_k, v_i)$ defines the probability of sampling node u_j given the node v_i and type t_k. Similarly, we can speed up Eq. (9) by approximating the expectation with the Monte-Carlo mean. According to the derivations of importance sampling, we conclude that the optimal sampler to minimize the variance, then we get an approximately optimal solution:

$$q^*(u_j|t_k, v_i) = \frac{p(u_j|t_k, v_i)}{\sum_{u_j \in nei[v_i, t_k]} p(u_j|t_k, v_i)} \propto a_{t_k}(v_i, u_j) \qquad (10)$$

where $a_{t_k}(v_i, u_j) = \frac{A_{t_k}(v_i, u_j)}{\sqrt{D_{t_k}(v_i)D_{t_k}(u_j)}}$ is defined as a re-normalized laplace for sampling and $A_{t_k}(v_i, u_j)$ denotes the entry in the adjacency matrix of type t_k.

For each node v_i and type t_k, we calculate the unnormalized scores $a_{t_k}(v_i, u_j)$ and preprocess them by the alias method [14], which provides constant complexity of sampling and linear preprocessing time. The schema-based attention and alias method based importance sampling both have constant complexity rather than being proportional to the number of neighbors, making the whole algorithm scalability. The overall feed-forward propagation is shown in Algorithm 1.

The time complexity of our approach is $O(KC^LH)$, where K denotes the number of types, C is the sample number for each layer, L denotes the

---

Algorithm 1. The overall feed-forward propagation.

1: **Input:** The target pair $\langle u, i \rangle$, the number of network layers L, sample number C_l for layer l, HIN $G = (V, E)$, the alias table $q*$
2: **Output:** The optimization criterion $\mathcal{L}$
3: Initialize the receptive field graph $S = \langle u, i \rangle$.
4: **for** $l \in [L-1, ..., 0]$ **do**
5:     **for** v_i is an unvisited node in S **do**
6:         Calculate $p_\theta^l(t_k|v_i)$ by Equation (8)
7:         **for** each $t_k \in nei[v_i]$ **do**
8:             Sample $C_l p_\theta(t_k|v_i)$ neighbors by alias method over $q^*(u_j|t_k, v_i)$ by Equation (10)
9:             Add v_i and all neighbors u_j to graph S
10:         Mark the node v_i as visited status
11: Initial $\mathbf{h}^0(v_i)$ by Equation (4) for all $v_i \in S$
12: **for** $l \in [0, ..., L-1]$ **do**
13:     Estimate $\mathbf{m}^l(v_i)$ by Equation (7)
14:     Calculate $\mathbf{h}^{l+1}(v_i)$ for next layer by Equation 3
15: Calculate the predicted rating score and loss $\mathcal{L}$

---

number of layers, and H is the dimension size. Compared to the GraphSAGE with the time complexity of $O(C^L H)$, the multiplier K results from the computation schema-based attention, and the importance sampling costs $O(C)$ by alias method instead of $O(D)$ where D is the degree of nodes.

3 Experiments

In this section, we will demonstrate the effectiveness of our approach by performing experiments on five real datasets compared to the state-of-the-art recommendation methods.

3.1 Experiment Setup

Datasets. We use four public datasets, including *Douban Movie*, *Douban Book*, *Small Yelp* and *Full Yelp*, and one private dataset *Ali* in our experiments. We show the statistics for each dataset in Table 1.

Table 1. Statistics of heterogeneous graphs for each dataset. We omit the statistical details of *Small Yelp* here, because it is a subset of *Full Yelp* and has the same meta graph schema of *Full Yelp* dataset. Please find the statistical details of *Small Yelp* in this paper [12].

Relation (A-B)	#A	#B	#A-B	Avg Degrees of A/B	Relation (A-B)	#A	#B	#A-B	Avg Degrees of A/B
Douban Movie (node:37,595, edge: 1,714,941, Sparsity:1.21%)					Douban Book (node: 48,055, edge:1,026,046, Sparsity: 0.44%)				
User-Movie	13,367	12,677	1,068,278	79.9/84.3	User-Book	13,024	22,347	792,026	60.8/35.4
User-User	2,440	2,294	4,085	1.7/1.8	User-User	12,748	12,748	169,150	13.3/13.3
User-Group	13,337	2,753	570,047	42.7/207.1	Book-Author	21,907	10,805	21,905	1.0/2.0
Movie-Director	10,179	2,449	11,276	1.1/4.6	Book-Publisher	21,773	1,815	21,773	1.0/11.9
Movie-Actor	11,718	6,311	33,587	2.9/5.3	Book-Year	21,192	64	21,192	1.0/331.1
Movie-Type	12,678	38	27,668	2.2/728.1					
Full Yelp (node: 1,970,659, edge: 15,186,279, Sparsity: 0.04%)					Ali (node: 49,042,421, edge: 398,302,071, Sparsity:0.017‰)				
User-Item	1,518,169	188,593	5,996,996	4.0/31.8	User-Item-Bought	28,016,674	8,512,543	136,322,333	4.9/16.0
User-User	716,849	716,861	13,237,194	18.5/18.5	User-Item-AddCart	29,894,459	9,686,796	148,877,338	5.0/15.4
User-Stars	261,371	9	1,518,169	5.8/168685.5	User-Item-Collect	15,739,564	8,372,466	81,748,045	5.2/9.8
Item-City	188,583	1,110	188,583	1.0/169.9	User-User	227,647	227,647	271,690	1.2/1.2
Item-State	188,593	69	188,593	1.0/2733.2	Item-Category	10,358,571	11,516	10,358,571	1.0/899.5
Item-Category	188,052	1,305	739,022	3.9/566.3	Item-City	10,358,571	556	10,358,571	1.0/18630.5
Item-Stars	188,593	9	188,593	1.0/20954.7	Item-Shop	10,350,952	1,258,192	10,350,952	1.0/8.2
Small Yelp (node: 31,092, edge: 817,686, Sparsity: 0.84%)					Category-Category	14,571	159	14,571	1.0/91.6

The two datasets of Douban are made up with book ratings from book review website, while the datasets of Yelp and Ali come from the e-commerce domain. The *Small Yelp* dataset in some paper [18] is also called the *Yelp K200* dataset, because it has 200K User-Item pairs that need to be scored. The above five public datasets can be downloaded through the following link.[2,3]

---

[2] https://github.com/librahu/HIN-Datasets-for-Recommendation-and-Network-Embedding.

[3] https://www.yelp.com/dataset.

The *Ali* dataset are collected in a commercial e-business website.[4] Besides the properties of user and item, the edges includes various kinds of user behaviors, such as "add cart", "bought", "collect". We use the rating of buying as the predictive score.

All dataset use the User-Item rating as the prediction target. *Douban Movie*, *Douban Book* and *Small Yelp* are partitioned randomly into 70%/10%/20% training/validation/test splits. While Full Yelp dataset and Ali dataset contain too many records, we have to take a 79%/1%/20% division on them in order to avoid large validation time consumption in the process of training. During training, the models are only allowed to exploit the heterogeneous information network of the training part, which means the evaluation is inductive.

Competitors. We consider the following methods to compare:

- PMF [11]. Probabilistic matrix factorization is a graphical model by factorizing the rating matrix into two low-dimensional matrices.
- SoMF [10]. The social relations (User-User) are characterized by a social regularization term, and integrated into the basic matrix factorization model.
- DSR [19]. DSR is a HIN based recommendation with dual similarity regularization, which imposes the constraints on entities with high and low similarities simultaneously.
- HERec [12]. HERec is a state-of-the-art HIN based model which utilizes the personalized non-linear fusion function to predict the rating score[5].
- GraphSAGE [6]. GraphSAGE is a general inductive framework which generates embeddings by sampling uniformly and aggregating features from a node's local neighborhood.
- GraphIcpt [17]. GraphInception (GraphIcpt) introduces the concept of meta-path into the propagation on the heterogeneous graph, which groups the neighbors according to their node types with predefined meta-path[6].
- HGAN [16]. Heterogeneous graph attention network aggregates neighbors by fusing the schema-level and node-level attention by predefined meta-path[7].
- TS-Rec. It is our proposed HIN based recommendation model with type-sensitive sampling.

With regard to the implementation of our method *TS-Rec*, we use tensorflow as the main framework of automatic differentiation. Besides, we add two self-defined C++ operators in tensorflow, for sampling the neighbor by alias methods. The self-defined operators are paralleled by threads leading to scalability and efficiency. Besides, we overwrite the sparse matrix storage format by Compressed Sparse Row Format (CSR) [5] and some operating functions in tensorflow for supporting multi-type edges storage and better performance.

---

[4] https://www.taobao.com.
[5] https://github.com/librahu/HERec.
[6] https://github.com/zyz282994112/GraphInception.
[7] https://github.com/Jhy1993/HAN.

We adopt the early stopping for all competitors, which wait 20 steps until no progress of RMSE on the validation set. The latent dimension is set to be 32 and the batch size is set to be 64. We use l2-loss in the training process, and the weight is set to be $5e - 5$. Adam is adopted as the optimizer and the initial learning rate is $5e-4$. We adopt two layer's aggregation for graph neural networks for all competitors, and the sample size is 25 for each layer. All experiments are conducted on a machine with one GPU (NVIDIA Tesla P100 16G) and 96 CPUs (Intel Xeon Platinum 2.50GHz).

3.2   Recommendation Effectiveness

By comparing our proposed model (TS-Rec) against the other competitors, we report their performance in terms of the metrics in Table 2 and the convergence time in Table 3.

Table 2. Experimental results on three datasets in terms of effectiveness. The symbol '-' denotes the competitor is not scalable on the datasets. The metrics on the test set are reported.

Dataset	Douban Movie		Douban Book		Small Yelp		Full Yelp		Ali	
Metrics	MAE	RMSE	MAE	RMSE	MAE	RMSE	MAE	RMSE	MAE	RMSE
PMF	0.5741	0.7641	0.5774	0.7414	1.0791	1.4816	1.0249	1.3498	1.1700	1.6112
SoMF	0.5817	0.7680	0.5756	0.7302	1.0373	1.3782	–	–	–	–
DSR	0.5681	0.7225	0.5740	0.7206	0.9098	1.1208	–	–	–	–
HERec	0.5515	0.7053	0.5602	0.6919	0.8475	1.0917	–	–	–	–
GraphSAGE	0.5529	0.7112	0.5661	0.7022	0.8281	1.0487	0.9781	1.2516	0.9328	1.1802
GraphIcpt	0.5491	0.7170	0.5569	0.7004	0.8139	1.0377	–	–	–	–
HGAN	0.5511	0.7034	0.5578	0.6911	0.8134	1.0335	–	–	–	–
TS-Rec	**0.5473**	**0.6966**	**0.5485**	**0.6766**	**0.7977**	**1.0269**	**0.9521**	**1.2221**	**0.9168**	**1.1585**

Among these baselines, HIN based methods (DSR, HERec, GraphInception) perform better than traditional MF based methods (PMF and SoMF), which indicates the usefulness of the heterogeneous information. PMF only utilizes the implicit feedback and SoMF use additional user-user information. Other competitors utilize the complete heterogeneous information network. One main drawback of the traditional HIN based recommendation is that, they use human expertise to select candidate meta-path by prior knowledge, which limits the combinatorial generalization of HIN.

The GNN based methods (GraphSAGE, GraphInception, HGAN) have comparable performance to the best HIN based model on Douban dataset. Specifically, they perform better than HERec significantly on Yelp dataset, which verifies that the predefined meta-path may not capture the combinatorial generalization of HIN.

However, the GNN based methods could not handle the trade-off between the power-law of types and the power-law of nodes on the setting of heterogeneous graphs. GraphSAGE sampling the neighbors uniformly and has shortest per-batch training time. However, GraphSAGE has small chance to sample the

Table 3. Convergence statistic on all datasets. The training ends until the early stopping. The number of epochs until convergence is abbreviated to '#Epochs', and 'Average' denotes the average time cost per epoch.

Statistic	Total time cost			#Eposhs	Average
Dataset	Small Yelp	Full Yelp	Ali		
HERec	33 min	–	–	–	–
HGAN	20 min	–	–	–	–
GraphSAGE	4 min	38 min	18.9 h	4.7	4.0 h
TS-Rec	50 s	21 min	14.3 h	2.2	6.5 h

minority of types, which leads to longer convergence time than the type-sensitive sampling and poorer performance. HGAN is type-sensitive by incorporating the attention mechanism into the aggregation, but the time cost is impractice for recommendation owing to the lack of sampling. GraphInception groups the partial neighbors by the predefined meta-path, which also limits the performance.

In Table 3, TS-Rec outperforms the state-of-the-art methods on convergence time. GraphSAGE has small chance to sample the minority of types, which results in invalid sampling instance and slower convergence. The reason why TS-Rec has the faster convergence time than GraphSAGE is that our method needs less step and GraphSAGE needs more steps to converge.

In brief, TS-Rec achieves the best performance and convergence time on all datasets. On the one hand, the schema-based attention outputs the distribution over types and makes the sampling type-sensitive. The attention mechanism and GNN keep the capacity of combinatorial generalization over the information of types. On the other hand, nodes with the same type usually have redundant information for recommendation, thus we use the importance sampling inside types with alias method to reduce the complexity significantly.

4   Conclusion

In this paper, we proposed a HIN based recommendation with type-sensitive sampling to handle the heterogeneous and power-law graphs. The type-sensitive sampling outputs the distribution over types by schema-based attention which makes a balance between two aspects of power law on heterogeneous graph, and it reduces the complexity by importance sampling with the alias method. Empirical studies on real-world and industrial datasets demonstrate the effectiveness of the proposed algorithm in the setting of heterogeneous and power-law graphs for recommendation tasks, and our approach achieves the fastest convergence time while outperforming the state-of-the-art competitors.

Acknowledgement. This work was partially supported by the National Key Research and Development Plan of China (No. 2019YFB2102100), NSFC under Grant No. 61832001, Alibaba-PKU Joint Program, and Zhejiang Lab (No. 2019KB0AB06).

References

1. Bai, J., Zhou, C., et al.: Personalized bundle list recommendation. In: Proceedings of the 2019 WWW (2019)
2. Battaglia, P.W., Hamrick, J.B., et al.: Relational inductive biases, deep learning, and graph networks. arXiv preprint arXiv:1806.01261 (2018)
3. Chen, J., Ma, T., Xiao, C.: Fastgcn: fast learning with graph convolutional networks via importance sampling. In: International Conference on Learning Representations (ICLR) (2018)
4. Gilmer, J., Schoenholz, S.S., Riley, P.F., Vinyals, O., Dahl, G.E.: Neural message passing for quantum chemistry. In: Proceedings of the 34th International Conference on Machine Learning, vol. 70, pp. 1263–1272. JMLR. org (2017)
5. Greathouse, J.L., Daga, M.: Efficient sparse matrix-vector multiplication on GPUs using the CSR storage format. In: Proceedings of the International Conference for High Performance Computing, Networking, Storage and Analysis, pp. 769–780. IEEE Press (2014)
6. Hamilton, W., Ying, Z., Leskovec, J.: Inductive representation learning on large graphs. In: Advances in NIPS, pp. 1024–1034 (2017)
7. Huang, W., Zhang, T., Rong, Y., Huang, J.: Adaptive sampling towards fast graph representation learning. In: Advances in NIPS, pp. 4563–4572 (2018)
8. Jiang, H., Song, Y., Wang, C., Zhang, M., Sun, Y.: Semi-supervised learning over heterogeneous information networks by ensemble of meta-graph guided random walks. In: IJCAI, pp. 1944–1950 (2017)
9. Li, Z., Zhenpeng, L., et al.: Addgraph: anomaly detection in dynamic graph using attention-based temporal GCN. In: IJCAI (2018)
10. Ma, H., Zhou, D., Liu, C., Lyu, M.R., King, I.: Recommender systems with social regularization. In: Proceedings of WSDM, pp. 287–296. ACM (2011)
11. Mnih, A., Salakhutdinov, R.R.: Probabilistic matrix factorization. In: Advances in NIPS, pp. 1257–1264 (2008)
12. Shi, C., Hu, B., Zhao, W.X., Philip, S.Y.: Heterogeneous information network embedding for recommendation. IEEE Trans. Knowl. Data Eng. **31**(2), 357–370 (2019)
13. Sun, Y., Han, J.: Mining heterogeneous information networks: a structural analysis approach. ACM SIGKDD Explor. Newsl. **14**(2), 20–28 (2013)
14. Walker, A.J.: New fast method for generating discrete random numbers with arbitrary frequency distributions. Electron. Lett. **10**(8), 127–128 (1974)
15. Wang, J., et al.: Billion-scale commodity embedding for e-commerce recommendation in Alibaba. In: Proceedings of the 24th ACM SIGKDD. ACM (2018)
16. Wang, X., Ji, H., et al.: Heterogeneous graph attention network. In: The World Wide Web Conference, pp. 2022–2032 (2019)
17. Zhang, Y., Xiong, Y., et al.: Deep collective classification in heterogeneous information networks. In: Proceedings of the 2018 WWW (2018)
18. Zhao, H., et al.: Meta-graph based recommendation fusion over heterogeneous information networks. In: Proceedings of the 23rd ACM SIGKDD. ACM (2017)
19. Zheng, J., Liu, J., Shi, C., Zhuang, F., Li, J., Wu, B.: Recommendation in heterogeneous information network via dual similarity regularization. Int. J. Data Sci. Anal. **3**(1), 35–48 (2016). https://doi.org/10.1007/s41060-016-0031-0
20. Zhou, C., Bai, J., et al.: Atrank: an attention-based user behavior modeling framework for recommendation. In: AAAI (2018)
21. Zhou, J., Cui, G., Zhang, Z., Yang, C., Liu, Z., Sun, M.: Graph neural networks: a review of methods and applications. arXiv preprint arXiv:1812.08434 (2018)

Adaptive Loading Plan Decision Based upon Limited Transport Capacity

Jiaye Liu[1], Jiali Mao[1(✉)], Jiajun Liao[1], Yuanhang Ma[1], Ye Guo[2], Huiqi Hu[1], Aoying Zhou[1], and Cheqing Jin[1]

[1] East China Normal University, Shanghai, China
{5118451030,51195100011}@stu.ecnu.edu.cn,
{jlmao,hqhu,ayzhou,cqjin}@dase.ecnu.edu.cn,
yhma.dev@outlook.com
[2] Jing Chuang Zhi Hui (Shanghai) Logistics Technology., LTD, Shanghai, China
guoye@jczh56.com

Abstract. Cargo distribution is one of most critical issues for steel logistics industry, whose core task is to determine cargo loading plan for each truck. Due to cargos far outnumber available transport capacity in steel logistics industry, traditional policies treat all cargos equally and distribute them to each arrived trucks with the aim of maximizing the load for each truck. However, they ignore timely delivering high-priority cargos, which causes a great loss to the profit of the steel enterprise. In this paper, we first bring forward a data-driven cargo loading plan decision framework based on the target of high-priority cargo delivery maximization, called as *ALPD*. To be specific, through analyzing historical steel logistics data, some significant limiting rules related to loading plan decision process are extracted. Then a two-step online decision mechanism is designed to achieve optimal cargo loading plan decision in each time period. It consists of *genetic algorithm*-based loading plan generation and *breadth first traversal*-based loading plan path searching. Furthermore, *adaptive time window* based solution is introduced to address the issue of low decision efficiency brought by uneven distribution of number of arrived trucks within different time periods. Extensive experimental results on real steel logistics data generated from *Rizhao Steel*'s logistics platform validate the effectiveness and practicality of our proposal.

Keywords: Steel logistics · Cargo loading plan · Searching · Adaptive time window

1 Introduction

Due of the characteristic of long-distance and heavy-haul in steel product transportation, only heavy truck satisfies this requirement in road transportation. But the number of heavy trucks is limited and has uneven distribution in different regions. Driven by limited number of heavy trucks and excess capacity, the steel

© Springer Nature Switzerland AG 2020
Y. Nah et al. (Eds.): DASFAA 2020, LNCS 12114, pp. 685–697, 2020.
https://doi.org/10.1007/978-3-030-59419-0_42

enterprises generally face serious problems like truck overloading and order over-
due. It's direct results include high cost and low efficiency of steel logistics, and
even bad effects on the development of social economy. It was reported that on
Oct.10, 2019, a bridge on national highway 312 in *Wuxi* city, *Jiangsu* province,
China, was collapsed because the load of a lorry transporting steel products was
far beyond the weight that bridge withstanded[1]. One important reason under-
lies that phenomenon of heavy overloading is lack of proper cargo loading plans.
Traditional policy attends to determine cargo loading list for each truck based
upon the consideration of maximizing the load of each truck. For steel industry,
the cargos far outnumber available transport capacity, which inevitably results
in the occurrence of overloading.

In order to boost profits for steel logistics platform, the cargo loading plan
decision mechanism is required to solve the unbalance issue between the number of available trucks and the demand of cargo delivery. Numerous researches
[8,16] on task-resource assignment in collaborative logistics field have emerged
in last few years. But due of their distinct scenarios from steel logistics, their
models cannot be applied for our proposed decision problem. With the transformation of steel logistics informationization, there have emerged many logistics
platform and produced huge amount of data. Take a logistics platform (*JCZH* for
short) of *Rizhao Steel*[2] as example, *JCZH* needs to complete the task of matching cargos of over 150 tons to trucks each minute, which generates tremendous
amount of data hourly. It provides us an opportunity to come up with appropriate decision-making on cargo loading plan. Based upon the analysis on historical
data generated from *JCZH*, we found that it is important to timely transporting cargos of urgent delivery demands, which can avoid the economic losses and
bring more profits to logistics platform. In view of that, we present a two-phase
data-driven cargo loading plan decision framework with the aim of maximizing
delivery proportion of high-priority cargos, called as *ALPD*.

At off-line analysis phase, massive analysis is implemented on historical logistics data to extract some limiting rules related to loading plan decision process,
e.g. transportation flow direction and variety of cargo, etc. At online decision
phase, to maximize delivery proportion of high-priority cargos, *genetic algorithm* is leveraged to generate K cargo loading plan candidates for each truck.
Then *breadth first traversal*-based loading plan path searching is implemented
to obtain optimal cargo loading list for all trucks arrived in each time window.
Additionally, adaptive time window model is introduced to relieve the issue of
low decision efficiency within some time periods brought by uneven distribution
of number of trucks. To summarize, this paper makes the following contributions: (1) We first propose a data-driven cargo loading plan decision framework,
called as *ALPD*, to tackle the unbalance issue between transport capacity supply
and cargo delivery demand in steel logistics industry. (2) Based on *adaptive time*

---

[1] http://www.xinhuanet.com/local/2019-10/10/c_1125089169.html.
[2] https://www.rizhaosteel.com/en/.

window model, we put forward a two-step online decision approach consisting of loading plan candidates generation and loading plan path searching, with the aim of maximizing delivery proportion of high-priority cargos. (3) We compare our proposal with other congeneric approaches by conducting substantial experiments on real data sets generated by *JCZH*. Experimental results validate the effectiveness and practicality of our proposal.

The remainder of this paper is organized as follows. Section 2 reviews the most related work. The preliminary concepts are introduced in Sect. 3 and the detail of *ALPD* framework is outlined in Sect. 4. Section 5 presents the experimental results, followed by conclusions in Sect. 6.

2   Related Work

Cargo loading plan decision issue can be viewed as a task-resource assignment problem (*TRA* for short) for transportation resources. It has been extensively studied by academe and industries [4,8,16,18,20]. Epstein and Lin studied the *Empty Container Repositioning* problem in the ocean logistics industry [4,8]. Yuan and Xu tried to solve *TRA* problem by forecasting stochastic demands in grey deep neural network model [16,18]. Recently, quite a few researches applied reinforcement learning method to solve *TRA* problem in the applications such as inventory optimization and fleet management [3,9]. Although the above mentioned models are powerful and practical, their different scenarios from steel logistics limit their usage in our proposed problem.

With broad application of car-hailing platform, there have numerous researches devoted to tackle its core issue of order dispatching [1,2,11,13,17,19], which can be also viewed as a *TRA* problem. Tong et al.studied the online minimum bipartite matching problem based on the prediction of the spatiotemporal distribution of tasks and workers [10–13]. Chen et al. built the matching model by taking into account three kinds of participant characters such as attributes, requirements and supplements [2]. Zhang et al. proposed a taxi-orders dispatch model which considered three kinds of participant characters [19]. Zhe and Wang presented optimizing models concerning long-term efficiency using reinforcement learning method [15,17]. Besides, graph convolution network was used for ride-hailing demand forecasting to improve the long-term efficiency [5]. Wang et al. used adaptive matching batch window, which has a remarkable effect in coping with dynamic scenarios [14]. However, the aforementioned methods are not applicable for our proposed cargo loading plan decision problem, due to specific limited rules and optimization goal in steel logistics industry.

3   Problem Formulation

Our goal is to maximize delivery proportion of high-priority cargos, denoted as E_{dp}. If there are N pieces of cargos to be distributed to M trucks, we represent the loading plan decision result as a matrix $LPD = \begin{pmatrix} d_{11} & \cdots & d_{1M} \\ \cdots & d_{ij} & \cdots \\ d_{N1} & \cdots & d_{NM} \end{pmatrix}$, where $1 \leq i \leq N$, $1 \leq j \leq M$, and $d_{ij} = \begin{cases} 1, & \text{if cargo } C_i \text{ is distributed to truck } T_j \\ 0, & \text{if cargo } C_i \text{ is not distributed to truck } T_j \end{cases}$

Noted that, in actual scenario, a trucker prefers to transport the cargos that have his familiar transportation flow direction and may bring him more profits. So, the acceptance of a cargo loading plan by a trucker is largely depended on whether it matches the transport preference of that trucker.

In addition, a truck can load multiple pieces of cargos while a piece of cargo is only loaded on one truck, which imposes the following constraint: $\forall i, \sum_{j=1}^{M} d_{ij} \leq 1$. A cargo loading list of a truck T_j, denoted as LP, can be represented as $LP_j = (d_{1j}, \ldots, d_{ij}, \ldots, d_{Nj})$. To maximize the profit of the logistics platform, such factors as weight and priority of cargos shall be taken into account. Let w_i and p_i respectively denote weight and priority of a cargo C_i $(1 \leq i \leq N)$, we can formulate the delivery proportion of high-priority cargos of one truck T_j with maximum load l_j as $E_{dp}^j = \frac{\sum_{i=1}^{N} p_i \cdot d_{ij} \cdot w_i}{\sum_{i=1}^{N} d_{ij} \cdot w_i}$.

Therefore, the total objective function of all trucks in a time window can be expressed as follows:

$$\begin{cases} \max E_{dp} = \max \sum_{j=1}^{M} E_{dp}^j \\ s.t. \lim_{\varepsilon \to 0^+} \left(\sum_{i=1}^{N} d_{ij} \cdot w_i \right) + \varepsilon = l_j \end{cases} \tag{1}$$

4   Overview of *ALPD* Framework

ALPD framework is comprised of two components: *off-line limiting rule extraction* and *online cargo loading plan decision*.

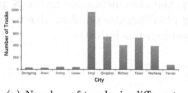
(a) Number of trucks in different cities

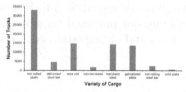
(b) Number of trucks for different varieties

Fig. 1. Histogram of Number of Trucks

4.1 Off-line Limiting Rule Extraction

Data sets for analysis and experiment are from *Rizhao steel*, which is a steel production company from *Shandong* province, China, with an annual output of 13 million tons of steel production capacity. We have collected for three months of data from multiple sources including *loading plan, outbound cargo, cargo order* and *inventory of warehouse*, etc. Through analyzing, it is found that (i) important influencing factors (e.g., the weight of cargo, the delivery date of the order, and the ratio of overall weight of chosen high-priority cargos to that of all chosen cargos in each loading list) are consistent with our optimization objective; (ii) the generation of a cargo loading plan is limited by some factors that reflect truckers' preference, which determine the probability of a trucker accepting one cargo loading plan decision. Figure 1 illustrates histogram of number of trucks as relative to city and variety of cargos. We can see that truckers have their respective favourite transportation destination city, as shown in Fig. 1(a). This is due to that they easily take the waybill for the return trip in those cities. Likewise, each trucker prefers to transport a few varieties of steel products because he tends to benefit more from those logistics orders, as illustrated in Fig. 1(b). On the basis of the extracted limiting rules related to decision process, we attend to group all trucks into various classes in term of truckers' transport preferences.

4.2 Online Cargo Loading Plan Decision

Based upon off-line analyzing results from historical steel logistics data, we put forward a two-step online decision approach (*OnLPD* for short) with the goal of maximizing delivery proportion of high-priority cargos.

Cargo Loading Plan Candidate Generation. Essentially, the issue of loading plan generation can be regarded as a combinatorial optimization problem of multi-objective optimization, whose fundamental problem is the same as the multiple knapsack problem.

Algorithm 1: Loading Plan Candidate Generation for Each Truck

---

Input: Truck T_j; Maximum Load l_j; List of Cargos $CL_j = (C_{j1}, C_{j2}, ...C_{jN})$;

Size of Candidate set K; Number of Iterations t

Output: Loading Plan Candidate Set LPC_j

1 //Initialize
2 **for** $i \leftarrow 1, K$ **do**
3   **while** $Weight(LP_i) < l_j$ **do**
4          $LP_i \leftarrow Random(C_{jl}$ in $CL_j)$
5   //Randomly select cargos from CL_j and ensure unique C_{jl} in LP_i
6      $LPC_j \leftarrow LPC_j \cup LP_i$ //append LP_i into LPC_j
7 **for** $p \leftarrow 1, t$ **do**
8   //Crossover
9   **for** $q \leftarrow 1, K$ **do**
10   Randomly select two loading plans LP_α and LP_β from LPC_j;
11   Exchange cargos between LP_α and LP_β to get $LP_\alpha^{'}$ and $LP_\beta^{'}$;
12   Insert them into LPC_j;
13   //Mutation
14   **for** $q \leftarrow 1, K$ **do**
15         $LP_j \leftarrow Random(LP$ in $LPC_j)$ //Randomly select loading plan from LPC_j $LP_j^{'} \leftarrow Mutation(LP_j)$
16   //Change some cargos of LP_j through Random(C_{jl} in CL_j)
17         $LP_j \leftarrow Max(LP_j, LP_j^{'})$ //Select one with larger E_{dp}^j from LP_j and $LP_j^{'}$
18   //Update
19     $LPC_j \leftarrow$ pick top K of LP_i in LPC_j
20 **return** LPC_j

In this paper, we treat trucks arrived within a short time as a micro-batch in a time window by using sliding window model. To generate a cargo loading plan LP_j for any arrived truck T_j in a time window of size W, we first sort the cargos by their delivery priorities in descending order, and screen out high-priority cargos to be delivered. Then we attempt to filter cargos according to each trucker's preference, and guarantee the weights of the cargos are in conformity with the maximum load of each truck. To maximize delivery proportion of high-priority cargos, the priority of chosen cargos is needed to be as high as possible (e.g., the cargos that the delivery date is approaching) and the total weight of chosen cargos is required to be as heavy as possible.

Algorithm 1 illustrates the generation of loading plan candidates for each truck using genetic algorithm. A loading plan is taken as a gene, each cargo in a loading plan is taken as the chromosome of this gene, and the candidate set of a car's loading plan is taken as a population. We initialize the loading plan candidate set with loading plans consisting of cargos through randomly selecting them from cargo list CL_j (at lines 1–4). Subsequently, we randomly select two loading plans, i.e. LP_α and LP_β, from LPC_j, and exchange some of their cargos to derive LP'_α as well as LP'_β, and meanwhile insert them into LPC_j (at lines 6–7). By the mutation component, we constantly change the original cargos of LP_j and the rest part of cargos in CL_j, and append the offspring with larger E^j_{dp} into LPC_j (at lines 8–11). By iterative execution, a group of loading plans are generated for truck T_j, we choose top K plans with highest E_{dp} value as a candidate set of loading plans LPC_j for truck T_j. It can be expressed as $LPC_j = \{LP^1_j, LP^2_j, \ldots, LP^K_j\}$, here K is a specified threshold to represent the size of the candidate set (at lines 12–13).

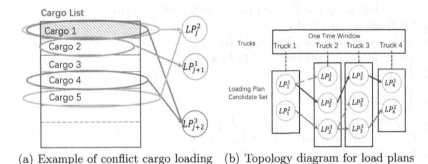

(a) Example of conflict cargo loading   (b) Topology diagram for load plans
plans

Fig. 2. Illustration of loading plan path searching

Figure 2(a) illustrates an example of cargo loading plan candidates generation. We can see that a candidate set of two loading plans is generated, i.e. LP^2_j and LP^3_{j+2}. There may be cases where some cargos are chosen repeatedly into different trucks' loading plans in a time window. As shown in Fig. 2(a), the first cargo is simultaneously chosen into LP^2_j and LP^3_{j+2}. It results in many loading plans cannot being implemented at the same time, hence LP^2_j and LP^3_{j+2} cannot exist in the same loading plan decision LPD. This case is in line with our definition for d_{ij}. To that end, we introduce a topology structure to assist in determining cargo loading plans for trucks.

Algorithm 2: Loading Plan Path Searching

Input: Truck Set $TS = (T_1, T_2, ... T_N)$,
         Loading Plan Candidate Set $LPCS = (LPC_1, LPC_2, ... LPC_N)$
Output: loading plan decision result matrix $LPD_{K \times N}$

1 Initialize $LPD_{K \times N}$;
2 Path list $PL \leftarrow \{\}$;
3 //The first truck's load plans are linked to paths
4 **for** $i \leftarrow 1, K$ **do**
5   //Each loading plan in LPC_1 is viewed as a starting node of path P_i
6      $P_i \leftarrow LPC_{1i}$;
7      $PL \leftarrow PL \cup P_i$;

8 //The follow-up truck's load plans are linked to paths
9 **for** truck T_j in TS **do**
10   **for** load plan LP_j^l in LPC_j **do**
11   **if** $non\text{-}conflicting(P_i, LP_j^l)$ **then**
12              $P_{new} \leftarrow P_i \cup LP_j^l$;
13      $PL \leftarrow PL \cup P_{new}$;

14 //Output the path with largest E_{dp}
15 $P_{best} \leftarrow$ the path with largest E_{dp} in PL ;
16 **for** *Any position ps corresponding to* LP_j *in* P_{best} **do**
17      $LPD_{ps} \leftarrow 1$;
18 **return** $LPD_{K \times N}$

Loading Plan Path Searching. We regard each loading plan LP_j^l ($1 \leq l \leq K$) as a node. In a time window, we try to search a group of nodes (i.e., a loading plan path) from the candidate sets of loading plans generated for various trucks. Algorithm 2 details the whole procedure of loading plan path searching. When traversed nodes are not conflict with each other, we add an edge between a pair of them (i.e., loading plans without common selected cargos), as shown in Fig. 2(b). To avoid the appearance of ring structure in the diagram which may influence on subsequent searching, we restrict an edge can be only added between two nodes respectively representing loading plans of two successively arrived trucks T_j and T_{j+1} (at lines 8–12). To ensure satisfying the constraint $\forall i, \sum_{j=1}^{M} d_{ij} \leq 1$, we attend to store the path $P_i = \{(LP_1^1 \ LP_2^1 \ ... \ LP_M^K), (.), ...\}$ instead of all the information for the topology structure. Finally, LPD has the largest E_{dp} in the set PL (at line 14).

***Adaptive Time Window*-Based *OnLPD*.** During the implementation of *OnLPD* on real data at *JCZH*, we find that number of trucks arrived in different time windows is different, as shown in Fig. 3. When numerous trucks arrive in a time window, it may lead to low decision efficiency and hence bring bad experience to the truckers. To address this issue, we adaptively determine the size of each time window based upon adaptive time window model. Further, we present an adaptive solution, called as $OnLPD_{ATW}$. To be specific, we attempt to determine the size of time window by constantly calculating the value of E_{dp} with the most feasible loading plans in current time window (denoted as W_{curr}). When the value of E_{dp} reaches the specified threshold thr_E (here, the value of thr_E is smaller than the average number of trucks arrived in all the time windows), or the value of W_{curr} is larger than W, the window slides forward.

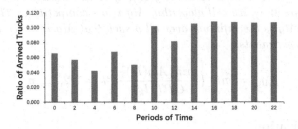

Fig. 3. Quantities distribution of trucks in different time horizons

5 Empirical Evaluation

In this section, we conduct extensive experiments to assess the effectiveness and efficiency of our proposal. First, to evaluate the effectiveness of $OnLPD$ and $OnLPD_{ATW}$, we utilize *competitive ratio* as evaluation metrics and compare our proposal with baseline approaches on real data from *JCZH*. Then, we compare the efficiency of $OnLPD$ and $OnLPD_{ATW}$.

All codes, written in Python, are conducted on a PC with Intel Core CPU 2.6 GHz Intel i7 processor and 16 GB RAM. The operating system is Windows 10. Unless mentioned otherwise, the parameters are set below, $K = 5$, $W = 2$ and $thr_E = 2$, which are obtained through choosing the average by repeating 10 experiments. The number of cycles of the genetic algorithm is controlled to be less than 10, and the mean value of the gene is set as 0.9.

5.1 Comparative Approach

We compared our proposed approaches ($OnLPD$ and $OnLPD_{ATW}$) with the following algorithms.

- **Hungarian algorithm(OPT):** it is leveraged as the optimum baseline algorithm, which can find optimal matching result in the off-line scenario in polynomial time [6].
- **Greedy algorithm(GR):** it determines a cargo loading plan according to the delivery sequence of cargos when each truck arrives, but ignores efficient usage of all trucks within a long time period.
- **Genetic algorithm(GA):** it has been widely used to solve combinatorial optimization problem. We use genetic algorithms in [7] as baseline algorithm, and change some details to fit our problem scenario.

5.2   Evaluation Metric

We utilize *competitive ratio* as evaluation metric to assess the performance of online cargo loading plan decision algorithm, which is widely used to measure the performance of an online algorithm in the condition of uncertain demands.

Definition 1. *Competitive Ratio* ρ: *Given an algorithm Alg, for any experimental instance I_k with a result $Alg(I_k)$, if it can always ensure $Alg(I_k) \leq \rho \cdot OPT(I_k)$, $\rho \varepsilon [0,1]$, we call algorithm Alg is $\rho - competitive$. The competitive ratio ρ_{Alg} of Alg is the infimum over all ρ such that Alg is $\rho - competitive$, i.e. in k-round experiments.*

$$\rho_{Alg} = min \left\{ \frac{min\{Alg(I_k)\}}{OPT(I_k)} , k = 1, 2, \ldots, n \right\} \tag{2}$$

5.3   Visualization

We first present verify the usefulness of our proposal through real user scenarios and present the visualized result.

As shown in Fig. 4(a), the information of truck and its corresponding loading plan (highlighted in red box) can be viewed through the user interface. In the first time window, our proposal first obtains cargo loading plan candidates for 4 trucks, and then determines the loading plans (highlighted in red box) with high-priority cargos (current time and delivery date of orders are marked with purple boxes) for them, as illustrated in Fig. 4(b). Likewise, Fig. 4(c) illustrates the cargo loading plans generated for another 9 trucks within subsequent time window. Owing to the advantage of loading plan path searching base upon adaptive time window model, our proposal can effectively generate optimal loading plans for different numbers of trucks in various time windows. Noted that the window sizes of two time windows are different, the former takes 32 s while the latter takes 29 s.

5.4   Quantitative Results

After running all the compared methods online for one day, we report the following evaluation metrics. The results show that $OnLPD$ and $OnLPD_{ATW}$ perform significantly better than the other two methods (GR and GA) in terms of the total objective function of all trucks in a time window E_{dp} and competitive ratio ρ, as illustrated in Fig. 5(a) and (b). On the comparative result of E_{dp}, $OnLPD_{ATW}$ is almost identical to $OnLPD$, and E_{dp}s of both approaches are close to that of OPT method, as shown in Fig. 5(a). This is due to that our proposals aim to maximize delivery proportion of high-priority cargos by optimizing the value of E_{dp} in each time window. In addition, on the *competitive ratio* (ρ) metric, $OnLPD_{ATW}$ almost obtains the same ρ value with $OnLPD$ as the number of trucks continuously increases. More important, they have constant competitive ratio (keep high ρ of 0.8) which are closed to competitive ratio of OPT method, as shown in Fig. 5(b). This also verify the rationality of our

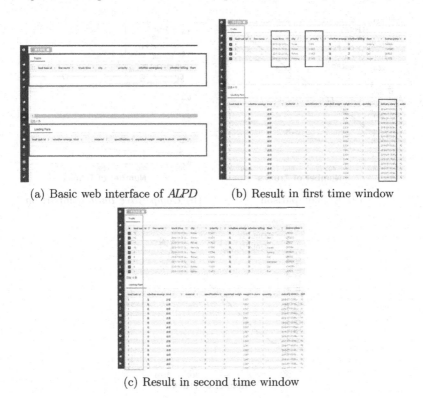

(a) Basic web interface of *ALPD*   (b) Result in first time window

(c) Result in second time window

Fig. 4. An illustration of cargo loading plan decision (Color figure online)

proposed online cargo loading plan decision approach. On the premise of insufficient transport capacity supply, our proposal can guarantee timely delivery of high-priority cargos by loading plan decision-making.

Finally, we compare the response time of $OnLPD$ and $OnLPD_{ATW}$, as illustrated in Fig. 5(c). We can see that $OnLPD_{ATW}$ approach always takes less response time than $OnLPD$ as the number of trucks rises. This is because number of trucks in each window is different. For $OnLPD$ approach using fixed size of time window (e.g., 2 min), when more trucks arrive in a time window, many truckers can only wait for loading plan decision in next time window. It results in a longer response time of loading plan decision. By contrast, $OnLPD_{ATW}$ determines the size of time window based on searching the most feasible decision result with best E_{dp}, it can avoid the low decision efficiency issue.

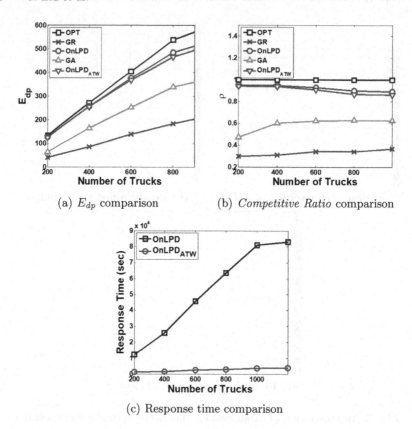

(a) E_{dp} comparison   (b) *Competitive Ratio* comparison

(c) Response time comparison

Fig. 5. Quantitative comparative results

6 Conclusion

In this work, we propose a *data-driven* cargo loading plan decision framework for steel logistics platform, called *ALPD*. It integrates off-line limiting rule extraction and online cargo loading plan decision. Based upon the off-line analysis on massive logistics data at *JCZH*, we derive valuable limiting rules related to the decision process. Further, on the basis of adaptive time window model, we provide a two-step decision approach consisting of loading plan candidates generation and loading plan path searching. We validate our proposal for performance by conducting extensive experiments on real logistics data at *JCZH*. In our future work, we will apply our cargo loading plan decision framework for more bulk logistics scenarios to improve efficiency.

Acknowledgements. The authors are very grateful to the editors and reviewers for their valuable comments and suggestions. This research was supported by the National Natural Science Foundation of China (NSFC) (Nos.U1911203, U1811264 and 61702423).

References

1. Besse, P.C., Guillouet, B., Loubes, J., Royer, F.: Destination prediction by trajectory distribution-based model. Trans. Intell. Transp. Syst. **19**(8), 2470–2481 (2018)
2. Chen, Y., Lv, P., Guo, D., Zhou, T., Xu, M.: A survey on task and participant matching in mobile crowd sensing. J. Comput. Sci. Technol. **33**(4), 768–791 (2018)
3. Chen, Y., et al.: Can sophisticated dispatching strategy acquired by reinforcement learning? In: AAMAS, pp. 1395–1403 (2019)
4. Epstein, R., et al.: A strategic empty container logistics optimization in a major shipping company. Interfaces **42**(1), 5–16 (2012)
5. Geng, X., et al.: Spatiotemporal multi-graph convolution network for ride-hailing demand forecasting. In: AAAI, pp. 3656–3663 (2019)
6. Kuhn, H.W.: The hungarian method for the assignment problem. Naval Res. Logist. Q. **2**(1–2), 83–97 (1955)
7. Li, M., Fan, S., Luo, A.: A partheno-genetic algorithm for combinatorial optimization. In: Pal, N.R., Kasabov, N., Mudi, R.K., Pal, S., Parui, S.K. (eds.) ICONIP 2004. LNCS, vol. 3316, pp. 224–229. Springer, Heidelberg (2004). https://doi.org/10.1007/978-3-540-30499-9_33
8. Li, X., Zhang, J., Bian, J., Tong, Y., Liu, T.: A cooperative multi-agent reinforcement learning framework for resource balancing in complex logistics network. In: AAMAS, pp. 980–988 (2019)
9. Lin, K., Zhao, R., Xu, Z., Zhou, J.: Efficient large-scale fleet management via multi-agent deep reinforcement learning. In: SIGKDD, pp. 1774–1783 (2018)
10. Tong, Y., et al.: The simpler the better: a unified approach to predicting original taxi demands based on large-scale online platforms. In: SIGKDD, pp. 1653–1662 (2017)
11. Tong, Y., She, J., Ding, B., Chen, L., Wo, T., Xu, K.: Online minimum matching in real-time spatial data: experiments and analysis. PVLDB **9**(12), 1053–1064 (2016)
12. Tong, Y., She, J., Ding, B., Wang, L., Chen, L.: Online mobile micro-task allocation in spatial crowdsourcing. In: ICDE, pp. 49–60 (2016)
13. Tong, Y., et al.: Flexible online task assignment in real-time spatial data. PVLDB **10**(11), 1334–1345 (2017)
14. Wang, Y., Tong, Y., Long, C., Xu, P., Xu, K., Lv, W.: Adaptive dynamic bipartite graph matching: a reinforcement learning approach. In: ICDE, pp. 1478–1489 (2019)
15. Wang, Z., Qin, Z., Tang, X., Ye, J., Zhu, H.: Deep reinforcement learning with knowledge transfer for online rides order dispatching. In: ICDM, pp. 617–626 (2018)
16. Xu, X., Hao, J., Yu, L., Deng, Y.: Fuzzy optimal allocation model for task-resource assignment problem in a collaborative logistics network. IEEE Trans. Fuzzy Syst. **27**(5), 1112–1125 (2019)
17. Xu, Z., et al.: Large-scale order dispatch in on-demand ride-hailing platforms: a learning and planning approach. In: SIGKDD, pp. 905–913 (2018)
18. Yuan, W., Chen, J., Cao, J., Jin, Z.: Forecast of logistics demand based on grey deep neural network model. In: ICMLC, pp. 251–256 (2018)
19. Zhang, L., et al.: A taxi order dispatch model based on combinatorial optimization. In: SIGKDD, pp. 2151–2159 (2017)
20. Zhou, C., Li, H., Liu, W., Stephen, A., Lee, L.H., Chew, E.P.: Challenges and opportunities in integration of simulation and optimization in maritime logistics. In: WSC, pp. 2897–2908 (2018)

Intention-Based Destination Recommendation
in Navigation Systems

Shuncheng Liu[1], Guanglin Cong[1], Bolong Zheng[2], Yan Zhao[3], Kai Zheng[1],
and Han Su[1(✉)]

[1] University of Electronic Science and Technology of China, Chengdu, China
{scliu,glcong,zhengkai,hansu}@uestc.edu.cn
[2] Huazhong University of Science and Technology, Wuhan, China
blzheng@hust.edu.cn
[3] Soochow University, Suzhou, China
yanzhao@suda.edu.cn

Abstract. Taking natural languages as inputs has been widely used in applications. Since the navigation application, which one of the fundamental and most used applications, is usually used during driving, thus the navigation application to take natural language as inputs can reduce the risk of driving. In reality, people use different ways to express their moving intentions. For example, 'I want to go to the bank' and 'I want to cash check' both reveal that the user wants to go to the bank. So we propose a new navigation system to take natural languages as inputs and recommend destinations to users by detecting users' moving intentions. The navigation system firstly utilizes a wealth of check-in data to extract corresponding words, including stuff and actions, for different types of locations. Then the extracted information is used to detect users' moving intentions and recommend suitable destinations. We formalize this task as a problem of constructing a model to classify users' input sentences into location types, and finding proper destinations to recommend. For empirical study, we conduct extensive experiments based on real datasets to evaluate the performance and effectiveness of our navigation system.

Keywords: Navigation system · Destination recommendation · Intention detectection

1 Introduction

Navigation applications that find optimal routes in road networks are one of the fundamental and most used applications in a wide variety of domains. Over the past several decades, the problem of computing the optimal path has been extensively studied and many efficient techniques have been developed, while the way of how users interact with navigation applications does not change a lot in the past years. Most navigation applications require users to input a specific source and a specific destination.

However, there is a very obvious trend in the IT industry. Many widely used applications take natural languages as inputs. That is to say, in the past, users need to tap words and click buttons as input, but now users only need to say their needs, and the application can respond accordingly. For example, we can use Siri to make phone calls

Y. Nah et al. (Eds.): DASFAA 2020, LNCS 12114, pp. 698–710, 2020.
https://doi.org/10.1007/978-3-030-59419-0_43

by simply saying 'call Michael'. Navigation applications are widely used during vehicle driving. At this time, it is very dangerous for users to use their hands to type in words on cellphones. So nowadays, a few navigation applications have followed the trend and taken natural language as inputs. As shown in Fig. 1(a), when a user says 'I want to find a bank', these navigation applications will recommend several nearby banks to the user. In this case, the user has specified the destination, i.e., bank. Meanwhile, users may only describe what they will do next, e.g., 'I want to cash check', without specifying the destination. Obviously, if the user wants to cash check, she needs to go to the bank. But current navigation systems fail to realize that the user intends to go to the bank and recommend nearby banks to the user, shown in Fig. 1(b). Meanwhile, with the ever-growing usage of check-in applications, e.g., Twitter and Foursquare, in mobile devices, a plethora of user-generated check-in data became available. This data is particularly useful for producing descriptions of locations. For example, users usually mention that they cash check in banks. Thus by mining the check-in data, we can know that 'cash check' has a strong correlation to 'bank' and the user may intend to go to the bank. So in this paper, we propose an intention-based destination recommendation algorithm, which detects users' moving intentions and recommends appropriate destinations to users, used in navigation systems.

(a) Recommedation Result with a Specific Destination   (b) Recommedation Result without a Specific Destination

Fig. 1. Existing navigation systems

In order to achieve intention-based destination recommendation for the users, there are some challenges we have to overcome. The first one is how to measure the correlations between stuff, actions and locations. The second one is how to detect a user's moving intention and correctly recommend locations to users. To address these challenges, we propose a two-phase framework, i.e., off-line category dictionary constructing and online recommendation. The off-line category dictionary constructing is to train a classifier model, and use a semantic word extractor to output a category dictionary. With the user's natural sentence t_{input} and her current location loc as input, the online recommendation module detects the moving intention by utilizing the trained classifier. And then the online module recommends a list of destinations to return the user.

In summary, the contributions of this paper are as follows:

- We propose a new navigation system to take natural languages as inputs and recommend destinations to users by detecting users' moving intentions. We formalize the intention-based destination recommendation problem, and then propose efficient algorithms to solve them optimally.
- We propose an intention-based destination recommendation algorithm, which detects users' moving intentions and recommends appropriate destinations to users. To the best of our knowledge, this paper pioneers to recommend destinations without specifying.
- We conduct extensive experiments based on real datasets, and as demonstrated by the results, our navigation system can accurately recommend destinations to users.

The remainder of the paper is structured as follows. Section 2 introduces the preliminary concepts and the work-flow of the proposed system. The intention-based destination recommendation algorithm is described in Sect. 3. The experimental studies are presented in Sect. 4. We review the related work in Sect. 5. Section 6 concludes the paper.

2 Problem Statement

In this section, we introduce preliminary concepts, and formally define the problem. Table 1 shows the major notations used in the rest of the paper.

Table 1. Summarize of notations

Notation	Definition	Notation	Definition
l	A POI in the space	$l.s$	The significance of a POI l
c	A venue category	$\mathbb{L}$	A list of l
t	A check-in data	V	A word vector
λ	A weight of linear combination	$D_{\mathbb{C}}$	The category dictionary
$F_{threshold}$	The filtering threshold	loc	A user's current location
$t_{input}.c$	The predicted category of a t_{input}	$t.c$	The category of a check-in data t
$\mathbb{T}_c$	A set of t with the same category c	t_{input}	A user-entered natural sentence

2.1 Preliminary

Definition 1 (POI). *A point of interest l, or POI, is a specific point location that someone may find useful or interesting.*

Definition 2 (Check-in data). *A Check-in data t is a short text that people announce their arrival at a hotel, airport, hospital, bank, or event.*

Definition 3 (Category). *A category c is a text label e.g.,Bank, Museum, Hotel, which refers to the Venue Category provided by Foursquare.*

In this work, we have nine root categories with one hundred and fifty leaf categories related to Foursquare Venue Categories. $\mathbb{C}$ represents the set of one hundred fifty nine category values obtained by adding nine root categories and one hundred fifty leaf categories. The root category is denoted by $c(R)$ and the leaf category is denoted by $c(L)$. e.g., $c(R)$ is Shop or Service then $c(L)$ is Bank. A POI l belongs to one or more categories, so we use $l.C$ to indicate all categories to which the POI l belongs. Since a t is a text that occurs at a specific POI, it has category $t.c$ corresponding to POI l.

Definition 4 (Category Dictionary). *A category dictionary D_c is a dictionary containing all words and corresponding to a category c.*

A D_c refers to a category and the corresponding dictionary content. Each word has a weight indicating its relevance to the category. e.g.,$D_{c=Bank}$ is money(0.74), deposit(0.63), withdraw(0.61) and so on. We use $D_{\mathbb{C}}$ to represent all dictionary contents.

Definition 5 (User Input). *A user input t_{input} is a natural sentence, which indicating a user's moving intention.*

In navigation applications, the input may or may not specify the destination. Previously, the task of natural sentence recognition including destination words has been well studied, so this article focuses on natural sentences t_{input} that do not contain destination words. Since we need to predict natural sentences that do not contain the destination vocabulary, we use $t_{input}.c$ to represent the category of text predicted by the system. $t_{input}.c$ here is formally consistent with c above.

From an overall perspective, we use check-in data t to achieve category dictionary $D_{\mathbb{C}}$, and then use the results of $D_{\mathbb{C}}$ to recommend a list of destinations $\mathbb{L}$ related to a user when the user enters t_{input} and location loc.

2.2 System Overview

In this section, we present the workflow of the recommendation system. Figure 2 presents the overview of the proposed system, which basically consists of four parts:

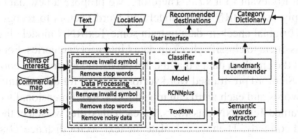

Fig. 2. System overview

data processing, classifier, landmark recommender and semantic words extractor. The system has two tasks.

The first task is an off-line task for constructing the category dictionary. The input of the whole task is the foursquare data set. First, we remove the illegal symbol, stop words and the noise data from the all check-in data t. Then we input the cleaned data into the classifier, tune and train the model, and save the best model parameters, the model outputs $t.c$ for each t. Next the semantic words extractor recognizes the correct predictions $(t,t.c)$, then aggregate them as $\mathbb{T}_c$ to extract keywords, finally generate the dictionary $D_{\mathbb{C}}$ as the output of the task.

The second task is an online task to recommend destinations. When the user enters t_{input} and loc, firstly the system has removed invalid symbols and stop words from the sentence t_{input}. Next, entering the cleaned t_{input} into the model trained by the off-line task. Then the model outputs the category of the natural sentence $t_{input}.c$. After that, the landmark recommender recommends the destinations $\mathbb{L}$ based on loc and $t_{input}.c$.

3   Destination Recommendation

In this section, we present our algorithm to explain how to extract user's moving intention based on natural sentences, including data clean algorithm, RCNN-plus classifier, and destination recommender. The user inputs a natural sentence t_{input} and the current location loc, then the system uses the above algorithms to output a recommended destination list $\mathbb{L}$. We will introduce these algorithms in the following subsections.

3.1   Data Clean Algorithm

With the ever-growing usage of check-in applications, e.g., Foursquare, a plethora of user-generated check-in data became available. The navigation system firstly utilizes a wealth of check-in data to extract corresponding words, including stuff and actions, for different types of locations. Textual content is the most frequently-used feature of check-in data, since users may mention semantic words while posting messages. However comment data may contain a lot of irrelevant characters including symbols, emotions, and URLs. Therefore, we need to do data preprocessing. We first remove irrelevant characters and the stop words. But we found that the check-in dataset contains a lot of noisy, which leads to a low accuracy classifying model. In other words, the t is ambiguous and irrelevant to it's $t.c$. Therefore, we propose a new data cleaning algorithm, we simply use TextRNN [6] model and word vectors to recognize noisy text. Firstly, we use original check-in data to train some TextRNN model classifiers, we use these rough classifiers to classify check-in data t into root categories. For each classifier, we filter out all the check-in data t that the TextRNN model predicts correctly, and prepare to use the Stanford pre-trained word vectors 'Glove' for data cleaning. The vectors contain 2.2M words and each word can transfer to a 300-dimensional word vector V.

After we trained TextRNN classifiers, for each classifier we select the check-in data t that predicted correctly. Then we calculate the text vector distance $Dis(V_t, V_c)$ based on Equation (1). In detail, the V_t represents the average of all word vectors contained in the t, similarly, the V_c represents the average of all word vectors contained in the $t.c$.

After that, we calculate the filtering threshold for each root category c using Equation (2). Each $F_{threshold}(c)$ represents how much noise the category c can tolerate. Next, we use $F_{threshold}(c)$ to clean all the original check-in data. According to Equation (1), we calculate the text vector distance between each t and $t.c$. When the distance value is greater than corresponding $F_{threshold}(c)$, the check-in data t should be deleted.

In summary, the algorithm filtered two parts of check-in data. The first part is misclassified check-in data, and the second part is check-in data that has a low correlation with its category. Therefore, the retained check-in data is of high quality.

$$Dis(V_t, V_c) = \|V_t - V_c\|_2 \qquad (1)$$

$$F_{threshold}(c) = \max_{\forall t.c \in c} Dis(V_t, V_c) \qquad (2)$$

3.2 TextRNN-RCNN Model

The RCNN [5] model is a text classifier with recurrent and convolutional neural network structure. It has a good effect on extracting key components in the text. The TextRNN [6] is a text classifier with recurrent structure. It has a good effect on obtaining context information. In order to leverage the power of both RCNN and TextRNN, we improve them and propose a joint model named *TextRNN-RCNN*.

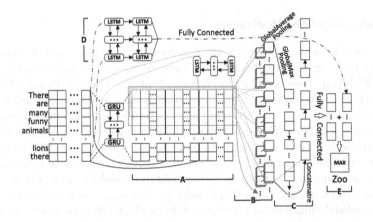

Fig. 3. TextRNN-RCNN model

For RCNN, it uses a bi-directional recurrent structure to capture the contextual information, and use the convolutional neural network to capture the semantic information of texts. [5] adds a max-pooling layer to capture the key component in the texts which can automatically judge the key features in text classification. Since the RCNN only uses one single max-pooling layer, under the 'without destination words' scenario the RCNN model is not comprehensive enough in feature selection. Besides, the capability of the model to capture semantic information features and the key components in the texts is insufficient. In order to improve the RCNN model, we modify the network structure and proposed a new model named RCNN-plus.

As Fig. 3 shown, firstly we delete the left context and right context structure to simplify the model. Then we modify the RCNN recurrent structure, we use the bidirectional GRU and bidirectional LSTM simultaneously as the bi-directional recurrent structure to capture the contextual information as much as possible. We concatenate the output of GRU, embedding layer and LSTM, shown in part A of Fig. 3. Besides, for convolutional Neural Network, we replace a single convolution kernel structure with a multi-convolution structure in RCNN-plus model, and use different convolution kernels to represent different word selection window widths, as shown in part B of Fig. 3. Convolution kernels of different colors represent windows of different lengths. The amount of convolution kernels depends on different text classification tasks. In order to capture key components in the texts with different sizes, shown in part C of Fig. 3, we perform maximum pooling and average pooling operations on all convolution kernels' results, while the RCNN model only uses a max-pooling layer to capture the key component. After that, we concatenate and input pooling results to the fully connected layer.

Since the above structure pays more attention to the key components in the texts essentially, we use the TextRNN to better obtain context information. As shown in part D of Fig. 2, we add a TextRNN to improve the generalization ability of the model, since it has a two-layer bidirectional LSTM structure. Then we use a threshold λ to linear combine two outputs and weigh them shown in part E of Fig. 2. Finally, we choose the maximum probability as the prediction.

3.3 Select Significant Destinations

In our application scenario, when the user provides a complete text, the model can calculate the corresponding POI category. Then the category needs to be replaced with certain POIs (e.g., Bank is replaced with Bank of America Financial Center). Therefore, we need to recommend proper destinations based on the selected category and the user's current location. It is a common sense that destinations have different significances. For instance, the White House is famous globally, but Pennsylvania Ave, where the White House is located, is only known to the locals in Washington DC. The selected destinations should have a high significance, so that users can find it easily. By regarding the users as authorities, destinations as hubs, and check-ins/visits as hyperlinks, we can leverage a HITS-like algorithm [16] to infer the significance of a destination $l.s$. The recommended destinations list $\mathbb{L}$ contains all POIs which belong to the certain category within a distance to the user's current location and are sorted by significance $l.s$.

4 Experiment

In this section, our experimental results are presented to evaluate the performance of model. This model is performed using Java and Python on Ubuntu 16.04. All the experiments are conducted on a computer with Intel Core i7-4770K(3.9GHz) CPU and 16GB memory.

4.1 Experimental Setup

Check-in Dataset: We use two check-in datasets (New York and Los Angeles) from Foursquare to carry out the experiments. The New York dataset contains 1,711,646 check-in texts, while the Los Angeles dataset contains 1,921,214 check-in texts.

POI Dataset: We have two POI databases. There are approximately 300 thousand POIs in New York and 500 thousand in Los Angeles.

Category Dataset: We have a category dataset. It contains nine root categories and 900 leaf categories. Among the 900 leaf categories, 150 leaf categories are tagged in a high frequency while the rest 750 leaf categories are tagged in a low frequency.

Baseline Methods: We compare our proposed TextRNN-RCNN method and text selecting method with the following algorithms:

- CNN-sentence [3] is a simple CNN with one layer of convolution on top of word vectors obtained from an unsupervised neural language model.
- RCNN [5] judges which words play key roles in text classification. It captures contextual information in texts with the recurrent structure and constructs the representation of text using a convolution neural network.
- TextRNN [6] assigns a separate LSTM layer for each task. Meanwhile, it introduces a bidirectional LSTM layer to capture the shared information for all tasks.
- TextAttBiRNN [10] is a simplified model of attention that is applicable to feedforward neural networks. The model demonstrates that it can solve the synthetic 'addition' and 'multiplication' long-term memory problems for sequence lengths which are both longer and more widely varying than the best-published results for these tasks.

Parameters Setting: The penalty for TextRNN-RCNN λ is set to 0.5 in our method and the number of RCNN-plus kernels is set to 5. In the following experiments, parameters are set to default values if unspecified.

4.2 Case Study

Before assessing the performance of the system, we perform a case study to show the effectiveness of intention-base destination recommendation. Figure 4.2 presents a case study where we input a sentence and get a POI recommendation. In this case, a user said 'I will deposit money' and the navigation system took the sentence as the input. Then the system removed invalid symbols and stop words of the sentence and got words 'deposit money'. After that, the system utilized the classifier to identify which category the words belonged to, i.e., category 'bank'. At last, the system recommended nearby POIs, which belongs to the category 'bank', to the users (Fig. 4).

Besides, we perform another case study to show the effectiveness of semantic words extraction. After training the check-in data, we can get a category dictionary that contains all words, including stuff and actions, corresponding to each category. Table 2 shows some categories, e.g.,'bank', 'zoo', 'bookstore', 'bar' and 'Asian restaurant', and their corresponding semantic words. For instance, for category 'bank', words such as 'deposit', 'money', 'credit', 'debit', 'withdraw'and 'ATM' have strong correlation to 'bank'.

Fig. 4. An example of intention-based destination recommendation navigation system

Table 2. Examples of the category dictionary

Category	Semantic words
Bank	Deposit,money,credit,debit,withdraw,ATM
Zoo	Animal,lion,monkey,feed,baboon,aquarium
Bookstore	Book,read,magazine,comic,section,literature
Bar	Cocktail,wine,whiskey,karaoke,beer,bartender
Asian restaurant	Noodle,dumpling,tofu,sushi,rice,tea

4.3   Performance Evaluation

In this subsection, we test the methods proposed previously by collecting data from foursquare. To make the evaluation purpose, we train different models to find categories of check-in data and user inputs t_{input}, and evaluate their performance by their accuracies. Besides, we do ablation experiments about λ to evaluate RCNN-plus and TextRNN. At last, we do another ablation experiments to evaluate the data-filtering method.

Model Accuracy. As discussed above, we train 80% of check-in data to generate a category classifier. After training, this classifier takes a check-in data t as its input, and then outputs the corresponding check-in category $t.c$ of t. We use the rest 20% of check-in data to test the effectiveness of the model. A good model should have a high accuracy of classifying test check-ins into correct check-in categories. So in this set of experiments, we calculate the accuracies of our algorithm (TextRNN-RCNN) with comparison methods (CNN-sentence, RCNN, TextRNN, TextAttBiRNN, RCNN-plus). Figure 5(a)–5(b) shows the accuracies of the root classifier and the accuracies of the leaf classifier. As shown in figures, TextRNN-RCNN outperforms all the comparison methods in both the root classifier and the leaf classifier, which demonstrates the superiority of our approach. Two TextRNN related methods, i.e., TextRNN and TextRNN-RCNN, have higher accuracies than others. It indicates that this model is more suitable in this application scenario since it is sensitive to obtain context information.

Effect of λ: For the joint model TextRNN+RCNN, RCNN-plus model pays more attention to key components in the texts essentially, and TextRNN has a good effect on obtaining context information. We use λ to linearly combine model RCNN-plus

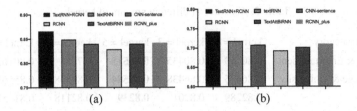

Fig. 5. (a)Accuracies of root classifier; (b) Accuracies of leaf classifier

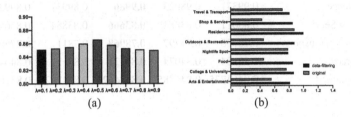

Fig. 6. (a)Accuracies of different λ; (b) Effectiveness of data filtering

and model TextRNN, and a bigger λ indicates a higher weight of TextRNN in the joint model. In this set of experiments, we tune λ from 0.1 to 0.9 with a step of 0.1. Figure 6(a) shows accuracies of different λ of the joint model. As shown in the figure, with the increase of λ, the accuracy of the joint model initially rises and subsequently falls. And the highest accuracy is obtained when λ is 0.5.

Effectiveness of Different Amounts of Kernels: RCNN-plus plays an important role in the joint model. The amount of the convolution kernels may affect the effectiveness of RCNN-plus, since the amount of the convolution kernel determines the amount of the sliding window. Specifically, the number of convolution kernels represents the incremental convolution kernel size. For example, kernel=3 means that the convolution kernel size from 1 to 3 are used simultaneously. In this set of experiments, we compare the accuracies of different amounts of kernels on root categories. As shown in Table 3, when kernel equals to 5, the RCNN-plus outperforms others in most cases. This is because with the rising of the kernel amounts, more continuous features can be extracted; while too many kernels are utilized, it may cause overfitting.

Effectiveness of Data Cleaning. We proposed a data clean algorithm in Sect. 3.1, that all experiments are conducted on cleaned data. So in this set of experiments, we evaluate the effectiveness of our data clean algorithm. We conduct a training process on both raw check-in data and cleaned check-in data and compare the accuracies of these two classifiers. Figure 6(b) shows the accuracies of leaf classifier. In this experiment, we can find that all the accuracies of cleaned data are higher than those of raw data. It indicates that our data clean algorithm is effective for this system.

Table 3. Accuracy of different amount of kernel

Root category	kernel = 1	kernel = 3	kernel = 5	kernel = 7	kernel = 9
Arts & Entertainment	0.80463	0.80435	**0.80578**	0.7988	0.80162
College & University	0.86475	0.86438	**0.86584**	0.86548	0.85563
Food	0.82289	0.8201	**0.8239**	0.82118	0.81852
Nightlife Spot	0.78391	**0.78785**	0.78023	0.77881	0.77361
Outdoors & Recreation	0.84325	**0.85297**	0.85158	0.84964	0.8527
Residence	0.99371	0.98023	**0.9986**	0.89937	0.89937
Shop & Service	0.83611	0.83884	**0.83666**	0.83884	0.83666
Travel & Transport	0.78736	0.77797	**0.78968**	0.7811	0.77594
Accuracy	0.83984	0.84074	**0.84121**	0.84	0.84048
Fine-grained	0.8464775	0.845385	**0.8484625**	0.8343025	0.83161375

5 Related Work

The problem referred to in this work is relevant to text processing and POI querying issues, including text classification, semantic location label mining, location recommendation and so on.

Text Classification. Given a set of texts, [5] uses recurrent structure to captures contextual information and uses convolutional neural network to constructs the representation of text. [6] proposes three multi-task architectures for RNN which learns to map the arbitrary text into semantic vector representations with both task-specific and shared layers. [6] uses a simple convolutional neural networks (CNN) structure trained on top of pre-trained word vectors for sentence-level classification tasks. On the contrary, in our work, an improved TextRNN-RCNN model is used for text classification tasks to improve accuracy.

Semantic Location Label Mining. Semantic place labeling is the process of defining a location with a category. Place labels have been proposed for automatically updating a person's status on social networking sites, such as [7], and automatically annotating check-ins, such as [15]. [4] develop a classifier that identifies place labels are based on the timing of visits to that place, nearby businesses, and simple demographics of the user. Several researchers use [9,13,14] nearby points of interests to characterize places. In addition, our focus is on mining semantic words corresponding to location label and we mine geographic categories from text features instead of using users' spatiotemporal features.

Location Recommendation. In different application scenarios, the system needs to recommend a series of locations according to the user's location and query category. [16] proposed a HITS-based model to infer users' travel experiences and the relative interest of a location. [2,12] build a recommender system by using multiple users' real-world location histories to recommend geographic locations. For users who need real-time recommendations, [1,8,11] typically recommend locations based on a user's real-time location.

6   Conclusions

In this paper, we have proposed a new navigation system based on check-in data, which can take natural language as inputs and recommend appropriate destinations to users. We propose a two-phase framework, i.e., off-line category dictionary constructing and online recommendation. The off-line category dictionary constructing is to train the classifier model, and use a semantic word extractor to output a category dictionary. The online recommendation module detects the moving intention by utilizing the trained classifier. And then, the online module recommends a list of destinations to return the user. We have conducted extensive experiments on real datasets. As demonstrated by the experimental results, in most cases, our navigation system can accurately recommend destinations to users.

Acknodgements. This work is partially supported by Natural Science Foundation of China (No. 61802054, 61972069, 61836007, 61832017, 61532018, 61902134), Alibaba Innovation Research (AIR) and the Fundamental Research Funds for the Central Universities (HUST: Grants No. 2019kfyXKJC021, 2019kfyXJJS091)

References

1. Abowd, G.D., Atkeson, C.G., Hong, J., Long, S., Kooper, R., Pinkerton, M.: Cyberguide: a mobile context-aware tour guide. Wirel. Netw. **3**(5), 421–433 (1997)
2. Horozov, T., Narasimhan, N., Vasudevan, V.: Using location for personalized poi recommendations in mobile environments. In: International Symposium on Applications and the Internet (SAINT 2006), p. 6-pp. IEEE (2006)
3. Kim, Y.: Convolutional neural networks for sentence classification. arXiv preprint arXiv:1408.5882 (2014)
4. Krumm, J., Rouhana, D.: Placer: semantic place labels from diary data. In: Proceedings of the 2013 ACM International Joint Conference on Pervasive and Ubiquitous Computing, pp. 163–172. ACM (2013)
5. Lai, S., Xu, L., Liu, K., Zhao, J.: Recurrent convolutional neural networks for text classification. In: Twenty-Ninth AAAI Conference on Artificial Intelligence (2015)
6. Liu, P., Qiu, X., Huang, X.: Recurrent neural network for text classification with multi-task learning. arXiv preprint arXiv:1605.05101 (2016)
7. Miluzzo, E., Lane, N.D., Eisenman, S.B., Campbell, A.T.: CenceMe – injecting sensing presence into social networking applications. In: Kortuem, G., Finney, J., Lea, R., Sundramoorthy, V. (eds.) EuroSSC 2007. LNCS, vol. 4793, pp. 1–28. Springer, Heidelberg (2007). https://doi.org/10.1007/978-3-540-75696-5_1
8. Park, M.-H., Hong, J.-H., Cho, S.-B.: Location-based recommendation system using Bayesian user's preference model in mobile devices. In: Indulska, J., Ma, J., Yang, L.T., Ungerer, T., Cao, J. (eds.) UIC 2007. LNCS, vol. 4611, pp. 1130–1139. Springer, Heidelberg (2007). https://doi.org/10.1007/978-3-540-73549-6_110
9. Phithakkitnukoon, S., Horanont, T., Di Lorenzo, G., Shibasaki, R., Ratti, C.: Activity-aware map: identifying human daily activity pattern using mobile phone data. In: Salah, A.A., Gevers, T., Sebe, N., Vinciarelli, A. (eds.) HBU 2010. LNCS, vol. 6219, pp. 14–25. Springer, Heidelberg (2010). https://doi.org/10.1007/978-3-642-14715-9_3
10. Raffel, C., Ellis, D.P.: Feed-forward networks with attention can solve some long-term memory problems. arXiv preprint arXiv:1512.08756 (2015)

11. Simon, R., Fröhlich, P.: A mobile application framework for the geospatial web. In: Proceedings of the 16th International Conference on World Wide Web, pp. 381–390. ACM (2007)

12. Takeuchi, Y., Sugimoto, M.: CityVoyager: an outdoor recommendation system based on user location history. In: Ma, J., Jin, H., Yang, L.T., Tsai, J.J.-P. (eds.) UIC 2006. LNCS, vol. 4159, pp. 625–636. Springer, Heidelberg (2006). https://doi.org/10.1007/11833529_64

13. Wolf, J., Guensler, R., Bachman, W.: Elimination of the travel diary: an experiment to derive trip purpose from GPS travel data. In: Transportation Research Board 80th Annual Meeting, pp. 7–11 (2001)

14. Xie, R., Luo, J., Yue, Y., Li, Q., Zou, X.: Pattern mining, semantic label identification and movement prediction using mobile phone data. In: Zhou, S., Zhang, S., Karypis, G. (eds.) ADMA 2012. LNCS (LNAI), vol. 7713, pp. 419–430. Springer, Heidelberg (2012). https://doi.org/10.1007/978-3-642-35527-1_35

15. Ye, M., Shou, D., Lee, W.-C., Yin, P., Janowicz, K.: On the semantic annotation of places in location-based social networks. In: Proceedings of the 17th ACM SIGKDD International Conference on Knowledge Discovery and Data Mining, pp. 520–528. ACM (2011)

16. Zheng, Y., Zhang, L., Xie, X., Ma, W.-Y.: Mining interesting locations and travel sequences from GPS trajectories. In: World Wide Web, pp. 791–800. ACM (2009)

Towards Accurate Retail Demand Forecasting Using Deep Neural Networks

Shanhe Liao, Jiaming Yin, and Weixiong Rao$^{(\boxtimes)}$

School of Software Engineering, Tongji University, Shanghai, China
{shhliao,14jiamingyin,wxrao}@tongji.edu.cn

Abstract. Accurate product sales forecasting, or known as demand forecasting, is important for retails to avoid either insufficient or excess inventory in product warehouse. Traditional works adopt either univariate time series models or multivariate time series models. Unfortunately, previous prediction methods frequently ignore the inherent structural information of product items such as the relations between product items and brands and the relations among various product items, and cannot perform accurate forecast. To this end, in this paper, we propose a deep learning-based prediction model, namely Structural Temporal Attention network (STANet), to adaptively capture the inherent inter-dependencies and temporal characteristics among product items. STANet uses the graph attention network and a variable-wise temporal attention to extract inter-dependencies among product items and to discover dynamic temporal characteristics, respectively. Evaluation on two real-world datasets validates that our model can achieve better results when compared with state-of-the-art methods.

Keywords: Demand forecasting · Multivariate time series · Graph attention network · Attention mechanism

1 Introduction

Demand forecasting aims to predict future product sales. Accurate forecasting in supermarkets is important to avoid either insufficient or excess inventory in product warehouses. Traditional works adopt either univariate time series models or multivariate time series models [19]. The univariate time series models, such as the autoregressive integrated moving average (ARIMA) [2], autoregression (AR), moving average (MA) and autoregressive moving average (ARMA), treat different product items separately. ARIMA is rather time consuming especially to process a large amount of product items. It is mainly because the model selection procedure, e.g., by Box-Jenkins methodology [2], is performed in a grid search manner. In addition, ARIMA assumes that the current value of time series is a linear combination of historical observations and a random noise. It is hard for ARIMA to capture non-linear relationships and inter-dependencies of different products. Some machine learning models can also be applied to solve

© Springer Nature Switzerland AG 2020
Y. Nah et al. (Eds.): DASFAA 2020, LNCS 12114, pp. 711–723, 2020.
https://doi.org/10.1007/978-3-030-59419-0_44

the demand forecasting problems, such as linear regression and linear support vector regression (SVR) [3]. Nonetheless, these machine learning models suffer from the similar weaknesses as ARIMA.

Consider that retail markets typically sell at least thousands of product items. Multivariate time series models are useful to overcome the issues of univariate time series models by taking into account inter-dependencies among product items. For example, as an extension of ARIMA, vector autoregression (VAR) [2, 12] can handle multivariate time series. However, the model capacity of VAR grows linearly over temporal window size and quadratically over the number of variables, making it hard to model thousands of products with a long history.

More recently, deep learning models have demonstrated outstanding performance in time series forecasting problems. There are basic recurrent neural network (RNN) models [7] and its variants including long short-term memory (LSTM) network [8,13] and gated recurrent unit (GRU) [4,6]. Besides, convolutional neural network (CNN) models [11,18] have shown outstanding performance by successfully extracting local and shift-invariant features in the fields of computer vision and time series modeling. The recent work LSTNet [10] combines CNN and GRU to perform multivariate time series forecasting. The LSTNet model uses CNN to extract short-term local dependency patterns among variables and uses a special recurrent-skip component to capture very long-term periodic patterns. However, it assumes that all variables in the multivariate time series have same periodicity, which is usually invalid for real datasets.

Beyond aforementioned weakness, the prediction methods above ignore the inherent structural information of product times, such as the relations between product items and brands, and the relations among various product items (which may share the same multi-level categories). All multivariate time series prediction models above simply try to learn the implicit dependencies among variables.

Our work is motivated by a previous work [5] which segments a set of product items into several clusters with help of a so-called *product tree*. This tree structure takes product categories as internal nodes and product items as leaf nodes. We extend the product tree by incorporating product brands and then construct a product graph structure. This structure explicitly represents the structural information of product items. Figure 1 illustrates an example of the graph structure of four product items. We can easily find that the brand *Master Kong* has three products, which belong to two different subcategories. Consider that a customer prefers the brand *Master Kong* and recently bought a product item *Master Kong Jasmine Tea*. It is reasonable to infer that he will try another product item *Master Kong Black Tea*, especially when *Master Kong Black Tea* involves a sale promotion campaign. Without the structural information as prior, previous methods either treat all product items equally or have to implicitly infer the inherent relationship but at the cost of accuracy loss.

To overcome the issues above, we propose a new deep-neural-network (DNN)-based retail demand prediction model in a multivariate time series forecasting manner, called Structural Temporal Attention Network (STANet). This network incorporates both the product graph structure (see Fig. 1) and temporal

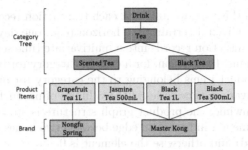

Fig. 1. An example of product structural graph with categories, items and brands.

characteristics of product items. In particular, we note that the inter-dependencies of products and temporal dependencies (e.g., temporal periodicity) may change over time. Thus, we leverage attention mechanism [15] to deal with these variations. In this way, STANet assigns various weights with respect to different inputs involving the variations. Based on graph attention network (GAT) [16], GRU, and a special variable-wise temporal attention, STANet performs better than existing methods. As a summary, we make the following contributions.

- We perform adequate analysis on real datasets to motivate our approach, including product structural inter-dependencies and temporal dependencies.
- We leverage GAT to capture the product structural inter-dependencies and GRU to capture temporal patterns. Moreover, a variable-wise temporal attention mechanism is adopted on the hidden states of GRU to deal with diverse temporal characteristics for different products. Thus, the two attention mechanisms, i.e., GAT and variable-wise temporal attention, can work together to comfortably learn the product structural inter-dependencies and temporal dependencies.
- Evaluations on two real-world sales datasets show that STANet achieves the best results compared with several state-of-the-art methods. Ablation experiments validate that the structural information of products and individualized processing for different items indeed improve forecasting accuracy.

The rest of this paper is organized as follows. Section 2 gives the problem formulation, and Sect. 3 describes the proposed approach STANet. After that, Sect. 4 reports evaluation results on two real-world datasets. Finally, Sect. 5 concludes the paper.

2   Problem Formulation

We consider a data set of transaction records in a supermarket. Each transaction record contains 3 fields: transaction timestamp, item ID, and amount of sold items. With help of the item ID, we can find a list of product categories (in our dataset, each product item is with a list of 4-level categories) and an associated product brand. In this way, we augment each transaction record by totally 8 (= 3 + 4 + 1) fields. Given a certain time horizon (e.g., one day or one week), we pre-process the transaction records into a multivariate time series of the volumes of sold product items. In addition, for a certain category (or brand), we sum the volumes of all product items belonging to the category (or brand). In this way, we have the multivariate time series of the volumes of product items, categories, and brands. Meanwhile, the product graph structure is stored in an adjacency matrix, where element 1 indicates an edge between two nodes (such as a product item and its brand) and otherwise the element is 0.

Formally, we denote the number of products by N_p and the total number of products, brands and categories by N. Given the augmented multivariate time series $X = [x_1, x_2, \ldots, x_T]$, where $x_t \in \mathbb{R}^{N \times 1}$, $t = 1, 2, \ldots, T$, and the adjacency matrix $M \in \mathbb{R}^{N \times N}$ that represents the product graph structure (for simplicity, we assume that the adjacency matrix M is static), we aim to predict future product sale volume x_{T+h} where h is the desirable horizon ahead of the current time stamp. The demand forecasting problem is equivalent to learning a function $f_M : \mathbb{R}^{N \times \tau} \to \mathbb{R}^{N \times 1}$. More specifically, to learn the function f_M, we use a time window of size τ to split training data into fixed length inputs $[x_t, x_{t+1}, \ldots, x_{t+\tau-1}]$ and corresponding labels $x_{t+\tau-1+h}$. Then the function f_M is learned from the inputs $[x_t, x_{t+1}, \ldots, x_{t+\tau-1}]$ to the labels $x_{t+\tau-1+h}$. Note that to evaluate the performance of the learned model, we only focus on these N_p product items, instead of the entire N product items, brands and categories.

3   Framework

In this section, we present the detail of the proposed model STANet. Figure 2 gives the framework of STANet.

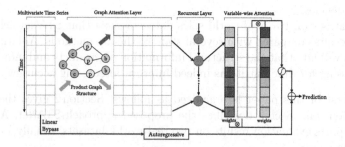

Fig. 2. The framework of STANet.

3.1   Graph Attention Component

For multivariate time series forecasting, one of the important tasks is to precisely capture the inter-dependencies between different variables. The challenge is that the inter-dependency weight between different variables may evolve over time, instead of a static weight. To explore inter-dependencies, we first build a graph neural network (GNN) on top of the product graph structure. Next, to learn dynamic inter-dependencies, we incorporate attention mechanism into the GNN, leading to a graph attention layer. In this way, we can assign various weights depending upon the input time series.

Given the input time series $X \in \mathbb{R}^{N \times \tau}$ and adjacency matrix $M \in \mathbb{R}^{N \times N}$, we use a multi-head graph attention layer to process X by time step. Formally, the graph hidden state of node i at time step t is given by

$$h_{gt}^i = \sigma \left(\frac{1}{K} \sum_{k=1}^{K} \sum_{j \in \mathcal{N}_i} \alpha_{ij}^k x_t^j W^k \right), \tag{1}$$

In the equation above, x_t^j is the sale volume of product item (or brand, or category) j at time step t, W^k is a learnable linear transformation in the graph attention layer to obtain sufficient expressive power, $\mathcal{N}_i$ refers to all the adjacent nodes of variable i in M, and K is the number of multi-head attentions and σ is an activation function. Here, we calculate the average of K multi-head attention mechanisms. Finally, the parameter α_{ij}^k is coefficients of attention mechanism computed by

$$\alpha_{ij}^k = \frac{exp \left(\text{LeakyReLU} \left(f_a \left(x_t^i W^k, x_t^j W^k \right) \right) \right)}{\sum_{\ell \in \mathcal{N}_i} exp \left(\text{LeakyReLU} \left(f_a \left(x_t^i W^k, x_t^\ell W^k \right) \right) \right)}, \tag{2}$$

where f_a is a scoring function to evaluate the relevance of two variables. It can be either a determinant function (e.g., cosine similarity) or a learnable component. In our model, it is a single-layer feed forward neural network.

Following the two equations above, for each time step of τ, given the input $x_t \in \mathbb{R}^{N \times 1}$ and $W^k \in \mathbb{R}^{1 \times F}$, we then generate the output $h_{gt} \in \mathbb{R}^{N \times F}$. After that, we flatten the output as $h'_{gt} \in \mathbb{R}^{NF \times 1}$. By repeating this step by τ times to process the input multivariate time series $X \in \mathbb{R}^{N \times \tau}$ and the matrix M, the final output of the graph attention component is $X^G \in \mathbb{R}^{NF \times \tau}$.

3.2   Recurrent Component

The 2nd component of **STANet** is a recurrent component to capture temporal patterns from X^G that is the output of the aforementioned graph attention component. Here, we use the gated recurrent unit (GRU) [6] as the recurrent layer to more capture long-term patterns, when compared with vanilla recurrent neural networks (RNN). The hidden state of recurrent units at time step t can be computed by the following equations:

$$r_t = \sigma \left(W_{xr} x_t^G + W_{hr} h_{t-1} + b_r \right), \tag{3}$$

$$z_t = \sigma \left(W_{xz} x_t^G + W_{hz} h_{t-1} + b_z \right), \tag{4}$$

$$c_t = tanh \left(W_{xc} x_t^G + W_{hc} \left(r_t \odot h_{t-1} \right) + b_c \right), \tag{5}$$

$$h_t = (1 - z_t) \odot h_{t-1} + z_t \odot c_t. \tag{6}$$

716   S. Liao et al.

Suppose the hidden size of GRU is d_r, then $x_t^G \in \mathbb{R}^{NF \times 1}$, $h_{t-1} \in \mathbb{R}^{d_r \times 1}$, W_{xr}, W_{xz} and $W_{xc} \in \mathbb{R}^{d_r \times NF}$, W_{hr}, W_{hz} and $W_{hc} \in \mathbb{R}^{d_r \times d_r}$, b_r, b_z and $b_c \in \mathbb{R}^{1 \times 1}$, r_t, z_t, c_t and $h_t \in \mathbb{R}^{d_r \times 1}$. The final output of the recurrent component is $X^R \in \mathbb{R}^{d_r \times \tau}$.

3.3   Variable-Wise Temporal Attention Component

After the graph attention component and recurrent component have successfully captured the inter-dependencies and basic temporal patterns, we then introduce a temporal attention into STANet to capture dynamic temporal pattern.

$$\alpha_{\tau-1} = f_a\left(H_{\tau-1}, h_{\tau-1}\right), \tag{7}$$

where $\alpha_{\tau-1} \in \mathbb{R}^{\tau \times 1}$, f_a is a scoring function and $h_{\tau-1}$ is the last hidden state of RNN, and $H_{\tau-1} = [h_0, h_1, \ldots, h_{\tau-1}]$ is a matrix stacking the hidden states of RNN.

Note that in real dataset, various product items may exhibit rather different temporal characteristics such as periodicity (that will be verified in Sect. 4.1). Thus, instead of using the same attention mechanism for all product items, we propose a variable-wise temporal attention mechanism to compute the attention coefficients for each variable independently as

$$\alpha_{\tau-1}^i = f_a\left(H_{\tau-1}^i, h_{\tau-1}^i\right). \tag{8}$$

Equation (8) is very similar to Eq. (7), except a superscript $i = 1, 2, \ldots, d_r$, indicating that the attention mechanism is calculated for a particular GRU hidden variable. In this way, our model could deal with different temporal characteristics such as periodicity for different product items (or brands, or categories). With the coefficients $\alpha_{\tau-1}^i$, we then compute a weighted context vector $c_{\tau-1}$, where each vector member $c_{\tau-1}^i$ is with respect to the ith hidden variable:

$$c_{\tau-1}^i = H_{\tau-1}^i \alpha_{\tau-1}^i, \tag{9}$$

where $H_{\tau-1}^i \in \mathbb{R}^{1 \times \tau}$ and $\alpha_{\tau-1}^i \in \mathbb{R}^{\tau \times 1}$. Given the context vector $c_{\tau-1} \in \mathbb{R}^{d_r \times 1}$ of all hidden variables, the final output of variable-wise temporal attention layer is the concatenation of the weighted context vector $c_{\tau-1}$ and last hidden state $h_{\tau-1}$, along with a linear projection operation. We then give the final output as

$$\hat{y}_{\tau-1}^D = W\left[c_{\tau-1}; h_{\tau-1}\right] + b, \tag{10}$$

where $W \in \mathbb{R}^{N \times 2d_r}$, $b \in \mathbb{R}^{1 \times 1}$ and $\hat{y}_{\tau-1}^D \in \mathbb{R}^{N \times 1}$.

3.4   Autoregressive Component

Finally, consider that the outputs scale of the neural network model is insensitive to its inputs scale, which may lead to inaccurate forecasting results for sudden changes. To this end, similar to LSTNet [10], we add an autoregressive component to capture the local trend. This component is a linear bypass that predicts future sales directly from origin input data to address the scale problem. This linear bypass will fit the historical data of all product items with a single linear layer as follows,

$$\widehat{y}_{\tau-1}^{L} = X'W_{ar} + b_{ar}, \tag{11}$$

where $X' \in \mathbb{R}^{N \times \tau'}$ is the latest τ' time steps' historical data of all product items, and $W_{ar} \in \mathbb{R}^{\tau' \times 1}$, $b_{ar} \in \mathbb{R}^{1 \times 1}$, $\tau' \leq \tau$.

The final prediction of STANet is then obtained by integrating the outputs of the neural network part and the autoregressive component using an automatically learned weight:

$$\widehat{y}_{\tau-1} = \widehat{y}_{\tau-1}^{D} + \widehat{y}_{\tau-1}^{L}, \tag{12}$$

where $\widehat{y}_{\tau-1}$ denotes the final prediction for horizon h ahead of the current time stamp using historical data of time window size τ.

4 Experiments and Evaluations

We first analyze two real-world datasets to motivate STANet, then compare STANet against 5 counterparts to show its superiority, and finally give an ablation study to study the effect of the components in STANet.

4.1 Datasets Description and Analysis

We use two real-world datasets collected from two medium size stores of a chain retail in Shandong Province, China. Table 1 summarizes the statistics.

Table 1. Statistics of two real data sets. T is the length of daily time step, P is the time interval, N_p, N_b and N_c are the numbers of product items, brands and categories, respectively, $N = N_p + N_b + N_c$, and sparsity means the proportion of empty values in the datasets.

Datasets	T	N_p	N_b	N_c	N	Sparsity
Dataset-1	572	1878	433	612	2923	53%
Dataset-2	833	1925	289	771	2985	39%

In Table 1, both dataset a non-trivial number of empty values, indicating that no sales for a certain product item within a time step. Such empty values make forecasting tasks rather challenging. Both datasets are split into training set (70%), validation set (15%) and testing set (15%) in chronological order. To explore the inter-dependencies and temporal characteristics of datasets, we give the following analyses.

Dynamic Inter-dependencies. We consider that two variables may have inter-dependencies if historical data of one can help forecast the other. Assuming two univariate time series $x = \{x_1, x_2, \ldots, x_T\}$, $y = \{y_1, y_2, \ldots, y_T\}$, and a specific time lag m, we model y as a regression of itself and x:

$$y_t = a_0 + a_1 y_{t-1} + \ldots + a_m y_{t-m} + b_1 x_{t-1} + \ldots + b_m x_{t-m}. \tag{13}$$

If y gets the best fitting when all $b_i = 0$ for $i = 1, 2, \ldots, m$, we believe x cannot help forecast y, in other words y has no dependency on x. To test whether x can help forecast y, we could leverage the Granger causality test [1], and for each lag m the result maybe differ. We use

$$GR_{x,y} = \frac{\text{number of } m \text{ where } x \text{ helps forecast } y}{\text{total number of } m} \tag{14}$$

to represent the importance of x to y, $GR_{x,y} \in [0, 1]$. We select one category (black tea) and two concrete product items (*Master Kong Black Tea* 500 mL and *Master Kong Black Tea* 1 L) to verify the inter-dependencies in Fig. 3.

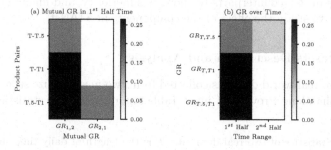

Fig. 3. GR among three variables, where $T.5$ represents *Black Tea* 500 mL, $T1$ is *Black Tea* 1 L and T is category of *Black Tea*. Time range is split into two parts in chronological order. (a) Mutual GR of three pairs in first half time. (b) Dynamic change of inter-dependencies over time.

As we can see from Fig. 3(a), $G_{T,T.5}$ and $G_{T,T1}$ are quite darker while adverse $G_{T.5,T}$ and $G_{T1,T}$ are almost white, means that historical data of the category *Black Tea* helps forecast sales of its children products, while children's history is of little importance to help forecast the volume of this category. $G_{T.5,T1}$ and $G_{T1,T.5}$ are in different degree of dark, which means they help each other when forecasting, but the importance is not equal. This is reasonable because a customer who purchased *Black Tea* 500 mL hardly purchases *Black Tea* 1 L immediately. There are competitive relationship between these two product items. Figure 3(b) shows that inter-dependency is dynamically changing over time. For example, history of *Black Tea* 500 mL helps forecast sale of *Black Tea* 1 L in the first half time, but almost has nothing to do with it in the second half time.

Diverse Temporal Characteristics. To verify that product items (or brands, or categories) have different temporal characteristics, we use Fast Fourier Transform (FFT) to plot periodogram [17] for the category *Black Tea* and its two children products in Fig. 4. A peak in periodogram indicates a periodicity candidate of original time series. As the figure shows in rectangle areas, the category only has two periodicity candidates, while its children product items have four candidates respectively. It is obvious that they have diverse temporal characteristics, which prevent us from applying the same attention to all variables.

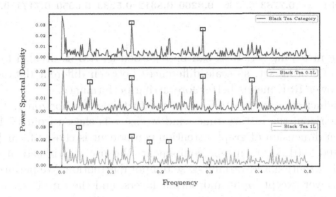

Fig. 4. Periodogram of one category and two product items. The rectangles highlight the points of periodicity candidates.

4.2 Experimental Setups and Results

Methods and Metric: We compare our model with five other methods. All models are as follows.

- AR [2]: a classic univariate time series modeling method.
- Ridge [14]: linear regression with $\ell 2$ regularization.
- LSVR [3]: linear SVR, another machine leaning method for regression.
- GRU [6]: Recurrent neural network using GRU cell.
- LSTNet [10]: State-of-the-art model with CNN, GRU and highway network.
- STANet: our model which leverages inherent product structural information.

For AR, Ridge and LSVR, we train a model for each product item, while for neural network models we use multivariate time series data as input directly.

We use root relative squared error (RSE) as the evaluation metric:

$$RSE = \frac{\sqrt{\sum_{i=1}^{N} \sum_{t=t_0}^{t_1} (y_{i,t} - \hat{y}_{i,t})^2}}{\sqrt{\sum_{i=1}^{N} \sum_{t=t_0}^{t_1} (y_{i,t} - mean(Y))^2}}, \tag{15}$$

where y and $\hat{y}$ are ground truth and predicted value respectively, t_0 and t_1 are start and end time of testing set, and $Y \in \mathbb{R}^{N \times (t_1 - t_0)}$ represents the matrix of

Table 2. RSE of six methods on two datasets with the horizon 1, 4, 7 and 14.

	Dataset-1				Dataset-2			
	1	4	7	14	1	4	7	14
AR [2]	0.8016	0.8206	0.8247	0.8331	0.5313	0.6656	0.7429	0.8393
Ridge [14]	0.7981	**0.8170**	0.8224	0.8327	0.5628	0.6806	0.7418	0.8322
LSVR [3]	3.2446	3.2370	3.2357	3.2397	7.8246	7.8412	7.8443	7.8514
GRU [6]	0.8535	0.8616	0.8448	0.8468	0.9067	0.9765	0.9310	0.9382
LSTNet [10]	0.7907	0.8203	0.8206	0.8316	0.7570	0.7543	0.8069	0.9131
STANet	**0.7783**	0.8186	**0.8200**	**0.8312**	**0.5233**	**0.6050**	**0.7171**	**0.7989**

all labels y in testing set. RSE can be regarded as RMSE divided by standard deviation of testing set, so scale differences between different datasets can be ignored. Lower RSE means better forecasting performance.

We conduct experiments to forecast product sales for horizon 1, 4, 7 and 14. For each horizon, the history time window size τ is chosen from $\{7, 14, 21, 28\}$. The hidden dimension of graph attention component is chosen from $\{2, 3, 4, 5\}$, and $\{50, 100, 200\}$ for recurrent component. We use multi-head attention for both graph attention and variable-wise temporal attention. We perform dropout after each layer except input and output layers, and the rate is set from 0.1 to 0.4. The Adam [9] algorithm is utilized to optimize the parameters of our model.

Table 2 provides the RSE of six methods on two datasets. Our proposed model STANet outperforms others except for horizon $h = 4$ on dataset-1. Nevertheless, the result of STANet even for horizon $h = 4$ on dataset-1 is very close to the best result by Ridge. For both datasets, the LSVR performs worst obviously among all six methods. In addition, the result of vanilla GRU is even worse than univariate model. It is mainly because not all products have strong inter-dependencies. Thus, adding unrelated data, or equivalently noise, would make the prediction worse.

In terms of the results of two datasets, we find that most methods perform better on dataset-2. It is mainly because the sparsity of dataset-2 is much lower than that of dataset-1. In addition, on dataset-1, LSTNet has better result than AR and Ridge except for horizon $h = 4$. Yet on dataset-2, the RSE of LSTNet is much higher than those of AR and Ridge. The reason is that the product items in dataset-2 have less inter-dependencies and more different temporal characteristics than those in dataset-1, so univariate models perform better. Instead, our model STANet leverages the attention mechanism and inherent products structural information to adaptively capture both structural and temporal dependencies, which make the model comfortably adapt to different datasets.

4.3   Ablation Study

To demonstrate the effect of each component, we give ablation study with seven variants for horizon $h = 1$ in Table 3. All the variants are as follows.

- STANet-CNN: use a CNN to substitute GAT component of STANet, which can seen as LSTNet with variable-wise temporal attention.
- STANet-oStructure: STANet without any product graph structural information. In this case neither GAT nor CNN is used.
- STANet-oCategory: STANet without product category information.
- STANet-oBrand: STANet without product brand information.
- STANet-oAttn: STANet without any temporal attention component.
- STANet-FixAttn: STANet using a fixed attention mechanism as Eq. (7) instead of the variable-wise temporal attention.
- STANet-oAR: STANet without autoregressive component.

Table 3. RSE of STANet and its seven variants on two datasets with horizon $h = 1$.

Methods	Dataset-1	Dataset-2
STANet-CNN	0.8230	0.5399
STANet-oStructure	0.8907	0.6179
STANet-oCategory	0.8256	0.5569
STANet-oBrand	0.8640	0.5471
STANet-oAttn	0.8155	0.5431
STANet-FixAttn	0.8105	0.5388
STANet-oAR	0.8327	0.8850
STANet	**0.7783**	**0.5233**

From Table 3, we have the following findings. Firstly, the product graph structural information plays a significant contribution to reduce forecasting errors. Secondly, explicitly using product category and brand information simultaneously with a GAT component achieves better result than using CNN to find product dependencies implicitly. Thirdly, as for the temporal part, STANet-FixAttn, using a fixed temporal attention, achieves better results than STANet-oAttn, but worse than STANet. These results indicate that the model capturing diverse temporal characteristics is quite necessary to demand forecasting. Finally, by comparing the results of STANet-oAR against other variants, we find that AR is the most important component for dataset-2 and yet the structural information for dataset-1. Such difference might be caused by the patterns that can be seen from data characteristics, such as sparsity.

5 Conclusions

In this paper, we propose a novel demand prediction model in a multivariate time series forecasting manner. The model integrates the components of GAT, GRU, variable-wise attention mechanism and auto-regressive to precisely capture the inherent product structural information and temporal periodicity for more accurate prediction. Our analytic results on two real-world datasets demonstrate that the two real datasets exhibit strong product structural information and temporal periodicity. The evaluation results validate that STANet outperforms five counterparts and seven variants. As for the future work, we plan to improve STANet and conduct more evaluation results on both online and offline transaction data.

Acknowledgement. This work is partially supported by National Natural Science Foundation of China (Grant No. 61772371 and No. 61972286). We also would like to thank anonymous reviewers for their valuable comments.

References

1. Arnold, A., Liu, Y., Abe, N.: Temporal causal modeling with graphical granger methods. In: Proceedings of the 13th ACM SIGKDD International Conference on Knowledge Discovery and Data Mining, pp. 66–75. ACM (2007)
2. Box, G.E., Jenkins, G.M., Reinsel, G.C., Ljung, G.M.: Time Series Analysis: Forecasting and Control. Wiley, New York (2015)
3. Cao, L.-J., Tay, F.E.H.: Support vector machine with adaptive parameters in financial time series forecasting. IEEE Trans. Neural Netw. **14**(6), 1506–1518 (2003)
4. Che, Z., Purushotham, S., Cho, K., Sontag, D., Liu, Y.: Recurrent neural networks for multivariate time series with missing values. Sci. Rep. **8**(1), 6085 (2018)
5. Chen, X., Huang, J.Z., Luo, J.: Purtreeclust: a purchase tree clustering algorithm for large-scale customer transaction data. In: 2016 IEEE 32nd International Conference on Data Engineering (ICDE), pp. 661–672. IEEE (2016)
6. Chung, J., Gulcehre, C., Cho, K., Bengio, Y.: Empirical evaluation of gated recurrent neural networks on sequence modeling. arXiv preprint arXiv:1412.3555 (2014)
7. Elman, J.L.: Finding structure in time. Cogn. Sci. **14**(2), 179–211 (1990)
8. Hochreiter, S., Schmidhuber, J.: Long short-term memory. Neural Comput. **9**(8), 1735–1780 (1997)
9. Kingma, D.P., Ba, J.: Adam: a method for stochastic optimization. arXiv preprint arXiv:1412.6980 (2014)
10. Lai, G., Chang, W.-C., Yang, Y., Liu, H.: Modeling long-and short-term temporal patterns with deep neural networks. In: The 41st International ACM SIGIR Conference on Research & Development in Information Retrieval, pp. 95–104. ACM (2018)
11. LeCun, Y., Bengio, Y., et al.: Convolutional networks for images, speech, and time series. In: The Handbook of Brain Theory and Neural Networks, vol. 3361, no. 10, p. 1995 (1995)
12. Lütkepohl, H.: New Introduction to Multiple Time Series Analysis. Springer Science & Business Media, Berlin, Heidelberg (2005). https://doi.org/10.1007/978-3-540-27752-1
13. Malhotra, P., Vig, L., Shroff, G., Agarwal, P.: Long short term memory networks for anomaly detection in time series. In: Proceedings, p. 89. Presses universitaires de Louvain (2015)
14. Seber, G.A., Lee, A.J.: Linear Regression Analysis, vol. 329. Wiley, New York (2012)
15. Vaswani, A., et al.: Attention is all you need. In: Advances in Neural Information Processing Systems, pp. 5998–6008 (2017)

16. Veličković, P., Cucurull, G., Casanova, A., Romero, A., Lio, P., Bengio, Y.: Graph attention networks. arXiv preprint arXiv:1710.10903 (2017)
17. Vlachos, M., Yu, P., Castelli, V.: On periodicity detection and structural periodic similarity. In: Proceedings of the 2005 SIAM International Conference on Data Mining, pp. 449–460. SIAM (2005)
18. Yang, J., Nguyen, M.N., San, P.P., Li, X.L., Krishnaswamy, S.: Deep convolutional neural networks on multichannel time series for human activity recognition. In: Twenty-Fourth International Joint Conference on Artificial Intelligence (2015)
19. Yin, J., et al.: Experimental study of multivariate time series forecasting models. In: Proceedings of the 28th ACM International Conference on Information and Knowledge Management, pp. 2833–2839 (2019)

Demo Papers

Demo Papers

AuthQX: Enabling Authenticated Query over Blockchain via Intel SGX

Shuaifeng Pang, Qifeng Shao, Zhao Zhang$^{(\boxtimes)}$, and Cheqing Jin

School of Data Science and Engineering,
East China Normal University, Shanghai, China
{sfpang,shao}@stu.ecnu.edu.cn, {zhzhang,cqjin}@dase.ecnu.edu.cn

Abstract. With the popularization of blockchain technology in traditional industries, though the desire for supporting various authenticated queries becomes more urgent, current blockchain platforms cannot offer sufficient means of achieving authenticated query for light clients, because Authenticated Data Structure (ADS) suffers from performance issues and state-of-the-art Trust Execution Environment (TEE) cannot deal with large-scale applications conveniently due to limited secure memory. In this study, we present a new query authentication scheme, named AuthQX, leveraging the commonly available trusted environment of Intel SGX. AuthQX organizes data hierarchically in trusted SGX enclave and untrusted memory to implement authenticated query cheaply.

Keywords: Blockchain · Authenticated query · MB-tree · Intel SGX

1 Introduction

As a kind of distributed ledger, blockchain has gained lots of attention and interest from public and academic communities. An application scenario of significant value blockchain supports is to achieve trusted data sharing among untrusted participants with no central authority. Due to heavy requirements on computing and storage resources, most blockchain platforms adopt *light clients*, targeted for portable devices like mobile phones, that forward queries to full nodes for execution and authenticate the integrity of the returned results, i.e., authenticated query. However, current systems have limited ability to back authenticated queries. Traditional methods to solve the problem of authenticated query, such as signature chaining [4] and Authenticated Data Structure (ADS) [1,3,6], are inadequate for the blockchain scenario since the former triggers tremendous signature computation overhead while ADS gives rise to the cost of verification objects (known as VO). Recently, the appearance of trusted hardware, like Intel Software Guard Extensions (SGX) [2] that provides a set of security enhancements within the processor, offers a promising direction of designing new query authentication schemes. With SGX, sensitive codes can be installed in a segment of trusted memory, named enclave, and run on untrusted machines while guaranteeing integrity and confidentiality.

This paper proposes a novel solution, AuthQX (for "**Auth**enticated **Q**uery via SG**X**"), for light clients. The authentication of VO is done by SGX so as to

© Springer Nature Switzerland AG 2020
Y. Nah et al. (Eds.): DASFAA 2020, LNCS 12114, pp. 727–731, 2020.
https://doi.org/10.1007/978-3-030-59419-0_45

diminish the network and computational overhead. We provide a hybrid index consisting of MB-tree and skip list to enable batch updates as well.

2   System Overview and Key Techniques

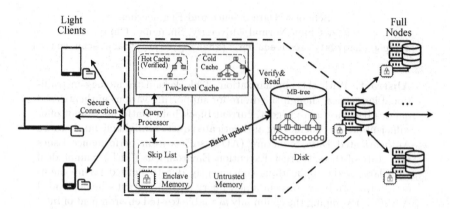

Fig. 1. AuthQX system architecture.

Figure 1 demonstrates our **AuthQX** system architecture that consists of several full nodes and light clients. When a light client desires to acquire information about transactions or states, it first connects to a full node that hosts an enclave and establishes secure communication. Query processor executes queries and verifies query results by VO inside enclave and returns results. The light client won't authenticate the query results since SGX can protect data confidentiality and code integrity from any malicious software. More specifically, our system contains three key components: (i) a query processor that responses to client queries; (ii) a hybrid index that enables authenticated queries and batch updates; (iii) a two-layer cache that buffers frequently-accessed MB-tree nodes in enclave memory and untrusted memory to reduce the cost of swapping pages and reading disk. Since updating any leaf node will cause digest propagation up to the root node, we allocates additional trusted memory to hold a skip list that buffers incoming blocks and periodically merge updates to MB-tree in batch. More technical details can be found in [5].

2.1   Query and Update

In our solution, the root node of MB-tree is always resident in enclave. Verified nodes are cached in enclave to shorten the verification path. Setting up skip list inside enclave enables batch update on MB-tree and diminishes hash computing overhead. We now details query and update procedures.

Authenticated Query on MB-tree. The query process is the same as the traditional one, during which, it accesses nodes from root to leaf, appends hashes of

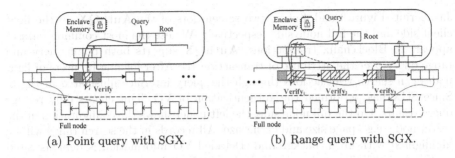

(a) Point query with SGX.   (b) Range query with SGX.

Fig. 2. Query and verify on a MB-tree with SGX enabled.

sibling nodes to VO and returns the query results. SGX performs the authentication of VO as suggested in Fig. 2. Since enclave may cache previously verified nodes, when computing Merkle path, the authenticating process can be early terminated once encountering a trusted node. The left and right boundaries of range query should be included in VO as well for further completeness authentication. As demonstrated in Fig. 2(b), the results of range query involve a set of verification paths but all end with verified nodes cached in enclave.

Batch Update. Our batch update strategy performs searching and digest computing only once for all keys belonging to the same leaf node, thus, cascading hash computing overhead is reduced dramatically. To find the leaf to which search key belongs, we start from the root node and advance down the tree using the given key. It will be split or merged once an internal node meets the rebalancing condition.

3   Demonstrations and Evaluations

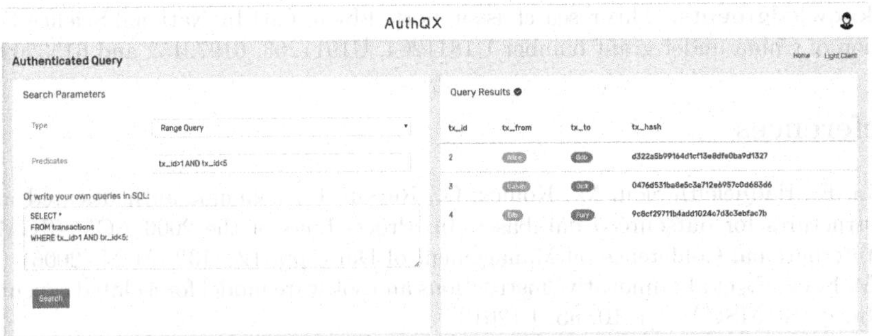

Fig. 3. The interface of light client. (Color figure online)

AuthQX backend is implemented in C++ and running on Ubuntu 16.04 OS with Intel SGX SDK and SGXSSL library enabled and its frontend is in

JavaScript. Figure 3 and 4 show two screenshots of the **AuthQX** on the light client side and the full node side respectively. We use the most common dataset appears in Blockchain, transactions. **AuthQX** suports both point query and range query over the attribute of transaction ID. After logging in, users of light nodes can search transactions through the query interface illustrated in Fig. 3. Since SGX has performed the authentication of VO, light clients only receive query results. On the full node side, the left panel shows the configuration of the server node, e.g., page size and cache size. All records in the search range will be highlighted with green background. Detailed VO information is demonstrated on Fig. 4 as well. Trusted nodes cached in SGX enclave are marked in red while unverified nodes are colored in green.

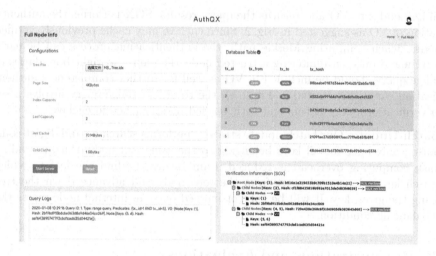

Fig. 4. The interface of full node. (Color figure online)

Acknowledgments. This research is supported by in part by National Science Foundation of China under grant number U1811264, U1911203, 61972152 and 61532021.

References

1. Li, F., Hadjieleftheriou, M., Kollios, G., Reyzin, L.: Dynamic authenticated index structures for outsourced databases. In: Proceedings of the 2006 ACM SIGMOD International Conference on Management of Data, pp. 121–132. ACM (2006)
2. McKeen, F., et al.: Innovative instructions and software model for isolated execution. In: HASP@ISCA, vol. 10, no. 1 (2013)
3. Merkle, R.C.: A certified digital signature. In: Brassard, G. (ed.) CRYPTO 1989. LNCS, vol. 435, pp. 218–238. Springer, New York (1990). https://doi.org/10.1007/0-387-34805-0_21
4. Pang, H., Tan, K.L.: Authenticating query results in edge computing. In: Proceedings of 20th International Conference on Data Engineering, pp. 560–571. IEEE (2004)

5. Shao, Q., Pang, S., Zhang, Z., Jin, C.: Authenticated range query using SGX for blockchain light clients. In: International Conference on Database Systems for Advanced Applications. Springer (2020)
6. Xu, C., Zhang, C., Xu, J.: vChain: enabling verifiable Boolean range queries over blockchain databases. In: Proceedings of the 2019 International Conference on Management of Data, pp. 141–158. ACM (2019)

SuperQuery: Single Query Access Technique for Heterogeneous DBMS

Philip Wootaek Shin, Kyujong Han, and Gibeom Kil[✉]

Data Streams Corporation, Seoul, South Korea
{wtshin,kjhan,gbkil}@datastreams.co.kr

Abstract. With the increasing interest in machine learning, data management has been receiving significant attention. We propose Super-Query, a big data virtualization technique that abstracts the physical elements of heterogeneous DBMS in various systems and integrates data into a single access channel. SuperQuery can integrate and collect data regardless of the location, shape, and structure of the data and enable rapid data analysis.

1 Introduction

The emergence of machine learning especially deep learning has made data management and data analysis important. Industries and government agencies are allocating significant resources to create data and analyze it, but more importantly they want to analyze and manage the data that they already own. However, since the data present not only in various systems but also stored in heterogeneous databases, it is not easy to integrate and analyze the data.

Big data integration differs from data integration on account of the volume, velocity, variety and veracity of data [1]. There are different methods like data warehouse, Data lake, Cloud, and data virtualization to integrate and manage data. Data virtualization provides data users a unified, abstracted and encapsulated view for querying and manipulating data stored in heterogeneous data storages [2]. We propose SuperQuery that can support analysts to easily preprocess big data. SuperQuery is based on data virtualization for different database management systems(DBMS) to access data with a single query and provide operations like aggregation, join, and search. SuperQuery will help big data processing, data integration and data extraction tasks.

2 System Architecture

There are three difficulties when accessing different DBMS with the single channel. First, a logical table (virtualized table) and a mapping of the physical data to the table are needed in SuperQuery to access data from the source. Second, data collection from various source systems is required. Third, there is a need for

Y. Nah et al. (Eds.): DASFAA 2020, LNCS 12114, pp. 732–735, 2020.
https://doi.org/10.1007/978-3-030-59419-0_46

a technology capable of performing high performance operations using collected physical data and meta information.

We solve these three difficulties in the following way. First, logical table/virtualized table and mapping of physical data can be collected from the MetaStream(Meta data management tool) [3]. Further, data collection from various source systems can be done using FACT (Relational Data base extracting engine) [4]. Lastly, the system was implemented to virtualize the physical data of various source systems and perform an operation with a single query by using Apache Spark [5].

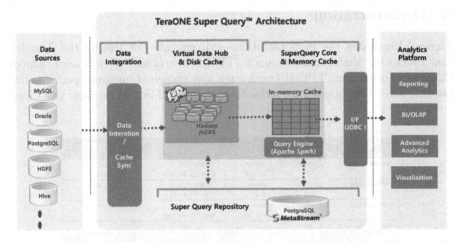

Fig. 1. System architecture of SuperQuery

Figure 1 shows the overall SuperQuery architecture. The Super Query Repository contains information such as job log, account management, and history that all common software solutions have. However, the difference is that meta information of table, column, etc. is brought and stored in the repository, and meta information is updated by user's need. These meta information play an important role in the logical table and memory cache implementation.

Suppose that SuperQuery collects information from multiple heterogeneous databases in a single query. If the desired information is stored in the cache, the requested information is collected from the cache and displayed to the user without the need for physical data integration. However, if it does not exist in the cache, the following process is performed. Super Query collects the requested information from several different data sources which us stored in two kinds of caches. The term cache is not a cache, which is a storage adjacent to the CPU, but is similar to the cache in terms of function. There are two types of cache implemented in SuperQuery: the first is the disk cache and the second the memory cache.

Hadoop Distributed File System (HDFS) [6] is deployed in a distributed environment in the form of a master-slave node. The master node processes the received information and the actual storage is distributed to slave nodes. This concept is similar to disk storage and is hence named as disk cache. However, to process query requests faster than this form, the system was implemented using Spark Apache's in memory processing to store. Spark uses a new model, called a resilient distributed dataset, which is stored in memory during computation without costly disk writes. Memory cache is implemented based on Spark Apache. In addition, this model follows Least Recently Used (LRU) rule similar to CPU cache. Super Query's API is implemented in JDBC to support programs that follow standard JDBC.

3   Demonstration

Figure 2 illustrates the demonstration environment. For testing the Transaction Processing Performance Council (TPC)'s benchmark [7] was used. Among different benchmarks, TPC-H benchmark was chosen to measure performance, because it consists of ad-hoc queries that are needed in real business situations.

1) Super Query

IP	OS	CPU Memory	Disk	Role
192.168.0.125	CentOS 7.5	24core 512GB	7200rpm SATA 3.0, 6.0 Gb/s	Master Node / Slave Node
192.168.0.126	CentOS 7.5	24core 512GB	7200rpm SATA 3.0, 6.0 Gb/s	Slave Node
192.168.0.127	CentOS 7.5	24core 512GB	7200rpm SATA 3.0, 6.0 Gb/s	Slave Node

2) Oracle Database

IP	OS	CPU Memory	Disk	Role
192.168.0.57	CentOS 6.9	48core 512GB	7200rpm SATA 3.0, 6.0 Gb/s	DB Server

Fig. 2. SuperQuery and Oracle database demonstration environment

In the test, for Oracle database, the time was measured to retrieve test queries from Oracle engine. For Super Query the time was calculated only when it was cached on either disk cache or memory cache. The collecting of data is done when there is modification in data, but most analysis of data can be done on disk cache(HDFS cache) or memory cache. The result of the HDFS cache is the source table information, which is a measure of execution time when it is cached in HDFS. The result of the memory cache is to measure the execution time with the source table information cached in memory. The data in the table was created with an authorized data generator provided by TPC-H, and a 10GB dataset and a 100GB dataset were generated respectively. 1 Node means that HDFS is constructed with a single node, and 3 nodes means that HDFS is constructed with 3 different nodes that can save data distributively.

As shown in Fig. 3, a single node SuperQuery outperforms in Memory Cache than Oracle data base. However depending on conditions, HDFS Cache performed equally or slightly slower than the Oracle database. For 3 node scenario,

Query		Oracle	Super Query (Single Node / 3 Nodes)		Remarks
			Memory Cache	HDFS Cache	
Query 01	10GB	1m 27s	20s / 9s	1m 43s / 50s	Top: 1 Node / Bottom: 3Nodes
	100GB	12m 26s	3m 10s / 1m 32s	14m 10s / 4m 54s	
Query 02	10GB	10s	7s / 4s	23s / 20s	
	100GB	29s	1m 5s / 17s	4m 41s / 1m 25s	
Query 03	10GB	41s	12s / 6s	1m 49s / 40s	
	100GB	4m 20s	2m 10s / 40s	15m 12s / 5m 28s	
Query 04	10GB	32s	6s / 4s	1m 35s / 37s	
	100GB	4m 8s	1m 20s / 27s	17m 31s / 5m 6s	
Query 05	10GB	40s	11s / 8s	1m 43s / 43s	
	100GB	5m 34s	2m 12s / 46s	16m 44s / 5m 46s	

Fig. 3. Results of queries for SuperQuery and Oracle database

HDFS cache performed 1.5 to 3 times faster than the single node scenario. Due to the nature of the big data ecosystem, increasing the nodes will increase the speed and it will scale out.

4 Summary

SuperQuery is a solution designed and constructed to cater to customers who want to analyze heterogeneous databases or different data source easily. From the user's point of view, the virtualization integration method that can integrate and collect data regardless of the location, form and structure of the data will be helpful for faster analysis than the existing methods.

Acknowledgement. This work was supported by National IT Industry Promotion Agency (NIPA) grants funded by the Korean Government (MSIT) (S0510-20-1001, Development of IoT and Big Data Infra Technology for Shipbuilding and Marine).

References

1. Dong, X.L., Srivastava, D.: Big data integration. Proc. VLDB Endow. **6**(11), 1188–1189 (2013)
2. van der Lans, R.: Data Virtualization for Business Intelligence Systems: Revolutionizing Data Integration for Data Warehouses, 1st edn. Morgan Kaufmann, Bulington (2012)
3. MetaStream. http://datastreams.co.kr/en/sub/prd/governance/metadata.asp. Accessed 23 Dec 2019
4. FACT. http://datastreams.co.kr/en/sub/prd/integration/etl.asp. Accessed 23 Dec 2019
5. Apache Spark. https://spark.apache.org/. Accessed 23 Dec 2019
6. HDFS. https://hadoop.apache.org/. Accessed 23 Dec 2019
7. TPC. http://www.tpc.org/information/about/abouttpc.asp. Accessed 23 Dec 2019

MDSE: Searching Multi-source Heterogeneous Material Data via Semantic Information Extraction

Jialing Liang, Peiquan Jin$^{(\boxtimes)}$, Lin Mu, Xin Hong, Linli Qi, and Shouhong Wan

University of Science and Technology of China, Hefei, China
jpq@ustc.edu.cn

Abstract. In this paper, we demonstrate MDSE, which provides effective information extraction and searching for multi-source heterogeneous materials data that are collected as XML documents. The major features of MDSE are: (1) We propose a transfer-learning-based approach to extract material information from non-textual material data, including images, videos, etc. (2) We present a heterogeneous-graph-based method to extract the semantic relationships among material data. (3) We build a search engine with both Google-like and tabular searching UIs to provide functional searching on integrated material data. After a brief introduction to the architecture and key technologies of MDSE, we present a case study to demonstrate the working process and the effectiveness of MDSE.

Keywords: Material · Information extraction · Search engine · Heterogeneous graph

1 Introduction

Material data are usually collected from multiple sources and represented by different data formats, yielding the heterogeneity of material data [1]. For example, a record of material data may include textual descriptions, tabular data, images, and videos. These heterogeneous material data introduce new challenges for material data utilization, e.g., how to extract semantic information from heterogeneous material data so that we can offer a unified view of multi-source material data. and how to find out the semantic relationships among multi-source material data.

In this paper, in order to address the above challenges, we develop a Web-based system named *MDSE (Material Data Search Engine)* that can extract and search semantic information and relationships from multi-source heterogeneous material data. The material data coming into MDSE are XML files consisting of heterogeneous data such as texts, images, tabular data, and videos. MDSE aims for extracting semantic information from those heterogeneous data inside each XML file and provide effective searching services for integrated data. To the best of our knowledge, MDSE is the first one that provides semantic information extraction and searching for multi-source heterogeneous material data. Briefly, MDSE has the following unique features:

© Springer Nature Switzerland AG 2020
Y. Nah et al. (Eds.): DASFAA 2020, LNCS 12114, pp. 736–740, 2020.
https://doi.org/10.1007/978-3-030-59419-0_47

(1) MDSE supports information extraction from multi-source heterogeneous material data, including textual data and non-textual data. Specially, we propose a transfer-learning method to extract information from non-textual material data.

(2) MDSE supports the extraction of semantic relationships among material entities. In particular, we propose to measure the semantic relationships among material entities using the heterogeneous graph and convolutional network.

(3) MDSE provides functional searching services, including a Google-like UI and a tabular searching UI, for integrated material information that are extracted from multi-source heterogeneous material data.

2 Architecture and Key Technologies of MDSE

Figure 1 shows the architecture of MDSE. It mainly includes an offline sub-system and an online sub-system. The offline sub-system consists of three modules, namely XML processing, material information extraction, and index construction. The module of XML processing is responsible to perform textual processing on the inputted XML document, including word segmentation, removing stop words, and stemming. The module of material information extraction is designed for extracting material semantics from heterogeneous data embedded in XML documents. The index-construction module constructs a typical inverted-file index for extracted information, which is represented by word vectors. The online sub-system consists of three modules including query processing, ranking, and user interface. The user interface provides Google-like as well as tabular interfaces for users. The query processing module is responsible for processing queries based on the index, and the ranking module sorts the results returned by the query processing and sends the ranked list to the user interface.

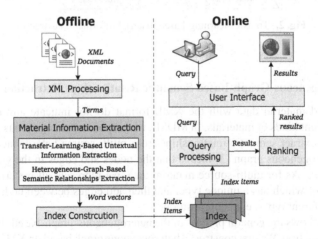

Fig. 1. Architecture of MDSE

2.1   Transfer-Learning-Based Non-textual Information Extraction

We propose a transfer learning [2] based non-textual information extraction from het-erogeneous material data. The general idea, as shown in Fig. 2, is to use the text learning model to extract information from non-textual data. Assume that we are to extract material information from an image embedded in an XML document, we can first collect the texts around the image. Combined with other texts such as the title and topic words of the XML document as well as the meta text of the image, we can prepare a set of texts to be learned for the target image. Next, we conduct a general text processing and modelling step toward the collected texts. We also perform the process of name-entity recognition and relation extraction on the processed texts. This can be done through some widely-used tools such as Stanford NLP tools (https://nlp.stanford.edu/software/). We formally transform the extracted entities and relations into word vectors, so that they can be integrated into the inverted index. With the above mechanism, we can transfer the knowledge of text information extraction to non-textual information extraction, so as to avoid the complex processing of images, videos, and other unstructured data. In addition, this transfer-learning method is extensible, meaning it can suit for new kinds of non-textual information extraction in the future.

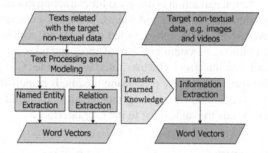

Fig. 2. Transfer-learning-based material information extraction

2.2   Heterogeneous-Graph-Based Semantic Relationships Extraction

The collected material data with the XML format do not indicate any relationships among different kinds of materials. In MDSE, we propose a heterogeneous-graph-based approach to extract semantic relationships from multi-source heterogeneous material data. A heterogeneous graph [3] means that the nodes and edges in the graph are with different types. As for multi-source material data, we represent all material entities as nodes, each of which has a unique type, and then add edges between nodes. Edges are also with different types, e.g., composition.

Figure 3 shows the general process of the heterogeneous-graph-based semantic relationships extraction. We first construct a heterogeneous graph based on XML documents. Then, we employ a graph convolutional network (GCN) [4] training, which aims to learn the influence of the connected nodes to a specific node. This can add the impacts of other nodes, e.g., the nodes connected to a node, into the node representation. This step consequently generates a node vector for each node. After that, we compute the node similarity between each two nodes in the graph. As a result, we use the similarity between two nodes to represent the semantic relationship between the two nodes. Such semantic relationships are finally integrated into node vectors, forming revised node vectors. A final node vector for a node not only represents the node attributes but also includes the semantic relationships between the node and other nodes. As each node represents a material entity, we can represent the semantic relationships among material entities through the heterogeneous graph-based approach.

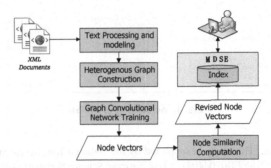

Fig. 3. Heterogeneous-graph-based relationship extraction for materials

3 Demonstration

Figure 4 shows a screenshot of the online system in *MDSE*. The offline system results in the index supporting the online search of material information. In the demonstration, we will input different keywords and show the results based on different ranking methods. The default ranking algorithm is based on the text-vector similarity [5]. As the vectors in MDSE have included information from heterogeneous material data as well as the semantic relationships of material data, our system can offer functional searching services for material information retrieval.

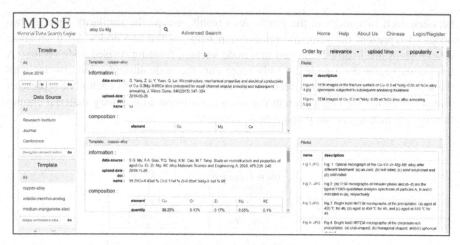

Fig. 4. Screenshot of MDSE

Acknowledgements. This study is supported by the National Key Research and Development Program of China (2018YFB0704404) and the National Science Foundation of China (61672479).

References

1. Zhou, J., Hong, X., Jin, P.: Information fusion for multi-source material data: progress and challenges. Appl. Sci. **9**(17), 3473 (2019)
2. Pan, S.J., Yang, Q.: A survey on transfer learning. IEEE Trans. Knowl. Data Eng. **22**(10), 1345–1359 (2010)
3. Wang, X., Ji, H., et al.: Heterogeneous graph attention network. In: WWW, pp. 2022–2032 (2019)
4. Hu, F., Zhu, Y., et al.: Hierarchical graph convolutional networks for semi-supervised node classification. In: IJCAI, pp. 4532–4539 (2019)
5. Jin, P., Li, X., Chen, H., Yue, L.: CT-rank: a time-aware ranking algorithm for web search. J. Convergence Inf. Technol. **5**(6), 99–111 (2010)

BigARM: A Big-Data-Driven Airport Resource Management Engine and Application Tools

Ka Ho Wong[1], Jiannong Cao[1], Yu Yang[1(✉)], Wengen Li[1], Jia Wang[1],
Zhongyu Yao[1], Suyan Xu[1], Esther Ahn Chian Ku[1], Chun On Wong[2],
and David Leung[2]

[1] The Hong Kong Polytechnic University, Hung Hom, Hong Kong
{khowong,jiannong.cao,wgcsli,esther.ku}@polyu.edu.hk
{csyyang,csjiawang}@comp.polyu.edu.hk
{frank7.yao,suyan.xu}@connect.polyu.hk
[2] Logistics and Supply Chain MultiTech R&D Centre, Pok Fu Lam, Hong Kong
{cowong,dleung}@lscm.hk

Abstract. Resource management becomes a critical issue in airport operation since passenger throughput grows rapidly but the fixed resources such as baggage carousels hardly increase. We propose a Big-data-driven Airport Resource Management (BigARM) engine and develop a suite of application tools for efficient resource utilization and achieving customer service excellence. Specifically, we apply BigARM to manage baggage carousels, which balances the overload carousels and reduces the planning and rescheduling workload for operators. With big data analytic techniques, BigARM accurately predicts the flight arrival time with features extracted from cross-domain data. Together with a multi-variable reinforcement learning allocation algorithm, BigARM makes intelligent allocation decisions for achieving baggage load balance. We demonstrate BigARM in generating full-day initial allocation plans and recommendations for the dynamic allocation adjustments and verify its effectiveness.

Keywords: Airport resource management · Inbound baggage handling · Big data analytics · Load balance

1 Introduction

Resources management is one of the most critical issues in airports [1]. In 2018, the Hong Kong International Airport handled more than $1,170$ daily flights and served over 72 million passengers, which increased almost 3 times in the past 20 years [2]. However, it only has twelve baggage carousels and no additional one has been added since the terminal opened in 1998. Allocating the best mix of flights per carousel with baggage load optimization is a crucial factor in providing excellent customer services. Traditional systems only consider incoming bag load against the number of carousels based on the flight schedule [3,4]. However,

© Springer Nature Switzerland AG 2020
Y. Nah et al. (Eds.): DASFAA 2020, LNCS 12114, pp. 741–744, 2020.
https://doi.org/10.1007/978-3-030-59419-0_48

there are various operation factors, such as flight arrival punctuality, baggage removal profile, weather conditions, etc., affecting the allocation [5]. These factors are highly dependent on each other and dynamically change over time making carousel allocation with a balanced baggage load become challenging.

We propose a Big-data-driven Airport Resource Management (BigARM) engine to manage the airport recourses with big data analytic techniques and develop a suite of application tools. We demonstrate BigARM in baggage carousel allocation, which aims to balance the overload carousels and reduce the workload of operators on planning and rescheduling the allocations. BigARM engine consists of three major components including big data collection and storage, flight arrival time prediction, and intelligent allocation decision making. Besides, we develop application tools together with a web-based graphical user interface for full-day allocation plan generation and real-time dynamic adjustments. The flight information, airfield operation data and weather conditions are automatically collected from an Airport Operator and stored into a MongoDB database. We design a data-driven approach to predict flight arrival time using features obtained by cross-domain data fusion, which makes BigARM aware of the dynamic change of flight status on time. A multi-variable reinforcement learning algorithm is proposed to make intelligent allocation decisions for achieving baggage load balance that is measured by the standard deviation of bag load across all carousels. Once flights' baggage load or predicted time of arrival has any update, BigARM will pop-up recommendations for operators to perform adjustments. Comparing to the estimated time of flight arrival (ETA) currently used by the airport, BigARM reduces the prediction error from 14.67 min to 8.86 min. Besides, BigARM achieves 36.61% and 33.65% improvement of baggage load balance in the initial allocation plan and day-end final plan respectively.

2   System Design

The system includes a BigARM engine and a suite of application tools as presented in Fig. 1. The BigARM engine consists of modules of the airport big data collection and storage, flight arrival time prediction, and intelligent allocation decision making for supporting the functions of application tools such as full-day allocation plan generation and recommendations of dynamic adjustment.

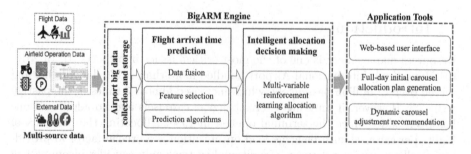

Fig. 1. Design of BigARM system.

2.1   Airport Data Collection and Storage

Cross-domain data on flights and airfield operations are collected from multiple entities owning the airport resource. Flight data contain the flight number, aircraft type, time of flight arrival, passenger count, baggage load, parking stand, etc. For airfield operation data, we have collected the on-belt-delivery time of first and last baggage for every flight, carousel allocation plan, allocation update transactions, etc. Since the airfield operation highly depends on the weather condition, we further collect the dew point, humidity, air pressure, temperature, visibility, wind direction, and wind speed from the Observatory. All data are collected at fixed time intervals and stored into a MongoDB database.

2.2   Flight Arrival Time Prediction

Flight delay is one of the key factors affecting the in-bound baggage handling. We perform feature engineering on flight information, real-time flight data, landing information, and weather conditions. After correlation analysis, 25 features that are statistically significant to the actual time of flight arrival are selected to predict the flight arrival time. To make better use of the information from cross-domain data, we perform feature level fusion by concatenation. Lastly, we train a Huber regression model to predict the flight arrival time which reduces the estimation error of ETA used by the airport from 14.67 min to 8.86 min.

2.3   Intelligent Allocation Decision Making

We design a multi-variable reinforcement learning algorithm to allocate the arrival bags to carousels such that the baggage load balance across all carousels is maximized. Each flight is regarded as a generator unit to generate allocation policies indicating how to select an allocation action under the current carousel status and maximize the accumulated rewards. To achieve the baggage load balance, the generator gets higher rewards when the standard deviation of the baggage load across all carousels becomes smaller. Overall, with the input of flight arrival time, baggage load and the predicted bags removal profile based on bags dwell time on the carousel, our proposed algorithm allocates the coming flights and achieves balanced baggage carousel throughput and utilization.

3   Demonstration

We demonstrate the full-day initial baggage carousel allocation plan generation and the recommendations of dynamic adjustment using BigARM and its application tools. The airport would schedule a full day carousel allocation plan the day before as a template for planning the next day operations. After login the BigARM system, the operator only needs to select a date and click a button named BigARM plan. BigARM will query the scheduled flight arrival time and its baggage load on that day from the database and run the allocation algorithm to generate a full day allocation plan as shown in Fig. 2. The color of each flight represents the baggage load from the lightest (green) to the heaviest (red). Besides reducing the time of generating the initial plan from hours to seconds, BigARM's initial plan achieves 36.61% more balanced than the one made by experienced operators in the airport.

 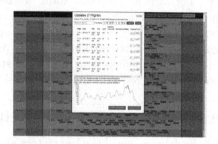

Fig. 2. Initial plan generation. (Color figure online)

Fig. 3. Recommendation of adjustments. (Color figure online)

In daily operations, operators have to frequently adjust the allocation due to flight delay and other factors. Once BigARM received the updated flight data, it will continuously predict the flight arrival time. If either flights' baggage load or predicted time of arrival has any update, the allocation algorithm will be triggered and pop-up recommendations for operators to perform the adjustment, which is presented in Fig. 3. Meanwhile, a load balance comparison curve are plotted out to highlight the impact of new allocation on baggage load balance for helping them make decisions. Comparing to the day-end plan made by the airport, BigARM achieves a 33.65% improvement in baggage load balance.

Acknowledgement. The work has been supported by the Innvoation and Technology Fund (ITP/024/18LP) and RGC General Research Fund (PolyU152199/17E). Thank Alan Lee, Patrick Yau, Gavin Lee, and Jiandong Li's effort in this work.

References

1. Budd, L., Stephen, I.: Air Transport Management: An International Perspective. Taylor & Francis (2016)
2. Hong Kong International Airport. Air Traffic Statistics (2018). https://www.hongkongairport.com/en/the-airport/hkia-at-a-glance/fact-figures.page
3. Malandri, C., Briccoli, M., Mantecchini, L., Paganelli, F.: A discrete event simulation model for inbound baggage handling. Transp. Res. Procedia **35**, 295–304 (2018)
4. Frey, M., Kiermaier, F., Kolisch, R.: Optimizing inbound baggage handling at airports. Transp. Sci. **51**(4), 1210–1225 (2017)
5. Yang, H., Morris, R., Păsăreanu, C.: Analysing the effect of uncertainty in airport surface operations. In: Companion Proceedings for the ISSTA/ECOOP 2018 Workshops, pp. 132–137 (2018)

S²AP: Sequential Senti-Weibo Analysis Platform

Shuo Wan, Bohan Li$^{(\boxtimes)}$, Anman Zhang, Wenhuan Wang, and Donghai Guan

Nanjing University of Aeronautics and Astronautics, Nanjing, China
{shuowan,bhli}@nuaa.edu.cn

Abstract. Microblogging sentiment analysis aims at exploring people's opinion on social networks such as Twitter and Weibo. Existing work mainly focus on the English corpus based on Distant Supervision, which ignores the noise data in corpus and internationalization. The field of Weibo sentiment analysis lacks a large-scale and complete corpus for application and evaluation. In this work, we formulate the problem of corpus construction into an Information Retrieval problem and construct a Weibo sentiment analysis corpus called Senti-weibo. We also release a weibo pre-processing toolkit in order to unify the pre-processing rules of Weibo text. Eventually, we apply these works to implement a real-time Weibo sentiment analysis platform: S²AP, which serves to analyze and track the sequential sentiment of Weibo topics.

Keywords: Corpus construction · Weibo sentiment analysis · Text pre-processing

1 Introduction

Weibo, the most popular microblogging social network in China, has 497 million monthly active users in September 2019. People can express their opinion about breaking news, current affairs, politics and other topics. These subjective data are rich in sentiment information [6], which brings great convenience to the research of sentiment analysis.

Sentiment analysis on Twitter has made significant progress in sentiment corpus. Most of the methods for constructing sentiment corpus use Distant Supervision [1], which labels the sentiment of Twitter with emotion symbols such as ":-)" and ":(". However, emotion symbols cannot fully represent the sentiment polarity of tweet, and corpus mentioned above are actually corpus with noise. Compared with Twitter sentiment analysis, the field of Weibo sentiment analysis lacks a large-scale and complete corpus.

Pre-processing for Twitter data is commonly used for denosing and dimensionality reduction. The experiment results of [2] show that appropriate text pre-processing methods can significantly enhance the classifier's performance. Weibo sentiment analysis is short of a unified rule for text cleaning. A unified pre-processing tool for training and online environment is indispensable. For

© Springer Nature Switzerland AG 2020
Y. Nah et al. (Eds.): DASFAA 2020, LNCS 12114, pp. 745–749, 2020.
https://doi.org/10.1007/978-3-030-59419-0_49

example, supposing that we have applied a couple of word segmentation tool and cleaning rule to preprocess Weibo dataset and train a corresponding classifier. Then if we want to make full use of this classifier in online environment, the segmentation tool and cleaning rules should be consistent with training environment. Otherwise, a large number of unknown words will be generated.

Our works are summarized below:

1. The problem of corpus construction is formulated into an Information Retrieval (IR) problem and we build a Weibo sentiment analysis corpus: Senti-weibo[1], a collection of 671,053 weibos (equals tweet of Twitter).
2. We unify weibo pre-processing rules and package it into a toolkit in Python: weibo-preprocess-toolkit[2], which is applied for weibo cleaning.
3. A Weibo sentiment analysis platform called S^2AP^3 is constructed to analyze the real-time sentiment of Weibo topics. A Weibo Topic Spider is designed to crawl real-time weibos for analysis and fresh sentiment weibos for corpus iteration. In particular, we take two topics as concrete examples and track the sentiment trend of them.

2   Corpus Iteration and Construction

Using emotion symbols or emojis to build corpus is actually equivalent to retrieving related sentiment tweets based on specified queries. The problem of corpus construction can be transformed into the problem of using spider to retrieve sentiment weibos from Weibo database, and improving the *Precision* and *Recall* of retrieved sentiment weibos.

Definition 1. *Given a database D (public dataset or Weibo server), the retrieved weibos as R(D). Precision is the sentiment relevant weibos of R(D), Recall is the sentiment weibos in R(D) retrieved from D:*

$$Precision = \frac{\#\,(sentiment\ weibos\ retrieved)}{\#\,(retrieved\ weibos)} = P\,(sentiment|R(D)) \qquad (1)$$

$$Recall = \frac{\#\,(sentiment\ weibos\ retrieved)}{\#\,(sentiment\ weibos)} = P\,(retrieved\ sentiment|D) \qquad (2)$$

In order to quickly initialize corpus, we firstly query sentiment weibos from public dataset with 40 typical emojis. The sentiment classifier trained on the initialized corpus with fastText [4] has the initial precision of 87.04%. The test dataset comes from COAE2014 [5] and we manually labeled 1,790 weibos.

Like most corpus [1,3] built with Distant Supervision, the initialized corpus also contains noise and slowly decays over time. From the perspective of IR, low *Precision* is responsible for noise in the recalled weibos, while the decay of corpus

---

[1] Available at: http://bit.ly/2IEzTw1.

[2] https://pypi.org/project/weibo-preprocess-toolkit.

[3] http://sentiweibo.top.

corresponds to low *Recall*. In order to construct corpus and train model with high performance, we need to optimize the *Precision* for high precision of sentiment classification model and improve the *Recall* for generalized performance.

Weibo Topic Spider is applied to retrieve sentiment weibos. We use spider with one emoji and its two synonyms words generated by word2vec as query to retrieve sentiment weibos from Weibo server. The combination of multiple features can improve the sentiment *Precision* of retrieved weibos, and the similarity between the features can ensure *Recall*. Spider continuously collects fresh sentiment weibos with higher sentiment consistency and accumulates them for corpus iteration. In the process of iteration, we divide the corpus into training set and verification set, and sample subset of training dataset to train the classifier. Multiple supervised learning algorithms are applied to train sentiment classifiers to verify the sentiment label of verification dataset, and then sift out the weibos whose classification results do not match the original label. After several rounds of iteration, the precision of classifier trained by fastText has increased by 3.07% (Fig. 1). Finally, we build a Weibo sentiment analysis Corpus: Senti-weibo.

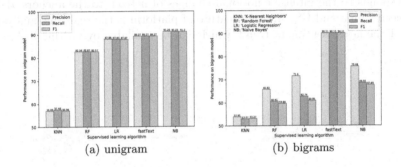

(a) unigram   (b) bigrams

Fig. 1. Multi-model evaluation of corpus

3   Platform Implementation and Demonstration

We design a demonstration platform called S²AP to present our works mentioned in each section. Figure 2 depicts the architecture of S²AP, which has two main function models: *offline scripts* and *front-end visualization.*

Offline scripts model encapsulates scripts such as spider, corpus iteration, and text pre-processing. Figure 3 shows the performance of the model caused by different pre-processing rules. We take six Chinese segmentation tools to cut test dataset, and apply the pre-trained model segmented by jieba[4] to classify each test dataset. Jieba is not the best Chinese segmentation tool, but the consistent rules in the training and test environment maximize the performance of model.

---

[4] https://github.com/fxsjy/jieba.

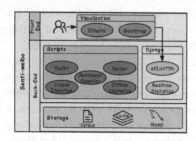

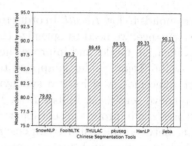

Fig. 2. Architecture of S^{2}AP   **Fig. 3.** Chinese segmentation tools

Front-end visualization provides interface to interact with back-end. Figure 4(a) represents the daily sentiment trend of Huawei from March to July. The negative sentiment of Huawei since May 16 accurately reflects the event that Huawei is added to the Entity List. Figure 4(b) reflects the sentiment trend of China-US trade from May 29 to May 30 in a five-minutes frequency, which corresponds to the intense emotional changes of netizen to the anchors' debate between China and US. Another feature of platform is the real-time analysis to reflect the latest sentiment trend, which can be available by visiting S^{2}AP.

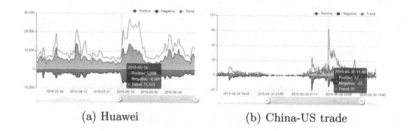

(a) Huawei   (b) China-US trade

Fig. 4. Sequential sentiment analysis

4   Conclusion

In this work, we aim at the construction and iteration of Chinese Weibo sentiment analysis corpus. More datasets can be constructed using Information Retrieval and we will continue to abstract the problem of corpus construction. The details of S^{2}AP can be available on GitHub[5], including introduction video, design of pre-processing tool, implement of web demonstration and others.

---

[5] https://github.com/wansho/senti-weibo.

References

1. Go, A., Bhayani, R., Huang, L.: Twitter sentiment classification using distant super-vision. CS224N Project Report, Stanford **1**(12), 2009 (2009)
2. Haddi, E., Liu, X., Shi, Y.: The role of text pre-processing in sentiment analysis. Procedia Comput. Sci. **17**, 26–32 (2013)
3. Iosifidis, V., Ntoutsi, E.: Large scale sentiment learning with limited labels. In: Proceedings of the 23rd ACM SIGKDD International Conference on Knowledge Discovery and Data Mining, pp. 1823–1832. ACM (2017)
4. Joulin, A., Grave, E., Bojanowski, P., Mikolov, T.: Bag of tricks for efficient text classification. arXiv preprint arXiv:1607.01759 (2016)
5. Yang, J., Weiran, X.U., Tan, S.: Task and data designing of sentiment sentence analysis evaluation in coae2014. J. Shanxi Univ. (2015)
6. Zhang, A., Li, B., Wan, S., Wang, K.: Cyberbullying detection with BiRNN and attention mechanism. In: International Conference on Machine Learning and Intelligent Communications, pp. 623–635. Springer (2019)

An Efficient Secondary Index for Spatial Data Based on LevelDB

Rui Xu, Zihao Liu, Huiqi Hu$^{(\boxtimes)}$, Weining Qian, and Aoying Zhou

School of Data Science and Engineering, East China Normal University,
Shanghai 200062, People's Republic of China
{51185100032,51185100023}@stu.ecnu.edu.cn,
{hqhu,wnqian,ayzhou}@dase.ecnu.edu.cn

Abstract. Spatial data has the characteristics of spatial location, unstructured, spatial relationships, massive data. However, the general commercial database itself is difficult to meet the requirements, it's non-trivial to add spatial expansion because spatial data in KVS has brought new challenges. First, the Key-Value database itself does not have a way to query key from its value. Second, we need to ensure both data consistency and timeliness of spatial data. To this end, we propose a secondary index based on LevelDB and R-tree, it supports two-dimensional data indexing and K-Nearest Neighbor algorithm querying. Further, we have optimized the query of a large amount of spatial data caused by the movement of objects. Finally, we conduct extensive experiments on real-world datasets which show our hierarchical index has small index and excellent query performance.

Keywords: Secondary index · R-tree · LevelDB

1 Introduction

Based on frequently accessed columns, the secondary index is used to promote database system query performance. In a relational database, a secondary index is usually built on a non-primary key column, pointing to the primary key. The steps of query via secondary index are as follows: (1) find the primary key from the non-primary key through the secondary index, (2) find the required data according to the primary key. Similarly in the spatial database, we use spatial coordinates value of the non-primary key to find its primary key via the secondary index. A spatial database is a database that is optimized for storing and querying data that represents objects defined in a geometric space. The secondary index makes it possible to store and query large spatial data and its performance directly affects the overall performance of spatial database. Otherwise, any queries for features require a "sequential scan" of each entry in the database.

At present, in scientific research such as geological surveys, urban planning and various life services based on location services, plenty of data related

© Springer Nature Switzerland AG 2020
Y. Nah et al. (Eds.): DASFAA 2020, LNCS 12114, pp. 750–754, 2020.
https://doi.org/10.1007/978-3-030-59419-0_50

to spatial two-dimensional coordinates are generated. However, the traditional relational database can't provide better support in the scenario where spatial data generates in high frequency. Meanwhile, the new NOSQL (Not only SQL) database lacks the design of the secondary index [3]. Take the short-distance delivery system for takeaway applications as an example, the issue of delayed deliveries is becoming more serious currently. To solve it, not only should the merchant notify the nearest delivery man as soon as possible, but the delivery clerk should also be tracked by real-time location. Our solution is as shown in Fig. 1, when a user places an order for delivery, LevelGIS will receive a request. And the LevelGIS will get all the couriers (value/candidate data-red dot) within five kilometers of the restaurant and return the nearest one (key) to user. If need, we can keep tracking by rider id (key). Compared to quadtree and k-d tree, MBR (Minimal Bounding Rectangle) of R-tree is very friendly to massive location-related data.

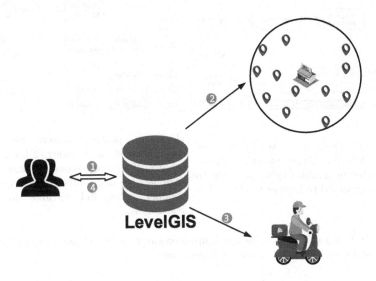

Fig. 1. Find the closest rider (key) from the coordinate (value) via LevelGIS. (Color figure online)

In this paper, we present a database system that combines LevelDB with R-tree [4], called as LevelGIS, to support 2D coordinate storage and index. Main contributions are summarized as follows:

- We propose a hierarchical index to promote data queries, design a corresponding secondary index for the memtable and L_0 (Level 0) individually, use a large R-tree to store the index of all the SSTable of L_1 and above, and finally form a three-layer structure of index.
- On the basis of LSM, LevelGIS implements a complete secondary index storage, query and maintenance solution. Meanwhile, it shows excellent read and write performance on two-dimensional coordinate date.

2   System Architecture and Key Techniques

Write and Delete. When performing a write operation, LevelDB first encapsulates "put" into "writebatch", and then directly dumps it into memory, finally performs persistence during merge phase, as we can see in Fig. 2(a). Likewise, LevelDB inserts a new value or a delete flag for "update" and "delete" instead of changing the original data immediately resulting in the key will have multiple values of different versions distinguished by sequence number. Considering the secondary index needs to maintain consistency with the primary index, Level-GIS has the characteristics of instant insertion and delayed update, as shown in Fig. 2(b).

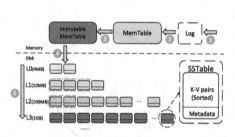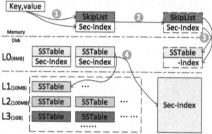

Steps of inserting a key-value pair: (1)the log file;(2)the memtable;(3)the immutable memtable;(4)dumping file to L_0;(5)compacted to further levels.

Insertion process:(1)Insert new data into skiplist and index;(2)Lock filled skiplist;(3)Dump secondary index and skiplist to disk directly;(4)Merge secondary index and SSTable downward separately.

Fig. 2. System architecture. The left figure shows architecture of the standard LevelDB. The right figure shows the LevelGIS architecture.

Read and Consistency. The query result is primary key due to the secondary index, but it can't guarantee whether the value is valid. Taking the 10NN query as an example, the query needs to traverse all the secondary indexes in parallel and verifies the sorted results after aggregation. The steps of verification are as follows: (1) the sequence number of the result needs to be judged whether it exceeds the specified sequence number by the current query; (2) the validation of query value needs to be checked via primary index. Validated results are put into queue. For the reason that secondary verification will greatly reduce the query efficiency of secondary indexes, LevelGIS maintains sequence number to reduce the number of verification.

Merge. Subsequently, the data structure of LevelDB is LSM tree (log-structured merge-tree). In terms of SSTable size and merge frequency, SSTables of L_0 have small size, the files are out of order externally, the merge frequency is high. On the contrary, SSTables above L_1 have large size, ordered and the merge frequency is low. So LevelGIS has hierarchical index like the right of Fig. 2(b). Each secondary index consists of an independent R-tree. Specifically, LevelGIS has a secondary index for L_0 in order to reduce merge frequency and improve query performance.

3 Demonstration

We conduct extensive experiments on real-world datasets to evaluate the trade-offs between different indexing techniques on various workloads. Similar dataset supported by Didi Chuxing can be found via GAIA open dataset [1]. There are one million entries of which key is 32B and of which value is 24B (six decimal places as a string). The comparison of experiment is on hierarchical index of R-tree [2], one index of R-tree and the origin non-secondary index LevelDB. As shown in Fig. 3(a), LevelGIS has better write performance than normal secondary-index which takes entire R-tree as a whole. Afterward, we select multiple percentages of candidate data, and then choose the closest from candidate data, it takes 28 ms for LevelGIS to select 0.1% of candidate data (key-8B) as shown in Fig. 3(b).

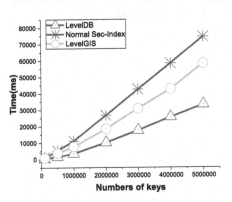

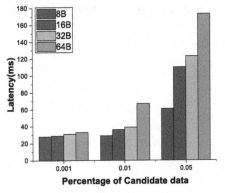

Insertion performance comparison..

Select candidate data on different key sizes.

Fig. 3. Experimental demonstration. The left picture shows the write performance and the right shows the read performance.

4 Conclusion

In this paper, we present a comprehensive study on LevelDB, exploring various design options to strike a good balance between query and write performance.

Instead of building R-tree as a whole index, we make it into hierarchical index. This performance improvement allows succinct tries to meet the requirements of a much wider range of real-world applications. Experimental results show that our LevelGIS outperforms state-of-the-art methods by a substantial margin.

Acknowledgments. This work is supported by National Key R&D Program of China (2018YFB1003404), Youth Program of National Science Foundation of China under grant number 61702189, Youth Science and Technology - Yang Fan Program of Shanghai under Grant Number 17YF1427800. Thanks to corresponding author Huiqi Hu.

References

1. Jin, J., et al.: CoRide: joint order dispatching and fleet management for multi-scale ride-hailing platforms. CoRR abs/1905.11353 (2019)
2. Luo, C., Carey, M.J.: LSM-based storage techniques: a survey. VLDB J. **29**(1), 393–418 (2020). https://doi.org/10.1007/s00778-019-00555-y
3. Qader, M.A., Cheng, S., Hristidis, V.: A comparative study of secondary indexing techniques in LSM-based NoSQL databases. In: Das, G., Jermaine, C.M., Bernstein, P.A. (eds.) Proceedings of the 2018 International Conference on Management of Data, SIGMOD Conference 2018, Houston, TX, USA, 10–15 June 2018, pp. 551–566. ACM (2018)
4. Sellis, T.K.: Review - the R*-Tree: an efficient and robust access method for points and rectangles. ACM SIGMOD Digit. Rev. **2**, 173–182 (2000)

A Trustworthy Evaluation System Based on Blockchain

Haokai Ji[1,2], Chundong Wang[1,2(✉)], Xu Jiao[1,2], Xiuliang Mo[1,2],
and Wenjun Yang[1,2]

[1] The Ministry of Education Key Laboratory, Tianjin University of Technology,
Tianjin 300384, China
michael3769@163.com
[2] Tianjin Key Laboratory of Intelligence Computing and Novel Software Technology,
Tianjin University of Technology, Tianjin 300384, China

Abstract. With the development of the Internet, online shopping, online movies, online music and other online services are greatly convenient for people's life. However, due to the widespread false information in these websites, the interests of users and privacy are threatened. To address this problem, we designed a trustworthy evaluation system based on blockchain named *TESB2*. *TESB2* uses blockchain technology to effectively protect the data and privacy of users. It effectively guides the evaluation behavior of users and ensures the fairness and credibility of evaluation information by combining the evaluation behavior of users with reputation rewards and punishments. And we innovatively propose a malicious user detection method by combining blockchain technology and machine learning algorithm. Finally, we showed *TESB2* through the movie scenario, and its performance was satisfactory.

Keywords: Trustworthy · Evaluation · Blockchain

1 Introduction

In recent years, with the development of the Internet, we have enjoyed great convenience through Internet shopping, listening to music, watching movies and so on. When people enjoy online services, they will check the ranking of products and product reviews written by other users. However, in order to pursue improper interests, some manufacturers use malicious users to inject a lot of false evaluation information into the scoring system. So as to improve the ranking and sales of their products, suppress the products of competitors, damage the interests of other manufacturers and normal users. The evaluation system of online service website is disturbed because of the lack of effective measures to reasonably restrict user behavior.

In order to solve the above problems, we propose a trustworthy evaluation system based on blockchain named *TESB2*. Blockchain technology has become a powerful tool to construct the monetary system and financial system [1,2].

Y. Nah et al. (Eds.): DASFAA 2020, LNCS 12114, pp. 755–759, 2020.
https://doi.org/10.1007/978-3-030-59419-0_51

Inspired by this, we first apply blockchain technology to the evaluation system of online services. *TESB2* has the following advantages: (1) It effectively protects the users' privacy, ensures the data not to be tampered, and protects the data security. (2) It can effectively reduce malicious users. The traditional evaluation system has malicious behavior because of less effective measures to reasonably restrict users. *TESB2* associates evaluation behavior with user interests, and guides users to consciously maintain the network ecological environment of the evaluation system. (3) It combines machine learning algorithm with blockchain technology to effectively detect malicious users, and uses decentralized blockchain technology to establish a high trust environment to ensure the credibility of data collected through user prosecutes. Using the advantage of high efficiency of centralized database operation, machine learning algorithm detection can improve the detection efficiency and reduce the false alarm rate.

2   System Design

Figure 1 is the main architecture of the system, which has three main function modules: *user credit evaluation module, data processing module* and *malicious user detection module*. *User credit evaluation module* is used to evaluate and constrain users' behavior. *Data processing module* is used to realize the data interactive flow of the system. The *malicious user detection module* combines machine learning and blockchain technology to detect malicious users. Through the system evaluation interface, users can evaluate and grade the products while browsing, and prosecute the users who make false and malicious evaluation. The administrator can view the blockchain data storage and personal data information of users.

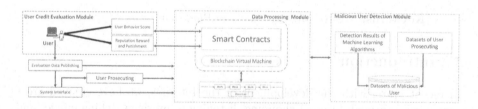

Fig. 1. The architecture of system.

2.1   User Credit Evaluation Module

This module mainly includes *user behavior score* and *credit reward and punishment*. User behavior score is a reasonable measure of user behavior and a standard for objective evaluation of user behavior. The increase or decrease of the score is managed by smart contract. When users conduct normal evaluation, or effectively prosecute malicious users, and maintain the fairness and credibility of the evaluation system, *TESB2* will increase its score according to the contract. On the contrary, if users are detected as malicious users, *TESB2* will reduce their score according to the contract. If a user has a high score, it means that it is a lofty reputation user, otherwise it is a dishonest user. Reputation reward and punishment are based on the reputation of users. If users are lofty reputation users, *TESB2* will give token rewards according to the contract. If users are dishonest users, *TESB2* will punish users according to the contract, such as reducing tokens or even restricting user behavior.

2.2 Data Processing Function Module

This function module is mainly responsible for the storage and interaction of the whole system data. The data of system includes the evaluation data published by users, the personal privacy data, the relevant data recording the user's behavior, malicious user detection data, etc. The data between each functional module is stored in the block, and interacted through the smart contract, which can effectively ensure the fair implementation of user reputation evaluation [3]. *TESB2* adopts the distributed P2P data storage environment built by the blockchain system. It is based on the popular open-source blockchain platform Ethereum, and uses the Ethereum client to build a decentralized network to support the intelligent contract interaction operation and information storage.

2.3 Malicious User Detection Module

This module uses machine learning algorithm to detect malicious users. In order to improve the detection efficiency, we build the detection function module on the centralized database, collect data from the blockchain, conduct data processing and detection in the centralized database, and then feed back the detection results to the blockchain, which is automatically processed by the smart contract. The detection algorithm of Naive Bayesian classification is used in this module. The classification efficiency of this algorithm is stable and robust. For details, please refer to [4].

   In order to improve the accuracy of detection and reduce the false alarm rate, we let lofty reputation users prosecute suspicious users. Because the prosecution behavior of users is combined with reputation rewards and punishments, and they should bear the corresponding rewards and punishments results, the user prosecuting data in the blockchain is reliable [5]. Finally, we can get the malicious user set by cross screening the suspicious user set and the data set detected by machine learning. Finally, the smart contract traces the hash address of the block to the corresponding malicious user.

3 Demonstration

This demo is based on the well-known public blockchain platform Ethereum as the development platform. Based on the current popular truffle framework, and the front end uses bootstrap framework to develop user pages. This demo can be utilized in multiple scenarios, not only for PC, but also for mobile. In this demo, we mainly show the evaluation system based on a movie scenario, using the commonly open data set MovieLens[1]. It has 20 million ratings and 465,000 tag applications applied to 27,000 movies by 138,000 users. It includes tag genome data with 15 million relevance scores across 1,129 tags. Figure 2(a) is the evaluation interface of the system. Users can browse the movie information, grade the movie, browse the evaluation information of other users, and prosecute the evaluation score of other users. Figure 2(b) is the user data viewing interface, from which you can see the user's behavior data. The administrator can also view the data records of the system. Figure 2(c) is the background data monitoring interface. The administrator can view the data records of the system, and can see the current block height, evaluation quantity and other information of the blockchain. The administrator can also monitor the block height, size, hash address, and generation time of each record data.

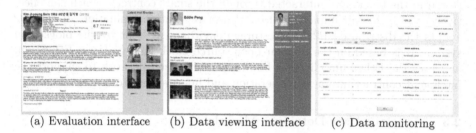

(a) Evaluation interface   (b) Data viewing interface   (c) Data monitoring

Fig. 2. System interface

4   Conclusion

The current evaluation system can not solve the problem of the authenticity and credibility of evaluation information. In this demonstration, we propose a good solution to protect the privacy interests and data security of merchants and users, save time, manpower and maintenance costs, and have a broad application prospect.

Acknowledgement. This work is supported by the General Project of Tianjin Municipal Science and Technology Commission under Grant (No. 15JCYBJC15600), the Major Project of Tianjin Municipal Science and Technology Commission under Grant (No. 15ZXDSGX00030), NSFC: The United Foundation of General Technology and Fundamental Research (No. U1536122).

---

[1] https://grouplens.org/datasets/movielens/.

References

1. Yuan, F., Wang, F.: Blockchain: the state of the art and future trends. Acta Automatica Sinica **42**(4), 481–494 (2016)
2. Zhu, L., Gao, F., et al.: Survey on privacy preserving techniques for blockchain technology. J. Comput. Res. Dev. **54**(10), 2170–2186 (2017)
3. Wang, J., Li, M., et al.: A blockchain based privacy-preserving incentive mechanism in crowdsensing applications. IEEE Access **6**, 17545–17556 (2018)
4. Wu, Z., Zhuang, Y., et al.: Shilling attack detection based on feature selection for recommendation systems. Acta Electronica Sinica **40**(8), 1687–1693 (2012)
5. Tso, R., Liu, Z., et al.: Distributed e-voting and e-bidding systems based on smart contract. Electronics **8**(4), 422 (2019)

An Interactive System for Knowledge Graph Search

Sinha Baivab[1], Xin Wang[2(✉)], Wei Jiang[1], Ju Ma[1], Huayi Zhan[1],
and Xueyan Zhong[2]

[1] Sichuan Changhong Electric Co., Ltd., Mianyang, China
{baivabsinha,wei.jiang,ju.ma,huayi.zhan}@changhong.com
[2] Southwest Petroleum University, Chengdu, China
xinwang.ed@gmail.com, zhongxueyan@sohu.com

Abstract. Recent years, knowledge graphs (KG) have experienced rapid growth since they contain enormous volume of facts about the real world, and become the source of various knowledge. It is hence highly desirable that the query-processing engine of a KG is capable of processing queries presented in natural language directly, though these natural language queries bring various ambiguities. In this paper, we present KGBot, an interactive system for searching information from knowledge graphs with natural language. KGBot has the following characteristics: it (1) understands queries issued with natural languages; (2) resolve query ambiguity via human-computer interaction; and (3) provides a graphical interface to interact with users.

1 Introduction

Recently, knowledge graphs have received tremendous attention in academia and industry, since they organize rich information with structured data, have become important resources for supporting open-domain question answering. However, with the volume of data presented and complex methods developed to retrieve the information, the task of providing a user-friendly query system for the casual users is challenging. Keyword-based search engines (*e.g.*, Google and Bing) have gained a huge success. Nevertheless, keyword searches may not be expressive enough to query structured data since keywords alone may not be able to well capture uses' search intention [8]. Structured query languages *e.g.*, SPARQL [4] and GRAPHQL [3] are developed for graph search; while they are too complicated for casual end users to write viable structured queries [10]. To facilitate search on knowledge graphs, practitioners advocate querying with natural language, since this offer users an intuitive way to express their search intention, rather than complicated structured query languages [1,7,9]. Nevertheless, due to the linguistic variability and ambiguity [5], it is extremely difficult to understand natural language questions precisely. Let us consider the below example.

Y. Nah et al. (Eds.): DASFAA 2020, LNCS 12114, pp. 760–765, 2020.
https://doi.org/10.1007/978-3-030-59419-0_52

Example 1. Consider a fraction of a knowledge graph depicted as graph G in Fig. 1(a). In the graph, different types of entities are denoted by different colored blocks and are classified as recipe name (denoted by R_i), *e.g.*, potato pie, potato omelette, etc., type of meal (denoted by T_i), *e.g.*, lunch, breakfast, etc., ingredients (denoted by I_i), *e.g.*, eggs, potatoes, chicken, etc., flavour (denoted by F_i), *e.g.*, sweet, salty, etc., cooking accessories (denoted by A_i), *e.g.*, curry power, scallion, etc., method of cooking (denoted by M_i), *e.g.*, baking, fried, etc. Furthermore, each recipe has additional properties, like recipe steps, time taken to prepare the dish and a brief introduction of the dish. These entities are linked with each other via different relations (denoted by *rel* : *i*) like type, taste, ingredients, etc. Now, suppose that one user wants to find *a baking recipe with ingredients eggs and potatoes* from the knowledge graph G. There may exist multiple dishes (nodes in G) that use potatoes and eggs (like R_1 and R_2) although they differ in flavour, time of meals, etc, as the answer of the ambiguous question. Note that, in our daily communication, when a person is facing an ambiguous question, he will try to make it clear by asking questions back to the questioner [2,10]. Likewise, in order to understand the user's intention better, a system needs to interact with users to eliminate ambiguities that appeared in questions. These interactions will help the system in understanding users better and return the most appropriate result to them.   □

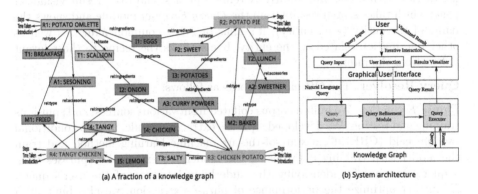

(a) A fraction of a knowledge graph   (b) System architecture

Fig. 1. Knowledge graph & system architecture

To implement this natural interactive process [5], we present KGBot, a novel system that takes natural language queries as input and effectively searches items for users via multiple round interactions. Its benefits are three folds: (1) since users know exactly what they want, letting users refine queries can achieve better accuracy; (2) knowledge graphs specify the structural relations among the mapping objects (entities/concepts/predicates) of the phrases in the question, this helps the formulation of queries; (3) the search process is combined with the query formulation, hence the overall system is less resource hungry and more cost-efficient. To the best of our knowledge, KGBot is among the first effort to search knowledge graphs with natural language queries and find the answer via multiple round iterations. It should also be remarked that recipe searching is just one application of this technique, one may apply the technique to find, for example, movies, places, persons, etc.

2   System Architecture

The architecture of KGBot is shown in Fig. 1(b). It consists of four modules. (1) *A Graphical User Interface* (GUI) provides a graphical interface to receive natural language queries asked by users, interact with users for question disambiguation and display query results to users. (2) *A Query Resolver* that performs preliminary analysis for queries, *e.g.*, key phrase extraction from natural language queries. (3) *A Query Refinement Module* (QRF) that fine tunes users' queries by asking questions back to them. (4) *A Query Executor* which uses refined queries to search the knowledge graph and returns final results to users. In the following, we elaborate each module in detail.

Graphical User Interface. The GUI module takes natural language queries as input, interacts with users and visualizes the final results. It consists of (1) a query input page, which accepts the natural language queries from users and sends queries to *Query Resolver*; (2) a user interaction page, which supports interaction with users, and iteratively refines queries; and (3) a result visualizer page; which takes search results from the *Query Executer* module and shows the name of the entity (*e.g.*, entity), it's alias or other common name, as well as other information related to the entity. It also links the recipe to the relevant videos from the internet.

Query Resolver. The QR module works as follows.

Phrase Extraction. We built a corpus collated from various sources, and split it as training and testing. We developed a phrase extraction model using a conditional random field (CRF) [6], a state-of-the-art machine learning method and train the model with training data. Using the model, we extract key phrases from a query sentence for identifying the underlying intention of the user's query. To further optimize the performance of phrase extraction, we also built up a dictionary to maintain common phrases (with occurrence frequency), by mining from query logs. In the run-time, the phrases detected from a query sentence with CRF model are further verified with the dictionary. Then high accuracy of phrase detection can be achieved.

Query Understanding. Given the phrases extracted, the query resolver needs to better understand the query based on the phrases. Since otherwise, there may exist an excessive search results, which leads to 1) low relevance of search results, and 2) high computational cost and long response time. These call for techniques for query understanding [8], which still remains an open problem. Considering the application domain (*e.g.*, recipe suggestion), we developed a viable method. Specifically, we have defined a set of phrase-phrase relationship, *e.g.*, the classifier relationship, the recipe relationship and the ingredient relationship, and stored them in a lookup table. We then use the relationship to formulate the queries, *i.e.*, based on the phrases extracted and appropriate relationship to finalize a structured representation Q, referred to as a pattern query.

Query Refinement Module. The QRF module interacts with users and gradually refines queries. The target of the interaction is towards eliminating ambiguity of a query. The module consists of two submodules: a submodule for asking questions to users; and a submodule for receiving and maintaining replies from users. The replies from users are then used to further refine queries. In QRF, we also use a query prefetching technique by exploiting the latency during the interaction, *i.e.*, we fetch necessary information from the knowledge graph or other tables during the time of user response at backend. Despite of benefiting from interaction, it will degrade users' experience if there are too many rounds of interactions. In light of this, QRF limits the number of interaction rounds to keep users engaging. We remark that the number of interaction rounds can be specified by system operator, depending on, *e.g.*, total response time, size of result set and analysis on users' feedback.

Query Executer. The Query Executer is responsible for generating the final answer for the user's query. The search procedure works in a graph pattern matching manner. More specifically, a structured representation Q of phrases detected along with constraints imposed by QRF are pushed onto the knowledge graph for matching evaluation. As will be demonstrated, owing to the constraints, the evaluation of graph pattern matching can be performed efficiently, and better still, the result set is often quite small, which facilitates users' inspection and understanding. As soon as the intended answer is returned, it is visualized and shown to the user using the GUI.

3   Demonstration Overview

The demonstration is to show the interaction with users and the result quality.

Setup. To demonstrate KGBot, we constructed a recipe knowledge graph which consists of 338192 recipes in total gathered from various sources. The record includes the recipe name with various attributes. The back-end of the system is implemented by Python and PHP, deployed on a virtual machine with 3.30 GHz CPU and 8 GB Memory.

Interacting with KGBot. We invite users to use the GUI, from issuing natural language queries to intuitive illustration of query results. (1) Query Input Page allows users to input natural language queries to find recipes. For example, a user issues a natural language query "please suggest me a baking recipe which contains potatoes and eggs", with text or audio (integrated with a Speech-to-Text module). (2) Through the Interaction Page, users can interact with KGBot by answering well selected questions for disambiguation. Like as shown in Fig. 2 (I), the system KGBot asks the user to refine the query via three questions. These questions will be asked using a drop-down list in the user interface, moreover, the system also aims at minimizing the number of questions to fully understand the user's intention. (3) The GUI provides intuitive ways to help users interpret query results. In particular, the GUI allows users to browse (a) all the final matches of Q. As an example, the query result of Q after refinement is shown in Fig. 2 (II).

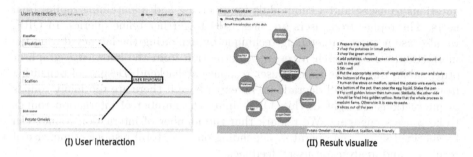

(I) User Interaction   (II) Result visualize

Fig. 2. User interaction and result visualize

Performance. The main overhead of the system on the server side comes from the phrase extraction, query formation and query evaluation. To develop a more efficient system, we have created a dictionary which contains the phrases related to the domain of recipe. For a better user experience, we load this phrase dictionary in memory, then once the system is initiated, the phrase detection works efficiently.

Summary. This demonstration aims to show the key idea and performance of a natural language Question-Answering system that employs user interaction to refine the query. The system is able to (1) understand natural language queries asked by a user, (2) effectively interact with users to remove ambiguities in the query, (3) find matches by integrating prefetch technique to utilize the interaction latency. These together convince us that the KGBot can serve as a promising tool for recipe search in real-life scenarios and this technique can be extended to other systems, for example, for searching movies, places, persons, etc as well.

References

1. Cheng, S., Xu, W., Mueller, K.: Colormapnd: a data-driven approach and tool for mapping multivariate data to color. IEEE Trans. Vis. Comput. Graph. **25**(2), 1361–1377 (2019)
2. Fader, A., Zettlemoyer, L., Etzioni, O.: Open question answering over curated and extracted knowledge bases. In: The 20th ACM SIGKDD International Conference on Knowledge Discovery and Data Mining, KDD 2014, New York, USA - 24–27 August 2014, pp. 1156–1165 (2014)

3. He, H., Singh, A.K.: Graphs-at-a-time: query language and access methods for graph databases. In: Proceedings of the ACM SIGMOD International Conference on Management of Data, SIGMOD 2008, Vancouver, BC, Canada, 10–12 June 2008, pp. 405–418 (2008)
4. Hogenboom, F., Milea, V., Frasincar, F., Kaymak, U.: RDF-GL: a SPARQL-based graphical query language for RDF. In: Chbeir, R., Badr, Y., Abraham, A., Hassanien, A.E. (eds.) Emergent Web Intelligence: Advanced Information Retrieval. AI&KP, pp. 87–116. Springer, London (2010). https://doi.org/10.1007/978-1-84996-074-8_4
5. Kaufmann, E., Bernstein, A.: Evaluating the usability of natural language query languages and interfaces to semantic web knowledge bases. J. Web Semant. **8**(4), 377–393 (2010)
6. Lafferty, J.D., McCallum, A., Pereira, F.C. N.: Conditional random fields: probabilistic models for segmenting and labeling sequence data. In: Proceedings of the Eighteenth International Conference on Machine Learning (ICML 2001), Williams College, Williamstown, MA, USA, 28 June–1 July 2001, pp. 282–289 (2001)
7. Li, F., Jagadish, H.V.: Constructing an interactive natural language interface for relational databases. PVLDB **8**(1), 73–84 (2014)
8. Pound, J., Hudek, A.K., Ilyas, I.F., Weddell, G.E.: Interpreting keyword queries over web knowledge bases. In: 21st ACM International Conference on Information and Knowledge Management, CIKM 2012, Maui, HI, USA, 29 October–02 November 2012, pp. 305–314 (2012)
9. Russell-Rose, T., Chamberlain, J., Shokraneh, F.: A visual approach to query formulation for systematic search. In: Proceedings of the 2019 Conference on Human Information Interaction and Retrieval, CHIIR 2019, Glasgow, Scotland, UK, 10–14 March 2019, pp. 379–383 (2019)
10. Zheng, W., Cheng, H., Zou, L., Yu, J.X., Zhao, K.: Natural language question/answering: Let users talk with the knowledge graph. In: Proceedings of the 2017 ACM on Conference on Information and Knowledge Management, CIKM 2017, Singapore, 06–10 November 2017 (2017)

STRATEGY: A Flexible Job-Shop Scheduling System for Large-Scale Complex Products

Zhiyu Liang, Hongzhi Wang$^{(\boxtimes)}$, and Jijia Yang

Harbin Institute of Technology, Harbin, China
{zyliang,wangzh,jijiayang}@hit.edu.cn

Abstract. Production scheduling plays an important role in manufacturing. With the rapid growth in product quantity and diversity, scheduling manually becomes increasingly difficult and inefficient. This attracts many researchers to develop systems and algorithms for automatic scheduling. However, existing solutions focus on standard flexible job-shop scheduling problem (FJSP) which requires the operations of each job to be totally ordered, while in reality they are usually partially sequential, resulting in a more general and complicated problem. To tackle this problem, we develop STRATEGY, a light-weight scheduling system with strong generality. In this paper, we describe the main features and key techniques of our system, and present the scenarios to be demonstrated.

Keywords: Scheduling system · Big data · Genetic algorithm

1 Introduction

Production scheduling is one of the most important tasks in manufacturing. Its target is to assign a series of manufacturing jobs to limited resources, such as equipments and workers, so that to minimize the makespan. Traditionally, the task is performed by experienced engineers. However, production is gradually evolving from traditional batch mode to large-scale customized manufacturing where every product is unique in processing, which will cause an explosion of the job and resource data in quantity and diversity. Thus, it becomes increasingly difficult to make a good scheduling plan manually, motivating people in both research and industry to tackle the large-scale scheduling problem automatically by developing algorithms and systems [3].

Generally, the production scheduling task can be formalized as a flexible job-shop scheduling problem (FJSP) [2], which has been proved to be NP-hard [3]. Over the last few decades, many optimization techniques have been proposed to solve FJSP and several tools are developed for scheduling, such as LEKIN [1] system and TORSCHE toolbox [5]. However, existing algorithms and tools can only handle FJSP when the operations of each jobs are totally sequential, which

© Springer Nature Switzerland AG 2020
Y. Nah et al. (Eds.): DASFAA 2020, LNCS 12114, pp. 766–770, 2020.
https://doi.org/10.1007/978-3-030-59419-0_53

is called standard FJSP, while in reality these operations are usually partially ordered. Since the partially ordered operations are mainly required by complex products such as assembly parts, we name the scheduling problem with this kind of operations as FJSP-CP (flexible job-shop scheduling problem for complex products). Apparently, FJSP-CP is an extension of standard FJSP which is more general and complicated.

To meet the requirement of large-scale customized production planning in reality, we develop STRATEGY, a light-weight flexible job-shop scheduling system for complex products. To the best of our knowledge, our system is the first one focusing on FJSP-CP. Compared with existing scheduling tools, our system has the following advantages.

- *Strong Generality.* STRATEGY is designed to deal with FJSP-CP, which is a superset of FJSP. It means that our system can also apply to any scheduling problem within the collections of FJSP, including single machine scheduling, parallel machines scheduling, flow-shop scheduling, flexible flow-shop scheduling, job-shop scheduling, and standard flexible job-shop scheduling.
- *Friendly UI.* Existing scheduling tools require the user to input all data of the scheduling task manually, which is inconvenient. Instead, utilizing our system, the user only needs to upload a structural task description file and input the hyper-parameters of the optimization algorithm, then the system will instantiate the scheduling problem automatically. The description file can be easily generated from the order planning and resource management system. The scheduling solution is depicted in Gantt Chart for visualization.
- *Light Weight.* Unlike existing systems that require deploying locally, our system is developed based on B/S architecture. Users can easily access the application only via a browser.

2 System Architecture and Implementation

The general structure of STRATEGY is shown in Fig. 1. The *Problem Solving* module receives the task description file and the hyper-parameters of the optimization algorithm input by the user, generates an instance of FJSP-CP from the file, and calls the algorithm from the library to solve the instantiated problem. The solution is delivered to the *Visualization* module and depicted in Gantt Chart. The system organizes all the data of a scheduling task, including the problem instance, the hyper-parameters and the solution as a scheduling project. The user can easily view and edit projects through the *Project Management* module, such as saving the current project, loading projects from the database and listing them through the *Visualization* module, searching for a project, viewing a project in detail via Gantt Chart, and deleting a project, etc.

STRATEGY is mainly implemented in Java and HTML based on B/S structure, following MVC pattern. We choose the well-known Apache Tomcat as Web server and adopt MongoDB to manage the project data.

3   Scheduling Algorithm

We adopt the widely used [3] genetic algorithm [4] to solve FJSP-CP because its powerful global searching, intrinsic parallelism and strong robustness can guarantee the effectiveness and efficiency.

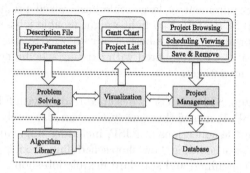

Fig. 1. System architecture

The most important procedure of genetic algorithm is the representation of the chromosome. Most existing methods are limited to jobs with totally ordered operations, which cannot apply to FJSP-CP. Thus, we propose a novel coding method that represents the order of operations in a DAG where the nodes are the operations and the edges reflect their sequence. Each directed edge is taken as a gene and the set of all genes forms a chromosome. The decoding is performed by mapping the edge set to a DAG and conducting topological sorting on it. The resulting operation series represents a scheduling solution.

Based on this coding strategy, we design a novel genetic algorithm for FJSP-CP. The key techniques of our algorithm is as follows.

Initial Population. Firstly, a basic individual is generated based on our coding method. Then, the algorithm conducts as many times random mutations as the population size on the basic individual to form the initial population. The mutation method is described below.

Fitness Function. The fitness score determines the probability an individual will be selected for reproduction. Because the goal of FJSP-CP is to minimum the makespan, i.e. the total working time of all the jobs, we define the fitness function as the reciprocal of the makespan.

Selection. After the experiment, we adopt the commonly used proportional selector in our algorithm for its contribution to efficient convergence.

Crossover. Basically, crossover is to exchange parts of genes between two selected individuals. However, in our case, if directed edges from two DAGs are exchanged, the intrinsic constrains on operation sequence could be broken, and it is computationally expensive to correct the children DAGs when they are in large scale. Thus we design an alternative reproduction method. For each pair of parents, the intersection of their edge sets are passed on to their offspring directly, and their own unique edges are inherited after being mutated. The order constrains are guaranteed by the mutation operator.

Mutation. We propose an edge mutation strategy to introduce gene diversity under the order constrains. For an edge $e_{a,b}$ from operation a to b, the algorithm selects an extra operation c and creates two new edges, $e_{a,c}$ and $e_{c,b}$ respectively, meeting that vertex v_c is not the predecessor of v_a or the successor of v_b. Then, the original $e_{a,b}$ is substituted by one of the two generated edges in probability if it doesn't exist in the chromosome.

Termination. The algorithm terminates after certain times of iterations, which is set as a hyper-parameter.

4   Demonstration Scenarios

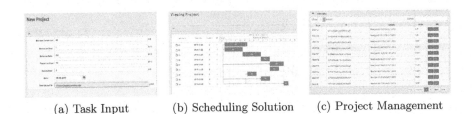

(a) Task Input   (b) Scheduling Solution   (c) Project Management

Fig. 2. System demonstration

We plan to demonstrate our STRATEGY in the following 3 parts.

- *Task Input.* Only task description file and hyper-parameters of the genetic algorithm are required to setup a new scheduling project. As shown in Fig. 2a
- *Problem Solving and Solution Visualization.* Once receiving the input, STRATEGY automatically creats an instance of FJSP-CP based on the description file and conducts the genetic algorithm on the optimization problem. The solution is visualized in Gantt Chart, as depicted in Fig. 2b
- *Project Management.* Our system provides friendly UI for users to manage their scheduling projects, such as listing, searching, viewing, saving and deleting, as Fig. 2c presents.

Acknowledgements. This paper was partially supported by NSFC grant U1866602, 61602129, 61772157, CCF-Huawei Database System Innovation Research Plan DBIR2019005B and Microsoft Research Asia.

References

1. Lekin-flexible job-shop scheduling system, October 2010. http://web-static.stern.nyu.edu/om/software/lekin/
2. Brucker, P., Schlie, R.: Job-shop scheduling with multi-purpose machines. Computing **45**(4), 369–375 (1990)
3. Chaudhry, I.A., Khan, A.: A research survey: review of flexible job shop scheduling techniques. Int. Trans. Oper. Res. (2015). https://doi.org/10.1111/itor.12199
4. Davis, L.: Handbook of Genetic Algorithms (1991)
5. Sucha, P., Kutil, M., Sojka, M., Hanzálek, Z.: TORSCHE scheduling toolbox for matlab. In: 2006 IEEE Conference on Computer Aided Control System Design, 2006 IEEE International Conference on Control Applications, 2006 IEEE International Symposium on Intelligent Control, pp. 1181–1186. IEEE (2006)

Federated Acoustic Model Optimization for Automatic Speech Recognition

Conghui Tan[1(✉)], Di Jiang[1], Huaxiao Mo[1], Jinhua Peng[1], Yongxin Tong[2],
Weiwei Zhao[1], Chaotao Chen[1], Rongzhong Lian[1], Yuanfeng Song[1],
and Qian Xu[1]

[1] AI Group, WeBank Co., Ltd., Shenzhen, China
{martintan,dijiang,vincentmo,kinvapeng,davezhao,chaotaochen,
ronlian,yfsong,qianxu}@webank.com
[2] BDBC, SKLSDE Lab and IRI, Beihang University, Beijing, China
yxtong@buaa.edu.cn

Abstract. Traditional Automatic Speech Recognition (ASR) systems
are usually trained with speech records centralized on the ASR vendor's
machines. However, with data regulations such as General Data Protection Regulation (GDPR) coming into force, sensitive data such as
speech records are not allowed to be utilized in such a centralized approach anymore. In this demonstration, we propose and show the method
of federated acoustic model optimization in order to solve this problem.
This demonstration does not only vividly show the underlying working
mechanisms of the proposed method but also provides an interface for
the user to customize its hyperparameters. With this demonstration, the
audience can experience the effect of federated learning in an interactive
fashion and we wish this demonstration would inspire more research on
GDPR-compliant ASR technologies.

Keywords: Automatic Speech Recognition · Federated learning

1 Introduction

Automatic Speech Recognition (ASR) is becoming the premise of a variety of
modern intelligent equipments. The performance of contemporary ASR systems
heavily rely on the robustness of acoustic models, which are conventionally
trained from speech records collected from diverse client scenarios in a centralized approach. However, due to the increasing awareness of data privacy
protection and the trends of strict data regulation such as the European Union
General Data Protection Regulation (GDPR) taking effect, collecting clients'
speech data and utilizing them in a centralized approach is becoming prohib-
ited. Therefore, in the new era of strict data privacy regulations, a new paradigm
is heavily needed to make it possible for the ASR vendors to consistently train
or refine their acoustic models.

The video of this paper can be found in https://youtu.be/H29PUN-xFxM.

© Springer Nature Switzerland AG 2020
Y. Nah et al. (Eds.): DASFAA 2020, LNCS 12114, pp. 771–774, 2020.
https://doi.org/10.1007/978-3-030-59419-0_54

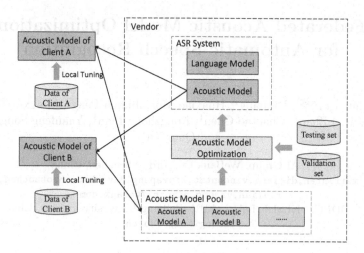

Fig. 1. System architecture

In this demonstration, we propose and demonstrate a novel paradigm of optimizing acoustic model for ASR vendors based on federated learning [2,6], which is a distributed training framework and enables the clients to collaboratively train a machine learning model without sharing the their private data. Through this demonstration, the audience would see that our proposed technique provides an effective approach for ASR vendors to refine their acoustic model in a privacy-preserving way and it potentially lay the foundation of developing more GDPR-compliant technologies for the ASR industry.

2   System Architecture

As shown in Fig. 1, our system consists of one vendor's central server and multiple clients. Private speech data is kept on the clients. The server maintains a global acoustic model, while each client has its local acoustic model, and server and clients will collaboratively optimize their models under the federated learning framework.

In each round of training, vendor will first distribute its acoustic model to all the clients, and each client optimize the received model on its local data. After summarizing the changes in the local model, only these changes are sent back to the vendor's server by using encrypted communication. After that, servers collects the updates from the clients, and tries to these updates to obtain a better global acoustic model. During the whole process, all the training data always stays on the clients' devices, and there is no worry for clients about leaking their local sensitive data to the vendor or the third-party.

Tuning on Client Data. To optimize acoustic model on the local data, transfer learning technique is adopted. More specially, we use model-based transfer learning with KLD-regularization [1], which aims at training a new model with lower training loss on the local data while keeping the KL divergence between the trained model and the vendor's original model not too large. In this way, we can obtain a new local model which is more adapted to the client's local setting but also robust to other scenarios. For the language model, it is also tuned on each client via transfer learning, but we will not synchronize it with the vendor via federated learning.

Acoustic Model Optimization. The tuned local models contains information of the extra local training data, and thus can be naturally be used for improving the global model. As a result, a new issue comes: how can we combine the updated models collected from different clients? Directly averaging them is a natural choice, but not good enough. Here we propose to merge the models by genetic algorithm [3]. In genetic algorithm, we need to maintain a set of candidate chromosomes, which are actually the encoded expressions for the models, and include all the collected models at initial stage. In each iteration of genetic algorithm, we generate new candidates by randomly mutating, crossovering and weighted-averaging the chromosomes in the candidate set. After that, we evaluate the performance of each candidate model by calculating its word error rate (WER) on the validation set, then the models with poor performances are eliminated from the set. After repeating such iterations for a certain time, we stop the algorithm and choose the candidate with lowest WER as the new global model.

3 System Implementation

The ASR system testbed is built upon the open source ASR toolkit Kaldi [4]. Without loss of generality, we utilize the Kaldi "chain" model as the acoustic model and the backoff n-gram model as the language model. The backoff n-gram language model is trained through the SRILM toolkit [5]. The whole system is deployed on multiple machines, one of which is the vendor node and the others are the client nodes. The hardware configuration of each machine is 314 GB memory, 72 Intel Core Processor (Xeon), Tesla K80 GPU and CentOS.

4 Performance Evaluation

We proceed to show a quantitative evaluation result of the proposed technique in a scenario of ten clients. The original vendor ASR system is trained with 10,000 h of speech data and each client has about 100 h of private speech data. Another 100 h of speech data is reserved for validation and testing. We compare the performances of the vendor's original acoustic model, the acoustic models tuned on clients' data, and the final optimized acoustic model of the vendor in terms of Word Error Rate (WER) on testing data. The experimental result is shown in Fig. 2. We observe that the optimized acoustic model significantly outperforms its courtparts, showing that the proposed technique can enable the ASR vendors to concisely refine its ASR system in the new era of strict data privacy regulations. In our demonstration, the audience is able to freely configure the experimental setting and observe the effects of changing the number of clients or their dataset.

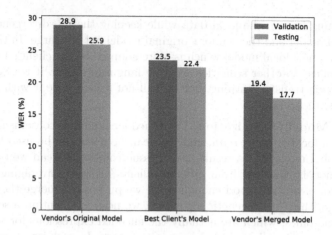

Fig. 2. Word Error Rates (WER) of each model

5   Conclusion

In this demonstration, we show a novel technique of federated acoustic model optimization, which solves the most critical problem faced by the ASR industry in the new era of strict data regulations. Through this demonstration, the audience will have a unique opportunity of experiencing how federated learning protects the clients' data privacy while provides the vendors sufficient information to further refine their acoustic models. We wish that the demonstration would shed light on developing better GDPR-compliant ASR technologies.

References

1. Huang, Y., Yu, D., Liu, C., Gong, Y.: Multi-accent deep neural network acoustic model with accent-specific top layer using the KLD-regularized model adaptation. In: Fifteenth Annual Conference of the International Speech Communication Association (2014)
2. Konečný, J., McMahan, H.B., Yu, F.X., Richtárik, P., Suresh, A.T., Bacon, D.: Federated learning: strategies for improving communication efficiency. arXiv preprint arXiv:1610.05492 (2016)
3. Mitchell, M.: An Introduction to Genetic Algorithms. MIT Press, Cambridge (1998)
4. Povey, D., et al.: The Kaldi speech recognition toolkit. In: IEEE 2011 Workshop on Automatic Speech Recognition and Understanding (2011)
5. Stolcke, A.: SRILM-an extensible language modeling toolkit. In: Seventh International Conference on Spoken Language Processing (2002)
6. Yang, Q., Liu, Y., Chen, T., Tong, Y.: Federated machine learning: concept and applications. ACM Trans. Intell. Syst. Technol. (TIST) **10**(2), 12 (2019)

EvsJSON: An Efficient Validator for Split JSON Documents

Bangjun He[1], Jie Zuo[1], Qiaoyan Feng[2(✉)], Guicai Xie[1], Ruiqi Qin[1],
Zihao Chen[1], and Lei Duan[1]

[1] School of Computer Science, Sichuan University, Chengdu, China
bangjun_he@163.com,
{zuojie,leiduan}@scu.edu.cn
[2] Dongguan Meteorological Bureau, Dongguan, China
dgqxt@163.com

Abstract. JSON is one of the most popular formats for publishing and exchanging data. In real application scenarios, due to the limitation in field length of data before storing in the database, a JSON document may be split into multiple documents if it is too long. In such case, the validation of the integrity and accuracy of documents is needed. However, this cannot be solved by existing methods. In this paper, we proposed a novel method to validate JSON documents characterized by being able to deal with split documents. Experiments demonstrated that the proposed method is efficient in validating large-scale JSON documents and performed better than the methods compared.

Keywords: JSON validation · Multiple documents · JSON schema

1 Introduction

As a lightweight and easy-to-use data-interchange format, JavaScript Object Notation (JSON) is frequently used for exchanging data. It flexibly organizes data such as records and arrays with a semi-structured model [1]. For example, from client to database, the general transmission of JSON data on the C/S system is shown in Fig. 1. Firstly, the user enters the data into a form. Then the data is received by the client and is encapsulated in the JSON format (i.e., JSON document). All documents must be validated before storing into the database. However long documents will be split into multiple ones due to the limitation in field length in database, which has made it a challenge for data validation.

Typically, existing JSON toolkits such as *evert-org* [2] and *json-schema* [3] contain two phases when validating JSON format, i.e., parsing JSON document and predefin JSON schema [4], where the corresponding JSON object and

This work was supported in part by the Guangdong Province Key Area R&D Program (2019B010940001), and the National Natural Science Foundation of China (61572332, 61972268).

© Springer Nature Switzerland AG 2020
Y. Nah et al. (Eds.): DASFAA 2020, LNCS 12114, pp. 775–779, 2020.
https://doi.org/10.1007/978-3-030-59419-0_55

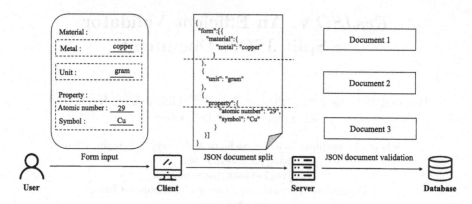

Fig. 1. JSON data transmission from client to database

schema object are obtained, respectively. By the predefined JSON schema, the structure of JSON document is restricted and its integrity can be checked [5]. By comparing the JSON object with schema object, the format of JSON document can be validated. However, the aforementioned toolkits fail to validate multiple JSON documents split from one long JSON document.

To fill this gap, we propose our method *EvsJSON* (an efficient validator for split JSON documents). The contributions of our work include:

- **Validation for split JSON documents**. We developed a novel and effective validation technique for JSON documents including the split JSON documents, considering the problem of field length limitation in database.
- **Efficient validation**. Compared with related methods, the proposed method performs better in efficiency with regards to the JSON document size.

2 Methodolody

The framework of *EvsJSON* is illustrated in Fig. 2(a). As is shown, *EvsJSON* comprises of four parts, and each one is detailed as follows:

Building Tree. A *JSON template* is first predefined, containing a batch of keys, specifying the data type of value corresponding to each key, and organizing all keys in a specific nested structure. Based on its nested structure, the template is transformed into a tree structure, called *JSON template tree*, where each node represents a key-value pair in the *JSON template* and each layer in the tree corresponds to a layer in the nested structure of the *JSON template*. An *JSON template tree* based on the JSON document in Fig. 1 is presented in Fig. 2(b).

Reading Document. A complete JSON document must be split into multiple ones if the document is too long in length. In this step, a reader is used to read JSON documents by reading data from stream. During this process, the integrity of each document will be checked. For documents split from a large one, all the

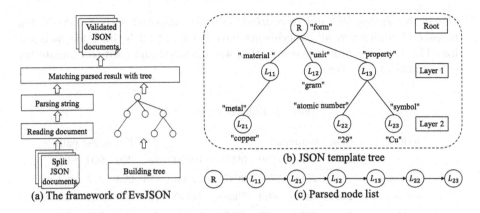

Fig. 2. Our proposed method

split parts will be concatenated into the original one. A specific ending symbol such as *EOF* is added at the end of every complete document.

Parsing Key-Value Pairs. A *jsonNode* is created based on the tree model, storing the information about each node itself as well as its parent node and child nodes. Then, by pre-order traversing, an empty node list $\mathcal{L}$ containing all tree nodes is obtained. According to the list, values corresponding to each node is parsed and extracted from each document. The information about all key-value pairs in JSON document is stored in a node list. The node list given in Fig. 2(c) is produced by traversing the *JSON template tree* in Fig. 2(b) in pre-order.

Matching Parsed Result with Tree. In this phase, each JSON document will be checked under three criteria and it will not be considered valid if: (1) any key in $\mathcal{L}$ does not appear in the JSON document; (2) the data type of the value corresponding to a key is not correct; (3) the structure of keys in the JSON document is not conformed to the order of $\mathcal{L}$. Specifically, each node will be taken out from $\mathcal{L}$ and if this node together with all its child nodes are verified to be valid, they will be removed from $\mathcal{L}$. By doing so, huge memory space can be saved when processing large document. Besides, once a node fails to pass the validation, its corresponding JSON document will be neglected at once without checking the remaining nodes. Anyway, the proposed method is able to find all mismatched nodes if necessary.

3 Experiments

We applied *EvsJSON* to a real-world JSON dataset from Booking website to test its performance in validating JSON data. The dataset comprising of customer reviews on hotels in Amsterdam was collected and provided by the Github project[1]. Four other JSON document parsing methods, i.e., *Evert-org* [2], *Json-*

---

[1] https://github.com/yipersevere/text-sentiment-classification-with-deep-neural-networks.

schema [3], *Justify* [6], and *Networknt* [7] were compared with *EvsJSON*. For the sake of sufficiency, all experiments have been run for 5 times for each setting. The average result for each setting was obtained and used to evaluate the performance of *EvsJSON*.

Table 1. Efficiency test (ms)

Method	# of Documents					# of Key-value pairs			
	100	500	1000	5000	10000	30	300	600	900
Evert-org	885	2818	4795	20479	29506	150	203	227	262
Json-schema	1200	3390	6092	23088	44240	419	490	528	550
Justify	1402	4333	7529	32696	62079	261	281	300	312
Networknt	865	2834	4608	20323	39148	253	262	265	272
EvsJSON	**834**	**2568**	**4561**	**20010**	**28913**	**87**	**115**	**143**	**155**

The number of JSON documents and the number of key-value pairs in one JSON document were tested to evaluate their impacts on *EvsJSON* in terms of efficiency. Table 1 shows the running time in milliseconds of each setting. The column of documents shows that, the total running time increases when the number of documents increases from 100 to 10000. However, *EvsJSON* costs the least time on the whole. Recall that, *EvsJSON* is characterized by validating split JSON documents. Since the documents in the original dataset are short in length, we produced the long documents by concatenating multiple short documents manually. And the number of key-value pairs in the processed documents vary from 30 to 900. As shown in the column of key-value pairs, *EvsJSON* tends to spend more time in processing data but it always spends the least time among all methods with number of key-value pairs increasing.

The above results indicate that *EvsJSON* is notably efficient in processing large-scale JSON documents as well as long JSON documents.

4 Conclusion

Validation for JSON documents is a key step in data processing. In this paper, we introduced *EvsJSON*, a novel and efficient method to validate JSON documents, especially the JSON documents split from a long one. Experiments on real-world datasets demonstrated that *EvsJSON* is better than other commonly used JSON document validating methods in efficiency.

References

1. Baazizi, M.A., Colazzo, D., Ghelli, G., Sartiani, C.: Schemas and types for JSON data: from theory to practice. In: SIGMOD Conference 2019, pp. 2060–2063 (2019)

2. Everit-org. https://github.com/everit-org/json-schema
3. Json-schema-validator. https://github.com/java-json-tools/json-schema-validator
4. Json-schema Homepage. http://json-schema.org
5. Pezoa, F., Reutter, J.L., Suárez, F., Ugarte, M., Vrgoc, D.: Foundations of JSON Schema. In: WWW 2016, pp. 263–273 (2016)
6. Justify. https://github.com/leadpony/justify
7. Networknt. https://github.com/networknt/json-schema-validator

GMDA: An Automatic Data Analysis System for Industrial Production

Zhiyu Liang, Hongzhi Wang$^{(\boxtimes)}$, Hao Zhang, and Hengyu Guo

Harbin Institute of Technology, Harbin, China
{zyliang,wangzh,haozhang,hyguo}@hit.edu.cn

Abstract. Data-driven method has shown many advantages over experience- and mechanism-based approaches in optimizing production. In this paper, we propose an AI-driven automatic data analysis system. The system is developed for small and medium-sized industrial enterprises who are lack of expertise on data analysis. To achieve this goal, we design a structural and understandable task description language for problem modeling, propose an supervised learning method for algorithm selecting and implement a random search algorithm for hyper-parameter optimization, which makes our system highly-automated and generic. We choose R language as the algorithm engine due to its powerful analysis performance. The system reliability is ensured by an interactive analysis mechanism. Examples show how our system can apply to representative analysis tasks in manufactory.

Keywords: Automatic · Interactive · Data analysis system

1 Introduction

Data-driven method has shown great advantages in optimizing production. Compared with traditional approaches which rely on numerous manual experience or complex mechanism models, data-driven approaches can mine potential rules and achieve accurate prediction for production improvement without any prior knowledge, thus becomes an importance part of industry 4.0. For instance, the Daimler Group introduces data analysis to production by deploying a comprehensive IBM SPSS Modeler [1] that increases productivity by about 25%.

While the benefits of data analysis for are obvious, the largest obstacle to its popularization is that the technicians of manufacturing lack of knowledge and experience on conducting data analysis on their own areas, including building data analysis model based on real-world problem, selecting appropriate algorithm to solve the model and tuning the hyper-parameter to achieve an admirable performance. Intuitively, a solution to tackling these problems is to adopt customized business solutions such as SPSS Modeler [1] and SAS [2], which is exactly what large international companies do. However, millions of small and medium manufacturing enterprises cannot afford such expensive data analytic service, nor do they demand for so many complicated functions provided by those softwares.

© Springer Nature Switzerland AG 2020
Y. Nah et al. (Eds.): DASFAA 2020, LNCS 12114, pp. 780–784, 2020.
https://doi.org/10.1007/978-3-030-59419-0_56

To fill this gap, we adopt AI techniques to conduct automatic data analysis [4] from demand to result demonstration, and develop an interactive data analytic system named GMDA [5]. To the best of our knowledge, our system is the first complete solution to meet the data analysis requirement of small and medium-sized manufactories. The main features of our system are as follows.

- *Generic.* We have done an comprehensively investigation on the representative analysis tasks in production and design a declarative language to describe the tasks to the system. Thus our system can process most of the data analysis tasks.
- *Highly Automated.* Once the task is declared, the system can automatically conduct end-to-end data analysis with minimum human effort.
- *Interactive.* Our system adopts an human-in-th-loop strategy for data analysis to maximally guarantee the quality of the result.

2   System Architecture and Key Techniques

The framework of GMDA is shown in Fig. 1. It consists of three major components, i.e., *Data Processor, Algorithm Selector* and *Parameter Optimizer.* Most components of GMDA are implemented in Java, while the analysis algorithm is performed by calling R language, taking full advantage of R in data analytics.

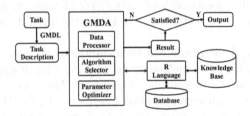

Fig. 1. System framework.

A real-world task is claimed by users through the interface, and represented by GMDL, a declarative task description language designed for non-expert users to write their data analysis requirements. GMDA receives the task description, interprets it and accomplishes the corresponding task with the following steps.

Initially, *Data Processor* loads the input data, of which each attribute is set a proper type automatically depending on its values. This strategy is typically time-saving for industrial data analysis where the data type is usually varied because the data could be generated from many different scenarios. Then, the meta-features of the input data are extracted and *Algorithm Selector* decides the most appropriate algorithm according to the task type and the meta-data. If necessary, *Data Processor* will preprocess the data, e.g. normalization, discretization and missing value imputation etc, to fit in the algorithm. Next, if the algorithm contains hyper-parameters, *Parameter Optimizer* searches for the optimum hyper-parameters. After that, GMDA runs the selected algorithm on the preprocessed data by calling R language, and returns the results. The results are finally evaluated by the user based on the actual demand. If the output is unsatisfactory, GMDA will adjust the hyper-parameters or even reselect the algorithm according to the user feedback until the outcome fills the bill.

The key techniques within our system are described below.

Task Description. We summarize representative data analysis tasks in manufacturing into four categories based on required analysis models, including association rule mining, classification, clustering and prediction. Then we design GMDL, a declarative language which describes the task requirements with a structural statement. The grammar of GMDL is as follows.

$COMMAND \rightarrow SETTING \quad COMMAND \mid SETTING$

$SETTING \rightarrow PROPERTY = "VALUE"$

where a *COMMAND* statement is used to describe the properties of a task and defined as at least two *SETTING* clauses separated by a space. Each *SETTING* clause declares the value of one of the properties of the described task and is independent with other clauses. For example, the task of forecasting the output of PCB product with the historical time-series data can be described by the following *COMMAND*.

$object = "prediction" \ sequential = "T" \ trainFile = "outputOfPCB.csv" \ target = "currentNumOfPCB"$

This statement consists of four *SETTING* clauses, declaring the analysis category is time series prediction, and assigning the input data and the attribute to be predicted.

The *PROPERTY* attribute can be instantiated by many keywords besides *object, sequential, trainFile* and *target* to guarantee its description ability. The details are listed in our system documentation. The user could interact with GMDA directly through command-line interface with GMDL, or through the GUI. In the later way, the system generates corresponding *COMMAND* automatically.

Algorithm Selection. We provide a list of candidate algorithms in GMDA system for various data analysis tasks. To support automatical analysis, we build a knowledge base for algorithm selection and propose an selection strategy based on similarity search. When a task can be conducted with multiple candidate algorithms, the system searches for the recorded dataset owning the most similar meta-features [3] with the dataset of the current task from the knowledge base, and chooses the best-performed algorithm of the retrieved dataset as the selection result. The knowledge base is established in advance by evaluating each candidate algorithm on a large number of different datasets with the accuracy as metric, and recording the features and best-performed algorithm of each dataset. It will be updated once a analysis result is accepted by the user, which instantly improves the intelligence of *Algorithm Selector*.

Hyper-Parameter Optimization. We design a heuristic search strategy for hyper-parameter tuning [5], and we have manually predefined the search space of each hyper-parameter in advance. Consequently, no extra efforts the user needs to pay for tuning the hyper-parameter. Even if the auto-tuning method fails in rare cases, the interactive analysis mechanism can ensure the resulting bad model is refused by the user, then the system will redo the analysis until achieving an acceptable result, which makes our system reliable and robust.

3   Demonstration Scenarios

We will take several representative data analysis tasks as examples to demonstrate our system.

Inventory Forecasting. The task requires to predict the product inventory over time based on the historical data. Our system models this task as a time-series prediction problem and solves it with ARIMA. The historical and predicted values are plotted in black and blue curves respectively, as Fig. 2a presents.

Automobile Evaluation. This task aims to learn the hidden preference of the customers on automobile configuration. GMDA fomulates this task as a rule mining problem and processes it by decision tree method. The main discovered rules are shown in Fig. 2b.

Tool Monitoring. The key of tool monitoring is to detect the fault based on its monitoring attributes. GMDA interprets this task as a general prediction problem. In this example, the system automatically selects multi-layer perceptron network from three candidates to predict the tool fault, resulting in an accuracy of 93%. Detection result is depicted in Fig. 2c.

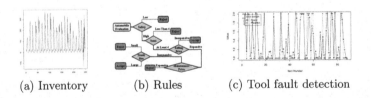

(a) Inventory   (b) Rules   (c) Tool fault detection

Fig. 2. Demonstration

Acknowledgements. This paper was partially supported by NSFC grant U1866602, 61602129, 61772157, CCF-Huawei Database System Innovation Research Plan DBIR2019005B and Microsoft Research Asia.

References

1. IBM SPSS Modeler. https://www.ibm.com/products/spss-modeler. Accessed 10 Jan 2020
2. SAS. https://www.sas.com/en_us/home.html. Accessed 10 Jan 2020
3. Castiello, C., Castellano, G., Fanelli, A.M.: Meta-data: characterization of input features for meta-learning. In: Torra, V., Narukawa, Y., Miyamoto, S. (eds.) MDAI 2005. LNCS (LNAI), vol. 3558, pp. 457–468. Springer, Heidelberg (2005). https://doi.org/10.1007/11526018_45
4. Hwang, Y., Tong, A., Choi, J.: Automatic construction of nonparametric relational regression models for multiple time series. In: International Conference on Machine Learning, pp. 3030–3039 (2016)
5. Zhang, H., Wang, H., Li, J., Gao, H.: A generic data analytics system for manufacturing production. Big Data Min. Anal. 1(2), 160–171 (2018)

An Intelligent Online Judge System for Programming Training

Yu Dong, Jingyang Hou, and Xuesong Lu[✉]

School of Data Science and Engineering, East China Normal University,
Shanghai, China
{51195100031,51195100033}@stu.ecnu.edu.cn,
xslu@dase.ecnu.edu.cn

Abstract. Online judge (OJ) systems are becoming increasingly popular in various applications such as programming training, competitive programming contests and even employee recruitment, mainly due to their ability of automatic evaluation of code submissions. In higher education, OJ systems have been extensively used in programming courses because the automatic evaluation feature can drastically reduce the grading workload of instructors and teaching assistants and thereby makes the class size scalable. However, in our teaching we feel that existing OJ systems should improve their ability on giving feedback to students and teachers, especially on code errors and knowledge states. The lack of such automatic feedback increases teachers' involvement and thus prevents college programming training from being more scalable. To tackle this challenge, we leverage historical student data obtained from our OJ system and implement two automated functions, namely, code error prediction and student knowledge tracing, using machine learning models. We demonstrate how students and teachers may benefit from the adoption of these two functions during programming training.

Keywords: Intelligent online judge · Error prediction · Knowledge tracing

1 Introduction

Today online judge (OJ) systems [3] have been widely adopted for programming education in the schools as well as in the MOOC courses. An OJ system usually consists of numerous programming problems, and authorized students may access some of the problems and submit code solutions using their preferred languages supported by the system. The prime function of an OJ system is the ability to automatically evaluate the submissions based on the comparison between the actual output of program execution and the expected output, given some predefined input. Each pair of input and output is named a "test case". Therefore for programming training, instructors and teaching assistants only need to design the problems with test cases, and OJ systems can take care of

© Springer Nature Switzerland AG 2020
Y. Nah et al. (Eds.): DASFAA 2020, LNCS 12114, pp. 785–789, 2020.
https://doi.org/10.1007/978-3-030-59419-0_57

the evaluation. This feature drastically reduces the grading workload of instructors and teaching assistants so that they can focus on the actual teaching, which enables learning at scale to some degree.

Despite the benefit of automatic evaluation, we notice two main issues of existing OJ systems in our teaching[1] that still lead to massive involvement of the teachers and thereby hinder the programming training to become more scalable. First, existing OJ systems offer next-to-zero feedback to students on the submissions of incorrect code. We refer to "incorrect code" as a piece of code that is compilable and generates wrong output for test cases. In this case, a student would only know her code has some errors but gain no additional hint on how to correct the code. The problem is even severer in the programming quiz, where students are not allowed to check the expected output of test cases[2] and have to search possible errors in a blinded manner. Second, existing OJ systems offer very little feedback to teachers on the knowledge states of students and the predictions of their future performance. As a result, teachers are not able to properly give personalized suggestions to students on improving their coding ability. These two issues have brought us relatively heavy workloads on helping the students correct solutions and improve programming skills.

In this demo work, we show two automated functions implemented in our OJ system that attempt to tackle the aforementioned challenges. First, we implement an error-prediction system that automatically predicts potential errors in the submitted code, which enables finer-grained feedback to the students and assists them on correction. Second, we implement a knowledge-tracing system that dynamically captures students' knowledge states and predicts their performance in future as they solve the problems, which facilitates personalized advising for the teachers. These two new functions would further reduce our tutoring workload, thereby scaling the programming training.

2 System Design

2.1 The Online Judge System

Our OJ system is built from OnlineJudge 2.0 open-sourced by Qingdao University[3]. The system has the separated frontend and backend, where the frontend website is built using Vue.js and the backend server is implemented using Django. In addition, the system has a module named JudgeServer, which runs a Judger Sandbox for automatic code evaluation. When receiving a code submission, the backend server sends the code in real time to the JudgeServer and obtains immediately the evaluation results. In order to generate the predictions on code errors and knowledge states, we add another module named ModelServer, which uses Flask to run the trained models and create an API for each model. After obtaining the evaluation results from the JudgeServer, the backend server calls the

---

[1] We teach C programming to first-year college students in a data science school.
[2] Otherwise students may fake the output.
[3] https://github.com/QingdaoU/OnlineJudge.

APIs by passing the required parameters and returns the model predictions to the frontend for rendering. The complete architecture of the system is shown in Fig. 1.

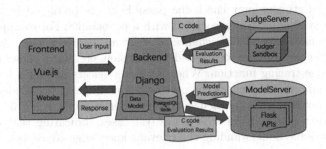

Fig. 1. The architecture of our OJ system.

2.2 The Models

We train two models for code error prediction and knowledge tracing, respectively, using historical student data in our OJ system. The error-prediction model is inspired by the work of *code2vec* [1] that attempts to predict semantic properties of code snippets. We build a similar model for the code-error prediction task using 1500 code snippets. In addition to the code embedding, we embed two more types of information in the model input, namely, the problem identity and the evaluation results on test cases. The output are the errors of the code labeled by us and each code may have multiple types of errors, that is, we build a multi-label classification model. In total, we identify 11 types of broad errors. The experimental results of the model are reported in Table 1. The knowledge-tracing model utilizes the idea of *performance factors analysis* [2]. We collect the submissions of 29 students in a C programming course, where each student has submitted code solutions to at least 100 problems. Each record of the model input represents a submission and has values pertaining to the knowledge concepts in the problem, the number of correct submissions and the number of incorrect submissions of the student in the past for each of the knowledge concepts. The output label is the ratio of the test cases that the current submission passes. Then we fit a linear regression model using the data and use it to predict the passing probabilities of new problems for each student. The experimental results are reported in Table 2.

Table 1. The error-prediction model.

Exact match	Precision	Recall	F1 score
0.587	0.681	0.669	0.674

Table 2. The knowledge-tracing model.

RMSE	MSE	MAE	R^2
0.410	0.168	0.371	0.131

3 Demonstration

In this section, we illustrate the usage of the two automated functions. Figure 2 shows the effect of the error-prediction function. In case of an incorrect code submission, a student may check the possible errors predicted by the system, where each type of error is associated with a probability. For example in Fig. 2, the student may have 99.65% chance to write an incorrect loop, and may also have 98.67% chance to use incorrect data precision, etc. Fig. 3 shows the effect of the knowledge-tracing function. When a teacher selects problems for an assignment or a quiz, she may check the passing probabilities of her students for each problem annotated with knowledge concepts. For example in Fig. 3, one student is predicted to pass only 11.28% of the test cases, indicating that the student may have trouble understanding the relevant knowledge concepts[4].

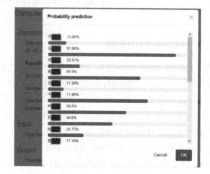

Fig. 2. Error prediction.   **Fig. 3.** Knowledge tracing.

4 Summary and Future Work

In this work, we show two intelligent functions implemented in our OJ system, namely, code error prediction and student knowledge tracing. The two functions would further reduce tutoring workloads of instructors and teaching assistants, and thereby enable more scalable programming training. In future we plan to refine our prediction models as more data are collected. Also we are designing an exercise-recommendation system based on the two functions, with the expectation of helping students identify personalized learning paths.

Acknowledgement. This work was partially supported by the grant from the National Natural Science Foundation of China (Grant No. U1811264).

---

[4] We partially anonymize the student names.

References

1. Alon, U., et al.: code2vec: learning distributed representations of code. In: Proceedings of the ACM on Programming Languages, vol. 3, no. POPL, pp. 1–29 (2019)
2. Pavlik Jr., P.I., Cen, H., Koedinger, K.R.: Performance factors analysis–a new alternative to knowledge tracing. In: Online Submission (2009)
3. Wasik, S., et al.: A survey on online judge systems and their applications. ACM Comput. Surv. (CSUR) 51(1), 1–34 (2018)

WTPST: Waiting Time Prediction for Steel Logistical Queuing Trucks

Wei Zhao[1], Jiali Mao[1(✉)], Shengcheng Cai[1], Peng Cai[1], Dai Sun[2],
Cheqing Jin[1], and Ye Guo[2]

[1] School of Data Science and Engineering, East China Normal University,
Shanghai, China
{52195100008,51195100041}@stu.ecnu.edu.cn,
{jimao,pcai,cqjin}@dase.ecnu.edu.cn
[2] Jing Chuang Zhi Hui (Shanghai) Logistics Technology, Shanghai, China
{sundai,guoye}@jczh56.com

Abstract. In the absence of reasonable queuing rules for trucks transporting steel raw materials, the trucks have to wait in long queues inside and outside the steel mill. It necessitates effective waiting time prediction method to help the managers to make better queuing rules and enhance the drivers' satisfaction. However, due to the particularity of steel logistic industry, few researches have conducted to tackle this issue. In transforming process of steel logistical informationization, huge amount of data has been generated in steel logistics platform, which offers us an opportunity to address this issue. This paper presents a waiting time prediction framework, called *WTPST*. Through analyzing the data from multiple sources including the in-plant and off-plant queuing information, in-plant trucks' unloading logs and cargo discharging operation capability data, some meaningful features related to the queuing waiting time are extracted. Based upon extracted features, a *Game-based* modeling mechanism is designed to proliferate predicting precision. We demonstrate that *WTPST* is capable of predicting the waiting time for each queuing truck, which enhances the efficiency of unloading in steel logistics. In addition, the comparison experimental results proves the prediction accuracy of *WTPST* outperforms the baseline approaches.

Keywords: Steel logistics · Queuing waiting time · Data fusion · Raw material · Machine learning

1 Introduction

Due to the absence of reasonable queuing rules, the raw material transporting trucks have to spend a long time from taking a queue number to unloading completion even if they arrive at the mill. Most of steel mills employ a rough rule of thumb by calculating the waiting time for off-plant queuing trucks in terms of the number of the trucks at the front of the line. They completely disregard the influence factors related to queuing like current unloading operation situation in the mill. This unavoidably results in long waiting queue and poor efficiency of the unloading, and hence prompts truck drivers' strong dissatisfactions. To this end, the first problem to be solved is to propose an effective waiting time prediction method to help the managers make proper queuing rules to enhance the drivers' service experience.

Queue waiting time prediction is a challenging problem and has been studied in many application areas. Various machine learning techniques have been applied to tackle it due to their strong learning ability for tremendous amount of data, such as K-Nearest Neighbor (K-NN) regression [1,2], and the other heuristic algorithms [3,4]. Nevertheless, such problem has not been carried out research in steel logistic field due to lacking of available data.

With the transformation of steel logistics information, huge amount of data has been generated and increased continuously in logistics platform. It offers us an opportunity to fuse and analyze data. We first build a large-scale data set on the waiting time prediction for raw material delivery trucks (*QWT-RMUP* for short) through fusing three real data sets from RiZhao Steel Holding Group. They record the time of each truck's complete queuing and unloading process (e.g., *QUEUE START TIME, ENTRY NOTICE TIME, ENTRY TIME, FINISH TIME*), waiting time, in-plant trucks unloading logs and warehouses working situation. Then, we need to tackle two challenges: (1) *How to generate valuable feature sets from multiple data sources?* and (2) *How to build a model which is suitable for QWT-RMUP data set?*.

2   System Overview

In this section, we present a waiting time prediction framework for steel logistics, called *WTPST*, as illustrated in Fig. 1. First, we fuse three real-world data sets from *RiZhao Steel Mill* logistics database, which include the off-plant queue, the in-plant trucks information and the cargo discharging operation capability of each warehouse, etc. This ensures the data that belongs to a same category contains all factors relative to its waiting time prediction. After that, we slice all data into two parts. One part is used for model selection and the other is used

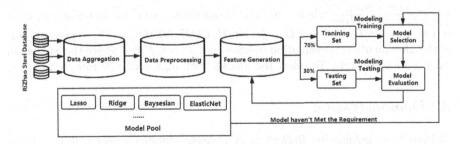

Fig. 1. *WTPST* framework

for model performance testing. Specifically, testing samples are used to measure the error between predicted waiting time and real waiting time using three evaluation indexes, i.e. the recall rates of error within half one hour ($EHOH$), error within one hour (EOH) and error within two hours (ETH). The process of feature generation and model building will be repeatedly executed until the model performance achieves satisfaction. Finally, the waiting time of each queuing trucks can be derived.

2.1   Problem Formulation

We formulate T_W as waiting time. Let $V(T_W)$ denote the waiting time of the raw material delivery truck generated by the predicted method, $R(T_W)$ denote the real waiting time of the raw material delivery truck, F_C denote the extracted features from QWT-$RMUP$ data set and M_i denote one model built by using any method in the model pool with hand-out cross validation (denoted as $HOCV$), i.e., $M_i \leftarrow HOCV(F_C, R(T_W))$. We aim to select a prediction method from the model pool which obtains a higher accuracy predicted waiting time, i.e., when satisfying $\min |R(T_W) - V(T_W)|$.

2.2   Feature Generation

Then the extracted information shall be converted into specific features by feature engineering and used for model training. Table 1 shows the final generated features for the model with a total of 324 dimensions.

2.3   Model Selection

Subsequently, to find out a model suitable for our QWT-$RMUP$ data set, we formulate the problem of model selection with a *Game-theory* based strategy as below: $Score(M_i) = \alpha * score^i_{EHOH} + \beta * score^i_{EOH} + \gamma * score^i_{ETH}$ where $score^i_{EHOH}$, $score^i_{EOH}$ and $score^i_{ETH}$ respectively represent the performance evaluation results of the model M_i according to three evaluation indexes, which include the recall of $EHOH$, EOH and ETH. $score^i_{EHOH}$, $score^i_{EOH}$ and $score^i_{ETH}$ are computed using the following formula: $score^i_j = \frac{\sum_{c \in C} P(recall_j|c)}{\sum_{k \in E} \sum_{c \in C} P(recall_k|c)}$ where c represents one category of raw material contained in category set C, and $j \in E = \{EHOH, EOH, ETH\}$. $recall_j$ represents one of the aforementioned evaluation indexes. $\alpha \in R, \beta \in R, \gamma \in R$ are the weights of each score separately. To select the models with high accuracy rate, $\alpha > \beta > \gamma$.

3   Demonstration

RiZhao Steel belongs to *RiZhao Steel Holding Group Company*. It is a super large steel joint enterprise in terms of steel production and transportation. Its logistics platform (called *JCZH*) generates about 0.5G logistics operational data

Table 1. Features generated by data analysis

Features description
One-hot category coding of the steel raw material
The amount of same-category trucks that are not called
The amount of same-category trucks that have been called but have not entered the mill
The amount of in-plant same-category trucks
The waiting time of 20 same-category trucks nearest to the current check-in truck that are not called
The waiting time of 20 same-category trucks nearest to the current check-in truck that are called but not entered the mill
The dwell time of in-plant 20 same-category trucks nearest to the current check-in truck

every day. When the raw material delivery trucks arrive at the mill, they rank to take their respective queue number at the parking place which is far away from the in-plant unloading warehouses. Only when their queue number are called, they can enter the mill for unloading. Figure 2 shows whole procedure of raw material delivery trucks queuing management application. First, the application starts when the truck checks in at the parking place, and the drivers need to fill in relevant information about goods. Then, the waiting time prediction module is activated and the status of that truck transforms from *Check in* to *Wait*. Accordingly, the waiting time of that truck is displayed on the interface, as shown in Fig. 2, the waiting time of the truck with plate number *LUL98732* is 12 min. Finally, the trucks will be called when its waiting time reduces to zero, and meanwhile *Qr code* will be showed to the driver for entering the steel mill.

Fig. 2. Illustration of queuing procedure

To evaluate the effectiveness of *WTPST*, we compared it with *DI-KNN* [1] and the original method adopted by *RiZhao Steel*. The latter is implemented simply according to calculating result of the formula $T_w = 60(min) * N$, where

794   W. Zhao et al.

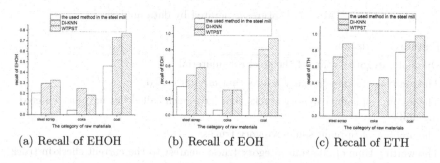

(a) Recall of EHOH   (b) Recall of EOH   (c) Recall of ETH

Fig. 3. Performance comparison results

T_w is the computed waiting time and N is the number of trucks in the front of queue. It is obviously unreasonable and has high error rate as compared to the real waiting time of the queuing trucks. The training data is generated from one week queuing data of the raw material trucks, which is about more than twenty thousands of records. The testing data is selected from the queuing data three days after the last day of the training data. We estimate the accuracy of *WTPST* according to the recall rates of *EHOH*, *EOH* and *ETH*. As shown in Fig. 3, we can see that all of the recall rates of our proposal are better than those of the other methods.

Acknowledgements. The authors are very grateful to the editors and reviewers for their valuable comments and suggestions. This work is supported by NSFC (No. 61702423, U1811264, U1911203).

References

1. Rahaman, M.S., Ren, Y., Hamilton, M., Salim, F.D.: Wait time prediction for airport taxis using weighted nearest neighbor regression. IEEE Access **6**, 74 660–74 672 (2018)
2. Bulut, M.F., Demirbas, M., Ferhatosmanoglu, H.: LineKing: coffee shop wait-time monitoring using smartphones. IEEE Trans. Mob. Comput. **14**(10), 2045–2058 (2014)
3. Bandi, C., Trichakis, N., Vayanos, P.: Robust multiclass queuing theory for wait time estimation in resource allocation systems. Manag. Sci. **65**(1), 152–187 (2018)
4. Vandenberghe, M., De Vuyst, S., Aghezzaf, E.-H., Bruneel, H.: Surgery sequencing to minimize the expected maximum waiting time of emergent patients. Eur. J. Oper. Res. **275**(3), 971–982 (2019)

A System for Risk Assessment of Privacy Disclosure

Zhihui Wang[1,2]($\boxtimes$), Siqin Li[1,2], Xuchen Zhou[1,2], Yu Wang[1,2], Wenbiao Xing[1,2], Yun Zhu[1,2], Zijing Tan[1,2], and Wei Wang[1,2]

[1] School of Computer Science, Fudan University, Shanghai, China
zhhwang@fudan.edu.cn
[2] Shanghai Key Laboratory of Data Science, Shanghai, China

Abstract. The wide spread and sharing of considerable information promotes the development of many industries such as the health care and so on. However, data owners should pay attention to the problem of privacy preservation during the sharing of data. A risk assessment system is presented in this paper. The system can assess the risk of privacy disclosure due to the sharing of data, and help data owners to evaluate whether it is safe or not to share the data.

Keywords: Risk assessment · Privacy preservation · Privacy disclosure

1 Introduction

With the rapid development of data storage and computing technologies, the amount of digital data in modern society is increasing rapidly. Various organizations generate and hold large amounts of data every day. By sharing and analyzing these data, the value of big data can be fully utilized to promote the development of many industries such as health care and so on. However, sharing these data may cause the disclosure of privacy. Data owners should make their efforts to protect private information [1].

Many research proposed to implement privacy preservation by perturbing data [2,3]. However, incorrect data may lead to misunderstanding, and accurate data are often necessary at most cases. Meanwhile, it is hard for data owners to assess the risk of privacy disclosure only through their personal intuitions. Therefore, a systematic management of data and a reasonable approach will be helpful to assess the risk of privacy disclosure.

According to the above considerations, a system for risk assessment of privacy disclosure is implemented in this paper. This system can provide data owners with the functions of data management and risk assessment of privacy disclosure during data sharing.

2 System Overview

In the system, there are two kinds of users (data owners and data users). Data owners hold and provide the data, while data users apply for the access to those

© Springer Nature Switzerland AG 2020
Y. Nah et al. (Eds.): DASFAA 2020, LNCS 12114, pp. 795–798, 2020.
https://doi.org/10.1007/978-3-030-59419-0_59

data. The risk assessment system of privacy disclosure is web-based, and it is divided into three parts: the front end, the back end and the middleware. The front end handles the user interface and acquires inputs from users. The back end implements data transfer with the database, and assesses the risk of privacy disclosure. The middleware takes charge of the interaction between the front end and the back end.

2.1 System Components

The system for risk assessment of privacy disclosure mainly includes the following components:

Management of Datasets. This component enables data owners to upload and manage their datasets. Moreover, data owners can set the status of their datasets. If the status of a dataset is set public, data users can search and apply for the access to it. Otherwise, the dataset is only visible for the corresponding data owner.

Management of Privacy Tags. This component implements the functions of adding, deleting and modifying the privacy tags for a single data attribute, the associated data attributes and the data operations. Privacy tags are represented as positive integers. The more the information might cause privacy disclosure, the greater the corresponding privacy tag is. Moreover, the privacy tags of associated attributes should be not lower than those of the corresponding separated ones, since the associated attributes generally contain more information than the separated ones. Besides, data owners can also set different privacy tags for various data operations which may be used by data users.

Submission of Data Requirement Applications. In the system, data users can customize their requirements for data and submit the corresponding data requirement applications. In a data requirement application, the data user need point out which data attributes are required and what kinds of operations will be performed on the data attributes. For example, a data requirement application may contain an attribute "age" and an operation "get average" in order to obtain the average age of a group of people.

Audit of Data Requirement Applications. Data owners can view the risk assessment results of privacy disclosure and audit the data requirement applications submitted by data users. According to the assessing results, they can determine whether it is safe enough or not to authorize the corresponding data users to access their required data.

3 Risk Assessment of Privacy Disclosure

The system is implemented with the approach, which is based on the matrix of data requirement, to assess the risk of privacy disclosure [4]. The procedure of risk assessment performs as follows:

1) Getting the data requirement and safety requirement: The system obtains the requirements from both data users and data owners. The data requirement U from a user are in the form of an application, which contains the required dataset, data attributes and data operations. The safety requirement R is the privacy tags of attributes, operations and associated attributes which are set by data owners in advance. For convenience, assume that p_f denotes the privacy tags of every single attribute, p_o denotes the privacy tags of operations, and p_r denotes the privacy tags of associated attributes.

2) Initializing the matrix of data requirement: Suppose that there are n attributes and m operations with respect to the dataset required by a data user. Let A denote an $m \times n$ matrix. Each element in the matrix is related to a specific operation on one attribute. The matrix A will be used to calculate the specific risk with respect to the user's data requirement application and the current risk interval of privacy disclosure for the required dataset. In this demonstration, the element a_{ij} of A is initialized as "1" for convenience.

3) Calculating the risk of privacy disclosure: The system traverses the matrix A row by row and if necessary modifies the value of its element a_{ij}, where a_{ij} is related to the j-th attribute and the i-th operation. If the j-th attribute and the i-th operation are in the data requirement application, the value of a_{ij} will be modified into $a_{ij} = p'_{f_j} \times p_{o_i}$, where p_{o_i} denotes the value of the i-th operation's privacy tag, and p'_{f_j} denotes the value of the j-th attribute's current privacy tag. If the i-th operation is only acted on the j-th attribute in the data requirement application, then p'_{f_j} is the privacy tag of the j-th attribute. Otherwise, the privacy tag p'_{f_j} is the largest value among the privacy tags of associated attributes F', where F' appears in the current data requirement with respect to the i-th operation and also contains the j-th attribute. Finally, the risk R_s of privacy disclosure is calculated by $R_s = \sum_{i=1}^{m} \sum_{j=1}^{n} a_{ij}$, where $a_{ij} \in A$.

4) Obtaining the current risk interval: Since the initialized matrix A means that the corresponding dataset is not yet authorized to access, the system considers the summation of elements in the initialized matrix A as the low-level risk R_l of privacy disclosure. That is, $R_l = m \times n$. The high-level risk R_h refers to the case that a data user is possible to get much more privacy information from the data requirement application. For handling this case, the system first modifies A's elements a_{kl} whose values are still 1 into $a_{kl} = p'_{f_l} \times p_{o_k}$, where p_{o_k} denotes the value of the k-th operation's privacy tag, and p'_{f_l} is the largest value among the privacy tags of associated attributes with respect to the k-th operation if there exists such associated attributes, otherwise p'_{f_l} is the privacy tag of the l-th attribute. Then, the system calculates the high-level risk $R_h = \sum_{i=1}^{m} \sum_{j=1}^{n} a_{ij}$, where $a_{ij} \in A$.

5) Converting into the risk level: In order to adjust the risk values measured on different scales to a common scale, the system first calculates the risk factor $p = \frac{R_s - R_l}{R_h - R_l}$, and then converts p into the risk level which is represented as an integer. The risk level denotes the risk assessment result of privacy disclosure. The larger the value of risk level, the higher the risk of privacy disclosure.

4   System Demonstration

The system is web-based and implemented by using Ajax, MySQL database, HTML and PHP, etc. The demonstration takes medical records from a hospital as an example. The data users submit their requests for assessing these data. As shown in Fig. 1, the demonstration will show the process of setting the privacy tags, describing the data requirement applications, and also the risk assessment and the auditing of privacy disclosure.

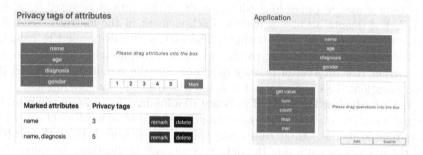

Fig. 1. System interface for demonstration.

5   Conclusion

This paper presents a risk assessment system for privacy disclosure of data sharing. By evaluating the data requirement applications submitted by data users, the system can provide data owners the ability to know in advance the potential risk of privacy disclosure. Therefore, data owners can take actions to preserve the data privacy.

Acknowledgments. This work is supported in part by Shanghai Science and Technology Development Fund (No. 16JC1400801), and National Natural Science Foundation of China (No. 61732004, No. U1636207 and No. 61572135).

References

1. OECD guidelines on the protection of privacy and transborder flows of personal data. http://www.oecd.org/sti/ieconomy/oecdguidelinesontheprotectionofprivacyandtransborderflowsofpersonaldata.htm
2. Dwork, C., Roth, A.: The algorithmic foundations of differential privacy. Found. Trends Theoret. Comput. Sci. **9**(3–4), 211–407 (2014)
3. Wang, T., et al.: Answering multi-dimensional analytical queries under local differential privacy. In: Proceedings of the SIGMOD Conference, pp. 159–176 (2019)
4. Zhou, X., Wang, Z., Wang, Y., Zhu, Y., Li, S., Wang, W.: A privacy leakage evaluation method in data openness based on matrix calculation. Comput. Appl. Softw. **37**(1), 298–303 (2020). (in Chinese)

Author Index

Printed in the United States
By Bookmasters